# 797,885 Books
are available to read at

www.ForgottenBooks.com

Forgotten Books' App
Available for mobile, tablet & eReader

ISBN 978-1-5284-0681-9
PIBN 10914407

This book is a reproduction of an important historical work. Forgotten Books uses state-of-the-art technology to digitally reconstruct the work, preserving the original format whilst repairing imperfections present in the aged copy. In rare cases, an imperfection in the original, such as a blemish or missing page, may be replicated in our edition. We do, however, repair the vast majority of imperfections successfully; any imperfections that remain are intentionally left to preserve the state of such historical works.

Forgotten Books is a registered trademark of FB &c Ltd.
Copyright © 2017 FB &c Ltd.
FB &c Ltd, Dalton House, 60 Windsor Avenue, London, SW19 2RR.
Company number 08720141. Registered in England and Wales.

For support please visit www.forgottenbooks.com

# 1 MONTH OF FREE READING

at

www.ForgottenBooks.com

By purchasing this book you are eligible for one month membership to ForgottenBooks.com, giving you unlimited access to our entire collection of over 700,000 titles via our web site and mobile apps.

To claim your free month visit:

www.forgottenbooks.com/free914407

\* Offer is valid for 45 days from date of purchase. Terms and conditions apply.

English
Français
Deutsche
Italiano
Español
Português

# www.forgottenbooks.com

**Mythology** Photography **Fiction**
Fishing Christianity **Art** Cooking
Essays **Buddhism** Freemasonry
Medicine **Biology** Music **Ancient Egypt** Evolution Carpentry Physics
Dance Geology **Mathematics** Fitness
Shakespeare **Folklore** Yoga Marketing
**Confidence** Immortality Biographies
Poetry **Psychology** Witchcraft
Electronics Chemistry History **Law**
Accounting **Philosophy** Anthropology
Alchemy Drama Quantum Mechanics
Atheism Sexual Health **Ancient History**
**Entrepreneurship** Languages Sport
Paleontology Needlework Islam
**Metaphysics** Investment Archaeology
Parenting Statistics Criminology
**Motivational**

U. S. DEPARTMENT OF COMMERCE
JESSE H. JONES, Secretary
COAST AND GEODETIC SURVEY
LEO OTIS COLBERT, Director

Special Publication No. 98
Revised (1940) Edition

# MANUAL OF HARMONIC ANALYSIS AND PREDICTION OF TIDES

BY

PAUL SCHUREMAN
Senior Mathematician

UNITED STATES
GOVERNMENT PRINTING OFFICE
WASHINGTON : 1941

For sale by the Superintendent of Documents, Washington, D. C. - - Price $1.00 (Buckram)

# PREFACE

This volume was designed primarily as a working manual for use in the United States Coast and Geodetic Survey and describes the procedure used in this office for the harmonic analysis and prediction of tides and tidal currents. It is based largely upon the works of Sir William Thomson, Prof. George H. Darwin, and Dr. Rollin A. Harris. In recent years there also has been considerable work done on this subject by Dr. A. T. Doodson, of the Tidal Institute of the University of Liverpool.

The first edition of the present work was published in 1924. In this revised edition there has been a rearrangement of the material in the first part of the volume to bring out more clearly the development of the tidal forces. Tables of astronomical data and other tables to facilitate the computations have been retained with a few revisions and additions and there has been added a list of symbols used in the work.

The collection of tidal harmonic constants for the world that appeared in the earlier edition has been omitted altogether because the work of maintaining such a list has now been taken over by the International Hydrographic Bureau at Monaco. These constants are now published in International Hydrographic Bureau Special Publication No. 26, which consists of a collection of loose sheets which permit the addition of new constants as they become available.

Special acknowledgment is due Walter B. Zerbe, associate mathematician of the Division of Tides and Currents, who reviewed the manuscript of this edition and offered many valuable suggestions.

|  | Page |
|---|---|
| Introduction | 1 |
| Historical statement | 1 |
| General explanation of tidal movement | 2 |
| Harmonic treatment of tidal data | 2 |
| Astronomical data | 3 |
| Degree of approximation | 8 |
| Development of tide-producing force | 10 |
| Fundamental formulas | 10 |
| Vertical component of force | 15 |
| Horizontal components of force | 26 |
| Equilibrium tide | 28 |
| Terms involving 4th power of moon's parallax | 34 |
| Solar tides | 39 |
| The $M_1$ tide | 41 |
| The $L_2$ tide | 43 |
| Lunisolar $K_1$ and $K_2$ tides | 44 |
| Meteorological and shallow-water tides | 46 |
| Analysis of observations | 49 |
| Harmonic constants | 49 |
| Observational data | 50 |
| Summations for analysis | 52 |
| Stencils | 53 |
| Secondary stencils | 57 |
| Fourier series | 62 |
| Augmenting factors | 71 |
| Phase lag or epoch | 75 |
| Inference of constants | 78 |
| Elimination | 84 |
| Long period constituents | 87 |
| Analysis of high and low waters | 100 |
| Forms used for analysis of tides | 104 |
| Analysis of tidal currents | 118 |
| Prediction of tides | 123 |
| Harmonic method | 123 |
| Tide-predicting machine | 126 |
| Forms used with tide-predicting machine | 143 |
| Prediction of tidal currents | 147 |
| Tables | 153 |
| Explanation of tables | 153 |
| 1. Fundamental astronomical data | 162 |
| 2. Harmonic constituents | 164 |
| 2a. Shallow-water constituents | 167 |
| 3. Latitude factors | 168 |
| 4. Mean longitude of lunar and solar elements | 170 |
| 5. Differences to adapt table 4 to any month, day, and hour | 172 |
| 6. Values of $I, \nu, \xi, \nu', $ and $2\nu''$ for each degree of $N$ | 173 |
| 7. Values of log $R_a$ for amplitude of constituent $L_2$ | 177 |
| 8. Values of $R$ for argument of constituent $L_2$ | 178 |
| 9. Values of log $Q_a$ for amplitude of constituent $M_1$ | 179 |
| 10. Values of $Q$ for argument of constituent $M_1$ | 180 |
| 11. Values of $u$ of equilibrium arguments for each degree of $N$ | 182 |
| 12. Values of log factor $F$ for each tenth degree of $I$ | 186 |
| 13. Values of $u$ and log factor $F$ for constituents $L_2$ and $M_1$ for the years 1900 to 2000 | 192 |
| 14. Node factor $f$ for middle of each year 1850 to 1999 | 199 |
| 15. Equilibrium argument $V_o+u$ for beginning of each year 1850 to 2000 | 204 |

## CONTENTS

Tables—Continued.

| | Page |
|---|---|
| 16. Differences to adapt table 15 to beginning of each calendar month | 212 |
| 17. Differences to adapt table 15 to beginning of each day of month | 213 |
| 18. Differences to adapt table 15 to beginning of each hour of day | 216 |
| 19. Products for Form 194 | 218 |
| 20. Augmenting factors | 228 |
| 21. Acceleration in epoch of $K_1$ due to $P_1$ | 229 |
| 22. Ratio of increase in amplitude of $K_1$ due to $P_1$ | 229 |
| 23. Acceleration in epoch of $S_2$ due to $K_2$ | 230 |
| 24. Ratio of increase in amplitude of $S_2$ due to $K_2$ | 230 |
| 25. Acceleration in epoch of $S_2$ due to $T_2$ | 231 |
| 26. Resultant amplitude $S_2$ due to $T_2$ | 232 |
| 27. Critical logarithms for Form 245 | 233 |
| 28. Constituent speed differences $(b-a)$ | 234 |
| 29. Elimination factors | 236 |
| 30. Products for Form 245 | 266 |
| 31. For construction of primary stencils | 268 |
| 32. Divisors for primary stencil sums | 288 |
| 33. For construction of secondary stencils | 299 |
| 34. Assignment of daily page sums for long-period constituents | 302 |
| 35. Products for Form 444 | 304 |
| 36. Angle differences for Form 445 | 306 |
| 37. Coast and Geodetic Survey tide-predicting machine No. 2—general gears | 307 |
| 38. Coast and Geodetic Survey tide-predicting machine No.2—constituent gears | 308 |
| 39. Synodic periods of constituents | 309 |
| 40. Day of common year corresponding to day of month | 309 |
| 41. Values of $h$ in formula $h=(1+r^2+2r \cos x)^{\frac{1}{2}}$ | 310 |
| 42. Values of $k$ in formula $k=\tan^{-1} \dfrac{r \sin x}{1+r \cos x}$ | 310 |
| Explanation of symbols | 311 |
| Index | 314 |

### ILLUSTRATIONS

| | Page |
|---|---|
| 1. Ecliptic, celestial equator, and moon's orbit | 6 |
| 2. Tide-producing force | 11 |
| 3. Celestial sphere | 16 |
| 4. Longitude relations | 19 |
| 5. Equilibrium tide with moon on equator | 29 |
| 6. Equilibrium tide with moon at maximum declination | 29 |
| 7. Constituent tide curve | 50 |
| 8. Phase relations | 77 |
| 9. Form 362, hourly heights | 105 |
| 10. Stencil for constituent M | 106 |
| 11. Application of stencil | 107 |
| 12. Form 142, stencil sums | 108 |
| 13. Computation of hourly means | 109 |
| 14. Form 244, computation of $V_0+u$ | 110 |
| 15. Form 244a, log $F$ and arguments for elimination | 111 |
| 16. Form 194, harmonic analysis | 112 |
| 17. Form 452, $R$, $\kappa$, and $\zeta$ from analysis and inference, diurnal tides | 115 |
| 18. Form 452, $R$, $\kappa$, and $\zeta$ from analysis and inference, semidiurnal tides | 116 |
| 19. Form 245, elimination | 117 |
| 20. Form 723, currents, harmonic comparison | 120 |
| 21. Coast and Geodetic Survey tide-predicting machine | 128 |
| 22. Tide-predicting machine, time side | 128 |
| 23. Tide-predicting machine, recording devices | 128 |
| 24. Tide-predicting machine, driving gears | 128 |
| 25. Tide-predicting machine, dial case from height side | 128 |
| 26. Tide-predicting machine, dial case from time side | 128 |
| 27. Tide-predicting machine, vertical driving shaft of middle section | 128 |
| 28. Tide-predicting machine, forward driving shaft of rear section | 128 |
| 29. Tide-predicting machine, rear end | 128 |
| 30. Tide-predicting machine, details of releasable gear | 128 |
| 31. Tide-predicting machine, details of constituent crank | 128 |
| 32. Form 444, standard harmonic constants for predictions | 143 |
| 33. Form 445, settings for tide-predicting machine | 145 |
| 34. Graphic solution of formulas (470) and (471) | 149 |

# MANUAL OF HARMONIC ANALYSIS AND PREDICTION OF TIDES

## INTRODUCTION

#### HISTORICAL STATEMENT

1. Sir William Thomson (Lord Kelvin) devised the method of reduction of tides by harmonic analysis about the year 1867. The principle upon which the system is based—which is that any periodic motion or oscillation can always be resolved into the sum of a series of simple harmonic motions—is said to have been discovered by Eudoxas as early as 356 B. C., when he explained the apparently irregular motions of the planets by combinations of uniform circular motions.[1] In the early part of the nineteenth century Laplace recognized the existence of partial tides that might be expressed by the cosine of an angle increasing uniformly with the time, and also applied the essential principles of the harmonic analysis to the reduction of high and low waters. Dr. Thomas Young suggested the importance of observing and analyzing the entire tidal curve rather than the high and low waters only. Sir George B. Airy also had an important part in laying the foundation for the harmonic analysis of the tides. To Sir William Thomson, however, we may give the credit for having placed the analysis on a practical basis.

2. In 1867 the British Association for the Advancement of Science appointed a committee for the purpose of promoting the extension, improvement, and harmonic analysis of tidal observations. The report on the subject was prepared by Sir William Thomson and was published in the Report of the British Association for the Advancement of Science in 1868. Supplementary reports were made from time to time by the tidal committee and published in subsequent reports of the British association. A few years later a committe, consisting of Profs. G. H. Darwin and J. C. Adams, drew up a very full report on the subject, which was published in the Report of the British Association for the Advancement of Science in 1883.

3. Among the American mathematicians who have had an important part in the development of this subject may be named Prof. William Ferrel and Dr. Rollin A. Harris, both of whom were associated with the U. S. Coast and Geodetic Survey. The Tidal Researches, by Professor Ferrel, was published in 1874, and additional articles on the harmonic analysis by the same author appeared from time to time in the annual reports of the Superintendent of the Coast and Geodetic Survey. The best known work of Doctor Harris is his Manual of Tides, which was published in several parts as appendices to the annual reports of the Superintendent of the Coast and Geodetic Survey. The subject of the harmonic analysis was treated principally in Part II of the Manual which appeared in 1897.

---

[1] Nautical Science, p. 279, by Charles Lane Poor.

2   U. S. COAST AND GEODETIC SURVEY

## GENERAL EXPLANATION OF TIDAL MOVEMENT

4. That the tidal movement results from the gravitational attraction of the moon and sun acting upon the rotating earth is now a well-established scientific fact. The movement includes both the vertical rise and fall of the tide and the horizontal flow of the tidal currents. It will be shown later that the tide-producing force due to this attraction, when taken in connection with the attraction between the particles of matter which constitute the earth, can be expressed by mathematical formulas based upon the well-known laws of gravitation.

5. Although the acting forces are well understood, the resultant tidal movement is exceedingly complicated because of the irregular distribution of land and water on the earth and the retarding effects of friction and inertia. Contrary to the popular idea of a progressive tidal wave following the moon around the earth, the basic tidal movement as evidenced by observations at numerous points along the shores of the oceans consists of a number of oscillating areas, the movement being somewhat similar to that in a pan of water that has been tilted. Such oscillations are technically known as stationary waves. The complex nature of the movement can be appreciated when consideration is given to the fact that such stationary waves may overlap or be superimposed upon each other and may be accompanied by a progressive wave movement.

6. Any basin of water has its natural free period of oscillation depending upon its size and depth. The usual formula for the period of oscillation in a rectangular tank of uniform depth is $2L/\sqrt{gd}$, in which $L$ is the length and $d$ the depth of the tank and $g$ is the acceleration of gravity. When a disturbing force is applied periodically at intervals corresponding to the free period of a body of water, it tends to build up an oscillation of much greater magnitude than would be possible with a single application of the force. The major tidal oscillations have periods approximating the half and the whole lunar day.

## HARMONIC TREATMENT OF TIDAL DATA

7. The harmonic analysis of tides is based upon an assumption that the rise and fall of the tide in any locality can be expressed mathematically by the sum of a series of harmonic terms having certain relations to astronomical conditions. A simple harmonic function is a quantity that varies as the cosine of an angle that increases uniformly with time. In the equation $y = A \cos at$, $y$ is an harmonic function of the angle $at$ in which $a$ is a constant and $t$ represents time as measured from some intitial epoch. The general equation for the height ($h$) of the tide at any time ($t$) may be written

$$h = H_0 + A \cos (at+\alpha) + B \cos (bt+\beta) + C \cos (ct+\gamma) + \text{etc.} \quad (1)$$

in which $H_0$ is the height of the mean water level above the datum used. Other symbols are explained in the following paragraph.

8. Each cosine term in equation (1) is known as a *constituent* or *component* tide. The coefficients $A$, $B$, $C$, etc. are the *amplitudes* of the constituents and are derived from observed tidal data in each locality. The expression in parentheses is a uniformly-varying angle and its value at any time is called its *phase*. Any constituent term has its maximum positive value when the phase of the angle is zero and a maximum negative value when the phase equals 180°, and the

term becomes zero when the phase equals 90° or 270°. The coefficient of $t$ represents the rate of change in the phase and is called the *speed* of the constituent and is usually expressed in degrees per hour. The time required for a constituent to pass through a complete cycle is known as its *period* and may be obtained by dividing 360° by its speed. The periods and corresponding speeds of the constituents are derived from astronomical data and are independent of the locality of the tide station. The symbols $\alpha$, $\beta$, $\gamma$, etc. refer to the initial phases of the constituent angles at the time when $t$ equals zero. The initial phases depend upon locality as well as the instant from which the time is reckoned and their values are derived from tidal observations. *Harmonic analysis* as applied to tides is the process by which the observed tidal data at any place are separated into a number of harmonic constituents. The quantities sought are known as *harmonic constants* and consist of the amplitudes and certain phase relations which will be more fully explained later. *Harmonic prediction* is accomplished by reuniting the elementary constituents in accordance with astronomical relations prevailing at the time for which the predictions are being made.

### ASTRONOMICAL DATA

**9.** In tidal work the only celestial bodies that need be considered are the moon and sun. Although every other celestial body whose gravitational influence reaches the earth creates a theoretical tide-producing force, the greater distance or smaller size of such body renders negligible any effect of this force upon the tides of the earth. In deriving mathematical expressions for the tide-producing forces of the moon and sun, the principal factors to be taken into consideration are the rotation of the earth, the revolution of the moon around the earth, the revolution of the earth around the sun, the inclination of the moon's orbit to the earth's equator, and the obliquity of the ecliptic. Numerical values pertaining to these factors will be found in table 1.

**10.** The earth rotates on its axis once each day. There are, however, several kinds of days—the sidereal day, the solar day, the lunar day, and the constituent day—depending upon the object used as a reference for the rotation. The *sidereal day* is defined by astronomers as the time required for the rotation of the earth with respect to the vernal equinox. Because of the precession of the equinox, this day differs slightly from the time of rotation with respect to a fixed star, the difference being less than the hundredth part of a second. The *solar day* and *lunar day* are respectively the times required for rotation with respect to the sun and moon. Since the motions of the earth and moon in their orbits are not uniform, the solar and lunar days vary a little in length and their average or mean values are taken as standard units of time. A *constituent day* is the time of the rotation of the earth with respect to a fictitious satellite representing one of the periodic elements in the tidal forces. It approximates in length the lunar or solar day and corresponds to the period of a diurnal constituent or twice the period of a semidiurnal constituent.

**11.** A *calendar day* is a mean solar day commencing at midnight. Such a calendar day is known also as a *civil day* to distinguish it from the *astronomical day* which commences at noon of the same date.

Prior to the year 1925, the astronomical day was in general use by astronomers for the recording of astronomical data, but beginning with the Ephemeris and Nautical Almanac published in 1925 the civil day has been adopted for the calculations. Each day of whatever kind may be divided into 24 equal parts known as hours which are qualified by the name of the kind of day of which they are a part, as *sideral hour, solar hour, lunar hour,* or *constituent hour.*

**12.** The moon revolves around the earth in an elliptical orbit. Although the average eccentricity of this orbit remains approximately constant for long periods of time, there are a number of perturbations in the moon's motion due, primarily, to the attractive force of the sun. Besides the revolution of the line of apsides and the regression of the nodes which take place more or less slowly, the principal inequalities in the moon's motion which affect the tides are the evection and variation. The evection depends upon the alternate increase and decrease of the eccentricity of the moon's orbit, which is always a maximum when the sun is passing the moon's line of apsides, and a minimum when the sun is at right angles to it. The variation inequality is due mainly to the tangential component of the disturbing force. The period of the revolution of the moon around the earth is called a month. The month is designated as sidereal, tropical, anomalistic, nodical, or synodical, according to whether the revolution is relative to a fixed star, the vernal equinox, the perigee, the ascending node, or the sun. The calendar month is a rough approximation to the synodical month.

**13.** It is customary to refer to the revolution of the earth around the sun, although it may be more accurately stated that they both revolve around their common center of gravity; but if we imagine the earth as fixed, the sun will describe an apparent path around the earth which is the same in size and form as the orbit of the earth around the sun, and the effect upon the tides would be the same. This orbit is an ellipse with an eccentricity that changes so slowly that it may be considered as practically constant. The period of the revolution of the earth around the sun is a year, but there are several kinds of years. The *sidereal year* is a revolution with respect to a fixed star, the *tropical year* is a revolution with respect to the vernal equinox, the *eclipse year* is a revolution with respect to the moon's ascending node, and the *anomalistic year* is a revolution with respect to the solar perigee.

**14.** A *calendar year* consists of an integral number of mean solar days and may be a *common year* of 365 days or a *leap year* of 366 days, these years being selected according to the calendars described below so that the average length will agree as nearly as practicable with the length of the tropical year which fixes the periodic changes in the seasons. The average length of the calendar year by the Julian calendar is exactly 365.25 days and by the Gregorian calendar 365.2425 days and these may be designated respectively as a *Julian year* and a *Gregorian year.*

**15.** The two principal kinds of calendars in use by most of the civilized world since the beginning of the Christian era are the Julian and the Gregorian calendars, the latter being the modern calendar in which the dates are sometimes referred to as "new style" to distinguish them from the dates of the older calendars. Prior to the year 45 B. C. there was more or less confusion in the calendars, intercalations of months and days being arbitrarily made by the priesthood and magistrates to bring the calendar into accord with the seasons and for other purposes.

16. The Julian calendar received its name from Julius Cæsar, who introduced it in the year 45 B. C. This calendar provided that the common year should consist of 365 days and every fourth year of 366 days, each year to begin on January 1. As proposed by Julius Cæsar, the 12 months beginning with January were to be alternately 31 days and 30 days in length with the exception that February should have only 29 days in the common years. When Augustus succeeded Julius Cæsar a few years later, he slightly modified this arrangement by transferring one day from February to the month of Sextilis, or August as it was then renamed, and also transferred the 31st day of September and November to October and December to avoid having three 31-day months in succession.

17. The Gregorian calendar received its name from Pope Gregory, who introduced it in the year 1582. It was immediately adopted by the Catholic countries but was not accepted by England until 1752. This calendar differs from the Julian calendar in having the century years not exactly divisible by 400 to consist of only 365 days, while in the Julian calendar every century year as well as every other year divisible by 4 is taken as a leap year with 366 days. For dates before Christ the year number must be diminished by 1 before testing its divisibility by 4 or 400 since the year 1 B. C. corresponds to the year 0 A. D. The Gregorian calendar will gain on the Julian calendar three days in each 400 years. When originally adopted, in order to adjust the Gregorian calendar so that the vernal equinox should fall upon March 21, as it had at the time of the Council of Nice in 325 A. D., 10 days were dropped and it was ordered that the day following October 4, 1582 of the Julian calendar should be designated as October 15, 1582 of the Gregorian calendar. This difference of 10 days between the dates of the two calendars continued until 1700, which was a leap year according to the Julian calendar and a common year by the Gregorian calendar. The difference between the two then became 11 days and in 1800 was increased to 12 days. Since 1900 the difference has been 13 days and will remain the same until the year 2100.

18. Dates of the Christian era prior to October 4, 1582, will, in general, conform to the Julian calendar. Since that time both calendars have been used. The Gregorian calendar was adopted in England by an act of Parliament passed in 1751, which provided that the day following September 2, 1752, should be called September 14, 1752, and also that the year 1752 and subsequent years should commence on the 1st day of January. Previous to this the legal year in England commenced on March 25. Except for this arbitrary beginning of the year, the old English calendar was the same as the Julian calendar. When Alaska was purchased from Russia by the United States, its calendar was altered by 11 days, one of these days being necessary because of the difference between the Asiatic and American dates when compared across the one hundred and eightieth meridian. Dates in the tables at the back of this volume refer to the Gregorian calendar.

19. The three great circles formed by the intersections of the planes of the earth's equator, the ecliptic, and the moon's orbit with the

celestial sphere are represented in figure 1. These circles intersect in six points, three of them being marked by symbols in the figure, namely, the *vernal equinox* ♈ at the intersection of the celestial equator and ecliptic, the ascending *lunar node* ☊ at the intersection of the ecliptic and the projection of the moon's orbit, and the *lunar intersection* $A$ at the intersection of the celestial equator and the projection of the moon's orbit. For brevity these three points are sometimes called respectively "the equinox," "the node," and "the intersection." The vernal equinox, although subject to a slow westward motion of about 50" per year, is generally taken as a fixed point of reference for the motion of other parts of the solar system. The moon's node has a westward motion of about 19° a year, which is sufficient to carry it entirely around a great circle in a little less than 19 years.

**20.** The angle $\omega$ between the ecliptic and the celestial equator is known as the obliquity of the ecliptic and has a nearly constant value of 23½°. The angle $i$ between the ecliptic and the plane of the moon's orbit is also constant with a value of about 5°.

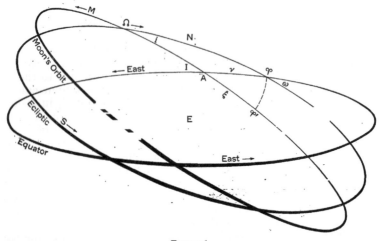

FIGURE 1.

The angle $I$ which measures the inclination of the moon's orbit to the celestial equator might appropriately be called the obliquity of the moon's orbit. Its magnitude changes with the position of the moon's node. When the moon's ascending node coincides with the vernal equinox, the angle $I$ equals the sum of $\omega$ and $i$, or about 28½°, and when the descending node coincides with the vernal equinox, the angle $I$ equals the difference between $\omega$ and $i$, or about 18½°. This variation in the obliquity of the moon's orbit with its period of approximately 18.6 years introduces an important inequality in the tidal movement which must be taken into account.

**21.** In the celestial sphere the terms "latitude" and "longitude" apply especially to measurements referred to the ecliptic and vernal equinox, but the terms may with propriety also be applied to measurements referred to other great circles and origins, provided they are sufficiently well defined to prevent any ambiguity. For example, we may say "longitude in the moon's orbit measured from the moon's node." Celestial longitude is always understood to be measured toward the east entirely around the circle. Longitude in the celestial equator reckoned from the vernal equinox is called right ascension, and the angular distance north or south of the celestial equator is called declination.

**22.** The true longitude of any point referred to any great circle in the celestial sphere may be defined as the arc of that circle intercepted between the accepted origin and the projection of the point on the circle, the measurement being always eastward from the origin to the projection of the point. The true longitude of any point will generally be different when referred to different circles, although reckoned from a common origin; and the longitude of a body moving at a uniform rate of speed in one great circle will not have a uniform rate of change when referred to another great circle.

**23.** The mean longitude of a body moving in a closed orbit and referred to any great circle may be defined as the longitude that would be attained by a point moving uniformly in the circle of reference at the same average angular velocity as that of the body and with the initial position of the point so taken that its mean longitude would be the same as the true longitude of the body at a certain selected position of that body in its orbit. With a common initial point, the mean longitude of a moving body will be the same in whatever circle it may be reckoned. Longitude in the ecliptic and in the celestial equator are usually reckoned from the vernal equinox ♈, which is common to both circles. In order to have an equivalent origin in the moon's orbit, we may lay off an arc ☊ ♈' (fig. 1) in the moon's orbit equal to ☊ ♈ in the ecliptic and for convenience call the point ♈' the referred equinox. The mean longitude of any body, if reckoned from either the equinox or the referred equinox, will be the same in any of the three orbits represented. This will, of course, not be the case for the true longitude.

**24.** Let us now examine more closely the spherical triangle ☊ ♈ $A$ in figure 1. The angles $\omega$ and $i$ are very nearly constant for long periods of time and have already been explained. The side ☊ ♈, usually designated by $N$, is the longitude of the moon's node and is undergoing a constant and practically uniform change due to the regression of the moon's nodes. This westward movement of the node, by which it is carried completely around the ecliptic in a period of approximately 18.6 years, causes a constant change in the form of the triangle, the elements of which are of considerable importance in the present discussion. The value of the angle $I$, the supplement of the angle ☊ $A$ ♈, has an important effect upon both the range and time of the tide, which will be noted later. The side $A$ ♈, designated by $\nu$, is the right ascension or longitude in the celestial equator of the intersection $A$. The arc designated by $\xi$ is equal to the side ☊ ♈ −side ☊ $A$ and is the longitude in the moon's orbit of the intersection $A$. Since the angles $i$ and $\omega$ are assumed to be constant, the values of $I$, $\nu$, and $\xi$ will depend directly upon $N$, the longitude of the moon's node, and may be readily obtained by the ordinary solution of the spherical triangle ☊ ♈ $A$. Table 6 give the values of $I$, $\nu$, and $\xi$ for each degree of $N$. In the computation of this table the value of $\omega$ for the beginning of the twentieth century was used. However, the secular change in the obliquity of the ecliptic is so slow that a difference of a century in the epoch taken as the basis of the computation would have resulted in differences of less than 0.02 of a degree in the tabular values. The table may therefore be used without material error for reductions pertaining to any modern time.

**25.** Looking again at figure 1, it will be noted that when the longitude of the moon's node is zero the value of the inclination $I$ will equal the sum of $\omega$ and $i$ and will be at its maximum. In this position the northern portion of the moon's orbit will be north of the ecliptic. When the longitude of the moon's node is 180°, the moon's orbit will be between the Equator and ecliptic, and the angle $I$ will be equal to angle $\omega$—angle $i$. The angle $I$ will be always positive and will vary from $\omega-i$ to $\omega+i$. When the longitude of the moon's node equals zero or 180°, the values of $\nu$ and $\xi$ will each be zero. For all positions of the moon's node north of the Equator as its longitude changes from 180 to 0°, $\nu$ and $\xi$ will have positive values, as indicated in the figure, these arcs being considered as positive when reckoned eastward from $\Upsilon$ and $\Upsilon'$, respectively. For all positions of the node south of the Equator, as the longitude changes from 360 to 180°, $\nu$ and $\xi$ will each be negative, since the intersection $A$ will then lay to the westward of $\Upsilon$ and $\Upsilon'$.

#### DEGREE OF APPROXIMATION

**26.** The problem of finding expressions for tidal forces and the equilibrium height of the tide in terms of time and place does not admit of a strict solution, but approximate expressions can be obtained which may be carried to as high an order of precision as desired. In ordinary numerical computations exact results are seldom obtained, the degree of precision depending upon the number of decimal places used in the computations, which, in turn, will be determined largely by the magnitude of the quantity sought. In general, the degree of approximation to the value of any quantity expressed numerically will be determined by the number of significant figures used. With a quantity represented by a single significant figure, the error may be as great as 33⅓ percent of the quantity itself, while the use of two significant figures will reduce the maximum error to less than 5 percent of the true value of the quantity. The large possible error in the first case renders it of little value, but in the latter case the approximation is sufficiently close to be useful when only rough results are necessary. The distance of the sun from the earth is popularly expressed by two significant figures as 93,000,000 miles.

**27.** With three or four significant figures fairly satisfactory approximations may be represented, and with a greater number very precise results may be expressed. For theoretical purposes the highest attainable precision is desirable, but for practical purposes, because of the increase in the labor without a corresponding increase in utility, it will be usually found advantageous to limit the degree of precision in accordance with the prevailing conditions.

**28.** Frequently a quantity that is to be used as a factor in an expression may be expanded into a series of terms. If the approximate value of such a series is near unity, terms which would affect the third decimal place, if expressed numerically, should usually be retained. The retention of the smaller terms will depend to some extent upon the labor involved since their rejection would not seriously affect the final results.

**29.** The formulas for the moon's true longitude and parallax on pages 19–20 are said to be given to the second order of approximation, a fraction of the first order being considered as one having an approximate value of 1/20 or 0.05, a fraction of the second order having an approximate value of $(0.05)^2$ or 0.0025, a fraction of third order having an approximate value of $(0.05)^3$ or 0.000125, etc. As these formulas provide important factors in the development of the equations representing the tide-producing forces, they determine to a large extent the degrees of precision to be expected in the results.

# DEVELOPMENT OF TIDE-PRODUCING FORCE

### FUNDAMENTAL FORMULAS

**30.** The tide-producing forces exerted by the moon and sun are similar in their action and mathematical expressions obtained for one may therefore by proper substitutions be adapted to the other. Because of the greater importance of the moon in its tide-producing effects, the following development will apply primarily to that body, the necessary changes to represent the solar tides being afterwards indicated.

**31.** The tide-producing force of the moon is that portion of its gravitational attraction which is effective in changing the water level on the earth's surface. This effective force is the difference between the attraction for the earth as a whole and the attraction for the different particles which constitute the yielding part of the earth's surface; or, if the entire earth were considered to be a plastic mass, the tide-producing force at any point within the mass would be the force that tended to change the position of a particle at that point relative to a particle at the center of the earth. That part of the earth's surface which is directly under the moon is nearer to that body than is the center of the earth and is therefore more strongly attracted since the force of gravity varies inversely as the square of the distance. For the same reason the center of the earth is more strongly attracted by the moon than is that part of the earth's surface which is turned away from the moon.

**32.** The tide-producing force, being the difference between the attraction for particles situated relatively near together, is small compared with the attraction itself. It may be interesting to note that, although the sun's attraction on the earth is nearly 200 times as great as that of the moon, its tide-producing force is less than one-half that of the moon. If the forces acting upon each particle of the earth were equal and parallel, no matter how great those forces might be, there would be no tendency to change the relative positions of those particles, and consequently there would be no tide-producing force.

**33.** The tide-producing force may be graphically represented as in figure 2.

Let $O$ = the center of the earth,
$C$ = the center of the moon,
$P$ = any point within or on the surface of the earth.

Then $OC$ will represent the direction of the attractive force of the moon upon a particle at the center of the earth and $PC$ the direction of the attractive force of the moon upon a particle at $P$. Now, let the magnitude of the moon's attraction at $P$ be represented by the length of the line $PC$. Then, since the attraction of gravitation varies inversely as the square of the distance, it is necessary, in order to represent the attraction at $O$ on the same scale, to take a line $CQ$ of such length that $CQ : CP = \overline{CP}^2 : \overline{CO}^2$.

**34.** The line $PQ$, joining $P$ and $Q$, will then represent the direction and magnitude of the resultant force that tends to disturb the position of $P$ relative to $O$, for it represents the difference between the force $PC$ and a force through $P$ equal and parallel to the force $QC$ which acts upon $O$. This last statement may be a little clearer to the reader if he will consider the force $PC$ as being resolved into a force $PD$ equal and parallel to $QC$, and the force $PQ$. The force $PD$, acting upon the particle at $P$, being equal and parallel to the force $QC$, acting upon a particle at $O$, will have no tendency to change the position of $P$ relative to $O$. The remaining force $PQ$ will tend to alter the position of $P$ relative to $O$ and is the tide-producing force of the moon at $P$. The force $PQ$ may be resolved into a vertical component $PR$, which tends to raise the water at $P$, and the horizontal component $PT$, which tends to move the water horizontally.

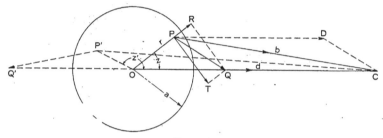

FIGURE 2.

**35.** If the point $P'$ is taken so that the distance $CP'$ is greater than the distance $CO$, the tide-producing force $P'Q'$ will be directed away from the moon. While at first sight this may appear paradoxical, it will be noted that the moon tends to separate $O$ from $P'$, but as $O$ is taken as the point of reference, this resulting force that tends to separate the points is considered as being applied at the point $P'$ only.

**36.** To express the tide-producing force by mathematical equations, refer to figure 2 and let

$r = OP$ = distance of particle $P$ from center of earth,
$b = PC$ = distance of particle $P$ from center of moon,
$d = OC$ = distance from center of earth to center of moon,
$z = COP$ = angle at center of earth between $OP$ and $OC$.

Also let

$M$ = mass of moon,
$E$ = mass of earth,
$a$ = mean radius of earth,
$\mu$ = attraction of gravitation between unit masses at unit distance.
$g$ = mean acceleration of gravity on earth's surface.

Since the force of gravitation varies directly as the mass and inversely as the square of the distance,

Attraction of moon for unit mass at point $O$ in direction $OC = \dfrac{\mu M}{d^2}$ (2)

Attraction of moon for unit mass at point $P$ in direction $PC = \dfrac{\mu M}{b^2}$ (3)

**37.** Let each of these forces be resolved into a vertical component along the radius $OP$ and a horizontal component perpendicular to the same in the plane $OPC$, and consider the direction from $O$ toward $P$ as positive for the vertical component and the direction corresponding to the azimuth of the moon as positive for the horizontal component. We then have from (2) and (3)

Attraction at $O$ in direction $O$ to $P = \dfrac{\mu M}{d^2} \cos z$ (4)

Attraction at $O$ perpendicular to $OP = \dfrac{\mu M}{d^2} \sin z$ (5)

Attraction at $P$ in direction $O$ to $P = \dfrac{\mu M}{b^2} \cos CPR$ (6)

Attraction at $P$ perpendicular to $OP = \dfrac{\mu M}{b^2} \sin CPR$ (7)

**38.** The tide-producing force of the moon at any point $P$ is measured by the difference between the attraction at $P$ and at the center of the earth. Letting

$F_v$ = vertical component of tide-producing force, and
$F_a$ = horizontal component in azimuth of moon,

and taking the differences between (6) and (4) and between (7) and (5), we obtain the following expressions for these component forces in terms of the unit $\mu$:

$$F_v / \mu = M\left(\dfrac{\cos CPR}{b^2} - \dfrac{\cos z}{d^2}\right) \quad (8)$$

$$F_a / \mu = M\left(\dfrac{\sin CPR}{b^2} - \dfrac{\sin z}{d^2}\right) \quad (9)$$

**39.** From the plane triangle $COP$ the following relations may be obtained:

$$b^2 = r^2 + d^2 - 2rd \cos z = d^2[1 - 2(r/d) \cos z + (r/d)^2] \quad (10)$$

$$\sin CPR = \sin CPO = (d/b) \sin z = \dfrac{\sin z}{[1 - 2(r/d) \cos z + (r/d)^2]^{\frac{1}{2}}} \quad (11)$$

$$\cos CPR = (1 - \sin^2 CPR)^{\frac{1}{2}} = \dfrac{\cos z - r/d}{[1 - 2(r/d) \cos z + (r/d)^2]^{\frac{1}{2}}} \quad (12)$$

**40.** In figure 2 it will be noted that the value of $z$, being reckoned in any plane from the line $OC$, may vary from zero to 180°, and also that the angle $CPR$ increases as $z$ increases within the same limits. Sin $z$ and sin $CPR$ will therefore always be positive. As the angle $OCP$ is always very small, the angle $CPR$ will differ by only a very small amount from the angle $z$ and will usually be in the same quadrant. In obtaining the square root for the numerator of (12) it was therefore necessary to use only that sign which would preserve this relationship. The denominators of (11) and (12) are to be considered as positive.

**41.** Substituting in equations (8) and (9) the equivalents for $b$, sin $CPR$, and cos $CPR$ from equations (10) to (12), the following basic formulas are obtained for the vertical and horizontal components of the tide-producing force at any point $P$ at $r$ distance from the center of the earth:

$$F_v/\mu = \frac{M}{d^2}\left[\frac{\cos z - r/d}{\{1-2(r/d)\cos z + (r/d)^2\}^{\frac{3}{2}}} - \cos z\right] \quad (13)$$

$$F_a/\mu = \frac{M}{d^2}\left[\frac{\sin z}{\{1-2(r/d)\cos z + (r/d)^2\}^{\frac{3}{2}}} - \sin z\right] \quad (14)$$

**42.** To express these forces in their relation to the mean acceleration of gravity on the earth's surface, represented by the symbol $g$, we have

$$g/\mu = E/a^2, \quad \text{or} \quad \mu/g = a^2/E \quad (15)$$

in which $E$ is the mass and $a$ is the mean radius of the earth. Substituting the above in formulas (13) and (14), we may write

$$F_v/g = (M/E)(a/d)^2\left[\frac{\cos z - r/d}{\{1-2(r/d)\cos z + (r/d)^2\}^{\frac{3}{2}}} = \cos z\right] \quad (16)$$

$$F_a/g = (M/E)(a/d)^2\left[\frac{\sin z}{\{1-2(r/d)\cos z + (r/d)^2\}^{\frac{3}{2}}} - \sin z\right] \quad (17)$$

**43.** Formulas (16) and (17) represent completely the vertical and horizontal components of the lunar tide-producing force at any point in the earth. If $r$ is taken equal to the mean radius $a$, the formulas will involve the constant ratio $M/E$ and two variable quantities— the angle $z$ which is the moon's zenith distance, and the ratio $a/d$ which is the sine of the moon's horizontal parallax in respect to the mean radius of the earth. Because of the smallness of the ratio $a/d$ it may also be taken as the parallax itself expressed as a fraction of a radian. The parallax is largest when the moon is in perigee and at this time the tide-producing force will reach its greatest magnitude. A more rapid change in the tidal force at any point on the earth's surface is caused by the continuous change in the zenith distance of the moon resulting from the earth's rotation. The vertical component attains its maximum value when $z$ equals zero, and the horizontal component has its maximum value when $z$ is a little less than 45°. Substituting numerical values in formulas (16) and (17) and in similar formulas for the tide-producing force of the sun, the following are obtained as the approximate extreme component forces when the moon and sun are nearest the earth:

Greatest $F_v/g = .144 \times 10^{-6}$ for moon, or $.054 \times 10^{-6}$ for sun $\quad (18)$

Greatest $F_a/g = .107 \times 10^{-6}$ for moon, or $.041 \times 10^{-6}$ for sun $\quad (19)$

The horizontal component of the tide-producing force may be measured by its deflection of the plumb line, the relation of this component to gravity as expressed by the above formula being the tangent of the angle of deflection. Under the most favorable conditions the greatest deflection due to the moon is about 0.022'' and the greatest deflection due to the sun is less than 0.009'' of arc.

**44.** To simplify the preceding formulas, the quantity involving the fractional exponent may be developed by Maclaurin's theorem into a series arranged according to the ascending powers of $r/d$, this being a small fraction with an approximate maximum value of 0.018. Thus

$$\frac{1}{\{1-2(r/d)\cos z+(r/d)^2\}^{\frac{1}{2}}} = 1+3\cos z\ (r/d)$$
$$+3/2\ (5\cos^2 z-1)(r/d)^2$$
$$+5/2\ (7\cos^3 z-3\cos z)(r/d)^3+\text{etc.} \qquad (20)$$

**45.** Substituting (20) in formulas (16) and (17) and neglecting the higher powers of $r/d$, we obtain the following formulas:

$$F_v/g = 3\ (M/E)(a/d)^2\ (\cos^2 z-1/3)\ (r/d)$$
$$+3/2\ (M/E)(a/d)^2\ (5\cos^3 z-3\cos z)\ (r/d)^2 \qquad (21)$$

$$F_a/g = 3/2\ (M/E)(a/d)^2\ (\sin 2z)\ (r/d)$$
$$+3/2\ (M/E)(a/d)^2\ \sin z\ (5\cos^2 z-1)\ (r/d)^2 \qquad (22)$$

**46.** If $r$, which represents the distance of the point of observation for the center of the earth, is replaced by the mean radius $a$, it will be noted that the first term of each of the above formulas involves the cube of the ratio $a/d$ while the second term involves the fourth power of this quantity. This ratio is essentially the moon's parallax expressed in the radian unit. These terms may now be written as separate formulas and for convenience of identification the digits "3" and "4" will be annexed to the formula symbol to represent respectively the terms involving the cube and fourth power of the parallax. Thus

$$F_{v3}/g = 3\ (M/E)(a/d)^3(\cos^2 z-1/3) \qquad (23)$$

$$F_{v4}/g = 3/2\ (M/E)(a/d)^4(5\cos^3 z-3\cos z) \qquad (24)$$

$$F_{a3}/g = 3/2\ (M/E)(a/d)^3\sin 2z \qquad (25)$$

$$F_{a4}/g = 3/2\ (M/E)(a/d)^4\sin z\ (5\cos^2 z-1) \qquad (26)$$

Formulas (23) and (25) involving the cube of the parallax represent the principal part of the tide-producing force. For the moon this is about 98 per cent of the whole and for the sun a higher percentage. The part of the tide-producing force represented by formulas (24) and (26) and involving the fourth power of the parallax is of very little practical importance but as a matter of theoretical interest will be later given further attention.

**47.** An examination of formulas (23) and (25) shows that the principal part of the tide-producing force is symmetrically distributed over the earth's surface with respect to a plane through the center of the earth and perpendicular to a line joining the centers of the earth and moon. The vertical component (23) has a maximum positive value when the zenith distance $z=0$ or $180°$ and a maximum negative value when $z=90°$, the maximum negative value being one-half as great as the maximum positive value. The vertical component becomes zero when $z=\cos^{-1}\pm\sqrt{1/3}$ (approx. 54.74° and 125.26°). The horizontal component (25) has its maximum value when $z=45°$ and an equal maximum negative value when $z=135°$. The horizontal component becomes zero when $z=0$, 90°, or 180.

**48.** If numerical values applicable to the mean parallax of the moon are substituted in (23) and (25), these component forces may be written

$$F_{v3}/g \text{ at mean parallax} = 0.000{,}000{,}167 \ (\cos^2 z - 1/3) \tag{27}$$

$$F_{a3}/g \text{ at mean parallax} = 0.000{,}000{,}084 \sin 2z \tag{28}$$

For the corresponding components of the solar tide-producing force, the numerical coefficients will be 0.46 times as great as those in the above formulas.

**49.** For the extreme values of the components represented by (23) and (25), with the moon and sun nearest the earth, the following may be obtained by suitable substitutions:

Greatest $F_{v3}/g = .140 \times 10^{-6}$ for moon, or $.054 \times 10^{-6}$ for sun (29)

Greatest $F_{a3}/g = .105 \times 10^{-6}$ for moon, or $.041 \times 10^{-6}$ for sun (30)

Comparing the above with (18) and (19), it will be noted that the maximum values of the lunar components involving the cube of the moon's parallax are only slightly less than the corresponding maximum values for the entire lunar force, while for the solar components the differences are too small to be shown with the number of decimal places used.

### VERTICAL COMPONENT OF FORCE

**50.** It is now proposed to expand into a series of harmonic terms formula (23) which represents the principal vertical component of the lunar tide-producing force. In figure 3 let $O$ represent the center of the earth and let projections on the celestial sphere be as follows:

$C$, the north pole
$IM'P'$, the earth's equator
$IM$, the moon's orbit
$M$, the position of the moon
$P$, the place of observation
$CMM'$, the hour circle of the moon
$CPP'$, the meridian of place of observation
$I$, the intersection of moon's orbit and equator

Also let

$I = $ angle $MIM' = $ inclination of moon's orbit to earth's equator
$t = $ arc $P'M'$ or angle $PCM = $ hour angle of moon
$X = IP' = $ longitude of $P$ measured in celestial equator from intersection $I$
$j = IM = $ longitude of moon in orbit reckoned from intersection $I$
$z = PM = $ zenith distance of moon
$D = M'M = $ declination of moon
$Y = P'P = $ latitude of $P$

The solution of a number of the spherical triangles represented in figure 3 will provide certain relations needed in the development of the formulas for the tide-producing force.

**51.** In spherical triangle $MCP$, the angle $C$ equals $t$ and the sides $MC$ and $PC$ are the complements of $D$ and $Y$, respectively. We may therefore write

$$\cos z = \sin Y \sin D + \cos Y \cos D \cos t \tag{31}$$

Substituting this value in formula (23), we obtain

$$\begin{aligned}
F_{v3}/g =\; & 3/2\ (M/E)(a/d)^3(1/2-3/2\sin^2 Y)(2/3-2\sin^2 D) \quad\text{---}\quad F_{v30}/g \\
& +3/2\ (M/E)(a/d)^3 \sin 2Y \sin 2D \cos t \quad\text{---}\quad F_{v31}/g \\
& +3/2\ (M/E)(a/d)^3 \cos^2 Y \cos^2 D \cos 2t \quad\text{---}\quad F_{v32}/g
\end{aligned} \tag{32}$$

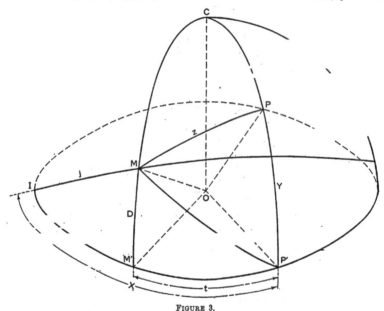

FIGURE 3.

**52.** In formula (32) the vertical component of the tide-producing force has been separated into three parts. The first term is independent of the rotation of the earth but is subject to variations arising from changes in declination and distance of the moon. It includes what are known as the *long-period constituents*, that is to say, constituents with periods somewhat longer than a day and in general a half month or longer. The second term involves the cosine of the hour angle ($t$) of the moon and this includes the *diurnal constituents* with periods approximating the lunar day. The last term involves the cosine of twice the hour angle of the moon and includes the *semidiurnal constituents* with periods approximating the half lunar day. The grouping of the tidal constituents according to their approximate periods affords an important classification in the further development of the tidal forces and these groups will be called *classes* or *species*. Symbols pertaining to a particular species are often identified by a subscript indicating the number of periods in a day, the subscript o being used for the long-period constituents. In formula (32) the individual terms are identified by the annexation of the species subscript to the general symbol for the formula.

**53.** As written, all of the three terms of formula (32) have the same coefficient $3/2$ $(M/E)$ $(a/d)^3$. In each case the latitude $(Y)$ factor has a maximum value of unity, this maximum being negative for the first term. For the long-period term $(F_{v30}/g)$, the latitude factor has a maximum positive value of $\frac{1}{2}$ at the equator, becomes zero in latitude 35.26° (approximately), and reaches a maximum negative value of $-1$ at the poles, the factor being the same for corresponding latitudes in both northern and southern hemispheres. For the diurnal term $(F_{v31}/g)$, the latitude factor is positive for the northern hemisphere and negative for the southern hemisphere. It has a maximum value of unity in latitude 45° and is zero at the equator and poles. For the semidiurnal terms $(F_{v32}/g)$, the latitude factor is always positive and has a maximum value of unity at the equator and equals zero at the poles.

**54.** For extreme values attainable for the declinational $(D)$ factors, consideration must be given to the greatest declination which can be reached by the tide-producing body. The periodic maximum declination reached by the moon in its 18.6 year node-cycle is 28.6° but this may be slightly increased by other inequalities in the moon's motion. The maximum declination for the sun, taken the same as the obliquity of the ecliptic, is 23.45°. The declination factor of the long-period term $(F_{v30}/g)$ has a maximum value of $2/3$ when the declination is zero. It diminishes with increasing north or south declination but must always remain positive because of the limits of the declination. For the diurnal term $(F_{v31}/g)$ the declinational factor has its greatest value when the declination is greatest. For the moon the maximum value of this factor is approximately 0.841 and for the sun 0.730. This factor is positive for the northern hemisphere and negative for the southern hemisphere. For the semidiurnal term $(F_{v32}/g)$ the declinational factor for both moon and sun is always positive and has a maximum value of unity at zero declination.

**55.** The greatest numerical values for the several terms of the vertical component of the tide-producing force as represented by formula (32) and applicable to the time when the moon and sun are nearest the earth, are as follows:

Greatest $F_{v30}/g = -.070 \times 10^{-6}$ for moon, or $-.027 \times 10^{-6}$ for sun (33)
Greatest $F_{v31}/g = \pm.088 \times 10^{-6}$ for moon, or $\pm.030 \times 10^{-6}$ for sun (34)
Greatest $F_{v32}/g = +.105 \times 10^{-6}$ for moon, or $+.041 \times 10^{-6}$ for sun (35)

For the long-period term (33) the greatest value applies to either pole and is negative. For the diurnal term (34) the greatest value applies in latitude 45° and may be positive or negative according to whether the latitude and declinational factors have the same or opposite signs. For the semidiurnal term (35) the greatest value applies to the equator and is positive.

**56.** Referring to formula (32), let $a/c$ equal the mean value of parallax $a/d$. Then $a/d$ may be replaced by its equivalent $(a/c)(c/d)$, in which the fraction $c/d$ expresses the relation between the true and the mean parallax. Also let $U=(M/E)(a/c)^3$, the numerical value of which will be found in table 1. Expressing separately the three terms of formula (32), we then have

$$F_{v30}/g = 3/2\ U\ (c/d)^3\ (1/2 - 3/2\sin^2 Y)(2/3 - 2\sin^2 D) \tag{36}$$
$$F_{v31}/g = 3/2\ U\ (c/d)^3\ \sin 2Y \sin 2D \cos t \tag{37}$$
$$F_{v32}/g = 3/2\ U\ (c/d)^3\ \cos^2 Y \cos^2 D \cos 2t \tag{38}$$

**57.** Referring to figure 3, the following relations may be obtained from the right spherical triangles $MIM'$ and $MP'M'$ and the oblique spherical triangle $MP'I$:

$$\sin D = \sin I \sin j \tag{39}$$
$$\cos D \cos t = \cos MP' \tag{40}$$
$$\cos MP' = \cos X \cos j + \sin X \sin j \cos I \tag{41}$$
$$\cos D \cos t = \cos X \cos j + \sin X \sin j \cos I$$
$$= \cos^2 \tfrac{1}{2}I \cos (X-j) + \sin^2 \tfrac{1}{2}I \cos (X+j) \tag{42}$$

**58.** Replacing the functions of $D$ and $t$ in formulas (36) to (38) by their equivalents derived from equations (39) and (42), there are obtained the following:

$$F_{v30}/g = 3/2\ U(c/d)^3(1/2 - 3/2\sin^2 Y) \times$$
$$[2/3 - \sin^2 I + \sin^2 I \cos 2j] \tag{43}$$
$$F_{v31}/g = 3/2\ U(c/d)^3 \sin 2Y \times$$
$$[\sin I \cos^2 \tfrac{1}{2}I \cos (X + 90° - 2j)$$
$$+ 1/2 \sin 2I \cos (X - 90°)$$
$$+ \sin I \sin^2 \tfrac{1}{2}I \cos (X - 90° + 2j)] \tag{44}$$
$$F_{v32}/g = 3/2\ U(c/d)^3 \cos^2 Y \times$$
$$[\cos^4 \tfrac{1}{2}I \cos (2X - 2j)$$
$$+ 1/2 \sin^2 I \cos 2X$$
$$+ \sin^4 \tfrac{1}{2}I \cos (2X + 2j)] \tag{45}$$

The above formulas involve the moon's actual distance $d$ and its true longitude $j$ as measured in its orbit from the intersection. While these are functions of time, they do not vary uniformly because of certain inequalities in the motion of the moon, and it is now desired to replace these quantities by elements that do change uniformly.

**59.** Referring to paragraphs 23–24 and to figure 1, it will be noted that longitude measured from intersection $A$ in the moon's orbit equals the longitude measured from the referred equinox $\Upsilon'$ less arc $\xi$, and longitude measured from intersection $A$ in the celestial equator equals the longitude measured from the equinox $\Upsilon$ less arc $\nu$.
Now let

$s'$ = true longitude of moon in orbit referred to equinox
$s$ = mean longitude of moon referred to equinox
$k$ = difference $(s' - s)$

Then
$$j = s' - \xi = s - \xi + k \tag{46}$$

**60.** In figure 4 let $S'$ and $P'$ be the points where the hour circles of the mean sun and place of observation intersect the celestial equator, $\Upsilon$ the vernal equinox, and $I$ the lunar intersection. Then $X$ will equal the arc $P'I$ and $\nu$ the arc $I\Upsilon$. Now let

$h$ = mean longitude of sun
$T$ = hour angle of mean sun

Then
$$X = T + h - \nu \tag{47}$$

**61.** Substituting the values of $j$ and $X$ from (46) and (47) in formulas (43) to (45), these may be written

$$F_{v30}/g = 3/2\ U(1/2 - 3/2\ \sin^2 Y) \times$$
$$[(c/d)^3(2/3 - \sin^2 I)$$
$$+ (c/d)^3 \sin^2 I \cos(2s - 2\xi + 2k)] \tag{48}$$

$$F_{v31}/g = 3/2\ U \sin 2Y \times$$
$$[(c/d)^3 \sin I \cos^2 \tfrac{1}{2}I \cos(T - 2s + h + 2\xi - \nu + 90° - 2k)$$
$$+ 1/2\ (c/d)^3 \sin 2I \cos(T + h - \nu - 90°)$$
$$+ (c/d)^3 \sin I \sin^2 \tfrac{1}{2}I \cos(T + 2s + h - 2\xi - \nu - 90° + 2k)] \tag{49}$$

$$F_{v32}/g = 3/2\ U \cos^2 Y \times$$
$$[(c/d)^3 \cos^4 \tfrac{1}{2}I \cos(2T - 2s + 2h + 2\xi - 2\nu - 2k)$$
$$+ 1/2\ (c/d)^3 \sin^2 I \cos(2T + 2h - 2\nu)$$
$$+ (c/d)^3 \sin^4 \tfrac{1}{2}I \cos(2T + 2s + 2h - 2\xi - 2\nu + 2k) \tag{50}$$

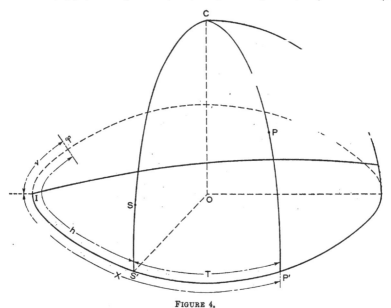

FIGURE 4.

Disregarding at this time the slow change in the function of $I$, the variable part of each term of the above formulas may be expressed in one of the following forms—$(c/d)^3$, $(c/d)^3 \cos A$, $(c/d)^3 \cos(A + 2k)$, or $(c/d)^3 \cos(A - 2k)$, in which $A$ includes all the elements of the variable angular function excepting the multiple of $k$.

**62.** The following equations for the motion of the moon were adapted from Godfrey's Elementary Treatise on the Lunar Theory:

$s'$ = true longitude of moon (in radians)
$= s$ _____ (mean longitude)
$+ 2e \sin(s - p) + 5/4\ e^2 \sin 2(s - p)$ _____ (elliptic inequality)
$+ 15/4\ me \sin(s - 2h + p)$ _____ (evectional inequality)
$+ 11/8\ m^2 \sin 2(s - h)$ _____ (variational inequality) (51)

$c/d =$ (true parallax of moon)/(mean parallax of moon)
$= $ unity
$+ e \cos (s-p) + e^2 \cos 2(s-p)$ ---------- (elliptic inequality)
$+ 15/8\ me \cos (s-2h+p)$ ---------- (evectional inequality)
$+ m^2 \cos 2(s-h)$ ------------ (variational inequality)   (52)

in which

$s' = $ true longitude of moon in orbit (referred to equinox)
$s = $ mean longitude of moon
$h = $ mean longitude of sun
$p = $ mean longitude of lunar perigee
$e = $ eccentricity of moon's orbit $= 0.0549$
$m = $ ratio of mean motion of sun to that of moon $= 0.0748$

The elements $e$ and $m$ are small fractions of the first order and the square of either or the product of both may be considered as being of the second order. In the following development the higher powers of these elements will be omitted.

**63.** Since $k$ has been taken as the difference between the true and the mean longitude of the moon, we may obtain from (51)

$$k = 2e \sin (s-p) + 5/4\ e^2 \sin 2(s-p) \\ + 15/4\ me \sin (s-2h+p) + 11/8\ m^2 \sin 2(s-h) \quad (53)$$

The value of $k$ is always small, its maximum value being about 0.137 radian. It may therefore be assumed without material error that the sine of $k$ or the sine of $2k$ is equal to the angle itself. Then

$$\sin 2k = 2k = 4e \sin (s-p) + 5/2\ e^2 \sin 2(s-p) \\ + 15/2\ me \sin (s-2h+p) + 11/4\ m^2 \sin 2(s-h) \quad (54)$$

$$\cos 2k = 1 - 2 \sin^2 k = 1 - 2k^2 \\ = 1 - 4e^2 + 4e^2 \cos 2(s-p) \quad (55)$$

terms smaller than those of the second order being omitted.

**64.** Cubing (52) and neglecting the smaller terms, we obtain

$$(c/d)^3 = 1 + 3/2\ e^2 + 3e \cos (s-p) + 9/2\ e^2 \cos 2(s-p) \\ + 45/8\ me \cos (s-2h+p) + 3\ m^2 \cos 2(s-h) \quad (56)$$

Multiplying (54) and (55) by (56)

$$(c/d)^3 \sin 2k = 4e \sin (s-p) + 17/2\ e^2 \sin 2(s-p) \\ + 15/2\ me \sin (s-2h+p) + 11/4\ m^2 \sin 2(s-h) \quad (57)$$

$$(c/d)^3 \cos 2k = 1 - 5/2\ e^2 + 3\ e \cos (s-p) + 17/2\ e^2 \cos 2(s-p) \\ + 45/8\ me \cos (s-2h+p) + 3\ m^2 \cos 2(s-h) \quad (58)$$

**65.** From (56), (57), and (58), we may obtain the following general expressions applicable to the further development of formulas (48) to (50). Negative coefficients have been avoided by the introduction of 180° in the angle when necessary.

$$(c/d)^3 \cos (A - 2k) = (c/d)^3 \cos 2k \cos A + (c/d)^3 \sin 2k \sin A \\ = (1 - 5/2\ e^2) \cos A \\ + 7/2\ e \cos (A-s+p) + 1/2 e \cos (A+s-p+180°) \\ + 17/2\ e^2 \cos (A-2s+2p) \\ + 105/16\ me \cos (A-s+2h-p) + 15/16\ me \cos (A+s-2h+p+180°) \\ + 23/8\ m^2 \cos (A-2s+2h) + 1/8\ m^2 \cos (A+2s-2h) \quad (59)$$

$(c/d)^3 \cos A = (1+3/2\ e^2) \cos A$
$+3/2\ e \cos (A-s+p)+3/2\ e \cos (A+s-p)$
$+9/4\ e^2 \cos (A-2s+2p)+9/4\ e^2 \cos (A+2s-2p)$
$+45/16\ me \cos (A-s+2h-p)+45/16\ me \cos (A+s-2h+p)$
$+3/2\ m^2 \cos (A-2s+2h)+3/2\ m^2 \cos (A+2s-2h)$ (60)

$(c/d)^3 \cos (A+2k) = (c/d)^3 \cos 2k \cos A - (c/d)^3 \sin 2k \sin A$
$= (1-5/2\ e^2) \cos A$
$+7/2\ e \cos (A+s-p)+1/2\ e \cos (A-s+p+180°)$
$+17/2\ e^2 \cos (A+2s-2p)$
$+105/16\ me \cos (A+s-2h+p)+15/16\ me \cos (A-s+2h-p+180°)$
$+23/8\ m^2 \cos (A+2s-2h)+1/8\ m^2 \cos (A-2s+2h)$ (61)

**66.** After suitable substitutions for $A$ have been made in the three preceding equations they are immediately applicable to the final expansion of the several terms in formulas (48) to (50), excepting the first term of (48) for which formula (56) may be used directly. Each term in the expanded formulas given below represents a constituent of the lunar tide-producing force and for convenience of reference is designated by the letter $A$ with a subscript. There are also given the generally recognized symbols for the principal constituents, and when such a symbol is enclosed in brackets it signifies that the term given only partially represents the constituent so named.

**67.** Formula for long-period constituents of vertical component of principal lunar tide-producing force:

$F_{v30}/g = 3/2\ U(1/2 - 3/2 \sin^2 Y) \times$

($A_1$)  [$(2/3 - \sin^2 I)\{(1+3/2\ e^2)$ _____ permanent term
($A_2$)  $+3\ e \cos (s-p)$ _____ Mm
($A_3$)  $+9/2\ e^2 \cos (2s-2p)$
($A_4$)  $+45/8\ me \cos (s-2h+p)$
($A_5$)  $+3\ m^2 \cos (2s-2h)\}$ _____ MSf
($A_6$)  $+ \sin^2 I\{(1-5/2\ e^2) \cos (2s-2\xi)$ _____ Mf
($A_7$)  $+7/2\ e \cos (3s-p-2\xi)$
($A_8$)  $+1/2\ e \cos (s+p+180°-2\xi)$
($A_9$)  $+17/2\ e^2 \cos (4s-2p-2\xi)$
($A_{10}$) $+105/16\ me \cos (3s-2h+p-2\xi)$
($A_{11}$) $+15/16\ me \cos (s+2h-p+180°-2\xi)$
($A_{12}$) $+23/8\ m^2 \cos (4s-2h-2\xi)$
($A_{13}$) $+1/8\ m^2 \cos (2h-2\xi)\}]$ (62)

**68.** Formula for diurnal constituents of vertical component of principal lunar tide-producing force:

$F_{v31}/g = 3/2\ U \sin 2Y \times$

($A_{14}$) [$\sin I \cos^2 1/2 I$
  $\{(1-5/2\ e^2) \cos (T-2s+h+90°+2\xi-\nu)$ ____ $O_1$
($A_{15}$) $+7/2\ e \cos (T-3s+h+p+90°+2\xi-\nu)$ ____ $Q_1$
($A_{16}$) $+1/2\ e \cos (T-s+h-p-90°+2\xi-\nu)$ _____ [$M_1$]
($A_{17}$) $+17/2\ e^2 \cos (T-4s+h+2p+90°+2\xi-\nu)$ __ $2Q_1$
($A_{18}$) $+105/16\ me \cos(T-3s+3h-p+90°+2\xi-\nu)$_ $\rho_1$
($A_{19}$) $+15/16\ me \cos (T-s-h+p-90°+2\xi-\nu)$
($A_{20}$) $+23/8\ m^2 \cos (T-4s+3h+90°+2\xi-\nu)$ ____ $\sigma_1$
($A_{21}$) $+1/8\ m^2 \cos (T-h+90°+2\xi-\nu)\}$

(Formula continued next page)

$$
\begin{aligned}
(A_{22})\quad &+\sin 2I\{(1/2+3/4\ e^2)\cos(T+h-90°-\nu)\text{------} &&[K_1]\\
(A_{23})\quad &+3/4\ e\cos(T-s+h+p-90°-\nu)\text{--------} &&[M_1]\\
(A_{24})\quad &+3/4\ e\cos(T+s+h-p-90°-\nu)\text{--------} &&J_1\\
(A_{25})\quad &+9/8\ e^2\cos(T-2s+h+2p-90°-\nu)\\
(A_{26})\quad &+9/8\ e^2\cos(T+2s+h-2p-90°-\nu)\\
(A_{27})\quad &+45/32\ me\cos(T-s+3h-p-90°-\nu)\text{-----} &&\chi_1\\
(A_{28})\quad &+45/32\ me\cos(T+s-h+p-90°-\nu)\text{-----} &&\theta_1\\
(A_{29})\quad &+3/4\ m^2\cos(T-2s+3h-90°-\nu)\text{--------} &&MP_1\\
(A_{30})\quad &+3/4\ m^2\cos(T+2s-h-90°-\nu)\}\text{--------} &&SO_1
\end{aligned}
$$

$$+\sin I\sin^2 \tfrac{1}{2}I$$

$$
\begin{aligned}
(A_{31})\quad &\{(1-5/2\ e^2)\cos(T+2s+h-90°-2\xi-\nu)\text{---} &&OO_1\\
(A_{32})\quad &+7/2\ e\cos(T+3s+h-p-90°-2\xi-\nu)\text{----} &&KQ_1\\
(A_{33})\quad &+1/2\ e\cos(T+s+h+p+90°-2\xi-\nu)\\
(A_{34})\quad &+17/2\ e^2\cos(T+4s+h-2p-90°-2\xi-\nu)\\
(A_{35})\quad &+105/16\ me\cos(T+3s-h+p-90°-2\xi-\nu)\\
(A_{36})\quad &+15/16\ me\cos(T+s+3h-p+90°-2\xi-\nu)\\
(A_{37})\quad &+23/8\ m^2\cos(T+4s-h-90°-2\xi-\nu)\\
(A_{38})\quad &+1/8\ m^2\cos(T+3h-90°-2\xi-\nu)\}] &&(63)
\end{aligned}
$$

**69.** Formula for semidiurnal constituents of vertical component of principal lunar tide-producing force:

$$F_{v32}/g = 3/2\ U\cos^2 Y \times$$

$$
\begin{aligned}
(A_{39})\quad &[\cos^4\tfrac{1}{2}I\{(1-5/2\ e^2)\cos(2T-2s+2h+2\xi-2\nu)\text{----} &&M_2\\
(A_{40})\quad &+7/2\ e\cos(2T-3s+2h+p+2\xi-2\nu)\text{-----} &&N_2\\
(A_{41})\quad &+1/2\ e\cos(2T-s+2h-p+180°+2\xi-2\nu)\text{-} &&[L_2]\\
(A_{42})\quad &+17/2\ e^2\cos(2T-4s+2h+2p+2\xi-2\nu)\text{---} &&2N_2\\
(A_{43})\quad &+105/16\ me\cos(2T-3s+4h-p+2\xi-2\nu)\text{-} &&\nu_2\\
(A_{44})\quad &+15/16\ me\cos(2T-s+p+180°+2\xi-2\nu)\text{-} &&\lambda_2\\
(A_{45})\quad &+23/8\ m^2\cos(2T-4s+4h+2\xi-2\nu)\text{-----} &&\mu_2\\
(A_{46})\quad &+1/8\ m^2\cos(2T+2\xi-2\nu)\}\\
(A_{47})\quad &+\sin^2 I\{(1/2+3/4\ e^2)\cos(2T+2h-2\nu)\text{-----------} &&[K_2]\\
(A_{48})\quad &+3/4\ e\cos(2T-s+2h+p-2\nu)\text{---------} &&[L_2]\\
(A_{49})\quad &+3/4\ e\cos(2T+s+2h-p-2\nu)\text{---------} &&KJ_2\\
(A_{50})\quad &+9/8\ e^2\cos(2T-2s+2h+2p-2\nu)\\
(A_{51})\quad &+9/8\ e^2\cos(2T+2s+2h-2p-2\nu)\\
(A_{52})\quad &+45/32\ me\cos(2T-s+4h-p-2\nu)\\
(A_{53})\quad &+45/32\ me\cos(2T+s+p-2\nu)\\
(A_{54})\quad &+3/4\ m^2\cos(2T-2s+4h-2\nu)\\
(A_{55})\quad &+3/4\ m^2\cos(2T+2s-2\nu)\}\\
(A_{56})\quad &+\sin^4\tfrac{1}{2}I\{(1-5/2\ e^2)\cos(2T+2s+2h-2\xi-2\nu)\\
(A_{57})\quad &+7/2\ e\cos(2T+3s+2h-p-2\xi-2\nu)\\
(A_{58})\quad &+1/2\ e\cos(2T+s+2h+p+180°-2\xi-2\nu)\\
(A_{59})\quad &+17/2\ e^2\cos(2T+4s+2h-2p-2\xi-2\nu)\\
(A_{60})\quad &+105/16\ me\cos(2T+3s+p-2\xi-2\nu)\\
(A_{61})\quad &+15/16\ me\cos(2T+s+4h-p+180°-2\xi-2\nu)\\
(A_{62})\quad &+23/8\ m^2\cos(2T+4s-2\xi-2\nu)\\
(A_{63})\quad &+1/8\ m^2\cos(2T+4h-2\xi-2\nu)\}] &&(64)
\end{aligned}
$$

**70.** *Arguments.*—Except for the slow changes in the values of $I$, $\xi$, and $\nu$ which result from the revolution of the moon's node, each term other than the permanent one in the three preceding formulas is an harmonic function of an angle that changes uniformly with time. This angle is known as the *argument* of the constituent, also as the *equilibrium argument* when obtained in connection with the development of the equilibrium tide. By analogy, the argument of the permanent term may be considered as zero, the cosine of zero being unity.

**71.** The argument serves to identify the constituent by determining its speed and period and fixing the times of the maxima and minima of the corresponding tidal force. It usually consists of two parts represented by the symbols $V$ and $u$. When referring to a particular instant of time such as the beginning of a series of observations, the $V$ is written with a subscript as $V_0$. The first part of the argument includes any constant and multiples of one or more of the following astronomical elements—$T$, the hour angle of the mean sun at the place of observation; $s$, the mean longitude of the moon; $h$ the mean longitude of the sun; and $p$, the longitude of the lunar perigee. The second part $u$ includes multiples of one or both of the elements $\xi$ and $\nu$, which are functions of the longitude of the moon's node and vary slowly between small positive and negative limits throughout a 19-year cycle. In a series of observations covering a year or less they are treated as constants with values pertaining to the middle of the series. They do not affect the average speed or period of the constituent. Their values corresponding to each degree of $N$, the longitude of the moon's node, are included in table 6, formulas for their computation being given on p. 156.

**72.** The hourly speed of a constituent may be obtained by adding the hourly speeds of the elements included in the $V$ of the argument. These elementary speeds will be found in table 1. The period of a constituent is obtained by dividing 360° by its speed. The approximate period is determined by the element of greatest speed contained in the argument. Thus, the hour angle $T$ has a speed of 15° per mean solar hour and all constituents with a single $T$ in their arguments have periods approximating one day, while constituents with arguments containing the multiple $2T$ have periods approximating the half day. Next to $T$, the element of greatest speed is $s$ the mean longitude of the moon, and long-period constituents with a single $s$ in their arguments will have periods approximating the month and with any multiple of $s$ the corresponding fraction of a month. The arguments and speeds of the constituents are listed in table 2. Numerical values of the arguments for the beginning of each calendar year from 1850 to 2000 are given in table 15 for constituents used in the Coast and Geodetic Survey tide-predicting machine. Tables 16 to 18 provide differences for referring these arguments to any day and hour of the year.

**73.** In order to visualize the arguments of the constituents depending primarily upon the rotation of the earth, some have found it convenient to conceive of a system of fictitious stars, or "astres fictifs" as they are sometimes called, which move at a uniform rate in the celestial equator, each constituent being represented by a separate star. Thus, for the principal lunar constituent we have the mean moon and for the principal solar constituent the mean sun, while the various inequalities in the motions of these bodies are served by imaginary stars which reach the meridian of the place of observation at times corresponding to the zero value of the constituent argument. For the diurnal constituents the argument equals the hour angle of the star but for the semidiurnal constituents the argument is double the hour angle of the star.

**74.** *Coefficients.*—The complete coefficient of each term of formulas (62) to (64) includes several important factors. First, the *basic factor* $U$, which equals the ratio of the mass of the moon to that of the earth multiplied by the cube of the mean parallax of the moon, is common to all of the terms. This together with the common numerical coefficient may be designated as the *general coefficient*. Next, the function involving the latitude $Y$ is known as the *latitude factor*, each formula having a different latitude factor. Following the latitude factor is a function of $I$, the inclination of the moon's orbit to the plane of the earth's equator, which may appropriately be called the *obliquity factor*, each factor applying to a group of terms. Lastly, we have an individual term coefficient which includes a numerical factor and involves the quantity $e$ or $m$. Since these factors are derived from the equations of elliptic motion, they will here be referred to as *elliptic factors*. The product of the elliptic factor by the mean value of the obliquity factor is known as the *mean constituent coefficient* $(C)$. Numerical values for these coefficients are given in table 2. Since all terms in any one of the formulas have the same general coefficient and latitude factor, their relative magnitudes will be proportional to their constituent coefficients. Terms of different formulas, however, have different latitude factors and their constituent coefficients are not directly comparable without taking into account the latitude of the place of observation.

**75.** The obliquity factors are subject to variations throughout an 18.6-year cycle because of the revolution of the moon's node. During this period the value of $I$ varies between the limits of $\omega-i$ and $\omega+i$, or from 18.3° to 28.6° approximately, and the functions of $I$ change accordingly. In order that tidal data pertaining to different years may be made comparable, it is necessary to adopt certain standard mean values for the obliquity factors to which results for different years may be reduced. While there are several systems of means which would serve equally well as standard values, the system adopted by Darwin in the early development of the harmonic analysis of tides has the sanction of long usage and is therefore followed. By the Darwin method, the mean for the obliquity factor is obtained from the product of the obliquity factor and the cosine of the elements $\xi$ and $\nu$ appearing in the argument. This may be expressed as the mean value of the product $J \cos u$, in which $J$ is the function of $I$ in the coefficient and $u$ the function of $\xi$ and $\nu$ in the argument. Since $u$ is relatively small and its cosine differs little from unity, the resulting mean will not differ greatly from the mean of $J$ alone or from the function of $I$ when given its mean value.

**76.** Using Darwin's system as described in section 6 of his paper on the Harmonic Analysis of Tidal Observations published in volume I of his collection of Scientific Papers (also in Report of the British Association for the Advancement of Science in 1883), the following mean values are obtained for the obliquity factors in formulas (62) to (64). These values were used in the computation of the corresponding constituent coefficients in table 2. The subscript $_0$ is here used to indicate the mean value of the function.

For terms $A_1$ to $A_5$ in formula (62)
$$[2/3-\sin^2 I]_0=(2/3-\sin^2 \omega)(1-3/2 \sin^2 i)=0.5021 \qquad (65)$$
For terms $A_6$ to $A_{13}$ in formula (62)
$$[\sin^2 I \cos 2\xi]_0=\sin^2 \omega \cos^4 \tfrac{1}{2}i=0.1578 \qquad (66)$$

For terms $A_{14}$ to $A_{21}$ in formula (63)
$[\sin I \cos^2 \tfrac{1}{2}I \cos (2\xi-\nu)]_0 = \sin \omega \cos^2 \tfrac{1}{2}\omega \cos^4 \tfrac{1}{2}i = 0.3800$ (67)
For terms $A_{22}$ to $A_{30}$ in formula (63)
$[\sin 2I \cos \nu]_0 = \sin 2\omega (1-3/2 \sin^2 i) = 0.7214$ (68)
For terms $A_{31}$ to $A_{38}$ in formula (63)
$[\sin I \sin^2 \tfrac{1}{2}I \cos (2\xi+\nu)]_0 = \sin \omega \sin^2 \tfrac{1}{2}\omega \cos^4 \tfrac{1}{2}i = 0.0164$ (69)
For terms $A_{39}$ to $A_{46}$ in formula (64)
$[\cos^4 \tfrac{1}{2}I \cos (2\xi-2\nu)]_0 = \cos^4 \tfrac{1}{2}\omega \cos^4 \tfrac{1}{2}i = 0.9154$ (70)
For terms $A_{47}$ to $A_{55}$ in formula (64)
$[\sin^2 I \cos 2\nu]_0 = \sin^2 \omega (1-3/2 \sin^2 i) = 0.1565$ (71)
For terms $A_{56}$ to $A_{63}$ in formula (64)
$[\sin^4 \tfrac{1}{2}I \cos (2\xi+2\nu)]_0 = \sin^4 \tfrac{1}{2}\omega \cos^4 \tfrac{1}{2}i = 0.0017$ (72)

**77.** The ratio obtained by dividing the true obliquity factor for any value of $I$ by its mean value may be called a *node factor* since it is a function of the longitude of the moon's node. The symbol generally used for the node factor is the small $f$. The node factor may be used with a mean constituent coefficient to obtain the true coefficient corresponding to a given longitude of the moon's node. Node factors for the several terms of formulas (62) to (64) may be expressed by the following ratios:

$f(A_1)$ to $f(A_5) = f(\text{Mm}) = (2/3 - \sin^2 I)/0.5021$ (73)
$f(A_6)$ to $f(A_{13}) = f(\text{Mf}) = \sin^2 I / 0.1578$ (74)
$f(A_{14})$ to $f(A_{21}) = f(O_1) = \sin I \cos^2 \tfrac{1}{2}I / 0.3800$ (75)
$f(A_{22})$ to $f(A_{30}) = f(J_1) = \sin 2I / 0.7214$ (76)
$f(A_{31})$ to $f(A_{38}) = f(OO_1) = \sin I \sin^2 \tfrac{1}{2}I / 0.0164$ (77)
$f(A_{39})$ to $f(A_{46}) = f(M_2) = \cos^4 \tfrac{1}{2}I / 0.9154$ (78)
$f(A_{47})$ to $f(A_{55}) = \sin^2 I / 0.1565$ (79)
$f(A_{56})$ to $f(A_{63}) = \sin^4 \tfrac{1}{2} I / 0.0017$ (80)

Node factors for the middle of each calendar year from 1850 to 1999 are given in table 14 for the constituents used in the Coast and Geodetic Survey tide-predicting machine. These include all the factors above excepting formulas (79) and (80). However, since formula (79) represents an increase of only about one per cent over formula (74), the tabular values for the latter are readily adapted to formula (79). Node factors change slowly and interpolations can be made in table 14 for any desired part of the year. For practical purposes, however, the values for the middle of the year are generally taken as constant for the entire year.

**78.** The reciprocal of the node factor is called the *reduction factor* and is usually represented by the capital $F$. Applied to tidal coefficients pertaining to any particular year, the reduction factors serve to reduce them to a uniform standard in order that they may be comparable. Logarithms of the reduction factors for every tenth of a degree of $I$ are given in table 12 for the constituents used on the tide-predicting machine of this office.

**79.** Formulas (62), (63), and (64), for the long-period, diurnal, and semidiurnal constituents of the vertical component of the tide-producing force may now be summarized as follows:

Let $E$ = constituent argument from table 2
$C$ = mean constituent coefficient from table 2
$f$ = node factor from table 14

Then

$$F_{v30}/g = 3/2 \; U(1/2 - 3/2 \sin^2 Y) \; \Sigma fC \cos E \tag{81}$$
$$F_{v31}/g = 3/2 \; U \sin 2Y \; \Sigma fC \cos E \tag{82}$$
$$F_{v32}/g = 3/2 \; U \cos^2 Y \; \Sigma fC \cos E \tag{83}$$

Latitude factors for each degree of $Y$ are given in table 3. The column symbol in this table is $Y$ with annexed letter and digits corresponding to those in the designation of the tidal forces. Thus, $Y_{v30}$ represents the latitude factor to be used with force $F_{v30}$, its value being equal to the function $(1/2 - 3/2 \sin^2 Y)$. Taking the numerical value for the basic factor $U$ from table 1, the general coefficient $3/2 \; U$ is found to be $0.8373 \times 10^{-7}$.

### HORIZONTAL COMPONENTS OF FORCE

**80.** The horizontal component of the principal part of the tide-producing force as expressed by formula (25), page 14, is in the direction of the azimuth of the tide-producing body. This component may be further resolved into a north-and-south and an east-and-west direction. In the following discussion the south and west will be considered as the positive directions for these components. Now let

$F_{s3}/g$ = south component of principal tide-producing force
$F_{w3}/g$ = west component of principal tide-producing force
$A$ = azimuth of moon reckoned from the south through the west.

From formula (25), we then have

$$F_{s3}/g = 3/2 \; (M/E)(a/d)^3 \sin 2z \cos A \tag{84}$$
$$F_{w3}/g = 3/2 \; (M/E)(a/d)^3 \sin 2z \sin A \tag{85}$$

**81.** Referring to figure 3, page 16, the angle $P'PM$ equals $A$, the azimuth of the moon. Now, keeping in mind that the angle $MPC$ is the supplement of $A$, the angle $PCM$ equals $t$, and the arcs $MC$ and $PC$ are the respective complements of $D$ and $Y$, we may obtain from the spherical triangle $MPC$ the following relations:

$$\sin z \cos A = -\cos Y \sin D + \sin Y \cos D \cos t \tag{86}$$
$$\sin z \sin A = \cos D \sin t \tag{87}$$

Multiplying each of the above equations by the value of $\cos z$ from formula (31), the following equations may be derived:

$$\begin{aligned}
\sin 2z \cos A &= 2 \sin z \cos z \cos A \\
&= 3/4 \sin 2Y \; (2/3 - 2 \sin^2 D) \\
&\quad - \cos 2Y \sin 2D \cos t \\
&\quad + 1/2 \sin 2Y \cos^2 D \cos 2t
\end{aligned} \tag{88}$$

$$\begin{aligned}
\sin 2z \sin A &= 2 \sin z \cos z \sin A \\
&= \sin Y \sin 2D \sin t \\
&\quad + \cos Y \cos^2 D \sin 2t
\end{aligned} \tag{89}$$

**82.** Substituting in (84) and (85) the quantities from equations (88) and (89), we have

$$\begin{aligned}
F_{s3}/g = \;& 9/8 \; (M/E)(a/d)^3 \sin 2Y \; (2/3 - 2 \sin^2 D) \text{........} & F_{s30}/g \\
& -3/2 \; (M/E)(a/d)^3 \cos 2Y \sin 2D \cos t \text{..........} & F_{s31}/g \\
& +3/4 \; (M/E)(a/d)^3 \sin 2Y \cos^2 D \cos 2t \text{..........} & F_{s32}/g
\end{aligned} \tag{90}$$

$$\begin{aligned}
F_{w3}/g = \;& 3/2 \; (M/E)(a/d)^3 \sin Y \sin 2D \sin t \text{............} & F_{w31}/g \\
& +3/2 \; (M/E)(a/d)^3 \cos Y \cos^2 D \sin 2t \text{...........} & F_{w32}/g
\end{aligned} \tag{91}$$

The south component is expressed by three terms representing respectively the long-period, diurnal, and semidiurnal constituents. For the west component there are only two terms—the diurnal and semidiurnal, there being no long-period constituents in the west component. Each term has been marked separately by a symbol with annexed digits analogous to those used for the vertical component to indicate the class to which the term belongs.

**83.** Comparing formula (90) for the south component with formula (32) for the vertical component, it will be noted that the same functions of $D$ and $t$ are involved in the corresponding terms of both formulas, and that the terms differ only in their numerical coefficient and the latitude factor. Allowing for these differences, summarized formulas analogous to those given for the vertical component (page 26) may be readily formed. In order to eliminate the negative sign of the coefficient of the middle term, 180° will be applied to the arguments of that term. With all symbols as before, we then have

$$F_{s30}/g = 9/8 \ U \sin 2Y \ \Sigma \ fC \cos E \tag{92}$$
$$F_{s31}/g = 3/2 \ U \cos 2Y \ \Sigma \ fC \cos (E+180°) \tag{93}$$
$$F_{s32}/g = 3/4 \ U \sin 2Y \ \Sigma \ fC \cos E \tag{94}$$

**84.** Comparing the two terms in formula (91) for the west component with the corresponding terms in formula (32) for the vertical component, it will be noted that the $D$ functions are the same but that in (91) the sine replaces the cosine for the functions of $t$. It may be shown that the corresponding development of these terms will be the same as for the vertical component except that in the developed series each argument will be represented by its sine instead of cosine. In order that the summarized formulas may be expressed in cosine functions, 90° will be subtracted from each argument. With the same symbols as before and allowing for differences in the latitude factors, we obtain

$$F_{w31}/g = 3/2 \ U \sin Y \ \Sigma \ fC \cos (E-90°) \tag{95}$$
$$F_{w32}/g = 3/2 \ U \cos Y \ \Sigma \ fC \cos (E-90°) \tag{96}$$

**85.** Formulas for the horizontal component of tide-producing force in any given direction may be derived as follows: Let $A$ equal the azimuth (measured from south through west) of given direction, and let $F_{a30}/g$, $F_{a31}/g$, and $F_{a32}/g$, respectively, represent the long-period, diurnal, and semidiurnal terms of the component in this direction. Then

$$F_{a30}/g = F_{s30}/g \times \cos A \tag{97}$$
$$F_{a31}/g = F_{s31}/g \times \cos A + F_{w31}/g \times \sin A \tag{98}$$
$$F_{a32}/g = F_{s32}/g \times \cos A + F_{w32}/g \times \sin A \tag{99}$$

As the long-period term has no west component, the summarized formula for the azimuth $A$ may be derived by simply introducing the factor $\cos A$ into the coefficient of formula (92). For the diurnal and semidiurnal terms it is necessary to combine the resolved elements from the south and west components.

**86.** Referring to formulas (93) to (96) and considering a single constituent in each species we obtain the following:

Diurnal constituent,

$$3/2\ UfC\ [\cos 2Y \cos A \cos (E+180°) + \sin Y \sin A \cos (E-90°)]$$
$$= 3/2\ UfC\ (-\cos 2Y \cos A \cos E + \sin Y \sin A \sin E)$$
$$= 3/2\ UfC\ P_1 \cos (E-X_1) \qquad (100)$$

in which

$$P_1 = (\cos^2 2Y \cos^2 A + \sin^2 Y \sin^2 A)^{\frac{1}{2}} \qquad (101)$$

$$X_1 = \tan^{-1} \frac{\sin Y \sin A}{-\cos 2Y \cos A} \qquad (102)$$

Semidiurnal constituent,

$$3/2\ UfC\ [\sin Y \cos Y \cos A \cos E + \cos Y \sin A \cos (E-90°)]$$
$$= 3/2\ UfC\ \cos Y\ (\sin Y \cos A \cos E + \sin A \sin E)$$
$$= 3/2\ UfC\ P_2 \cos (E-X_2) \qquad (103)$$

in which

$$P_2 = \cos Y\ (\sin^2 Y \cos^2 A + \sin^2 A)^{\frac{1}{2}} \qquad (104)$$

$$X_2 = \tan^{-1} \frac{\sin A}{\sin Y \cos A} \qquad (105)$$

**87.** Summarized formulas for the horizontal component of the tide-producing force in any direction $A$ may now be written as follows:

$$F_{a30}/g = 9/8\ U \sin 2Y \cos A\ \Sigma fC \cos E \qquad (106)$$

$$F_{a31}/g = 3/2\ UP_1\ \Sigma fC \cos (E-X_1) \qquad (107)$$

$$F_{a32}/g = 3/2\ UP_2\ \Sigma fC \cos (E-X_2) \qquad (108)$$

the values for $P_1$, $P_2$, $X_1$ and $X_2$ being obtained by formulas in the preceding paragraph. $P_1$ and $P_2$ are to be taken as positive and the following table will be found convenient in determining the proper quadrant for $X_1$ and $X_2$.

| A quadrant | North latitude | | South latitude | |
|---|---|---|---|---|
| | $X_1$ quadrant | $X_2$ quadrant | $X_1$ quadrant | $X_2$ quadrant |
| 1 | 2 or 1 | 1 | 3 or 4 | 2 |
| 2 | 1 or 2 | 2 | 4 or 3 | 1 |
| 3 | 4 or 3 | 3 | 1 or 2 | 4 |
| 4 | 3 or 4 | 4 | 2 or 1 | 3 |

For the $X_1$ quadrant the first value of each pair is applicable when the latitude does not exceed 45° north or south. Otherwise the second value is applicable.

## EQUILIBRIUM TIDE

**88.** The *equilibrium theory* of the tides is a hypothesis under which it is assumed that the waters covering the face of the earth instantly respond to the tide-producing forces of the moon and the sun and form a surface of equilibrium under the action of these forces. The theory disregards friction and inertia and the irregular distribution of the land masses of the earth. Although the actual tidal movement

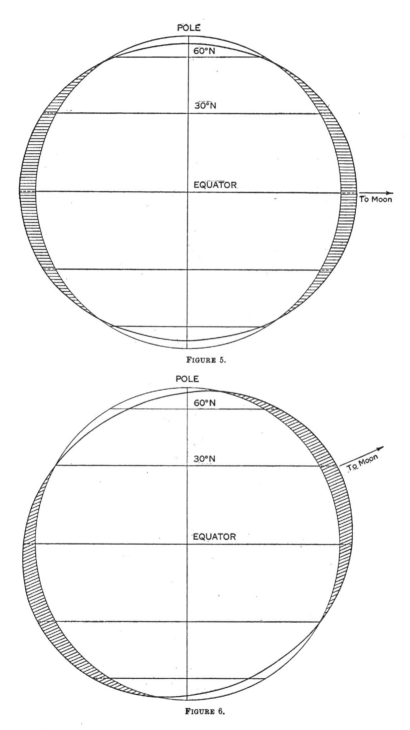

FIGURE 5.

FIGURE 6.

of nature does not even approximate to that which might be expected under the assumed conditions, the theory is of value as an aid in visualizing the distribution of the tidal forces over the surface of the earth. The theoretical tide formed under these conditions is known as the *equilibrium tide*, and sometimes as the *astronomical* or *gravitational* tide.

89. Under the equilibrium theory, the moon would tend to draw the earth into the shape of a prolate spheroid with the longest axis in line with the moon, thus producing one high water directly under the moon and another one on the opposite side of the earth with a low water belt extending entirely around the earth in a great circle midway between the high water points. It may be shown mathematically, however, that the total effect of the moon at its mean distance would be to raise the high water points about 14 inches above the mean surface of the earth and depress the low water belt about 7 inches below this surface, giving a maximum range of tide of about 21 inches. The corresponding range due to the sun is about 10 inches. Figures 5 and 6 illustrate on an exaggerated scale the theoretical disturbing effect of the moon on the earth. In the first figure the moon is assumed to be directly over the equator and in the last figure the moon is approximately at its greatest north declination.

90. With the moon over the equator (fig. 5), the range of the equilibrium tide will be at a maximum at the equator and diminish to zero at the poles and at any point there will be two high and low waters of equal range with each rotation of the earth. With the moon north or south of the equator (fig. 6), a declinational inequality is introduced and the two high and low waters of the day for any given latitude would no longer be equal except at the equator. This inequality would increase with the latitude and near the poles only one high and low water would occur with each rotation of the earth. Although latitude is an important factor in determining the range of the equilibrium tide, it is to be kept in mind that in the actual tide of nature the latitude of a place has no direct effect upon the rise and fall of the water.

91. A surface of equilibrium is a surface at every point of which the sum of the potentials of all the forces is a constant. On such a surface the resultant of all the forces at each point must be in the direction of the normal to the surface at that point. If the earth were a homogeneous mass with gravity as the only force acting, the surface of equilibrium would be that of a sphere. Each additional force will tend to disturb this spherical surface, and the total deformation will be represented by the sum of the disturbances of each of the forces acting separately. In the following investigation we need not be especially concerned with the more or less permanent deformation due to the centrifugal force of the earth's rotation, since we may assume that the disturbances of this spheriodal surface due to the tidal forces will not differ materially from the disturbances in a true spherical surface due to the same cause.

92. The potential at any point due to a force is the amount of work that would be required to move a unit of matter from that point, against the action of the force, to a position where the force is zero. This amount of work will be independent of the path along which the unit of matter is moved. If the force being considered is the gravity of the earth the potential at any point will be the amount

of work required to move a unit mass against the force of gravity from the point to an infinite distance from the earth's center. For the tide-producing force, the potential at any point will be measured by the amount of work necessary to move the unit of mass to the earth's center where this force is zero.

**93.** Referring to formula (21) for the vertical component of the tide-producing force, if the unit $g$ is replaced by the unit $\mu$ from equation (15), the formula may be written as follows:

$$F_v = \frac{3\mu M}{d^3}(\cos^2 z - 1/3)r + \frac{3\mu M}{2d^4}(5\cos^3 z - 3\cos z)r^2 \qquad (109)$$

**94.** Considering separately the tide-producing potential due to the two terms in the above formula, let the potential for the first term involving the cube of the moon's distance be represented by $V_3$ and the potential for the second term involving the 4th power of the moon's distance by $V_4$. In each case the work required to move a unit mass against the force through an infinitesimal distance $-dr$ toward the center of the earth is the product of the force by $-dr$, and the potential or total work required to move the particle to the center of the earth may be obtained by integrating between the limits $r$ and zero. Thus

$$V_3 = -\frac{3\mu M}{d^3}(\cos^2 z - 1/3)\int_r^0 r\, dr$$

$$= \frac{3\mu M}{2d^3}(\cos^2 z - 1/3)r^2 \qquad (110)$$

$$V_4 = -\frac{3\mu M}{2d^4}(5\cos^3 z - 3\cos z)\int_r^0 r^2\, dr$$

$$= \frac{\mu M}{2d^4}(5\cos^3 z - 3\cos z)r^3 \qquad (111)$$

**95.** At any instant of time the tide-producing potential at different points on the earth's surface will depend upon the zenith distance ($z$) of the moon and may be either positive or negative. It will now be shown that the average tide-producing potential for all points on the earth's surface, assuming it to be a sphere, is zero. Assume a series of right conical surfaces with common apex at center of earth and axis coinciding with the line joining centers of earth and moon, the angle between the generating line and the axis being $z$. These conical surfaces separated by infinitesimal angle $dz$ will cut the surface of the sphere into a series of equipotential rings, the surface area of any ring being equal to a $2\pi r^2 \sin z\, dz$. The average potential for the entire spherical surface may then be obtained by summing the products of the ring areas and corresponding potentials and dividing the sum by the total surface area of the sphere. Thus

$$\text{Average } V_3 = \frac{3\mu M r^2}{4d^3}\int_0^\pi (\cos^2 z - 1/3)\sin z\, dz$$

$$= \frac{3\mu M r^2}{4d^3}\left[-1/3\cos^3 z + 1/3\cos z\right]_0^\pi = 0 \qquad (112)$$

Average $V_4 = \dfrac{\mu M r^3}{4d^4} \displaystyle\int_0^\pi (5\cos^3 z - 3\cos z)\sin z\, dz$

$= \dfrac{\mu M r^3}{4d^4}\left[-5/4\cos^4 z + 3/2\cos^2 z\right]_0^\pi = 0 \qquad (113)$

**96.** Let $V_g$ represent the potential due to gravity at any point on the earth's surface. Since the force of gravity at any point on or above the earth's surface equals $\mu E/r^2$, the corresponding potential becomes

$$V_g = \mu E \int_r^\infty \frac{dr}{r^2} = \frac{\mu E}{r} \qquad (114)$$

If the earth is assumed to be a sphere with radius $a$, the gravitational potential at each point will equal $\mu E/a$, which may be taken as the average gravitational potential over the surface of the earth.

**97.** For a surface of equilibrium under the combined action of gravity and that part of the tide-producing force involving the cube of the moon's distance the sum of the corresponding potentials must be a constant, and since the average tide-producing potential for the entire surface of the earth is zero (par. 95), the constant will be the average gravitational potential or $\mu E/a$. Then from (110) and (114) we have

$$V_3 + V_g = \frac{3\mu M}{2d^3}(\cos^2 z - 1/3)r^2 + \frac{\mu E}{r} = \frac{\mu E}{a} \qquad (115)$$

Transposing and omitting common factor $\mu$, we may obtain

$$\frac{(r-a)a^2}{r^3} = 3/2(M/E)(a/d)^3(\cos^2 z - 1/3) \qquad (116)$$

Let

$$r = a + h \qquad (117)$$

so that $h$ represents the height of the equilibrium surface as referred to the undisturbed spherical surface of an equivalent sphere. Then

$$\frac{(r-a)a^2}{r^3} = \frac{ha^2}{(a+h)^3} = h/a - 3(h/a)^2 + 6(h/a)^3 - \text{etc.} \qquad (118)$$

As fraction $h/a$ is very small, its greatest value being less than 0.000001, the powers above the first may be neglected. Substituting in (116) and writing $h$ with subscript $_3$ to identify it with the principal tide-producing force, we have

$$h_3/a = 3/2\,(M/E)(a/d)^3(\cos^2 z - 1/3) \qquad (119)$$

**98.** Similarly, for a surface of equilibrium under the combined action of gravity and the part of the tide-producing force involving the 4th power of the moon's distance, we have from (111) and (114)

$$V_4 + V_g = \frac{\mu M}{2d^4}(5\cos^3 z - 3\cos z)r^3 + \frac{\mu E}{r} = \frac{\mu E}{a} \qquad (120)$$

$$\frac{(r-a)a^3}{r^4} = 1/2\ (M/E)(a/d)^4(5\cos^3 z - 3\cos z) \tag{121}$$

Letting $r = a + h_4$ and expanding the first member of the above formula, it becomes equal to $h_4/a$ after the rejection of the higher powers of this small fraction. The formula may then be written

$$h_4/a = 1/2\ (M/E)(a/d)^4(5\cos^3 z - 3\cos z) \tag{122}$$

**99.** Formulas (119) and (122) involving the cube and 4th power of the moon's parallax, respectively, represent the equilibrium heights of the tide due to the corresponding forces, the heights being expressed in respect to the mean radius ($a$) of the earth as the unit. In deriving these formulas the centrifugal force of the earth's rotation was disregarded and the resulting heights represent the disturbances in a true spherical surface due to the action of the tide-producing force. It may be inferred that in a condition of equilibrium the tidal forces would produce like disturbances in the spheroidal surface of the earth and the $h$ of the formulas may therefore be taken as being referred to the earth's surface as defined by the mean level of the sea.

**100.** The extreme limits of the equilibrium tide, applicable to the time when the tide-producing body is nearest the earth, may be obtained by substituting the proper numerical values in formulas (119) and (122). They are given below for both moon and sun.

From formula (119) involving the cube of parallax—
Greatest rise　$=1.46$ feet for moon, or $0.57$ foot for sun　(123)
Lowest fall　　$=0.73$ foot for moon, or $0.28$ foot for sun　(124)
Extreme range $=2.19$ feet for moon, or $0.85$ foot for sun.　(125)

From formula (122) involving the 4th power of parallax—
Greatest rise　$=0.026$ foot for moon, or $0.000025$ foot for sun　(126)
Lowest fall　　$=0.026$ foot for moon, or $0.000025$ foot for sun　(127)
Extreme range $=0.052$ foot for moon, or $0.00005$ foot for sun.　(128)

**101.** A comparison of formulas (23) and (119), the first expressing the relation of the vertical component of the principal tide-producing force to the acceleration of gravity ($g$) and the other the relation of the height of the corresponding equilibrium tide to the mean radius ($a$) of the earth, will show that they are identical with the single exception that the coefficient of the height formula is one-half that of the force formula. Therefore the development of the force formula into a series of harmonic constituents is immediately applicable in obtaining similar expressions for the equilibrium height of the tide. Using a notation for the height terms corresponding to that used for the force terms, let $h_{30}/a$, $h_{31}/a$, and $h_{32}/a$ represent, respectively, the long-period, diurnal, and semidiurnal terms of the equilibrium tide involving the cube of the moon's parallax. Then referring to formulas (81) to (83) we may write

$$h_{30}/a = 3/4\ U(1/2 - 3/2 \sin^2 Y)\ \Sigma\ fC \cos E \tag{129}$$
$$h_{31}/a = 3/4\ U \sin 2Y\ \Sigma\ fC \cos E \tag{130}$$
$$h_{32}/a = 3/4\ U \cos^2 Y\ \Sigma\ fC \cos E \tag{131}$$

the symbols having the same significance as in the preceding discussion of the tidal forces.

## TERMS INVOLVING 4TH POWER OF MOON'S PARALLAX

**102.** Formulas (24) and (26) represent the vertical and horizontal components of the part of the tide-producing force involving the 4th power of the moon's parallax. This part of the force constitutes only about 2 percent of the total tide-producing force of the moon and for brevity will be called the *lesser* force to distinguish it from the principal or primary part involving the cube of the parallax. The vertical component $F_{v4}/g$ has its maximum value when $z$ equals zero and, if numerical values pertaining to the moon and sun when nearest the earth are substituted in formula (24), the extreme values for this component are found to be $0.37 \times 10^{-8}$ for the moon and $0.35 \times 10^{-11}$ for the sun. The horizontal component $F_{a4}/g$ has its greatest value when $z$ equals about $31.09°$ and the substitution of numerical values in formula (26) gives the extreme value of this component as $0.26 \times 10^{-8}$ for the moon or $0.24 \times 10^{-11}$ for the sun.

**103.** Substituting in (24) the value of $\cos z$ from (31), the vertical component of the lesser force is expanded into four terms as follows:

$$\begin{aligned}
F_{v4}/g = &\ 15/4\ (M/E)(a/d)^4 \sin Y (\cos^2 Y - 2/5) \sin D (5 \cos^2 D - 2) & F_{v40}/g \\
& + 45/8\ (M/E)(a/d)^4 \cos Y (\cos^2 Y - 4/5) \cos D (5 \cos^2 D - 4) \cos t & F_{v41}/g \\
& + 45/4\ (M/E)(a/d)^4 \sin Y \cos^2 Y \sin D \cos^2 D \cos 2t & F_{v42}/g \\
& + 15/8\ (M/E)(a/d)^4 \cos^3 Y \cos^3 D \cos 3t & F_{v43}/g
\end{aligned}$$
(132)

These four terms represent, respectively, long-period, diurnal, semidiurnal, and terdiurnal constituents, according to the multiple of the hour angle $t$ involved in the term. Each term is followed by a symbol which is analogous to those used in the development of the principal force.

**104.** Each term in formula (132) may be further expanded by means of the relations given in formulas (39) and (42). Expressing these terms separately we have—

$$\begin{aligned}
F_{v40}/g = &\ 15/4\ (M/E)(a/d)^4 \sin Y(\cos^2 Y - 2/5) \times \\
& [3(\sin I - 5/4 \sin^3 I) \cos(j - 90°) \\
& + 5/4 \sin^3 I \cos(3j - 90°)]
\end{aligned}$$
(133)

$$\begin{aligned}
F_{v41}/g = &\ 45/8\ (M/E)(a/d)^4 \cos Y (\cos^2 Y - 4/5) \times \\
& [5/4 \sin^2 I \cos^2 \tfrac{1}{2} I \cos(X - 3j) \\
& + (1 - 10 \sin^2 \tfrac{1}{2} I + 15 \sin^4 \tfrac{1}{2} I) \cos^2 \tfrac{1}{2} I \cos(X - j) \\
& + (1 - 10 \cos^2 \tfrac{1}{2} I + 15 \cos^4 \tfrac{1}{2} I) \sin^2 \tfrac{1}{2} I \cos(X + j) \\
& + 5/4 \sin^2 I \sin^2 \tfrac{1}{2} I \cos(X + 3j)]
\end{aligned}$$
(134)

$$\begin{aligned}
F_{v42}/g = &\ 45/8\ (M/E)(a/d)^4 \sin Y \cos^2 Y \times \\
& [\sin I \cos^4 \tfrac{1}{2} I \cos(2X - 3j + 90°) \\
& + 3(\cos^2 \tfrac{1}{2} I - 2/3) \sin I \cos^2 \tfrac{1}{2} I \cos(2X - j - 90°) \\
& + 3(\cos^2 \tfrac{1}{2} I - 1/3) \sin I \sin^2 \tfrac{1}{2} I \cos(2X + j - 90°) \\
& + \sin I \sin^4 \tfrac{1}{2} I \cos(2X + 3j - 90°)]
\end{aligned}$$
(135)

$$\begin{aligned}
F_{v43}/g = &\ 15/8\ (M/E)(a/d)^4 \cos^3 Y \times \\
& [\cos^6 \tfrac{1}{2} I \cos(3X - 3j) \\
& + 3 \cos^4 \tfrac{1}{2} I \sin^2 \tfrac{1}{2} I \cos(3X - j) \\
& + 3 \cos^2 \tfrac{1}{2} I \sin^4 \tfrac{1}{2} I \cos(3X + j) \\
& + \sin^6 \tfrac{1}{2} I \cos(3X + 3j)]
\end{aligned}$$
(136)

**105.** If the common factor $(a/d)^4$ in formulas (133) to (136) is replaced by its equivalent $(a/c)^4 \times (c/d)^4$, these formulas may be de-

HARMONIC ANALYSIS AND PREDICTION OF TIDES 35

veloped into numerous constituent terms by a method similar to that already described in the development of the principal lunar force (paragraphs 59–69). In the following development constituents of very small magnitude are omitted. Those given are numbered consecutively with the constituent terms of the principal lunar force.

$F_{v40}/g = 15/4 \ (M/E)(a/c)^4 \sin Y (\cos^2 Y - 2/5) \times$

($A_{64}$)     $[(\sin I - 5/4 \sin^3 I)\{3(1+2e^2) \cos (s-90°-\xi)$
($A_{65}$)     $+9e \cos (2s-p-90°-\xi)$
($A_{66}$)     $+3e \cos (p-90°-\xi)\}$
($A_{67}$)     $+\sin^3 I\{5/4(1-6e^2) \cos (3s-90°-3\xi)$
($A_{68}$)     $+25/4 \ e \cos (4s-p-90°-3\xi)\}]$     (137)

$F_{v41}/g = 45/8 \ (M/E)(a/c)^4 \cos Y (\cos^2 Y - 4/5) \times$

($A_{69}$)     $[\sin^2 I \cos^2 \tfrac{1}{2}I\{5/4(1-6e^2) \cos (T-3s+h+3\xi-\nu)$
($A_{70}$)     $+25/4 \ e \cos (T-4s+h+p+3\xi-\nu)\}$
    $+(1-10 \sin^2 \tfrac{1}{2}I+15 \sin^4 \tfrac{1}{2}I) \cos^2 \tfrac{1}{2}I$
($A_{71}$)     $\{(1+2e^2) \cos (T-s+h+\xi-\nu)$    --------- [M$_1$]
($A_{72}$)     $+3e \cos (T-2s+h+p+\xi-\nu)$
($A_{73}$)     $+e \cos (T+h-p+\xi-\nu)\}$
    $+(1-10 \cos^2 \tfrac{1}{2}I+15 \cos^4 \tfrac{1}{2}I) \sin^2 \tfrac{1}{2}I$
($A_{74}$)     $\{(1+2e^2) \cos (T+s+h-\xi-\nu)$
($A_{75}$)     $+3e \cos (T+2s+h-p-\xi-\nu)\}]$     (138)

$F_{v42}/g = 45/8 \ (M/E)(a/c)^4 \sin Y \cos^2 Y \times$

($A_{76}$)     $[\sin I \cos^4 \tfrac{1}{2}I\{(1-6e^2) \cos (2T-3s+2h+90°+3\xi-2\nu)$
($A_{77}$)     $+5e \cos (2T-4s+2h+p+90°+3\xi-2\nu)$
($A_{78}$)     $+e \cos (2T-2s+2h-p-90°+3\xi-2\nu)\}$
    $+(\cos^2 \tfrac{1}{2}I - 2/3) \sin I \cos^2 \tfrac{1}{2}I$
($A_{79}$)     $\{3(1+2e^2) \cos (2T-s+2h-90°+\xi-2\nu)$
($A_{80}$)     $+9e \cos (2T-2s+2h+p-90°+\xi-2\nu)\}$
    $+(\cos^2 \tfrac{1}{2}I - 1/3) \sin I \sin^2 \tfrac{1}{2}I$
($A_{81}$)     $\{3(1+2e^2) \cos (2T+s+2h-90°-\xi-2\nu)\}]$     (139)

$F_{v43}/g = 15/8 \ (M/E)(a/c)^4 \cos^3 Y \times$

($A_{82}$)     $[\cos^6 \tfrac{1}{2}I\{(1-6e^2) \cos (3T-3s+3h+3\xi-3\nu)$ ----- M$_3$
($A_{83}$)     $+5e \cos (3T-4s+3h+p+3\xi-3\nu)$
($A_{84}$)     $+e \cos (3T-2s+3h-p+180°+3\xi-3\nu)$
($A_{85}$)     $+127/8 \ e^2 \cos (3T-5s+3h+2p+3\xi-3\nu)$
($A_{86}$)     $+75/8 \ me \cos (3T-4s+5h-p+3\xi-3\nu)\}$
    $+\cos^4 \tfrac{1}{2}I \sin^2 \tfrac{1}{2}I$
($A_{87}$)     $\{3(1+2e^2) \cos (3T-s+3h+\xi-3\nu)$
($A_{88}$)     $+9e \cos (3T-2s+3h+p+\xi-3\nu)\}]$     (140)

**106.** All of the constituent terms in formulas (137) to (140) are relatively unimportant but they are listed in table 1 because of their theoretical interest. The only one of these terms now used in the prediction of tides is ($A_{82}$) representing the constituent M$_3$ which has a speed exactly three-halves that of the principal lunar constituent M$_2$. Term ($A_{71}$) is of interest in having a speed exactly one-half that of M$_2$ and is sometimes called the true M$_1$ to distinguish it from the composite M$_1$ which is used in the prediction of tides and which will be described later.

**107.** For simplicity and the purposes of this publication, the mean values of the obliquity factors in the terms of the lesser tide-producing force will be taken as the values pertaining to the time when $I$ equals $\omega$ or 23.452°, excepting that for constituent $M_3$ and associated terms the mean has been obtained in accord with the system described in paragraph 75. The corresponding node factors (paragraph 77) may then be expressed by the following formulas in which the denominators are the accepted means of the obliquity factors:

$$f(A_{64}) \text{ to } f(A_{66}) = (\sin I - 5/4 \sin^3 I)/0.3192 \tag{141}$$

$$f(A_{67}) \text{ to } f(A_{68}) = \sin^3 I / 0.0630 \tag{142}$$

$$f(A_{69}) \text{ to } f(A_{70}) = \sin^2 I \cos^2\tfrac{1}{2} I / 0.1518 \tag{143}$$

$$f(A_{71}) \text{ to } f(A_{73}) = (1 - 10 \sin^2\tfrac{1}{2} I + 15 \sin^4\tfrac{1}{2} I) \cos^2\tfrac{1}{2} I / 0.5873 \tag{144}$$

$$f(A_{74}) \text{ to } f(A_{75}) = (1 - 10 \cos^2\tfrac{1}{2} I + 15 \cos^4\tfrac{1}{2} I) \sin^2\tfrac{1}{2} I / 0.2147 \tag{145}$$

$$f(A_{76}) \text{ to } f(A_{78}) = \sin I \cos^4\tfrac{1}{2} I / 0.3658 \tag{146}$$

$$f(A_{79}) \text{ to } f(A_{80}) = (\cos^2\tfrac{1}{2} I - 2/3) \sin I \cos^2\tfrac{1}{2} I / 0.1114 \tag{147}$$

$$f(A_{81}) = (\cos^2\tfrac{1}{2} I - 1/3) \sin I \sin^2\tfrac{1}{2} I / 0.0103 \tag{148}$$

$$f(A_{82}) \text{ to } f(A_{86}) = f(M_3) = \cos^6\tfrac{1}{2} I / 0.8758 \tag{149}$$

$$f(A_{87}) \text{ to } f(A_{88}) = \cos^4\tfrac{1}{2} I \sin^2\tfrac{1}{2} I / 0.0380 \tag{150}$$

Comparing formulas (149) and (78), it will be noted that the node factor for $M_3$ is equal to the node factor for $M_2$ raised to the 3/2 power. Computed values applicable to terms $A_{82}$ to $A_{86}$ are included in table 14 for years 1850 to 1999, inclusive.

**108.** For the tabulated constituent coefficients of the terms in formulas (137) to (140) there are included not only the elliptic and mean obliquity factors but also such other factors as may be necessary to permit the use of the general coefficient $(3/2\ U)$ of formulas (81) to (83) for the vertical component of the principal tide-producing force. The common coefficient $(M/E)(a/c)^4$ of formulas (137) to (140) is equal to $U$ multiplied by the parallax $a/c$, and the latter together with the necessary numerical factors is included in the constituent coefficients in table 2. Formulas (137) to (140) may then be summarized as follows:

$$F_{v40}/g = 3/2\ U \sin Y (\cos^2 Y - 2/5)\ \Sigma fC \cos E \tag{151}$$

$$F_{v41}/g = 3/2\ U \cos Y (\cos^2 Y - 4/5)\ \Sigma fC \cos E \tag{152}$$

$$F_{v42}/g = 3/2\ U \sin Y \cos^2 Y\ \Sigma fC \cos E \tag{153}$$

$$F_{v43}/g = 3/2\ U \cos^3 Y\ \Sigma fC \cos E \tag{154}$$

**109.** It is to be noted that in formulas (151), (152), and (153), the maximum value of the latitude factor in each is less than unity, being

0.4, 0.2754, and 0.3849, respectively, if the sign of the function is disregarded. In formula (154), as in the corresponding formulas for the principal tide-producing force, the maximum value of this factor is unity. In comparing the relative importance of the various constituents of the tide-producing force the latitude factor should be included with the mean coefficient. Attention is also called to the fact that the relative importance of the constituents involving the 4th power of the moon's parallax is greater in respect to the vertical component of the tide-producing force than in respect to the height of the equilibrium tide. In table 2 the mean coefficients are taken comparable in respect to the vertical component of the tide-producing force and the constituent coefficients pertaining to the lesser force are therefore 50 percent greater than they would be if taken comparable in respect to the equilibrium tide.

**110.** The south and west horizontal components of the lesser tide-producing force may be obtained by multiplying formula (26) by $\cos A$ and $\sin A$, respectively. Using the same system of notation as before, we then have

$$F_{s4}/g = 3/2 \ (M/E)(a/d)^4 \sin z \ (5 \cos^2 z - 1) \cos A \qquad (155)$$
$$F_{w4}/g = 3/2 \ (M/E)(a/d)^4 \sin z \ (5 \cos^2 z - 1) \sin A \qquad (156)$$

**111.** By means of the relations expressed in formulas (31), (86), and (87), the above component forces may be separated into long-period, diurnal, semidiurnal, and terdiurnal terms as follows:

South component,

$$F_{s40}/g = -15/4 \ (M/E)(a/d)^4 \cos Y(\cos^2 Y - 4/5) \sin D(5 \cos^2 D - 2) \quad (157)$$
$$F_{s41}/g = 45/8 \ (M/E)(a/d)^4 \sin Y(\cos^2 Y - 4/15) \cos D(5 \cos^2 D - 4) \cos t \quad (158)$$
$$F_{s42}/g = -45/4 \ (M/E)(a/d)^4 \cos Y(\cos^2 Y - 2/3) \sin D \cos^2 D \cos 2t \quad (159)$$
$$F_{s43}/g = 15/8 \ (M/E)(a/d)^4 \sin Y \cos^2 Y \cos^3 D \cos 3t \quad (160)$$

West component,

$$F_{w41}/g = 15/8 \ (M/E)(a/d)^4 (\cos^2 Y - 4/5) \cos D(5 \cos^2 D - 4) \sin t \quad (161)$$
$$F_{w42}/g = 15/4 \ (M/E)(a/d)^4 \sin 2Y \sin D \cos^2 D \sin 2t \quad (162)$$
$$F_{w43}/g = 15/8 \ (M/E)(a/d)^4 \cos^2 Y \cos^3 D \sin 3t \quad (163)$$

**112.** Comparing formulas (157) to (160) for the south component force with the corresponding terms of (132) for the vertical component, it will be noted that they differ only in the latitude factors and in sign for two of the terms. With adjustments for these differences the summarized formulas (151) to (154) are directly applicable for expressing the corresponding terms in the south component. Thus

$$F_{s40}/g = 3/2 \ U \cos Y(\cos^2 Y - 4/5) \ \Sigma fC \cos(E + 180°) \quad (164)$$
$$F_{s41}/g = 3/2 \ U \sin Y(\cos^2 Y - 4/15) \ \Sigma fC \cos E \quad (165)$$
$$F_{s42}/g = 3/2 \ U \cos Y(\cos^2 Y - 2/3) \ \Sigma fC \cos(E + 180°) \quad (166)$$
$$F_{s43}/g = 3/2 \ U \sin Y \cos^2 Y \ \Sigma fC \cos E \quad (167)$$

**113.** For the west component there is no long-period term. Comparing (161) to (163) with the corresponding terms of (132), it will be noted that the $t$-functions are expressed as sines instead of cosines but they may be changed to the latter by subtracting 90° from each argument. With this change and allowing for differences in the latitude factors and numerical coefficients, the summarized formulas for the west component will be similar to those for the vertical component and may be written as follows:

$$F_{u41}/g = 1/2 \ U \ (\cos^2 Y - 4/5) \ \Sigma fC \cos (E - 90°) \tag{168}$$
$$F_{w42}/g = 1/2 \ U \sin 2Y \ \Sigma fC \cos (E - 90°) \tag{169}$$
$$F_{w43}/g = 3/2 \ U \cos^2 Y \ \Sigma fC \cos (E - 90°) \tag{170}$$

**114.** To obtain the horizontal component of the lesser force in any direction, the same procedure may be followed as was used for the principal tide-producing force (paragraphs 85 to 87). With the same system of notation we then have

$$F_{a40}/g = 3/2 \ U \cos Y (\cos^2 Y - 4/5) \cos A \ \Sigma fC \cos (E + 180°) \tag{171}$$
$$F_{a41}/g = 3/2 \ U \ P_1 \ \Sigma fC \cos (E - X_1) \tag{172}$$
$$F_{a42}/g = 3/2 \ U \ P_2 \ \Sigma fC \cos (E - X_2) \tag{173}$$
$$F_{a43}/g = 3/2 \ U \ P_3 \ \Sigma fC \cos (E - X_3) \tag{174}$$

in which

$$P_1 = [\sin^2 Y (\cos^2 Y - 4/15)^2 \cos^2 A + 1/9 (\cos^2 Y - 4/5)^2 \sin^2 A]^{1/2} \tag{175}$$
$$P_2 = \cos Y [(\cos^2 Y - 2/3)^2 \cos^2 A + 4/9 \sin^2 Y \sin^2 A]^{1/2} \tag{176}$$
$$P_3 = \cos^2 Y (\sin^2 Y \cos^2 A + \sin^2 A)^{1/2} \tag{177}$$
$$X_1 = \tan^{-1} \frac{(\cos^2 Y - 4/5) \sin A}{3 \sin Y (\cos^2 Y - 4/15) \cos A} \tag{178}$$
$$X_2 = \tan^{-1} \frac{2 \sin Y \sin A}{-3 (\cos^2 Y - 2/3) \cos A} \tag{179}$$
$$X_3 = \tan^{-1} \frac{\sin A}{\sin Y \cos A} \tag{180}$$

The proper quadrants for $X_1$, $X_2$, and $X_3$ will be determined by the signs of the numerators and denominators in the above expressions, these signs being respectively the same as for the sine and cosine of the corresponding angles.

**115.** Comparing formula (122) for the equilibrium height of the tide due to the lesser tide-producing force with formula (24) for the vertical component of the force, it will be noted that they are the same with the exception that the numerical coefficient of the former is one-third that of the latter. With this change, the summarized formulas (151) to (154) for the vertical force may be used to express the corresponding equilibrium heights. Following the same system of notation as before, we have

$$h_{40}/a = 1/2 \ U \sin Y (\cos^2 Y - 2/5) \ \Sigma fC \cos E \tag{181}$$
$$h_{41}/a = 1/2 \ U \cos Y (\cos^2 Y - 4/5) \ \Sigma fC \cos E \tag{182}$$
$$h_{42}/a = 1/2 \ U \sin Y \cos^2 Y \ \Sigma fC \cos E \tag{183}$$
$$h_{43}/a = 1/2 \ U \cos^3 Y \ \Sigma fC \cos E \tag{184}$$

It is to be noted that the equilibrium height of the tide due to the principal tide-producing force when measured by the mean radius of the earth as a unit is one-half as great as the corresponding vertical component force referred to the mean acceleration of gravity as a unit, while the equilibrium height due to the lesser tide producing force similarly expressed is only one-third as great as the corresponding force. In table 2, the coefficients ($C$) of the constituents derived from the lesser force are made comparable with the others in respect to the vertical component force rather than in respect to the equilibrium height.

### SOLAR TIDES

**116.** Since the tide-producing force of the sun is similar in action to that of the moon, the formulas derived for the latter are applicable, with suitable substitutions, to the solar forces. Referring to formulas (62), (63), and (64), let $U$ be replaced by $U_1$ representing the product $(S/E)(a/c_1)^3$ in which $S$ is the mass of the sun and $(a/c_1)$ its mean parallax. Also replace $e$ by $e_1$, the eccentricity of the earth's orbit; $I$ by $\omega$, the obliquity of the ecliptic; $s$ by $h$, the mean longitude of the sun; and $p$ by $p_1$, the longitude of the solar perigee. For the solar forces the arcs $\xi$ and $\nu$ become zero and all terms representing the evectional and variational inequalities are omitted.

**117.** Making the changes indicated the solar constituents are now expressed in the following formulas. Each term is marked for identification by the letter $B$ with the same subscript used for the corresponding term in the lunar tide. The usual constituent symbol is also given for the more important terms. Using the same system of notation as before,

Solar $F_{v30}/g = 3/2\ U_1\ (1/2 - 3/2\ \sin^2 Y) \times$

| | | |
|---|---|---|
| $(B_1)$ | $[(2/3 - \sin^2 \omega)\{(1 + 3/2\ e^2_1)$ ---------permanent term | |
| $(B_2)$ | $+3\ e_1 \cos(h - p_1)$ | |
| $(B_3)$ | $+9/2\ e^2_1 \cos(2h - 2p_1)\}$ | |
| $(B_6)$ | $+\sin^2 \omega\{(1 - 5/2\ e^2_1) \cos 2h$ ----------------- | Ssa |
| $(B_7)$ | $+7/2\ e_1 \cos(3h - p_1)$ | |
| $(B_8)$ | $+1/2\ e_1 \cos(h + p_1 + 180°)$ | |
| $(B_9)$ | $+17/2\ e^2_1 \cos(4h - 2p_1)\}]$ | (185) |

Solar $F_{v31}/g = 3/2\ U_1 \sin 2Y \times$

| | | |
|---|---|---|
| $(B_{14})$ | $[\sin \omega \cos^2 \tfrac{1}{2}\omega\{(1 - 5/2\ e^2_1) \cos(T - h + 90°)$ ----- | $P_1$ |
| $(B_{15})$ | $+7/2\ e_1 \cos(T - 2h + p_1 + 90°)$ --------- | $\pi_1$ |
| $(B_{16})$ | $+1/2\ e_1 \cos(T - p_1 - 90°)$ | |
| $(B_{17})$ | $+17/2\ e^2_1 \cos(T - 3h + 2p_1 + 90°)\}$ | |
| $(B_{22})$ | $+\sin 2\omega\{(1/2 + 3/4\ e^2_1) \cos(T + h - 90°)$ -------- | [$K_1$] |
| $(B_{23})$ | $+3/4\ e_1 \cos(T + p_1 - 90°)$ | |
| $(B_{24})$ | $+3/4\ e_1 \cos(T + 2h - p_1 - 90°)$ -------- | $\psi_1$ |
| $(B_{25})$ | $+9/8\ e^2_1 \cos(T - h + 2p_1 - 90°)$ | |
| $(B_{26})$ | $+9/8\ e^2_1 \cos(T + 3h - 2p_1 - 90°)\}$ | |
| $(B_{31})$ | $+\sin \omega \sin^2 \tfrac{1}{2}\omega\{(1 - 5/2\ e^2_1) \cos(T + 3h - 90°)$ --- | $\phi_1$ |
| $(B_{32})$ | $+7/2\ e_1 \cos(T + 4h - p_1 - 90°)$ | |
| $(B_{33})$ | $+1/2\ e_1 \cos(T + 2h + p_1 + 90°)$ | |
| $(B_{34})$ | $+17/2\ e^2_1 \cos(T + 5h - 2p_1 - 90°)\}]$ | (186) |

Solar $F_{v32}/g = 3/2\ U_1 \cos^2 Y \times$

| | | |
|---|---|---|
| $(B_{39})$ | $[\cos^4 \tfrac{1}{2}\omega\{(1 - 5/2\ e^2_1) \cos(2T)$ ----------------- | $S_2$ |
| $(B_{40})$ | $+7/2\ e_1 \cos(2T - h + p_1)$ --------------- | $T_2$ |
| $(B_{41})$ | $+1/2\ e_1 \cos(2T + h - p_1 + 180°)$ -------- | $R_2$ |
| $(B_{42})$ | $+17/2\ e^2_1 \cos(2T - 2h + 2p_1)\}$ | |
| $(B_{47})$ | $+\sin^2 \omega\{(1/2 + 3/4\ e^2_1) \cos(2T + 2h)$ ---------- | [$K_2$] |
| $(B_{48})$ | $+3/4\ e_1 \cos(2T + h + p_1)$ | |
| $(B_{49})$ | $+3/4\ e_1 \cos(2T + 3h - p_1)$ | |

(Formula continued on next page)

$(B_{50})$ $\qquad +9/8\ e^2_1 \cos(2T+2p_1)$
$(B_{51})$ $\qquad +9/8\ e^2_1 \cos(2T+4h-2p_1)\}$
$(B_{56})$ $\quad +\sin^4 \tfrac{1}{2}\omega\{(1-5/2\ e^2_1)\cos(2T+4h)$
$(B_{57})$ $\qquad +7/2\ e_1 \cos(2T+5h-p_1)$
$(B_{58})$ $\qquad +1/2\ e_1 \cos(2T+3h+p_1+180°)$
$(B_{59})$ $\qquad +17/2\ e^2_1 \cos(2T+6h-2p_1)\}]$ $\hfill (187)$

118. The general coefficient for the solar tide-producing force differs from that of the lunar force in the basic factor. From the fundamental data in table 1, the ratio of $U_1/U$ is found to be 0.4602. This ratio, which will be designated as the *solar factor* with symbol $S'$, represents the theoretical relation between the principal solar and lunar tide-producing forces. In computing the constituent coefficients of the solar terms for use in table 2, the solar factor was included in order that the same general coefficient may be applicable to both lunar and solar terms. All of the summarized formulas involving the coefficients and arguments of table 2 are therefore applicable to both lunar and solar constituents. For the solar constituents, however, the node factor $(f)$ is always unity since $\omega$, the obliquity of the ecliptic, may be considered as a constant.

119. By substituting solar elements in formulas (137) to (140) the corresponding solar constituents pertaining to the 4th power of the sun's parallax are readily obtained. Since the theoretical magnitude of the lesser solar tide-producing force is less than 0.00002 part of the total tide-producing force of moon and sun, it is usually disregarded altogether. However, certain interest is attached to three of the constituents which are considered in connection with shallow water and meteorological tides (p. 46). These are constituents Sa, $S_1$, and $S_3$, corresponding respectively to terms $A_{64}$, $A_{71}$, and $A_{82}$ of the lunar series. They are listed in table 2 with reference letter $B$ and corresponding subscripts. Sa has a speed one-half that of constituent Ssa represented by term $B_6$ of formula (185). Its theoretical argument as derived from term $A_{64}$ contains the constant 90°, but being considered as a meteorological rather than an astronomical constituent, this constant is omitted from the argument. Constituents $S_1$ and $S_3$ have speeds respectively one-half and three-halves that of the principal solar constituent $S_2$.

120. The arguments of a number of the solar constituents include the element $p_1$ which represents the longitude of the solar perigee. As this changes less than 2° in a century, it may be considered as practically constant for the entire century. Referring to table 4 it will be noted that $p_1$ changes from 281.22° in 1900 to 282.94° in 2000. The value of 282° may therefore be adopted without material error for all work relating to the present century. With $p_1$ taken as a constant, it will be found that a number of terms in table 2 have the same speeds and may therefore be expected to merge into single constituents. Thus, constituents receiving contributions from more than one term are as follows: Sa from terms $B_2$, $B_3$, and $B_{64}$; Ssa from terms $B_3$ and $B_6$; $P_1$ from terms $B_{14}$ and $B_{25}$; $S_1$ from terms $B_{16}$, $B_{23}$, and $B_{71}$; $\psi_1$ from terms $B_{24}$ and $B_{33}$; $\phi_1$ from terms $B_{26}$ and $B_{31}$; $S_2$ from terms $B_{39}$ and $B_{50}$; and $R_2$ from terms $B_{41}$ and $B_{48}$. A few other solar terms also merge.

### THE M₁ TIDE

**121.** The separation of constituents from each other by the process of the analysis depends upon the differences in their speeds. Constituents with nearly equal speeds are not readily separated unless the analysis covers a very long series of observations but they tend to merge and form a single composite constituent. In formula (63), terms $A_{16}$ and $A_{23}$ have nearly equal speeds, one being a little less and the other a little greater than one-half the speed of the principal lunar constituent $M_2$. These two terms are usually considered as a single constituent and represented by the symbol $M_1$. Neglecting for the present the general coefficient and common latitude factor, the two terms may be written as follows:

$$\text{term } A_{16} = 1/2 \; e \sin I \cos^2 \tfrac{1}{2} I \cos (T - s + h - p - 90° + 2\xi - \nu) \quad (188)$$

$$\text{term } A_{23} = 3/2 \; e \sin I \cos I \cos (T - s + h + p - 90° - \nu) \quad (189)$$

The latter term, having a coefficient nearly three times as great as that of the first term, will predominate and determine the speed and period of the composite tide while the first term introduces certain inequalities in the coefficient and argument.

**122.** For brevity, let $A$ and $B$ represent the respective coefficients of terms $A_{16}$ and $A_{23}$ and let

$$\theta = T - s + h + p - 90° - \nu \quad (190)$$

Also let $P$ equal the mean longitude of the lunar perigee reckoned from the lunar intersection. Then

$$p = P + \xi \quad (191)$$

We then have

$$\text{term } A_{16} = A \cos (\theta - 2P)$$
$$= A \cos 2P \cos \theta + A \sin 2P \sin \theta \quad (192)$$

$$\text{term } A_{23} = B \cos \theta \quad (193)$$

$$M_1 = A_{16} + A_{23} = (A \cos 2P + B) \cos \theta + A \sin 2P \sin \theta$$
$$= (A^2 + 2AB \cos 2P + B^2)^{\tfrac{1}{2}} \cos \left[ \theta - \tan^{-1} \frac{A \sin 2P}{A \cos 2P + B} \right]$$
$$= \frac{e \sin I \cos^2 \tfrac{1}{2} I}{Q_a} \cos (T - s + h + p - 90° - \nu - Q_u) \quad (194)$$

in which

$$1/Q_a = \left[ 1/4 + 3/2 \frac{\cos I}{\cos^2 \tfrac{1}{2} I} \cos 2P + 9/4 \frac{\cos^2 I}{\cos^4 \tfrac{1}{2} I} \right]^{\tfrac{1}{2}} \quad (195)$$

$$Q_u = \tan^{-1} \frac{\sin 2P}{3 \cos I / \cos^2 \tfrac{1}{2} I + \cos 2P} \quad (196)$$

If $I$ is given its mean value corresponding to $\omega$, formula (195) may be reduced to the form

$$1/Q_a = (2.310 + 1.435 \cos 2P)^{\tfrac{1}{2}} \quad (197)$$

Values of $\log Q_a$ for each degree of $P$ based upon formula (197) are given in table 9.

**123.** The period of the composite constituent $M_1$ is very nearly an exact multiple of the period of the principal lunar constituent $M_2$, and for this reason the summations which are necessary for the analysis of the latter may be conveniently adapted to the analysis of the former. With other symbols as before, let

$$\theta = T - s + h - 90° + \xi - \nu \tag{198}$$

Terms $A_{16}$ and $A_{23}$ may then be combined as follows:

$$\text{term } A_{16} = A \cos(\theta - P)$$
$$= A \cos P \cos \theta + A \sin P \sin \theta \tag{199}$$

$$\text{term } A_{23} = B \cos(\theta + P)$$
$$= B \cos P \cos \theta - B \sin P \sin \theta \tag{200}$$

$$M_1 = A_{16} + A_{23} = (A+B) \cos P \cos \theta + (A-B) \sin P \sin \theta$$
$$= (A^2 + 2AB \cos 2P + B^2)^{\frac{1}{2}} \cos\left[\theta - \tan^{-1}\left(\frac{A-B}{A+B} \tan P\right)\right]$$
$$= \frac{e \sin I \cos^2 \tfrac{1}{2} I}{Q_a} \cos(T - s + h - 90° + \xi - \nu + Q) \tag{201}$$

in which

$$Q = \tan^{-1}\left(\frac{5 \cos I - 1}{7 \cos I + 1} \tan P\right) \tag{202}$$

If $I$ is given its mean value corresponding to $\omega$, formula (202) may be reduced to the following form which was used for computing the values of $Q$ in table 10.

$$\tan Q = 0.483 \tan P \tag{203}$$

**124.** Formulas (194) and (201) are the same except in the method of representing the argument. The elements $+p - Q_u$ in the first formula are replaced by $+\xi + Q$ in the latter, but it may be shown from (196) and (202) that

$$Q_u + Q = P = p - \xi \tag{204}$$

$$p - Q_u = \xi + Q \tag{205}$$

The complete arguments are therefore equal but in formula (201) the uniformly varying element $p$ has been transferred from the $V$ of the argument and included in the value of $Q$ where it is treated as a constant for a series of observations being analyzed. The speed of the argument as determined by the remaining part of the $V$ is then exactly one-half that of the principal constituent $M_2$ and with this assumption the summations for the latter may be adapted to the analysis of the former. It is to be noted, however, that the $u$ in this case has a progressive forward change of nearly 41° each year. The true average speed of this constituent is determined by the $V$ of formula (194) which includes the element $p$.

**125.** The obliquity factor for the composite $M_1$ constituent may be expressed by the formula $\sin I \cos^2 \tfrac{1}{2} I \times 1/Q_a$. According to the work of Darwin (Scientific Papers by Sir George H. Darwin, vol. 1, p. 39) the mean value of this factor is represented by the product $\sin \omega \cos^2 \tfrac{1}{2}\omega \cos^4 \tfrac{1}{2} i \times \sqrt{2.307}$, which equals $0.3800 \times 1.52$, or $0.5776$. When deriving the node-factor formula for $M_1$, Darwin inadvertently omitted the factor $\sqrt{2.307}$ and obtained the approximate equivalent of the following:

$$j(M_1) = \frac{\sin I \cos^2 \tfrac{1}{2} I}{\sin \omega \cos^2 \tfrac{1}{2} \omega \cos^4 \tfrac{1}{2} i} \times 1/Q_a = \frac{\sin I \cos^2 \tfrac{1}{2} I}{0.3800} \times 1/Q_a \qquad (206)$$

Comparing the above with formula (75), it will be noted that

$$f(M_1) = f(O_1) \times 1/Q_a \qquad (207)$$

Factors pertaining to constituent $M_1$ in tables 13 and 14 are based upon the above formulas.

126. Because of the omission of the factor $\sqrt{2.307}$ from formula (206), the node factors for $M_1$ which have been in general use since this system of tidal reductions was adopted are about 50 percent greater than was originally intended, while the reciprocal reduction factors are correspondingly too small. This constituent is relatively unimportant and no practical difficulties have resulted from the omission. The $M_1$ amplitudes as reduced from the observational data are comparable among themselves but should be increased by 50 percent to be on the same basis as the amplitudes of other constituents. The predicted tides have not been affected in the least since the node factors and reduction factors are reciprocal and compensating. The theoretical mean coefficient for this constituent with the factor $\sqrt{2.307}$ included is 0.0317; but in order that this coefficient may be adapted for use with the tabular node factors when computing tidal forces or the equilibrium height of the tide, the coefficient 0.0209 with the factor $\sqrt{2.307}$ excluded should be used.

127. Although $M_1$ is one of the relatively unimportant constituents and the error in the node factor has caused no serious difficulties, it may be questionable whether it should be perpetuated. It is obvious, however, that any change in the present procedure would lead to much confusion unless undertaken by general agreement among all the principal organizations engaged in tidal work. By making any change applicable to the analysis of all series of observations beginning after a certain specified date it would be possible to interpret the results on the basis of the period covered by the observations without the necessity of revising all previously published amplitudes for this constituent.

### THE $L_2$ TIDE

128. The composite $L_2$ constituent is formed by combining terms $A_{41}$ and $A_{48}$ of formula (64). Neglecting the general coefficient and common latitude factor these terms may be written

$$\text{term } A_{41} = 1/2\ e \cos^4 \tfrac{1}{2} I \cos (2T - s + 2h - p + 180° + 2\xi - 2\nu) \qquad (208)$$

$$\text{term } A_{48} = 3/4\ e \sin^2 I \cos (2T - s + 2h + p - 2\nu) \qquad (209)$$

A reference to table 2 will show that the mean coefficient of the first term is about four times as great as that of the latter term. The first term will therefore predominate and determine the speed of the composite constituent.

**129.** With other symbols as before, let $A$ and $B$ represent the respective coefficients of the two terms and $\theta$ the argument of the first term. We then have

$$A_{41}=A \cos \theta \tag{210}$$

$$A_{48}=B \cos (\theta+2P-180°)=-B \cos (\theta+2P) \tag{211}$$

$$L_2=A_{41}+A_{48}=(A-B \cos 2P) \cos \theta + B \sin 2P \sin \theta$$

$$=(A^2-2AB \cos 2P + B^2)^{\frac{1}{2}} \cos \left[\theta - \tan^{-1}\frac{B \sin 2P}{A-B \cos 2P}\right]$$

$$=1/2 \; e \frac{\cos^4 \frac{1}{2}I}{R_a} \cos (2T-s+2h-p+180°+2\xi-2\nu-R) \tag{212}$$

in which

$$1/R_a=(1-12 \tan^2 \tfrac{1}{2}I \cos 2P + 36 \tan^4 \tfrac{1}{2}I)^{\frac{1}{2}} \tag{213}$$

$$R=\tan^{-1}\frac{\sin 2P}{1/6 \cot^2 \tfrac{1}{2}I - \cos 2P} \tag{214}$$

Values of log $R_a$ and $R$ computed from the above formulas are given in tables 7 and 8, respectively.

**130.** The obliquity factor for the composite $L_2$ constituent may be expressed by the formula $\cos^4 \tfrac{1}{2}I \times 1/R_a$. The mean value of $1/R_a$ is approximately unity, and in accord with the Darwinian system the mean for the entire obliquity factor is taken as the product $\cos^4 \tfrac{1}{2}\omega \cos^4 \tfrac{1}{2}i$, which equals 0.9154 and is the same as the mean value of the obliquity factor for the principal constituent $M_2$. Multiplying this by the elliptic factor $\tfrac{1}{2}e$ gives 0.0251 as the mean constituent coefficient.

**131.** The node factor formula for constituent $L_2$ based upon the above mean for the obliquity factor is as follows:

$$f(L_2)=\frac{\cos^4 \tfrac{1}{2}I}{0.9145} \times 1/R_a = f(M_2) \times 1/R_a \tag{215}$$

Node factors for constituent $L_2$ based upon the above formula are included in table 14 for the middle of each year from 1850 to 1999, inclusive. The logarithms of the reciprocal reduction factors covering the period 1900 to 2000 are contained in table 13.

### LUNISOLAR $K_1$ AND $K_2$ TIDES

**132.** Lunar diurnal term $A_{22}$ of formula (63) and solar diurnal term $B_{22}$ of formula (186) have the same speed. Together they form the lunisolar $K_1$ constituent. Also, lunar semidiurnal term $A_{47}$ of formula (64) and solar semidiurnal term $B_{47}$ of formula (187) have speeds exactly twice that of constituent $K_1$ and together form the lunisolar $K_2$ constituent. In order that the solar terms may have the same general coefficient as the lunar terms, the solar factor $U_1/U$, which will be designated by the symbol $S'$, will be transferred from the general coefficient of the solar terms and included in the constituent coefficients. Then, neglecting the general coefficient and latitude factors common to the terms combined, we have the following formulas in which numerical values from table 1 have been substituted for constant quantities.

term $A_{22} = (1/2 + 3/4 e^2) \sin 2I \cos (T+h-90°-\nu)$
$= 0.5023 \sin 2I \cos (T+h-90°-\nu)$ (216)

term $B_{22} = (1/2 + 3/4 e_1^2) S' \sin 2\omega \cos (T+h-90°)$
$= 0.1681 \cos (T+h-90°)$ (217)

term $A_{47} = (1/2 + 3/4 e^2) \sin^2 I \cos (2T+2h-2\nu)$
$= 0.5023 \sin^2 I \cos (2T+2h-2\nu)$ (218)

term $B_{47} = (1/2 + 3/4 e_1^2) S' \sin^2 \omega \cos (2T+2h)$
$= 0.0365 \cos (T+2h)$ (219)

**133.** Taking first the diurnal terms, let $A$ represent the lunar coefficient $0.5023 \sin 2I$ and let $B$ represent the solar coefficient $0.1681$. We then have

$A_{22} = A \cos (T+h-90°-\nu)$
$= A \cos \nu \cos (T+h-90°) + A \sin \nu \sin (T+h-90°)$ (220)

$B_{22} = B \cos (T+h-90°)$ (221)

$K_1 = (A \cos \nu + B) \cos (T+h-90°) + A \sin \nu \sin (T+h-90°)$
$= (A^2 + 2AB \cos \nu + B^2)^{1/2} \cos \left[ T+h-90° - \tan^{-1} \dfrac{A \sin \nu}{A \cos \nu + B} \right]$
$= C_1 \cos (T+h-90°-\nu')$ (222)

in which

$C_1 = (A^2 + 2AB \cos \nu + B^2)^{\frac{1}{2}}$
$= (0.2523 \sin^2 2I + 0.1689 \sin 2I \cos \nu + 0.0283)^{\frac{1}{2}}$ (223)

$\nu' = \tan^{-1} \dfrac{A \sin \nu}{A \cos \nu + B} = \tan^{-1} \dfrac{\sin 2I \sin \nu}{\sin 2I \cos \nu + 0.3347}$ (224)

Values of $\nu'$ for each degree of $N$, which is the longitude of the moon's node, are included in table 6.

**134.** The obliquity factor for $K_1$ will be taken to include the entire coefficient $(A^2 + 2AB \cos \nu + B^2)^{\frac{1}{2}}$ and its mean value will be taken as the mean of the product $(A^2 + 2AB \cos \nu + B^2)^{\frac{1}{2}} \cos \nu'$.

From (224) we may obtain

$\cos \nu' = (A \cos \nu + B)/(A^2 + 2AB \cos \nu + B^2)^{\frac{1}{2}}$ (225)

Then for mean value of coefficient of $K_1$

$[(A^2 + 2AB \cos \nu + B^2)^{\frac{1}{2}} \cos \nu']_0 = [A \cos \nu + B]_0$
$= [0.5023 \sin 2I \cos \nu + 0.1681]_0 = 0.5305$ (226)

the numerical mean for $\sin 2I \cos \nu$ being obtained from formula (68). For the node factor of $K_1$ divide the coefficient of (222) by its mean value and obtain

$f(K_1) = (0.2523 \sin^2 2I + 0.1689 \sin 2I \cos \nu + 0.0283)^{\frac{1}{2}}/0.5305$
$= (0.8965 \sin^2 2I + 0.6001 \sin 2I \cos \nu + 0.1006)^{\frac{1}{2}}$ (227)

The node factors for the middle of each year 1850 to 1999 are included in table 14. Logarithms of the reciprocal reduction factors for each tenth of a degree of $I$ are given in table 12.

**135.** The semidiurnal terms $A_{47}$ and $B_{47}$ may be combined in a similar manner. Letting $A$ represent the lunar coefficient 0.5023 $\sin^2 I$ and $B$ the solar coefficient 0.0365, we have

$$A_{47} = A \cos (2T+2h-2\nu)$$
$$= A \cos 2\nu \cos (2T+2h) + A \sin 2\nu \sin (2T+2h) \qquad (228)$$

$$B_{47} = B \cos (2T+2h) \qquad (229)$$

$$\begin{aligned} K_2 &= (A \cos 2\nu + B) \cos (2T+2h) + A \sin 2\nu \sin (2T+2h) \\ &= (A^2 + 2AB \cos 2\nu + B^2)^{\frac{1}{2}} \cos \left[ 2T+2h - \tan^{-1} \frac{A \sin 2\nu}{A \cos 2\nu + B} \right] \\ &= C_2 \cos (2T+2h-2\nu'') \end{aligned} \qquad (230)$$

in which

$$\begin{aligned} C_2 &= (A^2 + 2AB \cos 2\nu + B^2)^{\frac{1}{2}} \\ &= (0.2523 \sin^4 I + 0.0367 \sin^2 I \cos 2\nu + 0.0013)^{\frac{1}{2}} \end{aligned} \qquad (231)$$

$$2\nu'' = \tan^{-1} \frac{A \sin 2\nu}{A \cos 2\nu + B} = \tan^{-1} \frac{\sin^2 I \sin 2\nu}{\sin^2 I \cos 2\nu + 0.0727} \qquad (232)$$

Values for $2\nu''$ for each degree of $N$ are included in table 6.

**136.** The obliquity factor for $K_2$ will be taken to include the entire coefficient $(A^2 + 2AB \cos 2\nu + B^2)^{\frac{1}{2}}$ and its mean value will be taken as the mean of the product $(A^2 + 2AB \cos 2\nu + B^2)^{\frac{1}{2}} \cos 2\nu''$. From (232)

$$\cos 2\nu'' = (A \cos 2\nu + B)/(A^2 + 2AB \cos 2\nu + B^2)^{\frac{1}{2}} \qquad (233)$$

Then for the mean value of coefficient of $K_2$

$$\begin{aligned} [(A^2 + 2AB \cos 2\nu + B^2)^{\frac{1}{2}} \cos 2\nu'']_0 &= [A \cos 2\nu + B]_0 \\ &= [0.5023 \sin^2 I \cos 2\nu + 0.0365]_0 = 0.1151 \end{aligned} \qquad (234)$$

the numerical mean for $\sin^2 I \cos 2\nu$ being obtained from formula (71). For the node factor of $K_2$ divide the coefficient of (230) by its mean value and obtain

$$\begin{aligned} f(K_2) &= (0.2523 \sin^4 I + 0.0367 \sin^2 I \cos 2\nu + 0.0013)^{\frac{1}{2}} / 0.1151 \\ &= (19.0444 \sin^4 I + 2.7702 \sin^2 I \cos 2\nu + 0.0981)^{\frac{1}{2}} \end{aligned} \qquad (235)$$

See table 14 for node factors and table 12 for reciprocal reduction factors.

### METEOROLOGICAL AND SHALLOW-WATER TIDES

**137.** In addition to the elementary constituents obtained from the development of the tide-producing forces of the moon and the sun, there are a number of harmonic terms that have their origin in meteorological changes or in shallow-water conditions. Variations in temperature, barometric pressure, and in the direction and force of the wind may be expected to cause fluctuations in the water level. Although in general such fluctuations are very irregular, there are some seasonal and daily variations which occur with a rough periodicity that admit of being expressed by harmonic terms. The meteorological constituents usually take into account in the tidal analysis are

Sa, Ssa, and $S_1$ with periods corresponding respectively to the tropical year, the half tropical year, and the solar day. These constituents are represented also by terms in the development of the tide-producing force of the sun but they are considered of greater importance as meteorological tides. Ssa occurs in the development of the principal solar force while Sa and $S_1$ would appear in a development involving the 4th power of the solar parallax (par. 119). In the analysis of tide observations both Sa and Ssa are usually found to have an appreciable affect on the water level. Constituent $S_1$ is relatively of little importance in its effect on the height of the tide but has been more noticeable in the velocity of off-shore tidal currents, probably as a result of periodic land and sea breezes.

138. The shallow-water constituents result from the fact that when a wave runs into shallow water its trough is retarded more than its crest and the wave loses its simple harmonic form. The shallow-water constituents are classified as overtides and compound tides, the overtide having a speed that is an exact multiple of one of the elementary constituents and the compound tide a speed that equals the sum or difference of the speeds of two or more elementary constituents.

139. The overtides were so named because of their analogy to the overtones in musical sounds and they may be considered as the higher harmonics of the fundamental tides. The only overtides usually taken into account in tidal work are the harmonics of the principal lunar and solar semidiurnal constituents $M_2$ and $S_2$, the lunar series being designated by the symbols $M_4$, $M_6$, and $M_8$, and the solar series by $S_4$, $S_6$, and $S_8$. The subscript indicates the number of periods in the constituent day. These overtides with their arguments and speeds are included in table 2a, the arguments and speeds being taken as exact multiples of those of the fundamental constituent. There are no theoretical expressions for the coefficients of the overtides but it is assumed that the amplitudes of the lunar series undergo variations due to changes in the longitude of the moon's node, which are analogous to those in the fundamental tide. The node factors for $M_4$, $M_6$, and $M_8$, respectively, are taken as the square, the cube, and the fourth power of the corresponding factor for $M_2$. For the solar terms this factor is always zero.

140. Compound tides were suggested by Helmholtz's theory of sound waves. Innumerable combinations are possible but the principal elementary constituents involved are $M_2$, $S_2$, $N_2$, $K_1$, and $O_1$. Table 2a includes the compound tides listed in International Hydrographic Bureau Special Publication No. 26, which is a compilation of the tidal harmonic constants for the world. The argument of a compound tide equals the sum or difference of the arguments of the elementary constituents of which it is compounded. The node factor is taken as the product of the node factors of the same constituents. Table 2a contains the arguments, speeds, and node factors of these tides.

141. Omitted from table 2a are a number of compound tides which have the same speeds as elementary constituents included in table 2. Thus, $2MS_2$, compounded by formula $2M_2-S_2$, has the same speed as constituent $\mu_2$ represented by term $A_{45}$ of formula (64). Considered as a compound tide there would be a small difference in the $u$ of the argument and also in the node factor. Since there is no practical way of separating the elementary constituent from the compound

tide of the same speed, this has been treated solely as an elementary constituent. Constituent MSf represented by term $A_5$ of formula (62) has the same speed as a compound tide of formula $S_2-M_2$. This constituent is relatively unimportant and it makes little difference whether treated as an elementary or a compound tide. Following the previous practice in this office it is treated in the harmonic analysis as a compound tide with corresponding argument and node factor. When included in the computation of tidal forces, however, the argument and node factor indicated in table 2 should be used.

## ANALYSIS OF OBSERVATIONS

### HARMONIC CONSTANTS

142. In the preceding discussion it has been shown that under the equilibrium theory the height of a theoretical tide at any place can be expressed mathematically by the sum of a number of harmonic terms involving certain astronomical data and the location of the place. It has also been pointed out that for obvious reasons the actual tide of nature does not conform to the theoretical equilibrium tide. However, the tide of nature can be conceived as being composed of the sum of a number of harmonic constituents having the same periods as those found in the tide-producing force. Although the complexity of the tidal movement is too great to permit a theoretical computation based upon astronomical conditions only, it is possible through the analysis of observational data at any place to obtain certain constants which can be introduced into the theoretical formulas and thus adapt them for the computation of the tide for any desired time.

143. In the formulas obtained for the height of the equilibrium tide each constituent term consists of the product of a coefficient by the cosine of an argument. For corresponding formulas expressing the actual height of the tide at any place, the entire theoretical coefficient including the latitude factor and the common general coefficient is replaced by a coefficient determined from an analysis of observational data for the station. This tidal coefficient, which is known as the *amplitude* of the constituent, is assumed to be subject to the same variations arising from changes in the longitude of the moon's node as the coefficient of the corresponding term in the equilibrium tide. The amplitude pertaining to any particular year is usually designated by the symbol $R$ while its mean value for an entire node period is represented by the symbol $H$. Amplitudes derived directly from an analysis of a limited series of observations must be multiplied by the reduction factor $F$ (par. 78) to obtain the mean amplitudes of the harmonic constants. For the prediction of tides, the mean amplitudes must be multiplied by the node factor $f$ (par. 77) to obtain the amplitudes pertaining to the year for which the predictions are to be made.

144. The phases of the constituents of the actual tide do not in general coincide with the phases of the corresponding constituents of the equilibrium tide but there may be lags varying from 0 to 360°. The interval between the high water phase of an equilibrium constituent and the following high water of the corresponding constituent in the actual tide is known as the *phase lag* or *epoch* of the constituent and is represented by the symbol $\kappa$ (kappa) which is expressed in angular measure. The amplitudes and epochs together are called harmonic constants and are the quantities sought in the harmonic analysis of tides. Each locality has a separate set of harmonic constants which can be derived only from observational data but which remain the same over a long period of time provided there are no

physical changes in the region that might affect the tidal conditions.

**145.** If we let $y_1$ equal the height of one of the tidal constituents as referred to mean sea level, it may be represented by the following formula:

$$y_1 = fH \cos (E-\kappa) = fH \cos (V+u-\kappa) \qquad (236)$$

The combination symbol $V+u$ is the equivalent of $E$ and represents the argument or phase of the equilibrium constituent.

**146.** Formula (236) is illustrated graphically in figure 7 by a cosine curve with amplitude $fH$. The horizontal line represents mean sea level and the vertical line through $T$ may be taken to indicate any instant of time under consideration. If the point $M$ represents the time when the constituent argument equals zero, the interval from $M$ to the following high water of the constituent will be the epoch $\kappa$. The interval from the preceding high water to $M$ is measured by the explement of $\kappa$ which may be expressed as $-\kappa$. The phase of the constituent argument at time $T$ is reckoned from $M$ and is expressed by the symbol $(V+u)$. The phase of the constit-

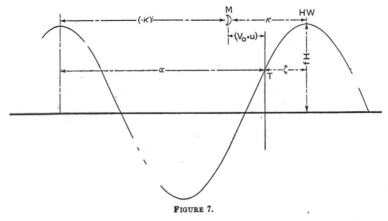

FIGURE 7.

uent itself at this time is reckoned from the preceding high water and therefore equals $(V+u-\kappa)$.

## OBSERVATIONAL DATA

**147.** The most satisfactory observational data for the harmonic analysis are from the record of an automatic tide gage that traces a continuous curve from which the height of the tide may be scaled at any desired interval of time. This record is usually tabulated to give the height of the tide at each solar hour of the series in the kind of time normally used at the place. It is important, however, that the time should be accurate and that the same system be used for the entire series of observations regardless of the fact that daylight saving time may have been adopted temporarily for other purposes during a portion of the year. When the continuous record from an automatic gage is not available, hourly heights of the tide as observed by other methods may be used. The record should be complete with each hour of the series represented. If a part of the record has been lost, the hiatus may be filled by interpolated values; or, if the gap is very extensive, the record may be broken up into shorter series which do not include the defective portion.

**148.** If hourly heights have not been observed but a record of high and low waters is available, an approximate evaluation of the more important constituents may be obtained by a special treatment. The results, however, are not nearly as satisfactory as those obtained from the hourly heights.

**149.** Although the hourly interval for the tabulated heights of the tide has usually been adopted as most convenient and practicable for the purposes of the harmonic analysis, a greater or less interval might be used. A shorter interval would cause a considerable increase in the amount of work without materially increasing the accuracy of the results for the constituents usually sought. However, if an attempt were made to analyze for the short period seiches a closer interval would be necessary. An interval greater than one hour would lessen the work of the analysis but would not be sufficient for the satisfactory development of the overtides.

**150.** In selecting the length of series of observations for the purpose of the analysis, consideration has been given to the fact that the procedure is most effective in separating two constituents from each other when the length of series is an exact multiple of the synodic period of these constituents. By synodic period is meant the interval between two consecutive conjunctions of like phases. Thus, if the speeds of the two constituents in degrees per solar hour are represented by $a$ and $b$, the synodic period will equal $360°/(a \sim b)$ hours. If there were only two constituents in the tide the best length of series could be easily fixed, but in the actual tide there are many constituents and the length of series most effective in one case may not be best adapted to another case. It is therefore necessary to adopt a length that is a compromise of the synodic periods involved, consideration being given to the relative importance of the different constituents.

**151.** Fortunately, the exact length of series is not of essential importance and for convenience all series may be taken to include an integral number of days. Theoretically, different lengths of series should be used in seeking different constituents, but practically it is more convenient to use the same length for all constituents, an exception being made in the case of a very short series. The longer the series of observations the less important is its exact length. Also the greater the number of synodic periods of any two constituents the more nearly complete will be their separation from each other. Constituents like $S_2$ and $K_2$ which have nearly equal speeds and a synodic period of about 6 months will require a series of not less than 6 months for a satisfactory separation. On the other hand, two constituents differing greatly in speed such as a diurnal and a semidiurnal constituent may have a synodic period that will not greatly exceed a day, and a moderately short series of observations will include a relatively large number of synodic periods. For this reason, when selecting the length of series no special consideration need be given to the effect of a diurnal and a semidiurnal constituent upon each other.

**152.** The following lengths of series have been selected as conforming approximately to multiples of synodic periods involving the more important constituents—14, 15, 29, 58, 87, 105, 134, 163, 192, 221, 250, 279, 297, 326, 355, and 369 days. The 369-day series is considered as a standard length to be used for the analysis whenever observations covering this period are available. This length conforms very closely with multiples of the synodic periods of practically all of the short-period constituents and is well adapted for the elimination of seasonal meteorological effects. When observations at any station are available for a number of years, it is desirable to have separate analyses made for different years in order that the results may be compared and serve as a check on each other. Although not essential, there are certain conveniences in having each such series commence on January 1 of the year, regardless of the fact that series of consecutive years may overlap by several days because the length of series is a little longer than the calendar year.

**153.** If the available observations cover a period less than 369 days, the next longest series listed above which is fully covered by the observations will usually be taken, any extra days of observations being rejected. However, if the observations lack only a few hours of being equal to the next greater length, it may be advantageous to extrapolate additional hourly heights to complete the larger series. The 29-day series is usually considered as a minimum standard for short series of observations. This is a little shorter than the synodical month and a little longer than the nodical, tropical, and anomalistic months. It is the minimum length for a satisfactory development of the more important constituents.

**154.** For observations of less than 29 days, but more than 14 days, provisions are made for an analysis of a 14-day series for the diurnal constituents and a 15-day series for the semidiurnal constituents, the first conforming to the synodic period of constituents $K_1$ and $O_1$, and the latter to the synodic period of $M_2$ and $S_2$. Through special treatment involving a comparison with another station, it is possible to utilize even shorter series of observations. This treatment is rarely required in case of tide observations but is useful in connection with tidal currents where observations may be limited to only a few days.

### SUMMATIONS FOR ANALYSIS

**155.** The first approximate separation of the constituents of the observed tide is accomplished by a system of summations, separate summations being made for all constituents with incommensurable periods. Designating the constituent sought by $A$, assume that the entire series of observations is divided into periods equal to the period of $A$ and each period is subdivided into a convenient number of equal parts, the subdivisions of each period being numbered consecutively beginning with zero at the initial instant of each period. All subdivisions of like numbers will then include the same phase of constituent $A$ but different phases for all other constituents with incommensurable speeds. The subdivisions will also include irregular variations arising from meteorological causes. By summing and averaging separately all heights corresponding to each of the numbered subdivisions over a sufficient length of time, the effects of constituents with incommensurable periods as well as the meteorological variations will be averaged out leaving intact constituent $A$ with its overtides.

**156.** The principle just described for separating constituent $A$ from the rest of the tide is applicable if the original periods into which the series of observations is divided are taken as some multiple of constituent $A$ period. In general practice, that multiple of the constituent period which is most nearly equal to the solar day is taken as the unit. This is the constituent day and includes one or more periods according to whether the constituent is diurnal, semidiurnal, etc. The constituent day is divided into 24 equal parts,the beginning of each part being numbered consecutively from 0 to 23 and these are known as constituent hours.

**157.** To carry out strictly the plan described above would require separate tabulations of the heights of the tide at different intervals for all constituents of incommensurable periods, a procedure involving an enormous amount of work. In actual practice the tabulated solar hourly heights are used for all of the summations, these heights being assigned to the nearest constituent hour. Corrections are afterwards applied to take account of any systematic error in this approximation.

**158.** There are two systems for the distribution and assignment of the solar hourly heights which differ slightly in detail. In the system ordinarily used and which is sometimes called the standard system, each solar hourly height is used once, and once only, by being assigned to its nearest constituent hour. By this system some constituent hours will be assigned two consecutive solar hourly heights or receive no assignment according to whether the constituent day is longer or shorter than the solar day. In the other system of distribution, each constituent hour receives one and only one solar hourly height necessitating the occasional rejection or double assignment of a solar hourly height. The difference in the results obtained from the two systems is practically negligible but the first system is generally used as it affords a quick method of checking the summations.

### STENCILS

**159.** The distribution of the tabulated solar hourly heights of the tide for the purpose of the harmonic analysis is conveniently accomplished by a system of stencils (fig. 10) which were devised by L. P. Shidy of the Coast and Geodetic Survey early in 1885 (Report of U. S. Coast and Geodetic Survey, 1893, vol. I, p. 108). Although the original construction of the stencils involves considerable work, they are serviceable for many years and have resulted in a very great saving of labor. These stencils are cut from the same forms which are used for the tabulation of the hourly heights of the tide and 106 sheets are required for the summation of a 369-day series of observations for a single constituent. Separate sets are provided for different constituents. Constituents with commensurable periods are included in a single summation and no stencils are required for constituents, $S_1$, $S_2$, $S_4$, etc.

**160.** The use of the stencils makes a standardized form for the tabulation of the hourly heights essential. This form (fig. 9) is a sheet 8 by 10½ inches, with spaces arranged for the tabulation of the 24 hourly heights of each day in a vertical column, with 7 days of record on each page. The hours of the day are numbered consecutively from $0^h$ at midnight to $23^h$ at 11 p. m. When the tabulated heights are entered, each day is indicated by its calendar date and also by a serial number commencing with 1 as the first day of series. The days on the stencil sheets are numbered serially to correspond with the tabulation sheets and may be used for any series regardless of the calendar dates.

**161.** The openings in the stencils are numbered to indicate the constituent hours that correspond most closely with the times of the height values showing through the openings when the stencil is applied to the tabulations. Openings applying to the same constituent hour are connected by a ruled line which clearly indicates to the eye the tabular heights which are to be summed together. For convenience in construction two stencil sheets are prepared for each page of tabulations, one sheet providing for the even constituent hours and the other sheet for the odd constituent hours.

**162.** The stencils are adapted for use with tabulations made in any kind of time provided the time used is uniform for the entire series of observations. For convenience the tabulations are usually made in the standard time of the place. The series to be analyzed, however, must commence with the zero hour of the day and this is also taken as the zero constituent hour for each constituent. Successive solar hours will fall either earlier or later than the corresponding constituent hour according to whether the constituent day is longer or shorter than the solar day.

**163.** For the construction of the stencils it is necessary to calculate the constituent hour that most nearly coincides with each solar hour of the series.

Let $a$=speed or rate of change in argument of constituent sought in degrees per solar hour.
$p$=number of constituent periods in constituent day; 1 for diurnal tides, 2 for semidiurnal tides, etc.
$sh$=number of solar hour reckoned from 0 at beginning of each solar day.
$shs$=number of solar hour reckoned from 0 at beginning of series.
$dos$=day of series counting from 1 as the first day.
$ch$=number of constituent hour reckoned from 0 at beginning of each constituent day.
$chs$=number of constituent hour reckoned from 0 at beginning of series.

Then

$$1 \text{ constituent period} = \frac{360}{a} \text{ solar hours.} \quad (237)$$

$$1 \text{ constituent day} = \frac{360p}{a} \text{ solar hours.} \quad (238)$$

$$1 \text{ constituent hour} = \frac{15p}{a} \text{ solar hours.} \quad (239)$$

$$1 \text{ solar hour} = \frac{a}{15p} \text{ constituent hours.} \quad (240)$$

Therefore,

$$(chs) = \frac{a}{15p}(shs) = \frac{a}{15p}[24\{(dos)-1\}+(sh)] \quad (241)$$

**164.** The above formula gives the constituent hour of the series $(chs)$ corresponding to any solar hour of the series $(shs)$. The observed heights of the tide being tabulated for the exact solar hours of the day, the $(shs)$ with which we are concerned will represent successive integers counting from 0 at the beginning of the series. The $(chs)$ as derived from the formula will generally be a mixed number. As it is desired to obtain the integral constituent hour corresponding most nearly with each solar hour, the $(chs)$ should be taken to the nearest integer by rejecting a fraction less than 0.5, or counting as an extra hour a fraction greater than 0.5, or adopting the usual rule for computations if the fraction is exactly 0.5. The constituent hour of the constituent day $(ch)$ required for the construction of the stencils may be obtained by rejecting multiples of 24 from the $(chs)$.

**165.** In the application of the above formula it will be found that the integral constituent hour will differ from the corresponding solar hour by a constant for a succession of solar hours, and then, with the difference changed by one, it will continue as a constant for another group of solar hours, etc. This fact is an aid in the preparation of a table of constituent hours corresponding to the solar hours of the series, as it renders it unnecessary to make an independent calculation for each hour. Instead of using the above formula for each value the time when the difference between the solar and constituent hours changes may be determined. The application of the differences to the solar hours will then give the desired constituent hours.

**166.** Formula (241) is true for any value of $(shs)$, whether integral or fractional. It represents the constituent time of any instant in the series of observations in terms of the solar time of that same instant, both kinds of time being reckoned from the beginning of the series as the zero hour. The difference between the constituent and the solar time of any instant may therefore be expressed by the following formula:

$$\text{Difference} = \frac{a}{15p}(shs) \sim (shs) = \frac{a \sim 15p}{15p}(shs) \qquad (242)$$

**167.** If the constituent day is shorter than the solar day, the speed $a$ will be greater than $15p$, and the constituent hour as reckoned from the beginning of the series will be greater than the solar hour of the same instant. If the constituent day is longer than the solar day the constituent hour at any instant will be less than the solar hour of the same instant. At the beginning of the series the difference between the constituent and solar time will be zero, but the difference will increase uniformly with the time of the series. As long as the difference does not exceed 0.5 of an hour the integral constituent hours will be designated by the same ordinals as the integral solar hours with which they most nearly coincide. Differences between 0.5 and 1.5 will be represented by the integer 1, differences between 1.5 and 2.5 by the integer 2, etc. If we let $d$ represent the integral difference, the time when the difference changes from $(d-1)$ to $d$, will be the time when the difference derived from formula (242) equals $(d-0.5)$. Substituting this in the formula, we may obtain

$$(shs) = \frac{15p}{a \sim 15p}(d-0.5) \qquad (243)$$

in which $(shs)$ represents the solar time when the integral difference between the constituent and solar time will change by one hour from $(d-1)$ to $d$. By substituting successively the integers 1, 2, 3, etc., for $d$ in the formula (243) the time of each change throughout the series may be obtained. The value of $(shs)$ thus obtained will generally be a mixed number; that is to say, the times of the changes will usually come between integral solar hours. The first integral solar hour after the change will be the one to which the new difference will apply if the usual system of distribution is to be adopted. In this case we are not concerned with the exact value of the fractional part of $(shs)$ but need note only the integral hours between which this value falls.

168. If, however, the second system of distribution should be desired, it should be noted whether the fractional part of $(shs)$ is greater or less than 0.5 hour. With a constituent day shorter than the solar day and the differences of formula (242) increasing positively, the application of the differences to the consecutive solar hours will result in the jumping or omission of a constituent hour at each change of difference. Under the second system of distribution each of the hours must be represented, and it will therefore be necessary in this case to apply two consecutive differences to the same solar hour to represent two consecutive constituent hours. The solar hour selected for this double use will be the one occurring nearest to the time of change of differences. If the fractional part of the $(shs)$ in (243) is less than 0.5 hour, the old and new differences will both be applied to the preceding integral solar hour; but if the fraction is greater than 0.5 hour the old and new differences will be applied to the integral solar hour following the change.

169. With a constituent day longer than the solar day and the differences of formula (242) increasing negatively, the application of the differences to the consecutive solar hours will result in two solar hours being assigned to the same constituent hour at each change of differences. Under the second system of distribution this must be avoided by the rejection of one of the solar hours. In this case the integral solar hour nearest the time of change will be rejected, since at the time of change the difference between the integral and the true difference is a maximum. Thus, if the fractional part of the $(shs)$, is less than 0.5 hour, the preceding solar hour will be rejected; but if the fraction is greater than 0.5 hour the next following solar hour will be rejected.

170. Table 31, computed from formula (243), gives the first solar hour of the group to which each difference applies when the usual system of distribution is adopted. Multiples of 24 have been rejected from the differences, since we are concerned only with the constituent hour of the constituent day rather than with the constituent hour of the series, and these differences may be applied directly to the solar hours of the day. For convenience equivalent positive and negative differences are given. By using the negative difference when it does not exceed the solar hour to which it is to be applied, and at other times using the positive difference, the necessity for adding or rejecting multiples of 24 hours from the results is avoided.

171. The tabulated solar hour is the integer hour that immediately follows the value for the $(shs)$ is formula (243). An asterisk (*) indicates that the fractional part of the $(shs)$ exceeds 0.5, and that the tabular hour is therefore the one nearest the exact value of $(shs)$. If the second system for the distribution of the hourly heights is adopted, the solar hours marked with the asterisk will be used with both old and new difference to represent two constituent hours, or will be rejected altogether according to whether the constituent day

is shorter or longer than the solar day. If the tabular hour is unmarked, the same rule of double use or rejection will apply to the untabulated solar hour immediately preceding the tabular unmarked hour. For the ordinary stencils no attention need be given to the asterisks. By the formula constituents with commensurable periods will have the same tabular values, and no distinction is made in the construction of the stencils. Thus, stencils for constituent M serve not only for $M_2$ but also for $M_3$, $M_4$, $M_6$, etc.

172. For the construction of a set of stencils for any constituent a preliminary set of the hourly height forms is prepared with days of series numbered consecutively beginning with 1 and each hourly height space numbered with its constituent hour as derived by the differences in table 31. The even and odd constituent hours are then transferred to separate sets of forms and the marked spaces cut out. In the Coast and Geodetic Survey this is done by a small machine with a punch operated by a hand lever. Spaces corresponding to the same constituent hour are connected by ruled lines which are numbered the same as the hours represented. Black ruling with red numbering is recommended, the red emphasizing the distinction between these numbers and the tabulated hourly heights which are to be summed.

173. When in use the stencils are placed one at a time on the sheets of tabulated heights, with days of series on stencils matching those on the tabulations, and all heights on the page corresponding to each constituent hour are then summed separately. For constituent S no stencils are necessary as the constituent hours in this case are identical with the solar hours. For constituents K, P, R, and T with speeds differing little from that of S, the lines joining the hourly spaces frequently become horizontal and the marginal sum previously obtained for constituent S becomes immediately available for the summation at hand. In these cases a hole in the margin of the stencil for the sum replaces the holes for the individual heights covered by the sum.

#### SECONDARY STENCILS

174. After the sums for certain principal constituents have been obtained by the stencils described in the preceding section, which for convenience will be called the primary stencils, the summations for other constituents may be abbreviated by the use of secondary stencils which are designed to regroup the hourly page sums already obtained for one constituent into new combinations conforming to the periods of other constituents. Certain irregularities are introduced by the process, but in a long series, such as 369 days, these are for the most part eliminated, and the resulting values for the harmonic constants compare favorably with those obtained by use of the primary stencils directly, the differences in the results obtained by the two methods being negligible. For short series the irregularities are less likely to be eliminated, and since the labor of summing for such a series is relatively small, the abbreviated form of summing is not recommended. As the length of series increases the saving in labor by the use of the secondary stencils increases, while the irregularities due to the short process tend to disappear. It is believed that the use of the secondary stencils will be found advantageous for all series more than 6 months in length.

**175.** In the primary summations there are obtained 24 sums for each page of tabulations, representing the 24 constituent hours of a constituent day. In general each sum will include 7 hourly heights, and the average interval between the first and last heights will be 6 constituent days. A few of the sums may, however, include a greater or less number of hourly heights within limits which may be a day greater or less than 6 constituent days.

**176.** Let the constituent for which summations have been made by use of the primary stencils be designated as $A$ and the constituent which is to be obtained by use of the secondary stencils as $B$. For convenience let it be first assumed that the heights included in the sums for constituent $A$ refer to the exact $A$-hours. This assumption is true for constituent S but only approximately true for the other constituents. It is now proposed to assign each hourly page sum obtained for constituent $A$ to the integral $B$-hour with which it most nearly coincides. Constituent $A$ and constituent $B$-hours separate at a uniform rate, and the proposed assignment will depend upon the relation of the hours on the middle day of each page of tabulations. The tabulated hourly heights on each full page of record run from zero (0) solar hour on the first day to the 23d solar hour on the seventh or last day of the page. The middle of the record on each such page is therefore at 11.5 solar hours on the fourth day, or 83.5 solar hours from the beginning of the page of record.

**177.** Let $a$ and $b$ represent the hourly speeds of the constituents $A$ and $B$, respectively, and $p$ and $p_1$ their respective subscripts, and let $n$ equal the number of the page of tabulation under consideration, beginning with number one as the first page.

The middle of page $n$ will then be

$$[168(n-1)+83.5] \text{ or } (168n-84.5) \text{ solar hours} \qquad (244)$$

from the beginning of the series.

Since one solar hour equals $a/15p$ constituent $A$-hours (formula 240), the middle of page $n$ will also correspond to

$$(168n-84.5)\frac{a}{15p} \text{constituent } A\text{-hours} \qquad (245)$$

from the beginning of the series.

As there are 24 constituent hours in each constituent day, the middle constituent $A$-day of each page will commence 12 constituent $A$-hours earlier than the time represented by the middle of the page, or at

$$[(168n-84.5)\frac{a}{15p}-12] \text{ constituent } A\text{-hours} \qquad (246)$$

from the beginning of the series.

**178.** The 24 integral constituent $A$-hours of the middle constituent day of the page will therefore be the integral constituent $A$-hours which immediately follow the time indicated by the last formula. The numerical value of this formula will usually be a mixed number. Let $f$ equal the fractional part, and let $m$ be an integer representing the number of any integral constituent hour according to its order in the middle constituent day of each page. For each page $m$ will have

HARMONIC ANALYSIS AND PREDICTION OF TIDES 59

successive values from 1 to 24. The integral constituent $A$-hours falling within the middle constituent day of each page of tabulations will then be represented by the general formula.

$$[(168n-84.5)\frac{a}{15p}-12-f+m] \text{ constituent } A\text{-hours} \quad (247)$$

from the beginning of the series.

**179.** The relation of the lengths of the constituent $A$- and constituent $B$-hours is given by the formula

$$1 \text{ constituent } A\text{-hour} = \frac{pb}{p_1 a} \text{constituent } B\text{-hours} \quad (248)$$

The constituent $B$-hour corresponding to the integral constituent $A$-hour of formula (247) is therefore

$$[(168n-84.5)\frac{a}{15p}-12-f+m]\frac{pb}{p_1 a} \text{constituent } B\text{-hours} \quad (249)$$

from the beginning of the series.

The last formula will, in general, represent a mixed number. The integral constituent $B$-hour to which the sum for the constituent $A$-hour is to be assigned will be the nearest integral number represented by this formula. Let $g$ be a fraction not greater than 0.5, which, applied either positively or negatively to the formula, will render it an integer.

**180.** The assignment of the hourly page sums for constituent $A$-hours to the constituent $B$-hours may now be represented as follows, multiples of 24 hours being rejected:

$$[(168n-84.5)\frac{a}{15p}-12-f+m-\text{multiple of } 24] \text{ constituent } A\text{-hour} \quad (250)$$

sum to be assigned to

$$[\{(168n-84.5)\frac{a}{15p}-12-f+m\}\frac{pb}{p_1 a}\pm g-\text{multiple of } 24] \text{ constituent } B\text{-hour.} \quad (251)$$

The difference between the constituent $A$-hour and the constituent $B$-hour to which the $A$-hour sum is to be assigned is

$$[\{(168n-84.5)\frac{a}{15p}-12-f+m\}\{\frac{pb}{p_1 a}-1\}\pm g-\text{multiple of } 24] \quad (252)$$

By means of the above formula table 33 has been prepared, giving the differences to be applied to the constituent $A$-hours of each page to obtain the constituent $B$-hours with which they most nearly coincide.

**181.** For the construction of secondary stencils the forms designated for the compilation of the stencil sums from the primary summations may be used. Because of the practical difficulties of constructing stencils with openings in adjacent line spaces it is desirable that the original compilation of the primary sums should be made so that each alternate line in the form for stencil sums is left vacant. As with the primary stencils, it will generally be found convenient to use two stencils for each page of the compiled primary sums, although in some cases it may be found desirable to use more than two stencils in order to separate more clearly the groups to be summed. The actual construction of the secondary stencils is similar to that of the primary stencils. A preliminary set of forms is filled out with constituent $B$-hours as derived by differences from table 33 applied to the constituent $A$-hours. The odd and even constituent $B$-hours are then transferred to separate forms and the spaces indicated cut out. The openings corresponding to the same constituent $B$-hour are connected with ruled lines and numbered to accord with the constituent hour represented. The page numbering corresponding to the page numbering on the compiled primary sums and referring to the pages of the original tabulated hourly heights is to be entered in the column provided near the left margin of the stencil.

**182.** In using the stencils each sheet is to be applied to the page of compiled primary sums having the same page numbering in the left-hand column as is given on the stencil. The primary sums applying to the same constituent $B$-hour are added and the results brought together in a stencil sum form, where the totals and means are obtained. A table of divisors for obtaining the means may be readily derived as follows: In a set of stencil sum forms corresponding to those used for the compilation of constituent $A$ primary sums the number of hourly heights included in each primary sum is entered in the space corresponding to that used for such primary sum. The secondary stencils for constituent $B$ are then applied and the sums of the numbers obtained and compiled in the same manner as that in which the constituent $B$ height sums are obtained. The divisors having been once obtained are applicable for all series of the same length.

**183.** In the analysis the means obtained by use of the secondary stencils may be treated as though obtained directly by the primary summations except that a special augmenting factor, to be discussed later, must be applied. The closeness of the agreement between the hourly means obtained by use of the secondary stencils and those obtained directly by use of primary stencils will depend to a large extent upon the relation of the speeds of constituents $A$ and $B$. The smaller the difference in the speeds the closer will be the agreement.

**184.** To determine the extreme difference in the time of an individual hourly height and of the $B$-hour to which it is assigned by the secondary stencils, let an assumed case be first considered in which the tabulated heights coincide exactly with the integral $A$-hours, and that on the middle day of the page of tabulated hourly heights one of the integral $B$-hours coincides exactly with an $A$-hour. At the corresponding $A$-hour, one $A$-day later, the $B$-hour will have increased by $24\,\dfrac{pb}{p_1a}$ constituent $B$-hours. Rejecting a multiple of 24 hours, this becomes $24\left(\dfrac{pb}{p_1a}-1\right)$, so that at the end of one $A$-day after the coincidence of integral hours of constituents $A$ and $B$ the constituent $A$ hourly height will differ in time from the integral constituent $B$-hour to which it is to be assigned by $24\left(\dfrac{pb}{p_1a}-1\right)$ constituent $B$-hours.

At the end of the third $A$-day this difference becomes $72\left(\dfrac{pb}{p_1a}-1\right)$ constituent $B$-hours. The same difference with opposite sign will apply to the third constituent day before the middle day of the page. Now, taking account of the fact that the $B$-hour on the middle day of the page may differ by an amount as great 0.5 of a $B$-hour from the integral $A$-hour, and that the integral $A$-hour may differ as much as 0.5 of a constituent $A$, or 0.5 $pb/p_1a$ of a constituent $B$ hour from the time of the actual observation of the solar hourly height, the extreme difference between the time of observation of an hourly height and the time represented by the $B$-hour with which this height is grouped by the secondary stencils may be represented by the formula

$$\pm\left[72\left(\dfrac{pb}{p_1a}\sim 1\right)+0.5\left(\dfrac{pb}{p_1a}+1\right)\right] \text{ constituent } B\text{-hours.} \quad (253)$$

The differences may be either positive or negative, and in a long series it may reasonably be expected that the number of positive and negative values will be approximately equal.

**185.** The above formula for the extreme difference furnishes a criterion by which to judge, to some extent, the reliability of the method. Testing the following schedule of constituents for which it is proposed to use the secondary stencils, the extreme differences as indicated are obtained. The differences are expressed in constituent $B$-hours and also in constituent $B$-degrees. It will be noted that one constituent hour is equivalent to a change of 15° in the phase of a diurnal constituent, 30° in the phase of a semidiurnal constituent, etc.

| Constituent $A$ | J | | S | | | | |
|---|---|---|---|---|---|---|---|
| Constituent $B$ | OO | 2SM | $K_1$ | $K_2$ | $R_2$ | $T_2$ | $P_1$ |
| Difference in hours | 3.58 | 1.36 | 1.20 | 1.20 | 1.10 | 1.10 | 1.20 |
| Difference in degrees | 54 | 41 | 18 | 36 | 33 | 33 | 18 |

| Constituent $A$ | | L | | 2MK | | |
|---|---|---|---|---|---|---|
| Constituent $B$ | MS | $\lambda_2$ | MK | MN | $\nu_2$ | $N_2$ |
| Difference in hours | 1.09 | 1.18 | 1.43 | 1.24 | 1.26 | 1.45 |
| Difference in degrees | 65 | 35 | 64 | 74 | 38 | 44 |

| Constituent $A$ | | | O | | |
|---|---|---|---|---|---|
| Constituent $B$ | $\mu_2$ | 2N | $\rho_1$ | Q | 2Q |
| Difference in hours | 1.21 | 1.02 | 3.42 | 3.79 | 6.58 |
| Difference in degrees | 36 | 31 | 51 | 57 | 99 |

**186.** In the ordinary primary summation the extreme difference between the time of the observation of a solar hourly height and the intregal constituent hour to which it is assigned is one-half of a constituent hour and, represented by constituent degrees, it is 7.5° for diurnal, 15° for semidiurnal, 22.5° for terdiurnal, 30° for quarter diurnal, 45° for sixth-diurnal, and 60° for eighth-diurnal constituents. By the above schedule it will be noted that the extreme difference exceeds 60° in only a few cases. The largest difference is 99° for constituent $2Q$ when based upon the primary summations for O. This is a small and unimportant constituent, and heretofore no analysis has been made for it, the value of its harmonic constants being inferred from those of constituent O. Although theoretically too small to justify a primary summation in general practice, the lesser work involved in the secondary summations may produce constants for this constituent which will be more satisfactory than the inferred constants.

**FOURIER SERIES**

**187.** A series involving only sines and cosines of whole multiples of a varying angle is generally known as the Fourier series. Such a series is of the form

$$h = H_0 + C_1 \cos \theta + C_2 \cos 2\theta + C_3 \cos 3\theta + \text{------} \\ + S_1 \sin \theta + S_2 \sin 2\theta + S_3 \sin 3\theta + \text{------} \quad (254)$$

It can be shown that by taking a sufficient number of terms the Fourier series may be made to represent any periodic function of $\theta$.

This series may be written also in the following form:

$$h = H_0 + A_1 \cos (\theta + \alpha_1) + A_2 \cos (2\theta + \alpha_2) + A_3 \cos (3\theta + \alpha_3) + \text{----} \quad (255)$$

in which

$$A_m = [C_m^2 + S_m^2]^{\frac{1}{2}} \text{ and } \alpha_m = -\tan^{-1} \frac{S_m}{C_m}$$

$m$ being the subscript of any term.

**188.** From the summations for any constituent 24 hourly means are obtained, these means being the approximate heights of the constituent tide at given intervals of time. These mean constituent hourly heights, together with the intermediate heights, may be represented by the Fourier series, in which

$H_0$ = mean value of the function corresponding to the height of mean sea level above the adopted datum.

$\theta$ = an angle that changes uniformly with time and completes a cycle of 360° in one constituent day. The values of $\theta$ corresponding to the 24 hourly means will be 0°, 15°, 30°, ------ 330°, and 345°.

Formula (254), or its equivalent (255), is the equation of a curve with the values of $\theta$ as the abscissæ and the corresponding values of $h$ as the ordinates. If the 24 constituent hourly means are plotted as ordinates corresponding to the values of 0°, 15°, 30°, ----- for $\theta$, it is possible to find values for $H_0$, $C_m$, and $S_m$, which when substituted in (255) will give the equation of a curve that will pass exactly through each of the 24 points representing these means.

**189.** In order to make the following discussion more general, let it be assumed that the period of $\theta$ has been divided into $n$ equal parts, and that the ordinate or value of $h$ pertaining to the beginning of each of those parts is known. Let $u$ equal the interval between these ordinates, then

$$n\,u = 2\pi, \text{ or } 360° \quad (256)$$

Let the given ordinates be $h_0$, $h_1$, $h_2$ ---- $h_{(n-1)}$ corresponding to the abscissae $o$, $u$, $2u$ ---- $(n-1)\,u$, respectively.

It is now proposed to show that the curve represented by the following Fourier series will pass through the $n$ points of which the ordinates are given:

$$h = H_0 + C_1 \cos \theta + C_2 \cos 2\theta + \text{\textemdash\textemdash\textemdash} C_k \cos k\theta$$
$$+ S_1 \sin \theta + S_2 \sin 2\theta + \text{\textemdash\textemdash\textemdash} S_l \sin l\theta$$
$$= H_0 + \sum_{m=1}^{m=k} C_m \cos m\theta + \sum_{m=1}^{m=l} S_m \sin m\theta \qquad (257)$$

in which the limit $k = \frac{n}{2}$ if $n$ is an even number, or $k = \frac{n-1}{2}$ if $n$ is an odd number; and the limit $l = \frac{n}{2} - 1$ if $n$ is even, or $\frac{n-1}{2}$ if $n$ is odd.

**190.** By substituting successively the coordinates of the $n$ given points in (257) we may obtain $n$ equations of the form

$$h_a = H_0 + \sum_{m=1}^{m=k} C_m \cos mau + \sum_{m=1}^{m=l} S_m \sin mau \qquad (258)$$

in which $a$ represents successively the integers 0 to $(n-1)$.

By the solution of these $n$ equations the values of $n$ unknown quantities may be obtained, including $H_0$ and the $(n-1)$ values for $C_m$ and $S_m$. It will be noted that the sum of the limits $k$ and $l$ of (257) or (258) equals $(n-1)$ for both even and odd values of $n$.

**191.** The reason for these limits is as follows:

A continued series $\Sigma\, C_m \cos m\, a\, u$ may be written

$$C_1 \cos a\, u + C_2 \cos 2\, a\, u + \text{\textemdash\textemdash} + C_n \cos n\, a\, u$$
$$+ C_{(n+1)} \cos (n+1)\, a\, u + C_{(n+2)} \cos (n+2)\, a\, u + \text{\textemdash\textemdash} + C_{2n} \cos 2n\, a\, u$$
$$+ C_{(2n+1)} \cos (2n+1)\, a\, u + C_{(2n+2)} \cos (2n+2)\, a\, u + \text{\textemdash\textemdash}$$
$$+ C_{3n} \cos 3n\, a\, u$$
$$+ \text{\textemdash\textemdash\textemdash\textemdash\textemdash\textemdash\textemdash\textemdash\textemdash\textemdash\textemdash\textemdash\textemdash\textemdash} \qquad (259)$$

Since $n\, u = 2\pi$ and $a$ is an integer, the above may be written

$$[C_1 + C_{(n+1)} + C_{(2n+1)} + \text{\textemdash\textemdash}]\cos a\, u$$
$$+ [C_2 + C_{(n+2)} + C_{(2n+2)} + \text{\textemdash\textemdash}]\cos 2\, a\, u$$
$$+ \text{\textemdash\textemdash\textemdash\textemdash\textemdash\textemdash\textemdash\textemdash\textemdash}$$
$$+ [C_{(n-1)} + C_{(2n-1)} + C_{(3n-1)} + \text{\textemdash\textemdash}]\cos (n-1)\, a\, u$$
$$+ [C_n + C_{2n} + C_{3n} + \text{\textemdash\textemdash}]\cos n\, a\, u \qquad (260)$$

Since $\cos n\, a\, u = \cos 2a\, \pi = 1$; $\cos (n-1)\, a\, u = \cos (2a\, \pi - a\, u) = \cos a\, u$; $\cos (n-2)\, a\, u = \cos 2\, a\, u$; etc., (260) may be written

$$[C_n + C_{2n} + C_{3n} + \text{\textemdash\textemdash}]\cos 0$$
$$+ [C_1 + C_{(n+1)} + C_{(2n+1)} + \text{\textemdash\textemdash}$$
$$+ C_{(n-1)} + C_{(2n-1)} + C_{(3n-1)} + \text{\textemdash\textemdash}]\cos a\, u$$
$$+ [C_2 + C_{(n+2)} + C_{(2n+2)} + \text{\textemdash\textemdash}$$
$$+ C_{(n-2)} + C_{(2n-2)} + C_{(3n-2)} + \text{\textemdash\textemdash}]\cos 2\, a\, u$$
$$\text{\textemdash\textemdash\textemdash\textemdash\textemdash\textemdash\textemdash\textemdash\textemdash\textemdash\textemdash\textemdash}$$
$$+ [C_k + C_{(n+k)} + C_{(2n+k)} + \text{\textemdash\textemdash}$$
$$+ C_{(n-k)} + C_{(2n-k)} + C_{(3n-k)} + \text{\textemdash\textemdash}]\cos k\, a\, u \qquad (261)$$

The first term of the above is a constant which will be included with the $H_0$ in the solution of (258). From an examination of (261) it is evident that the cosine terms will be completely represented when $k = \frac{n}{2}$, or $\frac{n-1}{2}$, according to whether $n$ is even or odd.

Similarly, the continued series $\Sigma\, S_m \sin m\, a\, u$ may be written

$$[S_n+S_{2n}+S_{3n}+\ \rule{2cm}{0.4pt}\ ] \sin 0$$
$$+[S_1+S_{(n+1)}+S_{(2n+1)}+\ \rule{1.5cm}{0.4pt}$$
$$-S_{(n-1)}-S_{(2n-1)}-S_{(3n-1)}-\ \rule{1.5cm}{0.4pt}\ ] \sin a\ u$$
$$+[S_2+S_{(n+2)}+S_{(2n+2)}+\ \rule{1.5cm}{0.4pt}$$
$$-S_{(n-2)}-S_{(2n-2)}-S_{(3n-2)}-\ \rule{1.5cm}{0.4pt}\ ] \sin 2\ a\ u$$
$$\rule{8cm}{0.4pt}$$
$$+[S_l+S_{(n+l)}+S_{(2n+l)}+$$
$$-S_{(n-l)}-S_{(2n-l)}-S_{(3n-l)}-\ \rule{1.5cm}{0.4pt}\ ] \sin l\ a\ u \qquad (262)$$

The first term in the above equals zero. The remaining terms will take complete account of the series $\Sigma\ S_m \sin m\ a\ u$, if $l=\frac{n}{2}-1$ when $n$ is even, or $\frac{n-1}{2}$ when $n$ is odd.

From the foregoing it is evident that the limit of $m$ will not exceed $\frac{n}{2}$.

**192.** If we let $u$ and $\alpha$ represent any angles with fixed values, $m$ and $p$ any integers with fixed values, and $a$ an integer having successive values from 0 to $(n-1)$, it may be shown that

$$\sum_{a=0}^{a=(n-1)} \sin(a\ m\ u+\alpha) = \frac{\sin \frac{1}{2} n\ m\ u}{\sin \frac{1}{2} m\ u} \sin [\tfrac{1}{2} (n-1)\ m\ u+\alpha] \qquad (263)$$

$$\sum_{a=0}^{a=(n-1)} \cos(a\ m\ u+\alpha) = \frac{\sin \frac{1}{2} n\ m\ u}{\sin \frac{1}{2} m\ u} \cos [\tfrac{1}{2} (n-1)\ m\ u+\alpha] \qquad (264)$$

$$\sum_{a=0}^{a=(n-1)} \sin a\ p\ u \sin a\ m\ u = \tfrac{1}{2}\frac{\sin \frac{1}{2} n\ (p-m)\ u \cos \frac{1}{2} (n-1)\ (p-m)\ u}{\sin \frac{1}{2} (p-m)\ u}$$
$$-\tfrac{1}{2}\frac{\sin \frac{1}{2} n\ (p+m)\ u \cos \frac{1}{2} (n-1)\ (p+m)\ u}{\sin \frac{1}{2} (p+m)\ u} \qquad (265)$$

$$\sum_{a=0}^{a=(n-1)} \cos a\ p\ u \cos a\ m\ u = \tfrac{1}{2}\frac{\sin \frac{1}{2} n\ (p-m)\ u \cos \frac{1}{2} (n-1)\ (p-m)\ u}{\sin \frac{1}{2} (p-m)\ u}$$
$$+\tfrac{1}{2}\frac{\sin \frac{1}{2} n\ (p+m)\ u \cos \frac{1}{2} (n-1)\ (p+m)\ u}{\sin \frac{1}{2} (p+m)\ u} \qquad (266)$$

$$\sum_{a=0}^{a=(n-1)} \sin a\ p\ u \cos a\ m\ u = \tfrac{1}{2}\frac{\sin \frac{1}{2} n\ (p-m)\ u \sin \frac{1}{2} (n-1)\ (p-m)\ u}{\sin \frac{1}{2} (p-m)\ u}$$
$$+\tfrac{1}{2}\frac{\sin \frac{1}{2} n\ (p+m)\ u \sin \frac{1}{2} (n-1)\ (p+m)\ u}{\sin \frac{1}{2} (p+m)\ u} \qquad (267)$$

**193.** If we let $\alpha=0$ and $u=\frac{2\pi}{n}$, or $n\ u=2\pi$, then formulas (263) to (267) may be written as follows:

$$\sum_{a=0}^{a=(n-1)} \sin a\ m\ u = \frac{\sin m\ \pi \sin\left(m\ \pi-\frac{m}{n}\pi\right)}{\sin \frac{m}{n}\pi} \qquad (268)$$

$$\sum_{a=0}^{a=(n-1)} \cos a\ m\ u = \frac{\sin m\ \pi \cos\left(m\ \pi-\frac{m}{n}\pi\right)}{\sin \frac{m}{n}\pi} \qquad (269)$$

$$\sum_{a=0}^{a=(n-1)} \sin a\, p\, u \sin a\, m\, u = \tfrac{1}{2} \frac{\sin (p-m)\,\pi \cos\left[(p-m)\,\pi - \frac{p-m}{n}\pi\right]}{\sin \frac{(p-m)}{n}\pi}$$

$$-\tfrac{1}{2} \frac{\sin (p+m)\,\pi \cos\left[(p+m)\,\pi - \frac{p+m}{n}\pi\right]}{\sin \frac{p+m}{n}\pi} \qquad (270)$$

$$\sum_{a=0}^{a=(n-1)} \cos a\, p\, u \cos a\, m\, u = \tfrac{1}{2} \frac{\sin (p-m)\,\pi \cos\left[(p-m)\,\pi - \frac{p-m}{n}\pi\right]}{\sin \frac{p-m}{n}\pi}$$

$$+\tfrac{1}{2} \frac{\sin (p+m)\,\pi \cos\left[(p+m)\,\pi - \frac{p+m}{n}\pi\right]}{\sin \frac{p+m}{n}\pi} \qquad (271)$$

$$\sum_{a=0}^{a=(n-1)} \sin a\, p\, u \cos a\, m\, u = \tfrac{1}{2} \frac{\sin (p-m)\,\pi \sin\left[(p-m)\,\pi - \frac{p-m}{n}\pi\right]}{\sin \frac{p-m}{n}\pi}$$

$$+\tfrac{1}{2} \frac{\sin (p+m)\,\pi \sin\left[(p+m)\,\pi - \frac{p+m}{n}\pi\right]}{\sin \frac{p+m}{n}\pi} \qquad (272)$$

**194.** If $p$ and $m$ are unequal integers and neither exceeds $\frac{n}{2}$, the above (268) to (272) become equal to zero. Thus,

$$\left. \begin{array}{l} \sum_{a=0}^{a=(n-1)} \sin a\, m\, u = 0 \\ \sum_{a=0}^{a=(n-1)} \cos a\, m\, u = 0 \\ \sum_{a=0}^{a=(n-1)} \sin a\, p\, u \sin a\, m\, u = 0 \\ \sum_{a=0}^{a=(n-1)} \cos a\, p\, u \cos a\, m\, u = 0 \\ \sum_{a=0}^{a=(n-1)} \sin a\, p\, u \cos a\, m\, u = 0 \end{array} \right\} \qquad (273)$$

**195.** If $p$ and $m$ are equal integers and do not exceed $\frac{n}{2}$, formulas (270), (271), and (272) will contain the indeterminate quantity $\frac{\sin (p-m)\pi}{\sin \frac{p-m}{n}\pi} = \frac{0}{0}$, and also when $p$ and $m$ each equal $\frac{n}{2}$, the indeterminate quantity $\frac{\sin (p+m)\pi}{\sin \frac{(p+m)}{n}\pi} = \frac{0}{0}$.

Evaluating these quantities we have

$$\left.\frac{\sin (p-m)\pi}{\sin \frac{p-m}{n}\pi}\right]_{(p-m)=0} = \left.\frac{\pi \cos (p-m)\pi}{\frac{\pi}{n} \cos \frac{p-m}{n}\pi}\right]_{(p-m)=0} = n \quad (274)$$

and

$$\left.\frac{\sin (p+m)\pi}{\sin \frac{p+m}{n}\pi}\right]_{(p+m)=n} = \left.\frac{\pi \cos (p+m)\pi}{\frac{\pi}{n} \cos \frac{p+m}{n}\pi}\right]_{(p+m)=n} = -n \quad (275)$$

In (275) it will be noted that when the integers $p$ and $m$ each equal $\frac{n}{2}$, $n$ must be an even number, and therefore $\cos n\pi$ is positive, while $\cos \pi$ is negative.

**196.** Assuming the condition that $p$ and $m$ are equal integers, each less than $\frac{n}{2}$, we have by substituting (274) in (270), (271), and (272),

$$\sum_{a=0}^{a=(n-1)} \sin a\, p\, u \sin a\, m\, u = \sum_{a=0}^{a=(n-1)} \sin^2 a\, m\, u = \tfrac{1}{2} n \quad (276)$$

$$\sum_{a=0}^{a=(n-1)} \cos a\, p\, u \cos a\, m\, u = \sum_{a=0}^{a=(n-1)} \cos^2 a\, m\, u = \tfrac{1}{2} n \quad (277)$$

$$\sum_{a=0}^{a=(n-1)} \sin a\, p\, u \cos a\, m\, u = \sum_{a=0}^{a=(n-1)} \sin a\, m\, u \cos a\, m\, u = 0 \quad (278)$$

**197.** Assuming the condition that $p$ and $m$ are each equal to $\frac{n}{2}$ we have by substituting (274) and (275) in (270), (271), and (272),

$$\sum_{a=0}^{a=(n-1)} \sin^2 a\, m\, u = \tfrac{1}{2} n + \tfrac{1}{2} n \cos \pi = 0 \quad (279)$$

$$\sum_{a=0}^{a=(n-1)} \cos^2 a\, m\, u = \tfrac{1}{2} n - \tfrac{1}{2} n \cos \pi = n \quad (280)$$

$$\sum_{a=0}^{a=(n-1)} \sin a\, m\, u \cos a\, m\, u = 0 \quad (281)$$

**198.** Returning now to the solution of (258), by substituting the successive values of $a$ from 0 to $(n-1)$, we have

$$\left.\begin{aligned}
h_0 &= H_0 + C_1 \cos 0 + C_2 \cos 0 + \text{\textemdash\textemdash} + C_k \cos 0 \\
    &\quad + S_1 \sin 0 + S_2 \sin 0 + \text{\textemdash\textemdash} + S_l \sin 0 \\
h_1 &= H_0 + C_1 \cos u + C_2 \cos 2u + \text{\textemdash\textemdash} + C_k \cos ku \\
    &\quad + S_1 \sin u + S_2 \sin 2u + \text{\textemdash\textemdash} + S_l \sin lu \\
h_2 &= H_0 + C_1 \cos 2u + C_2 \cos 4u + \text{\textemdash\textemdash} + C_k \cos 2ku \\
    &\quad + S_1 \sin 2u + S_2 \sin 4u + \text{\textemdash\textemdash} + S_l \sin 2lu \\
&\text{\textemdash\textemdash\textemdash\textemdash\textemdash\textemdash\textemdash\textemdash\textemdash\textemdash} \\
h_{(n-1)} &= H_0 + C_1 \cos (n-1)u + C_2 \cos 2(n-1)u + \text{\textemdash} \\
    &\quad + C_k \cos (n-1)ku \\
    &\quad + S_1 \sin (n-1)u + S_2 \sin 2(n-1)u + \text{\textemdash\textemdash} \\
    &\quad + S_l \sin (n-1)lu
\end{aligned}\right\} \quad (282)$$

**199.** To obtain value of $H_0$, add above equations

$$\sum_{a=0}^{a=(n-1)} h_a = n\, H_0$$

$$+ C_1 \sum_{a=0}^{a=(n-1)} \cos a\, u + C_2 \sum_{a=0}^{a=(n-1)} \cos 2\, a\, u + \text{--------} + C_k \sum_{a=0}^{a=(n-1)} \cos a\, k\, u$$

$$+ S_1 \sum_{a=0}^{a=(n-1)} \sin a\, u + S_2 \sum_{a=0}^{a=(n-1)} \sin 2\, a\, u + \text{--------} + S_l \sum_{a=0}^{a=(n-1)} \sin a\, l\, u$$

$$= n\, H_0 + \sum_{m=1}^{m=k} C_m \sum_{a=0}^{a=(n-1)} \cos a\, m\, u + \sum_{m=1}^{m=l} S_m \sum_{a=0}^{a=(n-1)} \sin a\, m\, u \qquad (283)$$

From (273), $\sum_{a=0}^{a=(n-1)} \cos a\, m\, u$ and $\sum_{a=0}^{a=(n-1)} \sin a\, m\, u$ each equals zero, since neither $k$ nor $l$, the maximum values of $m$ exceeds $\dfrac{n}{2}$

Therefore

$$\sum_{a=0}^{a=(n-1)} h_a = n\, H_0 \qquad (284)$$

and

$$H_0 = \frac{1}{n} \sum_{a=0}^{a=(n-1)} h_a \qquad (285)$$

**200.** To obtain the value of any coefficient $C$, such as $C_p$, multiply each equation of (282) by $\cos a\, p\, u$. Then

$h_0 \cos 0 = H_0 \cos 0$
  $\;+ C_1 \cos 0 + C_2 \cos 0 + \text{--------} + C_k \cos 0$
  $\;+ S_1 \sin 0 + S_2 \sin 0 + \text{--------} + S_l \sin 0$
$h_1 \cos p\, u = H_0 \cos p\, u$
  $\;+ C_1 \cos u \cos p\, u + C_2 \cos 2u \cos p\, u + \text{------} + C_k \cos k\, u \cos p\, u$
  $\;+ S_1 \sin u \cos p\, u + S_2 \sin 2u \cos p\, u + \text{--------} + S_l \sin l\, u \cos p\, u$
$h_2 \cos 2p\, u = H_0 \cos 2p\, u$
  $\;+ C_1 \cos 2u \cos 2p\, u + C_2 \cos 4u \cos 2p\, u + \text{--------}$
  $\;+ C_k \cos 2k\, u \cos 2p\, u$
  $\;+ S_1 \sin 2u \cos 2p\, u + S_2 \sin 4u \cos 2p\, u + \text{--------}$
  $\;+ S_l \sin 2l\, u \cos 2p\, u$

$\text{--------------------------------------------------}$

$h_{(n-1)} \cos (n-1)\, p\, u = H_0 \cos (n-1)\, p\, u$
  $\;+ C_1 \cos (n-1)\, u \cos (n-1)\, p\, u + C_2 \cos 2(n-1)\, u \cos (n-1)\, p\, u +$
  $\;+ C_k \cos (n-1)\, k\, u \cos (n-1)\, p\, u$
  $\;+ S_1 \sin (n-1)\, u \cos (n-1)\, p\, u + S_2 \sin 2(n-1)\, u \cos (n-1)\, p\, u + \text{--}$
  $\;+ S_l \sin (n-1)\, l\, u \cos (n-1)\, p\, u \qquad (286)$

Summing the above equations

$$\sum_{a=0}^{a=(n-1)} h_a \cos a\, p\, u = H_0 \sum_{a=0}^{a=(n-1)} \cos a\, p\, u$$

$$+ C_1 \sum_{a=0}^{a=(n-1)} \cos a\, u \cos a\, p\, u + S_1 \sum_{a=0}^{a=(n-1)} \sin a\, u \cos a\, p\, u \ldots$$

(Formula continued next page)

$$+ C_2 \sum_{a=0}^{a=(n-1)} \cos 2a\,u \cos a\,p\,u + S_2 \sum_{a=0}^{a=(n-1)} \sin 2a\,u \cos a\,p\,u$$

----

$$+ C_k \sum_{a=0}^{a=(n-1)} \cos a\,k\,u \cos a\,p\,u + S_l \sum_{a=0}^{a=(n-1)} \sin a\,l\,u \cos a\,p\,u$$

$$= H_0 \sum_{a=0}^{a=(n-1)} \cos a\,p\,u + \sum_{m=1}^{m=k} C_m \sum_{a=0}^{a=(n-1)} \cos a\,m\,u \cos a\,p\,u$$

$$+ \sum_{m=1}^{m=l} S_m \sum_{a=0}^{a=(n-1)} \sin a\,m\,u \cos a\,p\,u \qquad (287)$$

**201.** Examining the limits of (287), it will be noted by a reference to page 63 that $k$, the maximum value of $m$ for the $C$ terms is $\frac{n}{2}$ when $n$ is even and $\frac{n-1}{2}$ when $n$ is odd; also, that $l$ has a value of $\frac{n}{2}-1$ when $n$ is even and $\frac{n-1}{2}$ when $n$ is odd. The limits of $p$, which is a particular value of $m$, will, of course, be the same as those of $m$.

By (273) the quantity $\sum_{a=0}^{a=(n-1)} \cos a\,p\,u$ becomes zero for all the values of $p$, and the quantity $\sum_{a=0}^{a=(n-1)} \cos a\,m\,u \cos a\,p\,u$ becomes zero for all values of $m$ and $p$ except when $p$ equals $m$. By (273), (278) and (281) the quantity $\sum_{a=0}^{a=(n-1)} \sin a\,m\,u \cos a\,p\,u$ becomes zero for all values of $m$ and $p$.

Formula (287) may therefore be reduced to the form

$$\sum_{a=0}^{a=(n-1)} h_a \cos a\,p\,u = C_p \sum_{a=0}^{a=(n-1)} \cos^2 a\,p\,u \qquad (288)$$

For any value of $p$ less than $\frac{n}{2}$

$$\sum_{a=0}^{a=(n-1)} \cos^2 a\,p\,u = \tfrac{1}{2} n \qquad (277)$$

but when $p = \frac{n}{2}$, this quantity becomes equal to $n$ (280).

Therefore for all values of $p$ less than $\frac{n}{2}$

$$C_p = \frac{2}{n} \sum_{a=0}^{a=(n-1)} h_a \cos a\,p\,u \qquad (289)$$

but when $p$ is exactly $\frac{n}{2}$

$$C_p = \frac{1}{n} \sum_{a=0}^{a=(n-1)} h_a \cos a\,p\,u \qquad (290)$$

Since in tidal work $p$ is always taken less than $\frac{n}{2}$, we are not especially concerned with the latter formula.

**202.** To obtain the value of any coefficient $S$, such as $S_p$, multiply each equation of (282) by $\sin a\,p\,u$. Sum the resulting equations and obtain

$$\sum_{a=0}^{a=(n-1)} h_a \sin a\,p\,u = H_0 \sum_{a=0}^{a=(n-1)} \sin a\,p\,u$$

$$+ \sum_{m=1}^{m=k} C_m \sum_{a=0}^{a=(n-1)} \cos a\,m\,u \sin a\,p\,u$$

$$+ \sum_{m=1}^{m=l} S_m \sum_{a=0}^{a=(n-1)} \sin a\,m\,u \sin a\,p\,u \qquad (291)$$

By (273), (278), and (281) the quantities $\sum_{a=0}^{a=(n-1)} \sin a\,p\,u$ and $\sum_{a=0}^{a=(n-1)} \cos a\,m\,u \sin a\,p\,u$ are zero for all the values of $m$ and $p$; and $\sum_{a=0}^{a=(n-1)} \sin a\,m\,u \sin a\,p\,u$ becomes zero for all the values of $m$ and $p$ except when $m$ and $p$ are equal. In this case the limit of $l$ for $m$ and $p$ is less than $\frac{n}{2}$ and by (276), the quantity $\sum_{a=0}^{a=(n-1)} \sin^2 a\,p\,u = \frac{1}{2} n$. Therefore, formula (291) reduces to the form

$$\sum_{a=0}^{a=(n-1)} h_a \sin a\,p\,u = \frac{1}{2} n\, S_p \qquad (292)$$

and

$$S_p = \frac{2}{n} \sum_{a=0}^{a=(n-1)} h_a \sin a\,p\,u \qquad (293)$$

**203.** By substituting (285), (289), (290), and (293) in (257), the following equation of a curve, which will pass through the $n$ given points, will be obtained

$$h = \frac{1}{n} \sum_{a=0}^{a=(n-1)} h_a + \left[\frac{2}{n} \sum_{a=0}^{a=(n-1)} h_a \cos a\,u\right] \cos \theta$$

$$+ \left[\frac{2}{n} \sum_{a=0}^{a=(n-1)} h_a \sin a\,u\right] \sin \theta$$

$$+ \left[\frac{2}{n} \sum_{a=0}^{a=(n-1)} h_a \cos 2 a\,u\right] \cos 2\theta$$

$$+ \left[\frac{2}{n} \sum_{a=0}^{a=(n-1)} h_a \sin 2 a\,u\right] \sin 2\theta$$

-----

$$+ \left[\frac{2^*}{n} \sum_{a=0}^{a=(n-1)} h_a \cos k a\,u\right] \cos k\theta$$

$$+ \left[\frac{2}{n} \sum_{a=0}^{a=(n-1)} h_a \sin l a\,u\right] \sin l\theta \qquad (294)$$

*If $n$ is even and $k = \frac{n}{2}$, this fraction is $\frac{1}{n}$ instead of $\frac{2}{n}$.

204. Although by taking a sufficient number of terms the Fourier series may thus be made to represent a curve which will be exactly satisfied by the $n$ given ordinates, this is, in general, neither necessary nor desirable in tidal work, since it is known that the mean ordinates obtained from the summations of the hourly heights of the tide include many irregularities due to the imperfect elimination of the meteorological effects and also residual effects of constituents having periods incommensurable with that of the constituent sought. It is desirable to include only the terms of the series which represent the true periodic elements of the constituent. With series of observations of sufficient length, the coefficient of the other terms, if sought, will be found to approximate to zero.

205. The short-period constituents as derived from the equilibrium theory are, in general, either diurnal or semidiurnal. If the period of $\theta$ in formula (257) is taken to correspond to the constituent day, the diurnal constituents will be represented by the terms with coefficient $C_1$ and $S_1$, and the semidiurnal constituents by the terms with coefficients $C_2$ and $S_2$. For the long-period constituents, the period of $\theta$ may be taken to correspond to the constituent month or to the constituent year, in which case the coefficients $C_1$ and $S_1$ will refer to the monthly or annual constituents and the coefficients $C_2$ and $S_2$ to the semimonthly or semiannual constituents. For most of the constituents the coefficients $C_1$, $S_1$, $C_2$, and $S_2$ will be the only ones required, but for the tides depending upon the fourth power of the moon's parallax and for the overtides and the compound tides, other coefficients will be required. Terms beyond those with coefficients $C_8$ and $S_8$, for the overtides of the principal lunar constituent are not generally used in tidal work.

206. When it is known that certain periodic elements exist in a constituent tide and that the mean ordinates obtained from observations include accidental errors that are not periodic, it may be readily shown by the method known as the least square adjustment, using the observational equations represented by (258), that the most probable values of the constant $H_0$ and the coefficients $C_p$ and $S_p$ are the same as those given by formulas (285), (289), and (293), respectively.

207. Since in tidal work the value of $H_0$, which is the elevation of mean sea level above the datum of observations, is generally determined directly from the original tabulation of hourly heights, formula (285) is unnecessary except for checking purposes. Formulas (289) and (293) are used for obtaining the most probable values of the coefficients $C_p$ and $S_p$ from the hourly means obtained from the summations.

208. When 24 hourly means are used $n=24$ and $u=15°$, and the formulas may be written.

$$C_p = \frac{1}{12} \sum_{a=0}^{a=23} h_a \cos 15\,a\,p \qquad (295)$$

$$S_p = \frac{1}{12} \sum_{a=0}^{a=23} h_a \sin 15\,a\,p \qquad (296)$$

in which the angles are expressed in degrees.

If only 12 means are used, the formulas become

$$C_p = \frac{1}{6} \sum_{a=0}^{a=11} h_a \cos 30\,a\,p \qquad (297)$$

$$S_p = \frac{1}{6} \sum_{a=0}^{a=11} h_a \sin 30\ a\ p \qquad (298)$$

**209.** The upper part of Form 194 (fig. 16) is designed for the computation of the coefficients $C_p$ and $S_p$ in accordance with formulas (295) and (296) to take account of the 24 constituent hourly means.

It is now desired to express each constituent in the form

$$y = A \cos (p\ \theta + \alpha) \qquad (299)$$

or using a more specialized notation by

$$y = A \cos (p\ \theta - \zeta) \qquad (300)$$

By trigonometry

$$A \cos (p\ \theta - \zeta) = A \cos \zeta \cos p\ \theta + A \sin \zeta \sin p\ \theta \qquad (301)$$
$$= C_p \cos p\ \theta + S_p \sin p\ \theta$$

in which $\qquad C_p = A \cos \zeta \quad$ and $\quad S_p = A \sin \zeta \qquad (302)$

Therefore,

$$\tan \zeta = \frac{S_p}{C_p} \qquad (303)$$

and

$$A = \frac{C_p}{\cos \zeta} = \frac{S_p}{\sin \zeta} = \sqrt{C_p^2 + S_p^2} \qquad (304)$$

Substituting in formulas (303) and (304) the values of $C_p$ and $S_p$ from formulas (295) and (296), the corresponding values for $A$ and $\zeta$ may be obtained. Substituted in formula (300), these furnish an approximate representation of one of the tidal constituents, but a further processing is necessary in order to obtain the mean amplitude and epoch of the constituent.

### AUGMENTING FACTORS

**210.** In the usual summations with the primary stencils for all the short period constituents, except constituent S, the hourly ordinates which are summed in any single group are scattered more or less uniformly over a period from one-half of a constituent hour before to one-half of a constituent hour after the exact constituent hour which the group represents. Because of this the resulting mean will differ a little from the true mean ordinate that would be obtained if all the ordinates included were read on the exact constituent hour, as with constituent S, and the amplitude obtained will be less than the true amplitude of the constituent. The factor necessary to take account of this fact is called the augmenting factor.

**211.** Let any constituent be represented by the curve

$$y = A \cos (at + \alpha) \qquad (305)$$

in which
$\quad A =$ the true amplitude of the constituent
$\quad a =$ the speed of the constituent (degrees per solar hours)
$\quad t =$ variable time (expressed in solar hours)
$\quad \alpha =$ any constant.

The mean value of $y$ for a group of consecutive ordinates from $\tau/2$ hours before to $\tau/2$ hours after any given time $t$, $\tau$ being the number of solar hours covered by the group, is

$$\frac{A}{\tau}\int_{t-\tau/2}^{t+\tau/2} \cos(at+\alpha)dt = \left[\frac{180}{\pi}\frac{A}{a\tau}\sin(at+\alpha)\right]_{t-\tau/2}^{t+\tau/2}$$

$$= \frac{180}{\pi}\frac{A}{a\tau}\left[\sin\left(at+\alpha+\frac{a\tau}{2}\right)-\sin\left(at+\alpha-\frac{a\tau}{2}\right)\right]$$

$$= \frac{360}{\pi}\frac{A}{a\tau}\cos(at+\alpha)\sin\frac{a\tau}{2} = \frac{360}{\pi a\tau}\sin\frac{a\tau}{2} A \cos(at+\alpha) \quad (306)$$

**212.** Since the true value of $y$ at any time $t$, is equal to $A \cos(at+\alpha)$ by (305), it is evident that the relation of this true value to the mean value (306) for the group $\tau$ hours in length is

$$\frac{A \cos(at+\alpha)}{\frac{360}{\pi a\tau}\sin\frac{a\tau}{2} A \cos(at+\alpha)} = \frac{\pi a\tau}{360 \sin\frac{a\tau}{2}} \quad (307)$$

The quantity $\dfrac{\pi a\tau}{360 \sin\dfrac{a\tau}{2}}$ is the augmenting factor which is to be applied to the mean ordinate to obtain the true ordinate. In the use of this factor it is assumed that all the consecutive ordinates within the time $\tau/2$ hours before to $\tau/2$ hours after the given time have been used in obtaining the mean. This assumption is, of course, only approximately realized in the summation for any constituent, but the longer the series of observations the more nearly to the truth it approaches.

**213.** According to the usual summations with the primary stencils, the hourly heights included in a single group may be distributed over an interval from one-half of a constituent hour before to one-half of a constituent hour after the hour to be represented. In this case $\tau$ equals one constituent hour, or $\dfrac{15p}{a}$ solar hours.

Substituting this in (307), the

$$\text{augmenting factor} = \frac{\pi p}{24 \sin\dfrac{15p}{2}} \quad (308)$$

which is the formula generally adopted for the short-period constituents and is the one used in the calculation of the augmenting factors in Form 194. For the long-period constituents special factors are necessary which will be explained later.

**214.** If the second system of distribution of the hourly heights as described on page 53 is adopted, $\tau$ equals one solar hour and formula (307) becomes

$$\text{augmenting factor} = \frac{\pi a}{360 \sin\dfrac{a}{2}} \quad (309)$$

It will be noted that formula (308) depends upon the value of $p$ and therefore will be the same for all short period constituents (S excepted) with like subscripts. Formula (309) depends upon the speed $a$ of the constituent and will therefore be different for each constituent.

**215.** When the secondary stencils are used, the grouping of the ordinates is less simple than that provided by the primary stencils only. Let it be assumed that the series is of sufficient length so that the distribution of the ordinates is more or less uniform in accordance with the system adopted.

Suppose the original primary summations have been made for constituent $A$ with speed $a$ and that the secondary stencils have been used for constituent $B$ with speed $b$. Then let $p$ and $p'$ represent the subscripts of constituents $A$ and $B$, respectively.

The equation for constituent $B$ may be written

$$y = B \cos (bt + \beta) \tag{310}$$

**216.** In the primary summation for constituent $A$, the group of ordinates included in a single sum covers a period of one constituent $A$ hour or $\frac{15p}{a}$ solar hours. Expressed in time $t$, midway of this interval and representing the exact integral constituent $A$ hour to which the group applied, the average value of the $B$ ordinates included in such a group may be written

$$\frac{a}{15p} B \int_{t-\frac{15p}{a}}^{t+\frac{15p}{a}} \cos (bt + \beta) \, dt$$

$$= \frac{180}{\pi} \frac{a}{15pb} B \left[ \sin \left( bt + \beta + \frac{15pb}{2a} \right) - \sin \left( bt + \beta - \frac{15pb}{2a} \right) \right]$$

$$= \left( \frac{24}{\pi} \frac{a}{pb} \sin \frac{15pb}{2a} \right) B \cos (bt + \beta)$$

$$= F_1 B \cos (bt + \beta) \tag{311}$$

In which $F_1$, for brevity, is substituted for the coefficient $\frac{24}{\pi} \frac{a}{pb} \sin \frac{15pb}{2a}$ and gives the relation of the average $B$ ordinate included in the $A$ grouping to the true $B$ ordinate for the time $t$ represented by that group. The reciprocal of this coefficient will be that part of the augmenting factor necessary to take account of this primary grouping. If the primary summing has been for the constituent S, this coefficient may be taken as unity since the original S sums refer to the exact S hour.

**217.** When the secondary stencils are applied to the constituent $A$ group sums, the groups applying to an exact constituent $A$ hour at any time $t$ and represented by that time, will be distributed over an interval of a constituent $B$ hour, or $\frac{15p'}{b}$ solar hours.

For an integral constituent $B$ hour at any time $t$ within the middle day represented by a seven-day page of original tabulations the limits of this interval will be $\left( t - \frac{15p'}{2b} \right)$ and $\left( t + \frac{15p'}{2b} \right)$. For the same page of tabulations, letting $t$ represent the same time in the middle day, the limits of the group interval for the day following the middle one, are $\left(t+\frac{360p}{a}-\frac{15p'}{2b}\right)$ and $\left(t+\frac{360p}{a}+\frac{15p'}{2b}\right)$. If we let $n=-3, -2, -1, 0, +1, +2, +3$, respectively, for the seven successive days represented by a single page of original tabulations, the limits of the group interval for any day of the page may be represented by

$$\left(t+\frac{360pn}{a}-\frac{15p'}{2b}\right) \text{ and } \left(t+\frac{360pn}{a}+\frac{15p'}{2b}\right)$$

**218.** Formula (311) gives the mean value of the $B$ ordinate for grouping of the $A$ summations. The mean value of (311) obtained by combining the groups falling in any particular day of page of tabulations in the limits indicated above is

$$\frac{b}{15p'} F_1 B \int_{t+\frac{360pn}{a}-\frac{15p'}{2b}}^{t+\frac{360pn}{a}+\frac{15p'}{2b}} \cos(bt+\beta) \, dt$$

$$= \frac{180}{\pi} \frac{1}{15p'} F_1 B \left[ \sin\left(bt+\beta+\frac{360bpn}{a}+\frac{15p'}{2}\right) - \sin\left(bt+\beta+\frac{360bpn}{a}-\frac{15p'}{2}\right) \right]$$

$$= \left(\frac{24}{\pi}\frac{1}{p'} \sin \frac{15p'}{2}\right) F_1 B \cos\left(bt+\beta+\frac{360pn}{a}\right)$$

$$= F_1 F_2 B \cos\left(bt+\beta+\frac{360bpn}{a}\right) \quad (312)$$

if we put $F_2 = \frac{24}{\pi} \frac{1}{p'} \sin \frac{15p'}{2}$ for brevity.

**219.** Formula (312) represents the mean value of the $B$ ordinate for a particular day of the page record. The average value for the 7 days may be written

$$\tfrac{1}{7} F_1 F_2 B \sum_{n=-3}^{n=+3} \cos\left(t+\beta+\frac{360bpn}{a}\right)$$

$$= \tfrac{1}{7} F_1 F_2 B \left[ \cos(bt+\beta) \cos\left(-3\frac{360bp}{a}\right) - \sin(bt+\beta) \sin\left(-3\frac{360bp}{a}\right) \right.$$

$$+ \cos(bt+\beta) \cos\left(-2\frac{360bp}{a}\right) - \sin(bp+\beta) \sin\left(-2\frac{360bp}{a}\right)$$

$$+ \cos(bt+\beta) \cos\left(-1\frac{360bp}{a}\right) - \sin(bt+\beta) \sin\left(-1\frac{360bp}{a}\right)$$

$$+ \cos(bt+\beta) \cos 0 - \sin(bt+\beta) \sin 0$$

$$+ \cos(bt+\beta) \cos\left(\frac{360bp}{a}\right) - \sin(bt+\beta) \sin\left(\frac{360bp}{a}\right)$$

(Formula continued next page)

$$+\cos(bt+\beta)\cos\left(2\,\frac{360bp}{a}\right)-\sin(bt+\beta)\sin\left(2\,\frac{360bp}{a}\right)$$

$$+\cos(bt+\beta)\cos\left(3\,\frac{360bp}{a}\right)-\sin(bt+\beta)\sin\left(3\,\frac{360bp}{a}\right)\bigg]$$

$$=\tfrac{1}{7}F_1F_2B\left[1+2\cos\frac{360bp}{a}+2\cos 2\,\frac{360bp}{a}+2\cos 3\,\frac{360bp}{a}\right]\cos(bt+\beta)$$

$$=\tfrac{1}{7}F_1F_2B\left[2\,\frac{\sin 2\,\frac{360bp}{a}\cos\frac{3}{2}\,\frac{360bp}{a}}{\sin\tfrac{1}{2}\frac{360bp}{a}}-1\right]\cos(bt+\beta)$$

$$=\tfrac{1}{7}F_1F_2B\left[\frac{\sin\frac{1260bp}{a}}{\sin\frac{180bp}{a}}\right]\cos(bt+\beta). \qquad (313)$$

**220.** Replacing the equivalents of $F_1$ and $F_2$ in (313), the average value of the $B$ ordinate as obtained by the secondary summations may be written

$$\left[\frac{24a}{\pi pb}\sin\frac{15bp}{2a}\right]\left[\frac{24}{\pi p'}\sin\frac{15p'}{2}\right]\left[\frac{\sin\frac{1260bp}{a}}{7\sin\frac{180bp}{a}}\right]B\cos(bt+\beta) \qquad (314)$$

Since the true ordinate of constituent $B$ at any time $t$ is equal to $B\cos(bt+\beta)$, the reciprocal of the bracketed coefficient will be the augmenting factor necessary to reduce the $B$ ordinate as obtained from the summations to their true values.

This augmenting factor may be written

$$\left[\frac{\pi bp}{24a\sin\frac{15bp}{2a}}\right]\left[\frac{\pi p'}{24\sin\frac{15p'}{2}}\right]\left[\frac{7\sin\frac{180bp}{a}}{\sin\frac{1260bp}{a}}\right] \qquad (315)$$

The first factor of the above is to be omitted if the primary summations are for constituent S. It will be noted that the middle factor is the same as the augmenting factor that would be used if constituent $B$ had been subjected to the primary summations.

### PHASE LAG OR EPOCH

**221.** The phase lag or epoch of a tidal constituent, which is represented by the Greek kappa ($\kappa$), is the difference between the phase of the observed constituent and the phase of its argument at the same time. This difference remains approximately constant for any constituent in a particular locality. The phase of a constituent argument for any time may be obtained from the argument formula in table 2 by making suitable substitutions for the astronomical elements. The argument itself is represented by the general symbol $(V+u)$ or $E$ and its phase or value pertaining to an initial instant of time, such as the beginning of a series of observations, is expressed by $(V_0+u)$. Referring to formula (300), since $\theta$ is reckoned from the beginning of the series, the angular quantity $(-\zeta)$ is the corresponding phase of the observed constituent at this time. The phase lag may therefore be expressed by the following general formula:

$$\kappa = V_0 + u - (-\zeta) = V_0 + u + \zeta \qquad (316)$$

**222.** Since the argument formulas of all short-period constituents contain some multiple of the hour angle $(T)$ of the mean sun, the arguments themselves will have different values in different longitudes at the same instant of time. If $p$ equals the coefficient of $T$ or the subscript of the constituent and $L$ equals the longitude of the place in degrees reckoned west from Greenwich, $L$ being considered as negative for east longitude, the relation between the local and Greenwich argument for any constituent may be expressed as follows:

$$\text{local } (V+u) = \text{Greenwich } (V+u) - pL \qquad (317)$$

**223.** Also, since the absolute time of the beginning of a day or the beginning of a year depends upon the time meridian used in the locality, the initial instant taken for the beginning of a series of observations may differ in different localities even though expressed in the same clock time of the same calendar day. If we let $S$ equal the longitude of the time meridian in degrees, positive for west and negative for east, the same meridian expressed in hours becomes $S/15$. Letting $a$ equal the speed or hourly rate of change in the constituent argument, the difference in argument due to the difference in the absolute beginning of the series becomes $aS/15$, and the relation between the local and Greenwich argument due to this difference may be expressed as follows:

$$\text{local } (V_0+u) = \text{Greenwich } (V_0+u) - pL + aS/15 \qquad (318)$$

In the above formula the local and Greenwich $(V_0+u)$ pertain to the same clock time but not the same absolute time unless both clocks are set for the meridian of Greenwich.

**224.** Values of $(V_0+u)$ for the meridian of Greenwich at the beginning of each calendar year 1850 to 2000 are given in table 15 for all constituents represented in the Coast and Geodetic Survey tide-predicting machine. Tables 16 to 18 provide differences for referring the arguments to other days and hours of the year. In the preparation of table 15 that portion of the argument included in the $u$ was treated as a constant with a value pertaining to the middle of the calendar year. If the Greenwich $(V_0+u)$ with its corrections is substituted for the local $(V_0+u)$ in formula (316), we obtain

$$\kappa = \text{Greenwich } (V_0+u) - pL + aS/15 + \zeta \qquad (319)$$

**225.** The phase lag designated by $\kappa$ is sometimes called the local epoch to distinguish it from certain modified forms which may be used for special purposes. In the preparation of the harmonic constants for predictions it is convenient to combine the longitude and time meridian corrections with the local epoch to form a modified epoch designated by $k'$ or by the small $g$. The relation of the modified epoch to the local epoch may then be expressed by the following formula:

$$\kappa' \text{ or } g = \kappa + pL - aS/15 = \text{Greenwich } (V_0+u) + \zeta \quad (320)$$

**226.** The phases of the same tidal constituent in different parts of the world are not directly comparable through their local epochs since these involve the longitude of the locality. For such a comparison it is desirable to have a Greenwich epoch that is independent of both longitude and time meridian. Such an epoch may be designated by the capital $G$ and its relation to the corresponding local epoch expressed as follows:

$$\text{Greenwich epoch } (G) = \kappa + pL = \text{Greenwich } (V_0+u) + aS/15 + \zeta \quad (321)$$

**227.** The angle $\kappa$ may be graphically represented by figures 7 and 8. In figure 7, we have a simple representation of a single con-

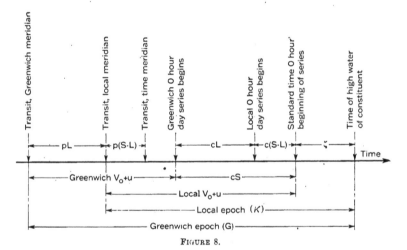

FIGURE 8.

stituent. In this figure changes in the phase or angle are measured along the horizontal line, positive change toward the right and negative change toward the left. The full vertical line indicates the beginning of the series, at which time the angle $p$ $\theta$, or $at$, equals 0. At the left of this vertical line, the symbol of a moon ($M$) indicates the zero value of the equilibrium argument that precedes the beginning of the series. For the principal lunar or solar constituent, this will be simultaneous with a transit of the mean moon (modified by longitude of moon's node) or of the mean sun, and for other short-period constituents with the transit of a fictitious star representing such constituent (p. 23). At the point represented by this moon, the angle $(V+u)$ has a value of zero. This angle increases to the right, and at the beginning of the series has a value represented by $(V_0+u)$, which may be readily computed for the beginning of any series. This interval from $M$ to the time of occurrence of the first following constituent high water is the epoch $\kappa$. This represents the lag or difference between the actual constituent high water at any place and the theoretical time as determined by the equilibrium theory. The distance from the beginning of the series to the following high water is the $\zeta$ of formula (300), which is determined directly from the analysis of the observations. From the figure it is evident that the $\kappa$ is the sum of $(V_0+u)$ and $\zeta$, and also that it is independent of the time of the beginning of the series.

**228.** Figure 8 gives a more detailed representation of the epoch of a constituent. In this figure the horizontal line represents changes in time. Distances along this line will be proportional to the changes in the angle of any single constituent, but since each constituent has a different speed equal distances along this line will not represent equal angles for different constituents. The time between the events may be converted into an equivalent constituent angle by multiplying by the speed of the constituent. The figure is to some extent self-explanatory. The word "transit" signifies the transit of the fictitious moon representing any constituent and also the time when the equilibrium argument of that constituent has a zero value. For all short-period constituents the time of such zero value will depend upon the longitude of the place of observation as well as upon absolute time. For long-period constituents the zero values are independent of the longitude of the place of observation, and the "transits" over the several meridians may be considered as occurring simultaneously, which is equivalent to taking the coefficient $p$ equal to zero. The figure illustrates the relation between the Greenwich $(V_0+u)$ calculated for the meridian of Greenwich and referring to standard Greenwich time and local $(V_0+u)$ referring to the meridian of observation and the actual time of the beginning of the observations.

### INFERENCE OF CONSTANTS

**229.** Under the conditions assumed for the equilibrium theory the amplitudes of the constituents could be computed directly by means of the coefficient formulas without the necessity of securing tidal observations, and the phases would correspond with the equilibrium arguments of the constituents. Under the conditions that actually exist it has been found from observations that the amplitudes of the constituents of a similar type at any place, although differing greatly from their theoretical values, have a relation that, in general, agrees fairly closely with the relations of their theoretical coefficients. It has also been ascertained from the results obtained from observations that the difference in the epochs or lags of the constituents have a relation conforming, in general, with the relation of the differences in their speeds. This last relation is based upon an assumption that the ages of the inequalities due to the disturbing influence of other constituents of a similar type are equal when expressed in time.

**230.** If the mean amplitudes, epochs, and speeds of several constituents $A$, $B$, $C$, are represented by $H(A)$, $H(B)$, $H(C)$, $\kappa(A)$, $\kappa(B)$, $\kappa(C)$, and $a$, $b$, $c$, respectively, the above relations may be expressed by the following formulas:

$$H(B) = \frac{\text{mean coefficient of } B}{\text{mean coefficient of } A} H(A) \tag{322}$$

$$\kappa(C)-\kappa(A)=\frac{c-a}{b-a}[\kappa(B)-\kappa(A)] \qquad (323)$$

or,

$$\kappa(C)=\kappa(A)+\frac{c-a}{b-a}[\kappa(B)-\kappa(A)] \qquad (324)$$

By formula (322) the amplitude of a constituent ($B$) may be inferred from the known amplitude of a constituent ($A$), and by formula (324) the epoch of a constituent ($C$) may be inferred from the known epochs of constituents ($A$) and ($B$).

**231.** These formulas have, however, certain limitations. They are not applicable to shallow water and meteorological constituents, nor are they adapted to the determination of a diurnal constituent from a semidiurnal constituent or of a semidiurnal constituent from a diurnal constituent. The results obtained by the application of the formulas to tides of similar type may be considered only as rough approximations to the truth. They may, however, be preferable to the values obtained for certain constituents when the series of observations is short.

**232.** By substituting the mean values of the coefficients and the speeds from table 2 the following special formulas may be derived from the general formulas (322) and (324)

*Diurnal constituents*

$H(J_1) = 0.079\ H(O_1);\quad \kappa(J_1) = \kappa(K_1)+0.496\ [\kappa(K_1)-\kappa(O_1)]$ (325)
$H(M_1) = 0.071\ H(O_1);\quad \kappa(M_1) = \kappa(K_1)-0.500\ [\kappa(K_1)-\kappa(O_1)]$ (326)
$H(OO) = 0.043\ H(O_1);\quad \kappa(OO) = \kappa(K_1)+1.000\ [\kappa(K_1)-\kappa(O_1)]$ (327)
$H(P_1) = 0.331\ H(K_1);\quad \kappa(P_1) = \kappa(K_1)-0.075\ [\kappa(K_1)-\kappa(O_1)]$ (328)
$H(Q_1) = 0.194\ H(O_1);\quad \kappa(Q_1) = \kappa(K_1)-1.496\ [\kappa(K_1)-\kappa(O_1)]$ (329)
$H(2Q) = 0.026\ H(O_1);\quad \kappa(2Q) = \kappa(K_1)-1.992\ [\kappa(K_1)-\kappa(O_1)]$ (330)
$H(\rho_1) = 0.038\ H(O_1);\quad \kappa(\rho_1) = \kappa(K_1)-1.429\ [\kappa(K_1)-\kappa(O_1)]$ (331)

*Semidiurnal constituents*

$H(K_2) = 0.272\ H(S_2)\ ;\quad \kappa(K_2) = \kappa(S_2)\ +0.081\ [\kappa(S_2)\ -\kappa(M_2)]$ (332)
$H(L_2) = 0.028\ H(M_2);\quad \kappa(L_2) = \kappa(S_2)\ -0.464\ [\kappa(S_2)\ -\kappa(M_2)]$ (333)
$\qquad\quad = 0.143\ H(N_2)\ ;\qquad\qquad = \kappa(M_2)+1.000\ [\kappa(M_2)-\kappa(N_2)]$ (334)
$H(N_2) = 0.194\ H(M_2);\quad \kappa(N_2) = \kappa(S_2)\ -1.536\ [\kappa(S_2)\ -\kappa(M_2)]$ (335)
$H(2N) = 0.026\ H(M_2);\quad \kappa(2N) = \kappa(S_2)\ -2.072\ [\kappa(S_2)\ -\kappa(M_2)]$ (336)
$\qquad\quad = 0.133\ H(N_2);\qquad\qquad = \kappa(M_2)-2.000\ [\kappa(M_2)-\kappa(N_2)]$ (337)
$H(R_2) = 0.008\ H(S_2);\quad \kappa(R_2) = \kappa(S_2)\ +0.040\ [\kappa(S_2)\ -\kappa(M_2)]$ (338)
$H(T_2) = 0.059\ H(S_2);\quad \kappa(T_2) = \kappa(S_2)\ -0.040\ [\kappa(S_2)\ -\kappa(M_2)]$ (339)
$H(\lambda_2) = 0.007\ H(M_2);\quad \kappa(\lambda_2) = \kappa(S_2)\ -0.536\ [\kappa(S_2)\ -\kappa(M_2)]$ (340)
$H(\mu_2) = 0.024\ H(M_2);\quad \kappa(\mu_2) = \kappa(S_2)\ -2.000\ [\kappa(S_2)\ -\kappa(M_2)]$ (341)
$H(\nu_2) = 0.038\ H(M_2);\quad \kappa(\nu_2) = \kappa(S_2)\ -1.464\ [\kappa(S_2)\ -\kappa(M_2)]$ (342)
$\qquad\quad = 0.194\ H(N_2);\qquad\qquad = \kappa(M_2)-0.866\ [\kappa(M_2)-\kappa(N_2)]$ (343)

**233.** In order to test the reliability of the results obtained by inference as above, 60 stations representing various types of tide in different parts of the world where the harmonic constants had been determined from observations were selected and a comparison was made between the values for certain constants as obtained by inference and by observations. The tests were applied to the diurnal constituents

$M_1$, $P_1$, and $Q_1$, and to the semidiurnal constituents $K_2$, $L_2$, and $\nu_2$, and formulas (326), (328), (329), (332), (333), and (342) were used for the purpose. The following results were obtained for the differences between values as obtained from inference and from observations. The average gross difference is the average difference without regard to the signs of the individual items, and the average net difference takes into account these signs so that a positive difference may offset a negative difference in the mean. The last two lines in the table show the percentage of cases in which the differences were less than 0.05 and 0.10 foot, respectively, for the amplitudes, and less than 10° and 20°, respectively, for the epochs.

|  | $M_1$ amplitude | $M_1$ epoch | $P_1$ amplitude | $P_1$ epoch | $Q_1$ amplitude | $Q_1$ epoch |
|---|---|---|---|---|---|---|
|  | Ft. | Deg. | Ft. | Deg. | Ft. | Deg. |
| Maximum difference | 0.05 | 149 | 0.27 | 49 | 0.05 | 105 |
| Average gross difference | .02 | 31 | .03 | 8 | .01 | 14 |
| Average net difference | .01 | 1 | .01 | 3 | .00 | 0 |
|  | % | % | % | % | % | % |
| Differences less than 0.05 foot or 10° | 93 | 37 | 85 | 76 | 96 | 58 |
| Differences less than 0.10 foot or 20° | 100 | 57 | 92 | 92 | 100 | 82 |

|  | $K_2$ amplitude | $K_2$ epoch | $L_2$ amplitude | $L_2$ epoch | $\nu_2$ amplitude | $\nu_2$ epoch |
|---|---|---|---|---|---|---|
|  | Ft. | Deg. | Ft. | Deg. | Ft. | Deg. |
| Maximum difference | 0.28 | 51 | 1.09 | 104 | 0.28 | 53 |
| Average gross difference | .02 | 9 | .09 | 25 | .04 | 14 |
| Average net difference | .00 | 5 | .08 | 4 | .02 | 4 |
|  | % | % | % | % | % | % |
| Differences less than 0.05 foot or 10° | 87 | 65 | 58 | 20 | 71 | 48 |
| Differences less than 0.10 foot or 20° | 97 | 93 | 78 | 44 | 88 | 83 |

By using formulas (334) and (343) for $L_2$ and $\nu_2$ the results are slightly improved, the average net differences for the amplitude and epoch of $L_2$ becoming 0.07 foot and 3°, respectively, the difference for the epoch of $\nu_2$ becoming 2°, while the average net difference for the amplitude of $\nu_2$ remains unchanged.

**234.** Although there is a fairly good agreement indicated by the average differences, it is evident that the inferred constants, especially the epochs, cannot be depended upon for a high degree of refinement. It may be stated, however, that for constituents with very small amplitudes the epochs determined from actual observations may be equally unreliable. This becomes evident when results from different years of observations are compared. Fortunately, the large discrepancies in epochs are found only in constituents of small amplitude and are therefore of little practical importance.

**235.** Constituent $\mu_2$ as determined by inference is relatively unimportant. However, this constituent has the same period as the compound tide $2MS_2$ and when obtained directly from the analysis of observations frequently differs considerably from the inferred $\mu_2$ both in amplitude and epoch. The inferred values for this constituent cannot therefore be considered as very satisfactory.

**236.** Prior to the elimination process described in the next section, certain preliminary corrections are applied to the amplitudes and epochs of constituents $S_2$ and $K_1$ because of the disturbing effects of $K_2$ and $T_2$ on the former and $P_1$ on the latter. In a short series of observations these effects may be considerable because of the small differences in the periods of the constituents involved.

**237.** Let

$$y_1 = A \cos (at+\alpha) \qquad (344)$$

and

$$y_2 = B \cos (bt+\beta) \qquad (345)$$

represent two constituents, the first being the principal or predominating constituent and the latter a secondary constituent whose effect is to modify the amplitude and epoch of the principal constituent. The resultant tide will then be represented by

$$y = y_1 + y_2 = A \cos (at+\alpha) + B \cos (bt+\beta) \qquad (346)$$

Values of $t$ which will render (344) a maximum must satisfy the derived equation

$$Aa \sin (at+\alpha) = 0 \qquad (347)$$

and the values of $t$ which will render (346) a maximum must satisfy the equation

$$Aa \sin (at+\alpha) + Bb \sin (bt+\beta) = 0 \qquad (348)$$

For a maximum of (344)

$$t = \frac{2n\pi - \alpha}{a} \qquad (349)$$

in which $n$ is any integer.

**238.** Let $\dfrac{\theta}{a}$ = the acceleration in the principal constituent $A$ due to the disturbing constituent $B$. Then for a maximum of (346)

$$t = \frac{2n\pi - \alpha - \theta}{a} \qquad (350)$$

This value of $t$ must satisfy equation (348), therefore we have

$$Aa \sin (2n\pi - \theta) + Bb \sin \left[\frac{b}{a}(2n\pi - \theta - \alpha) + \beta\right]$$

$$= -Aa \sin \theta + Bb \sin \left[\frac{b-a}{a}(2n\pi - \theta - \alpha) + \beta - \alpha - \theta\right] = 0 \qquad (351)$$

At the time of this maximum, when

$$t = \frac{2n\pi - \alpha - \theta}{a},$$

the phase of constituent $A$ will equal

$$(2n\pi - \alpha - \theta) + \alpha$$

and the phase of constituent $B$ will equal

$$\frac{b}{a}(2n\pi - \alpha - \theta) + \beta$$

Let $\phi$ = phase of constituent $B$ − phase of constituent $A$ at this time. Then

$$\phi = \frac{b-a}{a}(2n\pi - \alpha - \theta) + \beta - \alpha \qquad (352)$$

Substituting the above in (351)

$$-Aa \sin \theta + Bb \sin (\phi - \theta)$$
$$= -Aa \sin \theta + Bb \sin \phi \cos \theta - Bb \cos \phi \sin \theta$$
$$= -(Aa + Bb \cos \phi) \sin \theta + Bb \sin \phi \cos \theta = 0 \qquad (353)$$

Then

$$\tan \theta = \frac{Bb \sin \phi}{Aa + Bb \cos \phi} \qquad (354)$$

**239.** For the resultant amplitude at the time of this maximum substitute the values of $t$ from (350), in (346), and we have

$$y = A \cos (2n \pi - \theta) + B \cos \left[\frac{b}{a}(2n \pi - \theta - \alpha) + \beta\right]$$
$$= A \cos \theta + B \cos \left[\frac{b-a}{a}(2n \pi - \theta - \alpha) + \beta - \alpha - \theta\right]$$
$$= A \cos \theta + B \cos (\phi - \theta) \qquad (355)$$
$$= A \cos \theta + B \cos \phi \cos \theta + B \sin \phi \sin \theta$$
$$= (A + B \cos \phi) \cos \theta + B \sin \phi \sin \theta$$
$$= \sqrt{A^2 + B^2 + 2AB \cos \phi} \; \cos \left(\theta - \tan^{-1} \frac{B \sin \phi}{A + B \cos \phi}\right)$$

**240.** From (354)

$$\theta = \tan^{-1} \frac{B \sin \phi}{A \frac{a}{b} + B \cos \phi} = \tan^{-1} \frac{\sin \phi}{\frac{Aa}{Bb} + \cos \phi} \qquad (356)$$

In the special cases under consideration the ratio $\frac{a}{b}$ is near unity, and the difference between $\theta$ and $\tan^{-1} \frac{B \sin \phi}{A + B \cos \phi}$ is therefore very small, so that the cosine may be taken as unity.

The resultant amplitude may therefore be expressed by

$$\sqrt{A^2 + B^2 + 2AB \cos \phi} = A \sqrt{1 + \frac{B^2}{A^2} + 2\frac{B}{A} \cos \phi} \qquad (357)$$

The true amplitude of the constituent sought being A, the resultant amplitude must be divided by the factor

$$\sqrt{1 + \frac{B^2}{A^2} + 2\frac{B}{A} \cos \phi} \qquad (358)$$

in order to correct for the influence of the disturbing constituent.

**241.** The corrections for acceleration and amplitude as indicated by formulas (356) and (358) may to advantage be applied to the constants for constituent $K_1$ for an approximate elimination of the effects of constituent $P_1$ and to the constants for $S_2$ for an approximate elimination of the effects of constituents $K_2$ and $T_2$. By taking the relations of the theoretical coefficients for the ratios $\frac{B}{A}$ and the differences in the equilibrium arguments as the approximate equivalents of the phase differences represented by $\phi$, tables may be prepared giving the acceleration and resultant amplitudes with the arguments referring to certain solar elements.

Thus, from table 2, the following values may be obtained.

|  | $\frac{B}{A}$ | $\frac{Aa}{Bb}$ | $\phi$ |
|---|---|---|---|
| Effect of $P_1$ on $K_1$ | 0.33086 | 3.03904 | $-2h+\nu'+180°$. |
| Effect of $K_2$ on $S_2$ | 0.27213 | 3.66469 | $2h-2\nu''$. |
| Effect of $T_2$ on $S_2$ | 0.05881 | 17.02813 | $-h+p_1$. |

Substituting the above in (356) and (358) we have
Effect of $P_1$ on $K_1$

$$\text{Acceleration} = \tan^{-1} \frac{\sin(2h-\nu')}{3.0390 - \cos(2h-\nu')} \qquad (359)$$

$$\text{Resultant amplitude} = 0.813\sqrt{1.6767 - \cos(2h-\nu')} \qquad (360)$$

Effect of $K_2$ on $S_2$

$$\text{Acceleration} = \tan^{-1} \frac{\sin(2h-2\nu'')}{3.6647 + \cos(2h-2\nu'')} \qquad (361)$$

$$\text{Resultant amplitude} = 0.738\sqrt{1.9734 + \cos(2h-2\nu'')} \qquad (362)$$

Effect of $T_2$ on $S_2$

$$\text{Acceleration} = \tan^{-1} \frac{-\sin(h-p_1)}{17.0281 + \cos(h-p_1)} \qquad (363)$$

$$\text{Resultant amplitude} = 0.343\sqrt{8.5318 + \cos(h-p_1)} \qquad (364)$$

242. The above formulas give the accelerations and resulting amplitudes for any individual high water. For the correction of the constants derived from a series covering many high waters it is necessary to take averages covering the period of observations. Tables 21 to 26 give such average values for different lengths of series, the argument in each case referring to the beginning of the series.

In the preceding formulas the mean values of the coefficients were taken to obtain the ratios $\frac{A}{B}$. To take account of the longitude of the moon's node, the node factor should be introduced. If the mean coefficients are indicated by the subscript $o$, formulas (356) and (358) may be written

$$\text{Acceleration} = \tan^{-1} \frac{\sin \phi}{\frac{f(A)A_o a}{f(B)B_o b} + \cos \phi} \qquad (365)$$

$$\text{Resultant amplitude} = \sqrt{1 + \left(\frac{f(B)B_o}{f(A)A_o}\right)^2 + 2\frac{f(B)B_o}{f(A)A_o} \cos \phi} \qquad (366)$$

243. In the cases under consideration the ratio $\frac{f(A)}{f(B)}$ will not differ greatly from unity, the ratio $\frac{A_o a}{B_o b}$ will be rather large compared with $\cos \phi$, which can never exceed unity, and the acceleration itself is relatively small. Because of these conditions the following may be taken as the approximate equivalent of (365):

$A \sin \alpha' \cos \alpha = A' \sin \alpha' \cos \alpha' - \Sigma F_b \cos \{\frac{1}{2}(b-a)\tau + \beta\} \sin \alpha'$ (381)
$A \cos \alpha' \sin \alpha = A' \sin \alpha' \cos \alpha' - \Sigma F_b \sin \{\frac{1}{2}(b-a)\tau + \beta\} \cos \alpha'$ (382)

Subtracting (382) from (381)

$$A \sin (\alpha' - \alpha) = \Sigma F_b \sin \{\frac{1}{2}(b-a)\tau + \beta - \alpha'\} \qquad (383)$$

Multiplying (379) and (380) by $\cos \alpha'$ and $\sin \alpha'$, respectively,

$A \cos \alpha' \cos \alpha = A' \cos^2 \alpha' - \Sigma F_b \cos \{\frac{1}{2}(b-a)\tau + \beta\} \cos \alpha'$ (384)
$A \sin \alpha' \sin \alpha = A' \sin^2 \alpha' - \Sigma F_b \sin \{\frac{1}{2}(b-a)\tau + \beta\} \sin \alpha'$ (385)

Taking the sum of (384) and (385)

$$A \cos (\alpha' - \alpha) = A' - \Sigma F_b \cos \{\frac{1}{2}(b-a)\tau + \beta - \alpha'\} \qquad (386)$$

Dividing (383) by (386)

$$\tan (\alpha' - \alpha) = \frac{\Sigma F_b \sin \{\frac{1}{2}(b-a)\tau + \beta - \alpha'\}}{A' - \Sigma F_b \cos \{\frac{1}{2}(b-a)\tau + \beta - \alpha'\}} \qquad (387)$$

From (386)

$$A = \frac{A' - \Sigma F_b \cos \{\frac{1}{2}(b-a)\tau + \beta - \alpha'\}}{\cos (\alpha' - \alpha)} \qquad (388)$$

**249.** Substituting the value $F_b$ from (376) and the equivalents $R'(A)$, $R(A)$, $R(B)$, $-\zeta'(A) - \zeta(A)$, and $-\zeta(B)$ for $A'$, $A$, $B$, $\alpha'$, $\alpha$, and $\beta$, respectively, we have by (387) and (388)

$$\tan [\zeta(A) - \zeta'(A)] = \frac{\Sigma \frac{180}{\pi} \frac{\sin \frac{1}{2}(b-a)\tau}{\frac{1}{2}(b-a)\tau} R(B) \sin \{\frac{1}{2}(b-a)\tau - \zeta(B) + \zeta'(A)\}}{R'(A) - \Sigma \frac{180}{\pi} \frac{\sin \frac{1}{2}(b-a)\tau}{\frac{1}{2}(b-a)\tau} R(B) \cos \{\frac{1}{2}(b-a)\tau - \zeta(B) + \zeta'(A)\}} \qquad (389)$$

$$R(A) = \frac{R'(A) - \Sigma \frac{180}{\pi} \frac{\sin \frac{1}{2}(b-a)\tau}{\frac{1}{2}(b-a)\tau} R(B) \cos \{\frac{1}{2}(b-a)\tau - \zeta(B) + \zeta'(A)\}}{\cos [\zeta(A) - \zeta'(A)]} \qquad (390)$$

**250.** Formula (389) gives an expression for obtaining the difference to be applied to the uneliminated $\zeta'(A)$ in order to obtain the true $\zeta(A)$, and formula (390) gives an expression for obtaining the true amplitude $R(A)$. These formulas cannot, however, be rigorously applied, because the true values of $R(B)$ and $\zeta(B)$ of the disturbing constituents are, in general, not known, but very satisfactory results may be obtained by using the approximate values of $R(B)$ and $\zeta(B)$ derived from the analysis or by inference.

By a series of successive approximations, using each time in the formulas the newly eliminated values for the disturbing constituents, any desired degree of refinement may be obtained, but the first approximation is usually sufficient and all that is justified because of the greater irregularities existing from other causes.

**251.** Form 245 (fig. 19) provides for the computations necessary in applying formulas (389) and (390). In these formulas the factors represented by $\frac{180}{\pi} \frac{\sin \frac{1}{2}(b-a)\tau}{\frac{1}{2}(b-a)\tau}$ and the angles represented by

$\frac{1}{2}(b-a)\tau$ will depend upon the length of series; but for any given length of series they will be constant for all times and places. Table 29 has been computed to give these quantities for different lengths of series. The factor as directly obtained may be either positive or negative, but for convenience the tabular values are all given as positive, and when the factor as directly obtained is negative the angle has been modified by $\pm 180°$ in order to compensate for the change of sign in the factor and permit the tabular values to be used directly in formulas (389) and (390).

**252.** An examination of formulas (389) and (390) will show that the disturbing effect of one constituent upon another will depend largely upon the magnitude of the fraction $\frac{\sin \frac{1}{2}(b-a)\tau}{\frac{1}{2}(b-a)\tau}$. Assuming that $b$ is not equal to $a$, this fraction and the disturbing effect it represents will approach zero as the length of series $\tau$ approaches in value $\frac{360°}{(b-a)}$, or any multiple thereof, or, in other words, as $\tau$ approaches in length any multiple of the synodic period of constituents $A$ and $B$. Also, since the numerator of the fraction can never exceed unity, while the denominator may be increased indefinitely, the value of the fraction will, in general, be diminished by increasing the length of series and will approach zero as $\tau$ approaches infinity. The greater the difference $(b-a)$ between the speeds of the two constituents the less will be their disturbing effects upon each other. For this reason the effects upon each other of the diurnal and semidiurnal constituents are usually considered as negligible.

**253.** The quantities $R(B)$ and $\zeta(B)$ of formulas (389) and (390) refer to the true amplitudes and epochs of the disturbing constituents. These true values being in general unknown when the elimination process is to be applied, it is desirable that there should be used in the formulas the closest approximation to such values as are obtainable. If the series of observations covers a period of a year or more, the amplitudes and epochs as directly obtained from the analysis may be considered sufficiently close approximations for use in the formulas. For short series of observations, however, the values as directly obtained for the amplitudes and epochs of some of the constituents may be so far from the true values that they are entirely unserviceable for use in the formulas. In such cases inferred values for the disturbing constituents should be used.

### LONG-PERIOD CONSTITUENTS

**254.** The preceding discussions have been especially applicable to the reduction of the short-period constituents—those having a period of a constituent day or less. They are the constituents that determine the daily or semidaily rise and fall of the tide. Consideration will now be given to the long-period tides which affect the mean level of the water from day to day, but which have practically little or no effect upon the times of the high and low waters. There are five such long-period constituents that are usually treated in works on harmonic analysis—the lunar fortnightly Mf, the lunisolar synodic fortnightly MSf, the lunar monthly Mm, the solar semiannual Ssa, and the solar annual Sa. The first three are usually too small to be of practical importance, but the last two, depending largely upon meteorological conditions, often have an appreciable effect upon the mean daily level of the water.

**255.** To obtain the long-period constituents, methods similar to those adopted for the short-period constituents with certain modifications may be used. For the fortnightly and monthly constituents the constituent month may be divided into 24 equal parts, analogous to the 24 constituent hours of the day. Similarly, for the semiannual and annual constituents the constituent year may be divided into 24 equal parts, although it will often be found more convenient to divide the year into 12 parts to correspond approximately with the 12 calendar months.

**256.** Instead of distributing the individual hourly heights, as for the short-period constituents, a considerable amount of labor can be saved by using the daily sums of these heights. The mean of each sum is to be considered as applying to the middle instant of the period from 0 hour to 23d hour; that is, at the 11.5 hour of the day. If the constituent month or year is divided into 24 equal parts, the instants separating the groups may be numbered consecutively, like the hours, from 0 to 23, with the 0 instant of the first group taken at the exact beginning of the series. A table may now be prepared (table 34) which will show to which division each daily sum, or mean, of the series must be assigned.

**257.** Letting

$a=$ the hourly speed of any constituent, in degrees.
$p=1$ when applied to a monthly or an annual constituent, and
$p=2$ when applied to a fortnightly or a semiannual constituent.
$d=$ day of series.
$s=$ solar hour of day.

Then

$$1 \text{ constituent period} = \frac{360}{a} \text{ solar hours} \qquad (391)$$

and

$$1 \text{ constituent month} = \frac{360p}{a} \text{ solar hours} \qquad (392)$$

also

$$1 \text{ constituent year} = \frac{360p}{a} \text{ solar hours} \qquad (393)$$

Dividing the constituent month or year into 24 equal parts, the length of

$$1 \text{ constituent division} = \frac{15p}{a} \text{ solar hours} \qquad (394)$$

Therefore, to express the time of any solar hour in units of the constituent divisions to which the solar hourly heights are to be assigned, the solar hour should be multiplied by the factor $a/15p$.

Thus,

$$\text{Constituent division} = \frac{a}{15p} \text{ (solar hour of series)}$$

$$= \frac{a}{15p}[24(d-1)+s]$$

$$= \frac{a}{15p}[24(d-1)+11.5] \qquad (395)$$

since in using the daily sums, the solar hour of the day to which each such sum applies will always be 11.5 hour.

By substituting the speeds of the constituents from table 2 the following numerical values are obtained for the coefficient $\frac{a}{15p}$:

Mf, 0.036,601,10; MSf, 0.033,863,19; Mm, 0.036,291,65;
Sa and Ssa, 0.002,737,91.

By using the appropriate coefficient and substituting successively the numerals corresponding to the day of series ($d$), the corresponding value of the constituent division to which each daily sum is to be assigned may be readily obtained. The value of such division as obtained directly from the formula will usually be a mixed number. For table 34 the nearest integral number, less any multiple of 24, is used.

**258.** The distribution of the daily sums for the analysis of the long-period constituents may be conveniently accomplished by copying such sums in Form 142 (fig. 12), taking the constituent divisions as the equivalents of the constituent hours and using table 34 to determine the division or hour to which each sum should be assigned. The total sum and mean for each division may then be readily obtained. These means can then be treated as the hourly means of the short-period tides according to the processes outlined in Form 194 (fig. 16) with such modifications as will now be described.

**259.** In using the daily means as ordinates of a long-period constituent consideration must be given to the residual effects of any of the short-period constituents upon such means and steps taken to clear the means of these effects when necessary. Constituent $S_2$ with a period commensurate with the solar day, may be considered as being completely eliminated from each daily mean. Constituents $K_1$ and $K_2$ are very nearly eliminated because the K day is very nearly equal to the solar day. Other short-period constituents may affect the daily means to a greater or less extent, depending largely upon their amplitudes. Of these the principal ones are constituents $M_2$, $N_2$, and $O_1$. In the distribution and grouping of the daily means for the analysis of the several long-period constituents the disturbing effects of the short-period constituents just enumerated, excepting the effects of $M_2$ upon MSf, will be greatly reduced, and in a series covering several years may be practically eliminated. Because the period of MSf is the same as the synodic period of $M_2$ and $S_2$ there will always remain a residual effect of the constituent $M_2$ in the constituent MSf sums of the daily means, no matter how long the series, which must be removed by a special process.

**260.** Let the equation of one of the short-period constituents be

$$y = A \cos (at + \alpha) \qquad (396)$$

Letting $d$=day of series, the values of $t$ for the hours 0 to 23 of $d$ day will be

$$24(d-1),\ 24(d-1)+1,\ 24(d-1)+2,\ \ldots\ 24(d-1)+23.$$

Substituting these values for $t$ in (396) and designating the corresponding values of the ordinate $y$ as $y_0,\ y_1,\ y_2\ \ldots\ y_{23}$ the following are obtained:

$$y_0 = A \cos [24(d-1)a+\alpha]$$
$$y_1 = A \cos [24(d-1)a+\alpha+a]$$
$$y_2 = A \cos [24(d-1)a+\alpha+2a]$$
$$\text{-----------------}$$
$$y_{23} = A \cos [24(d-1)a+\alpha+23a]$$
(397)

Representing the mean of these 24 ordinates for $d$ day by $y_d$, we have

$$y_d = \frac{1}{24} A \cos \{24(d-1)a+\alpha\} [1+\cos a+\cos 2a+\text{----}+\cos 23a]$$
$$- \frac{1}{24} A \sin \{24(d-1)a+\alpha\}[\sin a+\sin 2a+\text{----}+\sin 23a]$$
$$= \frac{1}{24} A \frac{\sin 12a}{\sin \tfrac{1}{2}a} \left[ \cos \{24(d-1)a+\alpha\} \cos \frac{23}{2} a \right.$$
$$\left. - \sin \{24(d-1)a+\alpha\} \sin \frac{23}{2} a \right]$$
$$= \frac{1}{24} A \frac{\sin 12a}{\sin \tfrac{1}{2}a} \cos \{24(d-1)a+\alpha+11.5a\} \qquad (398)$$

**261.** Formula (398), representing the average value of the constituent $A$ ordinates contained in the daily mean for $d$ day, is the correction or clearance that must be subtracted from the mean for that day in order to eliminate the effects of $A$. It will be noted that if we let $A$ represent any of the solar constituents, $S_1$, $S_2$, $S_3$, $S_4$, etc., the factor $\sin 12a$, and consequently the entire formula, becomes zero for all values of $d$. By formula (398) clearances for each of the disturbing short-period constituents for each day of series may be computed and these clearances then applied individually to the daily means, or, if first multiplied by the factor 24, to the daily sums.

**262.** The labor involved in making independent calculations for the clearance of the effect of each short-period constituent for each day of series would be considerable, but this may be avoided to a large extent by means of a tide-computing machine.

If we let $t=$ time reckoned in mean solar hours from the beginning of the series, then for any value of $y_d$, which must apply to the 11.5 hour of $d$ day,

$$t = 24(d-1)+11.5$$

and

$$at = 24(d-1)a+11.5a \qquad (399)$$

If the above equivalent is substituted in (398) and $y_d$ replaced by $y_a$, we have

$$y_a = \frac{1}{24} A \frac{\sin 12a}{\sin \tfrac{1}{2}a} \cos (at+\alpha) \qquad (400)$$

which represents a continuous function of $t$; and for any value of $t$ corresponding to the 11.5 hour of $d$ day the corresponding value of $y_a$ will be $y_d$. This formula is the same as that for the short-period constituent $A$, except that it includes the factor $\dfrac{1}{24}\dfrac{\sin 12a}{\sin \frac{1}{2}a}$ in the coefficient. The speed $a$ is a known constant and the values of $A$ and $\alpha$ are presumed to have already been determined from the harmonic analysis of the short-period constituents. Similarly, the disturbing effects of other short-period constituents may be represented by

$$y_b = \frac{1}{24} B \frac{\sin 12b}{\sin \frac{1}{2}b} \cos (bt+\beta)$$

$$y_c = \frac{1}{24} C \frac{\sin 12c}{\sin \frac{1}{2}c} \cos (ct+\gamma) \tag{401}$$

etc.

The combined disturbing effect of all the short-period constituents may, therefore, be represented by the equation

$$y = y_a + y_b + \text{etc.} = \frac{1}{24} A \frac{\sin 12a}{\sin \frac{1}{2}a} \cos (at+\alpha)$$
$$+ \frac{1}{24} B \frac{\sin 12b}{\sin \frac{1}{2}b} \cos (bt+\beta) + \text{etc.} \tag{402}$$

**263.** This formula is adapted to use on the tide-computing machine. With the constituent cranks set in accordance with the coefficients and initial epochs of the above formula, the machine will indicate the values of $y$ corresponding to successive values of $t$. The values of $y$ desired for the clearances are those which correspond to $t$ at the 11.5 hour on each day. Thus, the clearance for each successive day of series may be read directly from the dials of the machine. In practice, it may be found more convenient to use the daily sums rather than the daily means for the analysis. In this case the coefficients of the terms of (402) should be multiplied by the factor 24 before being used in the tide-computing machine.

**264.** Assuming that all the daily sums are used in the analysis, the augmenting factor represented by formula (308) which is used for the short-period constituent is also applicable to the long-period constituents, with $p$ representing the number of constituent periods in a constituent month or year. Thus, for Mm and Sa, $p$ equals 1, and for Mf, MSf, and Ssa, $p$ equals 2. For the long-period constituents a further correction or augmenting factor is necessary, because the mean or sum of the 24 hourly heights of the day is used to represent the single ordinate at the 11.5 hour of the day.

**265.** If we let formula (396) be the equation of the long-period constituent sought, formula (400) will give the mean value of the 24 ordinates of the day which, in the grouping for the analysis, is taken as representing the 11.5 hour of the day or the $t_d$ hour of the series. Since the true constituent ordinate for this hour should be $A \cos (at_d + \alpha)$, it is evident that an augmenting factor of $24 \dfrac{\sin \frac{1}{2}a}{\sin 12a}$ must be applied to the mean ordinates as derived from the sum of the 24 hourly heights of the day in order to reduce the means to the 11.5 hour of each day.

**266.** The complete augmenting factor for the long-period constituents, the year or month being represented by 24 means, will be obtained by combining the above factor with that given in formula (308). Thus

$$\text{augmenting factor} = \frac{\pi p}{24 \sin \frac{15p}{2}} \times \frac{24 \sin \frac{1}{2}a}{\sin 12a} \qquad (403)$$

If the year or month is represented by only 12 means as when monthly means are used in evaluating Sa and Ssa, the formula becomes

$$\text{augmenting factor} = \frac{\pi p}{12 \sin 15p} \times \frac{24 \sin \frac{1}{2}a}{\sin 12a} \qquad (404)$$

Values obtained from these formulas are given in table 20.

**267.** The following method of reducing the long-period tides, which conforms to the system outlined by Sir George H. Darwin, differs to some extent from that just described. In this discussion it is assumed that a series of 365 days is used. Let the entire tide due to the five long-period constituents already named be represented by the equation

$$y = A \cos(at+\alpha) + B \cos(bt+\beta) + C \cos(at+\gamma) \qquad (405)$$
$$+ D \cos(dt+\delta) + E \cos(et+\epsilon)$$

**268.** For convenience in this discussion let $t$ be reckoned from the 11.5th solar hour of the first day of series instead of the midnight beginning that day. Every value of $t$ to which the daily means refer will then be either 0 or a multiple of 24.

Let $A'$, $B'$, $C'$, $D'$, and $E'$, equal
$A \cos \alpha$, $B \cos \beta$, $C \cos \gamma$, $D \cos \delta$, and $E \cos \epsilon$, respectively, and
$A''$, $B''$, $C''$, $D''$, and $E''$, equal
$-A \sin \alpha$, $-B \sin \beta$, $-C \sin \gamma$, $-D \sin \delta$, and $-E \sin \epsilon$, respectively. (406)

Then formula (405) may be written

$$y = A' \cos at + B' \cos bt + C' \cos ct + D' \cos dt + E' \cos et$$
$$+ A'' \sin at + B'' \sin bt + C'' \sin ct + D'' \sin dt + E'' \sin et \qquad (407)$$

In the above equation there are 10 unknown quantities, $A'$, $A''$, $B'$, $B''$, etc., for which values are sought in order to obtain from them the amplitudes and epochs of the five long-period constituents. The most probable values of these quantities may be found by the least square adjustment.

**269.** Let $y_1, y_2, \ldots y_{365}$ represent the daily means for a 365 day series, as obtained from observations. If we let $n$ be any day of the series, the value of $t$ to which that mean applies will be $24(n-1)$. By substituting in formula (407) the successive values of $y$ and the values of $t$ to which they correspond, 365 observational equations are formed as follows:

HARMONIC ANALYSIS AND PREDICTION OF TIDES 93

$$\left.\begin{array}{l} y_1 = A' \cos 0 + B' \cos 0 + \ldots \\ \phantom{y_1 =} + A'' \sin 0 + B'' \sin 0 + \ldots \\ y_2 = A' \cos 24a + B' \cos 24b + \ldots \\ \phantom{y_2 =} + A'' \sin 24a + B'' \sin 24b + \ldots \\ \overline{\phantom{xxxxxxxxxxxxxxxxxxxxxxxxxxxxxxxxxxxxxxx}} \\ y_{365} = A' \cos 24 \times 364a + B' \cos 24 \times 364b + \ldots \\ \phantom{y_{365} =} + A'' \sin 24 \times 364a + B'' \sin 24 \times 364b + \ldots \end{array}\right\} \quad (408)$$

**270.** A normal equation is now formed for each unknown quantity by multiplying each observational equation by the coefficient of the unknown quantity in that equation and adding the results. Thus, for the unknown quantity $A'$, we have

$$\left.\begin{array}{l} y_1 \cos 0 = A' \cos^2 0 + B' \cos 0 \cos 0 + \ldots \\ \phantom{y_1 \cos 0 =} + A'' \sin 0 \cos 0 + B'' \sin 0 \cos 0 + \ldots \\ y_2 \cos 24a = A' \cos^2 24a + B' \cos 24b \cos 24a + \ldots \\ \phantom{y_2 \cos 24a =} + A'' \sin 24a \cos 24a + B'' \sin 24b \cos 24a + \ldots \\ \overline{\phantom{xxxxxxxxxxxxxxxxxxxxxxxxxxxxxxxxxxxxxxxx}} \\ y_{365} \cos (24 \times 364a) = A' \cos^2 (24 \times 364a) \\ \phantom{y_{365}} + B' \cos (24 \times 364b) \cos (24 \times 364a) + \ldots \\ \phantom{y_{365}} + A'' \sin (24 \times 364a) \cos (24 \times 364a) \\ \phantom{y_{365}} + B'' \sin (24 \times 364b) \cos (24 \times 364a) + \ldots \end{array}\right\} \quad (409)$$

Summing

$$\sum_{n=1}^{n=365} y_n \cos 24(n-1)a = A' \sum_{n=1}^{n=365} \cos^2 24(n-1)a$$

$$+ A'' \sum_{n=1}^{n=365} \sin 24(n-1)a \cos 24(n-1)a$$

$$+ B' \sum_{n=1}^{n=365} \cos 24(n-1)b \cos 24(n-1)a$$

$$+ B'' \sum_{n=1}^{n=365} \sin 24(n-1)b \cos 24(n-1)a$$

$$+ C' \sum_{n=1}^{n=365} \cos 24(n-1)c \cos 24(n-1)a$$

$$+ C'' \sum_{n=1}^{n=365} \sin 24(n-1)c \cos 24(n-1)a$$

$$+ D' \sum_{n=1}^{n=365} \cos 24(n-1)d \cos 24(n-1)a$$

$$+ D'' \sum_{n=1}^{n=365} \sin 24(n-1)d \cos 24(n-1)a$$

$$+ E' \sum_{n=1}^{n=365} \cos 24(n-1)e \cos 24(n-1)a$$

$$+ E'' \sum_{n=1}^{n=365} \sin 24(n-1)e \cos 24(n-1)a \quad (410)$$

which is the normal equation for the unknown quantity $A'$.

**271.** In a similar manner we have for the normal equation for the quantity $A''$

$\Sigma\, y_n \sin 24(n-1)a$

$= A' \Sigma \cos 24(n-1)a \sin 24(n-1)a + A'' \Sigma \sin^2 24(n-1)a$

$+ B' \Sigma \cos 24(n-1)b \sin 24(n-1)a + B'' \Sigma \sin 24(n-1)b \sin 24(n-1)a$

$+ C' \Sigma \cos 24(n-1)c \sin 24(n-1)a + C'' \Sigma \sin 24(n-1)c \sin 24(n-1)a$

$+ D' \Sigma \cos 24(n-1)d \sin 24(n-1)a + D'' \Sigma \sin 24(n-1)d \sin 24(n-1)a$

$+ E' \Sigma \cos 24(n-1)e \sin 24(n-1)a + E'' \Sigma \sin 24(n-1)e \sin 24(n-1)a$

(411)

the limits of $n$ being the same as before.
Normal equations of forms similar to (410) and (411) are easily obtained for the other unknown quantities.

**272.** By changing the notation of formulas (265) to (267) the following relations may be derived:

$$\sum_{n=1}^{n=365} \cos^2 24(n-1)a = \tfrac{1}{2}n + \tfrac{1}{2}\frac{\sin 24na \cos 24(n-1)a}{\sin 24a}$$

$$= 182\tfrac{1}{2} + \tfrac{1}{2}\frac{\sin 8760a \cos 8736a}{\sin 24a} \qquad (412)$$

$$\sum_{n=1}^{n=365} \sin^2 24(n-1)a = \tfrac{1}{2}n - \tfrac{1}{2}\frac{\sin 24na \cos 24(n-1)a}{\sin 24a}$$

$$= 182\tfrac{1}{2} - \tfrac{1}{2}\frac{\sin 8760a \cos 8736a}{\sin 24a} \qquad (413)$$

$$\sum_{n=1}^{n=365} \cos 24(n-1)b \cos 24(n-1)a$$

$$= \tfrac{1}{2}\frac{\sin 12n(b-a) \cos 12(n-1)(b-a)}{\sin 12(b-a)}$$

$$+ \tfrac{1}{2}\frac{\sin 12n(b+a) \cos 12(n-1)(b+a)}{\sin 12(b+a)}$$

$$= \tfrac{1}{2}\frac{\sin 4380(b-a) \cos 4368(b-a)}{\sin 12(b-a)}$$

$$+ \tfrac{1}{2}\frac{\sin 4380(b+a) \cos 4368(b+a)}{\sin 12(b+a)} \qquad (414)$$

$$\sum_{n=1}^{n=365} \sin 24(n-1)b \sin 24(n-1)a$$

$$= \tfrac{1}{2}\frac{\sin 12n(b-a) \cos 12(n-1)(b-a)}{\sin 12(b-a)}$$

$$- \tfrac{1}{2}\frac{\sin 12n(b+a) \cos 12(n-1)(b+a)}{\sin 12(b+a)}$$

$$= \tfrac{1}{2}\frac{\sin 4380(b-a) \cos 4368(b-a)}{\sin 12(b-a)}$$

$$- \tfrac{1}{2}\frac{\sin 4380(b+a) \cos 4368(b+a)}{\sin 12(b+a)} \qquad (415)$$

$$\sum_{n=1}^{n=365} \sin 24(n-1)b \cos 24(n-1)a$$

$$= \tfrac{1}{2} \frac{\sin 12n(b-a) \sin 12(n-1)(b-a)}{\sin 12(b-a)}$$

$$+ \tfrac{1}{2} \frac{\sin 12n(b+a) \sin 12(n-1)(b+a)}{\sin 12(b+a)}$$

$$= \tfrac{1}{2} \frac{\sin 4380(b-a) \sin 4368(b-a)}{\sin 12(b-a)}$$

$$+ \tfrac{1}{2} \frac{\sin 4380(b+a) \sin 4368(b+a)}{\sin 12(b+a)} \qquad (416)$$

**273.** By substituting in (412) to (416) the numerical values of $a$, $b$, etc., from table 2, the corresponding equivalents for these expressions are obtained. These, in turn, may be substituted in (410), (411), and similar equations for the other unknown quantities to obtain the 10 normal equations given below. In preparing these equations the symbols $a$, $b$, $c$, $d$, and $e$ are taken, respectively, as the speeds of constituents Mm, Mf, MSf, Sa, and Ssa.

$$\sum_{n=1}^{n=365} y_n \cos 24(n-1)a$$
$$= 183.05A' + 0.72B' + 0.76C' + 4.88D' + 4.96E'$$
$$+ 2.14A'' + 4.29B'' + 5.04C''' - 0.34D'' - 0.70E'''$$

$$\sum_{n=1}^{n=365} y_n \sin 24(n-1)a$$
$$= 2.14A' - 4.15B' - 4.90C' + 3.80D' + 3.88E'$$
$$+ 181.95A'' + 1.01B'' + 1.06C''' + 0.34D'' + 0.68E'''$$
$$\qquad (417a)$$

$$\sum_{n=1}^{n=365} y_n \cos 24(n-1)b$$
$$= 0.72A' + 183.17B' + 0.56C' - 1.50D' - 1.51E'$$
$$- 4.15A'' + 0.88B'' + 0.92C''' - 0.09D'' - 0.18E'''$$

$$\sum_{n=1}^{n=365} y_n \sin 24(n-1)b$$
$$= 4.29A' + 0.88B' + 0.92C' + 3.05D' + 3.06E'$$
$$+ 1.01A'' + 181.83B'' - 0.80C''' - 0.08D'' - 0.17E'''$$
$$\qquad (417b)$$

$$\sum_{n=1}^{n=365} y_n \cos 24(n-1)c$$
$$= 0.76A' + 0.56B' + 183.19C' - 1.68D' - 1.70E'$$
$$- 4.90A'' + 0.92B'' + 0.97C''' - 0.11D'' - 0.21E'''$$

$$\sum_{n=1}^{n=365} y_n \sin 24(n-1)c$$
$$= 5.04A' + 0.92B' + 0.97C' + 3.24D' + 3.25E'$$
$$+ 1.06A'' - 0.80B'' + 181.81C''' - 0.10D'' - 0.20E'''$$
$$\qquad (417c)$$

$$\sum_{n=1}^{n=365} y_n \cos 24(n-1)d$$
$$= 4.88A' - 1.50B' - 1.68C' + 182.38D' - 0.24E'$$
$$+ 3.80A'' + 3.05B'' + 3.24C''' + 0.00D'' + 0.01E'''$$

$$\sum_{n=1}^{n=365} y_n \sin 24(n-1)d$$
$$= -0.34A' - 0.09B' - 0.11C' + 0.00D' + 0.00E'$$
$$+ 0.34A'' - 0.08B'' - 0.10C''' + 182.62D'' + 0.00E'''$$
$$\qquad (417d)$$

$$\sum_{n=1}^{n=365} y_n \cos 24(n-1)e$$
$$= 4.96A' - 1.51B' - 1.70C' - 0.24D' + 182.38E'$$
$$+ 3.88A'' + 3.06B'' + 3.25C'' + 0.00D'' + 0.00E''$$

$$\sum_{n=1}^{n=365} y_n \sin 24(n-1)e$$  (417e)
$$= -0.70A' - 0.18B' - 0.21C' + 0.01D' + 0.00E'$$
$$+ 0.68A'' - 0.17B'' - 0.20C'' + 0.00D'' + 182.62E''$$

**274.** The numerical value of the first member of each of the above normal equations is obtained from the observations by taking the sum of the product of each daily mean by the cosine or sine of the angle indicated. The solution of the equations give the values of $A'$, $A''$, $B'$, $B''$, etc., from which the corresponding values of quantities $A$ and $\alpha$, $B$ and $\beta$, etc., of formula (405) are readily obtained, since

$$A = \sqrt{(A')^2 + (A'')^2} \text{ and } \alpha = \tan^{-1} \frac{-A''}{A'}.$$

In calculating the corrected epoch, it must be kept in mind that the $t$ in this reduction is referred to the 11.5 hour of the first day of series instead of the preceding midnight.

**275.** Before solving equations (417), if the daily means have not already been cleared of the effects of the short-period constituents, it will be necessary to apply corrections to the first member of each of these equations in order to make the clearances.

The disturbance in a single daily mean due to the presence of a short-period constituent is represented by equation (398). Introducing the subscript $s$ to distinguish the symbols pertaining to the short-period constituents, the disturbance in the daily mean of the $n^{th}$ day of series due to the presence of the short-period constituent $A_s$ may be written

$$[y_s]_n = \frac{1}{24} A_s \frac{\sin 12a_s}{\sin \frac{1}{2}a_s} \cos \{24(n-1)a_s + 11.5a_s + \alpha_s\} \quad (418)$$

The disturbances in the products of the daily means by

$$\cos 24(n-1)a \text{ and } \sin 24(n-1)a$$

may therefore be written

$$[y_s]_n \cos 24(n-1)a$$

$$= \frac{1}{24} A_s \frac{\sin 12a_s}{\sin \frac{1}{2}a_s} \tfrac{1}{2} [\cos \{24(n-1)(a_s+a) + 11.5a_s + \alpha_s\}$$
$$+ \cos \{24(n-1)(a_s-a) + 11.5a_s + \alpha_s\}] \quad (419)$$

and

$$[y_s]_n \sin 24(n-1)a$$

$$= \frac{1}{24} A_s \frac{\sin 12a_s}{\sin \frac{1}{2}a_s} \tfrac{1}{2} [\sin \{24(n-1)(a_s+a) + 11.5a_s + \alpha_s\}$$
$$- \sin \{24(n-1)(a_s-a) + 11.5a_s + \alpha_s\}] \quad (420)$$

HARMONIC ANALYSIS AND PREDICTION OF TIDES 97

**276.** Then, referring to formulas (263) and (264)

$$\sum_{n=1}^{n=365} [y_s]_n \cos 24(n-1)a =$$

$$\frac{1}{48} A_s \frac{\sin 12a_s}{\sin \frac{1}{2}a_s} \left[ \frac{\sin 12 \times 365(a_s+a)}{\sin 12(a_s+a)} \cos \{12 \times 364(a_s+a) + 11.5a_s + \alpha_s\} \right.$$

$$\left. + \frac{\sin 12 \times 365(a_s-a)}{\sin 12(a_s-a)} \cos \{12 \times 364(a_s-a) + 11.5a_s + \alpha_s\} \right] \quad (421)$$

and

$$\sum_{n=1}^{n=365} [y_s]_n \sin 24(n-1)a =$$

$$\frac{1}{48} A_s \frac{\sin 12a_s}{\sin \frac{1}{2}a_s} \left[ \frac{\sin 12 \times 365(a_s+a)}{\sin 12(a_s+a)} \sin \{12 \times 364(a_s+a) + 11.5a_s + \alpha_s\} \right.$$

$$\left. - \frac{\sin 12 \times 365(a_s-a)}{\sin 12(a_s-a)} \sin \{12 \times 364(a_s-a) + 11.5a_s + \alpha_s\} \right] \quad (422)$$

Now let
$$A'_s = A_s \cos \alpha_s$$
and  $\quad\quad\quad\quad\quad\quad\quad\quad\quad\quad\quad\quad\quad\quad\quad\quad$ (423)
$$A''_s = -A_s \sin \alpha_s$$

then (421) and (422) may be reduced as follows:

$$\sum_{n=1}^{n=365} [y_s]_n \cos 24(n-1)a$$

$$= \frac{1}{48} \frac{\sin 12a_s}{\sin \frac{1}{2}a_s} \left[ \frac{\sin 12 \times 365(a_s+a)}{\sin 12(a_s+a)} \cos \{12 \times 364(a_s+a) + 11.5a_s\} \right.$$

$$\left. + \frac{\sin 12 \times 365(a_s-a)}{\sin 12(a_s-a)} \cos \{12 \times 364(a_s-a) + 11.5a_s\} \right] A'_s$$

$$+ \frac{1}{48} \frac{\sin 12a_s}{\sin \frac{1}{2}a_s} \left[ \frac{\sin 12 \times 365(a_s+a)}{\sin 12(a_s+a)} \sin \{12 \times 364(a_s+a) + 11.5a_s\} \right.$$

$$\left. + \frac{\sin 12 \times 365(a_s-a)}{\sin 12(a_s-a)} \sin \{12 \times 364(a_s-a) + 11.5a_s\} \right] A''_s \quad (424)$$

and

$$\sum_{n=1}^{n=365} [y_s]_n \sin 24(n-1)a$$

$$= \frac{1}{48} \frac{\sin 12a_s}{\sin \frac{1}{2}a_s} \left[ \frac{\sin 12 \times 365(a_s+a)}{\sin 12(a_s+a)} \sin \{12 \times 364(a_s+a) + 11.5a_s\} \right.$$

$$\left. - \frac{\sin 12 \times 365(a_s-a)}{\sin 12(a_s-a)} \sin \{12 \times 364(a_s-a) + 11.5a_s\} \right] A'_s$$

$$- \frac{1}{48} \frac{\sin 12a_s}{\sin \frac{1}{2}a_s} \left[ \frac{\sin 12 \times 365(a_s+a)}{\sin 12(a_s+a)} \cos \{12 \times 364(a_s+a) + 11.5a_s\} \right.$$

$$\left. - \frac{\sin 12 \times 365(a_s-a)}{\sin 12(a_s-a)} \cos \{12 \times 364(a_s-a) + 11.5a_s\} \right] A''_s \quad (425)$$

**277.** Formulas (424) and (425) represent the clearances for any long-period constituent $A$ due to any short-period constituent $A_s$. The first must be subtracted from terms corresponding to $\Sigma y_n \cos 24(n-1)a$ and the latter from terms corresponding to $\Sigma y_n \sin 24(n-1)a$ of formula (417) before solving the latter.

**278.** In (424) and (425) the coefficients of $A'_s$ and $A''_s$, which for brevity we may designate as $C'$, $C''$, $S'$, and $S''$, respectively, contain only values that are constant for all series and may therefore be computed once for all. Separate sets of such coefficients must, however, be computed for the effect of each short-period constituent upon each long-period constituent. In the usual reductions in which the effects of 3 short-period constituents upon 5 long-period constituents are considered, 15 sets of 4 coefficients each, or 60 coefficients in all, are required.

The coefficients are given in the following table:*

|  | Long-period constituents | | | | |
|---|---|---|---|---|---|
|  | Mm | Mf | MSf | Sa | Ssa |
| $M_2$ ($C'$) | −0.0556 | +0.0030 | +5.739 | −0.1041 | −0.1046 |
| ($C''$) | −0.1704 | −0.0377 | −2.923 | −0.0752 | −0.0755 |
| ($S'$) | −0.1708 | +0.0417 | −2.840 | −0.0018 | −0.0035 |
| ($S''$) | +0.0441 | +0.0105 | −5.727 | +0.0048 | +0.0096 |
| $N_2$ ($C'$) | −0.0588 | +0.0368 | +0.0294 | −0.0176 | −0.0176 |
| ($C''$) | −0.0776 | −0.2236 | −0.1938 | +0.0025 | +0.0025 |
| ($S'$) | −0.0206 | −0.1526 | −0.1221 | +0.0002 | +0.0004 |
| ($S''$) | +0.1138 | −0.0854 | −0.0808 | +0.0001 | +0.0002 |
| $O_1$ ($C'$) | −0.0648 | +0.0166 | +0.0157 | −0.1924 | −0.1934 |
| ($C''$) | −0.3476 | −0.0778 | −0.0816 | −0.1826 | −0.1831 |
| ($S'$) | −0.3452 | +0.0841 | +0.0875 | −0.0046 | −0.0093 |
| ($S''$) | +0.0405 | +0.0338 | +0.0331 | +0.0090 | +0.0180 |

In the above table the sign is so taken that the values are to be applied to the sums directly as indicated.

**279.** After the clearances have been applied and the normal equations (417) solved and the resulting amplitude and epoch obtained for each of the long-period constituents, the reductions will be completed in accordance with the processes already outlined, but it must be kept in mind that in this reduction the initial value of $t$ is taken to correspond to 11:30 a. m. on the first day of series. In obtaining the numerical values of such quantities as $\Sigma y_n \cos 24(n-1)a$ and $\Sigma y_n \sin 24(n-1)a$, in order to avoid the labor of separate multiplications for each day, the following abbreviations have been proposed by the British authorities. The values of $\cos 24(n-1)a$ and of $\sin 24(n-1)a$ are divided into 11 groups according as they fall nearest 0, 0.1, 0.2, 0.3, 0.4, 0.5, 0.6, 0.7, 0.8, 0.9, or 1.0. The daily values are then distributed into 11 corresponding groups, so that all values in one group will be multiplied by 0, another group by 0.1, etc. The $\cos 24(n-1)a$ and $\sin 24(n-1)a$ include negative as well as positive values. The former are taken into account by changing the sign of the daily mean to which the negative values apply.

**280.** As a part of the routine reductions of the tidal records from the principal tide stations it is the practice of the office to obtain the mean sea level for each calendar month. It is therefore desirable to

*From Scientific Papers by Sir George H. Darwin, Vol. I, p. 64.

have a method of using these means directly in the analysis for the annual and semiannual constituents, thus avoiding any special summation for the purpose. The period of the annual constituent is approximately the length of the Julian year, that is, 365.25 days. If this period is divided into 12 equal groups and the mean of the hourly heights for each group taken, these means represent the approximate height of the combined annual and semiannual constituents for the middle of each group, and the middle of the first group will be the initial point from which the zeta ($\zeta$) as obtained by the usual process is referred. As each group represents 30° of motion for the annual constituent, or 60° for the semiannual constituent, to refer this $\zeta$ to the actual beginning of the series of observations it will be necessary to apply a correction of 15° for the annual constituent or 30° for the semiannual constituent.

281. In obtaining the monthly means by calendar months the year is divided only approximately into 12 equal groups. The following table shows the difference between the middle of each group representing a calendar month and the middle of the corresponding group obtained by dividing the Julian year into 12 equal parts. It is to be noted that the hourly heights included in a monthly sum extend from 0 hour on the first day of the month to the 23d hour on the last day. The middle of the group as reckoned from the beginning of the month will therefore be 13.98 days, 14.48 days, 14.98 days, or 15.48 days, respectively, according to whether the month has 28, 29, 30, or 31 days.

| Month | Middle of group reckoned from beginning of year | | | Differences | |
|---|---|---|---|---|---|
| | Julian year | Common year | Leap year | Common year | Leap year |
| | *Days* | *Days* | *Days* | *Days* | *Days* |
| January | 15.22 | 15.48 | 15.48 | +0.26 | +0.26 |
| February | 45.66 | 44.98 | 45.48 | −0.68 | −0.18 |
| March | 76.09 | 74.48 | 75.48 | −1.61 | −0.61 |
| April | 106.53 | 104.98 | 105.98 | −1.55 | −0.55 |
| May | 136.97 | 135.48 | 136.48 | −1.49 | −0.49 |
| June | 167.41 | 165.98 | 166.98 | −1.43 | −0.43 |
| July | 197.84 | 196.48 | 197.48 | −1.36 | −0.36 |
| August | 228.28 | 227.48 | 228.48 | −0.80 | +0.20 |
| September | 258.72 | 257.98 | 258.98 | −0.74 | +0.26 |
| October | 289.16 | 288.48 | 289.48 | −0.68 | +0.32 |
| November | 319.59 | 318.98 | 319.98 | −0.61 | +0.39 |
| December | 350.03 | 349.48 | 350.48 | −0.55 | +0.45 |
| Sums | | | | −11.24 | −0.74 |
| Means | | | | −0.94 | −0.06 |
| Speed of Sa constituent per day = 0.9856°. | | | | ° | ° |
| Mean differences reduced to degrees of Sa | | | | −0.93 | −0.06 |
| Correction to $\zeta$ of Sa | | | | 14.07 | 14.94 |
| Correction to $\zeta$ of Ssa | | | | 28.14 | 29.88 |

282. From the above table it is evident that in the summation for the monthly means for a calendar year the middle of each group of a common year is on an average 0.93° earlier than the middle of the corresponding group when the Julian year is equally subdivided and the middle of each group of a leap year is on an average 0.06° earlier. Subtracting these values from 15°, the interval between the beginning of the observations and the middle of the first group of an equal subdivision, we have 14.07° and 14.94°, for common and leap years, respectively, as a correction to be applied to the $\zeta$ of Sa as

directly obtained, in order to refer the $\zeta$ to the 0 hour of the 1st day of January. For Ssa the corrections will be twice as great.

**283.** If the year commences on the first day of any month other than January, the corrections will differ a little from the above. Calculated in a manner similar to that above, the following table gives the correction to be applied to the $\zeta$ to refer to the first day of any month at which the series commences. The correction to the $\zeta$ of Ssa will be twice the tabular value for Sa.

| Observations commence— | Correction to $\zeta$ of Sa to refer to beginning of month | | Observations commence— | Correction to $\zeta$ of Sa to refer to beginning of month | |
|---|---|---|---|---|---|
| | Common year | Leap year | | Common year | Leap year |
| | ° | ° | | ° | ° |
| Jan. 1 | 14.07 | 14.94 | July 1 | 15.56 | 15.93 |
| Feb. 1 | 13.50 | 14.45 | Aug. 1 | 14.98 | 15.43 |
| Mar. 1 | 15.89 | 15.93 | Sept. 1 | 14.41 | 14.94 |
| Apr. 1 | 15.31 | 15.43 | Oct. 1 | 14.82 | 15.43 |
| May 1 | 15.72 | 15.93 | Nov. 1 | 14.24 | 14.94 |
| June 1 | 15.15 | 15.43 | Dec. 1 | 14.65 | 15.43 |

**284.** If the monthly means extend over many calendar years, it may be convenient to combine them for a single analysis. In this case the $(V_0+u)$ for January 1 may be taken as the average of the values for the beginning of each year included in the observations, and the correction to the $\zeta$ to refer to the beginning of the year will be a mean of the values given above for common and leap years, weighted in accordance with the number of each kind of year included. If only a few years of observations are available, it is better to analyze each year separately in order that the results may serve as a check on each other.

**285.** The augmenting factors to be used for constituents Sa and Ssa when derived from the monthly sea level values are based upon formula (404) in paragraph 266 and are as follows:

Sa 1.0115, logarithm 0.00497.
Ssa 1.0472, logarithm 0.02005.

### ANALYSIS OF HIGH AND LOW WATERS

**286.** The automatic tide gage, which furnishes a continuous record of the rise and fall of the tide, now being in general use, it is seldom necessary to rely only upon the high and low waters for an analysis. It may happen, however, that a record of high and low water observations is available for a more or less isolated locality where it has been impractical to secure continuous records. Such records, if they include all the high and low waters for a month or more may be utilized in determining approximate values of the principal harmonic constants, but the results are not as satisfactory as those obtained from an analysis of the hourly heights.

**287.** An elaborate mode of analysis of the high and low waters is contained in volume 1 of Scientific Papers, by Sir George H. Darwin. Other methods are given by Dr. R. A. Harris in his Manual of Tides. The process outlined below follows to some extent one of the methods of Doctor Harris, extending his treatment for the K and O to other constituents.

288. The lengths of series may be taken the same as the lengths used as the analysis of the hourly heights (see par. 152). It is sometimes convenient to divide a series, whatever its length, into periods of 29 days each. This permits a uniform method of procedure, and a comparison of the results from different series affords a check on the reliableness of the work.

289. The first process in this analysis consists in making the usual high and low water reductions, including the computation of the lunitidal intervals. Form 138 provides for this reduction. The times and heights of the high and low waters, together with the times of the moon's transits, are tabulated. For convenience the standard time of the place of observations may be used for the times of the high and low waters, and the Greenwich mean civil time of the moon's transits over the meridian of Greenwich may be used for the moon's transits. The interval between each transit and the following high and low water is then found, and the mean of all the high water intervals and the mean of all the low water intervals are then obtained separately. The true mean intervals between the time of the moon's transit over the local meridian and the time of the following high and low waters being desired, the means as directly obtained must be corrected to allow for any difference in the kind of time used for the transit of the moon and the time of the tides and also for the difference in time between the transit of the moon over the local meridian and the transit over the meridian to which the tabular values refer.

290. If the tide is of the semidiurnal type, the approximate amplitude and epoch for $M_2$ may be obtained directly from this high and low water reduction. On account of the presence of the other constituents the mean range from the high and low waters will always be a little larger than twice the amplitude of $M_2$. If the data are available for some other station in the general locality, the ratio of the $M_2$ amplitude to the mean range of tide at that station may be used in finding the $M_2$ amplitude from the mean range of tide at the station for which the results are sought. If this ratio cannot be obtained for any station in the general locality, the empirical ratio of 0.47 may be used with fairly satisfactory results. After the amplitude of $M_2$ has been thus obtained, it should be corrected for the longitude of the moon's node by factor $F$ from table 12.

291. The epoch of $M_2$ may be obtained from the corrected high and low water lunitidal intervals $HWI$, $LWI$ by the following formula:

$$M°_2 = \tfrac{1}{2}(HWI + LWI) \times 28.984 + 90° \tag{426}$$

In the above formula $HWI$ must be greater than $LWI$, 12.42 hours being added, if necessary, to the $HWI$ as directly obtained from the high and low water reductions.

292. The difference between the average duration of rise and fall of the tide at any place, where the tide is of the semidiurnal type, depends largely upon the constituent $M_4$. It is possible to obtain from the high and low waters a constituent with the speed of $M_4$ which, when used in the harmonic prediction of the tides, will cause the mean duration of rise and fall to be the same as that at the station. The effect of $M_4$ upon the mean duration of rise will depend chiefly upon the relation of its amplitude and epoch to the amplitude and epoch of the principal constituent $M_2$. By assuming an $M_4$ with epoch

such as to make the constituent symmetrically situated in regard to the maxima and minima of $M_2$, the amplitude necessary to account for the mean duration of rise of the tide may be readily calculated.

**293.** Let $DR$ = duration of rise of tide in hours as obtained from the lunitidal intervals,

$a$ = Hourly speed of $M_2$ = $28.°984$.
$M_2$ = Amplitude of $M_2$.
$M_2°$ = Epoch of $M_2$.
$M_4$ = Amplitude of $M_4$.
$M_4°$ = Epoch of $M_4$.

Then, for $M_4$ to be symmetrically situated with respect to the maxima and minima of $M_2$

$$M_4° = 2\ M_2° \pm 90° \tag{427}$$

in which the upper or lower sign is to be used according to whether $a(DR)$ is greater or less, respectively, than $180°$. Multiples of $360°$ may be added or rejected to obtain the result as a positive angle less than $360°$.

The equations of the constituents $M_2$ and $M_4$ may be written

$$y_1 = M_2 \cos(at + \alpha) \tag{428}$$

$$y_2 = M_4 \cos(2at + \beta) \tag{429}$$

and the resultant curve

$$y = M_2 \cos(at + \alpha) + M_4 \cos(2at + \beta) \tag{430}$$

**294.** Values of $t$ which will render (428) a maximum must satisfy the derived equation

$$aM_2 \sin(at + \alpha) = 0 \tag{431}$$

and for a maximum of (430) $t$ must satisfy the derived equation

$$aM_2 \sin(at + \alpha) + 2aM_4 \sin(2at + \beta) = 0 \tag{432}$$

For a maximum of (428)

$$t = \frac{2n\pi - \alpha}{a} \tag{433}$$

in which $n$ is any integer.

**295.** Let $\dfrac{\theta}{a}$ = the acceleration in the high waters of $M_2$ due to the presence of $M_4$. With the $M_4$ wave symmetrically situated with respect to the $M_2$ wave, $\dfrac{\theta}{a}$ will also equal the retardation in the low water of $M_2$, due to the presence of $M_4$, and $\dfrac{2\theta}{a}$ will equal the total amount by which the duration of rise of the tide has been diminished by $M_4$. If the duration of rise has been increased, $\theta$ will be negative. Then, for a maximum of (430)

$$t = \frac{2n\pi - \alpha - \theta}{a} \tag{434}$$

and this value of $t$ must satisfy equation (432).

**296.** Substituting in (432), we have

$$aM_2 \sin (2n\pi - \theta) + 2aM_4 \sin (4n\pi - 2\alpha + \beta - 2\theta) = -aM_2 \sin \theta - 2aM_4 \sin (2\theta + 2\alpha - \beta) = 0 \quad (435)$$

But

$$2\alpha - \beta = -2M_2° + M_4° \quad (436)$$

From (427)

$$-2M_2° + M_4° = \pm 90°$$

according to whether the duration of rise is greater or less than $\dfrac{180°}{a}$, or whether $\theta$ is negative or positive.
Then

$$2\alpha - \beta = \mp 90° \quad (437)$$

according to whether $\theta$ is positive or negative.
Substituting this in (435)

$$-aM_2 \sin \theta \pm 2aM_4 \cos 2\theta = 0 \quad (438)$$

and

$$\frac{M_4}{M_2} = \pm \tfrac{1}{2} \frac{\sin \theta}{\cos 2\theta} \quad (439)$$

the upper or lower sign being used according to whether $\theta$ is positive or negative. As under the assumed conditions $\theta$ must come within the limits $\pm 45°$, the ratio of $\dfrac{M_4}{M_2}$ as derived from (439) will always be positive.

**297.** The duration of rise of tide due solely to the constituent $M_2$ is $\dfrac{180°}{a}$.

The duration of rise as modified by the presence of the assumed $M_4$ is

$$DR = \frac{180°}{a} - \frac{2\theta}{a} \quad (440)$$

Therefore

$$\theta = \tfrac{1}{2}(180° - aDR) \quad (441)$$

Substituting the above in (439) we have

$$\frac{M_4}{M_2} = \pm \tfrac{1}{2} \frac{\sin (90° - \tfrac{1}{2}aDR)}{\cos (180° - aDR)} = \mp \tfrac{1}{2} \frac{\cos \tfrac{1}{2}aDR}{\cos aDR} \quad (442)$$

and

$$M_4 = \mp \tfrac{1}{2} \frac{\cos \tfrac{1}{2}aDR}{\cos aDR} M_2 \quad (443)$$

$M_4$ must be positive, and the sign of the above coefficient will depend upon whether $aDR$ is less or greater than $180°$.

**298.** The approximate constants for $S_2$, $N_2$, $K_1$, and $O_1$ may be obtained from the observed high and low waters as follows: Add to each low-water height the mean range of tide. Copy the high and modified low water heights into the form for hourly heights (form 362), always putting the values upon the nearest solar hour. Sum for the desired constituents, using the same stencils as are used for the regular analysis of the hourly heights. Account should be taken of the number of items entering into each sum and the mean for each constituent hour obtained. The 24 hourly means for each constituent are then to be analyzed in the usual manner.

**299.** The results obtained by this process are, of course, not as dependable as those obtained from a continuous record of hourly heights. The approximate results first obtained can, however, be improved by the following treatment if a tide-computing machine is available. Using the approximate constants as determined above for the principal constituents and inferred values for smaller constituents, set the machine for the beginning of the period of observations and find the predicted heights corresponding to the observed times of the high and low waters. Tabulate the differences between the observed and predicted heights for these times, using the hourly height form and entering the values according to the nearest solar hour. These differences are then to be summed and analyzed the same as the original observed heights. In this analysis of the residuals the constituent $M_2$ should be included. The results from the analysis of the residuals are then combined with the constants used for the setting of the predicting machine.

**300.** In making the combinations the following formulas may be used:

Let $A'$ and $\kappa'$ represent the first approximate values of the constants of any constituent.

$A''$ and $\kappa''$, the constants as obtained from the residuals.

$A$ and $\kappa$, the resultant constants sought.

Then

$$A = \sqrt{(A' \cos \kappa' + A'' \cos \kappa'')^2 + (A' \sin \kappa' + A'' \sin \kappa'')^2} \quad (444)$$

and

$$\kappa = \tan^{-1} \frac{A' \sin \kappa' + A'' \sin \kappa''}{A' \cos \kappa' + A'' \cos \kappa''} \quad (445)$$

#### FORMS USED FOR ANALYSIS OF TIDES

**301.** Forms used by the Coast and Geodetic Survey for the harmonic analysis of tide observations are shown in figures 9 to 19. A series of tide observations at Morro, California, covering the period February 13 to July 25, 1919, is taken as an example to illustrate the detail of the work.

**302.** *Form 362, Hourly heights* (fig. 9).—The hourly heights of the tide are first tabulated in form 362. Although the zero of the tide staff is usually taken as the height datum, any other fixed plane will serve this purpose. For practical convenience it is desirable that the datum be low enough to avoid negative tabulations but not so low as to cause the readings to be inconveniently large for summing.

**303.** The hours refer to mean solar time, which may be either local or standard, astronomical or civil, but standard civil time will generally be the most convenient to use. The series must commence with the zero (0) hour of the adopted time, and all vacancies in the record should be filled by interpolated values in order that each hour of the series may be represented by a tabulated height. It is the general practice to use brackets with interpolated values to distinguish them from the observed heights. The record for successive days of the series must be entered in successive columns of the form, and these columns are to be numbered consecutively, beginning with one (1) for the first day of the series.

**304.** The series analyzed should be one of the lengths indicated in paragraph 152. Series of observations very nearly equal to one of these standard lengths may be completed by the use of extrapolated hourly heights. If the observations cover a period of several years, the analysis for each year may be made separately, a comparison of the results affording an excellent check on the work.

**305.** The hourly heights on each page of form 362 are first summed horizontally and vertically. The total of the vertical sums must equal the total of the horizontal sums, and this page sum is entered in the lower right-hand corner of the page.

Form 362
DEPARTMENT OF COMMERCE
U. S. COAST AND GEODETIC SURVEY

**TIDES: HOURLY HEIGHTS**

Station: Morro, California. Year: 1919.
Chief of Party: E. B. Latham. Lat. 35° 22' N. Long. 120° 51' W.
Time Meridian: 120 W. Tide Gauge No. 107. Scale 1:9. Reduced to Staff.

| Month and Day. | mo. d. Feb. 13 | d. 14 | d. 15 | d. 16 | d. 17 | d. 18 | d. 19 | Horizontal Sum. |
|---|---|---|---|---|---|---|---|---|
| Day of Series | 1 | 2 | 3 | 4 | 5 | 6 | 7 | |
| Hour. | Feet. | Feet. | Feet. | Feet. | Feet | Feet. | Feet. | Feet. |
| 0 | 3.9 | 4.2 | 4.6 | 4.5 | 4.4 | 4.7 | 4.6 | 30.9 |
| 1 | 3.4 | 3.8 | 4.2 | 4.2 | 4.2 | 4.9 | 4.8 | 29.5 |
| 2 | 3.0 | 3.3 | 3.5 | 3.7 | 3.8 | 4.6 | 4.9 | 26.8 |
| 3 | 2.8 | 3.0 | 3.0 | 3.1 | 3.3 | 4.1 | 4.5 | 23.8 |
| 4 | 3.0 | 2.8 | 2.6 | 2.5 | 2.7 | 3.5 | 3.8 | 20.9 |
| 5 | 3.6 | 3.1 | 2.5 | 2.2 | 2.2 | 3.0 | 3.2 | 19.8 |
| 6 | 4.4 | 3.6 | 2.8 | 2.2 | 1.9 | 2.6 | 2.7 | 20.2 |
| 7 | 5.1 | 4.5 | 3.5 | 2.6 | 2.0 | 2.5 | 2.3 | 22.5 |
| 8 | 5.7 | 5.3 | 4.3 | 3.3 | 2.4 | 2.7 | 2.2 | 25.9 |
| 9 | 6.0 | 6.0 | 4.9 | 4.1 | 3.1 | 3.1 | 2.4 | 29.6 |
| 10 | 5.6 | 6.2 | 5.4 | 4.6 | 3.9 | 3.6 | 2.8 | 32.1 |
| 11 | 4.6 | 5.8 | 5.5 | 4.9 | 4.3 | 4.1 | 3.2 | 32.6 |
| Noon. | 3.9 | 5.1 | 5.1 | 4.8 | 4.4 | 4.5 | 3.6 | 31.4 |
| 13 | 3.4 | 4.3 | 4.4 | 4.3 | 4.2 | 4.5 | 3.8 | 28.9 |
| 14 | 2.6 | 3.4 | 3.5 | 3.6 | 3.7 | 4.3 | 3.8 | 24.9 |
| 15 | 1.9 | 2.6 | 2.8 | 2.9 | 3.1 | 3.8 | 3.6 | 20.7 |
| 16 | 1.2 | 2.0 | 2.2 | 2.2 | 2.6 | 3.2 | 3.2 | 16.6 |
| 17 | 1.0 | 1.6 | 1.7 | 1.6 | 2.1 | 2.7 | 2.8 | 13.5 |
| 18 | 1.3 | 1.6 | 1.5 | 1.3 | 1.9 | 2.4 | 2.5 | 12.5 |
| 19 | 2.3 | 2.2 | 1.8 | 1.4 | 1.9 | 2.3 | 2.3 | 14.2 |
| 20 | 3.2 | 3.1 | 2.6 | 2.0 | 2.3 | 2.5 | 2.4 | 18.1 |
| 21 | 4.0 | 3.9 | 3.4 | 2.8 | 3.0 | 3.0 | 2.9 | 23.0 |
| 22 | 4.3 | 4.5 | 4.1 | 3.6 | 3.8 | 3.6 | 3.7 | 27.6 |
| 23 | 4.5 | 4.7 | 4.5 | 4.1 | 4.4 | 4.2 | 4.2 | 30.6 |
| Sum. | 84.9 | 90.6 | 84.4 | 76.5 | 75.6 | 84.4 | 80.2 | 576.6 |

Sum for 29 days, 1 to 29 of — = Divisor=696; mean for 29 days=

FIGURE 9.

**306.** *Stencils* (figs. 10 and 11).—The first figure is a copy of the M stencil for the even hours of the first 7 days of the series, and the second figure illustrates the application of the same. This stencil being laid over the page of hourly heights shown in figure 9, the heights applying to each of the even constituent hours for this page show through the openings in the stencil, where they appear connected by diagonal lines, thus indicating each group to be summed.

**307.** For each constituent summation, excepting for S, there are provided two stencils for each page of tabulated hourly heights, one for the even constituent hours and the other for the odd constituent hours.

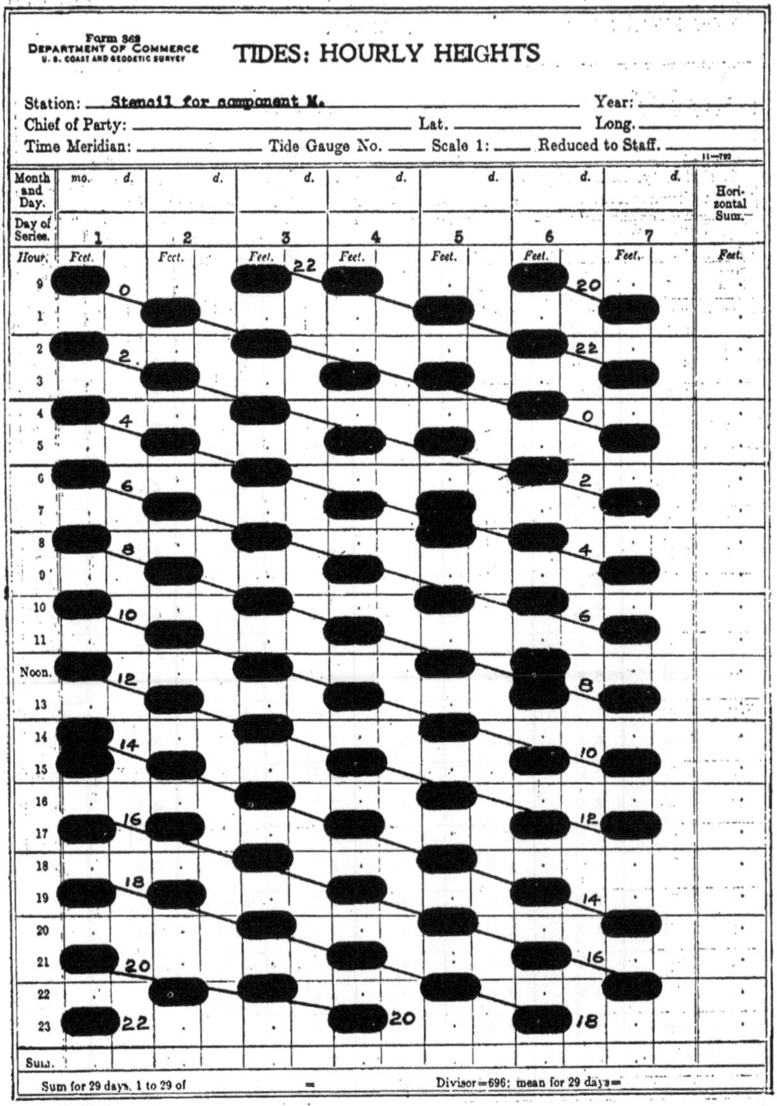

FIGURE 10.

HARMONIC ANALYSIS AND PREDICTION OF TIDES 107

The stencils are numbered with the days of series to which they apply, and special care must be taken to see that the days of series on each stencil correspond with the days of series on the page of tabulations with which it is used. For constituent S no stencils are necessary, as the constituent hours correspond to the solar hours of the tabulations and the horizontal sums from form 362 may be taken directly as the constituent hour sums.

**308.** *Form 142, Stencil sums* (figs. 12 and 13).—The sums for each constituent hour are entered in form 142, one line of the form being used for each page of the original tabulations. The total of the hour

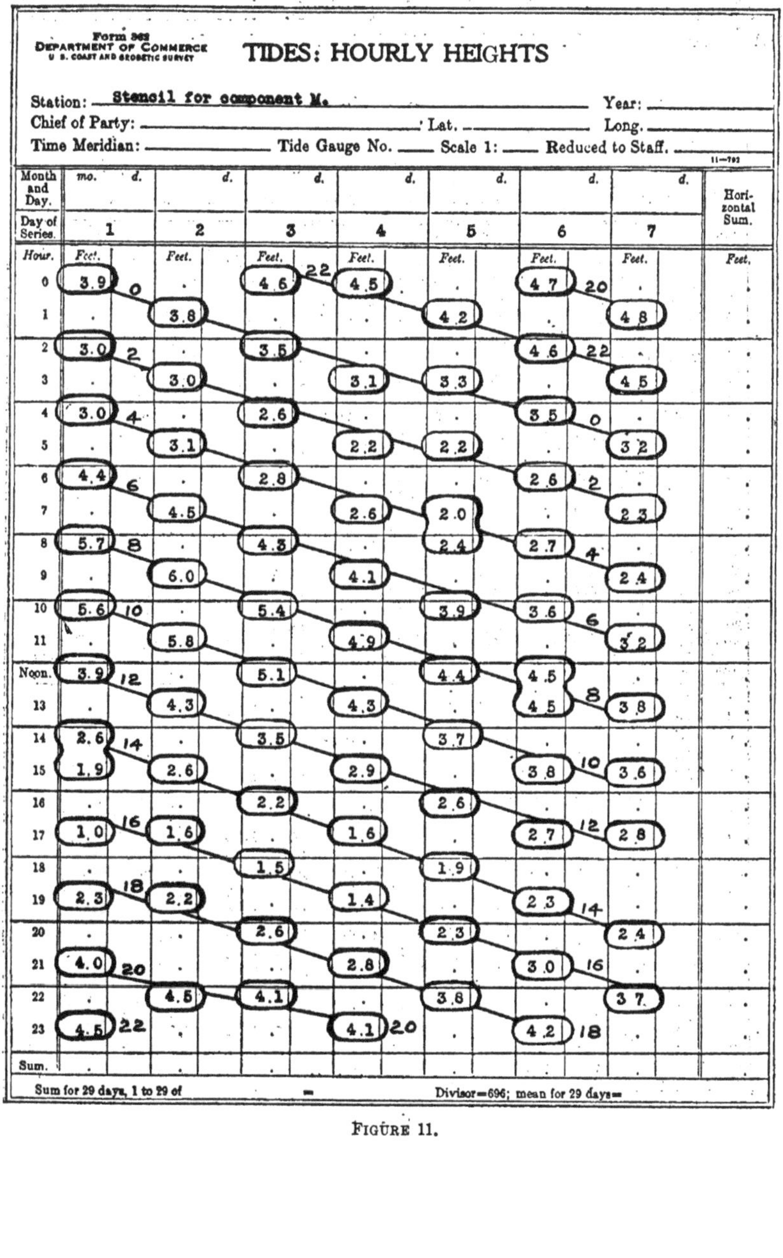

FIGURE 11.

sums in each line of the form must equal the corresponding page sum of the hourly heights in form 362, this serving as a check on the summation. After the summing of all the pages of the series has been completed for any constituent the totals for each constituent hour are obtained, the divisors from table 32 entered, and the constituent hourly means computed (fig. 13). These means should be carefully checked before proceeding with the analysis. Large errors can usually be detected by plotting the means.

309. *Form 244, Computation of* $(V_0+u)$ *(fig. 14).*—This form provides for the computation of the equilibrium arguments for the beginning of the series of observations, the computation being in accordance with formulas given in table 2. For the most part the form is self-explanatory. The values of the mean longitude of the

Form 142
DEPARTMENT OF COMMERCE
U. S. COAST AND GEODETIC SURVEY

**TIDES: STENCIL SUMS.**

Station: Morro, California. Lat.: 35° 22' N.
Component: "M". Length of series: 163. Series begins: 1919–Feb.–12–0 Long.: 120° 51' W.
Kind of time used: 120° W. Computed by Fred. A. Kummell, Dec. 9, 1920.

| Page. | 0ʰ | 1 | 2 | 3 | 4 | 5 | 6 | 7 | 8 | 9 | 10 | 11 |
|---|---|---|---|---|---|---|---|---|---|---|---|---|
| 1 | 24.3 | 20.6 | 17.9 | 16.9 | 21.0 | 23.0 | 28.0 | 31.9 | 39.2 | 34.8 | 31.9 | 27.4 |
| 2 | 21.8 | 17.5 | 14.4 | 13.6 | 11.2 | 12.3 | 14.6 | 20.5 | 21.7 | 23.9 | 24.7 | 24.3 |
| 3 | 19.7 | 16.9 | 11.0 | 9.6 | 12.2 | 17.7 | 28.1 | 24.9 | 27.4 | 27.6 | 29.8 | 21.1 |
| 4 | 26.4 | 18.0 | 17.3 | 17.3 | 22.7 | 22.6 | 26.0 | 34.3 | 36.5 | 41.2 | 33.3 | 28.5 |
| 5 | 21.5 | 21.4 | 17.9 | 18.2 | 16.3 | 19.9 | 24.9 | 29.9 | 37.7 | 34.8 | 33.2 | 29.6 |
| 6 | 20.3 | 16.8 | 15.8 | 12.1 | 12.5 | 15.1 | 21.0 | 21.4 | 23.6 | 24.6 | 25.0 | 28.1 |
| 7 | 23.1 | 16.1 | 13.3 | 13.1 | 15.6 | 23.7 | 28.6 | 33.9 | 30.1 | 34.9 | 27.6 | 23.6 |
| 8 | 25.5 | 23.0 | 21.4 | 21.6 | 20.8 | 23.5 | 27.2 | 29.7 | 43.5 | 36.4 | 32.6 | 27.5 |
| 9 | 20.9 | 18.5 | 15.2 | 12.1 | 11.3 | 13.6 | 18.1 | 26.3 | 26.8 | 28.6 | 28.0 | 29.6 |
| 10 | 16.9 | 13.2 | 10.2 | 8.7 | 11.5 | 15.5 | 18.3 | 21.5 | 24.4 | 25.3 | 28.7 | 24.3 |
| 11 | 18.6 | 15.0 | 12.5 | 15.7 | 17.2 | 23.2 | 29.5 | 41.1 | 36.7 | 34.4 | 27.8 | 24.0 |
| 12 | 24.5 | 25.5 | 20.4 | 20.7 | 21.0 | 24.6 | 32.1 | 31.7 | 32.7 | 36.5 | 31.6 | 25.6 |
| 13 | 25.7 | 17.3 | 13.2 | 10.0 | 11.9 | 12.5 | 16.2 | 20.3 | 24.3 | 30.2 | 30.3 | 24.4 |
| 14 | 16.7 | 12.6 | 8.3 | 8.7 | 9.5 | 14.3 | 19.0 | 23.4 | 30.2 | 27.4 | 27.2 | 26.2 |
| 15 | 19.0 | 16.1 | 16.3 | 15.7 | 20.1 | 26.5 | 37.7 | 37.7 | 40.2 | 39.3 | 35.6 | 29.6 |
| 16 | 29.6 | 22.8 | 21.6 | 22.5 | 25.7 | 31.7 | 31.9 | 34.9 | 35.8 | 38.1 | 28.3 | 27.3 |
| 17 | 22.9 | 18.6 | 14.5 | 11.1 | 10.5 | 12.3 | 15.3 | 19.0 | 25.4 | 24.5 | 24.9 | 23.6 |
| 18 | 15.4 | 10.0 | 6.2 | 3.2 | 4.9 | 10.2 | 16.0 | 24.5 | 24.5 | 25.1 | 25.4 | 27.6 |
| 19 | 16.7 | 15.4 | 13.1 | 15.3 | 19.8 | 29.4 | 31.8 | 35.8 | 38.3 | 37.9 | 38.7 | 29.3 |
| 20 | 27.6 | 21.0 | 19.8 | 20.4 | 28.1 | 29.9 | 31.5 | 36.1 | 36.4 | 39.0 | 38.9 | 22.8 |
| Sums. | 437.1 | 356.3 | 300.3 | 286.5 | 323.8 | 401.7 | 495.6 | 576.6 | 635.4 | 645.6 | 593.5 | 523.4 |

| Page. | 12 | 13 | 14 | 15 | 16 | 17 | 18 | 19 | 20 | 21 | 22 | 23 | |
|---|---|---|---|---|---|---|---|---|---|---|---|---|---|
| 1 | 22.7 | 21.1 | 17.5 | 13.5 | 14.5 | 17.5 | 17.9 | 26.2 | 26.2 | 27.7 | 26.9 | 28.0 | 576.6 |
| 2 | 25.8 | 17.6 | 17.4 | 18.9 | 21.5 | 28.8 | 32.1 | 35.1 | 36.5 | 35.9 | 35.8 | 26.6 | 552.5 |
| 3 | 17.5 | 17.5 | 14.7 | 15.4 | 21.1 | 23.5 | 29.2 | 33.3 | 35.3 | 39.5 | 30.0 | 24.6 | 547.6 |
| 4 | 23.2 | 20.8 | 12.9 | 9.0 | 7.1 | 7.9 | 14.0 | 20.8 | 22.2 | 24.9 | 25.6 | 25.3 | 538.0 |
| 5 | 27.5 | 20.2 | 16.9 | 15.5 | 16.4 | 23.8 | 25.3 | 30.3 | 27.9 | 29.0 | 34.0 | 25.4 | 597.5 |
| 6 | 23.8 | 23.0 | 20.0 | 24.2 | 23.1 | 25.2 | 27.5 | 28.8 | 39.4 | 38.2 | 28.7 | 24.4 | 562.6 |
| 7 | 19.5 | 15.5 | 17.8 | 14.6 | 15.5 | 20.2 | 29.8 | 31.4 | 33.7 | 33.1 | 30.6 | 29.0 | 574.3 |
| 8 | 22.4 | 19.4 | 12.3 | 8.8 | 7.4 | 10.7 | 14.9 | 19.5 | 24.1 | 26.6 | 32.0 | 27.3 | 558.1 |
| 9 | 22.1 | 18.1 | 15.4 | 14.0 | 17.1 | 19.1 | 24.3 | 29.3 | 36.0 | 29.0 | 28.0 | 24.6 | 528.2 |
| 10 | 23.2 | 22.6 | 26.3 | 22.0 | 24.4 | 25.9 | 28.7 | 33.8 | 28.1 | 30.7 | 26.4 | 24.5 | 536.1 |
| 11 | 19.4 | 18.2 | 11.8 | 12.5 | 13.1 | 16.8 | 25.3 | 26.6 | 29.1 | 29.0 | 30.4 | 22.9 | 550.8 |
| 12 | 21.5 | 13.3 | 8.4 | 4.4 | 4.4 | 8.7 | 15.0 | 21.0 | 25.5 | 31.9 | 28.1 | 27.0 | 536.3 |
| 13 | 21.7 | 18.8 | 16.5 | 17.6 | 17.6 | 21.6 | 26.9 | 36.1 | 35.5 | 36.4 | 30.2 | 27.0 | 542.2 |
| 14 | 28.6 | 23.6 | 27.4 | 26.4 | 26.0 | 28.8 | 36.8 | 32.5 | 31.5 | 28.2 | 28.0 | 25.0 | 566.3 |
| 15 | 20.3 | 16.4 | 12.9 | 10.5 | 14.6 | 21.5 | 22.0 | 25.7 | 27.8 | 31.1 | 25.4 | 22.6 | 584.6 |
| 16 | 20.9 | 14.9 | 10.9 | 6.1 | 4.8 | 9.9 | 15.9 | 23.2 | 30.1 | 33.0 | 30.6 | 28.9 | 580.2 |
| 17 | 21.3 | 22.2 | 17.1 | 16.8 | 17.9 | 26.7 | 27.6 | 36.0 | 33.5 | 36.5 | 35.4 | 32.1 | 545.7 |
| 18 | 22.2 | 21.1 | 21.5 | 23.6 | 31.2 | 31.8 | 30.4 | 31.7 | 31.3 | 28.2 | 27.6 | 21.1 | 514.4 |
| 19 | 19.6 | 15.5 | 12.5 | 11.3 | 10.0 | 12.2 | 19.9 | 26.1 | 28.6 | 24.1 | 22.5 | 19.7 | 542.5 |
| 20 | 20.4 | 15.2 | 7.8 | 3.6 | 3.0 | 7.2 | 17.1 | 22.2 | 27.4 | 29.8 | 34.3 | 27.7 | 558.1 |
| Sums | 443.6 | 375.0 | 318.0 | 288.9 | 310.7 | 388.8 | 481.6 | 569.6 | 613.7 | 620.6 | 590.7 | 513.9 | 11093.0 |

FIGURE 12.

moon ($s$), of the lunar perigee ($p$), of the sun ($h$), of the solar perigee ($p_1$), and of the moon's ascending node ($N$), may be obtained from table 4 for the beginning of any year between 1800 and 2000. The values for any year beyond these limits may be readily obtained by taking into account the rate of change in these elements as given in table 1. The corrections necessary in order to refer the elements to any desired month, day, and hour are given in table 5. As the tables refer to Greenwich mean civil time, the argument used in entering them should refer also to this kind of time, and in the lines for the beginning and middle of the series at the head of the form space is therefore provided for entering the equivalent Greenwich hour. Any change in the day may be avoided by using a negative Greenwich hour when necessary. For example; 1922, January 1, 0 hour, in the standard time of the meridian 15° east of Greenwich, may be written as 1922, January 1,—1 hour in Greenwich time, instead of 1921, December 31, 23 hour, as would otherwise be necessary. If a negative argument is used in table 5, the corresponding tabular value must be taken with its sign reversed. For the middle of the series the nearest integral hour is sufficient.

**310.** The values of $I$, $\nu$, $\xi$, $\nu'$, and $2\nu''$ are obtained for the middle of the series from table 6, using $N$ as the argument. If $N$ is between 180° and 360°, each of the last four quantities will be negative, but $I$

Form 148
DEPARTMENT OF COMMERCE
U. S. COAST AND GEODETIC SURVEY

**TIDES: STENCIL SUMS.**

Station: Morro, California. Lat.: 35° 22' N.

Component: M. Length of series: 163 Series begins: 1919 - Feb.-13-0 Long.: 120° 51' W.

Kind of time used: 120° W. Computed by Fred A. Kummell, Dec. 9, 1920.

| Page. | 0ʰ | 1 | 2 | 3 | 4 | 5 | 6 | 7 | 8 | 9 | 10 | 11 |
|---|---|---|---|---|---|---|---|---|---|---|---|---|
| 21 | 25.9 | 18.1 | 14.8 | 14.5 | 10.6 | 11.1 | 17.3 | 23.8 | 23.1 | 24.4 | 24.1 | 22.6 |
| 22 | 16.8 | 14.8 | 7.7 | 5.7 | 6.6 | 11.1 | 19.5 | 23.2 | 26.5 | 27.6 | 30.8 | 24.9 |
| 23 | 17.8 | 15.7 | 15.1 | 20.1 | 21.6 | 30.7 | 33.3 | 37.3 | 39.0 | 42.8 | 33.9 | 28.4 |
| 24 | 7.2 | 6.8 | 6.2 | 6.1 | 6.5 | 8.0 | 9.7 | 10.9 | 18.3 | 12.1 | 11.0 | 9.4 |
| Sums-21-24 | 67.7 | 55.4 | 43.8 | 46.4 | 45.3 | 60.9 | 79.8 | 95.2 | 106.9 | 106.9 | 99.8 | 85.3 |
| " 1-20 | 437.1 | 356.3 | 300.3 | 286.5 | 323.8 | 401.7 | 495.8 | 578.5 | 635.4 | 645.6 | 593.5 | 523.4 |
| Sums.- | 504.8 | 411.7 | 344.1 | 332.9 | 369.1 | 462.6 | 575.6 | 673.7 | 742.3 | 752.5 | 693.3 | 608.7 |
| Divisors.- | 164 | 163 | 162 | 165 | 164 | 163 | 163 | 163 | 164 | 165 | 163 | 162 |
| Means.- | 3.08 | 2.53 | 2.12 | 2.02 | 2.25 | 2.84 | 3.53 | 4.13 | 4.53 | 4.56 | 4.25 | 3.76 |

| Page. | 12ʰ | 13 | 14 | 15 | 16 | 17 | 18 | 19 | 20 | 21 | 22 | 23 | |
|---|---|---|---|---|---|---|---|---|---|---|---|---|---|
| 21 | 23.3 | 18.2 | 17.0 | 17.3 | 23.3 | 24.0 | 29.7 | 32.9 | 35.9 | 42.1 | 34.7 | 31.1 | 558.8 |
| 22 | 22.5 | 20.7 | 20.2 | 26.0 | 26.9 | 31.7 | 36.2 | 34.0 | 40.0 | 31.3 | 26.2 | 20.5 | 551.4 |
| 23 | 23.1 | 16.3 | 15.5 | 11.6 | 11.9 | 13.7 | 19.6 | 25.1 | 26.6 | 26.2 | 24.0 | 24.5 | 573.8 |
| 24 | 3.3 | 4.7 | 3.0 | 1.7 | 0.9 | 0.9 | 1.7 | 3.4 | 5.5 | 7.0 | 7.7 | 7.8 | 159.8 |
| Sums 21-24 | 72.2 | 59.9 | 55.7 | 56.6 | 63.0 | 70.3 | 86.2 | 95.4 | 108.0 | 106.6 | 92.6 | 83.9 | 1843.8 |
| " 1-20 | 443.6 | 375.0 | 318.0 | 288.9 | 310.7 | 388.8 | 481.6 | 569.6 | 613.7 | 620.6 | 590.7 | 513.9 | 11093.0 |
| Sums.- | 515.8 | 434.9 | 373.7 | 345.5 | 373.7 | 459.1 | 567.8 | 665.0 | 721.7 | 727.2 | 683.3 | 597.8 | 12936.8 |
| Divisors.- | 162 | 163 | 163 | 163 | 162 | 162 | 163 | 163 | 162 | 162 | 163 | 163 | |
| Means.- | 3.18 | 2.67 | 2.29 | 2.12 | 2.31 | 2.83 | 3.48 | 4.08 | 4.45 | 4.49 | 4.19 | 3.67 | |

FIGURE 13.

is always positive. Although table 6 is computed for the epoch, January 1, 1900, it is applicable without material error for any series of observations.

**311.** The values of $u$ of $L_2$ and $u$ of $M_1$, may be obtained from table 13 for any date between 1900 and 2000, inclusive, using the value of $N$ for interpolation. If the series falls beyond the limits of this table, the following formulas may be used:

$$u \text{ of } L_2 = 2\xi - 2\nu - R \text{ (par. 129)} \tag{446}$$

$$u \text{ of } M_1 = \xi - \nu + Q \text{ (par. 123)} \tag{447}$$

The values of $\xi$ and $\nu$ may be taken from form 244, the values of $R$ and $Q$ from tables 8 and 10, respectively, using the arguments $I$ and $P$ for the middle of the series.

Form 244.
DEPARTMENT OF COMMERCE,
U. S. COAST AND GEODETIC SURVEY.

**TIDES: Computation of $V_0+u$.**

Station Morro, California — Lat. 35° 22' N. — Long. 120° 51' W. 120.85 W. = $L$

Beginning of series 1919 (yr.) Feb. (mo.) 13 (d.) 0 (hr.) († Greenwich hr. 8) — Length of series 163 d. — Time mer. 120.00 W. = $S$

Middle of series 1919 (yr.) May (mo.) 5 (d.) 12 (hr.) († Greenwich hr. 20)

Compute all values to two decimal places. — Tables in Harmonic Analysis and Prediction of the Tide.

| | For the beginning of series. | | | | For the middle of series. | |
|---|---|---|---|---|---|---|
| | (1) = $s$ | (2) = $p$ | (3) = $h$ | (4) = $p_1$ | (5) = $p$ | (6) = $N$. |
| Table 4, for January 1 of year | 268.04 | 27.41 | 279.60 | 281.55 | 27.41 | 251.71 |
| Table 5, correction to 1st of month | 48.47 | 3.45 | 30.56 | 0.00 | 13.37 | −6.35 |
| Table 5, correction to day of month | 158.12 | 1.34 | 11.83 | 0.00 | 0.45 | −0.21 |
| Table 5, correction to Greenwich hr. | 4.39 | 0.04 | 0.33 | 0.00 | 0.09 | −0.04 |
| Sum = | (1) = 119.02 | (2) = 32.24 | (3) = 322.32 | (4) = 281.55 | (5) = 41.32 | (6) = 245.11 |

| | | | | |
|---|---|---|---|---|
| (7) = $I$ (Table 6) = 21.76<br>(8) = $\nu$ (Table 6) = −12.68<br>(9) = $\xi$ (Table 6) = −11.71<br>(10) = $\nu'$ (Table 6) = −8.53<br>(11) = $2\nu''$ (Table 6) = −16.63 | **$J_1$**<br>+ (19) = 138.25<br>− (8) = 12.68<br>$V_0+u$ = 150.93<br>---- = . | **$M_6$**<br>$V_0+u$ = 2 ($M_2$) = 140.52<br>**$M_8$**<br>$V_0+u$ = 2 ($M_2$) = 187.36 | **$R_2$**<br>(32) ± 180° = 220.77<br>+ (33) = −1.70<br>$V_0+u$ = 219.07<br>---- = . | **$(MK)_3$**<br>+ $M_2$ = 46.84<br>+ $K_1$ = 60.00<br>$V_0+u$ = 106.84<br>---- = . |
| (12) = P = (5) − (9) = 53.03<br>(13) = $u$ of $L_2$ ± 180° (Table 13) = 170.6<br>(14) = $u$ of $M_1$ + 90° (Table 13) = 123.6 | **$K_1$**<br>+ (17) = 51.47<br>− (10) = 8.53<br>$V_0+u$ = 60.00<br>---- = . | **$N_2$**<br>+ $M_2$ = 46.84<br>− (18) = −86.78<br>$V_0+u$ = −39.94<br>---- = 320.06 | **$S_1$**<br>$V_0+u$ = (15) ± 180°<br>---- = 179.15<br>**$S_2$**<br>$V_0+u$ = (33) = 358.30 | **$(2MK)_3$**<br>+ $M_2$ = 93.68<br>− $K_1$ = −60.00<br>$V_0+u$ = 33.68<br>---- = . |
| (15) = $S°-L°$ = −0.85<br>(16) = (3) + (15) = 321.47<br>(17) = (16) + 90° = 51.47 | **$K_2$**<br>+ (20) = 282.94<br>− (11) = 16.63<br>$V_0+u$ = 299.57<br>---- = . | **$(2N)_2$**<br>+ $N_2$ = 320.06<br>− (18) = −86.78<br>$V_0+u$ = 233.28<br>---- = . | **$S_4$**<br>$V_0+u$ = 2 (33) = 356.60<br>**$S_6$**<br>$V_0+u$ = 3 (33) = 354.90 | **$(MN)_4$**<br>+ $M_2$ = 46.84<br>+ $N_2$ = 320.06<br>$V_0+u$ = 366.90<br>---- = 6.90 |
| (18) = (1) − (2) = 86.78<br>(19) = (17) + (18) = 138.25<br>(20) = 2 (16) = 282.94<br>(21) = (1) + (2) = 151.26<br>(22) = (20) − (21) = 131.68 | **$L_2$**<br>+ (22) = 131.68<br>+ (13) = 170.6<br>$V_0+u$ = 302.28<br>---- = . | **$O_1$**<br>+ (31) = 334.15<br>− (30) = −351.46<br>$V_0+u$ = −17.31<br>---- = 342.69 | **$T_2$**<br>+ (33) = 358.30<br>− (32) = −40.77<br>$V_0+u$ = 317.53<br>---- = . | **$(MS)_4$**<br>+ $M_2$ = 46.84<br>+ $S_2$ = 358.30<br>$V_0+u$ = 405.14<br>---- = 45.14 |
| (23) = (16) − (1) = 202.45<br>(24) = (9) − (8) = 0.97<br>(25) = (23) + (24) = 203.42<br>(26) = 2 (25) = 46.84<br>(27) = (25) + (26) = 250.26 | **$M_1$**<br>+ (23) = 202.45<br>+ (14) = 123.6<br>$V_0+u$ = 326.05<br>---- = . | **$(OO)_1$**<br>+ (31) = 334.15<br>+ (30) = 351.46<br>$V_0+u$ = 685.61<br>---- = 325.61 | **$\lambda_2$**<br>(36) ± 180° = 180.24<br>− (18) = −86.78<br>$V_0+u$ = 93.46<br>---- = . | **$(2SM)_2$**<br>+ $S_4$ = 356.60<br>− $M_2$ = −46.84<br>$V_0+u$ = 309.76<br>---- = . |
| (28) = (1) − (9) = 130.73<br>(29) = 2 (28) = 261.46<br>(30) = (29) + 90° = 351.46 | **$M_2$**<br>$V_0+u$ = + (26) = 46.84<br>---- = . | **$P_1$**<br>(15) + 270° = 269.15<br>− (3) = −322.32<br>$V_0+u$ = −53.17<br>---- = 306.83 | **$\mu_2$**<br>+ $M_2$ = 46.84<br>+ (35) = 46.60<br>$V_0+u$ = 93.44<br>---- = . | **Mf**<br>$V_0+u$ = + (29) = 261.46<br>**MSf**<br>+ $S_2$ = 358.30<br>− $M_2$ = −46.84<br>$V_0+u$ = 311.46<br>---- = . |
| (31) = (16) − (8) = 334.15<br>(32) = (3) − (4) = 40.77<br>(33) = 2 (15) = −1.70 | **$M_3$**<br>$V_0+u$ = (27) ± 180°<br>---- = 70.26 | **$Q_1$**<br>+ $O_1$ = 342.69<br>− (18) = −86.78<br>$V_0+u$ = 255.91<br>---- = . | **$\nu_2$**<br>$\mu_2$ = 93.44<br>+ (18) = 86.78<br>$V_0+u$ = 180.22<br>---- = . | **Mm**<br>$V_0+u$ = + (18) = 86.78 |
| (34) = (3) − (1) = 203.30<br>(35) = 2 (34) = 46.60<br>(36) = (20) − (35) = 0.24<br>(37) = (3) − (2) = 290.08<br>(38) = 2 (37) = 220.16 | **$M_4$**<br>$V_0+u$ = 2 ($M_2$) = 93.68<br>---- = . | **$(2Q)_1$**<br>+ $Q_1$ = 255.91<br>− (18) = −86.78<br>$V_0+u$ = 169.13<br>---- = . | **$\rho_1$**<br>+ $Q_1$ = 255.91<br>+ (38) = 220.16<br>$V_0+u$ = 476.07<br>---- = 116.07 | **Sa**<br>$V_0+u$ = + (3) = 322.32<br>**Ssa**<br>$V_0+u$ = + 2 (3) = 284.64 |

† Greenwich hour = original hour + ($S°$ ÷ 15).

° Positive for West longitude; negative for East longitude.

FIGURE 14.

HARMONIC ANALYSIS AND PREDICTION OF TIDES    111

312. In finding the difference between the longitude of the time meridian ($S$) and the longitude of the place ($L$) consider west longitude as positive and east longitude as negative. In the ordinary use of form 244 it is assumed that civil time has been used in the tabulations of the observations. If, however, the original hourly heights as tabulated in form 362 are in accordance with astronomical time in which the 0 hour represents the noon of the corresponding civil day and the 12th hour the following midnight, form 244 will still be applicable if the longitude of the time meridian ($S$) is taken equal to the civil time meridian plus 180°. For example, if tabulations have been made in astronomical time for a locality where the civil time is based upon the meridian 15° E., the value for $S$ should be taken as $-15 + 180$, or 165°. If tabulations have been in Greenwich astronomical time, $S$ should be taken as 180°.

313. *Form 244a, Log F and arguments for elimination* (fig. 15).— Items (1) to (11) are compiled here for convenience of reference for

Form 244A
DEPARTMENT OF COMMERCE
U. S. COAST AND GEODETIC SURVEY

TIDES: Log $F$ and Arguments for Elimination

Station ...... Morro, California ......

Length of series ...... 163 ...... days. Series begins ...... 1919 Feb. 13 0

| Component | Log $F$ (4 dec.) | Component | Log $F$ (4 dec.) | Component | Log $F$ (4 dec.) |
|---|---|---|---|---|---|
| $J_1$ | 0.0201 | $M_1$ | 9.9726 | MK | 0.0091 |
| $K_1$ | 0.0160 | $N_2$, 2N | 9.9932 | 2MK | 0.0023 |
| $K_2$ | 0.0472 | $O_1$ | 0.0264 | MN | 9.9863 |
| $L_2$=Log $F(M_2)$+(7) | 9.9589 | OO | 0.0929 | MS, 2SM | 9.9932 |
| $M_1$=Log $F(O_1)$+(8) | 9.8856 | $P_1$ | 0.0000 | Mf | 0.0596 |
| $M_2$ | 9.9932 | $Q_1$, 2Q | 0.0264 | MSf | 9.9932 |
| $M_3$ | 9.9897 | $R_2$, $S_1$, $S_2$, $S_4$, $S_6$, $T_2$ | 0.0000 | Mm | 9.9772 |
| $M_4$ | 9.9863 | $\lambda_2$ $\mu_2$ $\nu_2$ | 9.9932 | Sa, Ssa | 0.0000 |
| $M_6$ | 9.9794 | $\rho_1$ | 0.0264 | | |

(1) = $N$ = item (6) from Form 244 = 245°.11 (2 dec.)
(2) = $I$ = item (7) from Form 244 = 21°.76 (2 dec.)
(3) = $P$ = item (12) from Form 244 = 53°.03 (2 dec.)

(4) = ($h-\frac{1}{2}\nu'$) = item (3) $-\frac{1}{2}$ item (10), from Form 244 = 327° (0 dec.)
(5) = ($h-\nu''$) = item (3) $-\frac{1}{2}$ item (11), from Form 244 = 331 (0 dec.)
(6) = ($h-p_1$) = item (3) – item (4), from Form 244 = 41 (0 dec.)

(7) = Log $R_a$ from Table 7 = 9.9657 (4 dec.)
(8) = Log $Q_a$ from Table 9 = 9.8592 (4 dec.)

(9) = Natural number from Log $F(K_1)$ = 1.038 (3 dec.)
(10) = Log $f(K_2)$ = 10 – Log $F(K_2)$ = 9.9528 (4 dec.)
(11) = Natural number $f(K_2)$ from (10) = 0.897 (3 dec.)

EXPLANATION.—For all tables see Special Publication No. 98. First fill in items (1) to (8). Then obtain values of log $F$ for all components excepting $L_2$ and $M_1$ from Table 12. Log $F(L_2)$ = log $F(M_2)$ + log $R_a$, and log $F(M_1)$ = log $F(O_1)$ + log $Q_a$. Items (9) to (11) are obtained after the rest of the form has been filled out.

FIGURE 15.

this and form 452. Items (1) to (6) are obtained from values given in form 244. Item (7) is obtained from table 7, using items (2) and (3) as arguments, and item (8) is obtained from table 9, using item (3) as argument. Items (9) to (11) are obtained after the rest of the form has been filled out.

**314.** The log $F$ for each of the listed constituents, except $L_2$ and $M_1$ and those for which the logarithm is given as zero, may be obtained from table 12, using item (2) as the argument. For constituents $L_2$ and $M_1$

$$\text{Log } F(L_2) = \log F(M_2) + \text{item (7)} \tag{448}$$

$$\text{Log } F(M_1) = \log F(O_1) + \text{item (8)} \tag{449}$$

If the tidal series analyzed was observed between the years 1900 and 2000, the log $F(L_2)$ and log $F(M_1)$ may be taken directly from

Form 194
DEPARTMENT OF COMMERCE
COAST AND GEODETIC SURVEY

**TIDES: HARMONIC ANALYSIS**

Station: Morro, California — Lat. 35° 22' N — Long. 120° 51' W

Component: M — Series begins (yr. m. d. h.) 1919 Feb. 13 0 — Length of series 163 (Days) — Time Mer. 120°

Hourly Means from Form 142.

| | | | | | | | | | | | | |
|---|---|---|---|---|---|---|---|---|---|---|---|---|
| (1) Hours 0 to 11 | 3.08 | 2.53 | 2.12 | 2.02 | 2.25 | 2.84 | 3.53 | 4.13 | 4.53 | 4.56 | 4.25 | 3.76 |
| (2) Hours 12 to 23 | 3.18 | 2.67 | 2.29 | 2.12 | 2.31 | 2.83 | 3.48 | 4.08 | 4.45 | 4.49 | 4.19 | 3.67 |
| (3)=(1)−(2) | −0.10 | −0.14 | −0.17 | −0.10 | −0.06 | 0.01 | 0.05 | 0.05 | 0.08 | 0.07 | 0.06 | 0.09 |
| (4)=Last 5 values of (3) reversed | 0.00 | 0.09 | 0.06 | 0.07 | 0.08 | 0.05 | 0.00 | | | | | |
| (5)=(1)+(2) | 6.26 | 5.20 | 4.41 | 4.14 | 4.56 | 5.67 | 7.01 | 8.21 | 8.98 | 9.05 | 8.44 | 7.43 |
| (6)=Last half of (5) | 7.01 | 8.21 | 8.98 | 9.05 | 8.44 | 7.43 | | | | | | |

| (7) | (8) | (9) | (10) | (11) | (12) | (13) | (14) | (15) | (16) | (17) | (18) | (19) | (20) | (21) |
|---|---|---|---|---|---|---|---|---|---|---|---|---|---|---|
| Sin. n×15° | (3)+(4) | (7)×(8) 12s₁ | Cos. n×15° | (3)−(4) | (10)×(11) 12c₁ | Sin. n×45° | (8)×(13) 12s₃ | Cos. n×45° | (11)×(15) 12c₃ | Sin. n×30° | (5)−(6) | (17)×(18) 12s₂ | Cos. n×30° | (18)×(20) 12c₂ |
| .000 | −0.10 | 0.000 | 1.000 | −0.10 | −0.100 | .000 | 0.000 | 1.000 | −0.100 | .000 | −0.75 | 0.000 | 1.000 | −0.750 |
| .259 | −0.05 | −0.013 | .966 | −0.23 | −0.222 | .707 | −0.035 | .707 | −0.163 | .500 | −3.01 | −1.505 | .866 | −2.607 |
| .500 | −0.11 | −0.055 | .866 | −0.23 | −0.199 | 1.000 | −0.110 | .000 | 0.000 | .866 | −4.57 | −3.958 | .500 | −2.285 |
| .707 | −0.03 | −0.021 | .707 | −0.17 | −0.120 | .707 | −0.021 | −.707 | +0.120 | 1.000 | −4.91 | −4.910 | .000 | 0.000 |
| .866 | 0.02 | 0.017 | .500 | −0.14 | −0.070 | .000 | 0.000 | −1.000 | +0.140 | .866 | −3.88 | −3.360 | −.500 | +1.940 |
| .966 | 0.06 | 0.058 | .259 | −0.04 | −0.010 | −.707 | −0.042 | −.707 | +0.028 | .500 | −1.76 | −0.880 | −.866 | +1.524 |
| 1.000 | 0.05 | 0.050 | .000 | 0.05 | 0.000 | −1.000 | −0.050 | .000 | 0.000 | | 12s₂= | −14.613 | 12c₂= | −2.178 |
| | 12s₁= | +0.036 | | 12c₁= | −0.721 | 12s₃= | −0.258 | 12c₃= | 0.025 | | | | | |

| (22) | (23) | (24) | (25) | (26) | (27) | (28) | (29) | (30) | (31) | (32) | (33) | (34) | (35) | (36) | (37) |
|---|---|---|---|---|---|---|---|---|---|---|---|---|---|---|---|
| Sin. n×90° | (18)×(22) 12s₆ | Cos. n×90° | (18)×(24) 12c₆ | (5)+(6) 1st half. | (5)+(6) 2d half. | (26)−(27) | Sin. n×60° | (28)×(29) 12s₄ | Cos. n×60° | (28)×(31) 12c₄ | (26)+(27) | Sin. n×120° | (33)×(34) 12s₈ | Cos. n×120° | (33)×(36) 12c₈ |
| 0 | 0.000 | +1 | −0.750 | 13.27 | 13.19 | 0.08 | .000 | 0.000 | 1.000 | 0.080 | 26.46 | .000 | 0.000 | 1.000 | 26.460 |
| +1 | −3.010 | 0 | 0.000 | 13.41 | 13.00 | 0.41 | .866 | 0.355 | .500 | 0.205 | 26.41 | .866 | 22.871 | −.500 | −13.205 |
| 0 | 0.000 | −1 | 4.570 | 13.39 | 13.10 | 0.29 | .866 | 0.251 | −.500 | −0.145 | 26.49 | −.866 | −22.940 | −.500 | −13.245 |
| −1 | 4.910 | 0 | 0.000 | | | | 12s₄= | 0.606 | 12c₄= | 0.140 | | 12s₈= | −0.000 | 12c₈= | 0.010 |
| 0 | 0.000 | +1 | −3.880 | | | | | | | | | | | | |
| +1 | −1.760 | 0 | 0.000 | | | | | | | | | | | | |
| 12s₆= | 0.140 | 12c₆= | −0.060 | | | | | | | | | | | | |

ζ′ is in 1st quadrant when we have +s and +c.
ζ′ is in 2d quadrant when we have +s and −c.
ζ′ is in 3d quadrant when we have −s and −c.
ζ′ is in 4th quadrant when we have −s and +c.

| Component | $M_1$ | $M_2$ | $M_3$ | $M_4$ | $M_6$ | $M_8$ |
|---|---|---|---|---|---|---|
| (38)=log. 12s | 8.55630 | 1.16474 | 9.41162 | 9.78247 | 9.14613 | 8.83885 |
| (39)=log. 12c | 9.85794 | 0.33806 | 8.39794 | 9.14613 | 8.77815 | 8.00000 |
| (40)=(38)−(39)=log. tan. ζ′ | 8.69836 | 0.82668 | 1.01368 | 0.63634 | 0.36798 | 0.83885 |
| (41)=ζ′, for beginning of the series | 177.14 | 261.52 | 275.53 | 76.99 | 113.20 | 278.25 |
| (42)=local $V_0+u$ (From Form 244) | 326.05 | 46.84 | 70.26 | 93.68 | 140.52 | 187.36 |
| (43)=(41)+(42)=κ′ | 143.19 | 308.36 | 345.79 | 170.67 | 253.72 | 105.61 |
| (44)=log. sin. ζ′, or | ------- | 9.99523 | 9.99797 | 9.98871 | 9.96338 | 9.99548 |
| (45)=log. cos. ζ′ | 9.99946 | ------- | ------- | ------- | ------- | ------- |
| (46)=(38)−(44), or (39)−(45) | 9.85848 | 1.16951 | 9.41365 | 9.79376 | 9.18275 | 8.84337 |
| (47)=log (augmenting factor ÷ 12) | 8.92206 | 8.92579 | 8.93204 | 8.94085 | 8.96642 | 9.00332 |
| (48)=log. (reciprocal of 12) for component S | 8.92082 | 8.92082 | 8.92082 | 8.92082 | 8.92082 | 8.92082 |
| (49)=(46)+(47), or (46)+(48)=log R′ | 8.78054 | 0.09530 | 8.34569 | 8.73461 | 8.14917 | 7.84669 |
| (50)=log factor F (From Form 244a) | 9.8856 | 9.9932 | 9.9897 | 9.9863 | 9.9794 | 9.9726 |
| (51)=(49)+(50)=log H′ | 8.66614 | 0.0885 | 8.33539 | 8.72091 | 8.12857 | 7.81929 |
| (52)=natural number from (49)=R′ | 0.060 | 1.245 | 0.022 | 0.054 | 0.014 | 0.007 |
| (53)=natural number from (51)=H′ | 0.046 | 1.226 | 0.022 | 0.053 | 0.013 | 0.007 |

FIGURE 16.

table 13, using the year of observations, together with item (1), as argument.

**315.** *Form 194, Harmonic analysis* (fig. 16).—This form is based, primarily, upon formulas (295), (296), (303), and (304) and is designed for the computations of the first approximate values of the epochs ($\kappa$) and the amplitudes ($H$) of the harmonic constants. Provisions are made for obtaining the diurnal, semidiurnal, terdiurnal, quarter-diurnal, sixth-diurnal and eighth-diurnal constituents, but only such items need be computed as are necessary for the particular constituents sought. For the principal lunar series $M_1$, $M_2$, $M_3$, $M_4$, $M_6$, and $M_8$, compute all items of the form. For the principal solar series $S_1$, $S_2$, $S_4$, and $S_6$, items (14), (16), (33), (35), and (37) may be omitted. For the lunisolar constituents $K_1$ and $K_2$, items (14), (16), and (23) to (37) may be omitted. For the diurnal constituents $J_1$, $O_1$, $OO$, $P_1$, $Q_1$, $2Q$, and $\rho_1$, items (5), (6), and (14) to (34) may be omitted. For the semidiurnal constituents $L_2$, $N_2$, $2N$, $R_2$, $T_2$, $\lambda_2$, $\mu_2$, $\nu_2$, and $2SM$, items (3), (4), (8) to (16), and (23) to (37) may be omitted. For ter-diurnal constituents MK and 2MK, items (5), (6), (9), (12), and (18) to (37) may be omitted. For quarter-diurnal constituents MN and MS, items (3), (4), (8) to (25), and (35) to (37) may be omitted. In the bottom portion of the form the symbol of the constituent is to be entered at the head of the column or columns indicated by the subscript corresponding to the number of constituent periods in a constituent day, the remaining columns being left blank.

**316.** The hourly means from form 142 (fig. 13) are entered as items (1) and (2) in regular order, beginning with the mean for 0 hour. Item (4) consists of the last five values of item (3) arranged in reverse order. Item (6) consists of the last six values of item (5) in their original order. For the computations of this form the following tables will be found convenient: table 19 of this publication for natural products, Vega's Logarithmic Tables for logarithms of linear quantities, and Bremiker's Funfstellige Logarithmen for logarithms of the trigonometrical functions. In the last table the angular arguments are given in degrees and decimals.

**317.** In choosing between items (44) and (45) the former should be used if the tabular value of (41) in the first quadrant is greater than 45° and the latter if this angle is less than 45°. In referring (41) to the proper quadrant it must be kept in mind that the signs of the natural numbers corresponding to (38) and (39) are respectively the signs of the sine and cosine of the required angles. Therefore (41) will be in the first quadrant if both $s$ and $c$ are positive, in the second quadrant if $s$ is positive and $c$ negative, in the third quadrant if both $s$ and $c$ are negative, and in the fourth quadrant if $s$ is negative and $c$ positive. In obtaining (49) use (46)+(47) for all constituents except S, and (46)+(48) for S. The log factor $F$ for item (50) may be obtained from form 244a.

**318.** Form 194 is designed for use when 24 constituent hourly means have been obtained and all the original hourly heights have been used in the summation. If in the summation for a constituent each constituent hour of the observation period received one and only one of the hourly heights, it will be necessary to take the log-augmenting factor from table 20 and add this to the sum of items (46) and (48) to obtain item (49), striking out item (47).

**319.** This form is also adapted for use with the long-period constituents. Assuming that the daily means have been cleared of the effects of the short-period constituents (p. 89), and that these means have been assorted into 24 groups to cover the constituent period, the 24 group means may then be entered in form 194 in place of the 24 hourly means used for the short-period constituents. Then, treating the constituents Mm and Sa the same as the diurnal tides and the constituents Mf, Msf, and Ssa as the semidiurnal tides, the form may be followed except that the log-augmenting factor must be taken from table 20 and then combined with items (46) and (48) to obtain item (49), striking out item (47).

**320.** To obtain Sa and Ssa from the monthly means of sea level, or tide level, the following process may be used: Enter the monthly means beginning with that for January in alternate spaces provided for the hourly means in form 194, placing the value for January in the space for the 0 hour. For convenience consider all the intermediate blank spaces as being filled with zero values and make the computations indicated by (3) to (12) and (18) to (21). Correct the coefficients of $s_1$ and $c_1$ from 12 to 6, at top and foot of columns (9), (12), (19), and (21). In bottom of form enter Sa in column having subscript 2 and Ssa in column with subscript 4 in order to obtain correct augmenting factors and strike out numerals indicating subscripts. For (38) and (39) take the logarithm of twice the values of $6s$ and $6c$ as obtained above. The $\zeta$'s as obtained from (40) must have the following corrections applied in order to refer them to 0 hour of the first day of January—common years, Sa correction$=+14.07°$, Ssa correction$=+28.14°$; leap years, Sa correction$=+14.94°$, Ssa correction$=+29.88°$. For convenience in recording the results it is suggested that the $\zeta$ as directly obtained from (40) be entered (in its proper quadrant) in the space just below the logarithm from which it is obtained, and that the $\zeta$ corrected to the first day of January be entered in the same line in the vacant column just to the right. The $V+u$, computed to the first day of January, may then be entered immediately under the corrected $\zeta$'s and the $\kappa'$ of (43) readily obtained. For (49) the combination (46)+(47) will be used.

**321.** *Form 452, R, $\kappa$, and $\zeta$ from analysis and inference* (figs. 17 and 18)—This form provides for certain computations preliminary to the regular elimination process. The constants for constituents $K_1$ and $S_2$ as obtained directly from form 194 may be improved by the application of corrections from tables 21 to 26; and constants for some of the smaller constituents, which have been poorly determined or not determined at all by the analysis, may be obtained by inference. If the series of observations is very short, the inferred values for the constants of some of the constituents may be better than the uneliminated values from form 194.

**322.** Form 452 is based upon paragraphs 229 to 243. It is designed to take account of the diurnal constituent on one side (fig. 17) and the semidiurnal constituents on the other side (fig. 18). The amplitudes and epochs indicated by the accent (') are to be taken from form 194 and the quantities indicated by the asterisk (*) from form 244 or 244a. If the series is less than 355 days, values for $S_1$ and 2SM may be omitted.

**323.** For all short series the values in columns (4) and (8) are to be computed in accordance with the equivalents and factors in columns

(3) and (7) respectively. If the series is 192 days or more in length, the κ of $M_1$, $P_1$, and $K_2$ for column (4), and the log $R$ of $M_1$, $P_1$, and $K_2$ for column (8) may be taken directly from form 194, and if the series is 355 days or more in length the κ and log $R$ of all the components for which analyses have been made may be taken directly from the same form. When a value is thus taken directly from the analysis, the corresponding equivalent in column (3) and factors in column (7) are to be crossed out.

**324.** The tabular values of items (12) and (13) for the diurnal constituents and items (14) to (18) for the semidiurnal constituents may be obtained from tables 21 to 26 or from plotted curves representing these tables, but for a series of 355 days or more in length the accelerations may be taken as zero and the resultant amplitude factors as unity.

Form 452
DEPARTMENT OF COMMERCE
U. S. COAST AND GEODETIC SURVEY

## TIDES: $R$, κ, AND ζ, FROM ANALYSIS AND INFERENCE.

*Station* Morro, California

Length of Series 163 days. Series begins 1919, February 13

DIURNAL COMPONENTS.

| Components. | From Analysis. $R'$ (1) Ft. (3 dec.) | From Analysis. ζ' (2) ° (2 dec.) | Components. | From Analysis and Inference. κ (3) Equivalent. | κ (4) ° (1 dec.) | $V_0+u.$* (5) ° (1 dec.) | ζ=(4)−(5) (6) ° (0 dec.) |
|---|---|---|---|---|---|---|---|
| $J_1$ | . | . | $J_1$ | $K_1°$+0.5×(14) | 116.7 | 150.9 | 326 |
| $K_1$ | 0.967 | 48.82 | $K_1$ | $(K_1°)'$ 108.8+(12) | 110.9 | 60.0 | 51 |
| $M_1$ | 0.060 | 177.14 | $M_1$ | $K_1°$−0.5×(14) | 105.1 | 326.0 | 139 |
| $O_1$ | 0.569 | 116.61 | $O_1$ | $(O_1°)'$ | 99.3 | 342.7 | 117 |
| OO | . | . | OO | $K_1°$+(14) | 122.5 | 325.6 | 157 |
| $P_1$ | 0.299 | 182.53 | $P_1$ | $K_1°$ | 110.9 | 306.8 | 164 |
| $Q_1$ | 0.107 | 210.95 | $Q_1$ | $K_1°$−1.5×(14) | 93.5 | 255.9 | 198 |
| 2Q | . | . | 2Q | $K_1°$−2.0×(14) | 87.7 | 169.1 | 279 |
| $S_1$ | . | . | $S_1$ | $(S_1°)'$ | -. | -. | - |
| $\rho_1$ | . | . | $\rho_1$ | $K_1°$−1.43×(14) | 94.3 | 116.1 | 338 |

(9)=P*= 53.03 (2 dec.); (10)=$F(K_1)$*= 1.038 (3 dec.); (11)=$(h-\frac{1}{2}\nu')$*= 327 (0 dec.)
(12)=acceleration in $K_1$ due to $P_1$=$F(K_1)$×Table 21= 2.1 (1 dec.)
(13)=resultant amplitude, $K_1$ and $P_1$=1+[$F(K_1)$ × (Table 22)]= 1.03 (2 dec.)
(14)=$(K_1°-O_1°)$= 11.6 (1 dec.)

*Explanation.*—Obtain from Form 194 the amplitudes and epochs indicated by the accent (′), and from Form 244 or 244a the quantities indicated by the asterisk (*).

If the series is less than 355 days, omit $S_1$. For all short series, the values in columns (4) and (8) will be computed in accordance with the Equivalents and Factors in columns (3) and (7), respectively; but if the series is 192 days or more in length, the κ of $M_1$ and $P_1$ for column (4), and the log. $R$ of $M_1$ and $P_1$ for column (8) may be taken directly from Form 194, and if the series is 355 days or more in length, the κ and log. $R$ of all the components for which analysis has been made may be taken directly from the same form. When a value is thus taken directly from the analysis, the corresponding equivalent in column (3) and factors in column (7) should be crossed out.

The tabular values for (12) and (13) may be obtained from Tables 21 and 22 in Special Publication No. 98 or from plotted curves representing this table. For a series of 355 days or more, (12)=0, and (13)=1.

Obtain the κ of $K_1$ by applying (12) to $(K_1°)'$ from Form 194, and use this corrected κ in computing (14). If the two angles in (14) differ by more than 180°, add 360° to the smaller before taking the difference, which may be either positive or negative.

In computing column (8) it will be noted that the corrected log. $R(K_1)$ is to be used when inferring $P_1$.

REMARKS:

| Components. | From Analysis and Inference. $R$ (7) Factors. | $R$ (8) (4 dec.) |
|---|---|---|
| $J_1$ | log. 0.079 | +8.8976 |
| | log. $H'(O_1)$ | +9.7814 |
| | log. $F(J_1)$* | −0.0201 |
| | log. $R(J_1)$ | 8.6589 |
| | $R(J_1)$ | |
| $K_1$ | log. $R'(K_1)$ | +9.9852 |
| | log. (13) | −0.0128 |
| | log. $R(K_1)$ | 9.9724 |
| | $R(K_1)$ | |
| $M_1$ | log. 0.071 | +8.8513 |
| | log. $R(O_1)$ | +9.7550 |
| | log. $Q_a$* | −9.8592 |
| | log. $R(M_1)$ | 8.7471 |
| | $R(M_1)$ | |
| $O_1$ | log. $R'(O_1)$ | 9.7550 |
| | $R(O_1)=R'(O_1)$ | |
| OO | log. 0.043 | +8.6335 |
| | log. $H'(O_1)$ | +9.7814 |
| | log. $F(OO)$* | −0.0929 |
| | log. $R(OO)$ | 8.3220 |
| | $R(OO)$ | |
| $P_1$ | log. 0.331 | +9.5198 |
| | log. $R(K_1)$ | +9.9724 |
| | log. $F(K_1)$* | +0.0160 |
| | log. $R(P_1)$ | 9.5082 |
| | $R(P_1)$ | |
| $Q_1$ | log. 0.194 | +9.2878 |
| | log. $R(O_1)$ | +9.7550 |
| | log. $R(Q_1)$ | 9.0428 |
| | $R(Q_1)$ | |
| 2Q | log. 0.026 | +8.4150 |
| | log. $R(O_1)$ | +9.7550 |
| | log. $R(2Q)$ | 8.1700 |
| | $R(2Q)$ | |
| $S_1$ | log. $R'(S_1)$ | --- |
| | $R(S_1)=R'(S_1)$ | |
| $\rho_1$ | log. 0.038 | +8.5798 |
| | log. $R(O_1)$ | +9.7550 |
| | log. $R(\rho_1)$ | 8.3348 |
| | $R(\rho_1)$ | |

FIGURE 17.

**325.** The $\kappa$'s of $K_1$ and $S_2$ are to be corrected by the accelerations as indicated before entering in column (4), and in computing item (14) for the diurnal constituents and (21) for the semidiurnal constituents the corrected $\kappa$'s are to be used. If the two angles in item (14) for the diurnal constituents, or in items (20) or (21) for the semidiurnal constituents, differ by more than 180°, the smaller angle should be increased by 360° before taking the difference, which may be either positive or negative. In computing column (8) it will be noted that the corrected log $R$'s of $K_1$ and $S_2$ are to be used in inferring other constituents depending upon them.

**326.** *Form 245, Elimination of component effects* (fig. 19).—This form is based upon formulas (389) and (390). One side of the form is designed for the elimination of the effects of the diurnal constituents upon each other and the other side for use with the semidiurnal constituents, the two sides being similar except for the listing of the

Figure 18.

HARMONIC ANALYSIS AND PREDICTION OF TIDES 117

constituents. The symbol $A$ represents the constituent to be cleared, and the symbol $B$ is the general designation for the disturbing constituents. The symbol applying to constituent $A$ is to be crossed out in column (1) and entered in column (8). The values for items (9) and (19) are to be taken from columns (1) and (2) of form 452.

**327.** For obtaining column (2) it will be found convenient to copy the logarithms of the $R$'s of $B$ from column (8) of form 452 on a horizontal strip of paper spaced the same as table 29. Applying this strip successively to the upper line of the tabular values for each con-

FIGURE 19.

stituent the logarithms of the resulting products for column (2) may be readily obtained. Similarly, for column (4), the ζ's of $B$ from column (6) of form 452 may be copied on a strip of paper and applied to the bottom line of the tabular values for each constituent and the differences obtained. The natural numbers for column (3) corresponding to the logarithms in column (2) can usually be obtained most expeditiously from table 27, this table giving the critical logarithm for each change of 0.001 in the corresponding natural number. If the logarithm is less than 6.6990, the natural number will be too small to appear in the third decimal place, and the effects of the corresponding constituent may be considered as nil. The products for columns (6) and (7) may be conveniently obtained from table 30. In column (8) the references to (6) and (7) are to the sums of these columns. The values of log $F(A)$ and $(V_0+u)$ for column (8) may be obtained from forms 244 and 244a.

**328.** In the use of this form it will be noted that the $R$'s and ζ's referring to constituent $B$ are to be the best known values whether derived from the analysis or by inference, but the $R'$ and $ζ'$ of constituent $A$, entered as items (9) and (19), respectively, must be the unmodified values as obtained directly by form 194.

### ANALYSIS OF TIDAL CURRENTS

**329.** Tidal currents are the periodic horizontal movements of the waters of the earth's surface. As they are caused by the same periodic forces that produce the vertical rise and fall of the tide, it is possible to represent these currents by harmonic expressions similar to those used for the tides. Constituents with the same periods as those contained in the tides are involved, but the current velocities take the place of the tidal heights. There are two general types of tidal currents, known as the reversing type and the rotary type.

**330.** In the reversing type the current flows alternately in opposite directions, the velocity increasing from zero at the time of turning to a maximum about 3 hours later and then diminishes to zero again, when it begins to flow in the opposite direction. By considering the velocities as positive in one direction and negative in the opposite direction, such a current may be expressed by a single harmonic series, such as

$$V = A \cos (at+\alpha) + B \cos (bt+\beta) + C \cos (ct+\gamma) + \text{etc.} \quad (450)$$

in which $V$=velocity of the current in the positive direction at any time $t$.

$A$, $B$, $C$, etc.=maximum velocities of current constituents.
$a$, $b$, $c$, etc.=speeds of constituents.
$\alpha$, $\beta$, $\gamma$, etc.=initial phases of constituents.

**331.** In the rotary type the direction of the current changes through all points of the compass, and the velocity, although varying in strength, seldom becomes zero. In the analysis of this type of current it is necessary to resolve the observed velocities in two directions at right angles to each other. For convenience the north and east directions are selected for this purpose, velocities toward the south and west being considered as negatives of these. For the harmonic representation of such currents it is, therefore, necessary to have two series—one for the north and the other for the east component.

**332.** For the analysis of either type of current the original hourly velocities or the resolved hourly velocities are tabulated in the same form used for the hourly heights of the tide. To avoid the inconvenience of negative readings in this tabulation, a constant, such as 3 knots, is added to all velocities. These hourly velocities are then summed with the same stencils that are used for the tides, and the hourly mean velocities are analyzed in the same manner as the hourly heights of the tide. The same forms are used for the currents, with the necessary modifications in the headings. The rotary currents will be represented by a double set of constants, one for the north components and the other for the east components.

**333.** For a 29-day series of observations, it is recommended that the analysis be made for the M series, the S series, and for $N_2$, $K_1$, and $O_1$. For longer series additional constituents may be included. In the analysis of current velocities, the harmonics of the higher degrees such as $M_4$ and $M_6$ may be expected to be of relatively greater magnitude than they are in the tides. From theoretical considerations it may also be shown that the magnitude of the diurnal constituents as compared with the semidiurnal constituents in a simple tidal oscillation is only about one-half as great in the current as in the tide. However, because of the complexity of the tidal and current movement, the actual relation between the various constituents as determined by the analysis is subject to wide variations. The constituent $S_1$, which is usually negligible in the tides, may be found to be of appreciable magnitude in offshore currents because of the effect of daily periodic land and sea breezes. However, as this constituent has a speed very nearly the same as that of $K_1$ it can be separated from the latter only by a long series of observations, preferably a year or more.

**334.** Form 723 (fig. 20) provides for the determination of harmonic constants from a series of current observations by comparison with corresponding constants from a tidal series covering the same period of time. This comparison is to be used if the series of observations is less than 29 days and may be used for longer series if desired. For the purpose of this comparison the hourly predicted heights at the tide station are usually to be preferred to actual observations since meteorological irregularities appearing in observed tides do not necessarily appear in a similar manner in the observed currents. In this work both currents and tides for the simultaneous period are to be summed for constituents M, S, N, K, and O; and the analysis is then carried through form 194. (Tides: Harmonic Analysis) to obtain the values of $R'$ and $\zeta'$ for each constituent. The harmonics $M_4$, $M_6$, and $M_8$ are to be obtained for the current series, but may be omitted in the tidal series.

**335.** Enter in Form 723 the accepted $H$ and $\kappa$ of the principal tidal constituents for the reference station and also the values of $R'$ and $\zeta'$ obtained from the analyses of the simultaneous series of tides and currents. The necessary calculations in the form are self-explanatory. The corrected velocity amplitude of each current constituent is obtained by a ratio on the assumption that for each constituent the relation of the corrected amplitude to the uncorrected amplitude is the same for both tide and current. The ratio derived for the con-

Form 723
DEPARTMENT OF COMMERCE
COAST AND GEODETIC SURVEY
Ed. Mar. 1935

## CURRENTS: HARMONIC COMPARISON

(A) Current Station No. 1, Bolivar Roads, Galveston Bay, Tex. Lat. 29° 20.8' N Long. 94° 46.1' W = L (A) 94.77°

(B) Tide Station Galveston, Tex. (predictions) Lat. 29° 18.7' N Long. 94° 47.5' W = L (B) 94.79°

Series begins Nov. 28, 1936 Length of Series 3 days

**Use one decimal for angles and three decimals for other quantities**

| Component | (1) Accepted H at (B) | (2) From Simultaneous Observations R' at (B) | (3) From Simultaneous Observations R' at (A) | (4) Ratio (1) ÷ (2) | (5) Corrected H at (A) (3) × (4) | (6) Accepted κ at (B) | (7) From Simultaneous Observations ζ' at (B) | (8) From Simultaneous Observations ζ' at (A) | (9) Difference (6) − (7) | (10) [L(B)−L(A)]p =[ 0.02 ]p | (11) Corrected κ at (A) (8)+(9)+(10) |
|---|---|---|---|---|---|---|---|---|---|---|---|
| | *Feet* | *Feet* | *Knots* | | *Knots* | *Degrees* | *Degrees* | *Degrees* | *Degrees* | *Degrees* | *Degrees* |
| $M_2$ | 0.271 | 0.382 | 0.859 | 0.709 | 0.609 | 117.7 | 101.3 | 94.4 | 16.4 | 0.0 | 110.8 |
| $M_4$ | ------- | ------- | 0.135 | do | 0.096 | ------- | ------- | 197.7 | $2M^{\circ}_2$ 32.8 | 0.1 | 230.6 |
| $M_6$ | ------- | ------- | 0.078 | do | 0.055 | ------- | ------- | 231.2 | $3M^{\circ}_2$ 49.2 | 0.1 | 280.5 |
| $M_8$ | ------- | ------- | 0.042 | do | 0.030 | ------- | ------- | 67.2 | $4M^{\circ}_2$ 65.6 | 0.2 | 133.0 |
| $S_2$ | 0.084 | 0.380 | 0.862 | 0.221 | 0.191 | 118.7 | 140.2 | 132.4 | −21.5 | 0.0 | 110.9 |
| $N_2$ | 0.066 | 0.365 | 0.814 | 0.181 | 0.147 | 97.1 | 73.2 | 67.1 | 23.9 | 0.0 | 91.0 |
| $K_1$ | 0.365 | 0.809 | 1.665 | 0.451 | 0.751 | 320.8 | 332.6 | 300.5 | −11.8 | 0.0 | 288.7 |
| $O_1$ | 0.343 | 0.868 | 1.868 | 0.395 | 0.738 | 311.3 | 284.4 | 256.0 | 26.9 | 0.0 | 282.9 |

Remarks: Direction of flood (positive velocities) 287° (true).

Direction of ebb (negative velocities) 111° (true).

FIGURE 20.

stituent $M_2$ is used also for the higher harmonics of M, this being considered more reliable than ratios determined directly from the much smaller amplitudes of these harmonics. The corrected epoch ($\kappa$) for each current constituent is calculated on the assumption that the difference between the corrected and uncorrected epoch is the same for tide and current. For convenience the zetas ($\zeta$) rather than the kappas from the simultaneous observations are used in the form and a longitude correction, column (10), is introduced to allow for this fact. Differences in column (9) for the higher harmonics of $M_2$ are derived from the difference for that constituent because of the uncertainty in the determination of epochs of constituents of very small amplitudes.

**336.** Short series of current observations are frequently taken at half-hourly intervals. As individual observations are somewhat rough, the utilization of the half-hourly observations will add materially to the accuracy of the results obtained from an analysis. Moreover, the closer spacing of the half-hourly values will give a better development of the higher harmonics of M which are of greater relative importance in the currents than in the tides. Special stencils have been prepared for the summation of these observations. Observations taken on the exact hour are tabulated in form 362 as usual, while observations on the half-hour are offset to the right on the intermediate lines. As the series of observations under consideration are short, provisions have been made for obtaining only the diurnal constituents $K_1$ and $O_1$; the semidiurnal constituents $M_2$, $S_2$, and $N_2$; and the higher harmonics of M.

**337.** For the diurnal constituents, the special stencils provide for the same distribution, with the inclusion of the half-hourly values, as is obtained with the standard stencils used for the hourly values only. Hourly means for the constituents are obtained and entered in form 194 and all subsequent computations are the same as those based upon the use of the standard stencils.

**338.** For the semidiurnal constituents $M_2$, $S_2$, and $N_2$, the semidiurnal period is divided into 24 parts. Special stencils for the constituents $M_2$ and $N_2$ provide for the distribution of the observed half-hourly velocities into the 24 groups indicated by this division. No stencil is required for the constituent $S_2$, the necessary grouping being accomplished by combining sums for afternoon observations with those for the forenoon observations of corresponding hours. Thus, the noon observations will be included with those taken at midnight, and the observations at 12:30 p. m. with those taken at 0:30 a. m.

**339.** The resulting means obtained for the semidiurnal constituents by the method described above are in reality half-hourly means, but in adapting form 194 for the analysis, these means may be entered in order in the spaces provided for the hourly means. Then, after doubling all subscripts in the form, the necessary computations may be carried out as indicated. Thus, all computations for the semidiurnal constituents will be made in the spaces originally designed for the diurnal constituents. The computations for all higher harmonics of even subscripts may be carried out in the same form using the spaces originally designed for the harmonics with subscripts one-half as great. In this adaptation of the form no provision is made for the computation of a harmonic of odd subscript which is here of relatively little importance. Other forms which are used in connection with the analysis will not be affected by the use of the special stencils for the half-hourly velocities.

340. Observations on the half-hour may also be analyzed separately from those on the exact hour, using the standard stencils for the summation. In this case the stencils are moved to the right one column and dropped one line, thus covering the hourly values and exposing those occurring on the half-hour. Allowance must be made for the difference of a half hour in the beginning of the series when computing the $(V_0+u)$'s in form 244. This may be conveniently done by assuming a time meridian a half-hour or $7\frac{1}{2}°$ westerly from the actual time meridian used so that the first half-hourly observation will correspond to the 0 hour of the assumed time meridian. The difference of 15 minutes for the middle of the series has a negligible effect in the computations and may be disregarded. In other respects the analysis is carried on in the same manner as the analysis for the hourly observations, and the results obtained afford a useful check on the latter.

# PREDICTION OF TIDES

## HARMONIC METHOD

**341.** The methods for the prediction of the tides may be classified as harmonic and nonharmonic. By the harmonic method the elementary constituent tides, represented by harmonic constants, are combined into a composite tide. By the nonharmonic method the predictions are made by applying to the times of the moon's transits and to the mean height of the tide systems of differences to take account of average conditions and various inequalities due to changes in the phase of the moon and in the declination and parallax of the moon and sun. Without the use of a predicting machine the harmonic method would involve too much labor to be of practical service, but with such a machine the harmonic method has many advantages over the nonharmonic systems and is now used exclusively by the Coast and Geodetic Survey in making predictions for the standard ports of this country.

**342.** The height of the tide at any time may be represented harmonically by the formula

$$h = H_0 + \Sigma f H \cos [at + (V_0 + u) - \kappa] \qquad (451)$$

in which
- $h$ = height of tide at any time $t$.
- $H_0$ = mean height of water level above datum used for prediction.
- $H$ = mean amplitude of any constituent $A$.
- $f$ = factor for reducing mean amplitude $H$ to year of prediction.
- $a$ = speed of constituent $A$.
- $t$ = time reckoned from some initial epoch such as beginning of year of predictions.
- $(V_0 + u)$ = value of equilibrium argument of constituent $A$ when $t = 0$.
- $\kappa$ = epoch of constituent $A$.

In the above formula all quantities except $h$ and $t$ may be considered as constants for any particular year and place, and when these constants are known the value of $h$, or the predicted height of the tide, may be computed for any value of $t$, or time. By comparing successive values of $h$ the heights of the high and low waters, together with the times of their occurrence, may be approximately determined. The harmonic method of predicting tides, therefore, consists essentially of the application of the above formula.

**343.** The exact value of $t$ for the times of high and low waters will be roots of the first derivative of formula (451) equated to zero, which may be written—

$$\frac{dh}{dt} = -\Sigma af H \sin [at + (V_0 + u) - \kappa] = 0 \qquad (452)$$

Although formula (452) cannot, in general, be solved by rigorous methods, it may be mechanically solved by a tide-predicting machine of the type used in the office of the Coast and Geodetic Survey.

**344.** The constant $H_0$ of formula (451) is the depression of the adopted datum below the mean level of the water at the place of prediction. For places on the open coast the mean water level is identical with mean sea level, but in the upper portions of tidal rivers that have an appreciable slope the mean water level may be somewhat higher than the mean sea level. The datum for the predictions may be more or less arbitrarily chosen but it is customary to use the low-water plane that has been adopted as the reference for the soundings on the hydrographic charts of the locality. For all places on the Atlantic and Gulf coasts of the United States, including Puerto Rico and the Atlantic coast of the Panama Canal Zone, this datum is mean low water. For the Pacific coast of the United States, Alaska, Hawaii and the Philippines, the datum is in general mean lower low water. For the rest of the world, the datum is in general mean low water springs, although there are many localities where somewhat lower planes are used. After the datum for any particular place has been adopted its relation to the mean water level may be readily obtained from simple nonharmonic reductions of the tides as observed in the locality. The value of $H_0$ thus determined is a constant that is available for future predictions at the stations.

**345.** The amplitude $H$ and the epoch $\kappa$ for each constituent tide to be included in the predictions are the harmonic constants determined by the analysis discussed in the preceding work. Each place will have its own set of harmonic constants, and when once determined will be available for all times, except as they may be slightly modified by a more accurate determination from a better series of observations or by changes in the physical conditions at the locality such as may occur from dredging, by the depositing of sediment, or by other causes.

**346.** The node factor $f$ (par. 77) is introduced in order to reduce the mean amplitude to the true amplitude depending upon the longitude of the moon's node. The factor $f$ for any single constituent, therefore, passes through a cycle of values. The change being slow, it is customary to take the value as of the middle of the year for which the predictions are being made and assume this as a constant for the entire year. The error resulting from this assumption is practically negligible. Each constituent has its own set of values for $f$, but these values are the same for all localities and have been compiled for convenient use in table 14 for the middle of each year from 1850 to 1999.

**347.** The quantity $a$ represents the angular speed of any constituent per unit of time. In the application of formulas (451) and (452) to the prediction of tides this is usually given in degrees per mean solar hour, the unit of $t$ being taken as the mean solar hour. The values of the speeds of the different constituents have been calculated from astronomical data by formulas derived from the development of the tide-producing force which has already been discussed. These speeds have been compiled in table 2 and are essentially constant for all times and places. The quantity $(V_0+u)$ is the value of the equilibrium argument of a constituent at the initial instant from which the value of $t$ is reckoned; that is, when $t$ equals zero. In the prediction of tides this initial epoch is usually taken at the midnight beginning the year for which the predictions are to be made. In strictness the $V$, or uniformly varying portion of the argument alone, refers to the initial epoch, while the $u$, or slow variation due to changes in the longitude of the moon's node, is taken as of the middle of the period of prediction and assumed to have this value as a constant for the entire period. The quantity $(V_0+u)$ is different for each constituent and is also different for each initial epoch and for different longitudes on the earth. In table 15 there have been compiled the values of this quantity for the beginning of each year from 1850 to 2000 for the the longitude of Greenwich. The values may be readily modified to adapt them to other initial epochs and other longitudes.

**348.** Let

$L=$ west longitude in degrees of station for which predictions are desired.

$S=$ west longitude in degrees of time meridian used at this station.

For east longitude, $L$ and $S$ will have negative values.

Now let

$p=0$ when referring to the long-period constituents.
   1 when referring to the diurnal constituents.
   2 when referring to the semidiurnal constituents, etc.

then $p$ will be the coefficient of the quantity $T$ in the equilibrium arguments. Now, $T$ is the hour angle of the mean sun and is the only quantity in these arguments that is a function of the longitude of the place of observation or of prediction. At any given instant of time the difference between the values of the hour angle $T$ at two stations will be equal to the difference in longitude of the stations. If, therefore, the value of the argument $(V_0+u)$ for any constituent at any given instant has been computed for the meridian of Greenwich, the correction to refer this argument for the same instant to a place in longitude $L°$ west of Greenwich will be $-pL$, the negative sign being necessary as the value of $T$ decreases as the west longitude increases.

**349.** The instant of time to which each of the tabular values of the Greenwich $(V_0+u)$'s of table 15 refers is the 0 hour of the Greenwich mean civil time at the beginning of a calendar year. In the predictions of the tides at any station it is desirable to take as the initial epoch the 0 hour of the standard or local time customarily used at that station. If, therefore, the longitude of the time meridian used is $S°$ west of Greenwich, the initial epoch of the predictions will usually be $S/15$ mean solar hours later than the instant to which the tabular Greenwich $(V_0+u)$'s are referred.

**350.** In formulas (451) and (452) the symbol $a$ is the general designation of the speed of any constituent; that is to say, it is the hourly rate of change in the argument. The difference in the argument due to a difference of $S/15$ hours in the initial epoch is therefore $aS/15$ degrees. The total correction to the tabular Greenwich $(V_0+u)$ of any year in order to obtain the local $(V_0+u)$ for a place in longitude $L°$ west at an initial epoch of 0 hours of time meridian $S°$ west at the beginning of the same calendar year is

$$\frac{aS}{15} - pL. \tag{453}$$

The general expression for the angles of (451) and (452) may now be written

$$at + (V_0 + u) - \kappa = at + \text{Greenwich } (V_0 + u) + \frac{aS}{15} - pL - \kappa \tag{454}$$

**351.** In order to avoid the necessity of applying the corrections for longitude and initial epoch to the Greenwich $(V_0+u)$'s for each year, these corrections may be applied once for all to the $\kappa$'s. Let

$$\frac{aS}{15} - pL - \kappa = -\kappa' \tag{455}$$

Then (454) may be written

$$at + (V_0 + u) - \kappa = at + \text{Greenwich } (V_0 + u) - \kappa' \tag{456}$$

Thus, by applying the corrections indicated in (455) to the $\kappa$'s for any station, a modified set of epochs is obtained. These will remain the same year after year and permit the direct use of the tabular Greenwich $(V_0+u)$'s in determining the actual constituent phases at the beginning of each calendar year.

**352.** Let

$$\text{Greenwich } (V_0 + u) - \kappa' = \alpha \tag{457}$$

then formulas (451) and (452) may be written

$$h = H_0 + \sum fH \cos (at + \alpha) \tag{458}$$

for height of tide at any time, and

$$\sum af H \sin (at + \alpha) = 0 \tag{459}$$

for times of high and low waters. Formula (458) may be easily solved for any single value of $t$, but for many values of $t$ as are necessary in the predictions of the tides for a year at any station the labor involved by an ordinary solution would be very great. Formula (459) can not, in general, be solved by rigorous methods. The invention of tide-predicting machines has rendered the solution of both formulas a comparatively simple matter.

### TIDE-PREDICTING MACHINE

**353.** The first tide-predicting machine was designed by Sir William Thomson (afterwards Lord Kelvin) and was made in 1873 under the auspices of the British Association for the Advancement of Science. This was an integrating machine designed to compute the height of the tide in accordance with formula (458). It provided for the summation of 10 of the principal constituents, and the resulting predicted heights were registered by a curve automatically traced by the machine. This machine is described in part I of Thomson and Tait's Natural Philosophy, edition of 1879. Several other tide-predicting machines designed upon the same general principles but providing for an increased number of constituents were afterwards constructed.

**354.** The first tide-predicting machine used in the United States was designed by William Ferrel, of the U. S. Coast and Geodetic Survey. This machine, which was completed in 1882, was based upon modified formulas and differed somewhat in design from any other machine that has ever been constructed. No curve was traced, but both the times and heights of the high and low waters were indicated directly by scales on the machine. The intermediate heights of the tide could be obtained only indirectly. A description of this machine is given in the report of the Coast and Geodetic Survey for the year 1883.

**355.** The first machine made to compute simultaneously the height of the tide and the times of high and low waters as represented by formulas (458) and (459), respectively, was designed and constructed in the office of the Coast and Geodetic Survey. It was completed in 1910 and is known as the United States Coast and Geodetic Survey tide-predicting machine No. 2. The machine sums simultaneously the terms of formulas (458) and (459) and registers successive heights of the tide by the movement of a pointer over a dial and also graphically by a curve automatically traced on a moving strip of paper. The times of high and low waters determined by the values of $t$ which satisfy equation (459) are indicated both by an automatic stopping of the machine and also by check marks on the graphic record.

**356.** The general appearance of the machine is illustrated by figure 21. It is about 11 feet long, 2 feet wide, and 6 feet high, and weighs approximately 2,500 pounds. The principal features are: First, the supporting framework; second, a system of gearing by means of which shafts representing the different constituents are made to rotate with angular speeds proportional to the actual speeds of the constituents; third, a system of cranks and sliding frames for obtaining harmonic motion; fourth, summation chains connecting the individual constituent elements, by means of which the sums of the harmonic terms of formulas (458) and (459) are transmitted to the recording devices; fifth, a system of dials and pointers for indicating in a convenient manner the height of the tide for successive instants of time and also the time of the high and low waters; sixth, a tide curve or graphic representation of the tide automatically constructed by the machine. The machine is designed to take account of the 37 constituents listed in table 38, including 32 short-period and 5 long-period constituents.

**357.** The heavy cast-iron base of the machine, which includes the operator's desk, has an extreme length of 11 feet and is 2 feet wide. This forms a very substantial foundation for the superstructure, increasing its stability and thereby diminishing errors that might result from a lack of rigidity in the fixed parts. On the left side of the desk is located the hand crank for applying the power (1, fig. 24), and under the desk are the primary gears for setting in motion the various parts of the machine. The superstructure is in three sections, each consisting of parallel hard-rolled brass plates held from 6 to 7 inches apart by brass bolts. Between these plates are located the shafts and gears that govern the motion of the different parts of the machine.

**358.** The front section, or dial case, rests upon the desk facing the operator and contains the apparatus for indicating and registering the results obtained by the machine. The middle section rests upon a depression in the base and contains the mechanism for the harmonic motions for the principal constituents $M_2$, $S_2$, $K_1$, $O_1$, $N_2$, and $M_4$. The rear section contains the mechanism for the harmonic motions for the remaining 31 constituents for which the machine provides.

**359.** The angular motions of the individual constituents, as indicated by the quantity $at$ in formulas (458) and (459), are represented in the machine by the rotation of short horizontal shafts having their bearings in the parallel plates. All of these constituent shafts are connected by a system of gearing with the hand crank at the left of the dial case and also with the time-registering dials, so that when the machine is in operation the motion of each of these shafts will be proportional to the speed $a$ of the corresponding constituent, and for any interval of time or increment in $t$ as indicated by the time dials the amount of angular motion in any constituent shaft will equal the increment in the product $at$ corresponding to that constituent.

**360.** Since the corresponding angles in formulas (458) and (459) are identical for all values of $t$, the motion provided by the gearing will be applicable alike to the solution of both formulas. The mechanism for the summation of the terms of formula (458) is situated on the side of the machine at the left of the operator, and for convenience this side of the machine is called the "height side" (fig. 21), and the mechanism for the summation of the terms of formula (459) is on the right-hand side of the machine, which is designated as the "time side" (fig. 22).

**361.** In table 37 are given the details of the general gearing from the hand-operating crank to the main vertical shafts, together with the details of all the gearing in the front section or dial case. It will be noted that $S-6$ (fig. 25) is the main vertical shaft of the dial case and is connected through the releasable gears to the hour hand, the minute hand, and the day dial, respectively. The releasable gears permit the adjustment of these indicators to any time desired. After an original adjustment is made so that the hour and minute hand will each read 0 at the same instant that the day dial indicates the beginning of a day, further adjustment will, in general, be unnecessary, as the gearing itself will cause the indicators to maintain a consistent relation throughout the year, and by use of the hand-operating crank the entire system may be made to indicate any time desired. The period of the hour-hand shaft is 24 dial hours, and the hand moves over a dial graduated accordingly (*3*, fig. 23). The minute-hand shaft, with a period of 1 dial hour, moves over a dial graduated into 60 minutes (*2*, fig. 23).

**362.** The day dial, which is about 10 inches in diameter, is graduated into 366 parts to represent the 366 days in a leap year. The names of the months and numerals to indicate every fifth day of each month are inscribed on the face of the dial. This dial is located just back of the front plate or face of the machine, in which there is an arc-shaped opening through which the graduations representing nearly two months are visible at any one time (*4*, fig. 23). The progress of the days as the machine is operated is indicated by the rotation of this dial past an index or pointer just below the opening (*6*, fig. 23). This pointer is secured to a short shaft which carries at its inner end a lever arm with a pin reaching under the lower edge of the day dial, against which it is pressed by a light spring. A portion of the edge of the dial equal to the angular distance from January $_1$ to February 28 is of a slightly larger radius, so that the pin pressing against it rises and throws the day pointer to the right one day when this portion has passed by. On

FIGURE 21.—COAST AND GEODETIC SURVEY TIDE-PREDICTING MACHINE.

FIGURE 22.—TIDE-PREDICTING MACHINE, TIME SIDE.

Special Publication No. 98

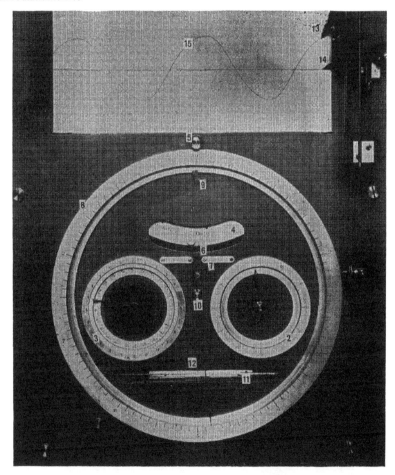

FIGURE 23.—TIDE-PREDICTING MACHINE, RECORDING DEVICES.

FIGURE 24.—TIDE-PREDICTING MACHINE, DRIVING GEARS.

Special Publication No. 98

FIGURE 25.—TIDE-PREDICTING MACHINE, DIAL CASE FROM HEIGHT SIDE.

Figure 26.—Tide-Predicting Machine, Dial Case From Time Side.

FIGURE 27.—TIDE-PREDICTING MACHINE, VERTICAL DRIVING SHAFT OF MIDDLE SECTION.

FIGURE 28.—TIDE-PREDICTING MACHINE, FORWARD DRIVING SHAFT OF REAR SECTION.

FIGURE 29.—TIDE-PREDICTING MACHINE, REAR END.

Special Publication No. 98

FIGURE 30.—TIDE-PREDICTING MACHINE, DETAILS OF RELEASABLE GEAR.

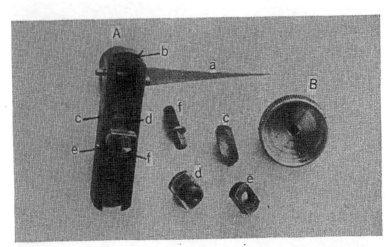

FIGURE 31.—TIDE-PREDICTING MACHINE, DETAILS OF CONSTITUENT CRANK.

the last day of December this pointer will move back one day to its original position.

**363.** On the same center with the day pointer there is a smaller index (7, fig. 23) which may be turned either to the right toward a plate inscribed "Common year," or to the left to a plate inscribed "Leap year." When this smaller index is turned toward the right, the day pointer is free to move in accordance with the change in radius of the edge of the dial. If the smaller index is turned toward the left, the day pointer is locked and must hold a fixed position throughout the year. For the prediction of the tides for two or more common years in succession the day dial must be set forward one day at the close of the year in order that the days of the succeeding year may be correctly registered. The day dial can be released for setting by the nut (5, fig. 23) immediately above the large dial ring. A slower movement of the day dial is provided by a releasable gear on the vertical shaft $S$–$6$ (fig. 25).

**364.** There are three main vertical shafts, $S$–$13$ (fig. 27), $S$–$14$ (fig. 28), and $S$–$16$ (fig. 29), to which are connected the gearing for the individual constituents. The period of rotation of each is 12 dial hours, and all move clockwise when viewed from above the machine. The connections between these main shafts and the individual constituent crankshafts are, in general, made by two pairs of bevel gears and an intermediate horizontal shaft, except that for the slow moving constituents Sa, Ssa, Mm, Mf, and MSf, a worm screw and wheel and a pair of spur gears are in each case substituted for a pair of bevel gears. In each case the gear on the main vertical shaft is releasable so that each crankshaft can be set independently.

**365.** Main shaft $S$–$13$ in the middle section of the machine drives 9 individual crankshafts representing 6 constituents, 3 of them being provided with two crankshafts each. These 6 constituents are $M_2$, $S_2$, $K_1$, $N_2$, $M_4$, and $O_1$, the first three having the double crankshafts. Main shaft $S$–$14$ at the front of the rear section of the machine drives 16 crankshafts representing one constituent each. These are $M_6$, MK, $S_4$, MN, $\nu_2$, $S_6$, $\mu_2$, and 2N in the upper range, and MS, $M_8$, $K_2$, 2MK, $L_2$, $M_3$, 2SM, and $P_1$ in the lower range. Main shaft $S$–$16$ at the back of the rear section drives 15 crankshafts. The constituents represented are OO, $\lambda_2$, $S_1$, $M_1$, $J_1$, Mm, and Ssa, in the upper range, and 2Q, $R_2$, $T_2$, $Q_1$, $\rho_1$, Mf, MSf, and Sa in the lower range.

**366.** For each of the five long-period constituents motion is communicated from the intermediate shaft by a worm screw and wheel to a small shaft on which is mounted a sliding spur gear. The latter engages a spur gear on the crankshaft, but may be easily disconnected by drawing out a pin on the time side of the machine, thus permitting the crankshaft to be turned freely when setting the machine.

**367.** *Gear speeds.*—The relative angular motion of each constituent crankshaft must correspond as nearly as possible to the theoretical speed of the constituent represented. The period of rotation of each of the three main vertical shafts being 12 dial hours, the angular motion of each of these shafts is 30° per dial hour. Table 38 contains the details of the gearing from the main vertical shafts to the individual crankshafts, the number of teeth in the different gears for each constituent being given in columns I, II, III, and IV. In designing the predicting machine it was necessary to find such values for these columns as would give gear speeds approximating as closely as possible with the theoretical speeds of the constituents. By comparing the gear speeds as obtained with the corresponding theoretical speeds it will be noted that the accumulated errors of the gears for an entire dial year for all the constituents are negligible in the prediction of the tides.

**368.** *Releasable gears.*—Releasable gears (52, fig. 27) on the main vertical shafts permit the independent adjustment of the time indicators and individual crankshafts. The details of these gears are illustrated in figure 30. A collar $C$, with a thread at its upper end and a flange at the bottom, is fastened to the shaft by means of three steel screws. The gear wheel $A$ fits closely upon this collar and rests upon the flange. It has sunk into its upper surface a recess $a$, which is filled by the flange of collar $B$. When in place, the latter is prevented from turning by a small steel screw reaching into a vertical groove $c$ in the collar $C$. The lower surface of collar $B$ is slightly dished, and the collar is split twice at right angles nearly to the top. When the milled nut $D$ is screwed down with a small pin wrench, the edge of the collar $B$ is pressed against the edge of the recess $a$ with such force as to make slipping practically impossible. When the nut is loosened, the gear may be turned independently of the main driving shaft. A small wrench (56, fig. 28) is used for setting these gears. Each of the three main driving shafts is provided with a clamp (55, fig. 28) to secure the shaft from turning when the nut of the releasable gear is being loosened or tightened.

**369.** *Constituent cranks.*—Secured to the ends of the constituent crank shafts, which projects through the brass plates on both sides of the machine, are brass cranks (40, fig. 25) which are provided for the constituent amplitudes. Those on the left or height side of the machine are designated as the constituent height cranks and are used for the coefficients of the cosine terms of formula (458), and those on the right or time side of the machine are designated as the constituent time cranks and are used for the coefficients of the sine terms of formula (459). The time crank on each constituent crank shaft is attached 90° in advance (in the direction of rotation) of the height crank on the same shaft. For the constituents Sa and Ssa no time cranks are provided, as the coefficients of the sine terms corresponding to these constituents are too small to be taken into account. The direction of rotation of each constituent crank shaft with its constituent cranks is clockwise when viewed from the time side of the machine and counterclockwise when viewed from the height side. The details of a constituent crank are shown in figure 31. The pointer $a$ is rigidly attached to the crank as an index for reading its position on a dial. In each crank there is a longitudinal groove $b$ with flanges in which a crank pin $d$ may be clamped in any desired position. The crank pin has a small rectangular block as a base which is designed to fit the groove in the crank, and through the center of the crank pin there is a threaded hole for the clamp screw $f$. Attached to the under side of the crank-pin block is a small spring $c$ that presses the block outward against the flanges of the groove, keeping it from slipping out of place when unclamped and at the same time permitting it to be moved along the groove when setting the machine. The crank pin may be securely fastened in any desired position by tightening up on the clamp screw, which, pressing against the small spring at the back, forces the crank-pin block outward against the flanges of the groove with sufficient pressure to prevent any slipping. A milled head wrench $B$ is used for tightening the clamp screw. A small rectangular block $e$ of hardened steel is fitted to turn freely upon the finely polished axle of the crank pin. This block is designed to fit into and slide along the slot of the constituent frame.

**370.** *Positive and negative direction.*—All the constituent crank shafts and cranks may be grouped into two ranges—those above the medial horizontal plane of the framework being in the upper range and those below this plane in the lower range. In the following discussion direction toward this medial plane is to be considered as negative and direction away from the plane as positive; that is to say, for all constituents in the upper range the positive direction will be upward and the negative direction downward, while for the constituents in the lower range the positive direction will be downward and the negative direction upward.

**371.** *Constituent dials.*—To indicate the angular positions of the constituent crank shafts, the pointer ($a$, fig. 31) moves around a dial ($41$, fig. 25) which is graduated in degrees. These dials are fastened to the frame of the machine back of the constituent cranks on both sides of the machine, those on the time side being graduated clockwise and those on the height side counterclockwise. These dials and pointers are so arranged that the angular position of a constituent crank shaft at any time will be the same whether read from the dial on the height side or from the dial on the time side of the machine, and at the zero reading for any constituent the height crank will be in a positive vertical position and the corresponding time crank in a horizontal position. At a reading of 90° the height crank will be horizontal and the time crank in a negative vertical position.

**372.** With the face of the machine registering the initial epoch, such as January 1, 0 hour, of any year, the value of $t$ then being taken as zero, each constituent crank shaft may be set, by means of its releasable gear, so that the dial readings will be equal to the $\alpha$ of the corresponding constituent as represented in formulas (458) and (459). If the machine is then put in operation, the dial readings will, for successive values of $t$, continuously correspond to the angle $(at+\alpha)$ of the formulas, as the gearing already described will provide for the increment $at$.

**373.** *Constituent sliding frames.*—For each constituent crank there is a light steel frame ($42$, fig. 25) fitted to slide vertically in grooves in a pair of angle pieces attached to the side plates of the machine. At the top of the frame there is a horizontal slot in which the crank pin slides. As the machine is operated the rotation of the crank shafts with their cranks cause each crank pin to move in the circumference of a circle, the radius of which depends upon the setting of the pin on the crank. This motion of the pin, acting in the horizontal slot of the sliding frame, imparts a vertical harmonic motion to that frame. The frame is in its zero position when the center horizontal line of the slot intersects the axis of the crank shaft. Positive motion is the direction away from the medial horizontal plane of the machine and negative motion is toward the medial plane. The displacement of each constituent height frame from its zero position will always equal the product of the amplitude setting of the crank pin by the cosine of the constituent dial reading, and the displacement of each constituent time frame will always equal the product of the amplitude setting by minus the sine of the constituent dial reading.

**374.** *Constituent pulleys.*—Each constituent frame is connected with a small movable pulley (*43*, fig. 25). For all constituents except $M_2$, $S_2$, $N_2$, $K_1$, $O_1$, and Sa on the height side and $M_2$, $S_2$, $N_2$, and $M_4$ on the time side this connection is by a single steel strip, so that the pulley has the same vertical motion as the corresponding frame.

**375.** *Doubling gears.*—Because of the very large amplitudes of some of the constituents two methods were used in order to keep the lengths of the cranks within practical limits. For $M_2$, $S_2$, and $K_1$ two sets of shafts and cranks were provided, so that the amplitudes of these constituents may be divided when necessary and a portion set on each. A further reduction in the length of the cranks for these and the other large constituents is accomplished by the use of doubling gears between the sliding frame and movable pulley. Two spur gears with the ratio of 1:2 (*48*, fig. 25) are arranged to turn together on the same axis. The smaller gear engages a rack (*46*) attached to the sliding frame and the larger gear engages a rack (*47*) attached to the constituent pulley. Each rack is held against its gear by a flange roller (*49*), and counterpoise weights are provided to take up the backlash in the gears. Through the action of these doubling gears any motion in the sliding frame causes a motion twice as great in the constituent pulley. Doubling gears are provided on the height side of the machine for constituents $M_2$, $S_2$, $N_2$, $K_1$, $O_1$, and Sa and on the time side for constituents $M_2$, $S_2$, $N_2$, and $M_4$.

**376.** *Scales for amplitude settings.*—The scales for setting the constituent amplitudes are attached to the frame of the machine and are, in general, graduated into units and tenths (*44*, fig. 25). The scales are arranged to read in a negative direction; that is, downward for the constituents of the upper range and upward for the constituents in the lower range. On a small adjustable plate (*45*) attached to each constituent pulley there is an index line which is set to read zero on the scale when the sliding frame is in its zero position. For setting the crank pins for the constituent amplitudes the cranks to be set are first turned to a negative vertical position. For the cranks on the height side of the machine this position corresponds to a dial reading of 180° and for the cranks on the time side to a reading of 90°.

**377.** The scales on the height side of the machine, which are used in setting the coefficients of formula (458), are graduated uniformly one-half inch to the unit. On the time side of the machine the scales are modified in order to automatically take account of the additional factor involving the speed of the constituent which appears in each of the coefficients of formula (459). Dividing the members of this formula by $m$, the speed of constituent $M_2$, it becomes

$$\sum \frac{a}{m} fH \sin (at+\alpha)=0 \qquad (460)$$

The modified scales are graduated 0.5 $a/m$ inch to the unit. The use of the modified scales on the time side of the machine permits both the height and time crank for any constituent to be set in accord with the factor $fH$ which is common to the coefficients of both formulas (458) and (459). There are also provided for special use on the time side of the machine unmodified scales graduated uniformly to read in a positive direction.

**378.** *Summation chains.*—The summations of the several cosine terms in formula (458) and of the several sine terms in formula (459) are carried on simultaneously by two chains, one (*27*, fig. 25) on the height side and the other (*28*, fig. 26) on the time side of the machine. The chains are of the chronometer fuse type, of tempered steel, and have 125 links per foot. The total length of the height chain is 27.6 feet and of the time chain 30.6 feet. A platinum point is attached to one of the links of the time chain 3.5 feet from its free end for an index.

**379.** Each of these chains is fastened at one end near the back part of the machine by a pair of adjusting screws (*53*, fig. 29, and *54*, fig. 22). From these adjusting screws each chain passes alternately downward under a constituent pulley of the lower range and upward over a constituent pulley of the upper range, spanning the space between the rear and middle section of the machine by two idler pulleys and continuing until every constituent pulley on each side of the machine is included in the system. The movable pulleys are so arranged that the direction of the chain in passing from one to another is always vertical and parallel to the direction of the motion of the sliding frames.

**380.** *Summation wheels.*—The free or movable end of each of the chains is attached to a threaded grooved wheel (*29, 30*, fig. 25), 12 inches in circumference and threaded to hold more than seven turns of the chain, or about 90 inches in all. These are called the height and time summation wheels. Each is mounted on a shaft that admits a small lateral motion, and by means of a fixed tooth attached to the framework of the machine and reaching into the threads of a screw fastened to the shaft the latter when rotating is forced into a screw motion with a pitch equal to that of the thread groove of the summation wheel; so that the path of the chain as it is wound or unwound from the summation wheel remains unchanged.

**381.** The height summation wheel (*29*, fig. 25) is located near the front edge of the middle section of the machine, where it receives the height summation chain directly from the nearest constituent pulley. The time summation pulley (*30*) is located inside the dial case near the lower left side, and three fixed pulleys are used to carry the time chain from the end constituent pulley to the summation wheel. Counterpoise weights are connected with the shafts containing the summation wheels in order to keep the summation chains taut.

**382.** When all of the sliding frames on either side of the machine are in their zero positions, the corresponding summation wheel is approximately half filled by turns of the summation chain. Any motion of a sliding frame in a positive direction will tend to unwind the chain from the wheel, and any motion in the negative direction will tend to slacken the chain so that it will be wound up by the counterpoise weight. With several of the sliding frames on either side of the machine moving simultaneously, the resultant motion, which is the algebraic sum of all, will be communicated to the summation wheel. The motion of the sliding frame being transmitted to the chain through a movable pulley, the motion of the free end of the chain must be twice as great as that in the pulley. The scale of the pulley motion is one-half inch to the unit of amplitude, and therefore the scale of the chain motion is 1 inch to the unit, and one complete rotation of the summation wheel represents a change of 12 units of amplitude.

**383.** The zero position of the height summation wheel is indicated by the conjunction of an index line (*50*, fig. 25) on the arm attached to the wheel and an index line (*51*, fig. 25) on a bracket attached to the framework of the machine just below the summation wheel, the wheel itself being approximately one-half filled with the summation chain. The length of the chain is adjusted so that the summation wheel will be in its zero position when all the sliding frames on the height side of machine are in their zero positions. It will be noted that the conjunction of the index lines will not alone determine the zero position of the wheel, since such conjunctions will occur at each turn of the wheel, while there is only one zero position, which is that taken when the constituent frames are set at zero.

**384.** The zero position of the time summation wheel is indicated by the conjunction of an index point (*11*, fig. 23) attached to the time summation chain and a fixed index (*12*, fig. 23) in the middle of the horizontal opening near the bottom of the dial case, and the length of the time summation chain is so adjusted that this conjunction will occur when all sliding frames on the time side of machine are in their zero positions.

**385.** *Predicted heights of the tide.*—When the machine is in operation, the sum of all the cosine terms of formula (458) included in the settings for a station will be transmitted through the height summation wheel to the face of the machine and there indicated in two ways—first by a pointer moving over a circular height scale (*8*, fig. 23) and second by the ordinates of a tide curve that is automatically traced on a roll of paper (*15*, fig. 23). The motion of the height summation wheel is transmitted by a gear ratio of 30:100 to a horizontal shaft which is located just back of the dial case. One complete rotation of this shaft represents 40 units in the height of the tide. From this shaft the motion is carried by two separate systems of gearing to the height pointer on the face of the machine and to the pen that traces the tide curve.

**386.** *Height scale.*—The height pointer is geared to make one complete revolution for a change of 40 units in the height of the tide. A height scale, with its circumference divided into 40 equal parts and each of these unit parts subdivided into tenths, provides for the direct registering of the sum of the cosine terms of formula (458) as communicated through the summation wheel. This scale has its zero graduation at the top and is graduated positively to the right and negatively to the left. The height pointer can easily be adjusted to any position by means of a small milled nut (*10*, fig. 23) at the end of its shaft. If it should be desired to refer the predicted heights to mean sea level, this pointer must be adjusted to read zero at the same time that the summation wheel is in its zero position; but if it is desired to refer to some other datum, the pointer will be adjusted according to the elevation of mean sea level above this datum. For the value of $h$ in formula (458) the pointer will be adjusted to a reading corresponding to the adopted value of $H_o$ at the time the summation wheel is in its zero position, then this value of $H_o$ will be automatically included with the sum of the cosine terms of that formula. As the machine is operated the height pointer will indicate the predicted height of the tide corresponding to the time shown on the time dials.

**387.** In order to increase the working scale of the machine when predicting tides with smaller ranges, two additional circular height scales are provided, one with the circle divided into 20 units and the other into 10 units, with the units subdivided into tenths. These scales may be easily removed or replaced on the machine, the scale in use being secured in place by a small button at the top (*9*, fig. 23). The 20-unit scale may be conveniently used when the extreme range of the predicted tide at any place is between 10 and 20 feet, and the 10-unit scale when the extreme range is less than 10 feet. If the 20-unit scale is to be used, the value of each coefficient of both the cosine and the sine terms must be doubled before setting the component cranks, and if the 10-unit scale is used these original coefficients must first be multiplied by 4 before setting the values in the machine. If the extreme tide is less than 4 feet, the 40-unit dial may be readily used as a 4-unit scale by considering the original unit graduations as tenths of units in the larger scale. In this case the coefficients of the cosine and sine terms of the formula must be multiplied by 10 before entering in the machine. The factor used for multiplying the coefficients to adapt them to the different height scales is called the working scale of the machine. Working scales of 1, 2, 4, and 10 are now in general use to take account of the different ranges of tide at the places for which predictions are made.

**388.** *Predicted times of the tide.*—Simultaneously with the summation of the cosine terms of formula (458) on the height side of the machine, the summation of the sine terms of formula (460), which was derived from formula (459), is being effected on the time side. Being concerned only with the time at which the sum of the sine terms is zero, no provision is made for registering the sum except at this time, which is indicated on the machine by the conjunction of the index point on the time chain and the fixed platinum index in the dial case. Near the time of a high water the index on the chain moves from right to left and near the time of a low water from left to right. The conjunction of the movable and fixed index is visible to the operator of the machine and he may note the corresponding dial readings for the time and height of the high or low water.

**389.** *Automatic stopping device.*—This device provides for automatically stopping the machine at each high and low water. Secured to the hand-crank shaft is a ratchet wheel and just above the ratchet wheel is a steel pawl (*25*, fig. 24) operated by an electromagnet (*26*) mounted under the desk top. The electric circuit for the electromagnet is closed by a contact spring that rests upon a hard-rubber cylinder (*31*, fig. 25) on the rear end of the shaft on which the time summation wheel is mounted. A small platinum plug in this rubber cylinder comes in contact with the spring, which is fitted with a fine motion adjustment, when the time summation chain registers zero. This closes the circuit and draws the pawl against the ratchet wheel, thereby automatically stopping the machine. The lateral screw motion of the shaft on which the rubber cylinder is mounted prevents the platinum plug from coming in contact with the spring on any revolution other than the one which brings the time chain to its zero position. The circuit is led through an insulated ring on the hub of the hand crank where a contact is kept closed by a spring. After the operator has noted the time and height readings of the high or low water he may easily break the circuit at the crank hub by a slight inward pressure against the crank handle, thus releasing the armature and pawl and permitting the machine to be turned forward to the next stop. By means of a small switch (*23*, fig. 24) just below the crank the circuit may be held open to prevent the automatic device from operating when so desired.

**390.** *Nonreversing ratchet.*—Upon the crank shaft, close to the bearing in the desk frame, there is a small ratchet wheel and above this there is a pawl (*24*, fig. 24) that is lifted away from the wheel by friction springs when the machine is being turned forward but which is instantly thrown into engagement when the crank is accidentally turned backward. By pushing in one of the small buttons (*22*, fig. 24) just above the crank the pawl is locked so that it cannot engage the ratchet, thus permitting the machine to be turned backward when desired. Pressure on another button releases the pawl.

**391.** *Tide curve.*—The tide curve which graphically represents the rise and fall of the predicted tide is automatically traced on a roll of paper by the machine at the same time that the results are being indicated on the dials. The curve is the resultant of a horizontal movement of the paper, corresponding to the passing of time, and a vertical movement of a fountain pen (*13*, fig. 23), corresponding to the rise and fall of the tide. The paper is 6 inches wide with about 380 feet to the roll, which is sufficient to include a little more than a full year of record of the predicted tides at a station. The paper should be about 0.0024 inch thick in order that the complete roll may be of a suitable size for use in the machine.

**392.** Within the dial case, near the upper right-hand corner, is a mandrel (*33*, fig. 25), which can be quickly removed and replaced. It is designed to hold the blank roll of paper, the latter being wound upon a wooden core especially designed to fit on the mandrel. At the bottom of the mandrel is an adjustable friction device to provide tension on the paper. From the blank roll the paper is led over an idler roller (*34*, fig. 25), mounted in the front plate of the dial case, then across the face of the machine for a distance of about 13 inches to a feed roller (*35*, fig. 25), then over the feed roller to the receiving roller (*36*, fig. 25), upon which it is wound.

**393.** The feed roller governs the motion of the paper across the face of the machine and is provided near each end with 12 fine needle points to prevent the paper from slipping. The feed roller is controlled by the main vertical shaft of the dial case through gearing of such ratio that the feed roller will turn at the same rate as the main vertical shaft; that is to say, one complete turn of the feed roller will represent 12 dial hours in time. The feed roller being 6 inches in circumference the paper will be moved forward at the rate of one-half inch to the dial hour. A ratchet and pawl (*37*, fig. 25) are so placed as to leave the paper at rest when the machine is turned backward. If desired, the paper feed can be thrown out of action altogether by turning a small milled head on the ratchet gear.

**394.** To provide for the winding up of the paper on the receiving roller there is a sprocket wheel (*38*, fig. 25) held by adjustable friction to the upper end of the feed roller. Fitted to the top of the receiving roller is a smaller sprocket which is driven by a chain from the feed-roller sprocket. The ratio of the sprockets is such as to force the receiving roller to wind up all the paper delivered by the feed roller, the tension on the paper being kept uniform by the friction device. To remove a completed roll of record the smaller sprocket is lifted from the receiving roller and a pin (*39*, fig. 25) at the back of the dial case is drawn out, releasing the upper bearing bracket. The bracket can then be raised and the receiving roller with its record removed. A similar bracket secured by a pin is provided for the removal of the mandrel on which the blank roll of paper is placed.

**395.** *Marigram gears.*—The pen that traces the tide curve is mounted in a carriage which is arranged to slide vertically on a pair of guiding rods and is controlled from a horizontal shaft at the back of the dial case. On this shaft there is mounted a set of three sliding change gears (*18*, fig. 26), which are designed to mesh, respectively, with three fixed gears mounted on a shaft just above. By sliding the change gears in different positions any one of them may be brought into mesh with its corresponding fixed gear. These gears provide for ratios of 1:1, 2:1, and 3:2, according to whether the innermost, the middle, or the outer gears are in mesh. At the outer end of the shaft containing the fixed gears is a thread-grooved wheel 4 inches in circumference (*19*, fig. 26), to which is attached one end of the pen-carriage chain (*20*, fig. 26). The chain is partly wound upon the wheel and from it passes through the dial case to the front of the machine, then upward over a pulley near the top to a counterpoise weight within the dial case. The pen carriage is secured to this chain by means of a clamp and can be adjusted to any desired position.

**396.** *Scale of tide curve.*—With a working scale of unity, the rotation of the height summation wheel, as transmitted through marigram gear ratio of 1:1 to the curve-line pen, will move the latter vertically 0.1 inch for each unit change in the sum of the harmonic terms and this may be taken as the basic or natural scale of the graphic record. This scale may be enlarged by the factor 3/2 or 2 through the use of one of the other gear ratios and may be further modified to any desired extent by the introduction of an arbitrary working scale factor. Letting $G$ equal the marigram gear ratio (1, 3/2, or 2) and $S$ equal the working scale factor applied to the amplitude settings, the vertical scale of the graphic record may be expressed as follows:

1 inch of graph represents $10/GS$ units of summation (461)

1 summation unit is represented by $GS/10$ inches in graph (462)

The scale ratio of the graph will differ with different units used in the predictions. Thus

Graph scale (amplitude settings in feet) $= GS/120$ (463)

Graph scale (amplitude settings in meters) $= GS/393.7$ (464)

Graph scale (amplitude settings in decimeters) $= GS/39.37$ (465)

**397.** In selecting the marigram gear ratio and scale factor for the predictions at any station, it is the general aim to secure as large a scale as possible while keeping the graph within the limits of the paper. Some consideration must be given also to the limits of the height dial scale and in some instances to the mechanical limits of the individual amplitude settings. The marigram gear ratio affects the graph only but the scale factor affects also the amplitude settings and the height dial readings. The extreme amplitude of the graphic record is limited by the width of the paper which extends 3 inches on either side of the medial line, but for mechanical reasons it is desirable in general to keep the record within a band 2½ inches on either side of the medial line. The following table suggests suitable scale, dial, and gear combinations for different tidal ranges and different current velocities. The tabular marigram scales are applicable only when the foot or knot has been used as the unit for machine settings. The marigram amplitude limits given in the last column are expressed in the same unit that is used in setting the machine regardless of what unit that may have been.

*Working scale, height dial, marigram gear, and scale*

| Tidal range limits | Current velocity limits | Working scale factor | Height dial | Marigram gear | Marigram scale | | Marigram amplitude limit |
|---|---|---|---|---|---|---|---|
| | | | | | Settings in feet | Settings in knots | |
| *Feet* | *Knots* | | | | *Ratio* | *Knots per inch* | *Units* |
| 0.0– 2.5 | 0.0– 1.0 | 10 | 4 | 2:1 | 1: 6 | 0.50 | 1.5 |
| 2.6– 3.5 | 1.1– 1.5 | 10 | 4 | 3:2 | 1: 8 | 0.67 | 2.0 |
| 3.6– 4.0 | 1.6– 2.0 | 10 | 4 | 1:1 | 1:12 | 1.00 | 3.0 |
| 4.1– 6.0 | 2.1– 3.0 | 4 | 10 | 2:1 | 1:15 | 1.25 | 3.7 |
| 6.1– 8.0 | 3.1– 4.0 | 4 | 10 | 3:2 | 1:20 | 1.67 | 5.0 |
| 8.1–10.0 | 4.1– 5.0 | 4 | 10 | 1:1 | 1:30 | 2.50 | 7.5 |
| 10.1–12.5 | 5.1– 6.0 | 2 | 20 | 2:1 | 1:30 | 2.50 | 7.5 |
| 12.6–16.5 | 6.1– 8.0 | 2 | 20 | 3:2 | 1:40 | 3.33 | 10.0 |
| 16.6–20.0 | 8.1–10.0 | 2 | 20 | 1:1 | 1:60 | 5.00 | 15.0 |
| 20.1–25.0 | 10.1–12.5 | 1 | 40 | 2:1 | 1:60 | 5.00 | 15.0 |
| 25.1–32.5 | 12.6–16.0 | 1 | 40 | 3:2 | 1:80 | 6.67 | 20.0 |
| 32.6– | 16.1– | 1 | 40 | 1:1 | 1:120 | 10.00 | 30.0 |

When height dial readings are not required, and amplitude settings are in feet, a convenient graph scale of 1:10 can be obtained by using any one of the following combinations; scale factor 12 with gear ratio 1:1, scale factor 8 with gear ratio 3:2, or scale factor 6 with gear ratio 2:1.

**398.** When the tide-predicting machine is used for the prediction of the tide-producing force, the graph scale to be adopted will depend upon the unit in which the force is to be expressed. Assume that the sum of all terms in the vertical component of the force (par. 79) is desired. Referring to paragraph 43, it will be noted that the extreme value of this component due to the combined action of moon and sun is approximately $0.2 \times 10^{-6}$ with the unit of force taken as $g$, the mean acceleration of gravity. In this case a convenient scale relation which will bring the graph within the desired limits on the paper is obtained by adopting a working scale factor of $6 \times 10^7$ with the marigram gear ratio of 2:1. With this combination 0.1 foot of graph ordinate will represent $10^{-7} g$ units of force. In practice the scale factor would be combined with the general coefficient common to all terms in the formulas.

**399.** *Pens.*—The curve-line pen (*13*, fig. 23) and the datum-line pen (*14*) are each of the ordinary fountain type. Each is fitted with a metal lock joint, so that it may be quickly removed and replaced in the same position, and is pressed against the paper by a light coil spring when in use. The curve-line pen is mounted in a swivel arm on a light carriage which slides vertically along two rods. The datum-line pen is mounted in a swivel arm that may be adjusted so that the mean sea-level line will be traced midway between the upper and lower edges of the paper.

**400.** *Hour-marking device.*—The arm for the datum-line pen is secured to the outer end of a shaft which carries two armatures, one for the upper and the other for the lower of two electromagnets (*17*, fig. 26). A spring keeps the armatures at equal distances from their respective electromagnets. The upper electromagnet is designed for indicating the hours on the datum line and is in a circuit that is opened and closed by a platinum-tipped contact spring resting upon the edge of an ivory disk in which are embedded, equally spaced, 24 narrow strips of platinum (*32*, fig. 25). The ivory disk is mounted on the shaft of the hour pointer, and as this rotates the platinum strips successively make an electric contact that throws the datum-line pen downward for an instant, making a corresponding jog in the datum line, the downward stroke of the pen indicating the exact hour. An extra strip of platinum placed close to the one representing the midnight hour causes a double jog for the beginning of each day, the downward stroke of the second jog indicating the zero hour.

**401.** *High and low water marking device.*—The lower electromagnet is in a circuit that is closed when the platinum index on the time chain (*11*, fig. 23) is in contact with the fixed platinum index (*12*); that is to say, at the times of high and low waters. When this contact is made, the electromagnet attracts the armature, which throws the datum-line pen upward, causing a corresponding upward jog in the datum line, and thus automatically marking the time of the high or low water. A small switch (*21*, fig. 24) just above the hand-crank shaft permits the cutting out of the current from the two electromagnets.

**402.** *Adjustment of machine.*—The adjustment of the machine should be tested at least once each year and at any other time when there is any reason for believing that a change may have taken place. The following adjustments are required.

**403.** *Height-chain adjustment.*—All amplitudes should be set at zero, so that the turning of each constituent crank shaft will produce no motion in the height chain. This should bring the summation wheel to its zero position, but on account of a certain amount of backlash and flexures in the machine this wheel may not be in an exact zero position even when the chain is in adjustment. Now, set a single constituent with a very small amplitude and operating the machine with the hand crank, note whether the index of the summation wheel oscillates equal distances on both sides of its zero position. If not, the chain should be adjusted by the adjusting nut at its fixed end at the back part of the machine.

**404.** *Time-chain adjustment.*—The adjustment of the time chain is similar to that of the height chain. The zero position is indicated by the conjunction of a small triangular-shaped index on the chain and a fixed platinum index in the middle of the horizontal opening in the dial face. A small amplitude being set on one of the constituent time cranks and the machine operated by the hand crank, the chain index should oscillate equal distances on both sides of the platinum point. If it does not, the necessary adjustment may be made at the fixed end of the chain.

**405.** *Hour-hand adjustment.*—This must be so adjusted that it will register the exact hour at the same instant the circuit for the electromagnet is closed for the hour mark on the marigram, which is indicated by a downward stroke of the datum-line pen. It is also necessary that the zero hour or beginning of the day shall correspond to the double hour mark on the marigram. This adjustment may be accomplished by moving the hour hand on its shaft after releasing its set screw. A finer adjustment may be effected by changing the position of the contact spring back of the dial face.

**406.** *Minute-hand adjustment.*—This is to be adjusted to read zero on the exact hour indicated by the hour hand and the closing of the electric circuit for the hour mark. The adjustment may be accomplished either by moving the minute hand on its shaft after releasing its set screw or by means of the releasable gears on the main vertical shaft of the dial case. The adjustments just described are those which need be made only occasionally. Other adjustments are taken into account each time the machine is set for a station.

**407.** *Setting predicting machine.*—The time indicators on the face of the machine are first set to represent the exact beginning of the period for which predictions are to be made, which will usually be 0 hour of January 1 of some year. The hour and minute hands should always be brought into place by the turning of the operating crank in order that the adjustment of these hands relative to the electromagnet circuit may not be affected. The date dial may, however, if desired, be set independently, using the binding nut just above the large dial ring for releasing and clamping. If only a small motion of the date dial is necessary, it is generally preferable to set it by the operating crank. The year index should be set to indicate the kind of year.

**408.** In the usual operation of the machine a ratchet prevents the operating crank from being turned backward, but this ratchet may be released when desired by pressing on a button in the side of the machine just above the crank. After the face of the machine has been thus set to register the beginning of the predictions the three main vertical shafts should be clamped to prevent them from turning.

**409.** *To set the height amplitudes.*—All the constituent cranks on the left or height side of the machine are first turned, by means of the releasable gears on the main vertical shafts, to a vertical position, the cranks of the upper range of constituents pointing downward and those in the lower range upward, in which position all angles will read 180°. For the long-period constituents the cranks can be more quickly brought to the vertical position by drawing out small knobs on the time side of the machine, thus disconnecting the gearing. The cranks are then turned by hand to the desired position and the knobs pushed back into place. The amplitudes may now be set according to the scales attached to the sides of the machine. The crank pin is unclamped by a small milled head wrench and is then moved along its groove until the index at the scale registers the amplitude setting given in Form 445, when it is clamped in this position. If no amplitude is given for any constituent, the corresponding crank must be set at zero.

**410.** *To set time amplitudes.*—The process is similar to that for the height amplitudes, the cranks on the time side of the machine being first turned to a vertical position with all angles reading 90°. The cranks are to be set with the same amplitudes as were used for the height side, the modified scales automatically taking account of the true differences in the amplitudes. For the constituents Sa and Ssa the amplitudes are set on the height side only.

**411. To set constituent angles.**—After the amplitudes have been set and checked on both sides of the machine the angles are set for the beginning of the period of predictions, these settings being given in Form 445. The angles may be set from either side of the machine, except for constituents Sa and Ssa, for which there are no dials on the time side, as the readings are the same for both sides. As each constituent angle is set its releasable gear is clamped to the main vertical shaft. After all the angles have been thus set the three main vertical shafts must be unclamped to permit them to turn.

**412. Changing height scale.**—There are three interchangeable height scales, known as the 40-foot, the 20-foot, and the 10-foot scale. The 40-foot ring may also be conveniently used as a 4-foot scale. The scale to be used for any station is indicated in Form 445. In removing a scale from the machine a small button at the top is turned to release the ring, which is then lifted slightly as it is being removed. The desired scale is then placed on the machine and secured in place by a button. Before removing or replacing the height scale it is desirable that the height pointer be set approximately 45° to the left of its zero position in order to interfere least with the removal or replacement of the scale.

**413. The datum or plane of reference.**—The hand-operating crank should be turned forward or backward until the index of the summation wheel on the height side of the machine indicates mean sea level. It must be kept in mind, however, that as the index lines may come in conjunction at each complete rotation of the summation wheel there is a possibility of being misled in regard to the mean sea-level position. When in doubt, the operating crank should be turned forward to obtain a number of conjunctions, the corresponding height dial reading for each being noted. The conjunction that corresponds most closely with the average of such height readings will be the one that applies to the true zero position. Each complete turn of the height summation wheel will cause a change in the height reading of 12 units, 6 units, or 3 units, respectively, according to whether the 40-unit, 20-unit, or 10-unit dial is used. The height hand, which can be released by the milled nut on the face of the machine, may now be set to the scale reading that corresponds to the height of mean sea level above the datum which has been adopted for the predictions, this value being given in Form 445.

**414. The marigram gear.**—There are three gear combinations, designated as the 1:1, 3:2, and 2:1 ratios. The gear ratio to be used for any station is indicated in Form 445. When it is necessary to change the gear ratio, the machine should be first turned to its mean sea-level position. The change is then effected by sliding the lower set of gears horizontally, being careful to hold the upper set with one hand to prevent it from turning when the gears are released. Before engaging the gears in their new ratios the counterpoise for the pen carriage should be brought to a position approximately midway between the limits of its range of motion. The 1:1 ratio is obtained by sliding the lower set of gears as far as possible toward the height side of the machine, thus engaging the innermost gears; the 3:2 ratio by moving these gears toward the time side until the outer gears are engaged, and the 2:1 ratio by engaging the middle gear of each set.

**415.** In setting up the machine for successive stations there is a mechanical advantage in making the necessary gear changes before setting the new amplitudes if the gear changes are in the order of 2:1, 3:2, 1:1, and after setting the amplitudes if the gear changes are in the reverse order. This precaution will lessen the chances of jamming the curve pen carriage and throwing the height chain off its pulleys when setting the amplitudes.

**416.** *Inserting paper roll.*—To place the paper on the machine, remove the mandril that is mounted within the dial case near the upper right-hand corner and slip the roll of paper over the mandril, the roll being so placed that the winding is clockwise when viewed from above and when on the machine the paper unwinds from the outer side of the roll. In placing the roll on the mandril care should be taken to see that the small projection on the base of the latter enters the cavity in the wooden core, so that the roll will fit flat against the base. After the mandril with the roll of paper has been returned to the machine and secured in place, the end of the paper is passed around a roller to the face of the machine, across the face, and over the feed roller at the left of the machine. The end is then inserted into the slit in the receiving roller, which is given a few turns to take up the slack paper and make it secure. Before passing the paper over the feeding roller and on the receiving roller these rollers should be released to permit them to turn independently, the release being effected by turning the small milled head on a ratchet stud gear near the base of the feeding roller and by lifting off from the top of the receiving roller the small knob holding the connecting chain. After the paper has been secured to the receiving roller these connections should be restored.

**417.** *Curve pen adjustment.*—With the machine in its mean sea-level position, the curve pen must be adjusted to bring the pen point on the mean sea-level line as drawn by the base-line pen. This adjustment may be effected by releasing the pen carriage from the operating chain and moving it to the desired position, where it is clamped in place by the binding screw.

**418.** *Verification of machine settings.*—Each step in the adjustment and setting of the machine should be carefully checked before proceeding with the next step. After the setting of the machine for any station has been completed an excellent check on the work is afforded, if the predictions for the same station for the preceding year are available, by turning the machine backward several days and then comparing the predicted tides with those previously obtained.

**419.** *Predicting.*—The datum and curve fountain pens are filled and put in place, the electric cut-out switch under the base of the machine closed, and the ratchet of the operating crank set to prevent the machine from being turned backward. If the predicted height of the tide for any given time is desired, the machine may be turned forward until the required time is registered on the time dials and the corresponding height read off of the height dial.

**420.** If the predicted high and low waters for the year are desired, the operating crank is turned forward until the machine is automatically stopped by the brake at a high or low water. To avoid the strain on the machine due to sudden stops, the operator should watch the small index on the time chain, and as this approaches the fixed index in the center of the opening on the face of the machine, turn the crank more slowly until the machine is stopped as the indexes come in contact with each other. The time and height may then be read directly from the dials on the face of the machine. The movement of the height pointer before the stopping of the machine and also the tide curve will clearly indicate whether the tide is a high or low water. After the tide has been recorded an inward pressure on the crank handle will release the brake and the machine can be turned forward to the next tide, the process being repeated until all the tides of the year have been predicted and recorded.

### FORMS USED WITH TIDE-PREDICTING MACHINE

**421.** *Form 444, standard harmonic constants for predictions* (fig. 32).—This form provides for the compilation of the harmonic con-

DEPARTMENT OF COMMERCE
U S COAST AND GEODETIC SURVEY
Form No. 444

TIDES / CURRENTS } STANDARD HARMONIC CONSTANTS FOR PREDICTION

STATION ........ Morro, California

Lat. 35 ° 22 ′ N.
Long. 120 ° 51 ′ W.
Long... 120.85° W.

| Component | H Amplitude | κ Epoch | A κ′−κ | B ×H | C 360°−κ′ | D −κ′ | REMARKS |
|---|---|---|---|---|---|---|---|
| | ft. | ° | ° | ft. | ° | ° | |
| $M_2$ | 1.227 | 309.5 | +9.8 | 4.91 | +41.7 | −318.3 | Time meridian 120 = 8.0 h. |
| $S_2$ | 0.320 | 304.3 | +1.7 | 1.28 | +54.0 | −306.0 | Extreme range 8.8 {ft. km. |
| $N_2$ | 0.260 | 284.5 | +14.2 | 1.04 | +61.3 | −298.7 | Dial 10 |
| $K_1$ | 1.001 | 111.2 | +0.5 | 4.00 | +248.3 | −111.7 | Marigram gear 3:12 |
| $M_4$ | 0.053 | 170.7 | +19.6 | 0.21 | +169.7 | −190.3 | Marigram scale 1:20 |
| $O_1$ | 0.608 | 99.1 | +9.3 | 2.43 | +251.6 | −108.4 | $A_0$ 2.40 ft. |
| | | | | | | | Permanent current kn. |
| $M_6$ | 0.013 | 253.7 | +29.5 | 0.05 | +76.8 | −283.2 | The DATUM is a plane ft. |
| $(MK)_3$ | | | | + | | − | below mean { low water springs lower low water |
| $S_6$ | 0.009 | 176.6 | +3.4 | 0.04 | +180.0 | −180.0 | |
| $(MN)_4$ | | | | + | | − | |
| $ν_2$ | 0.065 | 285.0 | +13.6 | 0.26 | +61.4 | −298.6 | |
| $S_4$ | 0.006 | 148.3 | +5.1 | 0.02 | +206.6 | −153.4 | |
| $μ_2$ | 0.025 | 174.2 | +18.0 | 0.10 | +167.8 | −192.2 | |
| $(2N)_2$ | (0.035 | 260.5) | +18.5 | 0.14 | +81.0 | −279.0 | |
| $(OO)_1$ | (0.026 | 123.3) | −8.3 | 0.10 | +245.0 | −115.0 | |
| $λ_2$ | (0.009 | 306.6) | +6.0 | 0.04 | +47.4 | −312.6 | |
| $S_1$ | | | | + | | − | |
| $M_1$ | 0.041 | 132.4 | +4.9 | 0.16 | +222.7 | −137.3 | |
| $J_1$ | (0.048 | 117.2) | −3.6 | 0.19 | +246.6 | −113.4 | |
| Mm | | | | + | | − | |
| Ssa | | | | + | | − | |
| Sa | | | | + | | − | |
| MSf | | | | + | | − | |
| Mf | | | | + | | − | |
| $ρ_1$ | (0.023 | 93.9) | +15.1 | 0.09 | +253.0 | −107.0 | |
| $Q_1$ | 0.107 | 98.9 | +13.7 | 0.43 | +247.4 | −112.6 | |
| $T_2$ | (0.019 | 304.5) | +2.0 | 0.08 | +53.5 | −305.5 | |
| $R_2$ | (0.003 | 304.1) | +1.4 | 0.01 | +54.5 | −305.5 | |
| $(2Q)_1$ | (0.016 | 87.1) | +18.0 | 0.06 | +254.9 | −105.1 | |
| $P_1$ | 0.274 | 107.7 | +1.2 | 1.10 | +251.1 | −108.9 | |
| $(2SM)_2$ | | | | | | | |
| $M_3$ | 0.022 | 345.8 | +14.7 | 0.09 | +359.5 | −0.5 | |
| $L_2$ | 0.050 | 307.5 | +5.5 | 0.20 | +47.0 | −313.0 | |
| $(2MK)_3$ | | | | | | | |
| $K_2$ | 0.103 | 289.8 | +1.0 | 0.41 | +69.2 | −290.8 | |
| $M_8$ | 0.007 | 105.6 | +39.3 | 0.03 | +215.1 | −144.9 | |
| $(MS)_4$ | | | | + | | − | |

Source of constants ....from observed hourly heights for 163 days beginning...........February 13, 1919.

Compiled by ...L.P.D... March 28, 1923..........   Verified by ..F.J.H... March 29, 1923..
                        (Date)                                                      (Date)

FIGURE 32.

stants for use in the prediction of the tides and also for certain permanent preliminary computations to adapt the constants for use with the U. S. Coast and Geodetic Survey tide-predicting machine No. 2. The form is used in a loose-leaf binder.

**422.** The constituents are listed in an order that conforms to the arrangement of the corresponding constituent shafts and cranks on the predicting machine. The accepted amplitudes and epochs are to be given in the columns provided for the purpose. At the bottom of the page a space is provided for indicating the source from which the constants were derived.

**423.** The column of Remarks provides for miscellaneous information pertaining to the predictions. This includes the kind of time in which the predictions are to be given, the approximate extreme range of tide at the place for determining the proper scale to be used, the height dial, the marigram gear, the marigram scale, and the datum to which the predicted heights are to be referred.

**424.** The extreme range may be estimated from the predictions for a preceding year or may be taken approximately as twice the sum of the amplitudes of the harmonic constants. The height dial, marigram gear, and marigram scale which are recommended for use with different extreme ranges are given in the table on page 138.

**425.** The principal hydrographic datums in general use are as follows: Mean low water for the Atlantic and Gulf coasts of the United States and Puerto Rico. Mean lower low water for the Pacific coast of the United States, Canada, and Alaska, and the Hawaiian and Philippine Islands. Approximate low water springs for the rest of the world, with a few exceptions. For use on the predicting machine the datum must be defined by its relation to the mean sea level, and this relation is usually determined from a reduction of the high and low waters.

**426.** Column A of Form 444 is designed for the differences by which the epochs of the constituents are adapted once for all for use with the unmodified Greenwich $(V_0+u)$'s of each year. These differences take account of the longitude of the station and also of the time meridian used for the predictions, and are computed by the formula

$$\kappa'-\kappa=pL-\frac{aS}{15} \qquad (466)$$

in which

$\kappa'-\kappa$=adapted epoch—true epoch.
$p$=subscript of constituent, which indicates number of periods in one constituent day. For the long-period constituents Mm, Ssa, Sa, MSf, and Mf, $p$ should be taken as zero.
$L$=longitude of station in degrees;+if west,—if east.
$a$=speed of constituent in degrees per solar hours.
$S$=longitude of time meridian in degrees;+if west,—if east.

The values of the products $\frac{aS}{15}$ for the principal time meridians may be taken from table 35. For any time meridian not given in the table the products may be obtained by direct multiplication, taking the values for the constituent speeds $(a)$ from table 2.

**427.** Column $B$ is designed for the reduction of the amplitudes to the working scale of the machine. The scale is unity when the 40-

HARMONIC ANALYSIS AND PREDICTION OF TIDES 145

foot height dial is used, 2 for the 20-foot height dial, 4 for the 10-foot height dial, and 10 for a 4-foot height dial. The working scale should be entered at the head of the column and used as a factor with the amplitudes in order to obtain the values for this column.

428. Columns $C$ and $D$ are designed to contain the adapted epochs in positive and negative forms which may be used additively with the Greenwich $(V_o+u)$'s. It will be found most convenient to compute column $D$ first, by applying the difference in column $A$ to the $\kappa$ in the preceding column and entering the result with the negative sign. If the direct application of the difference should give a negative result, this must be subtracted from 360° before entering in column $D$.

DEPARTMENT OF COMMERCE
U S. COAST AND GEODETIC SURVEY
Form No. 445

TIDES / CURRENTS } SETTINGS FOR TIDE PREDICTING MACHINE

STATION ___Morro, California___   YEAR 19_23_

| COMPO-NENT | AMPLI-TUDE SETTING | DIAL SETTING ||| | REMARKS |
|---|---|---|---|---|---|---|
| | | Jan. 1, 0ʰ | Feb. 1, 0ʰ | Dec.31,24ʰ | | |
| | ft. | ° | ° | ° | | |
| M₂ | 5.10 | 82.9 | 47.1 | 183.7 | | |
| S₂ | 1.30 | 54 | 54 | 54 | | Time Meridian 120 ___E.-W. |
| N₂ | 1.10 | 214 | 133 | 226 | | Dial 10 |
| K₁ | 3.55 | 255 | 286 | 255 | | Marigram gear 3:20 |
| M₄ | 0.20 | 252 | 180 | 94 | | Marigram scale 1:20 |
| O₁ | 2.00 | 287 | 221 | 28 | | |
| M₆ | 0.05 | 200 | 93 | 142 | | A₀ 2.40 ft. |
| (MK)₃ | --- | --- | --- | --- | | Permanent current ___kn. |
| S₄ | 0.05 | 180 | 180 | 180 | | |
| (MN)₄ | --- | --- | --- | --- | | |
| ν₂ | 0.25 | 33 | 6 | 323 | | |
| S₆ | --- | --- | --- | --- | | |
| μ₂ | 0.10 | 251 | 179 | 93 | | |
| (2N)₂ | 0.15 | 345 | 219 | 268 | | |
| (OO)₁ | 0.05 | 40 | 167 | 298 | | |
| λ₂ | 0.05 | 338 | 293 | 249 | | |
| S₁ | --- | --- | --- | --- | | |
| M₁ | 0.25 | 347 | 329 | 217 | | |
| J₁ | 0.15 | 140 | 216 | 228 | | |
| Mm | --- | --- | --- | --- | | |
| Ssa | --- | --- | --- | --- | | |
| Sa | --- | --- | --- | --- | | |
| MSf | --- | --- | --- | --- | | |
| Mf | --- | --- | --- | --- | | |
| ρ₁ | 0.05 | 219 | 162 | 150 | | |
| Q₁ | 0.35 | 35 | 284 | 47 | | |
| T₂ | 0.10 | 56 | 25 | 56 | | |
| R₂ | --- | --- | --- | --- | | |
| (2Q)₁ | 0.05 | 154 | 358 | 78 | | |
| P₁ | 1.10 | 242 | 211 | 242 | | |
| (2SM)₂ | --- | --- | --- | --- | | |
| M₃ | 0.10 | 61 | 7 | 32 | | |
| L₂ | 0.20 | 149 | 158 | 338 | | |
| (2MK)₃ | --- | --- | --- | --- | | |
| K₂ | 0.30 | 264 | 325 | 264 | | |
| M₈ | 0.05 | 20 | 237 | 63 | | |
| (MS)₄ | --- | --- | --- | --- | | |

Computed by __L.P.D.__ March 29, 1923   Verified by __F.J.H.__ March 29, 1923
Predicted by ___
Date ___

FIGURE 33.

The values for column $C$ may then be obtained by applying 360° to the negative values in column $D$.

**429.** *Form 445, settings for tide-predicting machine* (fig. 33).—This form is designed for the computations of the settings for the predicting machine for the beginning of each year of predictions. The forms are bound in books, a separate book being used for each year of predictions. This form is used in connection with Form 444, and for convenience the order of arrangement of the constituents is identical in the two forms. The name of the station, the time meridian, the height dial, marigram gear, marigram scale, and datum plane are copied directly from Form 444.

**430.** For the amplitude settings the amplitudes of column $B$ of Form 444 are multiplied by the factors $f$ from table 14 for the year for which the predictions are to be made. A convenient way to apply these factors is to prepare a strip of paper with the same vertical spacing as the lines on Form 444 and enter the factors $f$ for the required year on this strip. The strip may then be placed alongside of column $B$ of Form 444 and the multiplication be performed. The same strip will serve for every station for which predictions are to be made for the given year. It has been the recent practice to enter the amplitude settings to the nearest 0.05 foot as being sufficiently close for all practical purposes.

**431.** For the dial settings for January 1, 0 hour, the Greenwich equilibrium arguments of $(V_o+u)$'s from table 15 are to be applied, according to the indicated sign, to the angles of column $C$ or $D$ of Form 444, using the angle in column $D$ if it is less than the argument, otherwise using the angle in column $C$. For the application of the $(V_o+u)$'s a strip similar to that used for the factors $f$ should be prepared. The same strip will serve for all stations for the given year. For the dial settings it is customary to use whole degrees, except for constituent $M_2$, for which the setting is carried to the first decimal of a degree.

**432.** The settings for February 1 and December 31 are used for checking purposes to ascertain whether there has been any slipping of the gears during the operation of the machine. To obtain the dial settings for February 1, $0^h$, and December 31, $24^h$, prepare strips similar to those for the $f$'s and $(V_o+u)$'s. On one enter the angular motion of the constituents from January 1, $0^h$, to February 1, $0^h$; on a second and a third strip, the angular motion for February 1, $0^h$, to December 31, $24^h$, for a common and leap year, respectively. For checking purposes a fourth and fifth strip may contain the angular changes for a complete common and a complete leap year, respectively. The values for these strips may be obtained from table 36. These strips will be found more convenient if arranged with two columns each, one column containing the values in a positive form and the other column containing the equivalent negative value which is obtained by subtracting the first from 360°. These strips are good for all years, distinction being made between the common and leap years. By applying the first strip to the dial settings for January 1 the values for February 1 are readily obtained, and by applying the second or third strip to the latter settings those for the end of the year are obtained. The values obtained by applying the fourth or fifth strips to the settings for January 1 should also give the correct setting for the end of the year, and thus serve as a check. The angular changes for computing the settings for any day of the year may be obtained from tables 16 and 17.

### PREDICTION OF TIDAL CURRENTS

**433.** Since the tidal current velocities in any locality may be expressed by the sum of a series of harmonic terms involving the same periodic constituents that are found in the tides, the tide-predicting machine may be used for their prediction. For the currents, however, consideration must be given to the direction of flow, and in the use of the machine some particular direction must be assumed. At present the machine is used for the prediction of reversing currents in which the direction of the flood current is taken as positive and the maximum velocity in this direction corresponds to the high water of the predicted tide. The ebb current is then considered as having a negative velocity with its maximum corresponding to the low water of the predicted tide. Rotary currents may be predicted by taking the north and east components separately but the labor of obtaining the resultant velocities and directions from these components would be very great without a machine especially designed for the purpose. Predictions can, however, be made along the main axis of a rotary movement without serious difficulties. Formulas for referring the harmonic constants of the north and east components to any desired axis are given in Coast and Geodetic Survey Special Publication No. 215, Manual of Current Observations.

**434.** The harmonic constants for the prediction of current velocities are derived from current observations by an analysis similar to that used in obtaining the harmonic constants from tide observations. In the current harmonic constants, however, the amplitudes are expressed in a unit of velocity, usually the knot, instead of the linear unit that is used for the tidal harmonic constants. Forms 444 and 445 for the computation of the settings for the tide-predicting machine are applicable for the current predictions and the procedure in filling out these forms is essentially the same as described in paragraphs 421–432 for the tide predictions. The node factors $(f)$ and arguments $(V_o+u)$ are the same as for the tides. The height dial, marigram gear and scale suitable to the current velocity can be obtained from the table on page 138. Instead of a sea level elevation there should be entered in the column of "Remarks" the velocity of any permanent current along the axis in which the predictions are to be made. This velocity should be marked plus (+) or minus (−) according to whether the permanent current is in the flood or ebb direction.

**435.** The predicting machine is set with the current harmonic constants in the same manner as for the tidal harmonic constants. To take account of the permanent current the height summation wheel should be brought to its zero position and the height hand then set at a dial reading corresponding to the velocity of the permanent current, the hand being set to the right of the scale zero if the permanent current is in the flood direction and to the left if in the ebb direction. The hand crank should be then turned to bring the height hand to its zero position and the curve-pen set at the medial line of the paper, this line now representing zero velocity or slack water.

**436.** The operation of the machine for the prediction of the currents is similar to that for the prediction of the tides. The machine automatically stops at each maximum flood and ebb velocity and the corresponding times and velocities are then recorded, the flood velocities being read to the right and the ebb velocities to the left of the scale zero. In the prediction of the currents the times of slack water are also desired. These are indicated by the zero position of the recording hand as well as by the intersections of the curve and medial line in the graphic record. The velocity of the current at any intermediate time can be read directly from the height dial when the machine has been turned to the time desired and it may be also scaled from the graphic record.

**437.** Predictions of hydraulic currents in a strait, based upon the difference in the tidal head at the two entrances, may be made by means of harmonic constants derived from the tidal constants for the entrances. Differences in tidal range or in the times of the high and low waters at the two ends of a strait will cause the water surface at one end alternately to rise above and fall below that at the other end, thus creating a periodic reversing current in the strait. Theoretically, disregarding friction or inertia, the velocity of the current would vary as the square root of the difference in head, being zero when the surface is at the same level at both ends and reaching a maximum when the difference is greatest. Actually they will generally be a lag of some minutes in the response of the current movement to the difference in head which must be determined from observations.

**438.** Let the two ends of the strait be designated by $A$ and $B$, with the flow from $A$ toward $B$ considered as flood or positive and the flow in the opposite direction as ebb or negative. With the waterway receiving the tide from two sources, the application of the terms "flood" and "ebb" will be somewhat arbitrary, and care must be taken to indicate clearly the direction assumed for the flood movement. In the following discussion tidal constants pertaining to entrances $A$ and $B$ will be distinguished by subscripts $a$ and $b$, respectively, and those pertaining to the difference in tidal head by the subscript $d$. Since the usual constituent epochs known as "kappas" refer to the local meridian, it will be necessary for the purpose of comparison between places on different meridians to use the Greenwich epochs "$G$" (par. 226), these being independent of local time and longitude.

**439.** For any one constituent let $T$ represent time as expressed in degrees of the constituent reckoned from the phase zero of its Greenwich equilibrium argument. Also let $Y_a$ and $Y_b$ represent the height of the constituent tide for any time $T$ as referred to the mean level at locations $A$ and $B$, respectively; and let $Y_d$ equal the difference $(Y_a - Y_b)$. Formulas for heights and difference may now be written

$$Y_a = H_a \cos (T - G_a) \text{ for location "}A\text{"} \quad (467)$$

$$Y_b = H_b \cos (T - G_b) \text{ for location "}B\text{"} \quad (468)$$

$$Y_d = H_a \cos (T - G_a) - H_b \cos (T - G_b)$$
$$= (H_a \cos G_a - H_b \cos G_b) \cos T + (H_a \sin G_a - H_b \sin G_b) \sin T$$
$$= H_d \cos (T - G_d) \quad (469)$$

HARMONIC ANALYSIS AND PREDICTION OF TIDES  149

m which
$$H_d = [H_a^2 + H_b^2 - 2H_aH_b \cos(G_b - G_a)]^{\frac{1}{2}} \qquad (470)$$

$$G_d = \tan^{-1} \frac{H_a \sin G_a - H_b \sin G_b}{H_a \cos G_a - H_b \cos G_b} \qquad (471)$$

The proper quadrant for $G_d$ is determined by the signs of the numerator and denominator of the above fraction, these being the same, respectively, as for the sine and cosine of the angle. Formulas (470) and (471) may be solved graphically (fig. 34) by drawing from any point $C$ a line $CD$ to represent in length and direction $H_a$ and $G_a$, respectively; from the point $D$ a line $DE$ to represent in length and direction $H_b$ and $(G_b \pm 180°)$, respectively. The connecting line from

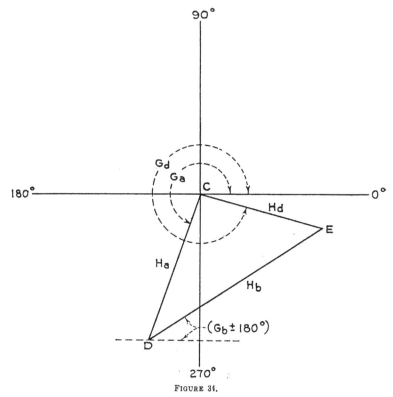

FIGURE 34.

$C$ to $E$ will represent by its length the amplitude $H_d$ and by its direction the epoch $G_d$.

**440.** Formulas (470) and (471) may be modified to adapt them for use with tables 41 and 42.
From (470) we may obtain

$$H_d/H_a = [1 + (H_b/H_a)^2 + 2(H_b/H_a) \cos(G_b - G_a \pm 180°)]^{\frac{1}{2}} \qquad (472)$$

or
$$H_d/H_b = [1 + (H_a/H_b)^2 + 2(H_a/H_b) \cos(G_a - G_b \pm 180°)]^{\frac{1}{2}} \qquad (473)$$

and from (471) we have

$$\text{Tan}(G_d - G_a) = \frac{(H_b/H_a) \sin(G_b - G_a \pm 180°)}{1 + (H_b/H_a) \cos(G_b - G_a \pm 180°)} \qquad (474)$$

or

$$\operatorname{Tan}\ (G_d - G_b \pm 180°) = \frac{(H_a/H_b)\ \sin\ (G_a - G_b \pm 180°)}{1 + (H_a/H_b)\ \cos\ (G_a - G_b \pm 180°)} \quad (475)$$

Formulas (472) and (474) are to be used when the ratio $H_b/H_a$ does not exceed unity. In this case take argument $r$ of the tables $=H_b/H_a$, and argument $x=(G_b-G_a\pm 180°)$. If the ratio $H_b/H_a$ exceeds unity use formulas (473) and (475) and take argument $r=H_a/H_b$ and argument $x=(G_a-G_b\pm 180°)$. The tabular values will give the ratios and angular differences represented in the first terms of the formulas. Therefore, in order to obtain the amplitude $H_d$, the tabular value from table 41 must be multiplied by $H_a$ if the ratio $H_b/H_a$ does not exceed unity, or by $H_b$ if this ratio does exceed unity. Also to obtain the epoch $G_d$, the tabular value from table 42 must be increased by $G_a$ if the ratio does not exceed unity or by $(G_b\pm 180°)$ if the ratio is greater than unity.

**441.** By the formulas given above separate computations are made for each of the tidal constituents. The values obtained for $H_d$ and $G_d$ are the corresponding amplitudes and Greenwich epochs in an harmonic expression for the continually changing difference in elevation of the water surface at the two entrances to the strait. When only a single time zone is involved, the small $g$'s or modified kappas $(\kappa')$ pertaining to that zone may be substituted for the Greenwich epochs $(G)$ in the formulas. For the prediction of the current, further modifications are necessary in the amplitudes to reduce to velocity units and in the epochs to allow for the lag in the response of the current to the changing difference in water level at the two entrances to the strait.

**442.** Since the velocity of an hydraulic current is theoretically proportional to the square root of the difference in head, we may write

$$(\text{Velocity})^2 = \text{constant}\ (C) \times \text{height difference} \quad (476)$$

If we now let $V$ equal the average velocity of the current at time of strength as determined from actual observations and assume that the corresponding difference in water level is 1.02 times the difference resulting from the principal constituent $M_2$, we may obtain an approximate value for the constant $(C)$ by the formula

$$C = V^2/(1.02 M_2) \quad (477)$$

in which $M_2$ is the amplitude of the constituent $M_2$ in the harmonic expression for the difference in head. The application of the factor $(C)$ to all the constituent amplitudes in this expression has the effect of changing the height units into units representing the square of the velocity of the resulting current.

**443.** The lag in the current is usually determined by a comparison of the times of strengths and slacks from actual observations with preliminary predictions of the corresponding phases based upon the harmonic constants derived by the method just described. This lag expressed in hours is multiplied successively by the speed of each constituent and the result applied to the preliminary epoch for that constituent.

**444.** In order that the magnitude of the constituent emplitudes may be adapted for use with the predicting machine, a scale factor $(S)$ is introduced. This factor, which depends upon the velocity of the current, is selected with the view of obtaining a reasonably large working scale without exceeding the limitations of the predicting machine. The following scale factors are suggested:

Average velocity of current at time of strength:  Scale factor
Less than 0.3 knot ---------------------------------------- 20
From 0.3 to 0.5 knot ------------------------------------- 10
From 0.5 to 1.0 knot -------------------------------------- 5
From 1.0 to 1.5 knots ------------------------------------- 3
From 1.5 to 2.0 knots ------------------------------------- 2
From 2.0 to 3.0 knots ------------------------------------- 1
From 3.0 to 4.0 knots ------------------------------------ 0.5
From 4.0 to 5.0 knots ----------------------------------- 0.25
From 5.0 to 10.0 knots ------------------------------------ 0.1

In practice, the scale factor is usually combined with the factor ($C$) and the product applied to each of the constituent amplitudes in the expression for difference in head.

**445.** Using the harmonic constants, modified in the manner described above, in the predicting machine, the resulting dial readings will represent the square of the current velocity. In order to avoid the necessity of extracting the square root of each individual reading, a square-root scale may be improvised and substituted for the regular height dial on the machine. From a consideration of the construction of this machine, it can be shown that with a scale factor of unity the angular position of a velocity graduation as measured in degrees from the zero point will be $9° \times (\text{velocity})^2$. Thus the 1-knot graduation will be spaced 9° from the zero, the 2-knot graduation at 36°, the 3-knot graduation at 81°, etc. For any scale factor ($S$), the formula for constructing the square-root scale becomes

$$\text{Angular distance from dial zero} = 9° \times S \times (\text{velocity})^2 \qquad (478)$$

**446.** To take account of any nontidal current not attributed to difference in head at the two entrances to the strait, a special graduation of the square-root scale is necessary. Let $V_o$ represent the nontidal current velocity, positive or negative according to whether it sets in the flood or ebb direction, and let $V$ represent the resultant velocity as indicated by any scale graduation, positive or negative according to whether it is flood or ebb. The angular distance of any scale graduation as measured from an initial point, usually marked by an arrow, may then be expressed by the following formula:

$$\text{Angle in degrees} = 9 \times S \times (V - V_o)^2 \qquad (479)$$

The required angle is to be measured to the right or to the left of the initial point according to whether the angle ($V - V_o$) is positive or negative. When setting the predicting machine the velocity pointer must be at the initial point marked by the arrow when the sum of the harmonic terms is zero.

**447.** In the graphic representation of the summation of the harmonic terms by the predicting machine, the scale of the marigram depends upon the marigram gear ratio as well as upon any scale factor which may have been introduced. With a gear ratio of unity, the scale of the marigram is 0.1 inch per unit of machine setting. In the summation for the hydraulic currents, the marigram read by a natural scale would indicate the square of the velocity. A special square-root reading scale for taking the velocities direct from the marigram may be prepared as follows: Let $Y=$ distance of any velocity graduation from zero of scale. Then

$$Y \text{ (in inches)} = 0.1 \times \text{(scale factor)} \times \text{(gear ratio)} \times \text{(velocity)}^2 \quad (480)$$

With the scale factor and gear ratio each unity, 1 knot of velocity would be represented by 0.1 inch on the marigram, 2 knots by 0.4 inch, 3 knots by 0.9 inch, etc. With a scale thus constructed the velocity of the tidal current may be taken directly from the marigram. Any nontidal current which is to be included may afterward be applied.

# TABLES

## EXPLANATION OF TABLES

Table 1. *Fundamental astronomical data.*—This table includes fundamental constants and formulas with references which form the basis for the computation of other tables contained in this volume. Because of the smallness of the solar and lunar parallax no distinction is made between the parallax and its sine. The eccentricity of the earth's orbit and the obliquity of the ecliptic are given for epoch January 1, 1900. The former changes about 0.000042 per century and the latter about 0.013 of a degree in a century. The values given may therefore be considered as applicable to the present century.

The formulas for longitude of both sun and moon are the same as used in the previous edition of this book and are from the work of Simon Newcomb. In a later work by Earnest W. Brown, slightly different values are obtained for the elements of the moon's orbit but the differences may be considered negligible in so far as the tidal work of the present century is concerned. In these formulas it will be noted that $T$ is the number of Julian centuries reckoned from Greenwich mean noon on December 31, 1899, of the Gregorian calendar which corresponds to December 19, 1899, by the Julian calendar. In the application of these formulas to early dates special care must be taken to make suitable allowances for the particular calendar in use at the time. See page 4 for information in regard to calendars.

Table 1 includes the numerical values of the mean longitude of the solar and lunar elements for the beginning of the century years 1600 to 2000 and also the rate of change in these longitudes as of January 1, 1900. As the variations in these rates are very small, they are applicable without material error for all modern times. This table includes also the principal astronomical periods depending on the solar and lunar elements with formulas showing how they are derived. In these formulas the longitude symbol is used to represent its own rate of change according to the unit in which the period is expressed.

Table 2. *Harmonic constituents.*—This table includes the arguments, speeds, and coefficients of the constituent harmonic terms obtained in the development of the tide-producing forces of the moon and sun. They are grouped with reference to the formulas of the text from which they are derived, the long-period constituents first, followed by the diurnal, semidiurnal, and terdiurnal terms. The reference numbers in the first column correspond to the numbered terms of the formulas of the text, the letter $A$ indicating a term in the lunar development and the letter $B$ a term in the solar development. In the second column the usual symbols are given for the principal constituents, parentheses being used when the term only partially represents the constituent.

For an explanation of the constituent argument $(E)$ see page 22. The argument consists of two parts—the $V$ which contains the uniformly changing elements and determines the speed and period of the constituent, and the $u$ which is a function of the moon's node with slow variations and which is treated as a constant for a limited series of observations. Because of the very small change in the element $p_1$ it may for practical purposes also be treated as a constant with a mean value of 282° for the present century.

The constituent speeds are obtained by adding the hourly rates of change in the elements appearing in the $V$ of the arguments. The hour angle ($T$) of the mean sun changes at the rate of 15° per hour. The hourly rate of change for each of the other elements will be found in table 1.

For an explanation of the constituent coefficients (C) see page 24. The coefficients of the solar terms include the solar factor $S'$ (paragraph 118), and coefficients of the lunar terms involving the 4th power of the moon's parallax include the factor $a/c$ (paragraph 108), in order that all terms may be comparable when used with the common basic factor $U$. It is to be noted that in the present system of coefficients for the terms of the principal tide-producing force there is included a factor "2" which was formerly incorporated in the general coefficient. For the terms involving the 4th power of the parallax there is a corresponding factor of "3" in order that all terms may be comparable in respect to the vertical component force.

In general the coefficients have been computed in accordance with the coefficient formulas of the text, but exceptions were made for the evectional and variational constituents $\rho_1$, $\nu_2$, $\lambda_2$, and $\mu_2$, the coefficients of which are based upon computations by Professor J. C. Adams who was associated with Darwin in the investigation of harmonic analysis and who carried the development of the lunar theory to a higher order of precision than is provided in this work. (See pp. 60-61 of the Report of British Association for the Advancement of Science for year 1883.)

The node factor ($f$) is explained on page 25. The last column of table 2 contains references to the formulas for the node factors of the various constituents.

Table 2a. *Shallow-water constituents.*—In this table there are listed the overtides and compound tides which are described on page 47.

Table 3. *Latitude factors.*—This table includes numerical values of the latitude ($Y$) functions which appear in the text as factors in formulas representing component tidal forces and the equilibrium height of the tide. The combination symbol at the head of each column is taken to suggest the formula to which it applies. Thus, the letters $v$, $s$, and $w$ refer respectively to the vertical, south, and west components of the force, the letter $v$ being applicable also to the formulas for the equilibrium height of the tide which have the same latitude factors as the corresponding terms in the vertical component of the force. The first numeral "3" or "4" indicates whether the formula is from the development of the principal force involving the cube of the parallax or from the development involving the 4th power of the parallax of the tide-producing body. The last digit "0," "1," "2," or "3" refers to the species of the constituents and indicates whether they are long-period, diurnal, semidiurnal, or terdiurnal. In several cases the same latitude factor is applicable to a number of different groups as indicated at the head of the column in the table.

HARMONIC ANALYSIS AND PREDICTION OF TIDES    155

The following formulas were used in computing the latitude factors. The maximum value (irrespective of sign) with corresponding latitude is also given for each function.

$Y_{v30} = (1/2 - 3/2 \sin^2 Y)$ _ _ _ _ _ _ maximum $-1$ when $Y = \pm 90°$.
$Y_{v31} = \sin 2Y$ _ _ _ _ _ _ _ _ _ _ _ _ _ maximum $\pm 1$ when $Y = \pm 45°$.
$Y_{s30}$, $Y_{s32}$, and $Y_{w42}$, same as $Y_{v31}$
$Y_{v32} = \cos^2 Y$ _ _ _ _ _ _ _ _ _ _ _ _ maximum $+1$ when $Y = 0$.
$Y_{w43}$ same as $Y_{v32}$
$Y_{s31} = \cos 2Y$ _ _ _ _ _ _ _ _ _ _ _ _ maximum $\pm 1$ when $Y = 0$ or $\pm 90°$.
$Y_{w31} = \sin Y$ _ _ _ _ _ _ _ _ _ _ _ _ _ maximum $\pm 1$ when $Y = \pm 90°$.
$Y_{w32} = \cos Y$ _ _ _ _ _ _ _ _ _ _ _ _ maximum $+1$ when $Y = 0$.
$Y_{v40} = \sin Y (\cos^2 Y - 2/5)$ _ _ _ maximum $\mp 0.4$ when $Y = \pm 90°$.
$Y_{v41} = \cos Y (\cos^2 Y - 4/5)$ _ _ _ maximum $-0.2754$ when $Y = \pm 58.91°$.
$Y_{s40}$ same as $Y_{v41}$
$Y_{v42} = \sin Y \cos^2 Y$ _ _ _ _ _ _ _ _ maximum $\pm 0.3849$ when $Y = \pm 35.26°$.
$Y_{s43}$ same as $Y_{v42}$
$Y_{v43} = \cos^3 Y$ _ _ _ _ _ _ _ _ _ _ _ maximum $+1$ when $Y = 0$.
$Y_{s41} = \sin Y (\cos^2 Y - 4/15)$ _ _ maximum $\mp 0.2667$ when $Y = \pm 90°$.
$Y_{s42} = \cos Y (\cos^2 Y - 2/3)$ _ _ _ maximum $-0.2095$ when $Y = \pm 61.87°$.
$Y_{w41} = (\cos^2 Y - 4/5)$ _ _ _ _ _ _ _ _ maximum $-0.8$ when $Y = \pm 90°$.

Table 4. *Mean longitude of lunar and solar elements.*—This table contains the mean longitude of the moon ($s$), of the lunar perigee ($p$), of the sun ($h$), of the solar perigee ($p_1$), and of the moon's ascending node ($N$), for January 1, 0 hour, Greenwich mean civil time, for each year from 1800 to 2000, the dates referring to the Gregorian calendar.

These values are readily derived from table 1, the rate of change in the mean longitude of the elements for the epoch January 1, 1900, being applicable without material error to any time within the two centuries 1800 to 2000 covered by table 4. The same rate of change may also be used, without introducing any errors of practical importance, to extend table 4 to dates beyond these limits. In extending the table, care should be taken to distinguish between the common and leap years, and for the earlier dates due consideration should be given to the kind of calendar in use. (See p. 4 for discussion of calendars.) It will be noted that each Julian century contains 36,525 days, while the common Gregorian century contains only 36,524 days; with an additional day every fourth century.

Table 5. *Differences to adapt table 4 to any month, day, and hour.*— These differences are derived from the daily and hourly rate of change of the elements as given in table 1, multiples of 360° being rejected when they occur. The table is prepared especially for common years, but is applicable to leap years by increasing the given date by one day if it is between March 1 and December 31, inclusive. The correction for the hour of the day refers to the Greenwich hour, and if the hour for which the elements are desired is expressed in another kind of time the equivalent Greenwich hour must be used for the table.

Table 6. *Values of $I$, $\nu$, $\xi$, $\nu'$, and $2\nu''$ for each degree of $N$.*—Referring to figure 1 (page 6), note that by construction arc $\Omega \Upsilon'$ equals arc $\Omega \Upsilon$. Then in the spherical triangle $\Omega \Upsilon A$, the three sides are $N$, $\nu$, and $(N-\xi)$, and the opposite angles are respectively $(180°-I)$, $i$, and $\omega$.

Therefore we have the following relations which may be used in computing the values of $I$, $\nu$, and $\xi$ in the table:

$$\cos I = \cos i \cos \omega - \sin i \sin \omega \cos N$$
$$= 0.91370 - 0.03569 \cos N$$

$$\tan \tfrac{1}{2}(N-\xi+\nu) = \frac{\cos \tfrac{1}{2}(\omega-i)}{\cos \tfrac{1}{2}(\omega+i)} \tan \tfrac{1}{2}N = 1.01883 \tan \tfrac{1}{2}N$$

$$\tan \tfrac{1}{2}(N-\xi-\nu) = \frac{\sin \tfrac{1}{2}(\omega-i)}{\sin \tfrac{1}{2}(\omega+i)} \tan \tfrac{1}{2}N = 0.64412 \tan \tfrac{1}{2}N$$

For the computation of $\nu'$ and $2\nu''$, formulas (224) and (232) on pages 45–46 may be used. The tabular values themselves were taken from the preceding edition of this work where they were based upon formulas differing slightly from those given here but any differences arising from the use of the latter may be considered as negligible.

Table 7. *Values of log $R_a$ for amplitude of constituent $L_2$.*—Values in this table are based upon formula (213) on page 44.

Table 8. *Values of $R$ for argument of constituent $L_2$.*—Values in this table are derived from formula (214) on page 44.

Table 9. *Values of log $Q_a$ for amplitude of constituent $M_1$.*—Values in this table are based upon formula (197) on page 41.

Table 10. *Values of $Q$ for argument of constituent $M_1$.*—Values in this table are derived from formula (203) on page 42.

Table 11. *Values of $u$ for equilibrium arguments.*—This table is based upon the $u$-formulas in table 2 and includes values for the principal lunar constituents for each degree of $N$. The $u$'s of $L_2$ and $M_1$, which are functions of both $N$ and $P$ are given separately in table 13 for the years 1900 to 2000.

Table 12. *Log factor $F$ for each degree of $I$.*—The factor $F$ is the reciprocal of the node factor $f$ to which references are given in table 2. The values in table 12 are based upon the formulas for these factors and are given for all the lunar constituents used in the tide-predicting machine, excepting values for $L_2$ and $M_1$ which are given separately in table 13.

Table 13. *Values of $u$ and log $F$ for $L_2$ and $M_1$.*—From a comparison of the $u$'s of constituents $L_2$, $M_1$, and $M_2$ in table 2, it will be noted that the following relations exist:

$$u \text{ of } L_2 = (u \text{ of } M_2) - R$$
$$u \text{ of } M_1 = \tfrac{1}{2}(u \text{ of } M_2) + Q$$

Also, the following relations may be derived from formula (215) on page 44 and formula (207) on page 43 since the factor $F$ is the reciprocal of the node factor $f$:

$$\log F(L_2) = \log F(M_2) + \log R_a$$
$$\log F(M_1) = \log F(O_1) + \log Q_a$$

The values for table 13 were computed by the above formulas, the component parts being taken from tables 7 to 12, inclusive. The values for log $F(M_1)$ in this table are in accord with Darwin's original formula from which a factor of approximately 1.5 was inadvertently omitted (see page 43).

Table 14. *Node factor f for middle of each year 1850 to 1999.*—The factor $f$ is the reciprocal of factor $F$. The values for the years 1850 to 1950 were taken directly from the Manual of Tides, by R. A. Harris, and the values for 1951 to 1999 were derived from tables 12 and 13.

Table 15. *Equilibrium argument $(V_o+u)$ for beginning of each year 1850 to 2000.*—The equilibrium argument is discussed on page 22. The tabular values are computed by the formulas for the argument in table 2, the $V_o$ referring to the value of $V$ on January 1, 0 hour Greenwich mean civil time, for each year, and the $u$ referring to the middle of the same calendar year; that is, Greenwich noon on July 2 in common years and the preceding midnight in leap years. The value of the $T$ of the formulas is 180° for each midnight, and the values of the other elements for the $V$ may be obtained from table 4. The $u$ of the argument may be obtained from tables 11 and 13 after the value of $N$ has been determined for the middle of each year from tables 4 and 5. In constructing table 15 the values for the years 1850 to 1950 were taken directly from the Manual of Tides, by R. A. Harris, and the values for the years 1951 to 2000 were computed as indicated above.

Tables 16, 17, and 18.—These tables give the differences to adapt table 15 to any month, day, and hour, and are computed from the hourly speeds of the constituents as given in table 2. The differences refer to the uniformly varying portion $V$ of the argument, it being assumed that for practical purposes the portion $u$ is constant for the entire year.

The approximate Greenwich $(V_o+u)$ for any desired Greenwich hour may be obtained by applying the appropriate differences from tables 16, 17, and 18 to the value for the first of January of the required year, as given in table 15. To refer this Greenwich $(V_o+u)$ to any local meridian, it is necessary to apply a further correction equal to the product of the longitude in degrees by the subscript of the constituent, which represents the number of periods in a constituent day. West longitude is to be considered as positive and east longitude as negative, and the subscripts of the long-period constituents are to be taken as zero. This correction is to be subtracted.

The $(V_o+u)$ obtained as above will, in general, differ by a small amount from the value as computed by Form 244, because in the former case the $u$ refers to the middle of the calendar year and in the latter case to the middle of the series of observations.

Table 19. *Products for Form 194.*—This is a multiplication table especially adapted for use with Form 194, the multipliers being the sines of multiples of 15°.

Table 20. *Augmenting factors.*—A discussion of augmenting factors is given on page 71. The tabular values for the short-period constituents are obtained by formulas (308) and (309) on page 72, and those for the long-period constituents by formulas (403) and (404) on page 92. For constituents $S_1$, $S_2$, etc. the augmenting factor is unity.

Tables 21 to 26.—These tables represent perturbations in $K_1$ and $S_2$ due to other constituents of nearly equal speeds. They are based upon formulas (359) to (364), inclusive, on page 83.

Table 27. *Critical logarithms for Form 245.*—This table was designed for quickly obtaining the natural numbers to three decimal places for column (3) of Form 245 from the logarithms of column (2). The logarithms are given for every change of 0.001 in the natural number. Each logarithm given in this table is derived from the natural number that is 0.0005 less than the tabular number to which it applies. Intermediate logarithms, therefore, apply to the same natural number as the preceding tabular logarithm. For example, logarithms less than 6.6990 apply to the natural number 0.000 and logarithms from 6.6990 to 7.1760 apply to the natural number 0.001, etc.

Table 28. *Constituent speed differences.*—The constituent speeds as given in table 2 were used in the computation of this table.

Table 29. *Elimination factors.*—These tables provide for certain constant factors in formulas (389) and (390). Separate tables for each length of series and different values for each term of the formulas are required. The tabular values are arranged in groups of three, determined as follows:

First value = logarithm of $\dfrac{180}{\pi} \dfrac{\sin \frac{1}{2}(b-a)\tau}{\frac{1}{2}(b-a)\tau}$.

Second value = natural number $\dfrac{180}{\pi} \dfrac{\sin \frac{1}{2}(b-a)\tau}{\frac{1}{2}(b-a)\tau}$ always taken as positive.

Third value = $\frac{1}{2}(b-a)\tau$, if $\dfrac{\sin \frac{1}{2}(b-a)\tau}{\frac{1}{2}(b-a)\tau}$ is positive,

or $\frac{1}{2}(b-a)\tau \pm 180$, if $\dfrac{\sin \frac{1}{2}(b-a)\tau}{\frac{1}{2}(b-a)\tau}$ is negative.

Table 30. *Products for Form 245.*—This table is designed for obtaining the products for columns (6) and (7) of Form 245.

Table 31. *For construction of primary stencils.*—This table gives the differences to be applied to the solar hours in order to obtain the constituent hours to which they most nearly coincide. Each difference applies to several successive solar hours, but for brevity only the first solar hour of each group to which the difference applies is given in the table.

An asterisk (*) indicates that the solar hour so marked is to be used twice or rejected according to whether the constituent speed is greater or less than $15p$, when in the summation it is desired to assign a single solar hour to each successive constituent hour. For the usual summations in which each solar hour height is assigned to the nearest constituent hour no attention need be given to the asterisk.

The table is computed by substituting successive integral values for $d$ in formula (243) and reducing the resulting solar hour of series ($shs$) to the corresponding day and hour. The solar hour to be tabulated is the integral hour that immediately follows the value of ($shs$) of the formula. If the fractional part of ($shs$) exceeds 0.5, the tabular solar hour is marked by an asterisk (*). The successive values of $d$, although used positively in formula (243), are to be considered as negative in the application of the table when the speed of the constitutent is less than $15p$. When the constituent speed is greater than $15p$, the difference is to be taken as positive. All tabular differences are brought within the limits +24 hours and −24 hours by rejecting multiples of ±24 hours when necessary, and for convenience in use all differences are given in both positive and negative forms.

The following example will illustrate the use of the table: To find constituent 2Q hours corresponding to solar hours 12 to 23 on 16th day of series. By the table we see that solar hour 12 of the 16th day of series is within the group beginning on solar hour 8 of the same day with the tabular difference of +19 or −5 hours, and that the difference changes by −1 hour on solar hours 15 and 21, the latter being marked by an asterisk. Applying the differences indicated, we have for these solar hours on the 16th day of series:

Solar hour____ 12, 13, 14*, 15, 16, 17, 18, 19, 20, 21*, 22, 23
Difference____ −5  −5  −5  −6  −6  −6  −6  −6  −6  −7  −7  −7

Constituent
2Q hour____  7,  8,  9*,  9, 10, 11, 12, 13, 14, 14*, 15, 16

In the results it will be noted that the constituent hours 9 and 14 are each represented by two solar hours. If it should be desired to limit the representation to a single solar hour each, the hours marked with the asterisk should be rejected.

To find constituent OO hours corresponding to solar hours 0 to 18 on the 22d day of series. The 0 hour of the 22d day is in the group beginning on solar hour 14 of the preceding day with the tabular difference of +14 or −10 hours, and changes of +1 hour in the differences occur on solar hours 3 and 17 of the 22d day. It will be noted that the hour 3 is marked by an asterisk. Applying the differences from the table as indicated, we have for the 22d day of series:

Solar hours__ 0,   1,   2,   3*,  4,   5,   6,   7,   8,   9,  10,  11,  12,  13,  14,  15,  16*, 17,  18
Differences__+14, +14, +14, +15, +15, +15, +15, +15, +15, −9,  −9,  −9,  −9,  −9,  −9,  −9,  −9,  −8,  −8

Constituent
OO hours__ 14, 15, 16, 18, 19, 20, 21, 22, 23, 0, 1, 2, 3, 4, 5, 6, 7, 9, 10

In the results it will be noted that constituent hours 17 and 8 are missing. If it is desired to have each of these hours represented also, the solar hours marked by asterisks will be used again. In this table the constituents have been arranged in accordance with the length of the constituent day.

Table 32. *Divisors for primary stencil sums.*—This table contains the number of solar hourly heights included in each constituent hour group for each of the standard length of series when all the hourly heights have been used in the summation.

Table 33. *For construction of secondary stencils.*—Constituent $A$ is the constituent for which the original primary summations have been made, and constituent $B$ is the constituent for which the sums are to be derived by the secondary stencils. The "Page" refers to the page of the original tabulations of the hourly heights in Form 362. The differences in this table were calculated by formula (252), and the corresponding "Constituent $A$ hours" from formula (250), $m$ being assigned successive values from 1 to 24 for each page of record. Special allowance was made for page 53 of the record to take account of the fact that in a 369-day series this page includes only 5 days of record. The sign of the difference is given at the top of the column. For K-P and R-T the positive sign is to be used for constituents K and R and the negative sign for constituents P and T.

For brevity all the 24 constituent hours for every page of record are not directly represented in the table. The difference for the omitted hours for any page should be taken numerically one greater than the difference for the given hours on that page. For an example, take the hours for page 2 for constituent OO as derived from constituent J. According to the table the difference for the constituent hours 10 to 3, inclusive, is 9 hours; therefore the difference for the omitted hours 4 to 9, inclusive, should be taken as 10 hours. For constituent 2Q as derived from constituent O the three differences usually required for each page are given in full.

The use of the table may be illustrated from the example above, as follows:

Page 2—

| J-hours | 0, | 1, | 2, | 3, | 4, | 5, | 6, | 7, | 8, | 9, | 10, | 11 |
|---|---|---|---|---|---|---|---|---|---|---|---|---|
| Difference | +9, | 9, | 9, | 9, | 10, | 10, | 10, | 10, | 10, | 10, | 9, | 9 |
| OO-hours | 9, | 10, | 11, | 12, | 14, | 15, | 16, | 17, | 18, | 19, | 19, | 20 |
| J-hours | 12, | 13, | 14, | 15, | 16, | 17, | 18, | 19, | 20, | 21, | 22, | 23 |
| Difference | +9, | 9, | 9, | 9, | 9, | 9, | 9, | 9, | 9, | 9, | 9, | 9 |
| OO-hours | 21, | 22, | 23, | 0, | 1, | 2, | 3, | 4, | 5, | 6, | 7, | 8 |

The period 24 hours should be added or subtracted when necessary in order that the resulting constituent hours may be between 0 and 23.

Table 34. *For summation of long-period constituents.*—This table is designed to show the assignment of the daily page sums of the hourly heights to the constituent divisions to which they most nearly correspond. The table is based upon formula (395). The constituent division to which each day of series is assigned is given in the left-hand column. For Mf, MSf, and Mm there will frequently occur two consecutive days which are to be assigned to the same constituent division. In such cases the day which most nearly corresponds to the constituent division is the only one given in the table, and this is marked by an asterisk (*). The missing day, whether it precedes or follows the one marked by the asterisk, is to be assigned to the same constituent division. For Sa a number of consecutive days of series are assigned to each constituent division. In the table there are given the first and last days of each group.

Table 35. *Products aS/15 for Form 444.*—This table contains the products of constituent speeds and time meridian longitudes for formula (466) which is used in obtaining values of $(\kappa'-\kappa)$ for column A of Form 444.

Table 36. *Angle differences for Form 445.*—This table gives the differences for obtaining and checking the dial settings for February 1 and December 31, as entered in Form 445. The differences are derived from tables 16 and 17.

Table 37. *Coast and Geodetic Survey tide-predicting machine No. 2— General gears.*—This table gives the details of the general gearing from the hand-operating crank to the main vertical shafts, together with the details of the gearing in the front section or dial case. In this table the gears and shafts are each numbered consecutively for convenience of reference, the gears being designated by the letter $G$ and the shafts by the letter $S$. In the second column are given the face of each bevel or spur gear and the diameter of each shaft. The next two columns contain the number of teeth and pitch of each bevel and spur gear. The pitch is the number of teeth per inch of diameter of the gear. The worm screw is equivalent to a gear of one tooth, as it requires a complete revolution of the screw to move the engaged wheel one tooth forward. The period of rotation of each shaft and gear is relative and refers to the time as indicated on the face of the machine, which for convenience is called dial time.

Table 38. *Coast and Geodetic Survey tide-predicting machine No. 2—Constituent gears.*—This table contains the details of the gearing from the main vertical shafts to the individual constituent cranks. Column I gives the number of teeth in the bevel gear on the main vertical shaft; column II, the number of teeth in the gear on the intermediate shaft that meshes with the gear on the vertical shaft; column III, the number of teeth in the gear on the intermediate shaft that meshes with the gear on the constituent crank shaft; and column IV, the number of teeth in the gear on the crank shaft.

For the long-period constituents the worm gear is taken as the equivalent of one tooth. For each of these constituents there is a short secondary shaft on which sliding gears are mounted, but the extra gears do not affect the speed of any of the crank shafts except that for constituent Sa in which case a ratio of 1:2 is introduced.

The crank-shaft speed per dial hour for each constituent is equal to $30° \times \frac{\text{column I}}{\text{column II}} \times \frac{\text{column III}}{\text{column IV}}$. For constituent Sa the product of both values appearing in each of the columns II and III is to be taken as the value for the column. The column of "Gear speed per dial hour" contains the speeds as computed by the above formula.

For comparison the table contains also the theoretical speed of each of the constituents and the accumulated error per year due to the difference between the theoretical and the gear speeds.

For convenience of reference the table includes also the maximum amplitude settings of the constituent cranks.

Table 39. *Synodic periods of constituents.*—This table is derived from table 28, the period represented by 360° being divided by the speed difference and the results reduced to days.

Table 40. *Day of year corresponding to any date.*—This table is convenient for obtaining the difference between any two dates and also in finding the middle of any series.

Table 41. *Values of h in formula* $h=(1+r^2+2r \cos x)^{\frac{1}{2}}$.—This table may be used with formulas (472) and (473) on page 149 to obtain constituent amplitudes for the prediction of hydraulic currents.

Table 42. *Values of k in formula* $k=\tan^{-1} \frac{r \sin x}{1+r \cos x}$.—This table may be used with formulas (474) and (475) on pages 149–150 to obtain constituent epochs for the prediction of hydraulic currents.

Table 1.—*Fundamental astronomical data*

| | |
|---|---|
| Mean distance, earth to sun | 92,897,416 miles [a] |
| Mean distance, earth to moon | 238,857 miles [a] |
| Equatorial radius of earth (Hayford's Spheroid of 1909) | 3,963.34 miles [a] |
| Polar radius of earth (Hayford's Spheroid of 1909) | 3,949.99 miles [a] |
| Mean radius of earth ($a$), (Intern. Ell.) 6,371,269 meters [b] | = 20,903,071 feet |
| | = 3,958.91 miles |
| Solar parallax (Paris Conference) 8.80″ [a] = | 0.000,042,66 radian |
| Lunar equatorial horizontal parallax (Brown) 57′ 2.70″ [a] = | 0.016,59 radian |
| Mean solar parallax in respect to mean radius ($a/c_1$) | 0.000,042,61 radian |
| Mean lunar parallax in respect to mean radius ($a/c$) | 0.016,57 radian |
| Eccentricity of earth's orbit ($e_1$), epoch Jan. 1, 1900 | 0.016,75 [c] |
| Eccentricity of moon's orbit ($e$) | 0.054,90 [d] |
| Obliquity of the ecliptic ($\omega$), epoch Jan. 1, 1900 | |
| 23° 27′ 8.26″ [c] = | 23.452° |
| Inclination of moon's orbit to plane of ecliptic ($i$) | |
| 5° 08′ 43.3546″ [d] = | 5.145° |
| Ratio of mass of sun to combined mass of earth and moon (Sitter) | 327,932 [b] |
| Ratio of mass of earth to mass of moon (Hinks) | 81.53 [b] |
| Mass of sun/mass of earth ($S/E$) | 331,954 |
| Mass of moon/mass of earth ($M/E$) | 0.012,27 |
| Solar coefficient $U_1 = (S/E)(a/c_1)^3$ | $.2569 \times 10^{-7}$ |
| Basic factor $U = (M/E)(a/c)^3$ | $.5582 \times 10^{-7}$ |
| Solar factor $S' = U_1/U$ | 0.4602 |

In the following formulas for longitude, $T$ represents the number of Julian centuries (36525 days) reckoned from Greenwich mean noon, December 31, 1899 (Gregorian Calendar).

Mean longitude of sun ($h$)
= 279° 41′ 48.04″ + 129,602,768.13″ $T$ + 1.089″ $T^2$ [c]

Longitude of solar perigee ($p_1$)
= 281° 13′ 15.0″ + 6,189.03″ $T$ + 1.63″ $T^2$ + 0.012″ $T^3$ [c]

Mean longitude of moon ($s$)
= 270° 26′ 14.72″ + (1336 rev. + 1,108,411.20″) $T$ + 9.09″ $T^2$ + 0.006,8″ $T^3$ [c]

Longitude of lunar perigee ($p$)
= 334° 19′ 40.87″ + (11 rev. + 392,515.94″) $T$ − 37.24″ $T^2$ − 0.045″ $T^3$ [c]

Longitude of moon's node ($N$)
= 259° 10′ 57.12″ − (5 rev. + 482,912.63″) $T$ + 7.58″ $T^2$ + 0.008″ $T^3$ [c]

Ratio of mean motion of sun to that of moon ($m$) .......... 0.074,804

[a] American Ephemeris and Nautical Almanac for year 1940, p. xx.
[b] Table of astronomical constants by W. de Sitter, published in Bulletin of the Astronomical Institutes of the Netherlands, Vol. VIII, No. 307, July 8, 1938, pp. 230–231.
[c] Astronomical Papers for the American Ephemeris, by Simon Newcomb: Vol. VI, pp. 9–10, and Vol. IX, pt. 1, p. 224.
[d] The Solar Parallax and Related Constants, by William Harkness, p. 140.

HARMONIC ANALYSIS AND PREDICTION OF TIDES 163

Table 1.—*Fundamental astronomical data*—Continued

MEAN LONGITUDE OF SOLAR AND LUNAR ELEMENTS FOR CENTURY YEARS

| Epoch, Gregorian calendar Greenwich mean civil time | Sun $h$ | Solar perigee $p_1$ | Moon $s$ | Lunar perigee $p$ | Moon's node $N$ |
|---|---|---|---|---|---|
| | ° | ° | ° | ° | ° |
| 1600, Jan. 1, 0 hour | 279.857 | 276.067 | 99.725 | 7.417 | 301.496 |
| 1700, Jan. 1, 0 hour | 280.624 | 277.784 | 47.604 | 116.501 | 167.343 |
| 1800, Jan. 1, 0 hour | 280.407 | 279.502 | 342.313 | 225.453 | 33.248 |
| 1900, Jan. 1, 0 hour | 280.190 | 281.221 | 277.026 | 334.384 | 259.156 |
| 2000, Jan. 1, 0 hour | 279.973 | 282.940 | 211.744 | 83.294 | 125.069 |

RATE OF CHANGE IN MEAN LONGITUDE OF SOLAR AND LUNAR ELEMENTS (EPOCH, JAN. 1, 1900)

| Elements | Per Julian century (36525 days) | Per common year (365 days) | Per solar day | Per solar hour |
|---|---|---|---|---|
| | ° | ° | ° | ° |
| Sun ($h$) | $100r+$ 0.769 | 359.761,28 | 0.985,647,3 | 0.041,068,64 |
| Solar perigee ($p_1$) | 1.719 | 0.017,18 | 0.000,047,1 | 0.000,001,96 |
| Moon ($s$) | $1336r+307.892$ | $13r+129.384,82$ | 13.176,396.8 | 0.549,016,53 |
| Lunar perigee ($p$) | $11r+109.032$ | 40.662,47 | 0.111,404,0 | 0,004,641.83 |
| Moon's node ($N$) | $-5r-134.142$ | $-19.328,19$ | $-0.052,953.9$ | $-0,002,206,41$ |

MEAN ASTRONOMICAL PERIODS
(*Symbols refer to rate of change in mean longitude*)

                                                                                                                      *Solar days*

Sidereal day, $360°/(360°+h)$ _____ 0.997,270
Lunar day, $360°/(360°+h-s)$ _____ 1.035,050

Nodical month, $360°/(s-N)$ _____ 27.212,220
Tropical month, $360°/s$ _____ 27.321,582
Anomalistic month, $360°/(s-p)$ _____ 27.554,550
Synodical month, $360°/(s-h)$ _____ 29.530,588
Moon's evectional period, $360°/(s-2h+p)$ _____ 31,811,939

Eclipse year, $360°/(h-N)$ _____ 346.620,0
Tropical year, $360°/h$ _____ 365.242,2
Anomalistic year, $360°/(h-p_1)$ _____ 365.259,6

Common year _____ 365.000,0
Mean Gregorian year _____ 365.242,5
Mean Julian year _____ 365.250,0
Leap year _____ 366.000,0

Evectional period in moon's parallax, $360°/(h-p)$ _____ 411.784,7

Revolution of lunar perigee, $360°/p$ _____ 8.85 Julian years
Revolution of moon's node, $360°/N$ _____ 18.61 Julian years

Revolution of solar perigee, $360°/p_1$ _____ 209 Julian centuries

## Table 2.—*Harmonic constituents*

| Ref. No. | Symbol | Argument (E) V | Argument (E) u | Speed per solar hour | Coefficient (C) | Factor-f formula |
|---|---|---|---|---|---|---|
| | | **LUNAR LONG-PERIOD TERMS, FORMULA (62)** | | | | |
| $A_1$ | | Zero (permanent term) | | zero° | 0.5044 | (73) |
| $A_2$ | Mm | $s-p$ | zero | 0.544,374,7 | 0.0827 | (73) |
| $A_3$ | | $2s-2p$ | zero | 1.088,749,4 | 0.0068 | (73) |
| $A_4$ | | $s-2h+p$ | zero | 0.471,521,1 | 0.0116 | (73) |
| $A_5$ | MSf | $2s-2h$ | zero | 1.015,895,8 | 0.0084 | (73) |
| $A_6$ | Mf | $2s$ | $-2\xi$ | 1.098,033,1 | 0.1566 | (74) |
| $A_7$ | | $3s-p$ | $-2\xi$ | 1.642,407,8 | 0.0303 | (74) |
| $A_8$ | | $s+p+180°$ | $-2\xi$ | 0.553,658,4 | 0.0043 | (74) |
| $A_9$ | | $4s-2p$ | $-2\xi$ | 2.186,782,5 | 0.0040 | (74) |
| $A_{10}$ | | $3s-2h+p$ | $-2\xi$ | 1.569,554,3 | 0.0043 | (74) |
| $A_{11}$ | | $s+2h-p+180°$ | $-2\xi$ | 0.626,512,0 | 0.0006 | (74) |
| $A_{12}$ | | $4s-2h$ | $-2\xi$ | 2.113,928,8 | 0.0025 | (74) |
| $A_{13}$ | | $2h$ | $-2\xi$ | 0.082,137,3 | 0.0001 | (74) |
| | | **LUNAR LONG-PERIOD TERMS, FORMULA (137)** | | | | |
| $A_{64}$ | | $s-90°$ | $-\xi$ | 0.549,016,5 | 0.0399 | (141) |
| $A_{65}$ | | $2s-p-90°$ | $-\xi$ | 1.093,391,2 | 0.0065 | (141) |
| $A_{66}$ | | $p-90°$ | $-\xi$ | 0.004,641,8 | 0.0022 | (141) |
| $A_{67}$ | | $3s-90°$ | $-3\xi$ | 1.647,049,6 | 0.0032 | (142) |
| $A_{68}$ | | $4s-p-90°$ | $-3\xi$ | 2.191,424,3 | 0.0009 | (142) |
| | | **SOLAR LONG-PERIOD TERMS, FORMULA (185)** | | | | |
| $B_1$ | | Zero (permanent term) | | zero | 0.2340 | unity |
| $B_2$ | | $h-p_1$ | zero | 0.041,066,7 | 0.0118 | unity |
| $B_3$ | | $2h-2p_1$ | zero | 0.082,133,4 | 0.0003 | unity |
| $B_6$ | Ssa | $2h$ | zero | 0.082,137,3 | 0.0728 | unity |
| $B_7$ | | $3h-p_1$ | zero | 0.123,204,0 | 0.0043 | unity |
| $B_8$ | | $h+p_1+180°$ | zero | 0.041,070,6 | 0.0006 | unity |
| $B_9$ | | $4h-2p_1$ | zero | 0.164,270,6 | 0.0002 | unity |
| | | **SOLAR LONG-PERIOD TERM, PARAGRAPH 119** | | | | |
| $B_{64}$ | Sa | $h$ | zero | 0.041,068,6 | | unity |
| | | **LUNAR DIURNAL TERMS, FORMULA (63)** | | | | |
| $A_{14}$ | $O_1$ | $T-2s+h+90°$ | $+2\xi-\nu$ | 13.943,035,6 | 0.3771 | (75) |
| $A_{15}$ | $Q_1$ | $T-3s+h+p+90°$ | $+2\xi-\nu$ | 13.398,660,9 | 0.0730 | (75) |
| $A_{16}$ | $(M_1)$ | $T-s+h-p+90°$ | $+2\xi-\nu$ | 14.487,410,3 | 0.0104 | (75) |
| $A_{17}$ | $2Q_1$ | $T-4s+h+2p+90°$ | $+2\xi-\nu$ | 12.854,286,2 | 0.0097 | (75) |
| $A_{18}$ | $\rho_1$ | $T-3s+3h-p+90°$ | $+2\xi-\nu$ | 13.471,514,5 | 0.0142 | (75) |
| $A_{19}$ | | $T-s-h+p-90°$ | $+2\xi-\nu$ | 14.414,556,7 | 0.0015 | (75) |
| $A_{20}$ | $\sigma_1$ | $T-4s+3h+90°$ | $+2\xi-\nu$ | 12.927,139,8 | 0.0061 | (75) |
| $A_{21}$ | | $T-h+90°$ | $+2\xi-\nu$ | 14.958,931,4 | 0.0003 | (75) |
| $A_{22}$ | $(K_1)$ | $T+h-90°$ | $-\nu$ | 15.041,068,6 | 0.3623 | (76) |
| $A_{23}$ | $(M_1)$ | $T-s+h+p-90°$ | $-\nu$ | 14.496,693,9 | 0.0297 | (76) |
| $A_{24}$ | $J_1$ | $T+s-h+p-90°$ | $-\nu$ | 15.585,443,3 | 0.0297 | (76) |
| $A_{25}$ | | $T-2s+h+2p-90°$ | $-\nu$ | 13.952,319,2 | 0.0024 | (76) |
| $A_{26}$ | | $T+2s+h-2p-90°$ | $-\nu$ | 16.129,818,0 | 0.0024 | (76) |
| $A_{27}$ | $\chi_1$ | $T-s+3h-p-90°$ | $-\nu$ | 14.569,547,6 | 0.0042 | (76) |
| $A_{28}$ | $\theta_1$ | $T+s-h+p-90°$ | $-\nu$ | 15.512,589,7 | 0.0042 | (76) |
| | | **LUNAR DIURNAL TERMS, FORMULA (63)** | | | | |
| $A_{29}$ | $MP_1$ | $T-2s+3h-90°$ | $-\nu$ | 14.025,172,9 | 0.0030 | (76) |
| $A_{30}$ | $SO_1$ | $T+2s-h-90°$ | $-\nu$ | 16.056,964,4 | 0.0030 | (76) |
| $A_{31}$ | $OO_1$ | $T+2s+h-90°$ | $-2\xi-\nu$ | 16.139,101,7 | 0.0163 | (77) |
| $A_{32}$ | $KQ_1$ | $T+3s-h-p-90°$ | $-2\xi-\nu$ | 16.683,476,4 | 0.0032 | (77) |
| $A_{33}$ | | $T+s+h+p-90°$ | $-2\xi-\nu$ | 15.594,727,0 | 0.0005 | (77) |
| $A_{34}$ | | $T+4s+h-2p-90°$ | $-2\xi-\nu$ | 17.227,851,1 | 0.0004 | (77) |
| $A_{35}$ | | $T+3s-h+p-90°$ | $-2\xi-\nu$ | 16.610,622,8 | 0.0004 | (77) |
| $A_{36}$ | | $T+s+3h-p+90°$ | $-2\xi-\nu$ | 15.667,580,6 | 0.0001 | (77) |
| $A_{37}$ | | $T+4s-h-90°$ | $-2\xi-\nu$ | 17.154,997,5 | 0.0003 | (77) |
| $A_{38}$ | | $T+3h-90°$ | $-2\xi-\nu$ | 15.123,205,9 | | (77) |

# HARMONIC ANALYSIS AND PREDICTION OF TIDES

Table 2.—*Harmonic constituents*—Continued

| Ref. No. | Symbol | Argument (E) V | Argument (E) u | Speed per solar hour | Coefficient (C) | Factor-f formula |
|---|---|---|---|---|---|---|
| | | LUNAR DIURNAL TERMS, FORMULA (138) | | | | |
| $A_{69}$ | | $T-3s+h$ | $+3\xi-\nu$ | 13.394,019,0 | 0.0116 | (143) |
| $A_{70}$ | | $T-4s+h+p$ | $+3\xi-\nu$ | 12.849,644,4 | 0.0032 | (143) |
| $A_{71}$ | $(M_1)$ | $T-s+h$ | $+\xi-\nu$ | 14.492,052,1 | 0.0367 | (144) |
| $A_{72}$ | | $T-2s+h+p$ | $+\xi-\nu$ | 13.947,677,4 | 0.0060 | (144) |
| $A_{73}$ | | $T+h-p$ | $+\xi-\nu$ | 15.036,426,8 | 0.0020 | (144) |
| $A_{74}$ | | $T+s+h$ | $-\xi-\nu$ | 15.590,085,2 | 0.0134 | (145) |
| $A_{75}$ | | $T+2s+h-p$ | $-\xi-\nu$ | 16.134,459,9 | 0.0022 | (145) |
| | | SOLAR DIURNAL TERMS, FORMULA (186) | | | | |
| $B_{14}$ | $P_1$ | $T-h+90°$ | zero | 14.958,931,4 | 0.1755 | unity |
| $B_{15}$ | $\pi_1$ | $T-2h+p_1+90°$ | zero | 14.917,864,7 | 0.0103 | unity |
| $B_{16}$ | | $T-p_1-90°$ | zero | 14.999,998,0 | 0.0015 | unity |
| $B_{17}$ | | $T-3h+2p_1+90°$ | zero | 14.876,798,0 | 0.0004 | unity |
| $B_{22}$ | $(K_1)$ | $T+h-90°$ | zero | 15.041,068,6 | 0.1681 | unity |
| $B_{23}$ | | $T+p_1-90°$ | zero | 15.000,002,0 | 0.0042 | unity |
| $B_{24}$ | $\psi_1$ | $T+2h-p_1-90°$ | zero | 15.082,135,3 | 0.0042 | unity |
| $B_{25}$ | | $T-h+2p_1-90°$ | zero | 14.958,935,4 | 0.0001 | unity |
| $B_{26}$ | | $T+3h-2p_1-90°$ | zero | 15.123,202,0 | 0.0001 | unity |
| $B_{31}$ | $\phi_1$ | $T+3h-90°$ | zero | 15.123,205,9 | 0.0076 | unity |
| $B_{32}$ | | $T+4h-p_1-90°$ | zero | 15.164,272,6 | 0.0004 | unity |
| $B_{33}$ | | $T+2h+p_1+90°$ | zero | 15.082,139,2 | 0.0001 | unity |
| $B_{34}$ | | $T+5h-2p_1-90°$ | zero | 15.205,339,3 | | unity |
| | | SOLAR DIURNAL TERM, PARAGRAPH 119 | | | | |
| $B_{71}$ | $S_1$ | $T$ | zero | 15.000,000,0 | | unity |
| | | COMBINATION DIURNAL TERMS, FORMULAS (194), (201), AND (222) | | | | |
| Note 1 | $M_1$ | $T-s+h+p-90°$ $T-s+h-90°$ | $-\nu-Q_u$ $+\xi-\nu+Q$ | 14.496,693,9 | 0.0209* | (206) |
| Note 2 | $K_1$ | $T+h-90°$ | $-\nu'$ | 15.041,068,6 | 0.5305 | (227) |
| | | LUNAR SEMIDIURNAL TERMS, FORMULA (64) | | | | |
| $A_{39}$ | $M_2$ | $2T-2s+2h$ | $+2\xi-2\nu$ | 28.984,104,2 | 0.9085 | (78) |
| $A_{40}$ | $N_2$ | $2T-3s+2h+p$ | $+2\xi-2\nu$ | 28.439,729,5 | 0.1759 | (78) |
| $A_{41}$ | $(L_2)$ | $2T-s+2h-p+180°$ | $+2\xi-2\nu$ | 29.528,478,9 | 0.0251 | (78) |
| $A_{42}$ | $2N_2$ | $2T-4s+2h+2p$ | $+2\xi-2\nu$ | 27.895,354,8 | 0.0235 | (78) |
| $A_{43}$ | $\nu_2$ | $2T-3s+4h-p$ | $+2\xi-2\nu$ | 28.512,583,1 | 0.0341 | (78) |
| $A_{44}$ | $\lambda_2$ | $2T-s+p+180°$ | $+2\xi-2\nu$ | 29.455,625,3 | 0.0066 | (78) |
| $A_{45}$ | $\mu_2$ | $2T-4s+4h$ | $+2\xi-2\nu$ | 27.968,208,4 | 0.0219 | (78) |
| $A_{46}$ | $(S_2)$ | $2T$ | $+2\xi-2\nu$ | 30.000,000,0 | 0.0006 | (78) |
| $A_{47}$ | $(K_2)$ | $2T+2h$ | $-2\nu$ | 30.082,137,3 | 0.0786 | (79) |
| $A_{48}$ | $(L_2)$ | $2T-s+2h+p$ | $-2\nu$ | 29.537,762,6 | 0.0064 | (79) |
| $A_{49}$ | $KJ_2$ | $2T+s+2h-p$ | $-2\nu$ | 30.626,512,0 | 0.0064 | (79) |
| $A_{50}$ | | $2T-2s+2h+2p$ | $-2\nu$ | 28.993,387,9 | 0.0005 | (79) |
| $A_{51}$ | | $2T+2s+2h-2p$ | $-2\nu$ | 31.170,886,7 | 0.0005 | (79) |
| $A_{52}$ | | $2T-s+4h-p$ | $-2\nu$ | 29.610,616,2 | 0.0009 | (79) |
| $A_{53}$ | | $2T+s+p$ | $-2\nu$ | 30.553,658,4 | 0.0009 | (79) |
| $A_{54}$ | | $2T-2s+4h$ | $-2\nu$ | 29.066,241,5 | 0.0007 | (79) |
| $A_{55}$ | | $2T+2s$ | $-2\nu$ | 31.098,033,1 | 0.0007 | (79) |
| $A_{56}$ | | $2T+2s+2h$ | $-2\xi-2\nu$ | 31.180,170,3 | 0.0017 | (80) |
| $A_{57}$ | | $2T+3s+2h-p$ | $-2\xi-2\nu$ | 31.724,545,0 | 0.0003 | (80) |
| $A_{58}$ | | $2T+s+2h+p+180°$ | $-2\xi-2\nu$ | 30.635,795,6 | | (80) |
| $A_{59}$ | | $2T+4s+2h-2p$ | $-2\xi-2\nu$ | 32.268,919,7 | | (80) |
| $A_{60}$ | | $2T+3s+p$ | $-2\xi-2\nu$ | 31.651,691,4 | | (80) |
| $A_{61}$ | | $2T+s+4h-p+180°$ | $-2\xi-2\nu$ | 30.708,649,3 | | (80) |
| $A_{62}$ | | $2T+4s$ | $-2\xi-2\nu$ | 32.196,066,1 | | (80) |
| $A_{63}$ | | $2T+4h$ | $-2\xi-2\nu$ | 30.164,274,6 | | (80) |

*Adapted for use with tabular node factors, theoretical value is 0.0317. See p. 43.

Table 2.—*Harmonic constituents*—Continued

| Ref. No. | Symbol | Argument (E) V | Argument (E) u | Speed per solar hour | Coefficient (C) | Factor-f formula |
|---|---|---|---|---|---|---|
| | | LUNAR SEMIDIURNAL TERMS, FORMULA (139) | | | | |
| $A_{76}$ | | $2T-3s+2h+90°$ | $+3\xi-2\nu$ | 28.435,087,7 | 0.0223 | (146) |
| $A_{77}$ | | $2T-4s+2h+p+90°$ | $+3\xi-2\nu$ | 27.890,713,0 | 0.0062 | (146) |
| $A_{78}$ | | $2T-2s+2h-p-90°$ | $+3\xi-2\nu$ | 28.979,462,4 | 0.0012 | (146) |
| $A_{79}$ | | $2T-s+2h-90°$ | $+\xi-2\nu$ | 29.533,120,8 | 0.0209 | (147) |
| $A_{80}$ | | $2T-2s+2h+p-90°$ | $+\xi-2\nu$ | 28.988,746,0 | 0.0034 | (147) |
| $A_{81}$ | | $2T+s+2h-90°$ | $-\xi-2\nu$ | 30.631,153,8 | 0.0019 | (148) |
| | | SOLAR SEMIDIURNAL TERMS, FORMULA (187) | | | | |
| $B_{39}$ | $S_2$ | $2T$ | zero | 30.000,000,0 | 0.4227 | unity |
| $B_{40}$ | $T_2$ | $2T-h+p_1$ | zero | 29.958,933,3 | 0.0248 | unity |
| $B_{41}$ | $R_2$ | $2T+h-p_1+180°$ | zero | 30.041,066,7 | 0.0035 | unity |
| $B_{42}$ | | $2T-2h+2p_1$ | zero | 29.917,866,6 | 0.0010 | unity |
| $B_{47}$ | $(K_2)$ | $2T+2h$ | zero | 30.082,137,3 | 0.0365 | unity |
| $B_{48}$ | | $2T+h+p_1$ | zero | 30.041,070,6 | 0.0009 | unity |
| $B_{49}$ | | $2T+3h-p_1$ | zero | 30.123,204,0 | 0.0009 | unity |
| $B_{50}$ | | $2T+2p_1$ | zero | 30.000,003,9 | | unity |
| $B_{51}$ | | $2T+4h-2p_1$ | zero | 30.164,270,6 | | unity |
| $B_{56}$ | | $2T+4h$ | zero | 30.164,274,6 | 0.0008 | unity |
| $B_{57}$ | | $2T+5h-p_1$ | zero | 30.205,341,2 | | unity |
| $B_{58}$ | | $2T+3h+p_1+180°$ | zero | 30.123,207,9 | | unity |
| $B_{59}$ | | $2T+6h-2p_1$ | zero | 30.246,407,9 | | unity |
| | | COMBINATION SEMIDIURNAL TERMS, FORMULAS (212) AND (230) | | | | |
| Note 3 | $L_2$ | $2T-s+2h-p+180°$ | $+2\xi-2\nu-R$ | 29.528,478,9 | 0.0251 | (215) |
| Note 4 | $K_2$ | $2T+2h$ | $-2\nu''$ | 30.082,137,3 | 0.1151 | (235) |
| | | LUNAR TERDIURNAL TERMS, FORMULA (140) | | | | |
| $A_{82}$ | $M_3$ | $3T-3s+3h$ | $+3\xi-3\nu$ | 43.476,156,3 | 0.0178 | (149) |
| $A_{83}$ | | $3T-4s+3h+p$ | $+3\xi-3\nu$ | 42.931,781,6 | 0.0050 | (149) |
| $A_{84}$ | | $3T-2s+3h-p+180°$ | $+3\xi-3\nu$ | 44.020,531,0 | 0.0010 | (149) |
| $A_{85}$ | | $3T-5s+3h+2p$ | $+3\xi-3\nu$ | 42.387,406,9 | 0.0009 | (149) |
| $A_{86}$ | | $3T-4s+5h-p$ | $+3\xi-3\nu$ | 43.004,635,2 | 0.0007 | (149) |
| $A_{87}$ | | $3T-s+3h$ | $+\xi-3\nu$ | 44.574,189,4 | 0.0024 | (150) |
| $A_{88}$ | | $3T-2s+3h+p$ | $+\xi-3\nu$ | 44.029,814,7 | 0.0004 | (150) |
| | | SOLAR TERDIURNAL TERM, PARAGRAPH 119 | | | | |
| $B_{62}$ | $S_3$ | $3T$ | zero | 45.000,000,0 | | unity |

Note 1—Combines terms $A_{16}$ and $A_{23}$.
Note 2—Combines terms $A_{22}$ and $B_{22}$.
Note 3—Combines terms $A_{41}$ and $A_{48}$
Note 4—Combines terms $A_{47}$ and $B_{47}$.

Table 2a.—*Shallow-water constituents*

| Symbol | Argument | | | Speed | Factor-$f$ |
|---|---|---|---|---|---|
| | Origin | $V$ | $u$ | | |
| | | *Semidiurnal* | | degrees per h. | |
| $MNS_2$ | $M_2+N_2-S_2$ | $2T-5s+4h+p$ | $+4\xi-4\nu$ | 27. 423, 833, 7 | $f^2(M_2)$ |
| $2SM_2$ | $2S_2-M_2$ | $2T+2s-2h$ | $-2\xi+2\nu$ | 31. 015, 895, 8 | $f(M_2)$ |
| | | *Terdiurnal* | | | |
| $MK_3$ | $M_2+K_1$ | $3T-2s+3h-90°$ | $+2\xi-2\nu-\nu'$ | 44. 025, 172, 9 | $f(M_2) \times f(K_1)$ |
| $MK_3$ | $2M_2-K_1$ | $3T-4s+3h+90°$ | $+4\xi-4\nu+\nu'$ | 42. 927, 139, 8 | $f^2(M_2) \times f(K_1)$ |
| $SK_3$ | $S_2+K_1$ | $3T+h-90°$ | $-\nu'$ | 45. 041, 068, 6 | $f(K_1)$ |
| $SO_3$ | $S_2+O_1$ | $3T-2s+h+90°$ | $+2\xi-\nu$ | 43. 943, 035, 6 | $f(O_1)$ |
| | | *Quarter diurnal* | | | |
| $M_4$ | $2M_2$ | $4T-4s+4h$ | $+4\xi-4\nu$ | 57. 968, 208, 4 | $f^2(M_2)$ |
| $MS_4$ | $M_2+S_2$ | $4T-2s+2h$ | $+2\xi-2\nu$ | 58. 984, 104, 2 | $f^2(M_2)$ |
| $MN_4$ | $M_2+N_2$ | $4T-5s+4h+p$ | $+4\xi-4\nu$ | 57. 423, 833, 7 | $f^2(M_2)$ |
| $MK_4$ | $M_2+K_2$ | $4T-2s+4h$ | $+2\xi-2\nu-2\nu''$ | 59. 066, 241, 5 | $f(M_2) \times f(K_2)$ |
| $S_4$ | $2S_2$ | $4T$ | zero | 60. 000, 000, 0 | unity |
| | | *Sixth diurnal* | | | |
| $M_6$ | $3M_2$ | $6T-6s+6h$ | $+6\xi-6\nu$ | 86. 952, 312, 7 | $f^3(M_2)$ |
| $2MS_6$ | $2M_2+S_2$ | $6T-4s+4h$ | $+4\xi-4\nu$ | 87. 968, 208, 4 | $f^2(M_2)$ |
| $2MN_6$ | $2M_2+N_2$ | $6T-7s+6h+p$ | $+6\xi-6\nu$ | 86. 407, 938, 0 | $f^3(M_2)$ |
| $2SM_6$ | $2S_2+M_2$ | $6T-2s+2h$ | $+2\xi-2\nu$ | 88. 984, 104, 2 | $f(M_2)$ |
| $MSN_6$ | $M_2+S_2+N_2$ | $6T-5s+4h+p$ | $+4\xi-4\nu$ | 87. 423, 833, 7 | $f^2(M_2)$ |
| $S_6$ | $3S_2$ | $6T$ | zero | 90. 000, 000, 0 | unity |
| | | *Eighth diurnal* | | | |
| $M_8$ | $4M_2$ | $8T-8s+8h$ | $+8\xi-8\nu$ | 115. 936, 416, 9 | $f^4(M_2)$ |
| $3MS_8$ | $3M_2+S_2$ | $8T-6s+6h$ | $+6\xi-6\nu$ | 116. 952, 312, 7 | $f^3(M_2)$ |
| $2(MS)_8$ | $2M_2+2S_2$ | $8T-4s+4h$ | $+4\xi-4\nu$ | 117. 968, 208, 4 | $f^2(M_2)$ |
| $2MSN_8$ | $2M_2+S_2+N_2$ | $8T-7s+6h+p$ | $+6\xi-6\nu$ | 116. 407, 938, 0 | $f^3(M_2)$ |
| $S_8$ | $4S_2$ | $8T$ | zero | 120. 000, 000, 0 | unity |

Table 3.—**Latitude factors**

| Y | $Y_{v30}$ | $Y_{v31}$ $Y_{s30}$ $Y_{s32}$ $Y_{w42}$ | $Y_{v32}$ $Y_{w43}$ | $Y_{s31}$ | $Y_{w31}$ | $Y_{w32}$ | $Y_{v40}$ | $Y_{v41}$ $Y_{s40}$ | $Y_{v42}$ $Y_{s43}$ | $Y_{v43}$ | $Y_{s41}$ | $Y_{s42}$ | $Y_{w41}$ | Y |
|---|---|---|---|---|---|---|---|---|---|---|---|---|---|---|
| ° |  | * |  |  | * |  | * |  | * |  | * |  |  | ° |
| 0 | 0.500 | 0.000 | 1.000 | 1.000 | 0.000 | 1.000 | 0.000 | 0.200 | 0.000 | 1.000 | 0.000 | 0.333 | 0.200 | 0 |
| 1 | .500 | .035 | 1.000 | 0.999 | .017 | 1.000 | .010 | .200 | .017 | 1.000 | .013 | .333 | .200 | 1 |
| 2 | .498 | .070 | 0.999 | .998 | .035 | 0.999 | .021 | .199 | .035 | 0.998 | .026 | .332 | .199 | 2 |
| 3 | .496 | .105 | .997 | .995 | .052 | .999 | .031 | .197 | .052 | .996 | .038 | .330 | .197 | 3 |
| 4 | .493 | .139 | .995 | .990 | .070 | .998 | .041 | .195 | .069 | .993 | .051 | .328 | .195 | 4 |
| 5 | .489 | .174 | .992 | .985 | .087 | .996 | .052 | .192 | .086 | .989 | .063 | .324 | .192 | 5 |
| 6 | .484 | .208 | .989 | .978 | .105 | .995 | .062 | .188 | .103 | .984 | .076 | .321 | .189 | 6 |
| 7 | .478 | .242 | .985 | .970 | .122 | .993 | .071 | .184 | .120 | .978 | .088 | .316 | .185 | 7 |
| 8 | .471 | .276 | .981 | .961 | .139 | .990 | .081 | .179 | .136 | .971 | .099 | .311 | .181 | 8 |
| 9 | .463 | .309 | .976 | .951 | .156 | .988 | .090 | .173 | .153 | .964 | .111 | .305 | .176 | 9 |
| 10 | .455 | .342 | .970 | .940 | .174 | .985 | .099 | .167 | .168 | .955 | .122 | .299 | .170 | 10 |
| 11 | .445 | .375 | .964 | .927 | .191 | .982 | .108 | .161 | .184 | .946 | .133 | .291 | .164 | 11 |
| 12 | .435 | .407 | .957 | .914 | .208 | .978 | .116 | .153 | .199 | .936 | .143 | .284 | .157 | 12 |
| 13 | .424 | .438 | .949 | .899 | .225 | .974 | .124 | .146 | .214 | .925 | .154 | .275 | .149 | 13 |
| 14 | .412 | .469 | .941 | .883 | .242 | .970 | .131 | .137 | .228 | .914 | .163 | .267 | .141 | 14 |
| 15 | .400 | .500 | .933 | .866 | .259 | .966 | .138 | .128 | .241 | .901 | .172 | .257 | .133 | 15 |
| 16 | .386 | .530 | .924 | .848 | .276 | .961 | .144 | .119 | .255 | .888 | .181 | .247 | .124 | 16 |
| 17 | .372 | .559 | .915 | .829 | .292 | .956 | .150 | .110 | .267 | .875 | .189 | .237 | .115 | 17 |
| 18 | .357 | .588 | .905 | .809 | .309 | .951 | .156 | .099 | .280 | .860 | .197 | .226 | .105 | 18 |
| 19 | .341 | .616 | .894 | .788 | .326 | .946 | .161 | .089 | .291 | .845 | .204 | .215 | .094 | 19 |
| 20 | .325 | .643 | .883 | .766 | .342 | .940 | .165 | .078 | .302 | .830 | .211 | .203 | .083 | 20 |
| 21 | .307 | .669 | .872 | .743 | .358 | .934 | .169 | .067 | .312 | .814 | .217 | .191 | .072 | 21 |
| 22 | .290 | .695 | .860 | .719 | .375 | .927 | .172 | .055 | .322 | .797 | .222 | .179 | .060 | 22 |
| 23 | .271 | .719 | .847 | .695 | .391 | .921 | .175 | .044 | .331 | .780 | .227 | .166 | .047 | 23 |
| 24 | .252 | .743 | .835 | .669 | .407 | .914 | .177 | .032 | .339 | .762 | .231 | .153 | .035 | 24 |
| 25 | .232 | .766 | .821 | .643 | .423 | .906 | .178 | .019 | .347 | .744 | .234 | .140 | .021 | 25 |
| 26 | .212 | .788 | .808 | .616 | .438 | .899 | .179 | .007 | .354 | .726 | .237 | .127 | .008 | 26 |
| 27 | .191 | .809 | .794 | .588 | .454 | .891 | .179 | −.005 | .360 | .707 | .239 | .113 | −.006 | 27 |
| 28 | .169 | .829 | .780 | .559 | .469 | .883 | .178 | −.018 | .366 | .688 | .241 | .100 | −.020 | 28 |
| 29 | .147 | .848 | .765 | .530 | .485 | .875 | .177 | −.031 | .371 | .671 | .242 | .086 | −.035 | 29 |
| 30 | .125 | .866 | .750 | .500 | .500 | .866 | .175 | −.043 | .375 | .650 | .242 | .072 | −.050 | 30 |
| 31 | .102 | .883 | .735 | .469 | .515 | .857 | .172 | −.056 | .378 | .630 | .241 | .058 | −.065 | 31 |
| 32 | .079 | .899 | .719 | .438 | .530 | .848 | .169 | −.069 | .381 | .610 | .240 | .045 | −.081 | 32 |
| 33 | .055 | .914 | .703 | .407 | .545 | .839 | .165 | −.081 | .383 | .590 | .238 | .031 | −.097 | 33 |
| 34 | .031 | .927 | .687 | .375 | .559 | .829 | .161 | −.093 | .384 | .570 | .235 | .071 | −.113 | 34 |
| 35 | .007 | .940 | .671 | .342 | .574 | .819 | .155 | −.106 | .385 | .550 | .232 | .004 | −.129 | 35 |
| 36 | −.018 | .951 | .655 | .309 | .588 | .809 | .150 | −.118 | .385 | .530 | .228 | −.010 | −.145 | 36 |
| 37 | −.043 | .961 | .638 | .276 | .602 | .799 | .143 | −.130 | .384 | .509 | .223 | −.023 | −.162 | 37 |
| 38 | −.069 | .970 | .621 | .242 | .616 | .788 | .136 | −.141 | .382 | .489 | .218 | −.036 | −.179 | 38 |
| 39 | −.094 | .978 | .604 | .208 | .629 | .777 | .128 | −.152 | .380 | .469 | .212 | −.049 | −.196 | 39 |
| 40 | −.120 | .985 | .587 | .174 | .643 | .766 | .120 | −.163 | .377 | .450 | .206 | −.061 | −.213 | 40 |
| 41 | −.146 | .990 | .570 | .139 | .656 | .755 | .111 | −.174 | .374 | .430 | .199 | −.073 | −.230 | 41 |
| 42 | −.172 | .995 | .552 | .105 | .669 | .743 | .102 | −.184 | .370 | .410 | .191 | −.085 | −.248 | 42 |
| 43 | −.198 | .998 | .535 | .070 | .682 | .731 | .092 | −.194 | .365 | .391 | .183 | −.096 | −.265 | 43 |
| 44 | −.224 | .999 | .517 | .035 | .695 | .719 | .082 | −.203 | .359 | .372 | .174 | −.107 | −.283 | 44 |
| 45 | −.250 | 1.000 | .500 | .000 | .707 | .707 | .071 | −.212 | .354 | .354 | .165 | −.118 | −.300 | 45 |

*In these columns reverse signs for south latitude. Other values are applicable to either north or south latitude.

HARMONIC ANALYSIS AND PREDICTION OF TIDES 169

Table 3.—*Latitude factors—Continued*

| Y | $Y_{v30}$ | $Y_{v31}$ $Y_{s30}$ $Y_{s32}$ $Y_{w42}$ | $Y_{v32}$ $Y_{w43}$ | $Y_{s31}$ | $Y_{w31}$ | $Y_{w32}$ | $Y_{v40}$ | $Y_{v41}$ $Y_{s40}$ | $Y_{v42}$ $Y_{s43}$ | $Y_{v43}$ | $Y_{s41}$ | $Y_{s42}$ | $Y_{w41}$ | Y |
|---|---|---|---|---|---|---|---|---|---|---|---|---|---|---|
| ° | * | | | * | | | * | | * | | * | | | ° |
| 45 | −0.250 | 1.000 | 0.500 | 0.000 | 0.707 | 0.707 | 0.071 | −0.212 | 0.354 | 0.354 | 0.165 | −0.118 | −0.300 | 45 |
| 46 | −.276 | 0.999 | .483 | −.035 | .719 | .695 | .059 | −.221 | .347 | .335 | .155 | −.128 | −.317 | 46 |
| 47 | −.302 | .998 | .465 | −.070 | .731 | .682 | .048 | −.228 | .340 | .317 | .145 | −.137 | −.335 | 47 |
| 48 | −.328 | .995 | .448 | −.105 | .743 | .669 | .035 | −.236 | .333 | .300 | .135 | −.146 | −.352 | 48 |
| 49 | −.354 | .990 | .430 | −.139 | .755 | .656 | .023 | −.242 | .325 | .282 | .124 | −.155 | −.370 | 49 |
| 50 | −.380 | .985 | .413 | −.174 | .766 | .643 | .010 | −.249 | .317 | .266 | .112 | −.163 | −.387 | 50 |
| 51 | −.406 | .978 | .396 | −.208 | .777 | .629 | −.003 | −.254 | .308 | .249 | .101 | −.170 | −.404 | 51 |
| 52 | −.431 | .970 | .379 | −.242 | .788 | .616 | −.017 | −.259 | .299 | .233 | .089 | −.177 | −.421 | 52 |
| 53 | −.457 | .961 | .362 | −.276 | .799 | .602 | −.030 | −.263 | .289 | .218 | .076 | −.183 | −.438 | 53 |
| 54 | −.482 | .951 | .345 | −.309 | .809 | .588 | −.044 | −.267 | .280 | .203 | .064 | −.189 | −.455 | 54 |
| 55 | −.507 | .940 | .329 | −.342 | .819 | .574 | −.058 | −.270 | .269 | .189 | .051 | −.194 | −.471 | 55 |
| 56 | −.531 | .927 | .313 | −.375 | .829 | .559 | −.072 | −.272 | .259 | .175 | .038 | −.198 | −.487 | 56 |
| 57 | −.555 | .914 | .297 | −.407 | .839 | .545 | −.087 | −.274 | .249 | .162 | .025 | −.202 | −.503 | 57 |
| 58 | −.579 | .899 | .281 | −.438 | .848 | .530 | −.101 | −.275 | .238 | .149 | .012 | −.204 | −.519 | 58 |
| 59 | −.602 | .883 | .265 | −.469 | .857 | .515 | −.115 | −.275 | .227 | .137 | −.001 | −.207 | −.535 | 59 |
| 60 | −.625 | .866 | .250 | −.500 | .866 | .500 | −.130 | −.275 | .217 | .125 | −.014 | −.208 | −.550 | 60 |
| 61 | −.647 | .848 | .235 | −.530 | .875 | .485 | −.144 | −.274 | .206 | .114 | −.028 | −.209 | −.565 | 61 |
| 62 | −.669 | .829 | .220 | −.559 | .883 | .469 | −.158 | −.272 | .195 | .103 | −.041 | −.210 | −.580 | 62 |
| 63 | −.691 | .809 | .206 | −.588 | .891 | .454 | −.173 | −.270 | .184 | .094 | −.054 | −.209 | −.594 | 63 |
| 64 | −.712 | .788 | .192 | −.616 | .899 | .438 | −.187 | −.266 | .173 | .084 | −.067 | −.208 | −.608 | 64 |
| 65 | −.732 | .766 | .179 | −.643 | .906 | .423 | −.201 | −.263 | .162 | .075 | −.080 | −.206 | −.621 | 65 |
| 66 | −.752 | .743 | .165 | −.669 | .914 | .407 | −.214 | −.258 | .151 | .067 | −.092 | −.204 | −.635 | 66 |
| 67 | −.771 | .719 | .153 | −.695 | .921 | .391 | −.228 | −.253 | .141 | .060 | −.105 | −.201 | −.647 | 67 |
| 68 | −.790 | .695 | .140 | −.719 | .927 | .375 | −.241 | −.247 | .130 | .053 | −.117 | −.197 | −.660 | 68 |
| 69 | −.807 | .669 | .128 | −.743 | .934 | .358 | −.254 | −.241 | .120 | .046 | −.129 | −.193 | −.672 | 69 |
| 70 | −.825 | .643 | .117 | −.766 | .940 | .342 | −.266 | −.234 | .110 | .040 | −.141 | −.188 | −.683 | 70 |
| 71 | −.841 | .616 | .106 | −.788 | .946 | .326 | −.278 | −.226 | .100 | .035 | −.152 | −.183 | −.694 | 71 |
| 72 | −.857 | .588 | .095 | −.809 | .951 | .309 | −.290 | −.218 | .091 | .030 | −.163 | −.177 | −.705 | 72 |
| 73 | −.872 | .559 | .085 | −.829 | .956 | .292 | −.301 | −.209 | .082 | .025 | −.173 | −.170 | −.715 | 73 |
| 74 | −.886 | .530 | .076 | −.848 | .961 | .276 | −.311 | −.200 | .073 | .021 | −.183 | −.163 | −.724 | 74 |
| 75 | −.900 | .500 | .067 | −.866 | .966 | .259 | −.322 | −.190 | .065 | .017 | −.193 | −.155 | −.733 | 75 |
| 76 | −.912 | .469 | .059 | −.883 | .970 | .242 | −.331 | −.179 | .057 | .014 | −.202 | −.147 | −.741 | 76 |
| 77 | −.924 | .438 | .051 | −.899 | .974 | .225 | −.340 | −.169 | .049 | .011 | −.211 | −.139 | −.749 | 77 |
| 78 | −.935 | .407 | .043 | −.914 | .978 | .208 | −.349 | −.159 | .042 | .009 | −.219 | −.130 | −.757 | 78 |
| 79 | −.945 | .375 | .036 | −.927 | .982 | .191 | −.357 | −.146 | .036 | .007 | −.226 | −.120 | −.764 | 79 |
| 80 | −.955 | .342 | .030 | −.940 | .985 | .174 | −.364 | −.134 | .030 | .005 | −.233 | −.111 | −.770 | 80 |
| 81 | −.963 | .309 | .024 | −.951 | .988 | .156 | −.371 | −.121 | .024 | .004 | −.239 | −.100 | −.776 | 81 |
| 82 | −.971 | .276 | .019 | −.961 | .990 | .139 | −.377 | −.109 | .019 | .003 | −.245 | −.090 | −.781 | 82 |
| 83 | −.978 | .242 | .015 | −.970 | .993 | .122 | −.382 | −.096 | .015 | .002 | −.250 | −.079 | −.785 | 83 |
| 84 | −.984 | .208 | .011 | −.978 | .995 | .105 | −.387 | −.082 | .011 | .001 | −.254 | −.069 | −.789 | 84 |
| 85 | −.989 | .174 | .008 | −.985 | .996 | .087 | −.391 | −.069 | .008 | .001 | −.258 | −.057 | −.792 | 85 |
| 86 | −.993 | .139 | .005 | −.990 | .998 | .070 | −.394 | −.055 | .005 | .000 | −.261 | −.046 | −.795 | 86 |
| 87 | −.996 | .105 | .003 | −.995 | .999 | .052 | −.397 | −.042 | .003 | .000 | −.264 | −.035 | −.797 | 87 |
| 88 | −.998 | .070 | .001 | −.998 | .999 | .035 | −.399 | −.028 | .001 | .000 | −.265 | −.023 | −.799 | 88 |
| 89 | −1.000 | .035 | .000 | −.999 | 1.000 | .017 | −.400 | −.014 | .000 | .000 | −.266 | −.012 | −.800 | 89 |
| 90 | −1.000 | .000 | .000 | −1.000 | 1.000 | .000 | −.400 | .000 | .000 | .000 | −.267 | .000 | −.800 | 90 |

*In these columns reverse signs for south latitude. Other values are applicable to either north or south latitude.

Table 4.—*Mean longitude of lunar and solar elements at Jan. 1, 0 hour, Greenwich mean civil time, of each year from 1800 to 2000*

[$s$ = mean longitude of moon; $p$ = mean longitude lunar perigee; $h$ = mean longitude of sun; $p_1$ = mean longitude solar perigee; $N$ = longitude of moon's node]

| Year | s | p | h | $p_1$ | N | Year | s | p | h | $p_1$ | N |
|---|---|---|---|---|---|---|---|---|---|---|---|
| | ° | ° | ° | ° | ° | | ° | ° | ° | ° | ° |
| 1800 | 342.31 | 225.45 | 280.41 | 279.50 | 33.25 | 1852 | 28.44 | 181.24 | 279.82 | 280.40 | 107.55 |
| 1801 | 111.70 | 266.12 | 280.17 | 279.52 | 13.92 | 1853 | 171.00 | 222.02 | 280.57 | 280.41 | 88.16 |
| 1802 | 241.08 | 306.78 | 279.93 | 279.54 | 354.59 | 1854 | 330.38 | 262.68 | 280.33 | 280.43 | 68.84 |
| 1803 | 10.47 | 347.44 | 279.69 | 279.55 | 335.26 | 1855 | 69.77 | 303.34 | 280.09 | 280.45 | 49.51 |
| 1804 | 139.85 | 28.10 | 279.45 | 279.57 | 315.93 | 1856 | 199.15 | 344.00 | 279.85 | 280.46 | 30.18 |
| 1805 | 282.41 | 68.88 | 280.20 | 279.59 | 296.55 | 1857 | 341.72 | 24.78 | 280.60 | 280.48 | 10.80 |
| 1806 | 51.80 | 109.54 | 279.96 | 279.61 | 277.23 | 1858 | 111.10 | 65.44 | 280.36 | 280.50 | 351.47 |
| 1807 | 181.18 | 150.20 | 279.72 | 279.62 | 257.90 | 1859 | 240.49 | 106.10 | 280.12 | 280.52 | 332.14 |
| 1808 | 310.57 | 190.86 | 279.48 | 279.64 | 238.57 | 1860 | 9.87 | 146.77 | 279.88 | 280.53 | 312.81 |
| 1809 | 93.13 | 231.64 | 280.23 | 279.66 | 219.19 | 1861 | 152.43 | 187.54 | 280.63 | 280.55 | 293.43 |
| 1810 | 222.51 | 272.30 | 279.99 | 279.67 | 199.86 | 1862 | 281.82 | 228.20 | 280.39 | 280.57 | 274.10 |
| 1811 | 351.90 | 312.96 | 279.75 | 279.69 | 180.53 | 1863 | 51.20 | 268.87 | 280.15 | 280.58 | 254.78 |
| 1812 | 121.28 | 353.63 | 279.51 | 279.71 | 161.20 | 1864 | 180.59 | 309.53 | 279.91 | 280.60 | 235.45 |
| 1813 | 263.84 | 34.40 | 280.26 | 279.73 | 141.82 | 1865 | 323.15 | 350.30 | 280.66 | 280.62 | 216.07 |
| 1814 | 33.23 | 75.06 | 280.02 | 279.74 | 122.49 | 1866 | 92.53 | 30.96 | 280.42 | 280.64 | 196.74 |
| 1815 | 162.61 | 115.73 | 279.78 | 279.76 | 103.17 | 1867 | 221.92 | 71.63 | 280.18 | 280.65 | 177.41 |
| 1816 | 292.00 | 156.39 | 279.54 | 279.78 | 83.84 | 1868 | 351.30 | 112.29 | 279.94 | 280.67 | 158.08 |
| 1817 | 74.56 | 197.16 | 280.29 | 279.79 | 64.46 | 1869 | 133.86 | 153.06 | 280.69 | 280.69 | 138.70 |
| 1818 | 203.94 | 237.82 | 280.05 | 279.81 | 45.13 | 1870 | 263.25 | 193.73 | 280.45 | 280.71 | 119.37 |
| 1819 | 333.33 | 278.49 | 279.81 | 279.83 | 25.80 | 1871 | 32.63 | 234.39 | 280.21 | 280.72 | 100.04 |
| 1820 | 102.71 | 319.15 | 279.57 | 279.85 | 6.47 | 1872 | 162.02 | 275.05 | 279.97 | 280.74 | 80.72 |
| 1821 | 245.28 | 359.92 | 280.32 | 279.86 | 347.09 | 1873 | 304.58 | 315.83 | 280.72 | 280.76 | 61.34 |
| 1822 | 14.66 | 40.59 | 280.08 | 279.88 | 327.76 | 1874 | 73.96 | 356.49 | 280.48 | 280.77 | 42.01 |
| 1823 | 144.04 | 81.25 | 279.84 | 279.90 | 308.43 | 1875 | 203.35 | 37.15 | 280.24 | 280.79 | 22.68 |
| 1824 | 273.43 | 121.91 | 279.61 | 279.91 | 289.11 | 1876 | 332.73 | 77.81 | 280.01 | 280.81 | 3.35 |
| 1825 | 55.99 | 162.69 | 280.35 | 279.93 | 269.72 | 1877 | 115.29 | 118.59 | 280.75 | 280.83 | 343.97 |
| 1826 | 185.38 | 203.35 | 280.11 | 279.95 | 250.40 | 1878 | 244.68 | 159.25 | 280.51 | 280.84 | 324.64 |
| 1827 | 314.76 | 244.01 | 279.87 | 279.97 | 231.07 | 1879 | 14.06 | 199.91 | 280.27 | 280.86 | 305.31 |
| 1828 | 84.15 | 284.67 | 279.64 | 279.98 | 211.74 | 1880 | 143.45 | 240.58 | 280.04 | 280.88 | 285.98 |
| 1829 | 226.71 | 325.45 | 280.38 | 280.00 | 192.36 | 1881 | 286.01 | 281.35 | 280.78 | 280.89 | 266.60 |
| 1830 | 356.09 | 6.11 | 280.14 | 280.02 | 173.03 | 1882 | 55.39 | 322.01 | 280.54 | 280.91 | 247.28 |
| 1831 | 125.48 | 46.77 | 279.91 | 280.03 | 153.70 | 1883 | 184.78 | 2.67 | 280.31 | 280.93 | 227.95 |
| 1832 | 254.86 | 87.43 | 279.67 | 280.05 | 134.37 | 1884 | 314.16 | 43.34 | 280.07 | 280.95 | 208.62 |
| 1833 | 37.42 | 128.21 | 280.41 | 280.07 | 114.99 | 1885 | 96.72 | 84.11 | 280.81 | 280.96 | 189.24 |
| 1834 | 166.81 | 168.87 | 280.18 | 280.09 | 95.66 | 1886 | 226.11 | 124.77 | 280.57 | 280.98 | 169.91 |
| 1835 | 296.19 | 209.53 | 279.94 | 280.10 | 76.34 | 1887 | 355.49 | 165.44 | 280.34 | 281.00 | 150.58 |
| 1836 | 65.58 | 250.20 | 279.70 | 280.12 | 57.01 | 1888 | 124.88 | 206.10 | 280.10 | 281.01 | 131.25 |
| 1837 | 208.14 | 290.97 | 280.44 | 280.14 | 37.63 | 1889 | 267.44 | 246.87 | 280.84 | 281.03 | 111.87 |
| 1838 | 337.52 | 331.63 | 280.21 | 280.16 | 18.30 | 1890 | 36.82 | 287.54 | 280.61 | 281.05 | 92.54 |
| 1839 | 106.91 | 12.30 | 279.97 | 280.17 | 358.97 | 1891 | 166.21 | 328.20 | 280.37 | 281.07 | 73.22 |
| 1840 | 236.29 | 52.96 | 279.73 | 280.19 | 339.64 | 1892 | 295.59 | 8.86 | 280.13 | 281.08 | 53.89 |
| 1841 | 18.85 | 93.73 | 280.48 | 280.21 | 320.26 | 1893 | 78.16 | 49.63 | 280.87 | 281.10 | 34.51 |
| 1842 | 148.24 | 134.39 | 280.24 | 280.22 | 300.93 | 1894 | 207.54 | 90.30 | 280.64 | 281.12 | 15.18 |
| 1843 | 277.62 | 175.06 | 280.00 | 280.24 | 281.61 | 1895 | 336.93 | 130.96 | 280.40 | 281.13 | 355.85 |
| 1844 | 47.01 | 215.72 | 279.76 | 280.26 | 262.28 | 1896 | 106.31 | 171.62 | 280.16 | 281.15 | 336.52 |
| 1845 | 189.57 | 256.49 | 280.51 | 280.28 | 242.90 | 1897 | 248.87 | 212.40 | 290.91 | 281.17 | 317.14 |
| 1846 | 318.95 | 297.16 | 280.27 | 280.29 | 223.57 | 1898 | 18.26 | 253.06 | 280.67 | 281.19 | 297.81 |
| 1847 | 88.34 | 337.82 | 280.03 | 280.31 | 204.24 | 1899 | 147.64 | 293.72 | 280.43 | 281.20 | 278.48 |
| 1848 | 217.72 | 18.48 | 279.79 | 280.33 | 184.91 | | | | | | |
| 1849 | 0.28 | 59.26 | 280.54 | 280.34 | 165.53 | | | | | | |
| 1850 | 129.67 | 99.92 | 280.30 | 280.36 | 146.20 | | | | | | |
| 1851 | 259.05 | 140.58 | 280.06 | 280.38 | 126.87 | | | | | | |

# HARMONIC ANALYSIS AND PREDICTION OF TIDES 171

Table 4.—*Mean longitude of lunar and solar elements at Jan. 1, 0 hour, Greenwich mean civil time, of each year from 1800 to 2000*—Continued

| Year | s | p | h | $p_1$ | N | Year | s | p | h | $p_1$ | N |
|---|---|---|---|---|---|---|---|---|---|---|---|
|  | ° | ° | ° | ° | ° |  | ° | ° | ° | ° | ° |
| 1900 | 277.03 | 334.38 | 280.19 | 281.22 | 259.16 | 1952 | 323.15 | 290.16 | 279.60 | 282.12 | 333.45 |
| 1901 | 46.41 | 15.05 | 279.95 | 281.24 | 239.83 | 1953 | 105.72 | 330.94 | 280.35 | 282.13 | 314.07 |
| 1902 | 175.80 | 55.71 | 279.71 | 281.26 | 220.50 | 1954 | 235.10 | 11.60 | 280.11 | 282.15 | 294.75 |
| 1903 | 305.18 | 96.37 | 279.47 | 281.27 | 201.17 | 1955 | 4.49 | 52.26 | 279.87 | 282.17 | 275.42 |
| 1904 | 74.57 | 137.03 | 279.23 | 281.29 | 181.84 | 1956 | 133.87 | 92.92 | 279.63 | 282.18 | 256.09 |
| 1905 | 217.13 | 177.81 | 279.98 | 281.31 | 162.46 | 1957 | 276.43 | 133.70 | 280.38 | 282.20 | 236.71 |
| 1906 | 346.51 | 218.47 | 279.74 | 281.32 | 143.13 | 1958 | 45.82 | 174.36 | 280.14 | 282.22 | 217.38 |
| 1907 | 115.90 | 259.13 | 279.50 | 281.34 | 123.81 | 1959 | 175.20 | 215.02 | 279.90 | 282.24 | 198.05 |
| 1908 | 245.28 | 299.79 | 279.27 | 281.36 | 104.48 | 1960 | 304.59 | 255.69 | 279.67 | 282.25 | 178.72 |
| 1909 | 27.84 | 340.57 | 280.01 | 281.38 | 85.10 | 1961 | 87.15 | 296.46 | 280.41 | 282.27 | 159.34 |
| 1910 | 157.23 | 21.23 | 279.77 | 281.39 | 65.77 | 1962 | 216.53 | 337.12 | 280.17 | 282.29 | 140.01 |
| 1911 | 286.61 | 61.89 | 279.53 | 281.41 | 46.44 | 1963 | 345.92 | 17.78 | 279.93 | 282.30 | 120.69 |
| 1912 | 56.00 | 102.55 | 279.30 | 281.43 | 27.11 | 1964 | 115.30 | 58.45 | 279.70 | 282.32 | 101.36 |
| 1913 | 198.56 | 143.33 | 280.04 | 281.44 | 7.73 | 1965 | 257.86 | 99.22 | 280.44 | 282.34 | 81.98 |
| 1914 | 327.94 | 183.99 | 279.80 | 281.46 | 348.40 | 1966 | 27.25 | 139.88 | 280.20 | 282.36 | 62.65 |
| 1915 | 97.33 | 224.65 | 279.57 | 281.48 | 329.07 | 1967 | 156.63 | 180.54 | 279.97 | 282.37 | 43.32 |
| 1916 | 226.71 | 265.32 | 279.33 | 281.50 | 309.75 | 1968 | 286.02 | 221.21 | 279.73 | 282.39 | 23.99 |
| 1917 | 9.27 | 306.09 | 280.07 | 281.51 | 290.36 | 1969 | 68.58 | 261.98 | 280.47 | 282.41 | 4.61 |
| 1918 | 138.66 | 346.75 | 279.84 | 281.53 | 271.04 | 1970 | 197.96 | 302.64 | 280.24 | 282.42 | 345.28 |
| 1919 | 268.04 | 27.41 | 279.60 | 281.55 | 251.71 | 1971 | 327.35 | 343.31 | 280.00 | 282.44 | 325.95 |
| 1920 | 37.43 | 68.08 | 279.36 | 281.56 | 232.38 | 1972 | 96.73 | 23.97 | 279.76 | 282.46 | 306.63 |
| 1921 | 179.99 | 108.85 | 280.10 | 281.58 | 213.00 | 1973 | 239.29 | 64.74 | 280.50 | 282.48 | 287.24 |
| 1922 | 309.37 | 149.51 | 279.87 | 281.60 | 193.67 | 1974 | 8.68 | 105.40 | 280.27 | 282.49 | 267.92 |
| 1923 | 78.76 | 190.18 | 279.63 | 281.62 | 174.34 | 1975 | 138.06 | 146.07 | 280.03 | 282.51 | 248.59 |
| 1924 | 208.14 | 230.84 | 279.39 | 281.63 | 155.01 | 1976 | 267.45 | 186.73 | 279.79 | 282.53 | 229.26 |
| 1925 | 350.71 | 271.61 | 280.14 | 281.65 | 135.63 | 1977 | 50.01 | 227.50 | 280.54 | 282.54 | 209.88 |
| 1926 | 120.09 | 312.27 | 279.90 | 281.67 | 116.31 | 1978 | 179.40 | 268.17 | 280.30 | 282.56 | 190.55 |
| 1927 | 249.47 | 352.94 | 279.66 | 281.69 | 96.98 | 1979 | 308.78 | 308.83 | 280.06 | 282.58 | 171.22 |
| 1928 | 18.86 | 33.60 | 279.42 | 281.70 | 77.65 | 1980 | 78.16 | 349.49 | 279.82 | 282.60 | 151.89 |
| 1929 | 161.42 | 74.37 | 280.17 | 281.72 | 58.27 | 1981 | 220.73 | 30.26 | 280.57 | 282.61 | 132.51 |
| 1930 | 290.81 | 115.03 | 279.93 | 281.74 | 38.94 | 1982 | 350.11 | 70.93 | 280.33 | 282.63 | 113.19 |
| 1931 | 60.19 | 155.70 | 279.69 | 281.75 | 19.61 | 1983 | 119.50 | 111.59 | 280.09 | 282.65 | 93.86 |
| 1932 | 189.58 | 196.36 | 279.45 | 281.77 | 0.28 | 1984 | 248.88 | 152.25 | 279.85 | 282.67 | 74.53 |
| 1933 | 332.14 | 237.13 | 280.20 | 281.79 | 340.90 | 1985 | 31.44 | 193.02 | 280.60 | 282.68 | 55.15 |
| 1934 | 101.52 | 277.80 | 279.96 | 281.81 | 321.57 | 1986 | 160.83 | 233.69 | 280.36 | 282.70 | 35.82 |
| 1935 | 230.91 | 318.46 | 279.72 | 281.82 | 302.25 | 1987 | 290.21 | 274.35 | 280.12 | 282.72 | 16.49 |
| 1936 | 0.29 | 359.12 | 279.48 | 281.84 | 282.92 | 1988 | 59.60 | 315.01 | 279.88 | 282.73 | 357.16 |
| 1937 | 142.85 | 39.89 | 280.23 | 281.86 | 263.54 | 1989 | 202.16 | 355.79 | 280.63 | 282.75 | 337.78 |
| 1938 | 272.24 | 80.56 | 279.99 | 281.87 | 244.21 | 1990 | 331.54 | 36.45 | 280.39 | 282.77 | 318.45 |
| 1939 | 41.62 | 121.22 | 279.75 | 281.89 | 224.88 | 1991 | 100.93 | 77.11 | 280.15 | 282.79 | 299.13 |
| 1940 | 171.01 | 161.88 | 279.51 | 281.91 | 205.55 | 1992 | 230.31 | 117.77 | 279.91 | 282.80 | 279.80 |
| 1941 | 313.57 | 202.65 | 280.26 | 281.93 | 186.17 | 1993 | 12.87 | 158.55 | 280.66 | 282.82 | 260.42 |
| 1942 | 82.95 | 243.32 | 280.02 | 281.94 | 166.84 | 1994 | 142.26 | 199.21 | 280.42 | 282.84 | 241.09 |
| 1943 | 212.34 | 283.98 | 279.78 | 281.96 | 147.51 | 1995 | 271.64 | 239.87 | 280.18 | 282.85 | 221.76 |
| 1944 | 341.72 | 324.64 | 279.54 | 281.98 | 128.19 | 1996 | 41.03 | 280.53 | 279.94 | 282.87 | 202.43 |
| 1945 | 124.28 | 5.42 | 280.29 | 281.99 | 108.80 | 1997 | 193.59 | 321.31 | 280.69 | 282.89 | 183.05 |
| 1946 | 253.67 | 46.08 | 280.05 | 282.01 | 89.48 | 1998 | 312.97 | 1.97 | 280.45 | 282.91 | 163.72 |
| 1947 | 23.05 | 86.74 | 279.81 | 282.03 | 70.15 | 1999 | 82.36 | 42.63 | 280.21 | 282.92 | 144.39 |
| 1948 | 152.44 | 127.40 | 279.57 | 282.05 | 50.82 | 2000 | 211.74 | 83.29 | 279.97 | 282.94 | 125.07 |
| 1949 | 295.00 | 168.18 | 280.32 | 282.06 | 31.44 |  |  |  |  |  |  |
| 1950 | 64.38 | 208.04 | 280.08 | 282.08 | 12.11 |  |  |  |  |  |  |
| 1951 | 193.77 | 249.50 | 279.84 | 282.10 | 352.78 |  |  |  |  |  |  |

Table 5.—*Differences to adapt table 4 to any month, day, and hour of Greenwich mean civil time*

### DIFFERENCES TO FIRST OF EACH CALENDAR MONTH OF COMMON YEARS [1]

| Month | s | p | h | p₁ | N | Month | s | p | h | p₁ | N |
|---|---|---|---|---|---|---|---|---|---|---|---|
| | ° | ° | ° | ° | ° | | ° | ° | ° | ° | ° |
| Jan. 1 | 0.00 | 0.00 | 0.00 | 0.00 | 0.00 | July 1 | 224.93 | 20.16 | 178.40 | 0.01 | −9.58 |
| Feb. 1 | 48.47 | 3.45 | 30.56 | 0.00 | −1.64 | Aug. 1 | 273.40 | 23.62 | 208.96 | 0.01 | −11.23 |
| Mar. 1 | 57.41 | 6.57 | 58.15 | 0.00 | −3.12 | Sept. 1 | 321.86 | 27.07 | 239.51 | 0.01 | −12.87 |
| Apr. 1 | 105.88 | 10.03 | 88.71 | 0.00 | −4.77 | Oct. 1 | 357.16 | 30.41 | 269.08 | 0.01 | −14.46 |
| May 1 | 141.17 | 13.37 | 118.28 | 0.01 | −6.35 | Nov. 1 | 45.62 | 33.87 | 299.64 | 0.01 | −16.10 |
| June 1 | 189.64 | 16.82 | 148.83 | 0.01 | −8.00 | Dec. 1 | 80.92 | 37.21 | 329.21 | 0.02 | −17.69 |

### DIFFERENCES TO BEGINNING OF EACH DAY OF MONTH FOR COMMON YEARS [1]

| Day | s | p | h | p₁ | N | Day | s | p | h | p₁ | N |
|---|---|---|---|---|---|---|---|---|---|---|---|
| | ° | ° | ° | ° | ° | | ° | ° | ° | ° | ° |
| 1 | 0.00 | 0.00 | 0.00 | 0.00 | 0.00 | 17 | 210.82 | 1.78 | 15.77 | 0.00 | −0.85 |
| 2 | 13.18 | 0.11 | 0.99 | 0.00 | −0.05 | 18 | 224.00 | 1.89 | 16.76 | 0.00 | −0.90 |
| 3 | 26.35 | 0.22 | 1.97 | 0.00 | −0.11 | 19 | 237.18 | 2.01 | 17.74 | 0.00 | −0.95 |
| 4 | 39.53 | 0.33 | 2.96 | 0.00 | −0.16 | 20 | 250.35 | 2.12 | 18.73 | 0.00 | −1.01 |
| 5 | 52.71 | 0.45 | 3.94 | 0.00 | −0.21 | 21 | 263.53 | 2.23 | 19.71 | 0.00 | −1.06 |
| 6 | 65.88 | 0.56 | 4.93 | 0.00 | −0.26 | 22 | 276.70 | 2.34 | 20.70 | 0.00 | −1.11 |
| 7 | 79.06 | 0.67 | 5.91 | 0.00 | −0.32 | 23 | 289.88 | 2.45 | 21.68 | 0.00 | −1.16 |
| 8 | 92.23 | 0.78 | 6.90 | 0.00 | −0.37 | 24 | 303.06 | 2.56 | 22.67 | 0.00 | −1.22 |
| 9 | 105.41 | 0.89 | 7.89 | 0.00 | −0.42 | 25 | 316.23 | 2.67 | 23.66 | 0.00 | −1.27 |
| 10 | 118.59 | 1.00 | 8.87 | 0.00 | −0.48 | 26 | 329.41 | 2.79 | 24.64 | 0.00 | −1.32 |
| 11 | 131.76 | 1.11 | 9.86 | 0.00 | −0.53 | 27 | 342.59 | 2.90 | 25.63 | 0.00 | −1.38 |
| 12 | 144.94 | 1.23 | 10.84 | 0.00 | −0.58 | 28 | 355.76 | 3.01 | 26.61 | 0.00 | −1.43 |
| 13 | 158.12 | 1.34 | 11.83 | 0.00 | −0.64 | 29 | 8.94 | 3.12 | 27.60 | 0.00 | −1.48 |
| 14 | 171.29 | 1.45 | 12.81 | 0.00 | −0.69 | 30 | 22.12 | 3.23 | 28.58 | 0.00 | −1.54 |
| 15 | 184.47 | 1.56 | 13.80 | 0.00 | −0.74 | 31 | 35.29 | 3.34 | 29.57 | 0.00 | −1.59 |
| 16 | 197.65 | 1.67 | 14.78 | 0.00 | −0.79 | 32 | 48.47 | 3.45 | 30.56 | 0.00 | −1.64 |

### DIFFERENCES TO BEGINNING OF EACH HOUR OF DAY, GREENWICH CIVIL TIME

| Hour | s | p | h | p₁ | N | Hour | s | p | h | p₁ | N |
|---|---|---|---|---|---|---|---|---|---|---|---|
| | ° | ° | ° | ° | ° | | ° | ° | ° | ° | ° |
| 0 | 0.00 | 0.00 | 0.00 | 0.00 | 0.00 | 12 | 6.59 | 0.06 | 0.49 | 0.00 | −0.03 |
| 1 | 0.55 | 0.00 | 0.04 | 0.00 | 0.00 | 13 | 7.14 | 0.06 | 0.53 | 0.00 | −0.03 |
| 2 | 1.10 | 0.01 | 0.08 | 0.00 | 0.00 | 14 | 7.69 | 0.06 | 0.57 | 0.00 | −0.03 |
| 3 | 1.65 | 0.01 | 0.12 | 0.00 | −0.01 | 15 | 8.24 | 0.07 | 0.62 | 0.00 | −0.03 |
| 4 | 2.20 | 0.02 | 0.16 | 0.00 | −0.01 | 16 | 8.78 | 0.07 | 0.66 | 0.00 | −0.04 |
| 5 | 2.75 | 0.02 | 0.21 | 0.00 | −0.01 | 17 | 9.33 | 0.08 | 0.70 | 0.00 | −0.04 |
| 6 | 3.29 | 0.03 | 0.25 | 0.00 | −0.01 | 18 | 9.88 | 0.08 | 0.74 | 0.00 | −0.04 |
| 7 | 3.84 | 0.03 | 0.29 | 0.00 | −0.02 | 19 | 10.43 | 0.09 | 0.78 | 0.00 | −0.04 |
| 8 | 4.39 | 0.04 | 0.33 | 0.00 | −0.02 | 20 | 10.98 | 0.09 | 0.82 | 0.00 | −0.04 |
| 9 | 4.94 | 0.04 | 0.37 | 0.00 | −0.02 | 21 | 11.53 | 0.10 | 0.86 | 0.00 | −0.05 |
| 10 | 5.49 | 0.05 | 0.41 | 0.00 | −0.02 | 22 | 12.08 | 0.10 | 0.90 | 0.00 | −0.05 |
| 11 | 6.04 | 0.05 | 0.45 | 0.00 | −0.02 | 23 | 12.63 | 0.11 | 0.94 | 0.00 | −0.05 |

[1] The table may also be used directly for dates between Jan. 1 and Feb. 29, inclusive, of leap years; but if the required date falls between Mar. 1 and Dec. 31, inclusive, of a leap year, the day of month should be increased by one before entering the table.

HARMONIC ANALYSIS AND PREDICTION OF TIDES

Table 6.—*Values of I, ν, ξ, ν', and 2ν'' for each degree of N.*

| N | Positive always | | Positive when $N$ is between 0 and 180°; negative when $N$ is between 180 and 360° | | | | | | | | N |
|---|---|---|---|---|---|---|---|---|---|---|---|
| | I | | ν | | ξ | | ν' | | 2ν'' | | |
| ° | ° | Diff. | ° | Diff. | ° | Diff. | ° | Diff. | ° | Diff. | ° |
| 0 | 28.60 | 0 | 0.00 | 19 | 0.00 | 17 | 0.00 | 13 | 0.00 | 28 | 360 |
| 1 | 28.60 | 0 | 0.19 | 19 | 0.17 | 17 | 0.13 | 14 | 0.28 | 29 | 359 |
| 2 | 28.60 | 1 | 0.38 | 18 | 0.34 | 17 | 0.27 | 13 | 0.57 | 28 | 358 |
| 3 | 28.59 | 0 | 0.56 | 19 | 0.51 | 16 | 0.40 | 14 | 0.85 | 29 | 357 |
| 4 | 28.59 | 1 | 0.75 | 19 | 0.67 | 17 | 0.54 | 13 | 1.14 | 28 | 356 |
| 5 | 28.58 | 0 | 0.94 | 18 | 0.84 | 17 | 0.67 | 13 | 1.42 | 28 | 355 |
| 6 | 28.58 | 1 | 1.12 | 19 | 1.01 | 17 | 0.80 | 14 | 1.70 | 29 | 354 |
| 7 | 28.57 | 1 | 1.31 | 19 | 1.18 | 17 | 0.94 | 13 | 1.99 | 28 | 353 |
| 8 | 28.56 | 1 | 1.50 | 18 | 1.35 | 16 | 1.07 | 13 | 2.27 | 28 | 352 |
| 9 | 28.55 | 2 | 1.68 | 19 | 1.51 | 17 | 1.20 | 14 | 2.55 | 28 | 351 |
| 10 | 28.53 | 1 | 1.87 | 18 | 1.68 | 17 | 1.34 | 13 | 2.83 | 28 | 350 |
| 11 | 28.52 | 2 | 2.05 | 19 | 1.85 | 17 | 1.47 | 13 | 3.11 | 28 | 349 |
| 12 | 28.50 | 1 | 2.24 | 18 | 2.02 | 16 | 1.60 | 13 | 3.39 | 28 | 348 |
| 13 | 28.49 | 2 | 2.42 | 19 | 2.18 | 17 | 1.73 | 13 | 3.67 | 28 | 347 |
| 14 | 28.47 | 2 | 2.61 | 18 | 2.35 | 16 | 1.86 | 13 | 3.95 | 28 | 346 |
| 15 | 28.45 | 2 | 2.79 | 19 | 2.51 | 17 | 1.99 | 13 | 4.23 | 28 | 345 |
| 16 | 28.43 | 2 | 2.98 | 18 | 2.68 | 16 | 2.12 | 13 | 4.51 | 27 | 344 |
| 17 | 28.41 | 2 | 3.16 | 18 | 2.84 | 17 | 2.25 | 13 | 4.78 | 28 | 343 |
| 18 | 28.39 | 3 | 3.34 | 18 | 3.01 | 16 | 2.38 | 13 | 5.06 | 27 | 342 |
| 19 | 28.36 | 2 | 3.52 | 18 | 3.17 | 17 | 2.51 | 13 | 5.33 | 27 | 341 |
| 20 | 28.34 | 3 | 3.70 | 18 | 3.34 | 16 | 2.64 | 13 | 5.60 | 27 | 340 |
| 21 | 28.31 | 2 | 3.88 | 18 | 3.50 | 16 | 2.77 | 13 | 5.87 | 27 | 339 |
| 22 | 28.29 | 3 | 4.06 | 18 | 3.66 | 16 | 2.90 | 13 | 6.14 | 27 | 338 |
| 23 | 28.26 | 3 | 4.24 | 18 | 3.82 | 16 | 3.03 | 12 | 6.41 | 27 | 337 |
| 24 | 28.23 | 3 | 4.42 | 18 | 3.98 | 16 | 3.15 | 13 | 6.68 | 26 | 336 |
| 25 | 28.20 | 4 | 4.60 | 18 | 4.14 | 16 | 3.28 | 12 | 6.94 | 27 | 335 |
| 26 | 28.16 | 3 | 4.78 | 17 | 4.30 | 16 | 3.40 | 13 | 7.21 | 26 | 334 |
| 27 | 28.13 | 4 | 4.95 | 18 | 4.46 | 16 | 3.53 | 12 | 7.47 | 26 | 333 |
| 28 | 28.09 | 3 | 5.13 | 17 | 4.62 | 16 | 3.65 | 13 | 7.73 | 26 | 332 |
| 29 | 28.06 | 4 | 5.30 | 18 | 4.78 | 16 | 3.78 | 12 | 7.99 | 26 | 331 |
| 30 | 28.02 | 4 | 5.48 | 17 | 4.94 | 16 | 3.90 | 12 | 8.25 | 25 | 330 |
| 31 | 27.98 | 4 | 5.65 | 17 | 5.10 | 15 | 4.02 | 12 | 8.50 | 25 | 329 |
| 32 | 27.94 | 4 | 5.82 | 17 | 5.25 | 16 | 4.14 | 12 | 8.75 | 25 | 328 |
| 33 | 27.90 | 4 | 5.99 | 17 | 5.41 | 15 | 4.26 | 12 | 9.00 | 25 | 327 |
| 34 | 27.86 | 4 | 6.16 | 17 | 5.56 | 15 | 4.38 | 12 | 9.25 | 25 | 326 |
| 35 | 27.82 | 5 | 6.33 | 17 | 5.71 | 15 | 4.50 | 12 | 9.50 | 24 | 325 |
| 36 | 27.77 | 4 | 6.50 | 16 | 5.86 | 15 | 4.62 | 12 | 9.74 | 24 | 324 |
| 37 | 27.73 | 5 | 6.66 | 17 | 6.01 | 15 | 4.74 | 11 | 9.98 | 24 | 323 |
| 38 | 27.68 | 5 | 6.83 | 16 | 6.16 | 15 | 4.85 | 12 | 10.22 | 24 | 322 |
| 39 | 27.63 | 5 | 6.99 | 16 | 6.31 | 15 | 4.97 | 11 | 10.46 | 23 | 321 |
| 40 | 27.58 | 5 | 7.15 | 16 | 6.46 | 15 | 5.08 | 11 | 10.69 | 24 | 320 |
| 41 | 27.53 | 5 | 7.31 | 16 | 6.61 | 14 | 5.19 | 11 | 10.93 | 23 | 319 |
| 42 | 27.48 | 5 | 7.47 | 16 | 6.75 | 15 | 5.30 | 11 | 11.16 | 22 | 318 |
| 43 | 27.43 | 5 | 7.63 | 16 | 6.90 | 14 | 5.41 | 11 | 11.38 | 22 | 317 |
| 44 | 27.38 | 6 | 7.79 | 15 | 7.04 | 14 | 5.52 | 11 | 11.60 | 22 | 316 |
| 45 | 27.32 | | 7.94 | | 7.18 | | 5.63 | | 11.82 | | 315 |

Table 6.—*Values of I, ν, ξ, ν', and 2ν'' for each degree of N*—Continued

| N | Positive always | | Positive when N is between 0 and 180°; negative when N is between 180 and 360° | | | | | | | | N |
|---|---|---|---|---|---|---|---|---|---|---|---|
| | I | Diff. | ν | Diff. | ξ | Diff. | ν' | Diff. | 2ν'' | Diff. | |
| ° | ° | | ° | | ° | | ° | | ° | | ° |
| 45 | 27.32 | 5 | 7.94 | 16 | 7.18 | 14 | 5.63 | 11 | 11.82 | 22 | 315 |
| 46 | 27.27 | 6 | 8.10 | 15 | 7.32 | 14 | 5.74 | 10 | 12.04 | 22 | 314 |
| 47 | 27.21 | 6 | 8.25 | 15 | 7.46 | 14 | 5.84 | 11 | 12.26 | 21 | 313 |
| 48 | 27.15 | 6 | 8.40 | 15 | 7.60 | 13 | 5.95 | 10 | 12.47 | 21 | 312 |
| 49 | 27.09 | 6 | 8.55 | 14 | 7.73 | 14 | 6.05 | 10 | 12.68 | 20 | 311 |
| 50 | 27.03 | 6 | 8.69 | 15 | 7.87 | 13 | 6.15 | 10 | 12.88 | 20 | 310 |
| 51 | 26.97 | 6 | 8.84 | 14 | 8.00 | 14 | 6.25 | 10 | 13.08 | 20 | 309 |
| 52 | 26.91 | 6 | 8.98 | 14 | 8.14 | 13 | 6.35 | 10 | 13.28 | 20 | 308 |
| 53 | 26.85 | 7 | 9.12 | 14 | 8.27 | 13 | 6.45 | 9 | 13.48 | 19 | 307 |
| 54 | 26.78 | 6 | 9.26 | 14 | 8.40 | 12 | 6.54 | 10 | 13.67 | 19 | 306 |
| 55 | 26.72 | 7 | 9.40 | 14 | 8.52 | 13 | 6.64 | 9 | 13.86 | 19 | 305 |
| 56 | 26.65 | 6 | 9.54 | 13 | 8.65 | 12 | 6.73 | 9 | 14.05 | 18 | 304 |
| 57 | 26.59 | 7 | 9.67 | 14 | 8.77 | 13 | 6.82 | 9 | 14.23 | 17 | 303 |
| 58 | 26.52 | 7 | 9.81 | 13 | 8.90 | 12 | 6.91 | 9 | 14.40 | 18 | 302 |
| 59 | 26.45 | 7 | 9.94 | 13 | 9.02 | 12 | 7.00 | 9 | 14.58 | 17 | 301 |
| 60 | 26.38 | 7 | 10.07 | 12 | 9.14 | 11 | 7.09 | 8 | 14.75 | 17 | 300 |
| 61 | 26.31 | 7 | 10.19 | 13 | 9.25 | 12 | 7.17 | 9 | 14.92 | 16 | 299 |
| 62 | 26.24 | 7 | 10.32 | 12 | 9.37 | 11 | 7.26 | 8 | 15.08 | 16 | 298 |
| 63 | 26.17 | 7 | 10.44 | 12 | 9.48 | 11 | 7.34 | 8 | 15.24 | 15 | 297 |
| 64 | 26.10 | 7 | 10.56 | 12 | 9.59 | 11 | 7.42 | 7 | 15.39 | 15 | 296 |
| 65 | 26.03 | 8 | 10.68 | 11 | 9.70 | 11 | 7.49 | 8 | 15.54 | 15 | 295 |
| 66 | 25.95 | 7 | 10.79 | 11 | 9.81 | 11 | 7.57 | 7 | 15.69 | 14 | 294 |
| 67 | 25.88 | 8 | 10.90 | 11 | 9.92 | 10 | 7.64 | 8 | 15.83 | 13 | 293 |
| 68 | 25.80 | 8 | 11.01 | 11 | 10.02 | 10 | 7.72 | 7 | 15.96 | 14 | 292 |
| 69 | 25.72 | 7 | 11.12 | 11 | 10.12 | 10 | 7.79 | 7 | 16.10 | 13 | 291 |
| 70 | 25.65 | 8 | 11.23 | 10 | 10.22 | 10 | 7.86 | 6 | 16.23 | 12 | 290 |
| 71 | 25.57 | 8 | 11.33 | 10 | 10.32 | 9 | 7.92 | 7 | 16.35 | 12 | 289 |
| 72 | 25.49 | 8 | 11.43 | 10 | 10.41 | 9 | 7.99 | 6 | 16.47 | 11 | 288 |
| 73 | 25.41 | 8 | 11.53 | 10 | 10.50 | 9 | 8.05 | 6 | 16.58 | 11 | 287 |
| 74 | 25.33 | 8 | 11.63 | 9 | 10.59 | 9 | 8.11 | 6 | 16.69 | 11 | 286 |
| 75 | 25.25 | 8 | 11.72 | 9 | 10.68 | 9 | 8.17 | 6 | 16.80 | 10 | 285 |
| 76 | 25.17 | 8 | 11.81 | 8 | 10.77 | 8 | 8.23 | 5 | 16.90 | 10 | 284 |
| 77 | 25.09 | 8 | 11.89 | 9 | 10.85 | 8 | 8.28 | 6 | 17.00 | 9 | 283 |
| 78 | 25.01 | 9 | 11.98 | 8 | 10.93 | 8 | 8.34 | 5 | 17.09 | 8 | 282 |
| 79 | 24.92 | 8 | 12.06 | 8 | 11.01 | 7 | 8.39 | 5 | 17.17 | 8 | 281 |
| 80 | 24.84 | 8 | 12.14 | 7 | 11.08 | 7 | 8.44 | 4 | 17.25 | 8 | 280 |
| 81 | 24.76 | 9 | 12.21 | 7 | 11.15 | 7 | 8.48 | 5 | 17.33 | 7 | 279 |
| 82 | 24.67 | 8 | 12.28 | 7 | 11.22 | 7 | 8.53 | 4 | 17.40 | 6 | 278 |
| 83 | 24.59 | 9 | 12.35 | 7 | 11.29 | 7 | 8.57 | 4 | 17.46 | 6 | 277 |
| 84 | 24.50 | 8 | 12.42 | 6 | 11.36 | 6 | 8.61 | 3 | 17.52 | 6 | 276 |
| 85 | 24.42 | 9 | 12.48 | 6 | 11.42 | 5 | 8.64 | 4 | 17.58 | 5 | 275 |
| 86 | 24.33 | 9 | 12.54 | 6 | 11.47 | 6 | 8.68 | 3 | 17.63 | 4 | 274 |
| 87 | 24.24 | 8 | 12.60 | 5 | 11.53 | 5 | 8.71 | 3 | 17.67 | 4 | 273 |
| 88 | 24.16 | 9 | 12.65 | 5 | 11.58 | 5 | 8.74 | 2 | 17.71 | 3 | 272 |
| 89 | 24.07 | 9 | 12.70 | 5 | 11.63 | 5 | 8.76 | 3 | 17.74 | 3 | 271 |
| 90 | 23.98 | | 12.75 | | 11.68 | | 8.79 | | 17.77 | | 270 |

Table 6.—*Values of I, ν, ξ, ν', and 2ν'' for each degree of N*—Continued

| N | Positive always | | Positive when N is between 0 and 180°; negative when N is between 180 and 360° | | | | | | | | N |
|---|---|---|---|---|---|---|---|---|---|---|---|
| | I | | ν | | ξ | | ν' | | 2ν'' | | |
| ° | ° | Diff. | ° | Diff. | ° | Diff. | ° | Diff. | ° | Diff. | ° |
| 90 | 23.98 | 9 | 12.75 | 4 | 11.68 | 4 | 8.79 | 2 | 17.77 | 2 | 270 |
| 91 | 23.89 | 9 | 12.79 | 4 | 11.72 | 4 | 8.81 | 2 | 17.79 | 2 | 269 |
| 92 | 23.80 | 8 | 12.83 | 4 | 11.76 | 4 | 8.83 | 2 | 17.81 | 1 | 268 |
| 93 | 23.72 | 9 | 12.87 | 3 | 11.80 | 3 | 8.85 | 1 | 17.82 | 1 | 267 |
| 94 | 23.63 | 9 | 12.90 | 3 | 11.83 | 3 | 8.86 | 1 | 17.83 | 0 | 266 |
| 95 | 23.54 | 9 | 12.93 | 2 | 11.86 | 3 | 8.87 | 1 | 17.83 | 1 | 265 |
| 96 | 23.45 | 9 | 12.95 | 2 | 11.89 | 3 | 8.88 | 1 | 17.82 | 1 | 264 |
| 97 | 23.36 | 9 | 12.97 | 2 | 11.92 | 2 | 8.89 | 1 | 17.81 | 2 | 263 |
| 98 | 23.27 | 9 | 12.99 | 2 | 11.94 | 1 | 8.90 | 0 | 17.79 | 2 | 262 |
| 99 | 23.18 | 9 | 13.01 | 1 | 11.95 | 1 | 8.90 | 1 | 17.77 | 3 | 261 |
| 100 | 23.09 | 9 | 13.02 | 0 | 11.96 | 1 | 8.89 | 0 | 17.74 | 3 | 260 |
| 101 | 23.00 | 9 | 13.02 | 0 | 11.97 | 1 | 8.89 | 1 | 17.71 | 4 | 259 |
| 102 | 22.91 | 9 | 13.02 | 0 | 11.98 | 0 | 8.88 | 1 | 17.67 | 5 | 258 |
| 103 | 22.82 | 9 | 13.02 | 1 | 11.98 | 0 | 8.87 | 1 | 17.62 | 5 | 257 |
| 104 | 22.73 | 9 | 13.01 | 1 | 11.98 | 1 | 8.86 | 2 | 17.57 | 6 | 256 |
| 105 | 22.64 | 9 | 13.00 | 1 | 11.97 | 1 | 8.84 | 2 | 17.51 | 6 | 255 |
| 106 | 22.55 | 9 | 12.99 | 2 | 11.96 | 1 | 8.82 | 2 | 17.45 | 7 | 254 |
| 107 | 22.46 | 9 | 12.97 | 2 | 11.95 | 2 | 8.80 | 2 | 17.38 | 8 | 253 |
| 108 | 22.37 | 9 | 12.95 | 3 | 11.93 | 2 | 8.78 | 3 | 17.30 | 8 | 252 |
| 109 | 22.28 | 8 | 12.92 | 3 | 11.91 | 2 | 8.75 | 3 | 17.22 | 8 | 251 |
| 110 | 22.20 | 9 | 12.89 | 4 | 11.89 | 3 | 8.72 | 3 | 17.14 | 9 | 250 |
| 111 | 22.11 | 9 | 12.85 | 4 | 11.86 | 3 | 8.69 | 4 | 17.05 | 10 | 249 |
| 112 | 22.02 | 9 | 12.81 | 4 | 11.83 | 4 | 8.65 | 4 | 16.95 | 11 | 248 |
| 113 | 21.93 | 9 | 12.77 | 5 | 11.79 | 4 | 8.61 | 4 | 16.84 | 11 | 247 |
| 114 | 21.84 | 9 | 12.72 | 5 | 11.75 | 5 | 8.57 | 5 | 16.73 | 11 | 246 |
| 115 | 21.75 | 8 | 12.67 | 6 | 11.70 | 5 | 8.52 | 4 | 16.62 | 12 | 245 |
| 116 | 21.67 | 9 | 12.61 | 6 | 11.65 | 5 | 8.48 | 5 | 16.50 | 13 | 244 |
| 117 | 21.58 | 8 | 12.55 | 7 | 11.60 | 6 | 8.43 | 6 | 16.37 | 13 | 243 |
| 118 | 21.50 | 9 | 12.48 | 7 | 11.54 | 6 | 8.37 | 6 | 16.24 | 14 | 242 |
| 119 | 21.41 | 9 | 12.41 | 8 | 11.48 | 7 | 8.31 | 6 | 16.10 | 14 | 241 |
| 120 | 21.32 | 8 | 12.33 | 8 | 11.41 | 7 | 8.25 | 6 | 15.96 | 15 | 240 |
| 121 | 21.24 | 9 | 12.25 | 8 | 11.34 | 8 | 8.19 | 6 | 15.81 | 15 | 239 |
| 122 | 21.15 | 8 | 12.17 | 9 | 11.26 | 8 | 8.13 | 7 | 15.66 | 16 | 238 |
| 123 | 21.07 | 8 | 12.08 | 10 | 11.18 | 8 | 8.06 | 7 | 15.50 | 17 | 237 |
| 124 | 20.99 | 8 | 11.98 | 10 | 11.10 | 9 | 7.99 | 8 | 15.33 | 17 | 236 |
| 125 | 20.91 | 9 | 11.88 | 10 | 11.01 | 9 | 7.91 | 8 | 15.16 | 17 | 235 |
| 126 | 20.82 | 8 | 11.78 | 11 | 10.92 | 10 | 7.83 | 8 | 14.99 | 18 | 234 |
| 127 | 20.74 | 8 | 11.67 | 12 | 10.82 | 10 | 7.75 | 8 | 14.81 | 19 | 233 |
| 128 | 20.66 | 8 | 11.55 | 12 | 10.72 | 11 | 7.67 | 9 | 14.62 | 19 | 232 |
| 129 | 20.58 | 7 | 11.43 | 12 | 10.61 | 11 | 7.58 | 9 | 14.43 | 20 | 231 |
| 130 | 20.51 | 8 | 11.31 | 13 | 10.50 | 12 | 7.49 | 9 | 14.23 | 20 | 230 |
| 131 | 20.43 | 8 | 11.18 | 13 | 10.38 | 12 | 7.40 | 10 | 14.03 | 20 | 229 |
| 132 | 20.35 | 7 | 11.05 | 14 | 10.26 | 13 | 7.30 | 10 | 13.83 | 21 | 228 |
| 133 | 20.28 | 8 | 10.91 | 14 | 10.13 | 13 | 7.20 | 10 | 13.62 | 22 | 227 |
| 134 | 20.20 | 7 | 10.77 | 15 | 10.00 | 13 | 7.10 | 10 | 13.40 | 22 | 226 |
| 135 | 20.13 | | 10.62 | | 9.87 | | 7.00 | | 13.18 | | 225 |

Table 6.—*Values of I, ν, ξ, ν′, and 2ν″ for each degree of N*—Continued

| N | Positive always | | Positive when N is between 0 and 180°; negative when N is between 180 and 360° | | | | | | | | N |
|---|---|---|---|---|---|---|---|---|---|---|---|
| | I | | ν | | ξ | | ν′ | | 2ν″ | | |
| ° | ° | Diff. | ° | Diff. | ° | Diff. | ° | Diff. | ° | Diff. | ° |
| 135 | 20.13 | 8 | 10.62 | 15 | 9.87 | 14 | 7.00 | 11 | 13.18 | 22 | 225 |
| 136 | 20.05 | 7 | 10.47 | 16 | 9.73 | 14 | 6.89 | 11 | 12.96 | 23 | 224 |
| 137 | 19.98 | 7 | 10.31 | 16 | 9.59 | 15 | 6.78 | 12 | 12.73 | 24 | 223 |
| 138 | 19.91 | 7 | 10.15 | 17 | 9.44 | 15 | 6.66 | 11 | 12.49 | 24 | 222 |
| 139 | 19.84 | 7 | 9.98 | 17 | 9.29 | 16 | 6.55 | 12 | 12.25 | 24 | 221 |
| 140 | 19.77 | 6 | 9.81 | 18 | 9.13 | 16 | 6.43 | 12 | 12.01 | 25 | 220 |
| 141 | 19.71 | 7 | 9.63 | 18 | 8.97 | 17 | 6.31 | 13 | 11.76 | 25 | 219 |
| 142 | 19.64 | 6 | 9.45 | 18 | 8.80 | 17 | 6.18 | 12 | 11.51 | 25 | 218 |
| 143 | 19.58 | 7 | 9.27 | 19 | 8.63 | 17 | 6.06 | 13 | 11.26 | 26 | 217 |
| 144 | 19.51 | 6 | 9.08 | 19 | 8.46 | 18 | 5.93 | 13 | 11.00 | 26 | 216 |
| 145 | 19.45 | 6 | 8.89 | 20 | 8.28 | 18 | 5.80 | 14 | 10.74 | 26 | 215 |
| 146 | 19.39 | 6 | 8.69 | 20 | 8.10 | 19 | 5.66 | 14 | 10.48 | 27 | 214 |
| 147 | 19.33 | 6 | 8.49 | 21 | 7.91 | 19 | 5.52 | 14 | 10.21 | 27 | 213 |
| 148 | 19.27 | 5 | 8.28 | 21 | 7.72 | 20 | 5.38 | 14 | 9.94 | 28 | 212 |
| 149 | 19.22 | 6 | 8.07 | 22 | 7.52 | 20 | 5.24 | 15 | 9.66 | 28 | 211 |
| 150 | 19.16 | 5 | 7.85 | 22 | 7.32 | 20 | 5.09 | 14 | 9.38 | 28 | 210 |
| 151 | 19.11 | 6 | 7.63 | 22 | 7.12 | 21 | 4.95 | 15 | 9.10 | 29 | 209 |
| 152 | 19.05 | 5 | 7.41 | 23 | 6.91 | 21 | 4.80 | 15 | 8.81 | 29 | 208 |
| 153 | 19.00 | 5 | 7.18 | 23 | 6.70 | 21 | 4.65 | 15 | 8.52 | 29 | 207 |
| 154 | 18.95 | 4 | 6.95 | 23 | 6.49 | 22 | 4.50 | 16 | 8.23 | 29 | 206 |
| 155 | 18.91 | 5 | 6.72 | 24 | 6.27 | 22 | 4.34 | 15 | 7.94 | 30 | 205 |
| 156 | 18.86 | 4 | 6.48 | 24 | 6.05 | 23 | 4.19 | 16 | 7.64 | 30 | 204 |
| 157 | 18.82 | 4 | 6.24 | 25 | 5.82 | 23 | 4.03 | 16 | 7.34 | 30 | 203 |
| 158 | 18.78 | 4 | 5.99 | 25 | 5.59 | 23 | 3.87 | 17 | 7.04 | 30 | 202 |
| 159 | 18.74 | 4 | 5.74 | 25 | 5.36 | 23 | 3.70 | 16 | 6.74 | 31 | 201 |
| 160 | 18.70 | 4 | 5.49 | 25 | 5.13 | 24 | 3.54 | 17 | 6.43 | 31 | 200 |
| 161 | 18.66 | 4 | 5.24 | 26 | 4.89 | 24 | 3.37 | 17 | 6.12 | 31 | 199 |
| 162 | 18.62 | 3 | 4.98 | 26 | 4.65 | 24 | 3.20 | 17 | 5.81 | 31 | 198 |
| 163 | 18.59 | 3 | 4.72 | 26 | 4.41 | 25 | 3.03 | 17 | 5.50 | 31 | 197 |
| 164 | 18.56 | 3 | 4.46 | 27 | 4.16 | 25 | 2.86 | 17 | 5.19 | 32 | 196 |
| 165 | 18.53 | 3 | 4.19 | 27 | 3.91 | 25 | 2.69 | 17 | 4.87 | 32 | 195 |
| 166 | 18.50 | 3 | 3.92 | 27 | 3.66 | 25 | 2.52 | 18 | 4.55 | 32 | 194 |
| 167 | 18.47 | 2 | 3.65 | 27 | 3.41 | 25 | 2.34 | 17 | 4.23 | 32 | 193 |
| 168 | 18.45 | 2 | 3.38 | 28 | 3.16 | 26 | 2.17 | 18 | 3.91 | 32 | 192 |
| 169 | 18.43 | 2 | 3.10 | 27 | 2.90 | 26 | 1.99 | 18 | 3.59 | 32 | 191 |
| 170 | 18.41 | 2 | 2.83 | 28 | 2.64 | 26 | 1.81 | 18 | 3.27 | 33 | 190 |
| 171 | 18.39 | 2 | 2.55 | 28 | 2.38 | 26 | 1.63 | 18 | 2.94 | 32 | 189 |
| 172 | 18.37 | 1 | 2.27 | 28 | 2.12 | 26 | 1.45 | 18 | 2.62 | 33 | 188 |
| 173 | 18.36 | 2 | 1.99 | 28 | 1.86 | 26 | 1.27 | 18 | 2.29 | 32 | 187 |
| 174 | 18.34 | 1 | 1.71 | 29 | 1.60 | 27 | 1.09 | 18 | 1.97 | 33 | 186 |
| 175 | 18.33 | 1 | 1.42 | 28 | 1.33 | 26 | 0.91 | 18 | 1.64 | 33 | 185 |
| 176 | 18.32 | 0 | 1.14 | 28 | 1.07 | 27 | 0.73 | 18 | 1.31 | 32 | 184 |
| 177 | 18.32 | 1 | 0.86 | 29 | 0.80 | 26 | 0.55 | 18 | 0.99 | 33 | 183 |
| 178 | 18.31 | 0 | 0.57 | 28 | 0.54 | 27 | 0.37 | 19 | 0.66 | 33 | 182 |
| 179 | 18.31 | 0 | 0.29 | 29 | 0.27 | 27 | 0.18 | 18 | 0.33 | 33 | 181 |
| 180 | 18.31 | | 0.00 | | 0.00 | | 0.00 | | 0.00 | | 180 |

HARMONIC ANALYSIS AND PREDICTION OF TIDES 177

Table 7.—Log $R_a$ for amplitude of constituent $L_2$

| P | I | 18° | 19° | 20° | 21° | 22° | 23° | 24° | 25° | 26° | 27° | 28° | 29° | I | P |
|---|---|---|---|---|---|---|---|---|---|---|---|---|---|---|---|
| ° | ° | | | | | | | | | | | | | ° | ° |
| 0 | 180 | 0.0708 | 0.0799 | 0.0897 | 0.1002 | 0.1117 | 0.1240 | 0.1373 | 0.1517 | 0.1674 | 0.1843 | 0.2027 | 0.2228 | 180 | 360 |
| 5 | 185 | 0.0695 | 0.0783 | 0.0878 | 0.0981 | 0.1092 | 0.1211 | 0.1340 | 0.1479 | 0.1628 | 0.1790 | 0.1966 | 0.2155 | 175 | 355 |
| 10 | 190 | 0.0654 | 0.0736 | 0.0824 | 0.0918 | 0.1019 | 0.1128 | 0.1244 | 0.1367 | 0.1500 | 0.1641 | 0.1792 | 0.1953 | 170 | 350 |
| 15 | 195 | 0.0590 | 0.0662 | 0.0739 | 0.0820 | 0.0906 | 0.0998 | 0.1095 | 0.1197 | 0.1305 | 0.1417 | 0.1535 | 0.1658 | 165 | 345 |
| 20 | 200 | 0.0506 | 0.0565 | 0.0628 | 0.0693 | 0.0762 | 0.0834 | 0.0908 | 0.0986 | 0.1065 | 0.1147 | 0.1230 | 0.1313 | 160 | 340 |
| 25 | 205 | 0.0407 | 0.0452 | 0.0498 | 0.0546 | 0.0596 | 0.0647 | 0.0698 | 0.0750 | 0.0802 | 0.0853 | 0.0904 | 0.0952 | 155 | 335 |
| 30 | 210 | 0.0297 | 0.0327 | 0.0357 | 0.0388 | 0.0418 | 0.0449 | 0.0478 | 0.0506 | 0.0533 | 0.0557 | 0.0578 | 0.0597 | 150 | 330 |
| 35 | 215 | 0.0182 | 0.0197 | 0.0212 | 0.0225 | 0.0238 | 0.0249 | 0.0258 | 0.0265 | 0.0269 | 0.0270 | 0.0268 | 0.0262 | 145 | 325 |
| 40 | 220 | 0.0065 | 0.0066 | 0.0066 | 0.0064 | 0.0060 | 0.0054 | 0.0045 | 0.0034 | 0.0019 | 0.0001 | 9.9979 | 9.9953 | 140 | 320 |
| 45 | 225 | 9.9951 | 9.9940 | 9.9926 | 9.9910 | 9.9891 | 9.9870 | 9.9846 | 9.9819 | 9.9789 | 9.9755 | 9.9717 | 9.9676 | 135 | 315 |
| 50 | 230 | 9.9843 | 9.9820 | 9.9794 | 9.9765 | 9.9734 | 9.9700 | 9.9663 | 9.9623 | 9.9580 | 9.9533 | 9.9483 | 9.9430 | 130 | 310 |
| 55 | 235 | 9.9743 | 9.9710 | 9.9673 | 9.9634 | 9.9592 | 9.9548 | 9.9500 | 9.9449 | 9.9395 | 9.9338 | 9.9278 | 9.9215 | 125 | 305 |
| 60 | 240 | 9.9653 | 9.9611 | 9.9566 | 9.9518 | 9.9467 | 9.9414 | 9.9357 | 9.9298 | 9.9235 | 9.9170 | 9.9102 | 9.9031 | 120 | 300 |
| 65 | 245 | 9.9575 | 9.9526 | 9.9473 | 9.9418 | 9.9360 | 9.9299 | 9.9236 | 9.9169 | 9.9100 | 9.9029 | 9.8954 | 9.8877 | 115 | 295 |
| 70 | 250 | 9.9510 | 9.9454 | 9.9396 | 9.9335 | 9.9271 | 9.9205 | 9.9136 | 9.9064 | 9.8990 | 9.8913 | 9.8834 | 9.8753 | 110 | 290 |
| 75 | 255 | 9.9458 | 9.9398 | 9.9336 | 9.9270 | 9.9202 | 9.9131 | 9.9058 | 9.8982 | 9.8904 | 9.8824 | 9.8742 | 9.8657 | 105 | 285 |
| 80 | 260 | 9.9421 | 9.9358 | 9.9292 | 9.9224 | 9.9152 | 9.9079 | 9.9003 | 9.8924 | 9.8844 | 9.8761 | 9.8676 | 9.8589 | 100 | 280 |
| 85 | 265 | 9.9399 | 9.9334 | 9.9266 | 9.9196 | 9.9123 | 9.9047 | 9.8969 | 9.8889 | 9.8807 | 9.8723 | 9.8636 | 9.8548 | 95 | 275 |
| 90 | 270 | 9.9391 | 9.9326 | 9.9257 | 9.9186 | 9.9113 | 9.9037 | 9.8958 | 9.8878 | 9.8795 | 9.8710 | 9.8623 | 9.8535 | 90 | 270 |

## Table 8.—Values of R for argument of constituent $L_2$

[Tabular values are positive when P is in first or third quadrants, negative when P is in second or fourth quadrants]

| I | P | 18° | 19° | 20° | 21° | 22° | 23° | 24° | 25° | 26° | 27° | 28° | 29° | I | P |
|---|---|---|---|---|---|---|---|---|---|---|---|---|---|---|---|
| ° | ° | | | | | | | | | | | | | ° | ° |
| 180 | 0 | 0.0 | 0.0 | 0.0 | 0.0 | 0.0 | 0.0 | 0.0 | 0.0 | 0.0 | 0.0 | 0.0 | 0.0 | 180 | 360 |
| 185 | 5 | 1.8 | 2.0 | 2.3 | 2.6 | 2.9 | 3.3 | 3.7 | 4.1 | 4.6 | 5.2 | 5.9 | 6.6 | 175 | 355 |
| 190 | 10 | 3.4 | 3.9 | 4.4 | 5.0 | 5.6 | 6.3 | 7.1 | 7.9 | 8.9 | 9.9 | 11.1 | 12.4 | 170 | 350 |
| 195 | 15 | 4.9 | 5.6 | 6.4 | 7.1 | 8.0 | 9.0 | 10.0 | 11.2 | 12.5 | 13.9 | 15.4 | 17.1 | 165 | 345 |
| 200 | 20 | 6.2 | 7.1 | 8.0 | 8.9 | 10.0 | 11.2 | 12.4 | 13.8 | 15.2 | 16.8 | 18.6 | 20.4 | 160 | 340 |
| 205 | 25 | 7.3 | 8.2 | 9.2 | 10.3 | 11.5 | 12.8 | 14.1 | 15.6 | 17.1 | 18.8 | 20.6 | 22.5 | 155 | 335 |
| 210 | 30 | 8.0 | 9.0 | 10.1 | 11.3 | 12.5 | 13.8 | 15.2 | 16.7 | 18.2 | 19.9 | 21.7 | 23.5 | 150 | 330 |
| 215 | 35 | 8.5 | 9.5 | 10.6 | 11.8 | 13.0 | 14.3 | 15.7 | 17.1 | 18.6 | 20.2 | 21.9 | 23.6 | 145 | 325 |
| 220 | 40 | 8.7 | 9.7 | 10.8 | 11.9 | 13.1 | 14.3 | 15.5 | 17.0 | 18.4 | 19.9 | 21.4 | 23.0 | 140 | 320 |
| 225 | 45 | 8.6 | 9.5 | 10.6 | 11.6 | 12.8 | 13.9 | 15.2 | 16.4 | 17.7 | 19.1 | 20.5 | 21.9 | 135 | 315 |
| 230 | 50 | 8.2 | 9.1 | 10.1 | 11.1 | 12.1 | 13.2 | 14.3 | 15.4 | 16.6 | 17.8 | 19.0 | 20.3 | 130 | 310 |
| 235 | 55 | 7.7 | 8.5 | 9.4 | 10.3 | 11.2 | 12.1 | 13.1 | 14.1 | 15.2 | 16.2 | 17.3 | 18.3 | 125 | 305 |
| 240 | 60 | 6.9 | 7.6 | 8.4 | 9.2 | 10.0 | 10.8 | 11.7 | 12.5 | 13.4 | 14.3 | 15.2 | 16.1 | 120 | 300 |
| 245 | 65 | 6.0 | 6.6 | 7.3 | 7.9 | 8.6 | 9.3 | 10.0 | 10.8 | 11.5 | 12.2 | 13.0 | 13.7 | 115 | 295 |
| 250 | 70 | 5.0 | 5.5 | 6.0 | 6.5 | 7.1 | 7.6 | 8.2 | 8.8 | 9.4 | 10.0 | 10.6 | 11.2 | 110 | 290 |
| 255 | 75 | 3.8 | 4.2 | 4.6 | 5.0 | 5.4 | 5.8 | 6.3 | 6.7 | 7.1 | 7.6 | 8.0 | 8.5 | 105 | 285 |
| 260 | 80 | 2.6 | 2.8 | 3.1 | 3.4 | 3.7 | 3.9 | 4.2 | 4.5 | 4.8 | 5.1 | 5.4 | 5.7 | 100 | 280 |
| 265 | 85 | 1.3 | 1.4 | 1.6 | 1.7 | 1.8 | 2.0 | 2.1 | 2.3 | 2.4 | 2.6 | 2.7 | 2.9 | 95 | 275 |
| 270 | 90 | 0.0 | 0.0 | 0.0 | 0.0 | 0.0 | 0.0 | 0.0 | 0.0 | 0.0 | 0.0 | 0.0 | 0.0 | 90 | 270 |

HARMONIC ANALYSIS AND PREDICTION OF TIDES 179

Table 9.—*Log $Q_a$ for amplitude of constituent* $M_1$

| P | | Log $Q_a$ | Diff. | P | | P | | Log $Q_a$ | Diff. | P | |
|---|---|---|---|---|---|---|---|---|---|---|---|
| ° | ° | | | ° | ° | ° | ° | | | ° | ° |
| 0 | 180 | 9.7133 | 0 | 180 | 360 | 45 | 225 | 9.8182 | 47 | 135 | 315 |
| 1 | 181 | 9.7133 | | 179 | 359 | 46 | 226 | 9.8229 | | 134 | 314 |
| 2 | 182 | 9.7135 | 2 | 178 | 358 | 47 | 227 | 9.8278 | 49 | 133 | 313 |
| 3 | 183 | 9.7137 | 2 | 177 | 357 | 48 | 228 | 9.8328 | 50 | 132 | 312 |
| | | | 4 | | | | | | 51 | | |
| 4 | 184 | 9.7141 | 4 | 176 | 356 | 49 | 229 | 9.8379 | 51 | 131 | 311 |
| 5 | 185 | 9.7145 | 6 | 175 | 355 | 50 | 230 | 9.8430 | 52 | 130 | 310 |
| 6 | 186 | 9.7151 | 7 | 174 | 354 | 51 | 231 | 9.8482 | 54 | 129 | 309 |
| 7 | 187 | 9.7158 | 7 | 173 | 353 | 52 | 232 | 9.8536 | 54 | 128 | 308 |
| 8 | 188 | 9.7165 | 9 | 172 | 352 | 53 | 233 | 9.8590 | 55 | 127 | 307 |
| 9 | 189 | 9.7174 | 10 | 171 | 351 | 54 | 234 | 9.8645 | 56 | 126 | 306 |
| 10 | 190 | 9.7184 | 10 | 170 | 350 | 55 | 235 | 9.8701 | 56 | 125 | 305 |
| 11 | 191 | 9.7194 | 12 | 169 | 349 | 56 | 236 | 9.8757 | 57 | 124 | 304 |
| 12 | 192 | 9.7206 | 13 | 168 | 348 | 57 | 237 | 9.8814 | 58 | 123 | 303 |
| 13 | 193 | 9.7219 | 13 | 167 | 347 | 58 | 238 | 9.8872 | 59 | 122 | 302 |
| 14 | 194 | 9.7232 | 15 | 166 | 346 | 59 | 239 | 9.8931 | 59 | 121 | 301 |
| 15 | 195 | 9.7247 | 16 | 165 | 345 | 60 | 240 | 9.8990 | 59 | 120 | 300 |
| 16 | 196 | 9.7263 | 17 | 164 | 344 | 61 | 241 | 9.9049 | 60 | 119 | 299 |
| 17 | 197 | 9.7280 | 18 | 163 | 343 | 62 | 242 | 9.9109 | 60 | 118 | 298 |
| 18 | 198 | 9.7298 | 19 | 162 | 342 | 63 | 243 | 9.9169 | 60 | 117 | 297 |
| 19 | 199 | 9.7317 | 20 | 161 | 341 | 64 | 244 | 9.9229 | 60 | 116 | 296 |
| 20 | 200 | 9.7337 | 21 | 160 | 340 | 65 | 245 | 9.9289 | 60 | 115 | 295 |
| 21 | 201 | 9.7358 | 22 | 159 | 339 | 66 | 246 | 9.9349 | 59 | 114 | 294 |
| 22 | 202 | 9.7380 | 23 | 158 | 338 | 67 | 247 | 9.9408 | 60 | 113 | 293 |
| 23 | 203 | 9.7403 | 24 | 157 | 337 | 68 | 248 | 9.9468 | 59 | 112 | 292 |
| 24 | 204 | 9.7427 | 25 | 156 | 336 | 69 | 249 | 9.9527 | 58 | 111 | 291 |
| 25 | 205 | 9.7452 | 27 | 155 | 335 | 70 | 250 | 9.9585 | 57 | 110 | 290 |
| 26 | 206 | 9.7479 | 27 | 154 | 334 | 71 | 251 | 9.9642 | 56 | 109 | 289 |
| 27 | 207 | 9.7506 | 28 | 153 | 333 | 72 | 252 | 9.9698 | 55 | 108 | 288 |
| 28 | 208 | 9.7534 | 30 | 152 | 332 | 73 | 253 | 9.9753 | 54 | 107 | 287 |
| 29 | 209 | 9.7564 | 31 | 151 | 331 | 74 | 254 | 9.9807 | 52 | 106 | 286 |
| 30 | 210 | 9.7595 | 31 | 150 | 330 | 75 | 255 | 9.9859 | 50 | 105 | 285 |
| 31 | 211 | 9.7626 | 33 | 149 | 329 | 76 | 256 | 9.9909 | 48 | 104 | 284 |
| 32 | 212 | 9.7659 | 34 | 148 | 328 | 77 | 257 | 9.9957 | 45 | 103 | 283 |
| 33 | 213 | 9.7693 | 35 | 147 | 327 | 78 | 258 | 0.0002 | 43 | 102 | 282 |
| 34 | 214 | 9.7728 | 36 | 146 | 326 | 79 | 259 | 0.0045 | 40 | 101 | 281 |
| 35 | 215 | 9.7764 | 37 | 145 | 325 | 80 | 260 | 0.0085 | 37 | 100 | 280 |
| 36 | 216 | 9.7801 | 38 | 144 | 324 | 81 | 261 | 0.0122 | 34 | 99 | 279 |
| 37 | 217 | 9.7839 | 39 | 143 | 323 | 82 | 262 | 0.0156 | 30 | 98 | 278 |
| 38 | 218 | 9.7878 | 40 | 142 | 322 | 83 | 263 | 0.0186 | 27 | 97 | 277 |
| 39 | 219 | 9.7918 | 42 | 141 | 321 | 84 | 264 | 0.0213 | 23 | 96 | 276 |
| 40 | 220 | 9.7960 | 42 | 140 | 320 | 85 | 265 | 0.0236 | 19 | 95 | 275 |
| 41 | 221 | 9.8002 | 43 | 139 | 319 | 86 | 266 | 0.0255 | 16 | 94 | 274 |
| 42 | 222 | 9.8045 | 45 | 138 | 318 | 87 | 267 | 0.0271 | 11 | 93 | 273 |
| 43 | 223 | 9.8090 | 46 | 137 | 317 | 88 | 268 | 0.0282 | 6 | 92 | 272 |
| 44 | 224 | 9.8136 | 46 | 136 | 316 | 89 | 269 | 0.0288 | 2 | 91 | 271 |
| 45 | 225 | 9.8182 | | 135 | 315 | 90 | 270 | 0.0290 | | 90 | 270 |

Table 10.—*Values of Q for argument of constituent* $M_1$

| P | Q | Diff. | P | Q | Diff. | P | Q | Diff. | P | Q | Diff. |
|---|---|---|---|---|---|---|---|---|---|---|---|
| ° | ° | | ° | ° | | ° | ° | | ° | ° | |
| 0 | 0.0 | 0.5 | 45 | 25.8 | 0.8 | 90 | 90.0 | 2.1 | 135 | 154.2 | 0.8 |
| 1 | 0.5 | 0.5 | 46 | 26.6 | 0.8 | 91 | 92.1 | 2.0 | 136 | 155.0 | 0.8 |
| 2 | 1.0 | 0.5 | 47 | 27.4 | 0.8 | 92 | 94.1 | 2.1 | 137 | 155.8 | 0.7 |
| 3 | 1.5 | 0.4 | 48 | 28.2 | 0.9 | 93 | 96.2 | 2.0 | 138 | 156.5 | 0.7 |
| 4 | 1.9 | 0.5 | 49 | 29.1 | 0.8 | 94 | 98.2 | 2.1 | 139 | 157.2 | 0.7 |
| 5 | 2.4 | 0.5 | 50 | 29.9 | 0.9 | 95 | 100.3 | 2.0 | 140 | 157.9 | 0.7 |
| 6 | 2.9 | 0.5 | 51 | 30.8 | 0.9 | 96 | 102.3 | 2.0 | 141 | 158.6 | 0.7 |
| 7 | 3.4 | 0.5 | 52 | 31.7 | 1.0 | 97 | 104.3 | 1.9 | 142 | 159.3 | 0.7 |
| 8 | 3.9 | 0.5 | 53 | 32.7 | 0.9 | 98 | 106.2 | 2.0 | 143 | 160.0 | 0.7 |
| 9 | 4.4 | 0.5 | 54 | 33.6 | 1.0 | 99 | 108.2 | 1.9 | 144 | 160.7 | 0.6 |
| 10 | 4.9 | 0.5 | 55 | 34.6 | 1.0 | 100 | 110.1 | 1.8 | 145 | 161.3 | 0.7 |
| 11 | 5.4 | 0.5 | 56 | 35.6 | 1.0 | 101 | 111.9 | 1.9 | 146 | 162.0 | 0.6 |
| 12 | 5.9 | 0.5 | 57 | 36.6 | 1.1 | 102 | 113.8 | 1.7 | 147 | 162.6 | 0.6 |
| 13 | 6.4 | 0.5 | 58 | 37.7 | 1.1 | 103 | 115.5 | 1.8 | 148 | 163.2 | 0.6 |
| 14 | 6.9 | 0.5 | 59 | 38.8 | 1.1 | 104 | 117.3 | 1.7 | 149 | 163.8 | 0.6 |
| 15 | 7.4 | 0.5 | 60 | 39.9 | 1.2 | 105 | 119.0 | 1.7 | 150 | 164.4 | 0.6 |
| 16 | 7.9 | 0.5 | 61 | 41.1 | 1.2 | 106 | 120.7 | 1.6 | 151 | 165.0 | 0.6 |
| 17 | 8.4 | 0.5 | 62 | 42.3 | 1.2 | 107 | 122.3 | 1.6 | 152 | 165.6 | 0.6 |
| 18 | 8.9 | 0.5 | 63 | 43.5 | 1.2 | 108 | 123.9 | 1.6 | 153 | 166.2 | 0.5 |
| 19 | 9.4 | 0.6 | 64 | 44.7 | 1.3 | 109 | 125.5 | 1.5 | 154 | 166.7 | 0.6 |
| 20 | 10.0 | 0.5 | 65 | 46.0 | 1.3 | 110 | 127.0 | 1.5 | 155 | 167.3 | 0.6 |
| 21 | 10.5 | 0.5 | 66 | 47.3 | 1.4 | 111 | 128.5 | 1.4 | 156 | 167.9 | 0.5 |
| 22 | 11.0 | 0.6 | 67 | 48.7 | 1.4 | 112 | 129.9 | 1.4 | 157 | 168.4 | 0.6 |
| 23 | 11.6 | 0.5 | 68 | 50.1 | 1.4 | 113 | 131.3 | 1.4 | 158 | 169.0 | 0.5 |
| 24 | 12.1 | 0.6 | 69 | 51.5 | 1.5 | 114 | 132.7 | 1.3 | 159 | 169.5 | 0.5 |
| 25 | 12.7 | 0.6 | 70 | 53.0 | 1.5 | 115 | 134.0 | 1.3 | 160 | 170.0 | 0.6 |
| 26 | 13.3 | 0.5 | 71 | 54.5 | 1.6 | 116 | 135.3 | 1.2 | 161 | 170.6 | 0.5 |
| 27 | 13.8 | 0.6 | 72 | 56.1 | 1.6 | 117 | 136.5 | 1.2 | 162 | 171.1 | 0.5 |
| 28 | 14.4 | 0.6 | 73 | 57.7 | 1.6 | 118 | 137.7 | 1.2 | 163 | 171.6 | 0.5 |
| 29 | 15.0 | 0.6 | 74 | 59.3 | 1.7 | 119 | 138.9 | 1.2 | 164 | 172.1 | 0.5 |
| 30 | 15.6 | 0.6 | 75 | 61.0 | 1.7 | 120 | 140.1 | 1.1 | 165 | 172.6 | 0.5 |
| 31 | 16.2 | 0.6 | 76 | 62.7 | 1.8 | 121 | 141.2 | 1.1 | 166 | 173.1 | 0.5 |
| 32 | 16.8 | 0.6 | 77 | 64.5 | 1.7 | 122 | 142.3 | 1.1 | 167 | 173.6 | 0.5 |
| 33 | 17.4 | 0.6 | 78 | 66.2 | 1.9 | 123 | 143.4 | 1.0 | 168 | 174.1 | 0.5 |
| 34 | 18.0 | 0.7 | 79 | 68.1 | 1.8 | 124 | 144.4 | 1.0 | 169 | 174.6 | 0.5 |
| 35 | 18.7 | 0.6 | 80 | 69.9 | 1.9 | 125 | 145.4 | 1.0 | 170 | 175.1 | 0.5 |
| 36 | 19.3 | 0.7 | 81 | 71.8 | 2.0 | 126 | 146.4 | 0.9 | 171 | 175.6 | 0.5 |
| 37 | 20.0 | 0.7 | 82 | 73.8 | 1.9 | 127 | 147.3 | 1.0 | 172 | 176.1 | 0.5 |
| 38 | 20.7 | 0.7 | 83 | 75.7 | 2.0 | 128 | 148.3 | 0.9 | 173 | 176.6 | 0.5 |
| 39 | 21.4 | 0.7 | 84 | 77.7 | 2.0 | 129 | 149.2 | 0.9 | 174 | 177.1 | 0.5 |
| 40 | 22.1 | 0.7 | 85 | 79.7 | 2.1 | 130 | 150.1 | 0.8 | 175 | 177.6 | 0.5 |
| 41 | 22.8 | 0.7 | 86 | 81.8 | 2.0 | 131 | 150.9 | 0.9 | 176 | 178.1 | 0.4 |
| 42 | 23.5 | 0.7 | 87 | 83.8 | 2.1 | 132 | 151.8 | 0.8 | 177 | 178.5 | 0.5 |
| 43 | 24.2 | 0.8 | 88 | 85.9 | 2.0 | 133 | 152.6 | 0.8 | 178 | 179.0 | 0.5 |
| 44 | 25.0 | 0.8 | 89 | 87.9 | 2.1 | 134 | 153.4 | 0.8 | 179 | 179.5 | 0.5 |
| 45 | 25.8 | | 90 | 90.0 | | 135 | 154.2 | | 180 | 180.0 | |

HARMONIC ANALYSIS AND PREDICTION OF TIDES  181

Table 10.—*Values of Q for argument of constituent* $M_1$—Continued

| P | Q | Diff. | P | Q | Diff. | P | Q | Diff. | P | Q | Diff. |
|---|---|---|---|---|---|---|---|---|---|---|---|
| ° | ° | | ° | ° | | ° | ° | | ° | ° | |
| 180 | 180.0 | 0.5 | 225 | 205.8 | 0.8 | 270 | 270.0 | 2.1 | 315 | 334.2 | 0.8 |
| 181 | 180.5 | 0.5 | 226 | 206.6 | 0.8 | 271 | 272.1 | 2.0 | 316 | 335.0 | 0.8 |
| 182 | 181.0 | 0.5 | 227 | 207.4 | 0.8 | 272 | 274.1 | 2.1 | 317 | 335.8 | 0.7 |
| 183 | 181.5 | 0.4 | 228 | 208.2 | 0.9 | 273 | 276.2 | 2.0 | 318 | 336.5 | 0.7 |
| 184 | 181.9 | 0.5 | 229 | 209.1 | 0.8 | 274 | 278.2 | 2.1 | 319 | 337.2 | 0.7 |
| 185 | 182.4 | 0.5 | 230 | 209.9 | 0.9 | 275 | 280.3 | 2.0 | 320 | 337.9 | 0.7 |
| 186 | 182.9 | 0.5 | 231 | 210.8 | 0.9 | 276 | 282.3 | 2.0 | 321 | 338.6 | 0.7 |
| 187 | 183.4 | 0.5 | 232 | 211.7 | 1.0 | 277 | 284.3 | 1.9 | 322 | 339.3 | 0.7 |
| 188 | 183.9 | 0.5 | 233 | 212.7 | 0.9 | 278 | 286.2 | 2.0 | 323 | 340.0 | 0.7 |
| 189 | 184.4 | 0.5 | 234 | 213.6 | 1.0 | 279 | 288.2 | 1.9 | 324 | 340.7 | 0.6 |
| 190 | 184.9 | 0.5 | 235 | 214.6 | 1.0 | 280 | 290.1 | 1.8 | 325 | 341.3 | 0.7 |
| 191 | 185.4 | 0.5 | 236 | 215.6 | 1.0 | 281 | 291.9 | 1.9 | 326 | 342.0 | 0.6 |
| 192 | 185.9 | 0.5 | 237 | 216.6 | 1.1 | 282 | 293.8 | 1.7 | 327 | 342.6 | 0.6 |
| 193 | 186.4 | 0.5 | 238 | 217.7 | 1.1 | 283 | 295.5 | 1.8 | 328 | 343.2 | 0.6 |
| 194 | 186.9 | 0.5 | 239 | 218.8 | 1.1 | 284 | 297.3 | 1.7 | 329 | 343.8 | 0.6 |
| 195 | 187.4 | 0.5 | 240 | 219.9 | 1.2 | 285 | 299.0 | 1.7 | 330 | 344.4 | 0.6 |
| 196 | 187.9 | 0.5 | 241 | 221.1 | 1.2 | 286 | 300.7 | 1.6 | 331 | 345.0 | 0.6 |
| 197 | 188.4 | 0.5 | 242 | 222.3 | 1.2 | 287 | 302.3 | 1.6 | 332 | 345.6 | 0.6 |
| 198 | 188.9 | 0.5 | 243 | 223.5 | 1.2 | 288 | 303.9 | 1.6 | 333 | 346.2 | 0.5 |
| 199 | 189.4 | 0.6 | 244 | 224.7 | 1.3 | 289 | 305.5 | 1.5 | 334 | 346.7 | 0.6 |
| 200 | 190.0 | 0.5 | 245 | 226.0 | 1.3 | 290 | 307.0 | 1.5 | 335 | 347.3 | 0.6 |
| 201 | 190.5 | 0.5 | 246 | 227.3 | 1.4 | 291 | 308.5 | 1.4 | 336 | 347.9 | 0.5 |
| 202 | 191.0 | 0.6 | 247 | 228.7 | 1.4 | 292 | 309.9 | 1.4 | 337 | 348.4 | 0.6 |
| 203 | 191.6 | 0.5 | 248 | 230.1 | 1.4 | 293 | 311.3 | 1.4 | 338 | 349.0 | 0.5 |
| 204 | 192.1 | 0.6 | 249 | 231.5 | 1.5 | 294 | 312.7 | 1.3 | 339 | 349.5 | 0.5 |
| 205 | 192.7 | 0.6 | 250 | 233.0 | 1.5 | 295 | 314.0 | 1.3 | 340 | 350.0 | 0.6 |
| 206 | 193.3 | 0.5 | 251 | 234.5 | 1.6 | 296 | 315.3 | 1.2 | 341 | 350.6 | 0.5 |
| 207 | 193.8 | 0.6 | 252 | 236.1 | 1.6 | 297 | 316.5 | 1.2 | 342 | 351.1 | 0.5 |
| 208 | 194.4 | 0.6 | 253 | 237.7 | 1.6 | 298 | 317.7 | 1.2 | 343 | 351.6 | 0.5 |
| 209 | 195.0 | 0.6 | 254 | 239.3 | 1.7 | 299 | 318.9 | 1.2 | 344 | 352.1 | 0.5 |
| 210 | 195.6 | 0.6 | 255 | 241.0 | 1.7 | 300 | 320.1 | 1.1 | 345 | 352.6 | 0.5 |
| 211 | 196.2 | 0.6 | 256 | 242.7 | 1.8 | 301 | 321.2 | 1.1 | 346 | 353.1 | 0.5 |
| 212 | 196.8 | 0.6 | 257 | 244.5 | 1.7 | 302 | 322.3 | 1.1 | 347 | 353.6 | 0.5 |
| 213 | 197.4 | 0.6 | 258 | 246.2 | 1.9 | 303 | 323.4 | 1.0 | 348 | 354.1 | 0.5 |
| 214 | 198.0 | 0.7 | 259 | 248.1 | 1.8 | 304 | 324.4 | 1.0 | 349 | 354.6 | 0.5 |
| 215 | 198.7 | 0.6 | 260 | 249.9 | 1.9 | 305 | 325.4 | 1.0 | 350 | 355.1 | 0.5 |
| 216 | 199.3 | 0.7 | 261 | 251.8 | 2.0 | 306 | 326.4 | 0.9 | 351 | 355.6 | 0.5 |
| 217 | 200.0 | 0.7 | 262 | 253.8 | 1.9 | 307 | 327.3 | 1.0 | 352 | 356.1 | 0.5 |
| 218 | 200.7 | 0.7 | 263 | 255.7 | 2.0 | 308 | 328.3 | 0.9 | 353 | 356.6 | 0.5 |
| 219 | 201.4 | 0.7 | 264 | 257.7 | 2.0 | 309 | 329.2 | 0.9 | 354 | 357.1 | 0.5 |
| 220 | 202.1 | 0.7 | 265 | 259.7 | 2.1 | 310 | 330.1 | 0.8 | 355 | 357.6 | 0.5 |
| 221 | 202.8 | 0.7 | 266 | 261.8 | 2.0 | 311 | 330.9 | 0.9 | 356 | 358.1 | 0.4 |
| 222 | 203.5 | 0.7 | 267 | 263.8 | 2.1 | 312 | 331.8 | 0.8 | 357 | 358.5 | 0.5 |
| 223 | 204.2 | 0.8 | 268 | 265.8 | 2.0 | 313 | 332.6 | 0.8 | 358 | 359.0 | 0.5 |
| 224 | 205.0 | 0.8 | 269 | 267.9 | 2.1 | 314 | 333.4 | 0.8 | 359 | 359.5 | 0.5 |
| 225 | 205.8 | | 270 | 270.0 | | 315 | 334.2 | | 360 | 360.0 | |

Table 11.—*Values of u for equilibrium arguments*

[Use sign at head of column when N is between 0 and 180°, reverse sign when N is between 180 and 360°]

| N | J$_1$ | K$_1$ | K$_2$ | M$_2$, N$_2$ 2N, MS λ, μ, ν | M$_3$ | M$_4$,MN | M$_6$ | M$_8$ | O$_1$, Q$_1$ 2Q, ρ | OO | MK | 2MK | Mf | N |
|---|---|---|---|---|---|---|---|---|---|---|---|---|---|---|
| ° | ° − | ° − | ° − | ° − | ° − | ° − | ° − | ° − | ° + | ° − | ° − | ° + | ° − | ° |
| 0 | 0.00 | 0.00 | 0.00 | 0.00 | 0.00 | 0.00 | 0.00 | 0.00 | 0.00 | 0.00 | 0.00 | 0.00 | 0.00 | 360 |
| 1 | 0.19 | 0.13 | 0.28 | 0.04 | 0.05 | 0.08 | 0.11 | 0.15 | 0.15 | 0.53 | 0.17 | 0.06 | 0.34 | 359 |
| 2 | 0.38 | 0.27 | 0.57 | 0.08 | 0.11 | 0.15 | 0.23 | 0.30 | 0.30 | 1.05 | 0.34 | 0.12 | 0.67 | 358 |
| 3 | 0.56 | 0.40 | 0.85 | 0.11 | 0.17 | 0.23 | 0.34 | 0.45 | 0.45 | 1.57 | 0.52 | 0.17 | 1.01 | 357 |
| 4 | 0.75 | 0.54 | 1.14 | 0.15 | 0.23 | 0.30 | 0.45 | 0.60 | 0.60 | 2.10 | 0.69 | 0.23 | 1.35 | 356 |
| 5 | 0.94 | 0.67 | 1.42 | 0.19 | 0.28 | 0.38 | 0.56 | 0.75 | 0.75 | 2.62 | 0.86 | 0.29 | 1.68 | 355 |
| 6 | 1.12 | 0.80 | 1.70 | 0.23 | 0.34 | 0.45 | 0.68 | 0.90 | 0.90 | 3.14 | 1.03 | 0.35 | 2.02 | 354 |
| 7 | 1.31 | 0.94 | 1.99 | 0.26 | 0.40 | 0.53 | 0.79 | 1.05 | 1.05 | 3.67 | 1.20 | 0.41 | 2.36 | 353 |
| 8 | 1.50 | 1.07 | 2.27 | 0.30 | 0.45 | 0.60 | 0.90 | 1.20 | 1.20 | 4.19 | 1.37 | 0.47 | 2.69 | 352 |
| 9 | 1.68 | 1.20 | 2.55 | 0.34 | 0.51 | 0.68 | 1.01 | 1.35 | 1.35 | 4.71 | 1.54 | 0.53 | 3.03 | 351 |
| 10 | 1.87 | 1.34 | 2.83 | 0.37 | 0.56 | 0.75 | 1.12 | 1.49 | 1.49 | 5.23 | 1.71 | 0.59 | 3.36 | 350 |
| 11 | 2.05 | 1.47 | 3.11 | 0.41 | 0.62 | 0.82 | 1.24 | 1.64 | 1.64 | 5.75 | 1.88 | 0.64 | 3.70 | 349 |
| 12 | 2.24 | 1.60 | 3.39 | 0.45 | 0.67 | 0.90 | 1.34 | 1.79 | 1.79 | 6.27 | 2.05 | 0.70 | 4.03 | 348 |
| 13 | 2.42 | 1.73 | 3.67 | 0.48 | 0.73 | 0.97 | 1.45 | 1.94 | 1.94 | 6.79 | 2.21 | 0.76 | 4.36 | 347 |
| 14 | 2.61 | 1.86 | 3.95 | 0.52 | 0.78 | 1.04 | 1.56 | 2.09 | 2.09 | 7.31 | 2.38 | 0.82 | 4.70 | 346 |
| 15 | 2.79 | 1.99 | 4.23 | 0.56 | 0.84 | 1.12 | 1.67 | 2.23 | 2.23 | 7.82 | 2.55 | 0.88 | 5.03 | 345 |
| 16 | 2.98 | 2.12 | 4.51 | 0.60 | 0.89 | 1.19 | 1.79 | 2.38 | 2.38 | 8.34 | 2.72 | 0.93 | 5.36 | 344 |
| 17 | 3.16 | 2.25 | 4.78 | 0.63 | 0.95 | 1.26 | 1.90 | 2.53 | 2.53 | 8.85 | 2.89 | 0.99 | 5.69 | 343 |
| 18 | 3.34 | 2.38 | 5.06 | 0.67 | 1.00 | 1.34 | 2.00 | 2.67 | 2.68 | 9.36 | 3.05 | 1.05 | 6.02 | 342 |
| 19 | 3.52 | 2.51 | 5.33 | 0.70 | 1.06 | 1.41 | 2.11 | 2.81 | 2.82 | 9.87 | 3.22 | 1.11 | 6.35 | 341 |
| 20 | 3.71 | 2.64 | 5.60 | 0.74 | 1.11 | 1.48 | 2.21 | 2.95 | 2.97 | 10.38 | 3.38 | 1.17 | 6.67 | 340 |
| 21 | 3.89 | 2.77 | 5.87 | 0.77 | 1.16 | 1.55 | 2.32 | 3.09 | 3.11 | 10.89 | 3.54 | 1.23 | 7.00 | 339 |
| 22 | 4.07 | 2.90 | 6.14 | 0.81 | 1.21 | 1.62 | 2.42 | 3.23 | 3.26 | 11.39 | 3.71 | 1.28 | 7.33 | 338 |
| 23 | 4.25 | 3.03 | 6.41 | 0.84 | 1.26 | 1.69 | 2.53 | 3.37 | 3.40 | 11.89 | 3.87 | 1.34 | 7.65 | 337 |
| 24 | 4.42 | 3.15 | 6.68 | 0.88 | 1.31 | 1.75 | 2.63 | 3.51 | 3.55 | 12.39 | 4.03 | 1.40 | 7.97 | 336 |
| 25 | 4.60 | 3.28 | 6.94 | 0.91 | 1.37 | 1.82 | 2.73 | 3.64 | 3.69 | 12.89 | 4.19 | 1.46 | 8.29 | 335 |
| 26 | 4.78 | 3.40 | 7.21 | 0.94 | 1.42 | 1.89 | 2.83 | 3.78 | 3.83 | 13.39 | 4.35 | 1.52 | 8.61 | 334 |
| 27 | 4.96 | 3.53 | 7.47 | 0.98 | 1.47 | 1.96 | 2.94 | 3.92 | 3.98 | 13.89 | 4.51 | 1.57 | 8.93 | 333 |
| 28 | 5.13 | 3.65 | 7.73 | 1.01 | 1.52 | 2.02 | 3.04 | 4.05 | 4.12 | 14.38 | 4.67 | 1.63 | 9.25 | 332 |
| 29 | 5.30 | 3.78 | 7.99 | 1.04 | 1.57 | 2.09 | 3.13 | 4.18 | 4.26 | 14.87 | 4.82 | 1.69 | 9.57 | 331 |
| 30 | 5.48 | 3.90 | 8.24 | 1.08 | 1.62 | 2.16 | 3.23 | 4.31 | 4.40 | 15.36 | 4.98 | 1.75 | 9.88 | 330 |
| 31 | 5.65 | 4.02 | 8.50 | 1.11 | 1.67 | 2.22 | 3.33 | 4.45 | 4.54 | 15.84 | 5.13 | 1.80 | 10.19 | 329 |
| 32 | 5.82 | 4.14 | 8.75 | 1.14 | 1.72 | 2.29 | 3.43 | 4.58 | 4.68 | 16.32 | 5.29 | 1.86 | 10.50 | 328 |
| 33 | 5.99 | 4.26 | 9.00 | 1.17 | 1.76 | 2.35 | 3.52 | 4.70 | 4.82 | 16.80 | 5.44 | 1.92 | 10.81 | 327 |
| 34 | 6.16 | 4.38 | 9.25 | 1.20 | 1.81 | 2.41 | 3.61 | 4.82 | 4.96 | 17.28 | 5.59 | 1.97 | 11.12 | 326 |
| 35 | 6.33 | 4.50 | 9.50 | 1.24 | 1.85 | 2.47 | 3.71 | 4.94 | 5.10 | 17.76 | 5.74 | 2.03 | 11.43 | 325 |
| 36 | 6.50 | 4.62 | 9.74 | 1.27 | 1.90 | 2.53 | 3.80 | 5.06 | 5.23 | 18.23 | 5.89 | 2.09 | 11.73 | 324 |
| 37 | 6.66 | 4.74 | 9.98 | 1.30 | 1.94 | 2.59 | 3.89 | 5.18 | 5.37 | 18.69 | 6.03 | 2.15 | 12.03 | 323 |
| 38 | 6.83 | 4.85 | 10.22 | 1.33 | 1.99 | 2.65 | 3.98 | 5.30 | 5.50 | 19.16 | 6.18 | 2.20 | 12.33 | 322 |
| 39 | 6.99 | 4.97 | 10.46 | 1.36 | 2.03 | 2.71 | 4.07 | 5.42 | 5.64 | 19.62 | 6.32 | 2.26 | 12.63 | 321 |
| 40 | 7.15 | 5.08 | 10.69 | 1.38 | 2.08 | 2.77 | 4.15 | 5.54 | 5.77 | 20.08 | 6.46 | 2.31 | 12.92 | 320 |
| 41 | 7.31 | 5.19 | 10.93 | 1.41 | 2.12 | 2.82 | 4.24 | 5.65 | 5.90 | 20.53 | 6.60 | 2.37 | 13.22 | 319 |
| 42 | 7.47 | 5.30 | 11.16 | 1.44 | 2.16 | 2.88 | 4.32 | 5.76 | 6.03 | 20.98 | 6.74 | 2.42 | 13.51 | 318 |
| 43 | 7.63 | 5.41 | 11.38 | 1.47 | 2.20 | 2.94 | 4.40 | 5.87 | 6.16 | 21.43 | 6.88 | 2.48 | 13.80 | 317 |
| 44 | 7.79 | 5.52 | 11.60 | 1.50 | 2.24 | 2.99 | 4.49 | 5.98 | 6.29 | 21.87 | 7.02 | 2.53 | 14.08 | 316 |
| 45 | 7.94 | 5.63 | 11.82 | 1.52 | 2.28 | 3.04 | 4.57 | 6.09 | 6.42 | 22.31 | 7.15 | 2.59 | 14.37 | 315 |

NOTE.—For L$_2$ and M$_1$ see Table 13; for 2SM and MSf, take u of M$_2$ with sign reversed; for P$_1$, R$_2$, S$_1$, S$_2$, S$_3$, S$_4$, T$_2$, Mm, Sa, and Ssa, take $u=0$.

HARMONIC ANALYSIS AND PREDICTION OF TIDES

Table 11.—*Values of u for equilibrium arguments*—Continued

[*Use sign at head of column when N is between 0 and 180°, reverse sign when N is between 180 and 360°*]

| N | $J_1$ | $K_1$ | $K_2$ | $M_2, N_2$ $2N, MS$ $\lambda, \mu, \nu$ | $M_3$ | $M_4, MN$ | $M_6$ | $M_8$ | $O_1, Q_1$ $2Q, \rho_1$ | OO | MK | 2MK | Mf | N |
|---|---|---|---|---|---|---|---|---|---|---|---|---|---|---|
| ° | ° | ° | ° | ° | ° | ° | ° | ° | ° + | ° | ° | ° + | ° | ° |
|   | −  | −  |    | −  | −  | −  |    |    |    |    |    |    | −  |    |
| 45 | 7.94 | 5.63 | 11.82 | 1.52 | 2.28 | 3.04 | 4.57 | 6.09 | 6.42 | 22.31 | 7.15 | 2.59 | 14.37 | 315 |
| 46 | 8.10 | 5.74 | 12.04 | 1.55 | 2.32 | 3.10 | 4.64 | 6.19 | 6.55 | 22.75 | 7.28 | 2.64 | 14.65 | 314 |
| 47 | 8.25 | 5.84 | 12.26 | 1.57 | 2.36 | 3.15 | 4.72 | 6.30 | 6.68 | 23.18 | 7.41 | 2.69 | 14.93 | 313 |
| 48 | 8.40 | 5.95 | 12.47 | 1.60 | 2.40 | 3.20 | 4.80 | 6.40 | 6.80 | 23.60 | 7.54 | 2.75 | 15.20 | 312 |
| 49 | 8.55 | 6.05 | 12.68 | 1.62 | 2.44 | 3.25 | 4.87 | 6.50 | 6.92 | 24.02 | 7.67 | 2.80 | 15.47 | 311 |
| 50 | 8.70 | 6.15 | 12.88 | 1.65 | 2.47 | 3.30 | 4.94 | 6.59 | 7.05 | 24.44 | 7.80 | 2.85 | 15.74 | 310 |
| 51 | 8.84 | 6.25 | 13.08 | 1.67 | 2.51 | 3.34 | 5.01 | 6.68 | 7.17 | 24.85 | 7.92 | 2.91 | 16.01 | 309 |
| 52 | 8.99 | 6.35 | 13.28 | 1.69 | 2.54 | 3.39 | 5.08 | 6.78 | 7.29 | 25.26 | 8.04 | 2.96 | 16.28 | 308 |
| 53 | 9.13 | 6.45 | 13.48 | 1.72 | 2.58 | 3.44 | 5.15 | 6.87 | 7.41 | 25.66 | 8.17 | 3.01 | 16.54 | 307 |
| 54 | 9.27 | 6.54 | 13.67 | 1.74 | 2.61 | 3.48 | 5.22 | 6.96 | 7.53 | 26.06 | 8.28 | 3.06 | 16.80 | 306 |
| 55 | 9.41 | 6.64 | 13.86 | 1.76 | 2.64 | 3.52 | 5.28 | 7.04 | 7.65 | 26.46 | 8.40 | 3.12 | 17.05 | 305 |
| 56 | 9.54 | 6.73 | 14.05 | 1.78 | 2.67 | 3.56 | 5.34 | 7.12 | 7.76 | 26.85 | 8.51 | 3.17 | 17.30 | 304 |
| 57 | 9.68 | 6.82 | 14.23 | 1.80 | 2.70 | 3.60 | 5.40 | 7.20 | 7.88 | 27.23 | 8.62 | 3.22 | 17.55 | 303 |
| 58 | 9.81 | 6.91 | 14.40 | 1.82 | 2.73 | 3.64 | 5.46 | 7.28 | 7.99 | 27.61 | 8.73 | 3.27 | 17.80 | 302 |
| 59 | 9.94 | 7.00 | 14.58 | 1.84 | 2.76 | 3.68 | 5.52 | 7.36 | 8.10 | 27.98 | 8.84 | 3.32 | 18.04 | 301 |
| 60 | 10.07 | 7.09 | 14.75 | 1.86 | 2.79 | 3.72 | 5.58 | 7.44 | 8.21 | 28.34 | 8.95 | 3.37 | 18.28 | 300 |
| 61 | 10.19 | 7.17 | 14.92 | 1.88 | 2.82 | 3.76 | 5.63 | 7.51 | 8.32 | 28.70 | 9.05 | 3.42 | 18.51 | 299 |
| 62 | 10.32 | 7.26 | 15.08 | 1.90 | 2.84 | 3.79 | 5.69 | 7.58 | 8.42 | 29.06 | 9.15 | 3.46 | 18.74 | 298 |
| 63 | 10.44 | 7.34 | 15.24 | 1.91 | 2.87 | 3.82 | 5.74 | 7.65 | 8.53 | 29.41 | 9.25 | 3.51 | 18.97 | 297 |
| 64 | 10.56 | 7.42 | 15.39 | 1.93 | 2.89 | 3.86 | 5.78 | 7.71 | 8.63 | 29.75 | 9.35 | 3.56 | 19.19 | 296 |
| 65 | 10.68 | 7.49 | 15.54 | 1.94 | 2.92 | 3.89 | 5.83 | 7.78 | 8.73 | 30.09 | 9.44 | 3.61 | 19.41 | 295 |
| 66 | 10.79 | 7.57 | 15.69 | 1.96 | 2.94 | 3.92 | 5.88 | 7.84 | 8.83 | 30.42 | 9.53 | 3.65 | 19.63 | 294 |
| 67 | 10.91 | 7.64 | 15.83 | 1.98 | 2.96 | 3.95 | 5.93 | 7.90 | 8.93 | 30.74 | 9.62 | 3.69 | 19.84 | 293 |
| 68 | 11.02 | 7.72 | 15.96 | 1.99 | 2.98 | 3.98 | 5.97 | 7.96 | 9.03 | 31.06 | 9.71 | 3.74 | 20.04 | 292 |
| 69 | 11.12 | 7.79 | 16.10 | 2.00 | 3.00 | 4.00 | 6.01 | 8.01 | 9.12 | 31.37 | 9.79 | 3.78 | 20.25 | 291 |
| 70 | 11.23 | 7.86 | 16.23 | 2.02 | 3.02 | 4.03 | 6.05 | 8.06 | 9.22 | 31.68 | 9.87 | 3.83 | 20.45 | 290 |
| 71 | 11.33 | 7.92 | 16.35 | 2.03 | 3.04 | 4.06 | 6.08 | 8.11 | 9.31 | 31.98 | 9.95 | 3.87 | 20.64 | 289 |
| 72 | 11.43 | 7.99 | 16.47 | 2.04 | 3.06 | 4.08 | 6.11 | 8.15 | 9.40 | 32.27 | 10.03 | 3.91 | 20.83 | 288 |
| 73 | 11.53 | 8.05 | 16.58 | 2.05 | 3.08 | 4.10 | 6.15 | 8.20 | 9.48 | 32.55 | 10.10 | 3.95 | 21.01 | 287 |
| 74 | 11.63 | 8.11 | 16.69 | 2.06 | 3.09 | 4.12 | 6.18 | 8.24 | 9.57 | 32.82 | 10.17 | 3.99 | 21.20 | 286 |
| 75 | 11.72 | 8.17 | 16.80 | 2.07 | 3.10 | 4.14 | 6.21 | 8.28 | 9.65 | 33.09 | 10.24 | 4.03 | 21.37 | 285 |
| 76 | 11.81 | 8.23 | 16.90 | 2.08 | 3.12 | 4.16 | 6.24 | 8.32 | 9.73 | 33.35 | 10.31 | 4.07 | 21.54 | 284 |
| 77 | 11.90 | 8.28 | 17.00 | 2.09 | 3.13 | 4.18 | 6.26 | 8.35 | 9.81 | 33.60 | 10.37 | 4.11 | 21.71 | 283 |
| 78 | 11.98 | 8.34 | 17.09 | 2.10 | 3.14 | 4.19 | 6.29 | 8.38 | 9.88 | 33.85 | 10.43 | 4.15 | 21.87 | 282 |
| 79 | 12.06 | 8.39 | 17.17 | 2.10 | 3.15 | 4.20 | 6.31 | 8.41 | 9.96 | 34.09 | 10.49 | 4.18 | 22.02 | 281 |
| 80 | 12.14 | 8.44 | 17.25 | 2.11 | 3.16 | 4.22 | 6.32 | 8.43 | 10.03 | 34.31 | 10.54 | 4.22 | 22.17 | 280 |
| 81 | 12.22 | 8.48 | 17.33 | 2.11 | 3.17 | 4.23 | 6.34 | 8.46 | 10.10 | 34.53 | 10.60 | 4.25 | 22.32 | 279 |
| 82 | 12.29 | 8.53 | 17.40 | 2.12 | 3.18 | 4.24 | 6.36 | 8.48 | 10.17 | 34.74 | 10.65 | 4.29 | 22.46 | 278 |
| 83 | 12.36 | 8.57 | 17.46 | 2.12 | 3.19 | 4.25 | 6.37 | 8.50 | 10.23 | 34.95 | 10.69 | 4.32 | 22.59 | 277 |
| 84 | 12.42 | 8.61 | 17.52 | 2.13 | 3.19 | 4.26 | 6.38 | 8.51 | 10.30 | 35.14 | 10.73 | 4.35 | 22.72 | 276 |
| 85 | 12.49 | 8.64 | 17.58 | 2.13 | 3.20 | 4.26 | 6.39 | 8.52 | 10.36 | 35.33 | 10.77 | 4.38 | 22.84 | 275 |
| 86 | 12.55 | 8.68 | 17.63 | 2.13 | 3.20 | 4.27 | 6.40 | 8.53 | 10.41 | 35.50 | 10.81 | 4.41 | 22.96 | 274 |
| 87 | 12.60 | 8.71 | 17.67 | 2.14 | 3.20 | 4.27 | 6.41 | 8.54 | 10.47 | 35.67 | 10.84 | 4.44 | 23.07 | 273 |
| 88 | 12.65 | 8.74 | 17.71 | 2.14 | 3.20 | 4.27 | 6.41 | 8.54 | 10.52 | 35.83 | 10.87 | 4.47 | 23.17 | 272 |
| 89 | 12.70 | 8.76 | 17.74 | 2.14 | 3.20 | 4.27 | 6.41 | 8.54 | 10.57 | 35.98 | 10.90 | 4.49 | 23.27 | 271 |
| 90 | 12.75 | 8.79 | 17.77 | 2.14 | 3.20 | 4.27 | 6.41 | 8.54 | 10.62 | 36.12 | 10.93 | 4.52 | 23.37 | 270 |

NOTE.—For $L_2$ and $M_1$ see Table 13; for 2SM and MSf, take *u* of $M_2$ with sign reversed; for $P_1$, $R_2$, $S_1$, $S_2$, $S_3$, $S_4$, $T_2$, Mm, Sa, and Ssa, take $u = 0$.

**Table 11.**—*Values of u for equilibrium arguments*—**Continued**

[*Use sign at head of column when N is between 0 and 180°, reverse sign when N is between 180 and 360°*]

| N | $J_1$ | $K_1$ | $K_2$ | $M_2, N_2$ $2N, MS$ $\lambda, \mu, \nu$ | $M_3$ | $M_4, MN$ | $M_6$ | $M_8$ | $O_1, Q_1$ $2Q, \rho_1$ | OO | MK | 2MK | Mf | N |
|---|---|---|---|---|---|---|---|---|---|---|---|---|---|---|
| ° | ° − | ° − | ° − | ° − | ° − | ° − | ° − | ° − | ° + | ° − | ° − | ° + | ° − | ° |
| 90 | 12.75 | 8.79 | 17.77 | 2.14 | 3.20 | 4.27 | 6.41 | 8.54 | 10.62 | 36.12 | 10.93 | 4.52 | 23.37 | 270 |
| 91 | 12.79 | 8.81 | 17.79 | 2.14 | 3.20 | 4.27 | 6.41 | 8.54 | 10.66 | 36.25 | 10.95 | 4.54 | 23.46 | 269 |
| 92 | 12.83 | 8.83 | 17.81 | 2.13 | 3.20 | 4.27 | 6.40 | 8.54 | 10.70 | 36.37 | 10.96 | 4.56 | 23.54 | 268 |
| 93 | 12.87 | 8.85 | 17.82 | 2.13 | 3.20 | 4.26 | 6.40 | 8.53 | 10.74 | 36.48 | 10.98 | 4.58 | 23.61 | 267 |
| 94 | 12.90 | 8.86 | 17.83 | 2.13 | 3.20 | 4.26 | 6.39 | 8.52 | 10.77 | 36.58 | 10.99 | 4.60 | 23.67 | 266 |
| 95 | 12.93 | 8.87 | 17.83 | 2.13 | 3.19 | 4.26 | 6.38 | 8.51 | 10.80 | 36.67 | 11.00 | 4.62 | 23.73 | 265 |
| 96 | 12.96 | 8.88 | 17.82 | 2.12 | 3.19 | 4.25 | 6.37 | 8.49 | 10.83 | 36.75 | 11.01 | 4.64 | 23.79 | 264 |
| 97 | 12.98 | 8.89 | 17.81 | 2.12 | 3.18 | 4.24 | 6.35 | 8.47 | 10.86 | 36.82 | 11.01 | 4.65 | 23.84 | 263 |
| 98 | 13.00 | 8.90 | 17.79 | 2.11 | 3.17 | 4.22 | 6.34 | 8.45 | 10.88 | 36.87 | 11.01 | 4.67 | 23.88 | 262 |
| 99 | 13.01 | 8.90 | 17.77 | 2.11 | 3.16 | 4.21 | 6.32 | 8.43 | 10.90 | 36.92 | 11.00 | 4.68 | 23.91 | 261 |
| 100 | 13.02 | 8.89 | 17.74 | 2.10 | 3.15 | 4.20 | 6.30 | 8.41 | 10.92 | 36.95 | 11.00 | 4.69 | 23.93 | 260 |
| 101 | 13.03 | 8.89 | 17.71 | 2.09 | 3.14 | 4.19 | 6.28 | 8.38 | 10.93 | 36.98 | 10.99 | 4.70 | 23.95 | 259 |
| 102 | 13.03 | 8.88 | 17.67 | 2.09 | 3.13 | 4.17 | 6.26 | 8.34 | 10.94 | 36.99 | 10.97 | 4.71 | 23.96 | 258 |
| 103 | 13.02 | 8.87 | 17.62 | 2.08 | 3.11 | 4.15 | 6.23 | 8.30 | 10.95 | 37.00 | 10.95 | 4.72 | 23.97 | 257 |
| 104 | 13.02 | 8.86 | 17.57 | 2.07 | 3.10 | 4.14 | 6.20 | 8.27 | 10.95 | 36.99 | 10.93 | 4.72 | 23.97 | 256 |
| 105 | 13.01 | 8.84 | 17.51 | 2.06 | 3.09 | 4.12 | 6.17 | 8.23 | 10.95 | 36.96 | 10.90 | 4.73 | 23.96 | 255 |
| 106 | 12.99 | 8.82 | 17.45 | 2.05 | 3.07 | 4.10 | 6.14 | 8.19 | 10.94 | 36.93 | 10.87 | 4.73 | 23.94 | 254 |
| 107 | 12.97 | 8.80 | 17.38 | 2.04 | 3.06 | 4.08 | 6.11 | 8.15 | 10.94 | 36.89 | 10.84 | 4.73 | 23.91 | 253 |
| 108 | 12.95 | 8.78 | 17.30 | 2.03 | 3.04 | 4.06 | 6.08 | 8.11 | 10.93 | 36.83 | 10.81 | 4.72 | 23.88 | 252 |
| 109 | 12.93 | 8.75 | 17.22 | 2.02 | 3.02 | 4.03 | 6.05 | 8.06 | 10.91 | 36.76 | 10.77 | 4.72 | 23.84 | 251 |
| 110 | 12.90 | 8.72 | 17.14 | 2.00 | 3.00 | 4.00 | 6.01 | 8.01 | 10.89 | 36.67 | 10.72 | 4.72 | 23.79 | 250 |
| 111 | 12.86 | 8.69 | 17.05 | 1.99 | 2.98 | 3.98 | 5.97 | 7.96 | 10.87 | 36.58 | 10.68 | 4.71 | 23.73 | 249 |
| 112 | 12.82 | 8.65 | 16.95 | 1.98 | 2.96 | 3.95 | 5.93 | 7.90 | 10.84 | 36.48 | 10.63 | 4.70 | 23.66 | 248 |
| 113 | 12.77 | 8.61 | 16.84 | 1.96 | 2.94 | 3.92 | 5.88 | 7.84 | 10.81 | 36.36 | 10.57 | 4.69 | 23.59 | 247 |
| 114 | 12.72 | 8.57 | 16.73 | 1.94 | 2.92 | 3.89 | 5.83 | 7.78 | 10.78 | 36.23 | 10.51 | 4.68 | 23.50 | 246 |
| 115 | 12.67 | 8.52 | 16.62 | 1.93 | 2.89 | 3.86 | 5.78 | 7.71 | 10.74 | 36.09 | 10.45 | 4.67 | 23.41 | 245 |
| 116 | 12.61 | 8.48 | 16.50 | 1.91 | 2.87 | 3.82 | 5.74 | 7.65 | 10.70 | 35.93 | 10.39 | 4.65 | 23.31 | 244 |
| 117 | 12.55 | 8.43 | 16.37 | 1.90 | 2.84 | 3.79 | 5.69 | 7.58 | 10.65 | 35.76 | 10.32 | 4.63 | 23.21 | 243 |
| 118 | 12.48 | 8.37 | 16.24 | 1.88 | 2.82 | 3.76 | 5.64 | 7.52 | 10.60 | 35.57 | 10.25 | 4.61 | 23.09 | 242 |
| 119 | 12.41 | 8.31 | 16.10 | 1.86 | 2.79 | 3.72 | 5.59 | 7.45 | 10.55 | 35.37 | 10.18 | 4.59 | 22.96 | 241 |
| 120 | 12.34 | 8.25 | 15.96 | 1.84 | 2.77 | 3.69 | 5.53 | 7.38 | 10.49 | 35.16 | 10.10 | 4.57 | 22.83 | 240 |
| 121 | 12.26 | 8.19 | 15.81 | 1.82 | 2.74 | 3.65 | 5.47 | 7.30 | 10.43 | 34.94 | 10.02 | 4.54 | 22.69 | 239 |
| 122 | 12.17 | 8.13 | 15.66 | 1.80 | 2.71 | 3.61 | 5.41 | 7.22 | 10.37 | 34.70 | 9.93 | 4.52 | 22.54 | 238 |
| 123 | 12.08 | 8.06 | 15.50 | 1.78 | 2.68 | 3.57 | 5.35 | 7.14 | 10.30 | 34.49 | 9.84 | 4.49 | 22.37 | 237 |
| 124 | 11.98 | 7.99 | 15.33 | 1.76 | 2.64 | 3.52 | 5.29 | 7.05 | 10.22 | 34.19 | 9.75 | 4.46 | 22.20 | 236 |
| 125 | 11.88 | 7.91 | 15.16 | 1.74 | 2.61 | 3.48 | 5.22 | 6.96 | 10.14 | 33.91 | 9.65 | 4.43 | 22.03 | 235 |
| 126 | 11.78 | 7.83 | 14.99 | 1.72 | 2.58 | 3.44 | 5.15 | 6.87 | 10.06 | 33.62 | 9.55 | 4.40 | 21.84 | 234 |
| 127 | 11.67 | 7.75 | 14.81 | 1.70 | 2.54 | 3.39 | 5.09 | 6.78 | 9.97 | 33.31 | 9.45 | 4.36 | 21.64 | 233 |
| 128 | 11.56 | 7.67 | 14.62 | 1.67 | 2.51 | 3.34 | 5.02 | 6.69 | 9.88 | 32.99 | 9.34 | 4.32 | 21.44 | 232 |
| 129 | 11.44 | 7.58 | 14.43 | 1.65 | 2.48 | 3.30 | 4.95 | 6.60 | 9.79 | 32.66 | 9.23 | 4.28 | 21.23 | 231 |
| 130 | 11.31 | 7.49 | 14.23 | 1.63 | 2.44 | 3.26 | 4.88 | 6.51 | 9.69 | 32.31 | 9.12 | 4.23 | 21.00 | 230 |
| 131 | 11.18 | 7.40 | 14.03 | 1.60 | 2.41 | 3.21 | 4.81 | 6.42 | 9.58 | 31.95 | 9.00 | 4.19 | 20.76 | 229 |
| 132 | 11.05 | 7.30 | 13.83 | 1.58 | 2.37 | 3.16 | 4.74 | 6.32 | 9.47 | 31.58 | 8.88 | 4.14 | 20.52 | 228 |
| 133 | 10.91 | 7.20 | 13.62 | 1.55 | 2.33 | 3.11 | 4.66 | 6.22 | 9.36 | 31.19 | 8.76 | 4.09 | 20.27 | 227 |
| 134 | 10.77 | 7.10 | 13.40 | 1.53 | 2.29 | 3.06 | 4.58 | 6.11 | 9.24 | 30.79 | 8.63 | 4.04 | 20.01 | 226 |
| 135 | 10.62 | 7.00 | 13.18 | 1.50 | 2.25 | 3.00 | 4.51 | 6.01 | 9.12 | 30.37 | 8.50 | 3.99 | 19.75 | 225 |

NOTE.—For $L_2$ and $M_1$ see table 13; for 2SM and MSf, take $u$ of $M_2$ with sign reversed; for $P_1$, $R_2$, $S_1$, $S_2$, $S_3$, $S_4$, $T_2$, Mm, Sa and Ssa, take $u=0$.

## Table 11.—Values of u for equilibrium arguments—Continued

[Use sign at head of column when $N$ is between 0 and 180°, reverse sign when $N$ is between 180 and 360°]

| $N$ | $J_1$ | $K_1$ | $K_2$ | $M_2, N_2$ $2N, MS$ $\lambda, \mu, \nu$ | $M_3$ | $M_4, MN$ | $M_6$ | $M_8$ | $O_1, Q_1$ $2Q, \rho$ | OO | MK | 2MK | Mf | $N$ |
|---|---|---|---|---|---|---|---|---|---|---|---|---|---|---|
| | ° | ° | ° | ° | ° | ° | ° | ° | °+ | ° | ° | °+ | ° | |
| | − | − | − | − | − | − | − | − | | − | − | | − | |
| 135 | 10.62 | 7.00 | 13.18 | 1.50 | 2.25 | 3.00 | 4.51 | 6.01 | 9.12 | 30.37 | 8.50 | 3.99 | 19.75 | 225 |
| 136 | 10.47 | 6.89 | 12.96 | 1.48 | 2.21 | 2.95 | 4.43 | 5.90 | 9.00 | 29.94 | 8.36 | 3.94 | 19.47 | 224 |
| 137 | 10.31 | 6.78 | 12.73 | 1.45 | 2.17 | 2.90 | 4.34 | 5.79 | 8.87 | 29.49 | 8.22 | 3.88 | 19.18 | 223 |
| 138 | 10.15 | 6.66 | 12.50 | 1.42 | 2.13 | 2.84 | 4.26 | 5.68 | 8.73 | 29.04 | 8.08 | 3.82 | 18.88 | 222 |
| 139 | 9.98 | 6.55 | 12.26 | 1.39 | 2.09 | 2.78 | 4.18 | 5.57 | 8.59 | 28.56 | 7.94 | 3.76 | 18.58 | 221 |
| 140 | 9.81 | 6.43 | 12.02 | 1.36 | 2.05 | 2.73 | 4.09 | 5.46 | 8.45 | 28.08 | 7.79 | 3.70 | 18.26 | 220 |
| 141 | 9.64 | 6.31 | 11.77 | 1.34 | 2.00 | 2.67 | 4.01 | 5.34 | 8.30 | 27.58 | 7.64 | 3.64 | 17.94 | 219 |
| 142 | 9.46 | 6.18 | 11.52 | 1.31 | 1.96 | 2.61 | 3.92 | 5.22 | 8.15 | 27.07 | 7.49 | 3.57 | 17.61 | 218 |
| 143 | 9.27 | 6.06 | 11.27 | 1.28 | 1.91 | 2.55 | 3.83 | 5.10 | 8.00 | 26.54 | 7.33 | 3.50 | 17.27 | 217 |
| 144 | 9.08 | 5.93 | 11.01 | 1.25 | 1.87 | 2.49 | 3.74 | 4.98 | 7.84 | 26.00 | 7.17 | 3.43 | 16.92 | 216 |
| 145 | 8.89 | 5.80 | 10.74 | 1.22 | 1.82 | 2.43 | 3.65 | 4.86 | 7.67 | 25.45 | 7.01 | 3.36 | 16.56 | 215 |
| 146 | 8.69 | 5.66 | 10.48 | 1.19 | 1.78 | 2.37 | 3.56 | 4.74 | 7.50 | 24.89 | 6.84 | 3.29 | 16.20 | 214 |
| 147 | 8.49 | 5.52 | 10.21 | 1.16 | 1.73 | 2.31 | 3.47 | 4.62 | 7.33 | 24.31 | 6.68 | 3.21 | 15.82 | 213 |
| 148 | 8.28 | 5.38 | 9.94 | 1.12 | 1.69 | 2.25 | 3.37 | 4.50 | 7.16 | 23.72 | 6.51 | 3.14 | 15.44 | 212 |
| 149 | 8.07 | 5.24 | 9.66 | 1.09 | 1.64 | 2.18 | 3.28 | 4.37 | 6.98 | 23.12 | 6.33 | 3.06 | 15.05 | 211 |
| 150 | 7.85 | 5.10 | 9.38 | 1.06 | 1.59 | 2.12 | 3.18 | 4.24 | 6.80 | 22.51 | 6.16 | 2.98 | 14.65 | 210 |
| 151 | 7.63 | 4.95 | 9.10 | 1.03 | 1.54 | 2.06 | 3.08 | 4.11 | 6.61 | 21.88 | 5.98 | 2.90 | 14.24 | 209 |
| 152 | 7.41 | 4.80 | 8.81 | 1.00 | 1.49 | 1.99 | 2.99 | 3.98 | 6.42 | 21.24 | 5.80 | 2.81 | 13.83 | 208 |
| 153 | 7.18 | 4.65 | 8.52 | 0.96 | 1.44 | 1.92 | 2.89 | 3.85 | 6.22 | 20.59 | 5.61 | 2.73 | 13.41 | 207 |
| 154 | 6.95 | 4.50 | 8.23 | 0.93 | 1.39 | 1.86 | 2.78 | 3.71 | 6.03 | 19.93 | 5.43 | 2.64 | 12.98 | 206 |
| 155 | 6.72 | 4.34 | 7.94 | 0.90 | 1.34 | 1.79 | 2.69 | 3.58 | 5.82 | 19.26 | 5.24 | 2.55 | 12.54 | 205 |
| 156 | 6.48 | 4.19 | 7.64 | 0.86 | 1.29 | 1.72 | 2.59 | 3.45 | 5.62 | 18.58 | 5.05 | 2.46 | 12.10 | 204 |
| 157 | 6.24 | 4.03 | 7.34 | 0.83 | 1.24 | 1.66 | 2.48 | 3.31 | 5.41 | 17.89 | 4.85 | 2.37 | 11.65 | 203 |
| 158 | 5.99 | 3.87 | 7.04 | 0.79 | 1.19 | 1.59 | 2.38 | 3.18 | 5.20 | 17.19 | 4.66 | 2.28 | 11.19 | 202 |
| 159 | 5.74 | 3.70 | 6.74 | 0.76 | 1.14 | 1.52 | 2.28 | 3.04 | 4.99 | 16.48 | 4.46 | 2.18 | 10.73 | 201 |
| 160 | 5.49 | 3.54 | 6.43 | 0.72 | 1.09 | 1.45 | 2.17 | 2.90 | 4.77 | 15.75 | 4.26 | 2.09 | 10.26 | 200 |
| 161 | 5.24 | 3.37 | 6.12 | 0.69 | 1.04 | 1.38 | 2.07 | 2.76 | 4.55 | 15.02 | 4.06 | 1.99 | 9.79 | 199 |
| 162 | 4.98 | 3.20 | 5.81 | 0.66 | 0.98 | 1.31 | 1.97 | 2.62 | 4.33 | 14.29 | 3.86 | 1.89 | 9.31 | 198 |
| 163 | 4.72 | 3.03 | 5.50 | 0.62 | 0.93 | 1.24 | 1.86 | 2.48 | 4.10 | 13.54 | 3.66 | 1.79 | 8.82 | 197 |
| 164 | 4.46 | 2.86 | 5.19 | 0.58 | 0.88 | 1.17 | 1.75 | 2.34 | 3.87 | 12.78 | 3.45 | 1.70 | 8.33 | 196 |
| 165 | 4.19 | 2.69 | 4.87 | 0.55 | 0.82 | 1.10 | 1.64 | 2.19 | 3.64 | 12.02 | 3.24 | 1.60 | 7.83 | 195 |
| 166 | 3.92 | 2.52 | 4.55 | 0.51 | 0.77 | 1.02 | 1.54 | 2.05 | 3.41 | 11.25 | 3.03 | 1.49 | 7.33 | 194 |
| 167 | 3.65 | 2.34 | 4.23 | 0.48 | 0.71 | 0.95 | 1.43 | 1.90 | 3.18 | 10.48 | 2.82 | 1.39 | 6.83 | 193 |
| 168 | 3.38 | 2.17 | 3.91 | 0.44 | 0.66 | 0.88 | 1.32 | 1.76 | 2.94 | 9.70 | 2.61 | 1.29 | 6.32 | 192 |
| 169 | 3.10 | 1.99 | 3.59 | 0.40 | 0.61 | 0.81 | 1.21 | 1.62 | 2.70 | 8.91 | 2.39 | 1.18 | 5.80 | 191 |
| 170 | 2.83 | 1.81 | 3.27 | 0.37 | 0.55 | 0.74 | 1.10 | 1.47 | 2.46 | 8.12 | 2.18 | 1.08 | 5.29 | 190 |
| 171 | 2.55 | 1.63 | 2.94 | 0.33 | 0.50 | 0.66 | 1.00 | 1.33 | 2.22 | 7.32 | 1.97 | 0.97 | 4.77 | 189 |
| 172 | 2.27 | 1.45 | 2.62 | 0.30 | 0.44 | 0.59 | 0.89 | 1.18 | 1.97 | 6.51 | 1.75 | 0.86 | 4.24 | 188 |
| 173 | 1.99 | 1.27 | 2.29 | 0.26 | 0.39 | 0.51 | 0.77 | 1.03 | 1.73 | 5.71 | 1.53 | 0.76 | 3.72 | 187 |
| 174 | 1.71 | 1.09 | 1.97 | 0.22 | 0.33 | 0.44 | 0.66 | 0.88 | 1.49 | 4.90 | 1.31 | 0.65 | 3.19 | 186 |
| 175 | 1.42 | 0.91 | 1.64 | 0.18 | 0.28 | 0.37 | 0.55 | 0.74 | 1.24 | 4.09 | 1.10 | 0.54 | 2.66 | 185 |
| 176 | 1.14 | 0.73 | 1.31 | 0.15 | 0.22 | 0.30 | 0.44 | 0.59 | 0.99 | 3.27 | 0.88 | 0.43 | 2.13 | 184 |
| 177 | 0.86 | 0.55 | 0.99 | 0.11 | 0.17 | 0.22 | 0.33 | 0.44 | 0.75 | 2.46 | 0.66 | 0.33 | 1.60 | 183 |
| 178 | 0.57 | 0.37 | 0.66 | 0.07 | 0.11 | 0.14 | 0.22 | 0.29 | 0.50 | 1.64 | 0.44 | 0.22 | 1.07 | 182 |
| 179 | 0.29 | 0.18 | 0.33 | 0.04 | 0.05 | 0.07 | 0.11 | 0.14 | 0.25 | 0.82 | 0.22 | 0.11 | 0.53 | 181 |
| 180 | 0.00 | 0.00 | 0.00 | 0.00 | 0.00 | 0.00 | 0.00 | 0.00 | 0.00 | 0.00 | 0.00 | 0.00 | 0.00 | 180 |

NOTE.—For $L_2$ and $M_1$ see Table 13; for 2SM and MSf, take $u$ of $M_2$ with sign reversed; for $P_1$, $R_2$, $S_1$, $S_2$, $S_3$, $S_4$, $T_2$, Mm, Sa, and Ssa, take $u=0$.

Table 12.—*Log factor F corresponding to every tenth of a degree of I*

| Constituent \ I | 18.3° | Diff. | 18.4° | Diff. | 18.5° | Diff. | 18.6° | Diff. | 18.7° | Diff. | 18.8° | Diff. |
|---|---|---|---|---|---|---|---|---|---|---|---|---|
| $J_1$ | 0.0827 | −21 | 0.0806 | −20 | 0.0786 | −20 | 0.0766 | −20 | 0.0746 | −19 | 0.0727 | −20 |
| $K_1$ | 0.0547 | −12 | 0.0535 | −12 | 0.0523 | −11 | 0.0512 | −12 | 0.0500 | −12 | 0.0488 | −11 |
| $K_2$ | 0.1263 | −22 | 0.1241 | −21 | 0.1220 | −21 | 0.1199 | −22 | 0.1177 | −22 | 0.1155 | −21 |
| $M_2^*$, $N_2$, $2N$ | 9.9839 | +2 | 9.9841 | +3 | 9.9844 | +2 | 9.9846 | +3 | 9.9849 | +2 | 9.9851 | +3 |
| $M_3$ | 9.9758 | +4 | 9.9762 | +4 | 9.9766 | +3 | 9.9769 | +4 | 9.9773 | +4 | 9.9777 | +3 |
| $M_4$, $MN$ | 9.9678 | +4 | 9.9682 | +5 | 9.9687 | +5 | 9.9692 | +5 | 9.9697 | +5 | 9.9702 | +5 |
| $M_6$ | 9.9516 | +8 | 9.9524 | +7 | 9.9531 | +7 | 9.9538 | +8 | 9.9546 | +7 | 9.9553 | +8 |
| $M_8$ | 9.9355 | +10 | 9.9365 | +10 | 9.9375 | +10 | 9.9385 | +10 | 9.9395 | +10 | 9.9405 | +10 |
| $O_1$, $Q_1$, $2Q$, $\rho_1$ | 0.0939 | −22 | 0.0917 | −21 | 0.0896 | −22 | 0.0874 | −21 | 0.0853 | −21 | 0.0832 | −21 |
| $OO$ | 0.3139 | −69 | 0.3070 | −70 | 0.3000 | −69 | 0.2931 | −68 | 0.2863 | −69 | 0.2794 | −68 |
| $MK$ | 0.0386 | −10 | 0.0376 | −9 | 0.0367 | −9 | 0.0358 | −9 | 0.0349 | −10 | 0.0339 | −9 |
| $2MK$ | 0.0224 | −7 | 0.0217 | −6 | 0.0211 | −7 | 0.0204 | −7 | 0.0197 | −7 | 0.0190 | −6 |
| $Mf$ | 0.2039 | −46 | 0.1993 | −45 | 0.1948 | −45 | 0.1903 | −45 | 0.1858 | −45 | 0.1813 | −44 |
| $Mm$ | 9.9465 | +8 | 9.9473 | +8 | 9.9481 | +8 | 9.9489 | +8 | 9.9497 | +8 | 9.9505 | +9 |

| Constituent \ I | 18.9° | Diff. | 19.0° | Diff. | 19.1° | Diff. | 19.2° | Diff. | 19.3° | Diff. | 19.4° | Diff. |
|---|---|---|---|---|---|---|---|---|---|---|---|---|
| $J_1$ | 0.0707 | −20 | 0.0687 | −19 | 0.0668 | −19 | 0.0649 | −19 | 0.0630 | −19 | 0.0611 | −19 |
| $K_1$ | 0.0477 | −12 | 0.0465 | −12 | 0.0453 | −11 | 0.0442 | −11 | 0.0431 | −12 | 0.0419 | −11 |
| $K_2$ | 0.1134 | −22 | 0.1112 | −22 | 0.1090 | −22 | 0.1068 | −23 | 0.1045 | −22 | 0.1023 | −22 |
| $M_2^*$, $N_2$, $2N$ | 9.9854 | +2 | 9.9856 | +3 | 9.9859 | +2 | 9.9861 | +3 | 9.9864 | +2 | 9.9866 | +3 |
| $M_3$ | 9.9780 | +4 | 9.9784 | +4 | 9.9788 | +4 | 9.9792 | +4 | 9.9796 | +4 | 9.9800 | +4 |
| $M_4$, $MN$ | 9.9707 | +5 | 9.9712 | +5 | 9.9717 | +6 | 9.9723 | +5 | 9.9728 | +5 | 9.9733 | +5 |
| $M_6$ | 9.9561 | +8 | 9.9569 | +7 | 9.9576 | +8 | 9.9584 | +8 | 9.9592 | +7 | 9.9599 | +8 |
| $M_8$ | 9.9415 | +10 | 9.9425 | +10 | 9.9435 | +10 | 9.9445 | +10 | 9.9455 | +11 | 9.9466 | +10 |
| $O_1$, $Q_1$, $2Q$, $\rho_1$ | 0.0811 | −21 | 0.0790 | −20 | 0.0770 | −21 | 0.0749 | −20 | 0.0729 | −20 | 0.0709 | −21 |
| $OO$ | 0.2726 | −67 | 0.2659 | −67 | 0.2592 | −67 | 0.2525 | −66 | 0.2459 | −66 | 0.2393 | −66 |
| $MK$ | 0.0330 | −9 | 0.0321 | −9 | 0.0312 | −9 | 0.0303 | −9 | 0.0294 | −9 | 0.0285 | −8 |
| $2MK$ | 0.0184 | −7 | 0.0177 | −6 | 0.0171 | −7 | 0.0164 | −6 | 0.0158 | −6 | 0.0152 | −6 |
| $Mf$ | 0.1769 | −44 | 0.1725 | −44 | 0.1681 | −44 | 0.1637 | −43 | 0.1594 | −43 | 0.1551 | −43 |
| $Mm$ | 9.9514 | +8 | 9.9522 | +8 | 9.9530 | +9 | 9.9539 | +8 | 9.9547 | +9 | 9.9556 | +8 |

| Constituent \ I | 19.5° | Diff. | 19.6° | Diff. | 19.7° | Diff. | 19.8° | Diff. | 19.9° | Diff. | 20.0° | Diff. |
|---|---|---|---|---|---|---|---|---|---|---|---|---|
| $J_1$ | 0.0592 | −18 | 0.0574 | −19 | 0.0555 | −18 | 0.0537 | −19 | 0.0518 | −18 | 0.0500 | −18 |
| $K_1$ | 0.0408 | −12 | 0.0396 | −11 | 0.0385 | −11 | 0.0374 | −12 | 0.0362 | −11 | 0.0351 | −11 |
| $K_2$ | 0.1001 | −23 | 0.0978 | −22 | 0.0956 | −23 | 0.0933 | −22 | 0.0911 | −23 | 0.0888 | −24 |
| $M_2^*$, $N_2$, $2N$ | 9.9869 | +3 | 9.9872 | +2 | 9.9874 | +3 | 9.9877 | +3 | 9.9880 | +2 | 9.9882 | +3 |
| $M_3$ | 9.9804 | +3 | 9.9807 | +4 | 9.9811 | +4 | 9.9815 | +4 | 9.9819 | +4 | 9.9823 | +4 |
| $M_4$, $MN$ | 9.9738 | +5 | 9.9743 | +6 | 9.9749 | +5 | 9.9754 | +5 | 9.9759 | +5 | 9.9764 | +6 |
| $M_6$ | 9.9607 | +8 | 9.9615 | +8 | 9.9623 | +8 | 9.9631 | +8 | 9.9639 | +8 | 9.9647 | +8 |
| $M_8$ | 9.9476 | +11 | 9.9487 | +10 | 9.9497 | +11 | 9.9508 | +10 | 9.9518 | +11 | 9.9529 | +11 |
| $O_1$, $Q_1$, $2Q$, $\rho_1$ | 0.0688 | −20 | 0.0668 | −20 | 0.0648 | −19 | 0.0629 | −20 | 0.0609 | −20 | 0.0589 | −19 |
| $OO$ | 0.2327 | −65 | 0.2262 | −65 | 0.2197 | −65 | 0.2132 | −64 | 0.2068 | −64 | 0.2004 | −64 |
| $MK$ | 0.0277 | −9 | 0.0268 | −9 | 0.0259 | −9 | 0.0250 | −8 | 0.0242 | −9 | 0.0233 | −8 |
| $2MK$ | 0.0146 | −6 | 0.0140 | −7 | 0.0133 | −6 | 0.0127 | −6 | 0.0121 | −6 | 0.0115 | −6 |
| $Mf$ | 0.1508 | −43 | 0.1465 | −42 | 0.1423 | −43 | 0.1380 | −42 | 0.1338 | −41 | 0.1297 | −42 |
| $Mm$ | 9.9564 | +9 | 9.9573 | +8 | 9.9581 | +9 | 9.9590 | +9 | 9.9599 | +9 | 9.9608 | +9 |

*Log F of $\lambda_1$, $\mu_2$, $\nu_2$, MS, 2SM, and MSf are each equal to log F of $M_2$.
Log F of $P_1$, $R_2$, $S_1$, $S_2$, $S_4$, $S_6$, $T_2$, Sa, and Ssa are each zero.
For log F of $L_2$ and $M_1$ see Table 13.

HARMONIC ANALYSIS AND PREDICTION OF TIDES 187

Table 12.—*Log factor F corresponding to every tenth of a degree of I—Con.*

| I / Constituent | 20.1° | Diff. | 20.2° | Diff. | 20.3° | Diff. | 20.4° | Diff. | 20.5° | Diff. | 20.6° | Diff. |
|---|---|---|---|---|---|---|---|---|---|---|---|---|
| $J_1$ | 0.0482 | −18 | 0.0464 | −17 | 0.0447 | −18 | 0.0429 | −18 | 0.0411 | −17 | 0.0394 | −17 |
| $K_1$ | 0.0340 | −11 | 0.0329 | −11 | 0.0318 | −11 | 0.0307 | −11 | 0.0296 | −11 | 0.0285 | −11 |
| $K_2$ | 0.0864 | −23 | 0.0841 | −23 | 0.0818 | −23 | 0.0795 | −24 | 0.0771 | −23 | 0.0748 | −23 |
| $M_2^*$, $N_2$, $2N$ | 9.9885 | +3 | 9.9888 | +2 | 9.9890 | +3 | 9.9893 | +3 | 9.9896 | +3 | 9.9899 | +2 |
| $M_3$ | 9.9827 | +4 | 9.9831 | +4 | 9.9835 | +5 | 9.9840 | +4 | 9.9844 | +4 | 9.9848 | +4 |
| $M_4$, $MN$ | 9.9770 | +5 | 9.9775 | +6 | 9.9781 | +5 | 9.9786 | +6 | 9.9792 | +5 | 9.9797 | +6 |
| $M_6$ | 9.9655 | +8 | 9.9663 | +8 | 9.9671 | +8 | 9.9679 | +8 | 9.9687 | +9 | 9.9696 | +8 |
| $M_8$ | 9.9540 | +10 | 9.9550 | +11 | 9.9561 | +11 | 9.9572 | +11 | 9.9583 | +11 | 9.9594 | +11 |
| $O_1$, $Q_1$, $2Q$, $\rho_1$ | 0.0570 | −19 | 0.0551 | −20 | 0.0531 | −19 | 0.0512 | −19 | 0.0493 | −18 | 0.0475 | −19 |
| $OO$ | 0.1940 | −63 | 0.1877 | −63 | 0.1814 | −63 | 0.1751 | −62 | 0.1689 | −62 | 0.1627 | −62 |
| $MK$ | 0.0225 | −8 | 0.0217 | −9 | 0.0208 | −8 | 0.0200 | −9 | 0.0191 | −8 | 0.0183 | −8 |
| $2MK$ | 0.0109 | −5 | 0.0104 | −6 | 0.0098 | −5 | 0.0093 | −6 | 0.0087 | −5 | 0.0082 | −6 |
| $Mf$ | 0.1255 | −41 | 0.1214 | −41 | 0.1173 | −41 | 0.1132 | −41 | 0.1091 | −40 | 0.1051 | −41 |
| $Mm$ | 9.9617 | +9 | 9.9626 | +9 | 9.9635 | +9 | 9.9644 | +9 | 9.9653 | +9 | 9.9662 | +9 |

| I / Constituent | 20.7° | Diff. | 20.8° | Diff. | 29.0° | Diff. | 21.0° | Diff. | 21.1° | Diff. | 21.2° | Diff. |
|---|---|---|---|---|---|---|---|---|---|---|---|---|
| $J_1$ | 0.0377 | −17 | 0.0360 | −17 | 0.0343 | −17 | 0.0326 | −17 | 0.0309 | −17 | 0.0292 | −16 |
| $K_1$ | 0.0274 | −11 | 0.0263 | −11 | 0.0252 | −11 | 0.0241 | −11 | 0.0230 | −11 | 0.0219 | −10 |
| $K_2$ | 0.0725 | −24 | 0.0701 | −23 | 0.0678 | −24 | 0.0654 | −24 | 0.0630 | −23 | 0.0607 | −24 |
| $M_2^*$, $N_2$, $2N$ | 9.9901 | +3 | 9.9904 | +3 | 9.9907 | +3 | 9.9910 | +2 | 9.9912 | +3 | 9.9915 | +3 |
| $M_3$ | 9.9852 | +4 | 9.9856 | +4 | 9.9860 | +4 | 9.9864 | +5 | 9.9869 | +4 | 9.9873 | +4 |
| $M_4$, $MN$ | 9.9803 | +5 | 9.9808 | +6 | 9.9814 | +5 | 9.9819 | +6 | 9.9825 | +6 | 9.9831 | +5 |
| $M_6$ | 9.9704 | +8 | 9.9712 | +8 | 9.9720 | +9 | 9.9729 | +8 | 9.9737 | +9 | 9.9746 | +8 |
| $M_8$ | 9.9605 | +11 | 9.9616 | +11 | 9.9627 | +12 | 9.9639 | +11 | 9.9650 | +11 | 9.9661 | +12 |
| $O_1$, $Q_1$, $2Q$, $\rho_1$ | 0.0456 | −19 | 0.0437 | −18 | 0.0419 | −19 | 0.0400 | −18 | 0.0382 | −18 | 0.0364 | −18 |
| $OO$ | 0.1565 | −61 | 0.1504 | −61 | 0.1443 | −61 | 0.1382 | −61 | 0.1321 | −60 | 0.1261 | −60 |
| $MK$ | 0.0175 | −8 | 0.0167 | −8 | 0.0159 | −8 | 0.0151 | −8 | 0.0143 | −8 | 0.0135 | −8 |
| $2MK$ | 0.0076 | −5 | 0.0071 | −6 | 0.0065 | −5 | 0.0060 | −5 | 0.0055 | −5 | 0.0050 | −5 |
| $Mf$ | 0.1010 | −40 | 0.0970 | −39 | 0.0931 | −40 | 0.0891 | −39 | 0.0852 | −40 | 0.0812 | −39 |
| $Mm$ | 9.9671 | +9 | 9.9680 | +10 | 9.9690 | +9 | 9.9699 | +10 | 9.9709 | +9 | 9.9718 | +10 |

| I / Constituent | 21.3° | Diff. | 21.4° | Diff. | 21.5° | Diff. | 21.6° | Diff. | 21.7° | Diff. | 21.8° | Diff. |
|---|---|---|---|---|---|---|---|---|---|---|---|---|
| $J_1$ | 0.0276 | −17 | 0.0259 | −16 | 0.0243 | −16 | 0.0227 | −16 | 0.0211 | −16 | 0.0195 | −16 |
| $K_1$ | 0.0209 | −11 | 0.0198 | −11 | 0.0187 | −10 | 0.0177 | −11 | 0.0166 | −10 | 0.0156 | −11 |
| $K_2$ | 0.0583 | −24 | 0.0559 | −25 | 0.0534 | −24 | 0.0510 | −24 | 0.0486 | −24 | 0.0462 | −24 |
| $M_2^*$, $N_2$, $2N$ | 9.9918 | +3 | 9.9921 | +3 | 9.9924 | +3 | 9.9927 | +3 | 9.9930 | +3 | 9.9933 | +3 |
| $M_3$ | 9.9877 | +5 | 9.9882 | +4 | 9.9886 | +4 | 9.9890 | +4 | 9.9894 | +5 | 9.9899 | +4 |
| $M_4$, $MN$ | 9.9836 | +6 | 9.9842 | +6 | 9.9848 | +6 | 9.9854 | +5 | 9.9859 | +6 | 9.9865 | +6 |
| $M_6$ | 9.9754 | +9 | 9.9763 | +9 | 9.9772 | +8 | 9.9780 | +9 | 9.9789 | +9 | 9.9798 | +8 |
| $M_8$ | 9.9673 | +11 | 9.9684 | +12 | 9.9696 | +11 | 9.9707 | +12 | 9.9719 | +11 | 9.9730 | +12 |
| $O_1$, $Q_1$, $2Q$, $\rho_1$ | 0.0346 | −18 | 0.0328 | −18 | 0.0310 | −18 | 0.0292 | −17 | 0.0275 | −18 | 0.0257 | −17 |
| $OO$ | 0.1201 | −60 | 0.1141 | −59 | 0.1082 | −59 | 0.1023 | −59 | 0.0964 | −58 | 0.0906 | −58 |
| $MK$ | 0.0127 | −8 | 0.0119 | −8 | 0.0111 | −8 | 0.0103 | −7 | 0.0096 | −8 | 0.0088 | −7 |
| $2MK$ | 0.0045 | −5 | 0.0040 | −5 | 0.0035 | −5 | 0.0030 | −5 | 0.0025 | −4 | 0.0021 | −5 |
| $Mf$ | 0.0773 | −38 | 0.0735 | −39 | 0.0696 | −38 | 0.0658 | −39 | 0.0619 | −38 | 0.0581 | −37 |
| $Mm$ | 9.9728 | +9 | 9.9737 | +10 | 9.9747 | +10 | 9.9757 | +10 | 9.9767 | +9 | 9.9776 | +10 |

*Log F of $\lambda_2$, $\mu_2$, $\nu_2$, SM, 2SM, and MSf are each equal to log F of $M_2$.
Log F of $P_1$, $R_2$, $S_1$, $S_2$, $S_4$, $S_6$, $T_2$, Sa, and Ssa are each zero.
For log F of $L_2$ and $M_1$ see Table 13.

Table 12.—*Log factor F corresponding to every tenth of a degree of I*—Con.

| Constituent / $I$ | 21.9° | Diff. | 22.0° | Diff. | 22.1° | Diff. | 22.2° | Diff. | 22.3° | Diff. | 22.4° | Diff. |
|---|---|---|---|---|---|---|---|---|---|---|---|---|
| $J_1$ | 0.0179 | −16 | 0.0163 | −15 | 0.0148 | −16 | 0.0132 | −15 | 0.0117 | −16 | 0.0101 | −15 |
| $K_1$ | 0.0145 | −10 | 0.0135 | −11 | 0.0124 | −10 | 0.0114 | −11 | 0.0103 | −10 | 0.0093 | −10 |
| $K_2$ | 0.0438 | −24 | 0.0414 | −24 | 0.0390 | −25 | 0.0365 | −24 | 0.0341 | −24 | 0.0317 | −24 |
| $M_2^*$, $N_2$, $2N$ | 9.9936 | +2 | 9.9938 | +3 | 9.9941 | +3 | 9.9944 | +3 | 9.9947 | +3 | 9.9950 | +3 |
| $M_3$ | 9.9903 | +5 | 9.9908 | +4 | 9.9912 | +5 | 9.9917 | +4 | 9.9921 | +5 | 9.9926 | +4 |
| $M_4$, MN | 9.9871 | +6 | 9.9877 | +6 | 9.9883 | +6 | 9.9889 | +6 | 9.9895 | +6 | 9.9901 | +6 |
| $M_6$ | 9.9806 | +9 | 9.9815 | +9 | 9.9824 | +9 | 9.9833 | +9 | 9.9842 | +9 | 9.9851 | +9 |
| $M_8$ | 9.9742 | +12 | 9.9754 | +12 | 9.9766 | +11 | 9.9777 | +12 | 9.9789 | +12 | 9.9801 | +12 |
| $O_1$, $Q_1$, $2Q$, $\rho_1$ | 0.0240 | −18 | 0.0222 | −17 | 0.0205 | −17 | 0.0188 | −17 | 0.0171 | −17 | 0.0154 | −17 |
| $OO$ | 0.0848 | −58 | 0.0790 | −58 | 0.0732 | −57 | 0.0675 | −57 | 0.0618 | −57 | 0.0561 | −57 |
| MK | 0.0081 | −8 | 0.0073 | −8 | 0.0065 | −7 | 0.0058 | −7 | 0.0051 | −8 | 0.0043 | −7 |
| 2MK | 0.0016 | −5 | 0.0011 | −4 | 0.0007 | −5 | 0.0002 | −4 | 9.9998 | −4 | 9.9994 | −4 |
| Mf | 0.0544 | −38 | 0.0506 | −37 | 0.0469 | −38 | 0.0431 | −37 | 0.0394 | −37 | 0.0357 | −36 |
| Mm | 9.9786 | +10 | 9.9796 | +10 | 9.9806 | +10 | 9.9816 | +11 | 9.9827 | +10 | 9.9837 | +10 |

| Constituent / $I$ | 22.5° | Diff. | 22.6° | Diff. | 22.7° | Diff. | 22.8° | Diff. | 22.9° | Diff. | 23.0° | Diff. |
|---|---|---|---|---|---|---|---|---|---|---|---|---|
| $J_1$ | 0.0086 | −15 | 0.0071 | −15 | 0.0056 | −15 | 0.0041 | −15 | 0.0026 | −14 | 0.0012 | −15 |
| $K_1$ | 0.0083 | −10 | 0.0073 | −10 | 0.0063 | −11 | 0.0052 | −10 | 0.0042 | −10 | 0.0032 | −10 |
| $K_2$ | 0.0293 | −25 | 0.0268 | −24 | 0.0244 | −25 | 0.0219 | −25 | 0.0194 | −24 | 0.0170 | −25 |
| $M_2^*$, $N_2$, $2N$ | 9.9953 | +3 | 9.9956 | +3 | 9.9959 | +3 | 9.9962 | +4 | 9.9966 | +3 | 9.9969 | +3 |
| $M_3$ | 9.9930 | +5 | 9.9935 | +4 | 9.9939 | +5 | 9.9944 | +4 | 9.9948 | +5 | 9.9953 | +5 |
| $M_4$, MN | 9.9907 | +6 | 9.9913 | +6 | 9.9919 | +6 | 9.9925 | +6 | 9.9931 | +6 | 9.9937 | +6 |
| $M_6$ | 9.9860 | +9 | 9.9869 | +9 | 9.9878 | +9 | 9.9887 | +10 | 9.9897 | +9 | 9.9906 | +9 |
| $M_8$ | 9.9813 | +12 | 9.9825 | +13 | 9.9838 | +12 | 9.9850 | +12 | 9.9862 | +12 | 9.9874 | +13 |
| $O_1$, $Q_1$, $2Q$, $\rho_1$ | 0.0137 | −17 | 0.0120 | −16 | 0.0104 | −17 | 0.0087 | −16 | 0.0071 | −17 | 0.0054 | −16 |
| $OO$ | 0.0504 | −56 | 0.0448 | −56 | 0.0392 | −56 | 0.0336 | −55 | 0.0281 | −55 | 0.0226 | −55 |
| MK | 0.0036 | −7 | 0.0029 | −7 | 0.0022 | −7 | 0.0015 | −7 | 0.0008 | −7 | 0.0001 | −7 |
| 2MK | 9.9990 | −5 | 9.9985 | −4 | 9.9981 | −4 | 9.9977 | −4 | 9.9973 | −4 | 9.9969 | −3 |
| Mf | 0.0321 | −37 | 0.0284 | −36 | 0.0248 | −36 | 0.0212 | −36 | 0.0176 | −36 | 0.0140 | −36 |
| Mm | 9.9847 | +10 | 9.9857 | +11 | 9.9868 | +10 | 9.9878 | +11 | 9.9889 | +10 | 9.9899 | +11 |

| Constituent / $I$ | 23.1° | Diff. | 23.2° | Diff. | 23.3° | Diff. | 23.4° | Diff. | 23.5° | Diff. | 23.6° | Diff. |
|---|---|---|---|---|---|---|---|---|---|---|---|---|
| $J_1$ | 9.9997 | −15 | 9.9982 | −14 | 9.9968 | −14 | 9.9954 | −14 | 9.9940 | −14 | 9.9926 | −14 |
| $K_1$ | 0.0022 | −10 | 0.0012 | −10 | 0.0002 | −10 | 9.9992 | −10 | 9.9982 | −9 | 9.9973 | −10 |
| $K_2$ | 0.0145 | −24 | 0.0121 | −25 | 0.0096 | −24 | 0.0072 | −25 | 0.0047 | −25 | 0.0022 | −24 |
| $M_2^*$, $N_2$, $2N$ | 9.9972 | +3 | 9.9975 | +3 | 9.9978 | +3 | 9.9981 | +3 | 9.9984 | +3 | 9.9987 | +4 |
| $M_3$ | 9.9958 | +5 | 9.9963 | +4 | 9.9967 | +5 | 9.9972 | +4 | 9.9976 | +5 | 9.9981 | +5 |
| $M_4$, MN | 9.9943 | +7 | 9.9950 | +6 | 9.9956 | +6 | 9.9962 | +6 | 9.9968 | +7 | 9.9975 | +6 |
| $M_6$ | 9.9915 | +9 | 9.9924 | +10 | 9.9934 | +9 | 9.9943 | +10 | 9.9953 | +9 | 9.9962 | +10 |
| $M_8$ | 9.9887 | +12 | 9.9899 | +13 | 9.9912 | +12 | 9.9924 | +13 | 9.9937 | +12 | 9.9949 | +13 |
| $O_1$, $Q_1$, $2Q$, $\rho_1$ | 0.0038 | −16 | 0.0022 | −16 | 0.0006 | −16 | 9.9990 | −16 | 9.9974 | −16 | 9.9958 | −16 |
| $OO$ | 0.0171 | −55 | 0.0116 | −54 | 0.0062 | −55 | 0.0007 | −54 | 9.9953 | −53 | 9.9900 | −54 |
| MK | 9.9994 | −7 | 9.9987 | −7 | 9.9980 | −6 | 9.9974 | −7 | 9.9967 | −7 | 9.9960 | −6 |
| 2MK | 9.9966 | −4 | 9.9962 | −4 | 9.9958 | −3 | 9.9955 | −4 | 9.9951 | −3 | 9.9948 | −4 |
| Mf | 0.0104 | −35 | 0.0069 | −35 | 0.0034 | −35 | 9.9999 | −35 | 9.9964 | −35 | 9.9929 | −35 |
| Mm | 9.9910 | +11 | 9.9921 | +10 | 9.9931 | +11 | 9.9942 | +11 | 9.9953 | +11 | 9.9964 | +11 |

*Log F of $\lambda_2$ $\mu_2$, $\nu_2$, MS, 2SM and MSf are each equal to log F of $M_2$.
Log F of $P_1$, $R_2$ $S_1$, $S_2$, $S_4$, $S_6$, $T_2$ Sa, and Ssa are each zero.
For log F of $L_2$ and $M_1$ see Table 13.

HARMONIC ANALYSIS AND PREDICTION OF TIDES 189

Table 12.—*Log factor F corresponding to every tenth of a degree of I*—Con.

| I  Constituent | 23.7° | Diff. | 23.8° | Diff. | 23.9° | Diff. | 24.0° | Diff. | 24.1° | Diff. | 24.2° | Diff. |
|---|---|---|---|---|---|---|---|---|---|---|---|---|
| $J_1$ | 9.9912 | −14 | 9.9898 | −14 | 9.9884 | −14 | 9.9870 | −13 | 9.9857 | −14 | 9.9843 | −13 |
| $K_1$ | 9.9963 | −9 | 9.9954 | −10 | 9.9944 | −10 | 9.9934 | −10 | 9.9924 | −9 | 9.9915 | −10 |
| $K_2$ | 9.9998 | −25 | 9.9973 | −25 | 9.9948 | −24 | 9.9924 | −25 | 9.9899 | −25 | 9.9874 | −24 |
| $M_2^*$, $N_2$, $2N$ | 9.9991 | +3 | 9.9994 | +3 | 9.9997 | +3 | 0.0000 | +3 | 0.0003 | +4 | 0.0007 | +3 |
| $M_3$ | 9.9986 | +5 | 9.9991 | +4 | 9.9995 | +5 | 0.0000 | +5 | 0.0005 | +5 | 0.0010 | +5 |
| $M_4$, MN | 9.9981 | +6 | 9.9987 | +7 | 9.9994 | +6 | 0.0000 | +7 | 0.0007 | +6 | 0.0013 | +7 |
| $M_6$ | 9.9972 | +9 | 9.9981 | +10 | 9.9991 | +10 | 0.0001 | +9 | 0.0010 | +10 | 0.0020 | +10 |
| $M_8$ | 9.9962 | +13 | 9.9975 | +13 | 9.9988 | +13 | 0.0001 | +12 | 0.0013 | +13 | 0.0026 | +13 |
| $O_1$, $Q_1$, $2Q$, $\rho_1$ | 9.9942 | −15 | 9.9927 | −16 | 9.9911 | −15 | 9.9896 | −16 | 9.9880 | −15 | 9.9865 | −15 |
| OO | 9.9846 | −53 | 9.9793 | −53 | 9.9740 | −53 | 9.9687 | −53 | 9.9634 | −52 | 9.9582 | −52 |
| MK | 9.9954 | −7 | 9.9947 | −6 | 9.9941 | −7 | 9.9934 | −7 | 9.9927 | −6 | 9.9921 | −6 |
| 2MK | 9.9944 | −3 | 9.9941 | −4 | 9.9937 | −3 | 9.9934 | −3 | 9.9931 | −3 | 9.9928 | −3 |
| Mf | 9.9894 | −34 | 9.9860 | −35 | 9.9825 | −34 | 9.9791 | −34 | 9.9757 | −33 | 9.9724 | −34 |
| Mm | 9.9975 | +11 | 9.9986 | +11 | 9.9997 | +12 | 0.0009 | +11 | 0.0020 | +11 | 0.0031 | +12 |

| I  Cotstituent | 24.3° | Diff. | 24.4° | Diff. | 24.5° | Diff. | 24.6° | Diff. | 24.7° | Diff. | 24.8° | Diff. |
|---|---|---|---|---|---|---|---|---|---|---|---|---|
| $J_1$ | 9.9830 | −14 | 9.9816 | −13 | 9.9803 | −13 | 9.9790 | −13 | 9.9777 | −13 | 9.9764 | −13 |
| $K_1$ | 9.9905 | −9 | 9.9896 | −9 | 9.9887 | −10 | 9.9877 | −9 | 9.9868 | −10 | 9.9858 | −9 |
| $K_2$ | 9.9850 | −25 | 9.9825 | −25 | 9.9800 | −24 | 9.9776 | −25 | 9.9751 | −25 | 9.9726 | −25 |
| $M_2^*$, $N_2$, $2N$ | 0.0010 | +3 | 0.0013 | +3 | 0.0016 | +4 | 0.0020 | +3 | 0.0023 | +3 | 0.0026 | +4 |
| $M_3$ | 0.0015 | +5 | 0.0020 | +5 | 0.0025 | +5 | 0.0030 | +5 | 0.0035 | +5 | 0.0040 | +5 |
| $M_4$, MN | 0.0020 | +6 | 0.0026 | +7 | 0.0033 | +6 | 0.0039 | +7 | 0.0046 | +7 | 0.0053 | +6 |
| $M_6$ | 0.0030 | +9 | 0.0039 | +10 | 0.0049 | +10 | 0.0059 | +10 | 0.0069 | +10 | 0.0079 | +10 |
| $M_8$ | 0.0039 | +14 | 0.0053 | +13 | 0.0066 | +13 | 0.0079 | +13 | 0.0092 | +13 | 0.0105 | +14 |
| $O_1$, $Q_1$, $2Q$, $\rho_1$ | 9.9850 | −15 | 9.9835 | −15 | 9.9820 | −15 | 9.9805 | −15 | 9.9790 | −15 | 9.9775 | −15 |
| OO | 9.9530 | −52 | 9.9478 | −52 | 9.9426 | −51 | 9.9375 | −51 | 9.9324 | −51 | 9.9273 | −51 |
| MK | 9.9915 | −6 | 9.9909 | −6 | 9.9903 | −6 | 9.9897 | −6 | 9.9891 | −6 | 9.9885 | −6 |
| 2MK | 9.9925 | −3 | 9.9922 | −3 | 9.9919 | −3 | 9.9916 | −2 | 9.9914 | −3 | 9.9911 | −2 |
| Mf | 9.9690 | −34 | 9.9656 | −33 | 9.9623 | −33 | 9.9590 | −33 | 9.9557 | −33 | 9.9524 | −33 |
| Mm | 0.0043 | +11 | 0.0054 | +12 | 0.0066 | +11 | 0.0077 | +12 | 0.0089 | +12 | 0.0101 | +11 |

| I  Constituent | 24.9° | Diff. | 25.0° | Diff. | 25.1° | Diff. | 25.2° | Diff. | 25.3° | Diff. | 25.4° | Diff. |
|---|---|---|---|---|---|---|---|---|---|---|---|---|
| $J_1$ | 9.9751 | −13 | 9.9738 | −12 | 9.9726 | −13 | 9.9713 | −12 | 9.9701 | −13 | 9.9688 | −12 |
| $K_1$ | 9.9849 | −9 | 9.9840 | −9 | 9.9831 | −9 | 9.9822 | −10 | 9.9812 | −9 | 9.9803 | −9 |
| $K_2$ | 9.9701 | −24 | 9.9677 | −25 | 9.9652 | −24 | 9.9628 | −25 | 9.9603 | −24 | 9.9579 | −25 |
| $M_2^*$, $N_2$, $2N$ | 0.0030 | +3 | 0.0033 | +3 | 0.0036 | +4 | 0.0040 | +3 | 0.0043 | +4 | 0.0047 | +3 |
| $M_3$ | 0.0045 | +5 | 0.0050 | +5 | 0.0055 | +5 | 0.0060 | +5 | 0.0065 | +5 | 0.0070 | +5 |
| $M_4$, MN | 0.0059 | +7 | 0.0066 | +7 | 0.0073 | +7 | 0.0080 | +6 | 0.0086 | +7 | 0.0093 | +7 |
| $M_6$ | 0.0089 | +10 | 0.0099 | +10 | 0.0109 | +10 | 0.0119 | +11 | 0.0130 | +10 | 0.0140 | +10 |
| $M_8$ | 0.0119 | +13 | 0.0132 | +14 | 0.0146 | +13 | 0.0159 | +14 | 0.0173 | +13 | 0.0186 | +14 |
| $O_1$, $Q_1$, $2Q$, $\rho_1$ | 9.9760 | −14 | 9.9746 | −15 | 9.9731 | −14 | 9.9717 | −15 | 9.9702 | −14 | 9.9688 | −14 |
| OO | 9.9222 | −51 | 9.9171 | −50 | 9.9121 | −50 | 9.9071 | −50 | 9.9021 | −50 | 9.8971 | −49 |
| MK | 9.9879 | −6 | 9.9873 | −6 | 9.9867 | −6 | 9.9861 | −5 | 9.9856 | −6 | 9.9850 | −6 |
| 2MK | 9.9909 | −3 | 9.9906 | −2 | 9.9904 | −3 | 9.9901 | −2 | 9.9899 | −2 | 9.9897 | −3 |
| Mf | 9.9491 | −32 | 9.9459 | −33 | 9.9426 | −32 | 9.9394 | −32 | 9.9362 | −32 | 9.9330 | −32 |
| Mm | 0.0112 | +12 | 0.0124 | +12 | 0.0136 | +12 | 0.0148 | +12 | 0.0160 | +12 | 0.0172 | +13 |

*Log $F$ of $\lambda_2$, $\mu_2$, $\nu_2$, MS, 2SM, and MSf are each equal to log $F$ of $M_2$.
Log $F$ of $P_1$, $R_2$, $S_1$, $S_2$, $S_4$, $S_6$, $T_2$, Sa, and Ssa are each zero.
For log $F$ of $L_2$ and $M_1$ see Table 13.

Table 12.—*Log factor F corresponding to every tenth of a degree of I*—Con.

| Constituent | 25.5° | Diff. | 25.6° | Diff. | 25.7° | Diff. | 25.8° | Diff. | 25.9° | Diff. | 26.0° | Diff. |
|---|---|---|---|---|---|---|---|---|---|---|---|---|
| $J_1$ | 9.9676 | −12 | 9.9664 | −12 | 9.9652 | −13 | 9.9639 | −12 | 9.9627 | −11 | 9.9616 | −12 |
| $K_1$ | 9.9794 | −9 | 9.9785 | −9 | 9.9776 | −8 | 9.9768 | −9 | 9.9759 | −9 | 9.9750 | −9 |
| $K_2$ | 9.9554 | −25 | 9.9529 | −25 | 9.9504 | −24 | 9.9480 | −25 | 9.9455 | −24 | 9.9431 | −25 |
| $M_2^*$, $N_2$, $2N$ | 0.0050 | +3 | 0.0053 | +4 | 0.0057 | +3 | 0.0060 | +4 | 0.0064 | +3 | 0.0067 | +4 |
| $M_3$ | 0.0075 | +5 | 0.0080 | +5 | 0.0085 | +6 | 0.0091 | +5 | 0.0096 | +5 | 0.0101 | +5 |
| $M_4$, MN | 0.0100 | +7 | 0.0107 | +7 | 0.0114 | +7 | 0.0121 | +7 | 0.0128 | +7 | 0.0135 | +7 |
| $M_6$ | 0.0150 | +10 | 0.0160 | +11 | 0.0171 | +10 | 0.0181 | +11 | 0.0192 | +10 | 0.0202 | +11 |
| $M_8$ | 0.0200 | +14 | 0.0214 | +14 | 0.0228 | +13 | 0.0241 | +14 | 0.0255 | +14 | 0.0269 | +14 |
| $O_1$, $Q_1$, $2Q$, $\rho_1$ | 9.9674 | −14 | 9.9660 | −14 | 9.9646 | −14 | 9.9632 | −14 | 9.9618 | −14 | 9.9604 | −14 |
| OO | 9.8922 | −49 | 9.8873 | −49 | 9.8824 | −49 | 9.8775 | −49 | 9.8726 | −49 | 9.8677 | −48 |
| MK | 9.9844 | −5 | 9.9839 | −6 | 9.9833 | −5 | 9.9828 | −5 | 9.9823 | −6 | 9.9817 | −5 |
| 2MK | 9.9894 | −2 | 9.9892 | −2 | 9.9890 | −2 | 9.9888 | −2 | 9.9886 | −1 | 9.9885 | −2 |
| Mf | 9.9298 | −32 | 9.9266 | −31 | 9.9235 | −32 | 9.9203 | −31 | 9.9172 | −31 | 9.9141 | −31 |
| Mm | 0.0185 | +12 | 0.0197 | +12 | 0.0209 | +13 | 0.0222 | +12 | 0.0234 | +13 | 0.0247 | +12 |

| Constituent | 26.1° | Diff. | 26.2° | Diff. | 26.3° | Diff. | 26.4° | Diff. | 26.5° | Diff. | 26.6° | Diff. |
|---|---|---|---|---|---|---|---|---|---|---|---|---|
| $J_1$ | 9.9604 | −12 | 9.9592 | −12 | 9.9580 | −11 | 9.9569 | −12 | 9.9557 | −11 | 9.9546 | −11 |
| $K_1$ | 9.9741 | −9 | 9.9732 | −8 | 9.9724 | −9 | 9.9715 | −9 | 9.9706 | −8 | 9.9698 | −9 |
| $K_2$ | 9.9406 | −24 | 9.9382 | −25 | 9.9357 | −24 | 9.9333 | −25 | 9.9308 | −24 | 9.9284 | −24 |
| $M_2^*$, $N_2$, $2N$ | 0.0071 | +3 | 0.0074 | +4 | 0.0078 | +3 | 0.0081 | +4 | 0.0085 | +4 | 0.0089 | +3 |
| $M_3$ | 0.0106 | +6 | 0.0112 | +5 | 0.0117 | +5 | 0.0122 | +6 | 0.0128 | +5 | 0.0133 | +5 |
| $M_4$, MN | 0.0142 | +7 | 0.0149 | +7 | 0.0156 | +7 | 0.0163 | +7 | 0.0170 | +7 | 0.0177 | +7 |
| $M_6$ | 0.0213 | +10 | 0.0223 | +11 | 0.0234 | +10 | 0.0244 | +11 | 0.0255 | +11 | 0.0266 | +11 |
| $M_8$ | 0.0283 | +15 | 0.0298 | +14 | 0.0312 | +14 | 0.0326 | +14 | 0.0340 | +14 | 0.0354 | +15 |
| $O_1$, $Q_1$, $2Q$, $\rho_1$ | 9.9590 | −13 | 9.9577 | −14 | 9.9563 | −14 | 9.9549 | −13 | 9.9536 | −13 | 9.9523 | −14 |
| OO | 9.8629 | −48 | 9.8581 | −48 | 9.8533 | −47 | 9.8486 | −48 | 9.8438 | −47 | 9.8391 | −47 |
| MK | 9.9812 | −5 | 9.9807 | −5 | 9.9802 | −5 | 9.9797 | −5 | 9.9792 | −5 | 9.9787 | −5 |
| 2MK | 9.9883 | −2 | 9.9881 | −1 | 9.9880 | −2 | 9.9878 | −1 | 9.9877 | −2 | 9.9875 | −1 |
| Mf | 9.9110 | −31 | 9.9079 | −31 | 9.9048 | −31 | 9.9017 | −30 | 9.8987 | −30 | 9.8957 | −31 |
| Mm | 0.0259 | +13 | 0.0272 | +13 | 0.0285 | +13 | 0.0298 | +12 | 0.0310 | +13 | 0.0323 | +13 |

| Constituent | 26.7° | Diff. | 26.8° | Diff. | 26.9° | Diff. | 27.0° | Diff. | 27.1° | Diff. | 27.2° | Diff. |
|---|---|---|---|---|---|---|---|---|---|---|---|---|
| $J_1$ | 9.9535 | −11 | 9.9524 | −12 | 9.9512 | −11 | 9.9501 | −11 | 9.9490 | −11 | 9.9479 | −10 |
| $K_1$ | 9.9689 | −8 | 9.9681 | −9 | 9.9672 | −8 | 9.9664 | −8 | 9.9656 | −9 | 9.9647 | −8 |
| $K_2$ | 9.9260 | −25 | 9.9235 | −24 | 9.9211 | −24 | 9.9187 | −25 | 9.9162 | −24 | 0.0138 | −24 |
| $M_2^*$, $N_2$, $2N$ | 0.0092 | +4 | 0.0096 | +3 | 0.0099 | +4 | 0.0103 | +4 | 0.0107 | +3 | 0.0110 | +4 |
| $M_3$ | 0.0138 | +6 | 0.0144 | +5 | 0.0149 | +6 | 0.0155 | +5 | 0.0160 | +6 | 0.0166 | +5 |
| $M_4$, MN | 0.0184 | +8 | 0.0192 | +7 | 0.0199 | +7 | 0.0206 | +7 | 0.0213 | +8 | 0.0221 | +7 |
| $M_6$ | 0.0277 | +10 | 0.0287 | +11 | 0.0298 | +11 | 0.0309 | +11 | 0.0320 | +11 | 0.0331 | +11 |
| $M_8$ | 0.0369 | +14 | 0.0383 | +15 | 0.0398 | +14 | 0.0412 | +15 | 0.0427 | +14 | 0.0441 | +15 |
| $O_1$, $Q_1$, $2Q$, $\rho_1$ | 9.9509 | −13 | 9.9496 | −13 | 9.9483 | −13 | 9.9470 | −13 | 9.9457 | −13 | 9.9444 | −13 |
| OO | 9.8344 | −47 | 9.8297 | −47 | 9.8250 | −47 | 9.8203 | −46 | 9.8157 | −46 | 9.8111 | −46 |
| MK | 9.9782 | −5 | 9.9777 | −5 | 9.9772 | −5 | 9.9767 | −5 | 9.9762 | −4 | 9.9758 | −5 |
| 2MK | 9.9874 | −2 | 9.9872 | −1 | 9.9871 | −1 | 9.9870 | −1 | 9.9869 | −1 | 9.9868 | −1 |
| Mf | 9.8926 | −30 | 9.8896 | −30 | 9.8866 | −29 | 9.8837 | −30 | 9.8807 | −30 | 9.8777 | −29 |
| Mm | 0.0336 | +14 | 0.0350 | +13 | 0.0363 | +13 | 0.0376 | +13 | 0.0389 | +14 | 0.0403 | +13 |

*Log F of $\lambda_2$, $\mu_2$, $\nu_2$, MS, 2SM, and MSf are each equal to log F of $M_2$.
Log F of $P_1$, $R_2$, $S_1$, $S_2$, $S_4$, $S_6$, $T_2$, Sa, and Ssa are each zero.
For log F of $L_2$ and $M_1$ see Table 13.

Table 12.—*Log factor F corresponding to every tenth of a degree of I*—Con.

| Constituent / I | 27.3° | Diff. | 27.4° | Diff. | 27.5° | Diff. | 27.6° | Diff. | 27.7° | Diff. |
|---|---|---|---|---|---|---|---|---|---|---|
| $J_1$ | 9.9469 | −11 | 9.9458 | −11 | 9.9447 | −10 | 9.9437 | −11 | 9.9426 | −10 |
| $K_1$ | 9.9639 | −8 | 9.9631 | −8 | 9.9623 | −8 | 9.9615 | −8 | 9.9607 | −8 |
| $K_2$ | 9.9114 | −24 | 9.9090 | −24 | 9.9066 | −24 | 9.9042 | −24 | 9.9018 | −24 |
| $M_2^*$, $N_2$, $2N$ | 0.0114 | +4 | 0.0118 | +3 | 0.0121 | +4 | 0.0125 | +4 | 0.0129 | +4 |
| $M_3$ | 0.0171 | +6 | 0.0177 | +5 | 0.0182 | +6 | 0.0188 | +5 | 0.0193 | +6 |
| $M_4$, MN | 0.0228 | +7 | 0.0235 | +8 | 0.0243 | +7 | 0.0250 | +8 | 0.0258 | +7 |
| $M_6$ | 0.0342 | +11 | 0.0353 | +11 | 0.0364 | +11 | 0.0375 | +12 | 0.0387 | +11 |
| $M_8$ | 0.0456 | +15 | 0.0471 | +15 | 0.0486 | +15 | 0.0501 | +14 | 0.0515 | +15 |
| $O_1$, $Q_1$, $2Q$, $\rho_1$ | 9.9431 | −13 | 9.9418 | −13 | 9.9405 | −12 | 9.9393 | −13 | 9.9380 | −12 |
| OO | 9.8065 | −46 | 9.8019 | −46 | 9.7973 | −45 | 9.7928 | −45 | 9.7883 | −45 |
| MK | 9.9753 | −4 | 9.9749 | −5 | 9.9744 | −4 | 9.9740 | −5 | 9.9735 | −4 |
| 2MK | 9.9867 | −1 | 9.9866 | −1 | 9.9865 | 0 | 9.9865 | −1 | 9.9864 | 0 |
| Mf | 9.8748 | −29 | 9.8719 | −30 | 9.8689 | −29 | 9.8660 | −29 | 9.8631 | −28 |
| Mm | 0.0416 | +14 | 0.0430 | +14 | 0.0444 | +13 | 0.0457 | +14 | 0.0471 | +14 |

| Constituent / I | 27.8° | Diff. | 27.9° | Diff. | 28.0° | Diff. | 28.1° | Diff. | 28.2° | Diff. |
|---|---|---|---|---|---|---|---|---|---|---|
| $J_1$ | 9.9416 | −11 | 9.9405 | −10 | 9.9395 | −10 | 9.9385 | −10 | 9.9375 | −10 |
| $K_1$ | 9.9599 | −9 | 9.9590 | −8 | 9.9582 | −8 | 9.9574 | −7 | 9.9567 | −8 |
| $K_2$ | 9.8994 | −24 | 9.8970 | −24 | 9.8946 | −24 | 9.8922 | −24 | 9.8898 | −24 |
| $M_2^*$, $N_2$, $2N$ | 0.0133 | +3 | 0.0136 | +4 | 0.0140 | +4 | 0.0144 | +4 | 0.0148 | +4 |
| $M_3$ | 0.0199 | +6 | 0.0205 | +5 | 0.0210 | +6 | 0.0216 | +6 | 0.0222 | +5 |
| $M_4$, MN | 0.0265 | +8 | 0.0273 | +7 | 0.0280 | +8 | 0.0288 | +7 | 0.0295 | +8 |
| $M_6$ | 0.0398 | +11 | 0.0409 | +11 | 0.0420 | +12 | 0.0432 | +11 | 0.0443 | +12 |
| $M_8$ | 0.0530 | +15 | 0.0545 | +16 | 0.0561 | +15 | 0.0576 | +15 | 0.0591 | +15 |
| $O_1$, $Q_1$, $2Q$, $\rho_1$ | 9.9368 | −13 | 9.9355 | −12 | 9.9343 | −13 | 9.9330 | −12 | 9.9318 | −12 |
| OO | 9.7838 | −45 | 9.7793 | −45 | 9.7748 | −45 | 9.7703 | −44 | 9.7659 | −44 |
| MK | 9.9731 | −4 | 9.9727 | −4 | 9.9723 | −5 | 9.9718 | −4 | 9.9714 | −4 |
| 2MK | 9.9864 | −1 | 9.9863 | 0 | 9.9863 | −1 | 9.9862 | 0 | 9.9862 | 0 |
| Mf | 9.8603 | −29 | 9.8574 | −29 | 9.8545 | −28 | 9.8517 | −28 | 9.8489 | −29 |
| Mm | 0.0485 | +14 | 0.0499 | +14 | 0.0513 | +14 | 0.0527 | +15 | 0.0542 | +14 |

| Constituent / I | 28.3° | Diff. | 28.4° | Diff. | 28.5° | Diff. | 28.6° |
|---|---|---|---|---|---|---|---|
| $J_1$ | 9.9365 | −10 | 9.9355 | −10 | 9.9345 | −10 | 9.9335 |
| $K_1$ | 9.9559 | −8 | 9.9551 | −8 | 9.9543 | −8 | 9.9535 |
| $K_2$ | 9.8874 | −24 | 9.8850 | −24 | 9.8826 | −23 | 9.8803 |
| $M_2^*$, $N_2$, $2N$ | 0.0152 | +3 | 0.0155 | +4 | 0.0159 | +4 | 0.0163 |
| $M_3$ | 0.0227 | +6 | 0.0233 | +6 | 0.0239 | +6 | 0.0245 |
| $M_4$, MN | 0.0303 | +8 | 0.0311 | +7 | 0.0318 | +8 | 0.0326 |
| $M_6$ | 0.0455 | +11 | 0.0466 | +12 | 0.0478 | +11 | 0.0489 |
| $M_8$ | 0.0606 | +15 | 0.0621 | +16 | 0.0637 | +15 | 0.0652 |
| $O_1$, $Q_1$, $2Q$, $\rho_1$ | 9.9306 | −12 | 9.9294 | −12 | 9.9282 | −12 | 9.9270 |
| OO | 9.7615 | −44 | 9.7571 | −44 | 9.7527 | −44 | 9.7483 |
| MK | 9.9710 | −4 | 9.9706 | −4 | 9.9702 | −4 | 9.9698 |
| 2MK | 9.9862 | 0 | 9.9862 | 0 | 9.9862 | 0 | 9.9862 |
| Mf | 9.8460 | −28 | 9.8432 | −28 | 9.8404 | −28 | 9.8376 |
| Mm | 0.0556 | +14 | 0.0570 | +14 | 0.0584 | +15 | 0.0599 |

*Log $F$ of $\lambda_2$, $\mu_2$ $\nu_2$, MS, and MSf are each equal to log $F$ of $M_2$.
Log $F$ of $P_1$, $R_2$, $S_1$, $S_2$, $S_4$, $S_6$, $T_2$, Sa, and Ssa are each zero.
For log $F$ of $L_2$ and $M_1$ see Table 13.

Table 13.—*Values of u and log F of $L_2$ and $M_1$ for years 1900 to 2000*

| Year | $N$ | $u$ of $L_2$ | Diff. | $u$ of $M_1$ | Diff. | Log $F$ ($L_2$) | Diff. | Log $F$ ($M_1$) | Diff. | $N$ |
|---|---|---|---|---|---|---|---|---|---|---|
| 1899 | 260° | +11.4° | 6.3 | 353.5° | 5.2 | 0.0964 | 161 | 9.7295 | 35 | 260° |
| 1900 | 255 | +5.1 | 6.4 | 358.7 | 5.0 | 0.1125 | 73 | 9.7260 | 78 | 255 |
|  | 250 | −1.3 | 4.9 | 3.7 | 5.1 | 0.1052 | 259 | 9.7338 | 189 | 250 |
|  | 245 | −6.2 | 2.9 | 8.8 | 5.4 | 0.0793 | 348 | 9.7527 | 297 | 245 |
|  | 240 | −9.1 | 0.9 | 14.2 | 6.2 | 0.0445 | 353 | 9.7824 | 404 | 240 |
| 1901 | 235 | −10.0 | 0.6 | 20.4 | 7.2 | 0.0092 | 313 | 9.8228 | 506 | 235 |
|  | 230 | −9.4 | 1.6 | 27.6 | 8.9 | 9.9779 | 250 | 9.8734 | 596 | 230 |
|  | 225 | −7.8 | 2.3 | 36.5 | 11.4 | 9.9529 | 182 | 9.9330 | 647 | 225 |
| 1902 | 220 | −5.5 | 2.5 | 47.9 | 14.6 | 9.9347 | 114 | 9.9977 | 609 | 220 |
|  | 215 | −3.0 | 2.6 | 62.5 | 18.0 | 9.9233 | 50 | 0.0586 | 412 | 215 |
|  | 210 | −0.4 | 2.6 | 80.5 | 19.4 | 9.9183 | 10 | 0.0998 | 58 | 210 |
|  | 205 | +2.2 | 2.3 | 99.9 | 17.7 | 9.9193 | 61 | 0.1056 | 276 | 205 |
| 1903 | 200 | +4.5 | 2.0 | 117.6 | 14.4 | 9.9254 | 115 | 0.0780 | 497 | 200 |
|  | 195 | +6.5 | 1.5 | 132.0 | 11.0 | 9.9369 | 154 | 0.0283 | 523 | 195 |
|  | 190 | +8.0 | 0.8 | 143.0 | 8.6 | 9.9523 | 190 | 9.9760 | 485 | 190 |
|  | 185 | +8.8 | 0.1 | 151.6 | 7.0 | 9.9713 | 213 | 9.9275 | 414 | 185 |
| 1904 | 180 | +8.9 | 0.7 | 158.6 | 5.8 | 9.9926 | 222 | 9.8861 | 336 | 180 |
|  | 175 | +8.2 | 1.8 | 164.4 | 5.1 | 0.0148 | 209 | 9.8525 | 256 | 175 |
|  | 170 | +6.4 | 2.8 | 169.5 | 4.6 | 0.0357 | 166 | 9.8269 | 178 | 170 |
|  | 165 | +3.6 | 3.5 | 174.1 | 4.5 | 0.0523 | 93 | 9.8091 | 100 | 165 |
| 1905 | 160 | +0.1 | 3.8 | 178.6 | 4.5 | 0.0616 | 5 | 9.7991 | 21 | 160 |
|  | 155 | −3.7 | 3.6 | 183.1 | 4.6 | 0.0611 | 109 | 9.7970 | 63 | 155 |
|  | 150 | −7.3 | 2.8 | 187.7 | 5.1 | 0.0502 | 195 | 9.8033 | 152 | 150 |
|  | 145 | −10.1 | 1.7 | 192.8 | 5.8 | 0.0307 | 249 | 9.8185 | 248 | 145 |
| 1906 | 140 | −11.8 | 0.3 | 198.6 | 7.0 | 0.0058 | 268 | 9.8433 | 349 | 140 |
|  | 135 | −12.1 | 0.8 | 205.6 | 8.8 | 9.9790 | 254 | 9.8782 | 446 | 135 |
|  | 130 | −11.3 | 1.9 | 214.4 | 11.5 | 9.9536 | 216 | 9.9228 | 516 | 130 |
|  | 125 | −9.4 | 2.7 | 225.9 | 15.4 | 9.9320 | 161 | 9.9744 | 494 | 125 |
| 1907 | 120 | −6.7 | 3.3 | 241.3 | 19.4 | 9.9159 | 90 | 0.0238 | 283 | 120 |
|  | 115 | −3.4 | 3.7 | 260.7 | 21.1 | 9.9069 | 11 | 0.0521 | 124 | 115 |
|  | 110 | +0.3 | 3.9 | 281.8 | 19.0 | 9.9058 | 79 | 0.0397 | 514 | 110 |
|  | 105 | +4.2 | 3.8 | 300.8 | 15.0 | 9.9137 | 177 | 9.9883 | 688 | 105 |
| 1908 | 100 | +8.0 | 3.2 | 315.8 | 11.3 | 9.9314 | 282 | 9.9195 | 683 | 100 |
|  | 95 | +11.2 | 2.2 | 327.1 | 8.7 | 9.9596 | 390 | 9.8512 | 591 | 95 |
|  | 90 | +13.4 | 0.4 | 335.8 | 7.2 | 9.9986 | 487 | 9.7921 | 472 | 90 |
| 1909 | 85 | +13.8 | 2.5 | 343.0 | 6.1 | 0.0473 | 536 | 9.7449 | 346 | 85 |
|  | 80 | +11.3 | 6.0 | 349.1 | 5.5 | 0.1009 | 463 | 9.7103 | 220 | 80 |
|  | 75 | +5.3 | 9.2 | 354.6 | 5.4 | 0.1472 | 191 | 9.6883 | 94 | 75 |
|  | 70 | −3.9 | 9.3 | 0.0 | 5.5 | 0.1663 | 209 | 9.6789 | 33 | 70 |
| 1910 | 65 | −13.2 | 6.2 | 5.5 | 5.8 | 0.1454 | 509 | 9.6822 | 164 | 65 |
|  | 60 | −19.4 | 2.1 | 11.3 | 6.7 | 0.0945 | 602 | 9.6986 | 299 | 60 |
|  | 55 | −21.5 | 1.1 | 18.0 | 8.2 | 0.0343 | 556 | 9.7285 | 437 | 55 |
|  | 50 | −20.4 | 3.4 | 26.2 | 10.5 | 9.9787 | 449 | 9.7722 | 564 | 50 |
| 1911 | 45 | −17.0 | 4.8 | 36.7 | 14.1 | 9.9338 | 324 | 9.8286 | 635 | 45 |
|  | 40 | −12.2 | 5.8 | 50.8 | 19.0 | 9.9014 | 191 | 9.8921 | 540 | 40 |
|  | 35 | −6.4 | 6.2 | 69.8 | 22.8 | 9.8823 | 59 | 9.9461 | 165 | 35 |
|  | 30 | −0.2 | 6.3 | 92.6 | 22.2 | 9.8764 | 77 | 9.9626 | 343 | 30 |
| 1912 | 25 | +6.1 | 6.0 | 114.8 | 17.7 | 9.8841 | 219 | 9.9283 | 643 | 25 |
|  | 20 | +12.1 | 5.3 | 132.5 | 13.1 | 9.9060 | 368 | 9.8640 | 681 | 20 |
|  | 15 | +17.4 | 3.9 | 145.6 | 9.9 | 9.9428 | 526 | 9.7959 | 585 | 15 |
|  | 10 | +21.3 | 1.4 | 155.5 | 7.7 | 9.9954 | 680 | 9.7374 | 450 | 10 |
| 1913 | 5 | +22.7 | 2.8 | 163.2 | 6.6 | 0.0634 | 780 | 9.6924 | 308 | 5 |
|  | 0 | +19.9 | 8.9 | 169.8 | 5.9 | 0.1414 | 675 | 9.6616 | 169 | 0 |
|  | 355 | +11.0 | 13.6 | 175.7 | 5.6 | 0.2089 | 187 | 9.6447 | 33 | 355 |
|  | 350 | −2.6 | 11.9 | 181.3 | 5.7 | 0.2276 | 450 | 9.6414 | 101 | 350 |
| 1914 | 345 | −14.5 | 6.0 | 187.0 | 6.1 | 0.1826 | 754 | 9.6515 | 240 | 345 |
|  | 340 | −20.5 | 0.7 | 193.1 | 6.9 | 0.1072 | 738 | 9.6755 | 378 | 340 |
|  | 335 | −21.2 | 2.5 | 200.0 | 8.5 | 0.0334 | 604 | 9.7133 | 518 | 335 |
|  | 330 | −18.7 |  | 208.5 |  | 9.9730 |  | 9.7651 |  | 330 |

HARMONIC ANALYSIS AND PREDICTION OF TIDES 193

Table 13.—*Values of u and log F of* $L_2$ *and* $M_1$ *for years 1900 to 2000*—Con.

| Year | N | u of $L_2$ | Diff. | u of $M_1$ | Diff. | Log F ($L_2$) | Diff. | Log F ($M_1$) | Diff. | N |
|---|---|---|---|---|---|---|---|---|---|---|
| | ° | ° | | ° | | | | | | ° |
| 1914 | 330 | −18.7 | 4.4 | 208.5 | 10.9 | 9.9730 | 447 | 9.7651 | 643 | 330 |
| 1915 | 325 | −14.3 | 5.5 | 219.4 | 14.6 | 9.9283 | 295 | 9.8294 | 699 | 325 |
| | 320 | −8.8 | 5.9 | 234.0 | 19.5 | 9.8988 | 153 | 9.8993 | 574 | 320 |
| | 315 | −2.9 | 5.9 | 253.5 | 22.8 | 9.8835 | 20 | 9.9567 | 174 | 315 |
| | 310 | +3.0 | 5.6 | 276.3 | 21.6 | 9.8815 | 108 | 9.9741 | 316 | 310 |
| 1916 | 305 | +8.6 | 5.0 | 297.9 | 17.0 | 9.8923 | 230 | 9.9425 | 581 | 305 |
| | 300 | +13.6 | 3.8 | 314.9 | 12.5 | 9.9153 | 346 | 9.8844 | 600 | 300 |
| | 295 | +17.4 | 2.2 | 327.4 | 9.4 | 9.9499 | 449 | 9.8244 | 506 | 295 |
| 1917 | 290 | +19.6 | 0.2 | 336.8 | 7.5 | 9.9948 | 510 | 9.7738 | 370 | 290 |
| | 285 | +19.4 | 3.2 | 344.3 | 6.3 | 0.0458 | 496 | 9.7368 | 239 | 285 |
| | 280 | +16.1 | 6.5 | 350.6 | 5.6 | 0.0954 | 351 | 9.7129 | 111 | 280 |
| | 275 | +9.6 | 7.9 | 356.2 | 5.2 | 0.1305 | 63 | 9.7018 | 15 | 275 |
| 1918 | 270 | +1.7 | 7.1 | 1.4 | 5.2 | 0.1368 | 227 | 9.7033 | 134 | 270 |
| | 265 | −5.4 | 4.4 | 6.6 | 5.5 | 0.1141 | 401 | 9.7167 | 252 | 265 |
| | 260 | −9.8 | 1.8 | 12.1 | 6.0 | 0.0740 | 440 | 9.7419 | 370 | 260 |
| | 255 | −11.6 | 0.4 | 18.1 | 7.1 | 0.0300 | 398 | 9.7789 | 486 | 255 |
| 1919 | 250 | −11.2 | 1.8 | 25.2 | 8.6 | 9.9902 | 320 | 9.8275 | 594 | 250 |
| | 245 | −9.4 | 2.6 | 33.8 | 11.2 | 9.9582 | 234 | 9.8869 | 670 | 245 |
| | 240 | −6.8 | 3.1 | 45.0 | 14.5 | 9.9348 | 148 | 9.9539 | 664 | 240 |
| | 235 | −3.7 | 3.2 | 59.5 | 18.2 | 9.9200 | 67 | 0.0203 | 493 | 235 |
| 1920 | 230 | −0.5 | 3.1 | 77.7 | 20.2 | 9.9133 | 6 | 0.0696 | 134 | 230 |
| | 225 | +2.6 | 2.8 | 97.9 | 18.8 | 9.9139 | 71 | 0.0830 | 250 | 225 |
| | 220 | +5.4 | 2.2 | 116.7 | 15.1 | 9.9210 | 128 | 0.0580 | 467 | 220 |
| | 215 | +7.6 | 1.6 | 131.8 | 11.5 | 9.9338 | 174 | 0.0113 | 510 | 215 |
| 1921 | 210 | +9.2 | 0.9 | 143.3 | 8.9 | 9.9512 | 209 | 9.9603 | 462 | 210 |
| | 205 | +10.1 | 0.0 | 152.2 | 7.0 | 9.9721 | 227 | 9.9141 | 381 | 205 |
| | 200 | +10.1 | 1.1 | 159.2 | 5.9 | 9.9948 | 224 | 9.8760 | 294 | 200 |
| | 195 | +9.0 | 2.0 | 165.1 | 5.0 | 0.0172 | 196 | 9.8466 | 209 | 195 |
| 1922 | 190 | +7.0 | 2.9 | 170.1 | 4.7 | 0.0368 | 141 | 9.8257 | 130 | 190 |
| | 185 | +4.1 | 3.5 | 174.8 | 4.4 | 0.0509 | 63 | 9.8127 | 54 | 185 |
| | 180 | +0.6 | 3.5 | 179.2 | 4.3 | 0.0572 | 27 | 9.8073 | 22 | 180 |
| | 175 | −2.9 | 3.2 | 183.5 | 4.5 | 0.0545 | 113 | 9.8095 | 97 | 175 |
| 1923 | 170 | −6.1 | 2.4 | 188.0 | 4.9 | 0.0432 | 179 | 9.8192 | 176 | 170 |
| | 165 | −8.5 | 1.4 | 192.9 | 5.6 | 0.0253 | 218 | 9.8368 | 259 | 165 |
| | 160 | −9.9 | 0.4 | 198.5 | 6.5 | 0.0035 | 231 | 9.8627 | 347 | 160 |
| 1924 | 155 | −10.3 | 0.5 | 205.0 | 8.2 | 9.9804 | 220 | 9.8974 | 433 | 155 |
| | 150 | −9.8 | 1.4 | 213.2 | 10.4 | 9.9584 | 191 | 9.9407 | 499 | 150 |
| | 145 | −8.4 | 2.1 | 223.6 | 13.9 | 9.9393 | 148 | 9.9906 | 498 | 145 |
| | 140 | −6.3 | 2.6 | 237.5 | 17.6 | 9.9245 | 94 | 0.0404 | 351 | 140 |
| 1925 | 135 | −3.7 | 3.0 | 255.1 | 20.2 | 9.9151 | 30 | 0.0755 | 9 | 135 |
| | 130 | −0.7 | 3.3 | 275.3 | 19.3 | 9.9121 | 42 | 0.0764 | 387 | 130 |
| | 125 | +2.6 | 3.2 | 294.6 | 15.9 | 9.9163 | 121 | 0.0377 | 631 | 125 |
| | 120 | +5.8 | 2.9 | 310.5 | 12.2 | 9.9284 | 207 | 9.9746 | 683 | 120 |
| 1926 | 115 | +8.7 | 2.2 | 322.7 | 9.4 | 9.9491 | 296 | 9.9063 | 626 | 115 |
| | 110 | +10.9 | 0.9 | 332.1 | 7.6 | 9.9787 | 380 | 9.8437 | 525 | 110 |
| | 105 | +11.8 | 0.9 | 339.7 | 6.3 | 0.0167 | 440 | 9.7912 | 409 | 105 |
| | 100 | +10.9 | 3.6 | 346.0 | 5.7 | 0.0607 | 434 | 9.7503 | 290 | 100 |
| 1927 | 95 | +7.3 | 6.4 | 351.7 | 5.3 | 0.1041 | 306 | 9.7213 | 172 | 95 |
| | 90 | +0.9 | 8.1 | 357.0 | 5.2 | 0.1347 | 34 | 9.7041 | 52 | 90 |
| | 85 | −7.2 | 7.3 | 2.2 | 5.4 | 0.1381 | 276 | 9.6989 | 71 | 85 |
| | 80 | −14.5 | 4.4 | 7.6 | 6.0 | 0.1105 | 475 | 9.7060 | 200 | 80 |
| 1928 | 75 | −18.9 | 1.1 | 13.6 | 7.1 | 0.0630 | 525 | 9.7260 | 332 | 75 |
| | 70 | −20.0 | 1.6 | 20.7 | 8.7 | 0.0105 | 477 | 9.7592 | 466 | 70 |
| | 65 | −18.4 | 3.4 | 29.4 | 11.4 | 9.9628 | 383 | 9.8058 | 579 | 65 |
| | 60 | −15.0 | 4.7 | 40.8 | 15.5 | 9.9245 | 268 | 9.8637 | 609 | 60 |
| 1929 | 55 | −10.3 | 5.5 | 56.3 | 20.4 | 9.8977 | 146 | 9.9246 | 431 | 55 |
| | 50 | −4.8 | 5.9 | 76.7 | 23.0 | 9.8831 | 20 | 9.9677 | 13 | 50 |
| | 45 | +1.1 | 6.0 | 99.7 | 20.9 | 9.8811 | 112 | 9.9664 | 480 | 45 |
| | 40 | +7.1 | | 120.6 | | 9.8923 | | 9.9184 | | 40 |

Table 13.—*Values of u and log F of $L_2$ and $M_1$ for years 1900 to 2000*—Con.

| Year | N | u of $L_2$ | Diff. | u of $M_1$ | Diff. | Log $F(L_2)$ | Diff. | Log $F(M_1)$ | Diff. | N |
|---|---|---|---|---|---|---|---|---|---|---|
| 1929 | 40 | +7.1 | 5.7 | 120.6 | 16.1 | 9.8923 | 252 | 9.9184 | 686 | 40 |
| 1930 | 35 | +12.8 | 4.8 | 136.7 | 11.9 | 9.9175 | 401 | 9.8498 | 670 | 35 |
|      | 30 | +17.6 | 3.3 | 148.6 | 9.0  | 9.9576 | 558 | 9.7828 | 558 | 30 |
|      | 25 | +20.9 | 0.4 | 157.6 | 7.4  | 0.0134 | 706 | 9.7270 | 420 | 25 |
|      | 20 | +21.3 | 4.2 | 165.0 | 6.3  | 0.0840 | 774 | 9.6850 | 281 | 20 |
| 1931 | 15 | +17.1 | 10.4 | 171.3 | 5.7 | 0.1614 | 588 | 9.6569 | 141 | 15 |
|      | 10 | +6.7  | 13.9 | 177.0 | 5.6 | 0.2202 | 16  | 9.6428 | 7   | 10 |
|      | 5  | −7.2  | 10.6 | 182.6 | 5.8 | 0.2218 | 572 | 9.6421 | 129 | 5  |
| 1932 | 0   | −17.8 | 4.5 | 188.4 | 6.3 | 0.1646 | 781 | 9.6550 | 267 | 0   |
|      | 355 | −22.3 | 0.3 | 194.7 | 7.3 | 0.0865 | 720 | 9.6817 | 408 | 355 |
|      | 350 | −22.0 | 3.3 | 202.0 | 9.1 | 0.0145 | 576 | 9.7225 | 546 | 350 |
|      | 345 | −18.7 | 4.9 | 211.1 | 12.0 | 9.9569 | 412 | 9.7771 | 662 | 345 |
| 1933 | 340 | −13.8 | 5.9 | 223.1 | 16.2 | 9.9157 | 262 | 9.8433 | 677 | 340 |
|      | 335 | −7.9  | 6.3 | 239.3 | 21.0 | 9.8895 | 119 | 9.9110 | 468 | 335 |
|      | 330 | −1.6  | 6.2 | 260.3 | 23.2 | 9.8776 | 18  | 9.9578 | 6   | 330 |
|      | 325 | +4.6  | 5.9 | 283.5 | 20.4 | 9.8794 | 151 | 9.9572 | 449 | 325 |
| 1934 | 320 | +10.5 | 5.1 | 303.9 | 15.6 | 9.8945 | 282 | 9.9123 | 633 | 320 |
|      | 315 | +15.6 | 3.9 | 319.5 | 11.4 | 9.9227 | 410 | 9.8490 | 594 | 315 |
|      | 310 | +19.5 | 1.9 | 330.9 | 8.7  | 9.9637 | 523 | 9.7896 | 478 | 310 |
|      | 305 | +21.4 | 1.1 | 339.6 | 7.1  | 0.0160 | 592 | 9.7418 | 340 | 305 |
| 1935 | 300 | +20.3 | 4.9 | 346.7 | 6.1 | 0.0752 | 550 | 9.7078 | 204 | 300 |
|      | 295 | +15.4 | 8.5 | 352.8 | 5.5 | 0.1302 | 316 | 9.6874 | 71  | 295 |
|      | 290 | +6.9  | 9.4 | 358.3 | 5.4 | 0.1618 | 70  | 9.6803 | 55  | 290 |
|      | 285 | −2.5  | 7.1 | 3.7   | 5.4 | 0.1548 | 393 | 9.6858 | 181 | 285 |
| 1936 | 280 | −9.6  | 3.6 | 9.1   | 6.0 | 0.1155 | 530 | 9.7039 | 310 | 280 |
|      | 275 | −13.2 | 0.3 | 15.1  | 6.6 | 0.0625 | 496 | 9.7349 | 428 | 275 |
|      | 270 | −13.5 | 1.6 | 21.7  | 8.1 | 0.0129 | 418 | 9.7777 | 554 | 270 |
|      | 265 | −11.9 | 2.9 | 29.8  | 10.4 | 9.9711 | 313 | 9.8331 | 658 | 265 |
| 1937 | 260 | −9.0  | 3.6 | 40.2  | 13.6 | 9.9398 | 208 | 9.8989 | 701 | 260 |
|      | 255 | −5.4  | 3.8 | 53.8  | 17.7 | 9.9190 | 103 | 9.9690 | 595 | 255 |
|      | 250 | −1.6  | 3.8 | 71.5  | 20.7 | 9.9087 | 24  | 0.0285 | 270 | 250 |
|      | 245 | +2.2  | 3.4 | 92.2  | 20.3 | 9.9063 | 63  | 0.0555 | 160 | 245 |
| 1938 | 240 | +5.6  | 2.9 | 112.5 | 16.8 | 9.9126 | 136 | 0.0395 | 446 | 240 |
|      | 235 | +8.5  | 2.1 | 129.3 | 12.7 | 9.9262 | 196 | 9.9949 | 522 | 235 |
|      | 230 | +10.6 | 1.2 | 142.0 | 9.6  | 9.9458 | 240 | 9.9427 | 475 | 230 |
|      | 225 | +11.8 | 0.0 | 151.6 | 7.4  | 9.9698 | 262 | 9.8952 | 383 | 225 |
| 1939 | 220 | +11.8 | 1.1 | 159.0 | 6.2 | 9.9960 | 256 | 9.8569 | 284 | 220 |
|      | 215 | +10.7 | 2.4 | 165.2 | 5.3 | 0.0216 | 216 | 9.8285 | 186 | 215 |
|      | 210 | +8.3  | 3.3 | 170.5 | 4.7 | 0.0432 | 141 | 9.8099 | 97  | 210 |
| 1940 | 205 | +5.0  | 3.8 | 175.2 | 4.5 | 0.0573 | 44  | 9.8002 | 13  | 205 |
|      | 200 | +1.2  | 3.6 | 179.7 | 4.5 | 0.0617 | 56  | 9.7989 | 68  | 200 |
|      | 195 | −2.4  | 3.1 | 184.2 | 4.5 | 0.0561 | 139 | 9.8057 | 145 | 195 |
|      | 190 | −5.5  | 2.1 | 188.7 | 4.9 | 0.0422 | 194 | 9.8202 | 223 | 190 |
| 1941 | 185 | −7.6  | 1.2 | 193.6 | 5.5 | 0.0228 | 218 | 9.8425 | 301 | 185 |
|      | 180 | −8.8  | 0.3 | 199.1 | 6.5 | 0.0010 | 219 | 9.8726 | 381 | 180 |
|      | 175 | −9.1  | 0.6 | 205.6 | 7.9 | 9.9791 | 200 | 9.9107 | 457 | 175 |
|      | 170 | −8.5  | 1.3 | 213.5 | 10.1 | 9.9591 | 171 | 9.9564 | 513 | 170 |
| 1942 | 165 | −7.2  | 1.8 | 223.6 | 13.0 | 9.9420 | 130 | 0.0077 | 509 | 165 |
|      | 160 | −5.4  | 2.2 | 236.6 | 16.6 | 9.9290 | 83  | 0.0586 | 380 | 160 |
|      | 155 | −3.2  | 2.6 | 253.2 | 19.2 | 9.9207 | 31  | 0.0966 | 78  | 155 |
|      | 150 | −0.6  | 2.7 | 272.4 | 18.9 | 9.9176 | 28  | 0.1044 | 299 | 150 |
| 1943 | 145 | +2.1  | 2.6 | 291.3 | 16.0 | 9.9204 | 91  | 0.0745 | 563 | 145 |
|      | 140 | +4.7  | 2.4 | 307.3 | 12.5 | 9.9295 | 160 | 0.0182 | 652 | 140 |
|      | 135 | +7.1  | 1.9 | 319.8 | 9.8  | 9.9455 | 230 | 9.9530 | 624 | 135 |
|      | 130 | +9.0  | 1.0 | 329.6 | 7.7  | 9.9685 | 297 | 9.8906 | 542 | 130 |
| 1944 | 125 | +10.0 | 0.3 | 337.3 | 6.5 | 9.9982 | 349 | 9.8364 | 441 | 125 |
|      | 120 | +9.7  | 2.1 | 343.8 | 5.6 | 0.0331 | 365 | 9.7923 | 334 | 120 |
|      | 115 | +7.6  | 4.4 | 349.4 | 5.2 | 0.0696 | 306 | 9.7589 | 225 | 115 |
|      | 110 | +3.2  |     | 354.6 |     | 0.1002 |     | 9.7364 |     | 110 |

HARMONIC ANALYSIS AND PREDICTION OF TIDES 195

Table 13.—*Values of u and log F of $L_2$ and $M_1$ for years 1900 to 2000*—Con.

| Year | N | u of $L_2$ | Diff. | u of $M_1$ | Diff. | Log. $F$ ($L_2$) | Diff. | Log $F$ ($M_1$) | Diff. | N |
|---|---|---|---|---|---|---|---|---|---|---|
| | | ° | | ° | | | | | | |
| 1944 | 110 | +3.2 | 6.2 | 354.6 | 5.1 | 0.1002 | 147 | 9.7364 | 115 | 110 |
| 1945 | 105 | −3.0 | 6.6 | 359.7 | 5.1 | 0.1149 | 88 | 9.7249 | 3 | 105 |
| | 100 | −9.6 | 5.3 | 4.8 | 5.5 | 0.1061 | 304 | 9.7246 | 117 | 100 |
| | 95 | −14.9 | 2.8 | 10.3 | 6.2 | 0.0757 | 426 | 9.7363 | 240 | 95 |
| | 90 | −17.7 | 0.2 | 16.5 | 7.5 | 0.0331 | 446 | 9.7603 | 370 | 90 |
| 1946 | 85 | −17.9 | 1.8 | 24.0 | 9.5 | 9.9885 | 397 | 9.7973 | 494 | 85 |
| | 80 | −16.1 | 3.4 | 33.5 | 12.6 | 9.9488 | 325 | 9.8467 | 583 | 80 |
| | 75 | −12.7 | 4.5 | 46.1 | 17.0 | 9.9163 | 200 | 9.9050 | 553 | 75 |
| 1947 | 70 | −8.2 | 5.2 | 63.1 | 21.5 | 9.8963 | 99 | 9.9603 | 279 | 70 |
| | 65 | −3.0 | 5.6 | 84.6 | 22.6 | 9.8864 | 20 | 9.9882 | 205 | 65 |
| | 60 | +2.6 | 5.5 | 107.2 | 19.0 | 9.8884 | 146 | 9.9677 | 589 | 60 |
| | 55 | +8.1 | 5.1 | 126.2 | 14.5 | 9.9030 | 282 | 9.9088 | 706 | 55 |
| 1948 | 50 | +13.2 | 4.2 | 140.7 | 10.7 | 9.9312 | 428 | 9.8382 | 648 | 50 |
| | 45 | +17.4 | 2.4 | 151.4 | 8.4 | 9.9740 | 579 | 9.7734 | 526 | 45 |
| | 40 | +19.8 | 0.8 | 159.8 | 6.9 | 0.0319 | 706 | 9.7208 | 388 | 40 |
| | 35 | +19.0 | 5.6 | 166.7 | 6.0 | 0.1025 | 724 | 9.6820 | 250 | 35 |
| 1949 | 30 | +13.4 | 11.4 | 172.7 | 5.7 | 0.1749 | 447 | 9.6570 | 115 | 30 |
| | 25 | +2.0 | 13.2 | 178.4 | 5.6 | 0.2196 | 150 | 9.6455 | 20 | 25 |
| | 20 | −11.2 | 8.9 | 184.0 | 5.8 | 0.2046 | 639 | 9.6475 | 154 | 20 |
| | 15 | −20.1 | 3.0 | 189.8 | 6.6 | 0.1407 | 764 | 9.6629 | 295 | 15 |
| 1950 | 10 | −23.1 | 1.1 | 196.4 | 7.7 | 0.0643 | 680 | 9.6924 | 437 | 10 |
| | 5 | −22.0 | 3.8 | 204.1 | 9.8 | 9.9963 | 532 | 9.7361 | 572 | 5 |
| | 0 | −18.2 | 5.3 | 213.9 | 13.1 | 9.9431 | 377 | 9.7933 | 670 | 0 |
| | 355 | −12.9 | 6.1 | 227.0 | 17.7 | 9.9054 | 227 | 9.8603 | 634 | 355 |
| 1951 | 350 | −6.8 | 6.5 | 244.7 | 22.3 | 9.8827 | 84 | 9.9237 | 335 | 350 |
| | 345 | −0.3 | 6.4 | 267.0 | 22.9 | 9.8743 | 54 | 9.9572 | 174 | 345 |
| | 340 | +6.1 | 6.0 | 289.9 | 19.1 | 9.8797 | 193 | 9.9398 | 555 | 340 |
| | 335 | +12.1 | 5.2 | 309.0 | 14.2 | 9.8990 | 332 | 9.8843 | 652 | 335 |
| 1952 | 330 | +17.3 | 3.8 | 323.2 | 10.6 | 9.9322 | 471 | 9.8191 | 579 | 330 |
| | 325 | +21.1 | 1.5 | 333.8 | 8.2 | 9.9793 | 597 | 9.7612 | 450 | 325 |
| | 320 | +22.6 | 2.1 | 342.0 | 6.7 | 0.0390 | 666 | 9.7162 | 310 | 320 |
| | 315 | +20.5 | 6.7 | 348.7 | 6.0 | 0.1056 | 583 | 9.6852 | 172 | 315 |
| 1953 | 310 | +13.8 | 10.6 | 354.7 | 5.5 | 0.1639 | 242 | 9.6680 | 38 | 310 |
| | 305 | +3.2 | 10.4 | 0.2 | 5.5 | 0.1881 | 238 | 9.6642 | 91 | 305 |
| | 300 | −7.2 | 6.4 | 5.7 | 5.7 | 0.1643 | 558 | 9.6733 | 225 | 300 |
| | 295 | −13.6 | 2.3 | 11.4 | 6.3 | 0.1085 | 600 | 9.6958 | 349 | 295 |
| 1954 | 290 | −15.9 | 0.9 | 17.7 | 7.4 | 0.0485 | 537 | 9.7307 | 484 | 290 |
| | 285 | −15.0 | 2.8 | 25.1 | 9.3 | 9.9948 | 420 | 9.7791 | 608 | 285 |
| | 280 | −12.2 | 3.8 | 34.4 | 12.2 | 9.9528 | 298 | 9.8399 | 703 | 280 |
| 1955 | 275 | −8.4 | 4.4 | 46.6 | 16.1 | 9.9230 | 175 | 9.9102 | 674 | 275 |
| | 270 | −4.0 | 4.5 | 62.7 | 20.2 | 9.9055 | 68 | 9.9776 | 449 | 270 |
| | 265 | +0.5 | 4.3 | 82.9 | 21.8 | 9.8987 | 33 | 0.0225 | 12 | 265 |
| | 260 | +4.8 | 3.7 | 104.7 | 19.0 | 9.9020 | 124 | 0.0237 | 379 | 260 |
| 1956 | 255 | +8.5 | 3.0 | 123.7 | 14.6 | 9.9144 | 204 | 9.9858 | 538 | 255 |
| | 250 | +11.5 | 1.9 | 138.3 | 10.9 | 9.9348 | 267 | 9.9320 | 516 | 250 |
| | 245 | +13.4 | 0.6 | 149.2 | 8.4 | 9.9615 | 308 | 9.8804 | 424 | 245 |
| | 240 | +14.0 | 0.9 | 157.6 | 6.7 | 9.9923 | 315 | 9.8380 | 312 | 240 |
| 1957 | 235 | +13.1 | 2.5 | 164.3 | 5.6 | 0.0238 | 273 | 9.8068 | 202 | 235 |
| | 230 | +10.6 | 3.9 | 169.9 | 5.1 | 0.0511 | 182 | 9.7866 | 97 | 230 |
| | 225 | +6.7 | 4.4 | 175.0 | 4.7 | 0.0693 | 53 | 9.7769 | 1 | 225 |
| | 220 | +2.3 | 4.3 | 179.7 | 4.6 | 0.0746 | 77 | 9.7770 | 92 | 220 |
| 1958 | 215 | −2.0 | 3.4 | 184.3 | 4.7 | 0.0669 | 177 | 9.7862 | 180 | 215 |
| | 210 | −5.4 | 2.1 | 189.0 | 5.1 | 0.0492 | 232 | 9.8042 | 266 | 210 |
| | 205 | −7.5 | 1.0 | 194.1 | 5.6 | 0.0260 | 246 | 9.8308 | 350 | 205 |
| | 200 | −8.5 | 0.0 | 199.7 | 6.6 | 0.0014 | 233 | 9.8658 | 432 | 200 |
| 1959 | 195 | −8.5 | 0.9 | 206.3 | 8.1 | 9.9781 | 202 | 9.9090 | 506 | 195 |
| | 190 | −7.6 | 1.3 | 214.4 | 10.2 | 9.9579 | 162 | 9.9596 | 555 | 190 |
| | 185 | −6.3 | 1.9 | 224.6 | 13.1 | 9.9417 | 118 | 0.0151 | 541 | 185 |
| | 180 | −4.4 | | 237.7 | | 9.9299 | | 0.0692 | | 180 |

Table 13.— *Values of u and log F of* $L_2$ *and* $M_1$ *for years 1900 to 2000*—Con.

| Year | N | u of $L_2$ | Diff. | u of $M_1$ | Diff. | Log $F$ ($L_2$) | Diff. | Log $F$ ($M_1$) | Diff. | N |
|---|---|---|---|---|---|---|---|---|---|---|
| | | ° | | ° | | | | | | |
| 1959 | 180 | −4.4 | 2.1 | 237.7 | 16.4 | 9.9299 | 71 | 0.0692 | 404 | 180 |
| 1960 | 175 | −2.3 | 2.3 | 254.1 | 18.6 | 9.9228 | 21 | 0.1096 | 107 | 175 |
| | 170 | 0.0 | 2.3 | 272.7 | 18.3 | 9.9207 | 30 | 0.1203 | 252 | 170 |
| | 165 | +2.3 | 2.2 | 291.0 | 15.6 | 9.9237 | 83 | 0.0951 | 507 | 165 |
| | 160 | +4.5 | 2.0 | 306.6 | 12.3 | 9.9320 | 137 | 0.0444 | 603 | 160 |
| 1961 | 155 | +6.5 | 1.6 | 318.9 | 9.6 | 9.9457 | 191 | 9.9841 | 589 | 155 |
| | 150 | +8.1 | 0.8 | 328.5 | 7.6 | 9.9648 | 242 | 9.9252 | 523 | 150 |
| | 145 | +8.9 | 0.2 | 336.1 | 6.4 | 9.9890 | 280 | 9.8729 | 435 | 145 |
| 1962 | 140 | +8.7 | 1.5 | 342.5 | 5.6 | 0.0170 | 295 | 9.8294 | 344 | 140 |
| | 135 | +7.2 | 3.1 | 348.1 | 5.1 | 0.0465 | 263 | 9.7950 | 247 | 135 |
| | 130 | +4.1 | 4.7 | 353.2 | 4.8 | 0.0728 | 166 | 9.7703 | 149 | 130 |
| | 125 | −0.6 | 5.4 | 358.0 | 4.9 | 0.0894 | 8 | 9.7554 | 47 | 125 |
| 1963 | 120 | −6.0 | 5.0 | 2.9 | 5.1 | 0.0902 | 169 | 9.7507 | 58 | 120 |
| | 115 | −11.0 | 3.5 | 8.0 | 5.6 | 0.0733 | 305 | 9.7565 | 170 | 115 |
| | 110 | −14.5 | 1.5 | 13.6 | 6.6 | 0.0428 | 370 | 9.7735 | 290 | 110 |
| | 105 | −16.0 | 0.5 | 20.2 | 8.1 | 0.0058 | 368 | 9.8025 | 412 | 105 |
| 1964 | 100 | −15.5 | 2.1 | 28.3 | 10.5 | 9.9690 | 320 | 9.8437 | 510 | 100 |
| | 95 | −13.4 | 3.3 | 38.8 | 14.1 | 9.9370 | 245 | 9.8956 | 567 | 95 |
| | 90 | −10.1 | 4.2 | 52.9 | 18.7 | 9.9125 | 156 | 9.9523 | 465 | 90 |
| | 85 | −5.9 | 4.8 | 71.6 | 22.1 | 9.8969 | 48 | 9.9988 | 61 | 85 |
| 1965 | 80 | −1.1 | 5.1 | 93.7 | 21.2 | 9.8921 | 60 | 0.0049 | 392 | 80 |
| | 75 | +4.0 | 5.0 | 114.9 | 17.0 | 9.8981 | 180 | 9.9657 | 666 | 75 |
| | 70 | +9.0 | 4.3 | 131.9 | 12.8 | 9.9161 | 309 | 9.8991 | 702 | 70 |
| | 65 | +13.3 | 3.2 | 144.7 | 9.6 | 9.9470 | 447 | 9.8289 | 615 | 65 |
| 1966 | 60 | +16.5 | 1.4 | 154.3 | 7.7 | 9.9917 | 581 | 9.7674 | 489 | 60 |
| | 55 | +17.9 | 2.0 | 162.0 | 6.5 | 0.0498 | 675 | 9.7185 | 353 | 55 |
| | 50 | +15.9 | 7.0 | 168.5 | 5.8 | 0.1173 | 622 | 9.6832 | 218 | 50 |
| | 45 | +8.9 | 11.5 | 174.3 | 5.5 | 0.1795 | 274 | 9.6614 | 85 | 45 |
| 1967 | 40 | −2.6 | 11.6 | 179.8 | 5.6 | 0.2069 | 283 | 9.6529 | 48 | 40 |
| | 35 | −14.2 | 7.0 | 185.4 | 6.0 | 0.1786 | 651 | 9.6577 | 182 | 35 |
| | 30 | −21.2 | 1.9 | 191.4 | 6.8 | 0.1135 | 717 | 9.6759 | 322 | 30 |
| | 25 | −23.1 | 1.8 | 198.2 | 8.2 | 0.0418 | 624 | 9.7081 | 464 | 25 |
| 1968 | 20 | −21.3 | 4.1 | 206.4 | 10.6 | 9.9794 | 484 | 9.7545 | 592 | 20 |
| | 15 | −17.2 | 5.4 | 217.0 | 14.3 | 9.9310 | 335 | 9.8137 | 666 | 15 |
| | 10 | −11.8 | 6.2 | 231.3 | 19.2 | 9.8975 | 190 | 9.8803 | 567 | 10 |
| | 5 | −5.6 | 6.5 | 250.5 | 23.0 | 9.8785 | 50 | 9.9370 | 185 | 5 |
| 1969 | 0 | +0.9 | 6.5 | 273.5 | 22.2 | 9.8735 | 91 | 9.9555 | 328 | 0 |
| | 355 | +7.4 | 6.1 | 295.7 | 17.7 | 9.8826 | 232 | 9.9227 | 624 | 355 |
| | 350 | +13.5 | 5.1 | 313.4 | 13.0 | 9.9058 | 379 | 9.8603 | 657 | 350 |
| 1970 | 345 | +18.6 | 3.6 | 326.4 | 9.8 | 9.9437 | 528 | 9.7946 | 558 | 345 |
| | 340 | +22.2 | 0.9 | 336.2 | 7.7 | 9.9965 | 663 | 9.7388 | 423 | 340 |
| | 335 | +23.1 | 3.3 | 343.9 | 6.5 | 0.0628 | 727 | 9.6965 | 282 | 335 |
| | 330 | +19.8 | 8.7 | 350.4 | 5.9 | 0.1355 | 583 | 9.6683 | 143 | 330 |
| 1971 | 325 | +11.1 | 12.2 | 356.3 | 5.5 | 0.1938 | 121 | 9.6540 | 9 | 325 |
| | 320 | −1.1 | 10.6 | 1.8 | 5.6 | 0.2059 | 416 | 9.6531 | 125 | 320 |
| | 315 | −11.7 | 5.4 | 7.4 | 6.0 | 0.1643 | 669 | 9.6656 | 256 | 315 |
| | 310 | −17.1 | 0.9 | 13.4 | 6.8 | 0.0974 | 661 | 9.6912 | 393 | 310 |
| 1972 | 305 | −18.0 | 2.1 | 20.2 | 8.2 | 0.0313 | 546 | 9.7305 | 528 | 305 |
| | 300 | −15.9 | 3.7 | 28.4 | 10.4 | 9.9767 | 408 | 9.7833 | 649 | 300 |
| | 295 | −12.2 | 4.7 | 38.8 | 14.0 | 9.9359 | 271 | 9.8482 | 709 | 295 |
| | 290 | −7.5 | 5.1 | 52.8 | 18.5 | 9.9088 | 143 | 9.9191 | 611 | 290 |
| 1973 | 285 | −2.4 | 5.1 | 71.3 | 22.0 | 9.8945 | 26 | 9.9802 | 256 | 285 |
| | 280 | +2.7 | 4.6 | 93.3 | 21.6 | 9.8919 | 86 | 0.0058 | 224 | 280 |
| | 275 | +7.3 | 4.0 | 114.9 | 17.1 | 9.9005 | 183 | 9.9834 | 508 | 275 |
| | 270 | +11.3 | 3.1 | 132.0 | 12.8 | 9.9188 | 274 | 9.9326 | 566 | 270 |
| 1974 | 265 | +14.4 | 1.6 | 144.8 | 9.7 | 9.9462 | 343 | 9.8760 | 491 | 265 |
| | 260 | +16.0 | 0.1 | 154.5 | 7.6 | 9.9805 | 380 | 9.8269 | 374 | 260 |
| | 255 | +15.9 | 2.2 | 162.1 | 6.2 | 0.0185 | 364 | 9.7895 | 253 | 255 |
| | 250 | +13.7 | | 168.3 | | 0.0549 | | 9.7642 | | 250 |

HARMONIC ANALYSIS AND PREDICTION OF TIDES 197

Table 13.—*Values of u and log F of $L_2$ and $M_1$ for years 1900 to 2000*—Con.

| Year | N | u of $L_2$ | Diff. | u of $M_1$ | Diff. | Log $F$ ($L_2$) | Diff. | Log $F$ ($M_1$) | Diff. | N |
|---|---|---|---|---|---|---|---|---|---|---|
| | | ° | | ° | | | | | | |
| 1974 | 250 | +13.7 | 4.1 | 168.3 | 5.5 | 0.0549 | 274 | 9.7642 | 134 | 250 |
| 1975 | 245 | +9.6 | 5.4 | 173.8 | 5.0 | 0.0823 | 116 | 9.7508 | 22 | 245 |
| | 240 | +4.2 | 5.3 | 178.8 | 4.8 | 0.0939 | 68 | 9.7486 | 82 | 240 |
| | 235 | −1.1 | 4.2 | 183.6 | 4.9 | 0.0871 | 211 | 9.7568 | 183 | 235 |
| | 230 | −5.3 | 2.7 | 188.5 | 5.3 | 0.0660 | 286 | 9.7751 | 282 | 230 |
| 1976 | 225 | −8.0 | 1.0 | 193.8 | 5.8 | 0.0374 | 300 | 9.8033 | 378 | 225 |
| | 220 | −9.0 | 0.2 | 199.6 | 6.8 | 0.0074 | 273 | 9.8411 | 471 | 220 |
| | 215 | −8.8 | 1.1 | 206.4 | 8.3 | 9.9801 | 228 | 9.8882 | 554 | 215 |
| | 210 | −7.7 | 1.8 | 214.7 | 10.5 | 9.9573 | 174 | 9.9436 | 606 | 210 |
| 1977 | 205 | −5.9 | 2.2 | 225.2 | 13.5 | 9.9399 | 118 | 0.0042 | 589 | 205 |
| | 200 | −3.7 | 2.3 | 238.7 | 16.8 | 9.9281 | 63 | 0.0631 | 439 | 200 |
| | 195 | −1.4 | 2.4 | 255.5 | 18.7 | 9.9218 | 10 | 0.1070 | 129 | 195 |
| 1978 | 190 | +1.0 | 2.2 | 274.2 | 18.1 | 9.9208 | 40 | 0.1199 | 228 | 190 |
| | 185 | +3.2 | 2.0 | 292.3 | 15.3 | 9.9248 | 89 | 0.0971 | 469 | 185 |
| | 180 | +5.2 | 1.7 | 307.6 | 11.9 | 9.9337 | 135 | 0.0502 | 553 | 180 |
| | 175 | +6.9 | 1.2 | 319.5 | 9.3 | 9.9472 | 177 | 9.9949 | 536 | 175 |
| 1979 | 170 | +8.1 | 0.5 | 328.8 | 7.4 | 9.9649 | 213 | 9.9413 | 477 | 170 |
| | 165 | +8.6 | 0.3 | 336.2 | 6.2 | 9.9862 | 237 | 9.8936 | 398 | 165 |
| | 160 | +8.3 | 1.4 | 342.4 | 5.4 | 0.0099 | 241 | 9.8538 | 317 | 160 |
| | 155 | +6.9 | 2.6 | 347.8 | 4.9 | 0.0340 | 213 | 9.8221 | 232 | 155 |
| 1980 | 150 | +4.3 | 3.7 | 352.7 | 4.6 | 0.0553 | 143 | 9.7989 | 147 | 150 |
| | 145 | +0.6 | 4.3 | 357.3 | 4.6 | 0.0696 | 33 | 9.7842 | 59 | 145 |
| | 140 | −3.7 | 4.3 | 1.9 | 4.8 | 0.0729 | 99 | 9.7783 | 33 | 140 |
| | 135 | −8.0 | 3.4 | 6.7 | 5.2 | 0.0630 | 214 | 9.7816 | 131 | 135 |
| 1981 | 130 | −11.4 | 2.0 | 11.9 | 6.0 | 0.0416 | 287 | 9.7947 | 237 | 130 |
| | 125 | −13.4 | 0.4 | 17.9 | 7.1 | 0.0129 | 312 | 9.8184 | 349 | 125 |
| | 120 | −13.8 | 1.0 | 25.0 | 9.0 | 9.9817 | 293 | 9.8533 | 457 | 120 |
| | 115 | −12.8 | 2.3 | 34.0 | 12.0 | 9.9524 | 246 | 9.8990 | 533 | 115 |
| 1982 | 110 | −10.5 | 3.2 | 46.0 | 16.0 | 9.9278 | 177 | 9.9523 | 508 | 110 |
| | 105 | −7.3 | 3.9 | 62.0 | 20.1 | 9.9101 | 95 | 0.0031 | 276 | 105 |
| | 100 | −3.4 | 4.3 | 82.1 | 21.7 | 9.9006 | 2 | 0.0307 | 161 | 100 |
| | 95 | +0.9 | 4.4 | 103.8 | 19.1 | 9.9004 | 100 | 0.0146 | 549 | 95 |
| 1983 | 90 | +5.3 | 4.2 | 122.9 | 14.7 | 9.9104 | 211 | 9.9597 | 703 | 90 |
| | 85 | +9.5 | 3.5 | 137.6 | 11.1 | 9.9315 | 330 | 9.8894 | 676 | 85 |
| | 80 | +13.0 | 2.3 | 148.7 | 8.6 | 9.9645 | 454 | 9.8218 | 572 | 80 |
| | 75 | +15.3 | 0.1 | 157.3 | 7.0 | 0.0099 | 560 | 9.7646 | 445 | 75 |
| 1984 | 70 | +15.4 | 3.4 | 164.3 | 6.1 | 0.0659 | 604 | 9.7201 | 312 | 70 |
| | 65 | +12.0 | 7.8 | 170.4 | 5.6 | 0.1263 | 478 | 9.6889 | 183 | 65 |
| | 60 | +4.2 | 10.8 | 176.0 | 5.4 | 0.1741 | 98 | 9.6706 | 54 | 60 |
| 1985 | 55 | −6.6 | 9.6 | 181.4 | 5.6 | 0.1839 | 365 | 9.6652 | 77 | 55 |
| | 50 | −16.2 | 5.3 | 187.0 | 6.2 | 0.1474 | 625 | 9.6729 | 212 | 50 |
| | 45 | −21.5 | 0.9 | 193.2 | 7.1 | 0.0849 | 650 | 9.6941 | 350 | 45 |
| | 40 | −22.4 | 2.3 | 200.3 | 8.7 | 0.0199 | 560 | 9.7291 | 490 | 40 |
| 1986 | 35 | −20.1 | 4.2 | 209.0 | 11.4 | 9.9639 | 429 | 9.7781 | 614 | 35 |
| | 30 | −15.9 | 5.5 | 220.4 | 15.7 | 9.9210 | 290 | 9.8395 | 640 | 30 |
| | 25 | −10.4 | 6.2 | 236.1 | 20.5 | 9.8920 | 152 | 9.9035 | 471 | 25 |
| | 20 | −4.2 | 6.5 | 256.6 | 23.4 | 9.8768 | 14 | 9.9506 | 18 | 20 |
| 1987 | 15 | +2.3 | 6.3 | 280.0 | 21.0 | 9.8754 | 126 | 9.9524 | 456 | 15 |
| | 10 | +8.6 | 5.9 | 301.0 | 16.3 | 9.8880 | 270 | 9.9068 | 668 | 10 |
| | 5 | +14.5 | 4.9 | 317.3 | 11.9 | 9.9150 | 422 | 9.8400 | 650 | 5 |
| | 0 | +19.4 | 3.2 | 329.2 | 9.1 | 9.9572 | 579 | 9.7750 | 535 | 0 |
| 1988 | 355 | +22.6 | 0.1 | 338.3 | 7.3 | 0.0151 | 717 | 9.7215 | 395 | 355 |
| | 350 | +22.7 | 4.6 | 345.6 | 6.3 | 0.0868 | 761 | 9.6820 | 255 | 350 |
| | 345 | +18.1 | 10.7 | 351.9 | 5.8 | 0.1629 | 536 | 9.6565 | 114 | 345 |
| | 340 | +7.4 | 13.4 | 357.7 | 5.6 | 0.2165 | 39 | 9.6451 | 18 | 340 |
| 1989 | 335 | −6.0 | 9.8 | 3.3 | 5.7 | 0.2126 | 573 | 9.6469 | 154 | 335 |
| | 330 | −15.8 | 4.0 | 9.0 | 6.2 | 0.1553 | 744 | 9.6623 | 289 | 330 |
| | 325 | −19.8 | 0.4 | 15.2 | 7.3 | 0.0809 | 677 | 9.6912 | 429 | 325 |
| | 320 | −19.4 | | 22.5 | | 0.0132 | | 9.7341 | | 320 |

198    U. S. COAST AND GEODETIC SURVEY

Table 13.—*Values of u and log F of $L_2$ and $M_1$ for years 1900 to 2000*—Con.

| Year | N | u of $L_2$ | Diff. | u of $M_1$ | Diff. | Log $F(L_2)$ | Diff. | Log $F(M_1)$ | Diff. | N |
|---|---|---|---|---|---|---|---|---|---|---|
|      |     | °     |     | °     |      |        |     |        |     |     |
| 1989 | 320 | −19.4 |     | 22.5  |      | 0.0132 |     | 9.7341 |     | 320 |
|      |     |       | 3.0 |       | 9.0  |        | 537 |        | 565 |     |
| 1990 | 315 | −16.4 |     | 31.5  |      | 9.9595 |     | 9.7906 |     | 315 |
|      | 310 | −11.8 | 4.6 | 43.2  | 11.7 | 9.9209 | 386 | 9.8582 | 676 | 310 |
|      | 305 | −6.5  | 5.3 | 59.0  | 15.8 | 9.8967 | .242| 9.9276 | 694 | 305 |
|      | 300 | −0.9  | 5.6 | 79.5  | 20.5 | 9.8859 | 108 | 9.9776 | 500 | 300 |
|      |     |       | 5.5 |       | 22.7 |        | 17  |        | 51  |     |
| 1991 | 295 | +4.6  |     | 102.2 |      | 9.8876 |     | 9.9827 |     | 295 |
|      | 290 | +9.7  | 5.1 | 122.6 | 20.4 | 9.9011 | 135 | 9.9433 | 394 | 290 |
|      | 285 | +14.0 | 4.3 | 138.2 | 15.6 | 9.9257 | 246 | 9.8848 | 585 | 285 |
|      | 280 | +17.0 | 3.0 | 149.7 | 11.5 | 9.9601 | 344 | 9.8283 | 565 | 280 |
|      |     |       | 1.3 |       | 8.9  |        | 428 |        | 464 |     |
| 1992 | 275 | +18.3 |     | 158.6 |      | 0.0029 |     | 9.7819 |     | 275 |
|      | 270 | +17.3 | 1.0 | 165.5 | 6.9  | 0.0479 | 450 | 9.7493 | 326 | 270 |
|      | 265 | +13.6 | 3.7 | 171.5 | 6.0  | 0.0887 | 408 | 9.7290 | 203 | 265 |
|      |     |       | 5.9 |       | 5.4  |        | 250 |        | 79  |     |
| 1993 | 260 | +7.7  |     | 176.9 |      | 0.1137 |     | 9.7211 |     | 260 |
|      | 255 | +0.8  | 6.9 | 182.0 | 5.1  | 0.1147 | 10  | 9.7249 | 38  | 255 |
|      | 250 | −5.0  | 5.8 | 187.0 | 5.0  | 0.0933 | 214 | 9.7400 | 151 | 250 |
|      | 245 | −8.6  | 3.6 | 192.4 | 5.4  | 0.0591 | 342 | 9.7662 | 262 | 245 |
|      |     |       | 1.6 |       | 5.9  |        | 373 |        | 371 |     |
| 1994 | 240 | −10.2 |     | 198.3 |      | 0.0218 |     | 9.8033 |     | 240 |
|      | 235 | −9.9  | 0.3 | 205.2 | 6.9  | 9.9878 | 340 | 9.8510 | 477 | 235 |
|      | 230 | −8.5  | 1.4 | 213.5 | 8.3  | 9.9598 | 280 | 9.9086 | 576 | 230 |
|      | 225 | −6.4  | 2.1 | 224.1 | 10.6 | 9.9389 | 209 | 9.9731 | 645 | 225 |
|      |     |       | 2.6 |       | 13.8 |        | 137 |        | 641 |     |
| 1995 | 220 | −3.8  |     | 237.9 |      | 9.9252 |     | 0.0372 |     | 220 |
|      | 215 | −1.1  | 2.7 | 255.1 | 17.2 | 9.9183 | 69  | 0.0868 | 496 | 215 |
|      | 210 | +1.6  | 2.7 | 274.5 | 19.4 | 9.9177 | 6   | 0.1041 | 173 | 210 |
|      | 205 | +4.1  | 2.5 | 293.1 | 18.6 | 9.9228 | 51  | 0.0839 | 202 | 205 |
|      |     |       | 2.1 |       | 15.4 |        | 102 |        | 444 |     |
| 1996 | 200 | +6.2  |     | 308.5 |      | 9.9330 |     | 0.0395 |     | 200 |
|      | 195 | +7.9  | 1.7 | 320.4 | 11.9 | 9.9477 | 147 | 9.9876 | 519 | 195 |
|      | 190 | +8.9  | 1.0 | 329.7 | 9.3  | 9.9661 | 184 | 9.9382 | 494 | 190 |
|      | 185 | +9.2  | 0.3 | 337.0 | 7.3  | 9.9870 | 209 | 9.8955 | 427 | 185 |
|      |     |       | 0.6 |       | 6.0  |        | 221 |        | 348 |     |
| 1997 | 180 | +8.6  |     | 343.0 |      | 0.0091 |     | 9.8607 |     | 180 |
|      | 175 | +7.0  | 1.6 | 348.3 | 5.3  | 0.0302 | 211 | 9.8339 | 268 | 175 |
|      | 170 | +4.5  | 2.5 | 353.0 | 4.7  | 0.0477 | 175 | 9.8148 | 191 | 170 |
|      | 165 | +1.2  | 3.3 | 357.5 | 4.5  | 0.0586 | 109 | 9.8035 | 113 | 165 |
|      |     |       | 3.7 |       | 4.4  |        | 19  |        | 36  |     |
| 1998 | 160 | −2.5  |     | 1.9   |      | 0.0605 |     | 9.7999 |     | 160 |
|      | 155 | −6.1  | 3.6 | 6.4   | 4.5  | 0.0524 | 81  | 9.8044 | 45  | 155 |
|      | 150 | −9.1  | 3.0 | 11.3  | 4.9  | 0.0355 | 169 | 9.8174 | 130 | 150 |
|      | 145 | −11.1 | 2.0 | 16.9  | 5.6  | 0.0124 | 231 | 9.8395 | 221 | 145 |
|      |     |       | 0.7 |       | 6.5  |        | 258 |        | 319 |     |
| 1999 | 140 | −11.8 |     | 23.4  |      | 9.9866 |     | 9.8714 |     | 140 |
|      | 135 | −11.4 | 0.4 | 31.5  | 8.1  | 9.9613 | 253 | 9.9130 | 416 | 135 |
|      | 130 | −9.9  | 1.5 | 42.0  | 10.5 | 9.9389 | 224 | 9.9627 | 497 | 130 |
|      |     |       | 2.4 |       | 13.9 |        | 174 |        | 514 |     |
| 2000 | 125 | −7.5  |     | 55.9  |      | 9.9215 |     | 0.0141 |     | 125 |
|      | 120 | −4.4  | 3.1 | 73.9  | 18.0 | 9.9103 | 112 | 0.0516 | 375 | 120 |
|      | 115 | −0.9  | 3.5 | 94.8  | 20.9 | 9.9066 | 37  | 0.0540 | 24  | 115 |
|      | 110 | +2.8  | 3.7 | 114.7 | 19.9 | 9.9111 | 45  | 0.0142 | 398 | 110 |

Table 14.—Node factor f for middle of each year, 1850 to 1999

| Constituent | 1850 | 1851 | 1852 | 1853 | 1854 | 1855 | 1856 | 1857 | 1858 | 1859 |
|---|---|---|---|---|---|---|---|---|---|---|
| $J_1$ | 0.892 | 0.948 | 1.007 | 1.061 | 1.105 | 1.138 | 1.158 | 1.165 | 1.160 | 1.141 |
| $K_1$ | 0.922 | 0.959 | 0.999 | 1.037 | 1.069 | 1.092 | 1.107 | 1.113 | 1.108 | 1.095 |
| $K_2$ | 0.816 | 0.887 | 0.977 | 1.075 | 1.168 | 1.246 | 1.298 | 1.317 | 1.302 | 1.254 |
| $L_2$ | 1.163 | 0.905 | 0.725 | 1.055 | 1.263 | 0.944 | 0.469 | 0.962 | 1.283 | 1.001 |
| $M_1$ | 1.023 | 1.675 | 1.974 | 1.559 | 1.118 | 1.860 | 2.348 | 1.872 | 1.177 | 1.776 |
| $M_2$*, $N_2$, $2N$, $\lambda_2$, $\mu_2$, $\nu_2$ | 1.027 | 1.017 | 1.005 | 0.993 | 0.981 | 0.972 | 0.966 | 0.963 | 0.965 | 0.971 |
| $M_3$ | 1.042 | 1.026 | 1.008 | 0.989 | 0.972 | 0.958 | 0.949 | 0.945 | 0.948 | 0.957 |
| $M_4$, $MN$ | 1.056 | 1.035 | 1.011 | 0.986 | 0.963 | 0.944 | 0.932 | 0.928 | 0.931 | 0.942 |
| $M_6$ | 1.085 | 1.053 | 1.016 | 0.978 | 0.944 | 0.918 | 0.900 | 0.894 | 0.899 | 0.915 |
| $M_8$ | 1.114 | 1.071 | 1.021 | 0.971 | 0.927 | 0.892 | 0.869 | 0.861 | 0.867 | 0.888 |
| $O_1$, $Q_1$, $2Q$, $\rho_1$ | 0.874 | 0.933 | 0.998 | 1.059 | 1.110 | 1.150 | 1.174 | 1.183 | 1.176 | 1.153 |
| $OO$ | 0.631 | 0.786 | 0.983 | 1.204 | 1.422 | 1.608 | 1.735 | 1.783 | 1.745 | 1.627 |
| $MK$ | 0.948 | 0.976 | 1.004 | 1.029 | 1.048 | 1.062 | 1.069 | 1.072 | 1.070 | 1.063 |
| $2MK$ | 0.974 | 0.993 | 1.010 | 1.022 | 1.029 | 1.032 | 1.032 | 1.032 | 1.032 | 1.032 |
| $Mf$ | 0.743 | 0.856 | 0.990 | 1.129 | 1.257 | 1.360 | 1.427 | 1.452 | 1.432 | 1.370 |
| $Mm$ | 1.094 | 1.059 | 1.016 | 0.973 | 0.933 | 0.900 | 0.879 | 0.871 | 0.878 | 0.897 |

| Constituent | 1860 | 1861 | 1862 | 1863 | 1864 | 1865 | 1866 | 1867 | 1868 | 1869 |
|---|---|---|---|---|---|---|---|---|---|---|
| $J_1$ | 1.110 | 1.066 | 1.013 | 0.955 | 0.898 | 0.852 | 0.829 | 0.832 | 0.863 | 0.912 |
| $K_1$ | 1.072 | 1.041 | 1.004 | 0.964 | 0.926 | 0.898 | 0.883 | 0.885 | 0.904 | 0.936 |
| $K_2$ | 1.179 | 1.086 | 0.988 | 0.897 | 0.823 | 0.773 | 0.749 | 0.753 | 0.784 | 0.840 |
| $L_2$ | 0.568 | 0.924 | 1.225 | 1.117 | 0.865 | 0.879 | 1.082 | 1.190 | 1.091 | 0.840 |
| $M_1$ | 2.227 | 1.792 | 1.046 | 1.260 | 1.680 | 1.609 | 1.164 | 0.812 | 1.189 | 1.731 |
| $M_2$*, $N_2$, $2N$, $\lambda_2$, $\mu_2$, $\nu_2$ | 0.980 | 0.991 | 1.004 | 1.016 | 1.026 | 1.034 | 1.038 | 1.037 | 1.032 | 1.024 |
| $M_3$ | 0.970 | 0.987 | 1.006 | 1.024 | 1.040 | 1.051 | 1.057 | 1.056 | 1.049 | 1.036 |
| $M_4$, $MN$ | 0.960 | 0.983 | 1.008 | 1.032 | 1.054 | 1.069 | 1.076 | 1.075 | 1.066 | 1.048 |
| $M_6$ | 0.941 | 0.974 | 1.011 | 1.049 | 1.081 | 1.105 | 1.117 | 1.115 | 1.100 | 1.073 |
| $M_8$ | 0.922 | 0.966 | 1.015 | 1.065 | 1.110 | 1.143 | 1.159 | 1.156 | 1.135 | 1.099 |
| $O_1$, $Q_1$, $2Q$, $\rho_1$ | 1.116 | 1.065 | 1.005 | 0.941 | 0.880 | 0.832 | 0.808 | 0.812 | 0.843 | 0.896 |
| $OO$ | 1.447 | 1.230 | 1.008 | 0.807 | 0.647 | 0.540 | 0.489 | 0.497 | 0.563 | 0.685 |
| $MK$ | 1.050 | 1.032 | 1.007 | 0.979 | 0.951 | 0.928 | 0.916 | 0.918 | 0.933 | 0.958 |
| $2MK$ | 1.029 | 1.023 | 1.011 | 0.995 | 0.976 | 0.960 | 0.950 | 0.952 | 0.964 | 0.981 |
| $Mf$ | 1.270 | 1.145 | 1.007 | 0.872 | 0.755 | 0.670 | 0.629 | 0.635 | 0.689 | 0.783 |
| $Mm$ | 0.928 | 0.968 | 1.011 | 1.054 | 1.091 | 1.117 | 1.130 | 1.128 | 1.111 | 1.082 |

| Constituent | 1870 | 1871 | 1872 | 1873 | 1874 | 1875 | 1876 | 1877 | 1878 | 1879 |
|---|---|---|---|---|---|---|---|---|---|---|
| $J_1$ | 0.971 | 1.028 | 1.079 | 1.120 | 1.147 | 1.162 | 1.164 | 1.154 | 1.130 | 1.094 |
| $K_1$ | 0.974 | 1.014 | 1.050 | 1.079 | 1.099 | 1.111 | 1.112 | 1.104 | 1.087 | 1.061 |
| $K_2$ | 0.920 | 1.014 | 1.112 | 1.201 | 1.269 | 1.309 | 1.315 | 1.287 | 1.227 | 1.144 |
| $L_2$ | 0.828 | 1.148 | 1.224 | 0.816 | 0.545 | 1.087 | 1.270 | 0.858 | 0.543 | 1.203 |
| $M_1$ | 1.811 | 1.300 | 1.185 | 2.004 | 2.286 | 1.656 | 1.227 | 1.998 | 2.269 | 1.645 |
| $M_2$*, $N_2$, $2N$, $\lambda_2$, $\mu_2$, $\nu_2$ | 1.013 | 1.000 | 0.988 | 0.977 | 0.969 | 0.964 | 0.963 | 0.967 | 0.974 | 0.984 |
| $M_3$ | 1.019 | 1.001 | 0.982 | 0.966 | 0.954 | 0.947 | 0.946 | 0.951 | 0.961 | 0.976 |
| $M_4$, $MN$ | 1.026 | 1.001 | 0.976 | 0.955 | 0.939 | 0.930 | 0.928 | 0.935 | 0.949 | 0.968 |
| $M_6$ | 1.039 | 1.001 | 0.965 | 0.933 | 0.910 | 0.896 | 0.894 | 0.904 | 0.924 | 0.953 |
| $M_8$ | 1.052 | 1.002 | 0.953 | 0.912 | 0.881 | 0.864 | 0.862 | 0.874 | 0.900 | 0.938 |
| $O_1$, $Q_1$, $2Q$, $\rho_1$ | 0.958 | 1.022 | 1.080 | 1.127 | 1.161 | 1.179 | 1.182 | 1.169 | 1.140 | 1.098 |
| $OO$ | 0.858 | 1.067 | 1.290 | 1.500 | 1.666 | 1.764 | 1.779 | 1.708 | 1.563 | 1.366 |
| $MK$ | 0.987 | 1.014 | 1.037 | 1.054 | 1.065 | 1.071 | 1.072 | 1.068 | 1.059 | 1.044 |
| $2MK$ | 1.000 | 1.015 | 1.025 | 1.030 | 1.032 | 1.032 | 1.032 | 1.032 | 1.031 | 1.027 |
| $Mf$ | 0.907 | 1.044 | 1.181 | 1.300 | 1.390 | 1.442 | 1.450 | 1.413 | 1.335 | 1.224 |
| $Mm$ | 1.043 | 1.000 | 0.957 | 0.919 | 0.891 | 0.874 | 0.872 | 0.884 | 0.908 | 0.943 |

*Factor $f$ of MS, 2SM, and MSf are each equal to factor $f$ of $M_2$.
Factor $f$ of $P_1$, $R_2$, $S_1$, $S_2$, $S_4$, $S_6$, $T_2$, Sa, and Ssa are each unity.

Table 14.—Node factor f for middle of each year, 1850 to 1999—Continued

| Constituent | 1880 | 1881 | 1882 | 1883 | 1884 | 1885 | 1886 | 1887 | 1888 | 1889 |
|---|---|---|---|---|---|---|---|---|---|---|
| $J_1$ | 1.047 | 0.991 | 0.932 | 0.878 | 0.840 | 0.827 | 0.841 | 0.880 | 0.934 | 0.994 |
| $K_1$ | 1.027 | 0.988 | 0.949 | 0.914 | 0.890 | 0.882 | 0.891 | 0.915 | 0.950 | 0.990 |
| $K_2$ | 1.048 | 0.951 | 0.866 | 0.800 | 0.760 | 0.748 | 0.762 | 0.803 | 0.869 | 0.955 |
| $L_2$ | 1.246 | 1.020 | 0.786 | 0.944 | 1.152 | 1.171 | 1.006 | 0.824 | 0.945 | 1.205 |
| $M_1$ | 1.046 | 1.528 | 1.824 | 1.529 | 0.970 | 0.877 | 1.364 | 1.721 | 1.593 | 1.075 |
| $M_2^*, N_2, 2N, \lambda_2, \mu_2, \nu_2$ | 0.996 | 1.009 | 1.020 | 1.030 | 1.036 | 1.038 | 1.036 | 1.029 | 1.020 | 1.008 |
| $M_3$ | 0.994 | 1.013 | 1.031 | 1.045 | 1.054 | 1.057 | 1.054 | 1.044 | 1.030 | 1.012 |
| $M_4, MN$ | 0.992 | 1.017 | 1.041 | 1.060 | 1.073 | 1.077 | 1.072 | 1.060 | 1.040 | 1.016 |
| $M_6$ | 0.988 | 1.026 | 1.062 | 1.092 | 1.111 | 1.118 | 1.111 | 1.091 | 1.061 | 1.024 |
| $M_8$ | 0.984 | 1.035 | 1.084 | 1.124 | 1.151 | 1.160 | 1.150 | 1.123 | 1.082 | 1.033 |
| $O_1, Q_1, 2Q, \rho_1$ | 1.043 | 0.980 | 0.916 | 0.860 | 0.820 | 0.806 | 0.821 | 0.862 | 0.919 | 0.983 |
| $OO$ | 1.144 | 0.926 | 0.739 | 0.599 | 0.513 | 0.486 | 0.516 | 0.604 | 0.746 | 0.936 |
| $MK$ | 1.023 | 0.997 | 0.968 | 0.941 | 0.922 | 0.915 | 0.922 | 0.942 | 0.969 | 0.998 |
| $2MK$ | 1.019 | 1.005 | 0.988 | 0.969 | 0.955 | 0.950 | 0.955 | 0.970 | 0.988 | 1.006 |
| $Mf$ | 1.092 | 0.953 | 0.823 | 0.717 | 0.649 | 0.626 | 0.651 | 0.721 | 0.828 | 0.959 |
| $Mm$ | 0.984 | 1.028 | 1.069 | 1.102 | 1.124 | 1.131 | 1.123 | 1.101 | 1.067 | 1.026 |

| Constituent | 1890 | 1891 | 1892 | 1893 | 1894 | 1895 | 1896 | 1897 | 1898 | 1899 |
|---|---|---|---|---|---|---|---|---|---|---|
| $J_1$ | 1.049 | 1.096 | 1.132 | 1.155 | 1.165 | 1.162 | 1.146 | 1.118 | 1.077 | 1.026 |
| $K_1$ | 1.028 | 1.062 | 1.088 | 1.105 | 1.112 | 1.110 | 1.099 | 1.078 | 1.054 | 1.012 |
| $K_2$ | 1.052 | 1.148 | 1.230 | 1.289 | 1.316 | 1.308 | 1.267 | 1.197 | 1.108 | 1.010 |
| $L_2$ | 1.153 | 0.709 | 0.683 | 1.185 | 1.219 | 0.704 | 0.607 | 1.141 | 1.229 | 0.897 |
| $M_1$ | 1.323 | 2.091 | 2.158 | 1.434 | 1.369 | 2.176 | 2.240 | 1.471 | 1.166 | 1.781 |
| $M_2^*, N_2, 2N, \lambda_2, \mu_2, \nu_2$ | 0.996 | 0.984 | 0.974 | 0.967 | 0.963 | 0.964 | 0.969 | 0.978 | 0.989 | 1.001 |
| $M_3$ | 0.993 | 0.976 | 0.961 | 0.950 | 0.946 | 0.947 | 0.954 | 0.967 | 0.983 | 1.002 |
| $M_4, MN$ | 0.991 | 0.968 | 0.948 | 0.934 | 0.928 | 0.930 | 0.939 | 0.956 | 0.977 | 1.002 |
| $M_6$ | 0.987 | 0.952 | 0.923 | 0.903 | 0.894 | 0.896 | 0.910 | 0.934 | 0.966 | 1.003 |
| $M_8$ | 0.982 | 0.936 | 0.898 | 0.873 | 0.861 | 0.864 | 0.882 | 0.913 | 0.955 | 1.004 |
| $O_1, Q_1, 2Q, \rho_1$ | 1.046 | 1.100 | 1.142 | 1.170 | 1.182 | 1.179 | 1.160 | 1.125 | 1.078 | 1.020 |
| $OO$ | 1.153 | 1.375 | 1.571 | 1.713 | 1.780 | 1.761 | 1.660 | 1.491 | 1.281 | 1.058 |
| $MK$ | 1.024 | 1.045 | 1.059 | 1.068 | 1.072 | 1.071 | 1.065 | 1.054 | 1.036 | 1.013 |
| $2MK$ | 1.019 | 1.028 | 1.031 | 1.032 | 1.032 | 1.032 | 1.032 | 1.030 | 1.025 | 1.014 |
| $Mf$ | 1.098 | 1.230 | 1.339 | 1.416 | 1.451 | 1.441 | 1.387 | 1.296 | 1.175 | 1.038 |
| $Mm$ | 0.983 | 0.941 | 0.907 | 0.883 | 0.872 | 0.875 | 0.892 | 0.921 | 0.958 | 1.001 |

| Constituent | 1900 | 1901 | 1902 | 1903 | 1904 | 1905 | 1906 | 1907 | 1908 | 1909 |
|---|---|---|---|---|---|---|---|---|---|---|
| $J_1$ | 0.968 | 0.910 | 0.861 | 0.832 | 0.829 | 0.854 | 0.900 | 0.957 | 1.016 | 1.069 |
| $K_1$ | 0.973 | 0.934 | 0.903 | 0.885 | 0.883 | 0.899 | 0.928 | 0.965 | 1.005 | 1.042 |
| $K_2$ | 0.916 | 0.838 | 0.782 | 0.752 | 0.750 | 0.774 | 0.825 | 0.900 | 0.992 | 1.090 |
| $L_2$ | 0.753 | 1.030 | 1.193 | 1.117 | 0.925 | 0.858 | 1.051 | 1.221 | 1.062 | 0.653 |
| $M_1$ | 1.902 | 1.399 | 0.858 | 1.069 | 1.507 | 1.643 | 1.340 | 0.946 | 1.479 | 2.112 |
| $M_2^*, N_2, 2N, \lambda_2, \mu_2, \nu_2$ | 1.013 | 1.024 | 1.032 | 1.037 | 1.038 | 1.034 | 1.026 | 1.016 | 1.003 | 0.991 |
| $M_3$ | 1.020 | 1.036 | 1.049 | 1.056 | 1.057 | 1.051 | 1.039 | 1.023 | 1.005 | 0.986 |
| $M_4, MN$ | 1.027 | 1.049 | 1.066 | 1.076 | 1.076 | 1.068 | 1.053 | 1.031 | 1.007 | 0.982 |
| $M_6$ | 1.040 | 1.074 | 1.101 | 1.115 | 1.117 | 1.104 | 1.080 | 1.047 | 1.010 | 0.973 |
| $M_8$ | 1.054 | 1.100 | 1.136 | 1.157 | 1.159 | 1.142 | 1.108 | 1.063 | 1.013 | 0.964 |
| $O_1, Q_1, 2Q, \rho_1$ | 0.956 | 0.893 | 0.842 | 0.811 | 0.808 | 0.834 | 0.882 | 0.944 | 1.008 | 1.068 |
| $OO$ | 0.850 | 0.679 | 0.550 | 0.496 | 0.490 | 0.543 | 0.652 | 0.814 | 1.017 | 1.240 |
| $MK$ | 0.986 | 0.957 | 0.933 | 0.918 | 0.916 | 0.929 | 0.952 | 0.980 | 1.008 | 1.033 |
| $2MK$ | 0.999 | 0.980 | 0.963 | 0.952 | 0.951 | 0.960 | 0.977 | 0.996 | 1.012 | 1.023 |
| $Mf$ | 0.901 | 0.779 | 0.686 | 0.634 | 0.630 | 0.673 | 0.759 | 0.877 | 1.012 | 1.151 |
| $Mm$ | 1.045 | 1.083 | 1.112 | 1.128 | 1.130 | 1.116 | 1.089 | 1.052 | 1.010 | 0.966 |

*Factor $f$ of MS, 2SM, and MSf are each equal to factor $f$ of $M_2$.
Factor $f$ of $P_1$, $R_2$, $S_1$, $S_2$, $S_4$, $S_6$, $T_2$, Sa, and Ssa are each unity.

HARMONIC ANALYSIS AND PREDICTION OF TIDES 201

Table 14.—*Node factor f for middle of each year, 1850 to 1999*—Continued

| Constituent | 1910 | 1911 | 1912 | 1913 | 1914 | 1915 | 1916 | 1917 | 1918 | 1919 |
|---|---|---|---|---|---|---|---|---|---|---|
| $J_1$ | 1.111 | 1.142 | 1.160 | 1.165 | 1.157 | 1.137 | 1.104 | 1.059 | 1.004 | 0.945 |
| $K_1$ | 1.073 | 1.096 | 1.109 | 1.113 | 1.107 | 1.092 | 1.067 | 1.035 | 0.997 | 0.958 |
| $K_2$ | 1.182 | 1.256 | 1.303 | 1.317 | 1.296 | 1.243 | 1.165 | 1.071 | 0.973 | 0.884 |
| $L_2$ | 0.834 | 1.246 | 1.135 | 0.561 | 0.729 | 1.221 | 1.172 | 0.761 | 0.780 | 1.118 |
| $M_1$ | 1.972 | 1.248 | 1.557 | 2.297 | 2.146 | 1.310 | 1.371 | 1.993 | 1.909 | 1.243 |
| $M_2^*$, $N_2$, 2N, $\lambda_2$, $\mu_2$, $\nu_2$ | 0.980 | 0.970 | 0.965 | 0.963 | 0.966 | 0.972 | 0.982 | 0.993 | 1.006 | 1.018 |
| $M_3$ | 0.969 | 0.956 | 0.948 | 0.945 | 0.949 | 0.958 | 0.972 | 0.990 | 1.009 | 1.027 |
| $M_4$, MN | 0.959 | 0.942 | 0.931 | 0.928 | 0.933 | 0.945 | 0.964 | 0.987 | 1.012 | 1.036 |
| $M_6$ | 0.940 | 0.914 | 0.898 | 0.894 | 0.901 | 0.918 | 0.946 | 0.980 | 1.017 | 1.054 |
| $M_8$ | 0.920 | 0.887 | 0.867 | 0.861 | 0.870 | 0.893 | 0.928 | 0.973 | 1.023 | 1.073 |
| $O_1$, $Q_1$, 2Q, $\rho_1$ | 1.118 | 1.154 | 1.176 | 1.183 | 1.173 | 1.148 | 1.109 | 1.056 | 0.995 | 0.931 |
| OO | 1.455 | 1.633 | 1.748 | 1.783 | 1.732 | 1.602 | 1.414 | 1.195 | 0.974 | 0.779 |
| MK | 1.051 | 1.063 | 1.070 | 1.072 | 1.069 | 1.061 | 1.048 | 1.028 | 1.003 | 0.975 |
| 2MK | 1.029 | 1.032 | 1.032 | 1.032 | 1.032 | 1.032 | 1.028 | 1.021 | 1.009 | 0.992 |
| Mf | 1.275 | 1.373 | 1.434 | 1.452 | 1.425 | 1.356 | 1.252 | 1.124 | 0.985 | 0.851 |
| Mm | 0.927 | 0.896 | 0.877 | 0.871 | 0.880 | 0.902 | 0.934 | 0.975 | 1.018 | 1.060 |

| Constituent | 1920 | 1921 | 1922 | 1923 | 1924 | 1925 | 1926 | 1927 | 1928 | 1929 |
|---|---|---|---|---|---|---|---|---|---|---|
| $J_1$ | 0.890 | 0.847 | 0.827 | 0.836 | 0.870 | 0.921 | 0.980 | 1.037 | 1.086 | 1.125 |
| $K_1$ | 0.921 | 0.894 | 0.882 | 0.887 | 0.909 | 0.942 | 0.981 | 1.020 | 1.055 | 1.083 |
| $K_2$ | 0.813 | 0.767 | 0.748 | 0.756 | 0.791 | 0.852 | 0.934 | 1.030 | 1.127 | 1.214 |
| $L_2$ | 1.198 | 1.034 | 0.870 | 0.932 | 1.133 | 1.199 | 0.963 | 0.669 | 0.975 | 1.270 |
| $M_1$ | 0.896 | 1.308 | 1.597 | 1.503 | 1.082 | 0.954 | 1.619 | 2.063 | 1.739 | 1.138 |
| $M_2^*$, $N_2$, 2N, $\lambda_2$, $\mu_2$, $\nu_2$ | 1.028 | 1.035 | 1.038 | 1.036 | 1.031 | 1.022 | 1.011 | 0.998 | 0.986 | 0.976 |
| $M_3$ | 1.042 | 1.052 | 1.057 | 1.055 | 1.047 | 1.034 | 1.016 | 0.998 | 0.979 | 0.964 |
| $M_4$, MN | 1.056 | 1.071 | 1.077 | 1.074 | 1.063 | 1.045 | 1.022 | 0.997 | 0.973 | 0.952 |
| $M_6$ | 1.086 | 1.108 | 1.118 | 1.114 | 1.096 | 1.068 | 1.033 | 0.995 | 0.959 | 0.929 |
| $M_8$ | 1.116 | 1.146 | 1.160 | 1.154 | 1.130 | 1.092 | 1.044 | 0.994 | 0.946 | 0.906 |
| $O_1$, $Q_1$, 2Q, $\rho_1$ | 0.871 | 0.827 | 0.806 | 0.815 | 0.850 | 0.905 | 0.968 | 1.032 | 1.088 | 1.134 |
| OO | 0.626 | 0.528 | 0.487 | 0.504 | 0.579 | 0.710 | 0.889 | 1.102 | 1.325 | 1.530 |
| MK | 0.947 | 0.925 | 0.915 | 0.920 | 0.937 | 0.963 | 0.992 | 1.018 | 1.040 | 1.056 |
| 2MK | 0.973 | 0.958 | 0.950 | 0.953 | 0.966 | 0.984 | 1.002 | 1.017 | 1.026 | 1.031 |
| Mf | 0.730 | 0.660 | 0.626 | 0.641 | 0.701 | 0.801 | 0.928 | 1.066 | 1.201 | 1.317 |
| Mm | 1.096 | 1.120 | 1.131 | 1.126 | 1.107 | 1.076 | 1.036 | 0.993 | 0.950 | 0.914 |

| Constituent | 1930 | 1931 | 1932 | 1933 | 1934 | 1935 | 1936 | 1937 | 1938 | 1939 |
|---|---|---|---|---|---|---|---|---|---|---|
| $J_1$ | 1.150 | 1.163 | 1.164 | 1.151 | 1.126 | 1.088 | 1.038 | 0.982 | 0.923 | 0.871 |
| $K_1$ | 1.102 | 1.112 | 1.112 | 1.102 | 1.083 | 1.056 | 1.021 | 0.982 | 0.943 | 0.909 |
| $K_2$ | 1.278 | 1.312 | 1.313 | 1.279 | 1.216 | 1.130 | 1.033 | 0.937 | 0.854 | 0.792 |
| $L_2$ | 1.022 | 0.471 | 0.873 | 1.270 | 1.078 | 0.636 | 0.859 | 1.190 | 1.162 | 0.933 |
| $M_1$ | 1.312 | 2.353 | 1.992 | 1.107 | 1.614 | 2.148 | 1.850 | 1.098 | 1.079 | 1.534 |
| $M_2^*$, $N_2$, 2N, $\lambda_2$, $\mu_2$, $\nu_2$ | 0.968 | 0.964 | 0.964 | 0.968 | 0.975 | 0.986 | 0.998 | 1.011 | 1.022 | 1.031 |
| $M_3$ | 0.952 | 0.946 | 0.946 | 0.952 | 0.963 | 0.979 | 0.997 | 1.016 | 1.033 | 1.047 |
| $M_4$, MN | 0.937 | 0.929 | 0.929 | 0.936 | 0.951 | 0.972 | 0.996 | 1.021 | 1.044 | 1.063 |
| $M_6$ | 0.907 | 0.895 | 0.895 | 0.906 | 0.928 | 0.958 | 0.994 | 1.032 | 1.067 | 1.096 |
| $M_8$ | 0.878 | 0.863 | 0.862 | 0.877 | 0.905 | 0.945 | 0.992 | 1.043 | 1.091 | 1.130 |
| $O_1$, $Q_1$, 2Q, $\rho_1$ | 1.165 | 1.181 | 1.181 | 1.165 | 1.134 | 1.090 | 1.034 | 0.970 | 0.907 | 0.852 |
| OO | 1.686 | 1.772 | 1.773 | 1.690 | 1.535 | 1.332 | 1.108 | 0.894 | 0.714 | 0.581 |
| MK | 1.066 | 1.071 | 1.071 | 1.067 | 1.057 | 1.041 | 1.019 | 0.992 | 0.964 | 0.938 |
| 2MK | 1.032 | 1.032 | 1.032 | 1.032 | 1.031 | 1.026 | 1.017 | 1.003 | 0.985 | 0.966 |
| Mf | 1.402 | 1.446 | 1.447 | 1.403 | 1.320 | 1.205 | 1.070 | 0.931 | 0.804 | 0.704 |
| Mm | 0.887 | 0.873 | 0.873 | 0.887 | 0.913 | 0.949 | 0.991 | 1.035 | 1.075 | 1.107 |

*Factor $f$ of MS, 2MS, and MSf are each equal to factor $f$ of $M_2$.
Factor $f$ of $P_1$, $R_2$, $S_1$, $S_2$, $S_4$, $S_6$, $T_2$, Sa, and Ssa are each unity.

Table 14.—*Node factor f for middle of each year, 1850 to 1999*—Continued

| Constituent | 1940 | 1941 | 1942 | 1943 | 1944 | 1945 | 1946 | 1947 | 1948 | 1949 |
|---|---|---|---|---|---|---|---|---|---|---|
| $J_1$ | 0.836 | 0.827 | 0.846 | 0.888 | 0.944 | 1.003 | 1.057 | 1.103 | 1.136 | 1.157 |
| $K_1$ | 0.888 | 0.882 | 0.894 | 0.920 | 0.956 | 0.996 | 1.034 | 1.067 | 1.091 | 1.107 |
| $K_2$ | 0.757 | 0.748 | 0.766 | 0.812 | 0.882 | 0.970 | 1.068 | 1.162 | 1.242 | 1.295 |
| $L_2$ | 0.860 | 1.021 | 1.180 | 1.144 | 0.876 | 0.748 | 1.091 | 1.255 | 0.894 | 0.482 |
| $M_1$ | 1.623 | 1.313 | 0.879 | 1.076 | 1.714 | 1.944 | 1.480 | 1.138 | 1.927 | 2.339 |
| $M_2^*$, $N_2$, $2N$, $\lambda_2$, $\mu_2$, $\nu_2$ | 1.036 | 1.038 | 1.035 | 1.028 | 1.018 | 1.006 | 0.994 | 0.982 | 0.972 | 0.966 |
| $M_3$ | 1.055 | 1.057 | 1.053 | 1.042 | 1.027 | 1.009 | 0.990 | 0.973 | 0.959 | 0.949 |
| $M_4$, MN | 1.074 | 1.077 | 1.071 | 1.057 | 1.036 | 1.012 | 0.987 | 0.964 | 0.945 | 0.933 |
| $M_6$ | 1.113 | 1.118 | 1.108 | 1.086 | 1.055 | 1.018 | 9.981 | 0.947 | 0.919 | 0.901 |
| $M_8$ | 1.154 | 1.160 | 1.147 | 1.117 | 1.074 | 1.025 | 0.975 | 0.929 | 0.894 | 0.870 |
| $O_1$, $Q_1$, $2Q$, $\rho_1$ | 0.816 | 0.806 | 0.826 | 0.870 | 0.929 | 0.994 | 1.055 | 1.107 | 1.147 | 1.173 |
| OO | 0.505 | 0.486 | 0.526 | 0.623 | 0.774 | 0.969 | 1.189 | 1.408 | 1.598 | 1.729 |
| MK | 0.920 | 0.915 | 0.925 | 0.946 | 0.974 | 1.002 | 1.028 | 1.047 | 1.061 | 1.069 |
| 2MK | 0.953 | 0.950 | 0.957 | 0.973 | 0.991 | 1.008 | 1.021 | 1.028 | 1.032 | 1.032 |
| Mf | 0.642 | 0.626 | 0.659 | 0.736 | 0.848 | 0.981 | 1.120 | 1.249 | 1.354 | 1.424 |
| Mm | 1.126 | 1.131 | 1.121 | 1.096 | 1.061 | 1.019 | 0.976 | 0.935 | 0.902 | 0.880 |

| Constituent | 1950 | 1951 | 1952 | 1953 | 1954 | 1955 | 1956 | 1957 | 1958 | 1959 |
|---|---|---|---|---|---|---|---|---|---|---|
| $J_1$ | 1.165 | 1.160 | 1.143 | 1.112 | 1.070 | 1.002 | 0.959 | 0.901 | 0.855 | 0.829 |
| $K_1$ | 1.113 | 1.109 | 1.096 | 1.074 | 1.043 | 1.001 | 0.966 | 0.929 | 0.900 | 0.883 |
| $K_2$ | 1.317 | 1.303 | 1.257 | 1.184 | 1.092 | 0.995 | 0.903 | 0.827 | 0.776 | 0.770 |
| $L_2$ | 1.074 | 1.330 | 1.014 | 0.653 | 1.001 | 1.260 | 1.112 | 0.867 | 0.915 | 1.115 |
| $M_1$ | 1.717 | 1.120 | 1.778 | 2.161 | 1.664 | 0.964 | 1.276 | 1.656 | 1.527 | 1.053 |
| $M_2^*$, $N_2$, $2N$, $\lambda_2$, $\mu_2$, $\nu_2$ | 0.963 | 0.965 | 0.970 | 0.982 | 0.990 | 1.003 | 1.015 | 1.026 | 1.033 | 1.038 |
| $M_3$ | 0.945 | 0.948 | 0.956 | 0.969 | 0.986 | 1.004 | 1.023 | 1.039 | 1.051 | 1.057 |
| $M_4$, MN | 0.928 | 0.931 | 0.941 | 0.959 | 0.981 | 1.006 | 1.031 | 1.052 | 1.068 | 1.076 |
| $M_6$ | 0.894 | 0.898 | 0.914 | 0.939 | 0.972 | 1.009 | 1.046 | 1.079 | 1.104 | 1.116 |
| $M_8$ | 0.861 | 0.867 | 0.887 | 0.920 | 0.962 | 1.012 | 1.062 | 1.107 | 1.141 | 1.158 |
| $O_1$, $Q_1$, $2Q$, $\rho_1$ | 1.183 | 1.177 | 1.155 | 1.119 | 1.069 | 1.010 | 0.945 | 0.884 | 0.835 | 0.808 |
| OO | 1.784 | 1.750 | 1.637 | 1.459 | 1.246 | 1.023 | 0.819 | 0.656 | 0.546 | 0.491 |
| MK | 1.072 | 1.070 | 1.063 | 1.051 | 1.033 | 1.009 | 0.981 | 0.953 | 0.930 | 0.916 |
| 2MK | 1.032 | 1.032 | 1.032 | 1.029 | 1.023 | 1.012 | 0.996 | 0.977 | 0.960 | 0.951 |
| Mf | 1.452 | 1.435 | 1.375 | 1.278 | 1.154 | 1.016 | 0.880 | 0.761 | 0.675 | 0.630 |
| Mm | 0.872 | 0.877 | 0.896 | 0.926 | 0.965 | 1.008 | 1.051 | 1.088 | 1.116 | 1.130 |

| Constituent | 1960 | 1961 | 1962 | 1963 | 1964 | 1965 | 1966 | 1967 | 1968 | 1969 |
|---|---|---|---|---|---|---|---|---|---|---|
| $J_1$ | 0.831 | 0.860 | 0.909 | 0.766 | 1.025 | 1.076 | 1.117 | 1.146 | 1.161 | 1.165 |
| $K_1$ | 0.885 | 0.903 | 0.934 | 0.972 | 1.011 | 1.048 | 1.077 | 1.098 | 1.110 | 1.113 |
| $K_2$ | 0.752 | 0.781 | 0.836 | 0.914 | 1.008 | 1.106 | 1.195 | 1.265 | 1.307 | 1.316 |
| $L_2$ | 1.199 | 1.081 | 0.849 | 0.893 | 1.200 | 1.237 | 0.838 | 0.690 | 1.185 | 1.310 |
| $M_1$ | 0.767 | 1.197 | 1.690 | 1.699 | 1.166 | 1.175 | 1.976 | 2.175 | 1.503 | 1.197 |
| $M_2^*$, $N_2$, $2N$, $\lambda_2$, $\mu_2$, $\nu_2$ | 1.037 | 1.033 | 1.024 | 1.014 | 1.001 | 0.989 | 0.978 | 0.969 | 0.964 | 0.963 |
| $M_3$ | 1.056 | 1.049 | 1.037 | 1.020 | 1.002 | 0.983 | 0.967 | 0.954 | 0.947 | 0.945 |
| $M_4$, MN | 1.076 | 1.066 | 1.050 | 1.027 | 1.003 | 0.978 | 0.956 | 0.940 | 0.930 | 0.928 |
| $M_6$ | 1.116 | 1.111 | 1.075 | 1.041 | 1.004 | 0.967 | 0.935 | 0.911 | 0.897 | 0.894 |
| $M_8$ | 1.157 | 1.137 | 1.102 | 1.055 | 1.005 | 0.956 | 0.914 | 0.883 | 0.865 | 0.861 |
| $O_1$, $Q_1$, $2Q$, $\rho_1$ | 0.810 | 0.840 | 0.891 | 0.954 | 1.018 | 1.076 | 1.124 | 1.159 | 1.178 | 1.182 |
| OO | 0.495 | 0.557 | 0.675 | 0.845 | 1.053 | 1.276 | 1.487 | 1.655 | 1.758 | 1.782 |
| MK | 0.917 | 0.932 | 0.956 | 0.985 | 1.013 | 1.036 | 1.053 | 1.064 | 1.071 | 1.072 |
| 2MK | 0.952 | 0.962 | 0.980 | 0.998 | 1.014 | 1.024 | 1.030 | 1.032 | 1.032 | 1.032 |
| Mf | 0.633 | 0.684 | 0.776 | 0.898 | 1.035 | 1.172 | 1.293 | 1.385 | 1.439 | 1.451 |
| Mm | 1.128 | 1.113 | 1.084 | 1.046 | 1.003 | 0.959 | 0.922 | 0.893 | 0.876 | 0.872 |

*Factor $f$ of MS, 2SM, and MSf are each equal to factor $f$ of $M_2$.
Factor $f$ of $P_1$, $R_2$, $S_1$, $S_2$, $S_4$, $S_6$, $T_2$, $Sa$, and Ssa are each unity.

Table 14.—*Node factor f for middle of each year, 1850 to 1999*—Continued

| Constituent | 1970 | 1971 | 1972 | 1973 | 1974 | 1975 | 1976 | 1977 | 1978 | 1979 |
|---|---|---|---|---|---|---|---|---|---|---|
| $J_1$ | 1.155 | 1.132 | 1.097 | 1.051 | 0.995 | 0.936 | 0.881 | 0.842 | 0.827 | 0.839 |
| $K_1$ | 1.105 | 1.088 | 1.063 | 1.029 | 0.991 | 0.951 | 0.916 | 0.891 | 0.882 | 0.890 |
| $K_2$ | 1.289 | 1.232 | 1.150 | 1.055 | 0.957 | 0.871 | 0.804 | 0.763 | 0.748 | 0.760 |
| $L_2$ | 0.882 | 0.668 | 1.118 | 1.270 | 1.014 | 0.808 | 0.988 | 1.179 | 1.169 | 0.994 |
| $M_1$ | 1.987 | 2.176 | 1.503 | 1.012 | 1.535 | 1.777 | 1.428 | 0.870 | 0.874 | 1.361 |
| $M_2^*$, $N_2$, $2N$, $\lambda_2$, $\mu_2$, $\nu_2$ | 0.966 | 0.973 | 0.983 | 0.995 | 1.008 | 1.020 | 1.029 | 1.035 | 1.038 | 1.036 |
| $M_3$ | 0.950 | 0.960 | 0.975 | 0.993 | 1.012 | 1.029 | 1.044 | 1.054 | 1.057 | 1.054 |
| $M_4$, $MN$ | 0.934 | 0.948 | 0.967 | 0.991 | 1.016 | 1.039 | 1.059 | 1.072 | 1.077 | 1.073 |
| $M_6$ | 0.903 | 0.922 | 0.951 | 0.986 | 1.024 | 1.060 | 1.090 | 1.110 | 1.118 | 1.112 |
| $M_8$ | 0.873 | 0.898 | 0.935 | 0.981 | 1.032 | 1.081 | 1.122 | 1.149 | 1.160 | 1.151 |
| $O_1$, $Q_1$, $2Q$, $\rho_1$ | 1.170 | 1.143 | 1.101 | 1.047 | 0.984 | 0.920 | 0.863 | 0.822 | 0.806 | 0.819 |
| $OO$ | 1.716 | 1.575 | 1.380 | 1.159 | 0.940 | 0.750 | 0.607 | 0.517 | 0.485 | 0.512 |
| $MK$ | 1.068 | 1.059 | 1.045 | 1.024 | 0.998 | 0.970 | 0.943 | 0.923 | 0.915 | 0.922 |
| $2MK$ | 1.032 | 1.031 | 1.028 | 1.020 | 1.006 | 0.989 | 0.970 | 0.956 | 0.950 | 0.955 |
| $Mf$ | 1.417 | 1.341 | 1.233 | 1.102 | 0.962 | 0.831 | 0.723 | 0.652 | 0.625 | 0.647 |
| $Mm$ | 0.882 | 0.906 | 0.940 | 0.982 | 1.025 | 1.067 | 1.100 | 1.123 | 1.131 | 1.124 |

| Constituent | 1980 | 1981 | 1982 | 1983 | 1984 | 1985 | 1986 | 1987 | 1988 | 1989 |
|---|---|---|---|---|---|---|---|---|---|---|
| $J_1$ | 0.877 | 0.930 | 0.989 | 1.045 | 1.093 | 1.130 | 1.153 | 1.164 | 1.163 | 1.148 |
| $K_1$ | 0.913 | 0.948 | 0.987 | 1.026 | 1.060 | 1.086 | 1.104 | 1.112 | 1.111 | 1.100 |
| $K_2$ | 0.799 | 0.864 | 0.949 | 1.045 | 1.142 | 1.226 | 1.285 | 1.315 | 1.310 | 1.270 |
| $L_2$ | 0.848 | 1.001 | 1.238 | 1.157 | 0.745 | 0.811 | 1.263 | 1.244 | 0.749 | 0.746 |
| $M_1$ | 1.656 | 1.468 | 0.974 | 1.323 | 2.050 | 2.032 | 1.292 | 1.367 | 2.142 | 2.122 |
| $M_2^*$, $N_2$, $2N$, $\lambda_2$, $\mu_2$, $\nu_2$ | 1.030 | 1.021 | 1.009 | 0.997 | 0.984 | 0.974 | 0.967 | 0.964 | 0.964 | 0.969 |
| $M_3$ | 1.045 | 1.031 | 1.013 | 0.994 | 0.977 | 0.962 | 0.951 | 0.946 | 0.947 | 0.954 |
| $M_4$, $MN$ | 1.061 | 1.042 | 1.018 | 0.993 | 0.969 | 0.949 | 0.935 | 0.928 | 0.930 | 0.939 |
| $M_6$ | 1.092 | 1.063 | 1.027 | 0.989 | 0.954 | 0.924 | 0.904 | 0.894 | 0.896 | 0.910 |
| $M_8$ | 1.125 | 1.085 | 1.036 | 0.986 | 0.939 | 0.901 | 0.874 | 0.862 | 0.864 | 0.881 |
| $O_1$, $Q_1$, $2Q$, $\rho_1$ | 0.858 | 0.915 | 0.979 | 1.041 | 1.096 | 1.140 | 1.168 | 1.182 | 1.180 | 1.161 |
| $OO$ | 0.596 | 0.735 | 0.921 | 1.137 | 1.361 | 1.560 | 1.706 | 1.778 | 1.766 | 1.668 |
| $MK$ | 0.941 | 0.967 | 0.996 | 1.022 | 1.043 | 1.058 | 1.068 | 1.072 | 1.071 | 1.065 |
| $2MK$ | 0.969 | 0.987 | 1.005 | 1.019 | 1.027 | 1.031 | 1.032 | 1.032 | 1.032 | 1.032 |
| $Mf$ | 0.715 | 0.820 | 0.949 | 1.088 | 1.221 | 1.333 | 1.412 | 1.450 | 1.443 | 1.392 |
| $Mm$ | 1.103 | 1.070 | 1.029 | 0.986 | 0.944 | 0.909 | 0.884 | 0.872 | 0.874 | 0.891 |

| Constituent | 1990 | 1991 | 1992 | 1993 | 1994 | 1995 | 1996 | 1997 | 1998 | 1999 |
|---|---|---|---|---|---|---|---|---|---|---|
| $J_1$ | 1.120 | 1.080 | 1.030 | 0.972 | 0.914 | 0.864 | 0.833 | 0.829 | 0.852 | 0.896 |
| $K_1$ | 1.079 | 1.051 | 1.015 | 0.976 | 0.937 | 0.905 | 0.886 | 0.883 | 0.897 | 0.926 |
| $K_2$ | 1.203 | 1.115 | 1.016 | 0.922 | 0.842 | 0.785 | 0.754 | 0.750 | 0.772 | 0.821 |
| $L_2$ | 1.216 | 1.248 | 0.898 | 0.801 | 1.077 | 1.208 | 1.107 | 0.921 | 0.893 | 1.096 |
| $M_1$ | 1.334 | 1.156 | 1.778 | 1.829 | 1.282 | 0.800 | 1.083 | 1.487 | 1.560 | 1.214 |
| $M_2^*$, $N_2$, $2N$, $\lambda_2$, $\mu_2$, $\nu_2$ | 0.977 | 0.988 | 1.000 | 1.013 | 1.024 | 1.032 | 1.037 | 1.038 | 1.034 | 1.027 |
| $M_3$ | 0.966 | 0.982 | 1.000 | 1.019 | 1.036 | 1.048 | 1.056 | 1.057 | 1.051 | 1.040 |
| $M_4$, $MN$ | 0.955 | 0.976 | 1.000 | 1.025 | 1.048 | 1.065 | 1.075 | 1.076 | 1.069 | 1.054 |
| $M_6$ | 0.932 | 0.964 | 1.000 | 1.038 | 1.072 | 1.099 | 1.115 | 1.117 | 1.105 | 1.082 |
| $M_8$ | 0.911 | 0.952 | 1.000 | 1.051 | 1.098 | 1.134 | 1.156 | 1.159 | 1.143 | 1.111 |
| $O_1$, $Q_1$, $2Q$, $\rho_1$ | 1.128 | 1.081 | 1.024 | 0.960 | 0.897 | 0.844 | 0.812 | 0.808 | 0.832 | 0.879 |
| $OO$ | 1.505 | 1.296 | 1.072 | 0.863 | 0.688 | 0.565 | 0.498 | 0.489 | 0.538 | 0.643 |
| $MK$ | 1.054 | 1.038 | 1.015 | 0.988 | 0.959 | 0.934 | 0.918 | 0.916 | 0.928 | 0.950 |
| $2MK$ | 1.030 | 1.025 | 1.015 | 1.000 | 0.982 | 0.964 | 0.952 | 0.951 | 0.959 | 0.976 |
| $Mf$ | 1.303 | 1.184 | 1.048 | 0.910 | 0.786 | 0.691 | 0.636 | 0.629 | 0.669 | 0.752 |
| $Mm$ | 0.918 | 0.956 | 0.998 | 1.042 | 1.081 | 1.110 | 1.128 | 1.130 | 1.117 | 1.091 |

*Factor $f$ of MS, 2SM, and MSf are each equal to factor $f$ of $M_2$.
Factor $f$ of $P_1$, $R_2$, $S_1$, $S_2$, $S_4$, $S_6$, $T_2$, Sa, and Ssa are each unity.

Table 15.—Equilibrium argument $(V_0+u)$ for meridian of Greenwich at beginning of each calendar year, 1850 to 2000

| Constituent | 1850 | 1851 | 1852 | 1853 | 1854 | 1855 | 1856 | 1857 | 1858 | 1859 | 1860 | 1861 | 1862 | 1863 | 1864 | 1865 | 1866 | 1867 | 1868 | 1869 |
|---|---|---|---|---|---|---|---|---|---|---|---|---|---|---|---|---|---|---|---|---|
| J₁ | 29.7 | 116.0 | 204.0 | 307.0 | 38.1 | 129.4 | 221.2 | 327.3 | 59.4 | 151.1 | 242.6 | 347.4 | 77.0 | 165.2 | 251.7 | 350.6 | 74.0 | 157.0 | 240.8 | 340.1 |
| K₁ | 3.5 | 1.6 | 0.9 | 5.2 | 3.3 | 5.0 | 7.1 | 10.4 | 12.8 | 14.9 | 16.7 | 18.9 | 19.3 | 18.7 | 17.0 | 15.2 | 11.7 | 8.0 | 4.6 | 3.1 |
| K₂ | 187.8 | 183.8 | 181.8 | 184.0 | 186.0 | 189.5 | 194.0 | 200.9 | 205.8 | 210.4 | 214.0 | 218.2 | 218.6 | 217.0 | 213.2 | 209.7 | 203.2 | 196.4 | 190.0 | 187.0 |
| L₂ | 155.7 | 350.6 | 161.5 | 330.9 | 177.7 | 26.0 | 194.6 | 353.3 | 204.2 | 54.4 | 229.0 | 26.0 | 229.0 | 72.2 | 259.4 | 62.5 | 249.2 | 86.0 | 283.9 | 100.4 |
| M₁ | 6.8 | 273.5 | 165.2 | 50.6 | 346.9 | 274.8 | 170.8 | 55.4 | 347.2 | 281.7 | 178.9 | 61.0 | 341.0 | 280.0 | 179.7 | 56.8 | 311.8 | 235.5 | 162.6 | 47.2 |
| M₂ | 299.8 | 40.1 | 140.6 | 217.0 | 318.0 | 59.3 | 160.6 | 237.7 | 339.2 | 80.6 | 181.8 | 258.5 | 359.3 | 99.8 | 200.2 | 276.0 | 16.0 | 116.1 | 216.2 | 292.0 |
| M₃ | 269.7 | 240.2 | 280.6 | 145.5 | 117.0 | 88.9 | 61.0 | 356.6 | 328.8 | 300.9 | 272.7 | 207.7 | 178.9 | 149.7 | 120.2 | 53.9 | 24.0 | 354.1 | 324.2 | 258.0 |
| M₄ | 239.6 | 80.2 | 281.3 | 74.1 | 276.1 | 118.5 | 321.3 | 115.4 | 318.4 | 161.2 | 3.6 | 156.9 | 358.5 | 199.6 | 40.3 | 191.9 | 32.1 | 232.2 | 72.3 | 224.0 |
| M₆ | 179.4 | 120.3 | 61.9 | 291.1 | 234.1 | 177.8 | 121.9 | 353.1 | 297.6 | 241.7 | 185.4 | 55.4 | 357.1 | 299.5 | 240.5 | 107.9 | 49.8 | 348.2 | 288.5 | 156.0 |
| M₈ | 119.2 | 160.4 | 202.6 | 148.1 | 192.1 | 237.0 | 282.5 | 230.9 | 276.7 | 322.3 | 7.3 | 313.9 | 357.1 | 39.3 | 80.7 | 23.8 | 64.1 | 88.1 | 144.7 | 88.0 |
| N₂ | 270.0 | 281.6 | 293.4 | 268.0 | 280.3 | 292.8 | 305.5 | 280.8 | 293.5 | 306.2 | 318.7 | 293.6 | 305.6 | 317.5 | 329.1 | 303.1 | 314.5 | 325.8 | 337.2 | 311.2 |
| 2N | 240.3 | 163.2 | 86.2 | 319.1 | 242.6 | 166.4 | 90.3 | 323.8 | 247.3 | 171.8 | 95.6 | 328.7 | 252.2 | 175.2 | 98.0 | 330.3 | 252.9 | 175.5 | 98.1 | 330.4 |
| O₁ | 299.9 | 42.6 | 143.8 | 218.5 | 317.7 | 56.3 | 154.6 | 115.4 | 63.7 | 162.3 | 56.8 | 236.0 | 335.9 | 148.7 | 179.5 | 258.2 | 3.6 | 109.3 | 214.4 | 292.7 |
| OO | 240.0 | 132.5 | 29.8 | 318.6 | 223.1 | 129.6 | 37.5 | 333.4 | 242.0 | 150.0 | 56.8 | 348.9 | 250.7 | 148.7 | 41.8 | 317.2 | 201.3 | 84.1 | 329.1 | 245.8 |
| P₁ | 349.7 | 349.9 | 350.2 | 349.4 | 349.7 | 349.9 | 350.2 | 349.4 | 349.6 | 349.9 | 350.1 | 349.4 | 349.6 | 349.8 | 350.1 | 349.3 | 349.6 | 349.8 | 350.1 | 349.3 |
| Q₁ | 270.1 | 284.1 | 296.6 | 269.5 | 280.0 | 289.9 | 299.4 | 277.0 | 279.8 | 289.9 | 299.2 | 271.1 | 228.7 | 294.7 | 308.5 | 285.4 | 302.5 | 319.0 | 335.4 | 311.9 |
| 2Q | 240.4 | 165.6 | 89.4 | 320.5 | 242.2 | 153.4 | 84.3 | 313.4 | 234.1 | 154.9 | 76.1 | 306.2 | 228.7 | 152.2 | 77.4 | 312.6 | 240.5 | 168.8 | 96.4 | 331.1 |
| R₂ | 179.9 | 179.7 | 179.4 | 180.2 | 179.9 | 179.6 | 179.4 | 180.1 | 179.8 | 179.6 | 179.3 | 180.1 | 179.8 | 179.6 | 179.3 | 180.0 | 179.8 | 179.5 | 179.3 | 180.0 |
| S₁ | 180.0 | 180.0 | 180.0 | 180.0 | 180.0 | 180.0 | 180.0 | 180.0 | 180.0 | 180.0 | 180.0 | 180.0 | 180.0 | 180.0 | 180.0 | 180.0 | 180.0 | 180.0 | 180.0 | 180.0 |
| S₃,₄,₆ | 0.0 | 0.0 | 0.0 | 0.0 | 0.0 | 0.0 | 0.0 | 0.0 | 0.0 | 0.0 | 0.0 | 0.0 | 0.0 | 0.0 | 0.0 | 0.0 | 0.0 | 0.0 | 0.0 | 0.0 |
| T₂ | 0.1 | 0.3 | 0.6 | 359.8 | 0.1 | 0.4 | 0.6 | 359.9 | 0.2 | 0.4 | 0.7 | 359.9 | 0.2 | 0.4 | 0.7 | 0.0 | 0.2 | 0.5 | 0.7 | 0.0 |
| λ₂ | 148.8 | 59.6 | 330.7 | 228.9 | 140.4 | 52.2 | 324.1 | 223.0 | 135.4 | 46.9 | 318.7 | 217.2 | 128.5 | 39.6 | 310.5 | 208.1 | 118.7 | 29.3 | 299.9 | 197.6 |
| μ₂ | 241.0 | 82.1 | 283.4 | 76.2 | 277.9 | 119.9 | 322.6 | 115.5 | 317.7 | 159.8 | 1.8 | 154.9 | 356.4 | 197.7 | 38.8 | 191.9 | 31.8 | 232.6 | 73.4 | 206.4 |
| ν₂ | 270.8 | 200.6 | 130.6 | 25.1 | 315.6 | 246.3 | 177.2 | 72.4 | 50.0 | 294.2 | 224.9 | 119.9 | 50.0 | 340.1 | 269.2 | 163.8 | 93.4 | 312.4 | 312.4 | 206.6 |
| ρ₁ | 270.9 | 203.1 | 133.8 | 26.6 | 315.3 | 243.4 | 171.1 | 62.0 | 349.6 | 277.3 | 205.4 | 97.3 | 26.7 | 317.2 | 249.2 | 146.1 | 80.9 | 16.2 | 310.7 | 207.2 |
| MK | 303.3 | 41.8 | 141.6 | 219.2 | 321.1 | 64.3 | 167.8 | 248.2 | 352.0 | 95.5 | 198.5 | 277.4 | 18.5 | 118.5 | 217.2 | 291.2 | 27.7 | 124.0 | 220.8 | 285.1 |
| 2MK | 236.1 | 78.6 | 280.4 | 71.9 | 272.8 | 113.5 | 314.9 | 105.5 | 305.6 | 146.2 | 346.9 | 138.1 | 339.2 | 181.0 | 23.3 | 176.7 | 20.4 | 224.2 | 67.7 | 220.9 |
| MN | 209.8 | 321.8 | 74.1 | 125.1 | 238.4 | 352.1 | 106.1 | 158.5 | 272.7 | 26.8 | 140.5 | 192.0 | 304.9 | 57.3 | 169.3 | 219.1 | 330.5 | 81.9 | 193.3 | 243.2 |
| MS | 299.8 | 40.1 | 140.6 | 217.0 | 318.0 | 59.3 | 160.6 | 237.7 | 339.2 | 80.6 | 181.8 | 258.5 | 359.3 | 99.8 | 200.5 | 276.0 | 16.0 | 116.1 | 216.2 | 292.0 |
| 2SM | 60.2 | 319.9 | 219.4 | 143.0 | 42.0 | 300.7 | 199.4 | 122.3 | 20.8 | 279.4 | 178.2 | 101.5 | 0.7 | 260.2 | 159.8 | 84.0 | 344.0 | 243.9 | 143.8 | 68.0 |
| Mf | 240.0 | 134.9 | 33.0 | 320.1 | 222.7 | 126.7 | 31.5 | 323.1 | 228.3 | 133.2 | 37.3 | 326.4 | 227.4 | 125.8 | 21.1 | 299.4 | 188.8 | 77.4 | 327.3 | 246.5 |
| MSf | 60.2 | 319.9 | 219.4 | 143.0 | 42.0 | 300.7 | 199.4 | 122.3 | 20.8 | 279.4 | 178.2 | 101.5 | 0.7 | 260.2 | 159.8 | 84.0 | 344.0 | 243.9 | 143.8 | 68.0 |
| Mm | 29.8 | 118.5 | 219.4 | 309.0 | 37.7 | 126.1 | 199.4 | 316.9 | 45.1 | 134.1 | 223.1 | 324.9 | 53.6 | 142.3 | 231.1 | 332.8 | 61.6 | 150.3 | 239.0 | 340.8 |
| Sa | 280.3 | 280.1 | 279.8 | 280.6 | 280.3 | 280.1 | 279.8 | 280.6 | 280.4 | 280.1 | 279.9 | 280.6 | 280.4 | 280.2 | 279.9 | 280.7 | 279.8 | 280.2 | 279.9 | 280.7 |
| Ssa | 200.6 | 200.1 | 199.6 | 201.1 | 200.7 | 200.2 | 199.7 | 201.2 | 200.7 | 200.2 | 199.8 | 201.3 | 200.8 | 200.3 | 199.8 | 201.3 | 200.8 | 200.4 | 199.9 | 201.4 |

HARMONIC ANALYSIS AND PREDICTION OF TIDES 205

Table 15.—*Equilibrium argument* ($V_0+u$) *for meridian of Greenwich at beginning of each calendar year, 1850 to 2000*—Con.

| Constituent | 1870 | 1871 | 1872 | 1873 | 1874 | 1875 | 1876 | 1877 | 1878 | 1879 | 1880 | 1881 | 1882 | 1883 | 1884 | 1885 | 1886 | 1887 | 1888 | 1889 |
|---|---|---|---|---|---|---|---|---|---|---|---|---|---|---|---|---|---|---|---|---|
| J₁ | 67.1 | 155.7 | 245.6 | 350.5 | 82.1 | 174.0 | 266.1 | 12.2 | 103.9 | 195.0 | 285.3 | 28.5 | 116.1 | 201.9 | 286.1 | 23.3 | 106.5 | 190.8 | 276.7 | 18.4 |
| K₁ | 1.7 | 1.4 | 2.0 | 4.4 | 6.3 | 8.5 | 10.9 | 14.1 | 16.2 | 17.7 | 18.6 | 19.7 | 18.6 | 16.5 | 13.4 | 10.7 | 7.1 | 4.0 | 1.9 | 2.0 |
| K₂ | 183.7 | 182.6 | 183.6 | 188.2 | 192.1 | 196.8 | 201.8 | 208.6 | 212.9 | 216.0 | 217.6 | 219.2 | 216.7 | 212.0 | 206.2 | 201.5 | 194.8 | 188.3 | 184.4 | 184.0 |
| L₂ | 271.0 | 100.6 | 307.8 | 139.1 | 296.9 | 121.8 | 334.7 | 170.6 | 331.4 | 151.8 | 359.2 | 189.8 | 8.8 | 184.5 | 17.7 | 204.5 | 38.9 | 219.8 | 35.4 | 219.6 |
| M₁ | 297.7 | 201.0 | 145.3 | 48.0 | 301.8 | 202.5 | 145.5 | 55.4 | 310.0 | 207.3 | 142.2 | 58.2 | 312.8 | 203.4 | 108.2 | 31.1 | 301.4 | 192.6 | 83.7 | 344.5 |
| M₂ | 32.4 | 133.0 | 233.9 | 310.6 | 51.9 | 153.3 | 254.4 | 331.8 | 73.2 | 174.4 | 275.3 | 351.6 | 92.1 | 192.4 | 292.5 | 8.2 | 108.3 | 208.3 | 308.6 | 24.7 |
| M₃ | 228.6 | 199.5 | 170.8 | 105.9 | 77.8 | 50.0 | 22.2 | 317.8 | 289.8 | 261.5 | 232.9 | 167.4 | 138.1 | 108.5 | 78.7 | 12.2 | 342.3 | 312.5 | 282.5 | 217.1 |
| M₄ | 64.8 | 266.0 | 107.8 | 261.2 | 103.8 | 306.6 | 149.0 | 303.7 | 146.4 | 348.7 | 190.6 | 343.2 | 184.2 | 24.7 | 225.4 | 16.3 | 216.4 | 56.7 | 257.2 | 49.4 |
| M₆ | 97.2 | 39.0 | 341.6 | 211.8 | 155.6 | 99.9 | 44.3 | 275.5 | 219.6 | 163.0 | 105.9 | 334.8 | 276.2 | 217.1 | 157.4 | 24.4 | 324.6 | 265.0 | 205.8 | 72.1 |
| M₈ | 129.6 | 172.1 | 215.5 | 162.4 | 207.5 | 253.2 | 299.1 | 247.4 | 292.7 | 337.4 | 21.2 | 326.5 | 8.3 | 49.4 | 89.9 | 32.6 | 72.8 | 113.3 | 154.4 | 98.8 |
| N₂ | 322.9 | 334.8 | 346.9 | 321.8 | 334.4 | 347.1 | 359.8 | 335.1 | 347.8 | 0.2 | 12.4 | 347.0 | 358.7 | 10.2 | 21.6 | 355.5 | 6.9 | 18.3 | 29.8 | 4.1 |
| 2N | 253.4 | 176.5 | 99.9 | 333.1 | 256.9 | 180.9 | 104.9 | 338.4 | 262.3 | 186.0 | 109.5 | 342.3 | 265.3 | 188.1 | 110.8 | 342.9 | 265.5 | 188.2 | 111.0 | 343.6 |
| O₁ | 34.8 | 135.6 | 235.2 | 308.8 | 47.3 | 148.5 | 243.6 | 316.4 | 54.7 | 153.5 | 252.8 | 327.8 | 69.4 | 172.5 | 277.2 | 357.5 | 103.1 | 207.6 | 310.7 | 28.9 |
| OO | 140.3 | 39.3 | 302.0 | 294.8 | 141.1 | 50.2 | 318.8 | 254.6 | 162.2 | 68.3 | 332.0 | 259.8 | 156.0 | 47.1 | 283.4 | 203.9 | 87.2 | 333.7 | 225.1 | 148.8 |
| P₁ | 349.6 | 349.8 | 350.0 | 349.3 | 349.5 | 349.8 | 350.0 | 349.2 | 349.5 | 349.7 | 350.0 | 349.2 | 349.5 | 349.7 | 349.9 | 349.2 | 349.4 | 349.7 | 349.9 | 349.2 |
| Q₁ | 325.3 | 337.3 | 348.3 | 320.0 | 329.8 | 339.3 | 348.7 | 319.6 | 329.3 | 339.3 | 350.0 | 323.2 | 336.0 | 350.4 | 6.4 | 344.8 | 1.7 | 17.6 | 31.9 | 345.7 |
| 2Q | 255.6 | 179.1 | 161.3 | 331.3 | 252.3 | 173.1 | 93.7 | 322.9 | 243.9 | 165.2 | 87.1 | 318.5 | 242.6 | 168.3 | 95.5 | 332.2 | 260.1 | 187.5 | 113.1 | 345.7 |
| R₂ | 179.7 | 179.5 | 180.0 | 180.0 | 179.7 | 179.4 | 179.2 | 179.9 | 179.7 | 179.4 | 179.2 | 179.9 | 179.6 | 179.4 | 179.1 | 179.8 | 179.6 | 179.3 | 179.1 | 179.8 |
| S₁ | 180.0 | 180.0 | 180.0 | 180.0 | 180.0 | 180.0 | 180.0 | 180.0 | 180.0 | 180.0 | 180.0 | 180.0 | 180.0 | 180.0 | 180.0 | 180.0 | 180.0 | 180.0 | 180.0 | 180.0 |
| S₂, ₄, ₆ | 0.0 | 0.0 | 0.0 | 0.0 | 0.0 | 0.0 | 0.0 | 0.0 | 0.0 | 0.0 | 0.0 | 0.0 | 0.0 | 0.0 | 0.0 | 0.0 | 0.0 | 0.0 | 0.0 | 0.0 |
| T₂ | 0.3 | 0.5 | 0.8 | 0.0 | 0.3 | 0.6 | 0.8 | 0.1 | 0.3 | 0.6 | 0.8 | 0.1 | 0.4 | 0.6 | 0.9 | 0.2 | 0.4 | 0.7 | 0.9 | 0.2 |
| λ₂ | 108.5 | 19.6 | 291.0 | 189.6 | 101.4 | 13.3 | 285.3 | 184.2 | 96.1 | 7.8 | 279.5 | 177.2 | 88.4 | 359.2 | 269.8 | 167.4 | 77.9 | 348.6 | 259.4 | 157.3 |
| μ₂ | 66.8 | 268.2 | 109.8 | 262.9 | 104.9 | 307.1 | 149.3 | 302.8 | 144.8 | 346.8 | 188.5 | 341.2 | 182.2 | 350.4 | 224.3 | 15.3 | 217.1 | 58.0 | 259.4 | 51.5 |
| ν₂ | 196.3 | 66.4 | 356.8 | 251.6 | 182.4 | 113.3 | 44.2 | 299.5 | 230.3 | 160.9 | 91.3 | 345.8 | 275.8 | 205.5 | 135.1 | 28.9 | 318.5 | 248.1 | 177.8 | 72.1 |
| ρ₁ | 138.8 | 358.1 | 249.8 | 358.1 | 177.8 | 105.5 | 33.1 | 294.1 | 221.8 | 140.0 | 68.9 | 322.0 | 253.7 | 185.7 | 119.8 | 18.2 | 313.3 | 247.4 | 177.9 | 74.2 |
| MK | 34.1 | 134.4 | 235.9 | 315.0 | 58.2 | 161.8 | 265.6 | 346.0 | 89.3 | 192.1 | 293.9 | 11.3 | 110.7 | 208.9 | 305.4 | 18.9 | 115.3 | 212.3 | 310.6 | 26.7 |
| 2MK | 63.1 | 264.6 | 105.7 | 256.8 | 97.5 | 298.4 | 138.7 | 289.6 | 130.1 | 331.0 | 172.0 | 323.2 | 165.5 | 8.2 | 211.5 | 9.5 | 209.3 | 52.7 | 255.3 | 47.5 |
| MN | 355.3 | 107.8 | 272.4 | 272.4 | 25.3 | 140.4 | 254.4 | 307.0 | 60.9 | 174.1 | 287.7 | 338.6 | 90.8 | 202.6 | 314.1 | 3.7 | 115.1 | 226.6 | 338.4 | 24.8 |
| MS | 32.4 | 133.0 | 233.9 | 310.6 | 51.9 | 153.3 | 254.4 | 331.8 | 73.2 | 174.4 | 275.3 | 351.6 | 92.1 | 192.4 | 292.5 | 8.2 | 108.3 | 208.3 | 308.6 | 24.7 |
| 2SM | 327.6 | 227.0 | 126.1 | 49.4 | 308.1 | 206.7 | 105.2 | 28.2 | 286.8 | 185.6 | 84.7 | 8.4 | 267.9 | 167.6 | 67.5 | 351.8 | 251.7 | 151.7 | 51.4 | 335.3 |
| Mf | 142.7 | 41.9 | 303.4 | 233.0 | 137.1 | 42.3 | 307.6 | 239.1 | 143.7 | 47.4 | 309.6 | 236.0 | 133.3 | 27.3 | 278.1 | 193.2 | 82.1 | 333.0 | 227.2 | 150.9 |
| MSf | 327.6 | 227.0 | 49.4 | 167.7 | 308.1 | 206.7 | 105.2 | 167.7 | 286.8 | 185.6 | 84.7 | 8.4 | 267.9 | 167.6 | 67.5 | 351.8 | 251.8 | 151.7 | 51.4 | 335.3 |
| Mm | 69.5 | 158.2 | 247.0 | 348.8 | 77.5 | 166.2 | 254.9 | 356.7 | 85.4 | 174.1 | 262.9 | 4.7 | 93.4 | 182.1 | 270.8 | 12.6 | 101.3 | 190.1 | 278.8 | 20.6 |
| Sa | 280.4 | 280.2 | 280.0 | 280.7 | 280.5 | 280.2 | 280.0 | 280.8 | 280.5 | 280.3 | 280.0 | 280.8 | 280.5 | 280.3 | 280.1 | 280.8 | 280.6 | 280.3 | 280.1 | 280.8 |
| Ssa | 200.9 | 200.4 | 200.0 | 201.4 | 201.0 | 200.5 | 200.0 | 201.5 | 201.0 | 200.6 | 200.1 | 201.5 | 201.1 | 200.6 | 200.1 | 201.6 | 201.2 | 200.7 | 200.2 | 201.7 |

246037—41——14

Table 15.—Equilibrium argument ($V_0 + u$) for meridian of Greenwich at beginning of each calendar year, 1850 to 2000—Con.

| Constituent | 1890 | 1891 | 1892 | 1893 | 1894 | 1895 | 1896 | 1897 | 1898 | 1899 | 1900 | 1901 | 1902 | 1903 | 1904 | 1905 | 1906 | 1907 | 1908 | 1909 |
|---|---|---|---|---|---|---|---|---|---|---|---|---|---|---|---|---|---|---|---|---|
| $J_1$ | 107.6 | 197.9 | 289.0 | 34.8 | 126.9 | 219.0 | 310.9 | 56.4 | 147.3 | 237.2 | 325.7 | 52.7 | 137.8 | 221.5 | 304.6 | 42.1 | 127.0 | 213.6 | 301.8 | 45.5 |
| $K_1$ | 2.0 | 3.0 | 4.6 | 7.6 | 9.9 | 12.2 | 14.4 | 17.3 | 18.6 | 19.2 | 18.9 | 17.5 | 14.9 | 11.6 | 7.8 | 5.3 | 2.6 | 0.9 | 0.4 | 1.8 |
| $K_2$ | 183.8 | 185.4 | 188.6 | 194.8 | 199.7 | 204.4 | 209.4 | 215.2 | 217.8 | 218.7 | 217.5 | 214.2 | 209.0 | 202.7 | 195.9 | 191.4 | 186.6 | 182.3 | 180.7 | 183.7 |
| $L_2$ | 65.8 | 260.3 | 55.9 | 240.3 | 94.0 | 295.3 | 86.6 | 268.4 | 118.4 | 317.1 | 127.1 | 309.0 | 147.1 | 345.0 | 174.1 | 339.6 | 162.5 | 1.2 | 204.8 | 17.5 |
| $M_1$ | 287.7 | 192.3 | 85.3 | 340.4 | 288.6 | 199.8 | 93.4 | 344.8 | 289.6 | 205.1 | 97.5 | 351.6 | 271.7 | 263.3 | 101.6 | 338.1 | 232.1 | 158.1 | 90.6 | 336.2 |
| $M_2$ | 125.4 | 226.4 | 327.6 | 44.5 | 146.0 | 247.4 | 348.9 | 65.8 | 166.8 | 267.7 | 8.3 | 108.7 | 208.9 | 309.0 | 49.0 | 124.5 | 224.9 | 325.3 | 65.8 | 142.2 |
| $M_3$ | 188.1 | 159.6 | 131.3 | 66.8 | 39.0 | 11.1 | 343.3 | 278.6 | 250.3 | 221.6 | 192.5 | 163.0 | 133.4 | 103.5 | 73.6 | 7.1 | 337.4 | 307.9 | 278.7 | 213.4 |
| $M_4$ | 250.8 | 92.8 | 295.1 | 89.0 | 291.9 | 134.9 | 337.7 | 131.5 | 333.7 | 175.4 | 16.6 | 217.4 | 57.3 | 258.0 | 98.1 | 249.4 | 89.8 | 290.5 | 131.6 | 284.5 |
| $M_6$ | 16.3 | 319.1 | 262.7 | 133.6 | 77.7 | 22.3 | 326.6 | 197.8 | 140.5 | 83.1 | 24.9 | 326.1 | 266.7 | 207.0 | 147.1 | 14.2 | 314.7 | 255.8 | 197.5 | 66.7 |
| $M_8$ | 141.7 | 185.5 | 230.2 | 178.1 | 223.9 | 269.8 | 315.4 | 263.0 | 307.4 | 350.8 | 33.2 | 74.8 | 115.6 | 156.0 | 196.1 | 138.9 | 179.6 | 221.0 | 263.3 | 209.0 |
| $N_2$ | 16.1 | 28.4 | 40.8 | 16.0 | 28.7 | 41.5 | 54.2 | 29.3 | 41.6 | 53.8 | 65.7 | 77.3 | 88.8 | 100.2 | 111.5 | 85.4 | 96.8 | 108.5 | 120.3 | 95.0 |
| $2N$ | 266.8 | 190.3 | 114.1 | 347.5 | 271.5 | 195.5 | 119.5 | 352.8 | 276.4 | 199.8 | 123.0 | 46.0 | 328.7 | 251.4 | 174.0 | 46.1 | 328.8 | 251.7 | 174.8 | 47.7 |
| $O_1$ | 127.2 | 226.5 | 325.2 | 38.2 | 136.4 | 234.5 | 332.7 | 45.8 | 144.8 | 244.5 | 345.2 | 87.4 | 191.2 | 296.3 | 42.0 | 122.0 | 226.0 | 329.5 | 69.5 | 144.0 |
| $OO$ | 49.4 | 313.2 | 219.4 | 154.4 | 62.8 | 331.5 | 239.7 | 174.1 | 79.4 | 342.0 | 240.9 | 135.2 | 24.3 | 178.2 | 152.0 | 63.5 | 311.8 | 205.1 | 103.2 | 32.5 |
| $P_1$ | 349.4 | 349.6 | 349.9 | 349.1 | 349.4 | 349.6 | 349.8 | 349.1 | 349.3 | 349.6 | 349.8 | 350.0 | 350.3 | 350.5 | 350.7 | 350.5 | 350.3 | 350.5 | 350.7 | 350.0 |
| $Q_1$ | 17.9 | 28.5 | 38.5 | 9.7 | 19.1 | 38.0 | 38.0 | 19.3 | 19.6 | 30.5 | 42.6 | 56.0 | 71.1 | 87.5 | 104.5 | 82.6 | 98.0 | 111.7 | 124.0 | 96.7 |
| $2Q$ | 268.6 | 190.5 | 111.8 | 341.2 | 261.9 | 182.5 | 103.3 | 352.8 | 254.4 | 176.6 | 99.9 | 24.7 | 311.0 | 238.6 | 167.0 | 43.3 | 329.9 | 254.9 | 178.5 | 49.4 |
| $R_2$ | 179.6 | 179.3 | 179.0 | 179.8 | 179.5 | 179.3 | 179.0 | 179.7 | 179.5 | 179.2 | 179.0 | 178.7 | 178.4 | 178.2 | 177.9 | 178.7 | 178.4 | 179.0 | 178.6 | 178.6 |
| $S_1$ | 180.0 | 180.0 | 180.0 | 180.0 | 180.0 | 180.0 | 180.0 | 180.0 | 180.0 | 180.0 | 180.0 | 180.0 | 180.0 | 180.0 | 180.0 | 180.0 | 180.0 | 180.0 | 180.0 | 180.0 |
| $S_{2,4,6}$ | 0.0 | 0.0 | 0.0 | 0.0 | 0.0 | 0.0 | 0.0 | 0.0 | 0.0 | 0.0 | 0.0 | 0.0 | 0.0 | 0.0 | 0.0 | 0.0 | 0.0 | 0.0 | 0.0 | 0.0 |
| $T_2$ | 0.4 | 0.7 | 1.0 | 0.2 | 0.5 | 0.7 | 1.0 | 0.3 | 0.5 | 0.8 | 1.0 | 1.3 | 1.6 | 1.8 | 2.1 | 1.3 | 1.6 | 1.8 | 2.1 | 1.4 |
| $\lambda_2$ | 68.6 | 340.1 | 251.8 | 150.6 | 62.5 | 334.5 | 246.5 | 145.2 | 56.8 | 328.2 | 239.6 | 150.3 | 61.0 | 331.6 | 242.2 | 139.7 | 50.4 | 321.5 | 232.4 | 130.6 |
| $\mu_2$ | 253.0 | 94.7 | 246.6 | 89.9 | 292.2 | 134.4 | 336.6 | 129.8 | 331.8 | 173.3 | 14.6 | 215.5 | 56.7 | 257.6 | 98.4 | 250.4 | 91.4 | 292.5 | 133.8 | 286.6 |
| $\nu_2$ | 2.3 | 92.7 | 223.4 | 118.5 | 49.4 | 340.4 | 271.2 | 166.3 | 96.8 | 27.2 | 317.3 | 247.1 | 176.8 | 106.4 | 35.9 | 289.8 | 219.4 | 149.2 | 79.3 | 333.8 |
| $\rho_1$ | 4.0 | 292.8 | 221.0 | 112.2 | 39.8 | 327.4 | 255.2 | 146.3 | 74.8 | 4.0 | 294.2 | 225.8 | 159.1 | 93.7 | 28.9 | 287.0 | 220.5 | 152.5 | 82.9 | 335.6 |
| $MK$ | 127.5 | 229.4 | 332.1 | 52.1 | 155.9 | 259.7 | 3.3 | 83.1 | 185.5 | 287.0 | 27.2 | 126.2 | 223.8 | 320.2 | 56.8 | 130.0 | 227.5 | 326.2 | 66.2 | 144.1 |
| $2MK$ | 248.8 | 89.4 | 290.5 | 81.4 | 282.7 | 122.6 | 323.3 | 114.2 | 315.0 | 156.1 | 357.1 | 199.9 | 42.9 | 246.4 | 90.3 | 244.2 | 87.7 | 289.6 | 131.3 | 282.7 |
| $MN$ | 141.6 | 254.7 | 8.4 | 60.5 | 174.7 | 288.5 | 43.0 | 95.0 | 208.5 | 321.5 | 74.0 | 186.0 | 297.7 | 49.2 | 160.5 | 210.1 | 321.8 | 73.8 | 186.2 | 297.2 |
| $MS$ | 125.4 | 226.4 | 327.6 | 44.5 | 146.0 | 247.4 | 348.9 | 65.8 | 166.8 | 267.7 | 8.3 | 108.7 | 208.9 | 309.0 | 49.0 | 124.7 | 224.9 | 325.3 | 65.8 | 142.2 |
| $2SM$ | 234.6 | 133.6 | 32.4 | 315.5 | 214.0 | 112.6 | 11.1 | 294.2 | 193.2 | 92.3 | 351.7 | 251.3 | 151.1 | 51.0 | 311.0 | 235.3 | 135.1 | 34.7 | 294.2 | 217.8 |
| $Mf$ | 51.1 | 313.3 | 217.1 | 148.1 | 53.2 | 318.5 | 223.5 | 154.2 | 57.3 | 313.8 | 217.8 | 113.9 | 6.6 | 256.4 | 145.0 | 60.8 | 312.9 | 208.3 | 106.9 | 34.3 |
| $MSf$ | 234.6 | 133.6 | 32.4 | 315.5 | 214.0 | 112.6 | 11.1 | 294.2 | 193.2 | 92.3 | 351.7 | 251.3 | 151.1 | 51.0 | 311.0 | 235.3 | 135.1 | 34.7 | 294.2 | 217.8 |
| $Mm$ | 109.3 | 198.0 | 286.7 | 154.4 | 117.2 | 206.0 | 294.7 | 36.5 | 125.2 | 213.9 | 302.6 | 31.4 | 120.1 | 208.8 | 297.5 | 39.3 | 128.0 | 216.8 | 305.5 | 47.3 |
| $Sa$ | 280.6 | 280.4 | 280.1 | 280.9 | 280.6 | 280.4 | 280.2 | 280.9 | 280.7 | 280.4 | 280.2 | 280.0 | 279.7 | 279.5 | 279.2 | 280.0 | 279.7 | 279.5 | 279.3 | 280.0 |
| $Ssa$ | 201.2 | 200.7 | 200.3 | 201.8 | 201.3 | 200.8 | 200.3 | 201.8 | 201.3 | 200.9 | 200.4 | 199.9 | 199.4 | 199.0 | 198.5 | 200.0 | 199.5 | 199.0 | 198.5 | 200.0 |

Table 15.—*Equilibrium argument* ($V_0+u$) *for meridian at Greenwich at beginning of each calendar year, 1850 to 2000*—Con.

| Constituent | 1910 | 1911 | 1912 | 1913 | 1914 | 1915 | 1916 | 1917 | 1918 | 1919 | 1920 | 1921 | 1922 | 1923 | 1924 | 1925 | 1926 | 1927 | 1928 | 1929 |
|---|---|---|---|---|---|---|---|---|---|---|---|---|---|---|---|---|---|---|---|---|
| J₁ | 136.2 | 227.6 | 319.5 | 65.6 | 157.7 | 249.5 | 340.8 | 85.4 | 174.8 | 262.7 | 349.0 | 87.6 | 170.9 | 253.9 | 337.9 | 77.4 | 164.7 | 253.6 | 343.7 | 88.7 |
| K₁ | 3.0 | 4.8 | 7.0 | 10.3 | 12.6 | 14.7 | 16.4 | 18.5 | 18.7 | 18.0 | 16.1 | 14.2 | 10.6 | 6.9 | 3.6 | 2.3 | 1.1 | 0.9 | 1.7 | 4.2 |
| K₂ | 185.5 | 189.1 | 193.0 | 200.6 | 205.6 | 210.0 | 213.4 | 217.4 | 217.5 | 215.5 | 211.4 | 207.7 | 201.6 | 194.3 | 188.1 | 185.3 | 182.4 | 181.6 | 182.9 | 187.1 |
| L₂ | 179.7 | 22.0 | 235.0 | 55.4 | 206.6 | 49.0 | 260.2 | 81.5 | 245.3 | 75.8 | 277.1 | 101.6 | 284.2 | 101.7 | 291.2 | 119.9 | 319.1 | 133.4 | 306.7 | 141.4 |
| M₁ | 229.6 | 146.2 | 92.1 | 343.4 | 237.0 | 149.2 | 96.3 | 350.1 | 242.0 | 142.6 | 78.4 | 344.0 | 236.0 | 124.6 | 25.1 | 309.9 | 226.5 | 120.0 | 15.3 | 292.0 |
| M₂ | 243.3 | 344.5 | 85.9 | 163.0 | 264.5 | 5.9 | 107.1 | 183.7 | 284.4 | 25.0 | 125.3 | 201.0 | 301.1 | 41.2 | 141.3 | 217.1 | 317.6 | 58.2 | 159.1 | 235.9 |
| M₃ | 184.9 | 156.8 | 128.9 | 64.5 | 36.7 | 8.8 | 340.6 | 275.5 | 246.7 | 217.5 | 187.9 | 121.6 | 91.7 | 61.7 | 31.9 | 325.7 | 296.3 | 267.3 | 238.7 | 173.8 |
| M₄ | 126.6 | 329.1 | 171.9 | 326.1 | 169.0 | 11.7 | 214.2 | 7.4 | 208.9 | 49.6 | 250.6 | 42.1 | 242.2 | 123.5 | 282.5 | 74.4 | 275.1 | 116.4 | 318.2 | 111.7 |
| M₆ | 9.9 | 313.6 | 257.8 | 129.1 | 73.5 | 17.6 | 321.2 | 191.1 | 133.8 | 74.9 | 15.8 | 243.2 | 183.3 | 123.5 | 63.8 | 291.4 | 232.6 | 174.6 | 117.3 | 347.6 |
| M₈ | 253.2 | 298.2 | 343.8 | 292.1 | 337.9 | 23.4 | 68.3 | 14.8 | 57.8 | 99.9 | 141.1 | 84.2 | 124.4 | 164.6 | 205.1 | 148.5 | 190.2 | 232.8 | 276.5 | 223.4 |
| N₂ | 107.3 | 119.8 | 132.5 | 107.8 | 120.5 | 133.2 | 145.7 | 183.7 (?) | 132.5 | 144.3 | 155.9 | 129.9 | 141.2 | 152.6 | 164.0 | 138.0 | 149.7 | 161.7 | 173.8 | 148.8 |
| 2N | 331.3 | 255.1 | 179.0 | 52.6 | 336.6 | 260.5 | 184.3 | 57.3 | 340.6 | 263.7 | 186.6 | 58.8 | 341.4 | 264.0 | 186.6 | 58.9 | 341.9 | 265.1 | 188.6 | 61.8 |
| O₁ | 243.1 | 341.6 | 79.9 | 152.6 | 250.8 | 349.0 | 87.7 | 161.5 | 261.6 | 2.9 | 105.7 | 184.6 | 290.1 | 35.8 | 140.7 | 218.8 | 320.6 | 61.2 | 160.7 | 234.2 |
| OO | 297.4 | 204.2 | 112.2 | 48.2 | 316.7 | 224.6 | 131.1 | 62.8 | 324.1 | 221.3 | 113.6 | 28.2 | 271.9 | 154.9 | 40.4 | 317.9 | 213.2 | 112.9 | 16.1 | 309.2 |
| P₁ | 350.2 | 350.5 | 350.7 | 350.0 | 350.8 | 350.4 | 350.7 | 349.9 | 349.9 | 350.1 | 350.3 | 349.9 | 350.1 | 350.2 | 350.6 | 349.9 | 350.1 | 350.3 | 350.6 | 349.8 |
| Q₁ | 107.1 | 116.9 | 126.4 | 97.4 | 106.8 | 116.4 | 126.3 | 98.8 | 109.7 | 122.3 | 136.3 | 113.5 | 130.2 | 147.2 | 163.4 | 139.7 | 152.8 | 164.6 | 175.4 | 147.1 |
| 2Q | 331.1 | 252.2 | 173.0 | 42.2 | 322.8 | 243.7 | 164.9 | 35.2 | 317.8 | 241.6 | 167.0 | 42.4 | 330.4 | 258.6 | 186.1 | 60.6 | 345.0 | 263.1 | 190.2 | 60.1 |
| R₂ | 178.4 | 178.1 | 177.9 | 178.6 | 178.6 | 178.1 | 177.8 | 178.6 | 178.3 | 178.0 | 177.8 | 178.5 | 178.3 | 178.0 | 177.8 | 178.5 | 178.2 | 178.0 | 177.7 | 178.4 |
| S₁ | 180.0 | 180.0 | 180.0 | 180.0 | 180.0 | 180.0 | 180.0 | 180.0 | 180.0 | 180.0 | 180.0 | 180.0 | 180.0 | 180.0 | 180.0 | 180.0 | 180.0 | 180.0 | 180.0 | 180.0 |
| S₂,₄,₆ | 0.0 | 0.0 | 0.0 | 0.0 | 0.0 | 0.0 | 0.0 | 0.0 | 0.0 | 0.0 | 0.0 | 0.0 | 0.0 | 0.0 | 0.0 | 0.0 | 0.0 | 0.0 | 0.0 | 0.0 |
| T₂ | 1.6 | 1.9 | 2.1 | 1.4 | 1.7 | 1.9 | 2.2 | 1.4 | 1.7 | 2.0 | 2.2 | 1.5 | 1.7 | 2.0 | 2.2 | 1.5 | 1.8 | 2.0 | 2.3 | 1.6 |
| λ | 42.2 | 314.0 | 225.9 | 124.8 | 36.8 | 308.7 | 220.5 | 118.9 | 30.2 | 301.2 | 212.1 | 109.7 | 20.3 | 290.8 | 201.5 | 99.2 | 10.1 | 281.3 | 192.8 | 91.3 |
| μ | 128.4 | 330.4 | 172.5 | 326.1 | 168.2 | 10.3 | 212.5 | 5.3 | 206.8 | 48.1 | 249.5 | 41.3 | 242.1 | 82.9 | 283.7 | 76.0 | 277.1 | 118.6 | 320.2 | 113.3 |
| ν₂ | 264.4 | 195.1 | 126.0 | 21.2 | 312.2 | 243.0 | 173.7 | 68.5 | 358.7 | 288.7 | 218.5 | 112.4 | 41.9 | 331.5 | 261.0 | 155.1 | 85.0 | 15.1 | 305.5 | 200.4 |
| ρ₁ | 264.2 | 119.9 | 119.0 | 10.8 | 298.4 | 226.2 | 154.2 | 46.3 | 335.8 | 266.6 | 198.8 | 96.0 | 30.9 | 326.1 | 260.5 | 156.7 | 88.0 | 18.1 | 305.1 | 198.7 |
| MK | 246.3 | 349.4 | 92.9 | 173.3 | 277.1 | 20.6 | 123.5 | 202.2 | 303.2 | 43.0 | 141.4 | 215.2 | 311.7 | 48.1 | 144.9 | 219.4 | 318.6 | 59.1 | 160.6 | 240.0 |
| 2MK | 123.6 | 324.3 | 164.9 | 315.8 | 156.4 | 357.0 | 197.7 | 348.9 | 190.2 | 32.0 | 234.5 | 27.7 | 231.6 | 75.4 | 278.2 | 72.0 | 274.0 | 115.5 | 316.5 | 107.6 |
| MN | 350.6 | 104.4 | 218.4 | 270.8 | 25.0 | 139.0 | 252.8 | 304.7 | 57.0 | 169.3 | 281.2 | 331.0 | 82.4 | 193.7 | 305.2 | 355.2 | 107.3 | 219.9 | 333.0 | 24.7 |
| MS | 243.3 | 344.5 | 85.9 | 163.0 | 264.5 | 5.9 | 107.1 | 183.7 | 284.4 | 25.0 | 125.3 | 201.0 | 301.1 | 41.2 | 141.3 | 217.1 | 317.6 | 58.2 | 159.1 | 235.9 |
| 2SM | 116.7 | 15.5 | 274.1 | 197.0 | 95.5 | 354.1 | 252.9 | 176.3 | 75.6 | 335.0 | 234.7 | 159.0 | 58.9 | 318.8 | 218.7 | 142.9 | 42.4 | 301.8 | 200.9 | 124.1 |
| Mf | 297.1 | 201.3 | 106.2 | 37.8 | 303.0 | 207.1 | 111.7 | 40.6 | 301.2 | 199.2 | 94.0 | 11.8 | 260.9 | 149.5 | 39.9 | 319.6 | 216.3 | 115.9 | 17.7 | 307.5 |
| MSf | 116.7 | 274.1 | 274.1 | 197.0 | 95.5 | 354.1 | 252.9 | 176.3 | 75.6 | 335.0 | 234.7 | 159.0 | 58.9 | 318.8 | 218.7 | 142.9 | 42.4 | 301.8 | 200.9 | 124.1 |
| Mm | 136.0 | 224.7 | 313.4 | 55.2 | 144.0 | 232.7 | 321.4 | 63.2 | 151.9 | 240.6 | 329.4 | 71.1 | 159.9 | 248.6 | 337.3 | 79.1 | 167.8 | 256.5 | 345.3 | 87.0 |
| Sa | 279.8 | 279.5 | 279.3 | 280.0 | 279.8 | 279.6 | 279.3 | 280.1 | 279.8 | 279.4 | 279.4 | 280.1 | 279.9 | 279.6 | 279.4 | 280.1 | 279.9 | 279.7 | 279.4 | 280.2 |
| Ssa | 199.6 | 199.1 | 198.6 | 200.1 | 199.6 | 199.1 | 198.7 | 200.2 | 199.7 | 199.3 | 198.7 | 200.2 | 199.7 | 199.3 | 198.8 | 200.3 | 199.8 | 199.3 | 198.8 | 200.3 |

Table 15.—*Equilibrium argument* ($V_o + u$) *for meridian of Greenwich at beginning of each calendar year, 1850 to 2000*—Con.

| Constituent | 1930 | 1931 | 1932 | 1933 | 1934 | 1935 | 1936 | 1937 | 1938 | 1939 | 1940 | 1941 | 1942 | 1943 | 1944 | 1945 | 1946 | 1947 | 1948 | 1949 |
|---|---|---|---|---|---|---|---|---|---|---|---|---|---|---|---|---|---|---|---|---|
| $J_1$ | 180.4 | 272.3 | 4.4 | 110.5 | 202.1 | 293.1 | 23.2 | 126.2 | 213.5 | 299.1 | 23.1 | 120.2 | 203.5 | 288.0 | 14.2 | 116.2 | 205.5 | 296.0 | 27.3 | 133.1 |
| $K_1$ | 6.1 | 8.4 | 10.7 | 14.0 | 15.9 | 17.4 | 18.2 | 19.1 | 17.9 | 15.6 | 12.4 | 9.6 | 6.7 | 3.1 | 1.2 | 1.4 | 1.6 | 2.7 | 4.4 | 7.4 |
| $K_2$ | 191.8 | 196.6 | 201.6 | 208.3 | 212.4 | 215.3 | 216.6 | 217.9 | 215.1 | 210.3 | 204.2 | 199.4 | 192.7 | 187.0 | 182.8 | 182.8 | 182.8 | 184.8 | 188.2 | 194.6 |
| $L_2$ | 352.2 | 170.1 | 330.8 | 167.9 | 18.8 | 201.7 | 6.0 | 193.1 | 35.9 | 227.4 | 44.3 | 215.2 | 49.6 | 249.0 | 82.0 | 240.2 | 64.4 | 271.8 | 118.7 | 270.7 |
| $M_1$ | 228.3 | 128.5 | 21.6 | 293.3 | 234.4 | 134.1 | 27.0 | 285.7 | 227.4 | 132.6 | 22.0 | 260.9 | 173.4 | 109.5 | 12.9 | 252.0 | 151.2 | 91.3 | 14.5 | 257.4 |
| $M_2$ | 337.2 | 78.6 | 180.1 | 257.1 | 358.7 | 99.6 | 200.5 | 276.8 | 17.2 | 117.5 | 217.6 | 293.2 | 33.3 | 133.4 | 233.8 | 309.9 | 50.6 | 151.6 | 252.8 | 329.8 |
| $M_3$ | 145.8 | 117.9 | 90.1 | 25.7 | 357.7 | 329.4 | 300.7 | 235.2 | 205.8 | 176.2 | 146.4 | 79.9 | 49.9 | 20.2 | 350.6 | 284.8 | 256.0 | 227.4 | 199.2 | 134.7 |
| $M_4$ | 314.3 | 157.2 | 0.2 | 154.3 | 356.6 | 199.2 | 41.0 | 193.6 | 34.4 | 234.9 | 75.1 | 226.5 | 66.6 | 266.9 | 107.5 | 259.8 | 101.3 | 303.2 | 145.7 | 299.6 |
| $M_6$ | 291.5 | 235.8 | 180.2 | 51.4 | 355.4 | 208.8 | 241.5 | 94.8 | 51.6 | 352.4 | 292.7 | 159.7 | 99.9 | 133.4 | 341.2 | 209.6 | 151.9 | 38.5 | 145.7 | 269.5 |
| $M_8$ | 268.7 | 314.4 | 0.3 | 308.6 | 353.8 | 38.4 | 82.0 | 27.1 | 68.9 | 109.8 | 150.3 | 92.9 | 133.2 | 173.8 | 215.0 | 159.5 | 202.5 | 246.5 | 291.3 | 239.3 |
| $N_2$ | 161.4 | 174.1 | 186.9 | 162.1 | 174.7 | 187.1 | 199.3 | 173.8 | 185.5 | 197.1 | 208.4 | 182.3 | 193.6 | 205.1 | 216.7 | 191.0 | 203.1 | 215.3 | 227.8 | 203.0 |
| $2N$ | 345.6 | 269.6 | 193.6 | 67.1 | 351.0 | 274.7 | 198.2 | 70.9 | 353.8 | 276.8 | 199.8 | 71.4 | 354.2 | 276.8 | 199.6 | 72.1 | 355.4 | 279.0 | 202.8 | 76.2 |
| $O_1$ | 332.6 | 70.8 | 168.9 | 241.7 | 361.0 | 78.7 | 178.4 | 253.6 | 355.5 | 98.8 | 203.6 | 284.0 | 29.5 | 133.6 | 236.6 | 312.6 | 52.7 | 151.9 | 250.6 | 323.5 |
| $OO$ | 216.6 | 124.9 | 33.6 | 329.2 | 236.7 | 142.4 | 45.7 | 332.9 | 228.3 | 118.6 | 4.2 | 274.6 | 158.2 | 45.4 | 297.5 | 222.0 | 123.2 | 27.4 | 293.9 | 229.1 |
| $P_1$ | 350.1 | 350.3 | 350.6 | 349.8 | 350.0 | 350.3 | 350.5 | 349.8 | 350.0 | 350.2 | 350.4 | 349.7 | 350.0 | 350.2 | 350.5 | 349.7 | 350.0 | 350.2 | 350.4 | 349.7 |
| $Q_1$ | 156.8 | 166.3 | 175.6 | 146.7 | 156.0 | 166.2 | 176.1 | 47.6 | 163.0 | 178.4 | 194.5 | 173.1 | 189.2 | 205.5 | 219.6 | 193.7 | 357.5 | 279.3 | 200.5 | 196.7 |
| $2Q$ | 341.0 | 261.8 | 182.4 | 51.7 | 332.6 | 254.0 | 176.1 | 47.6 | 332.0 | 258.0 | 185.4 | 62.1 | 350.2 | 277.1 | 202.5 | 74.9 | 357.5 | 279.3 | 200.5 | 69.9 |
| $R_2$ | 178.2 | 177.9 | 177.7 | 178.4 | 178.2 | 177.9 | 177.6 | 178.4 | 178.1 | 177.9 | 177.6 | 178.3 | 178.1 | 177.8 | 177.6 | 178.3 | 178.0 | 177.8 | 177.5 | 178.3 |
| $S_1$ | 180.0 | 180.0 | 180.0 | 180.0 | 180.0 | 180.0 | 180.0 | 180.0 | 180.0 | 180.0 | 180.0 | 180.0 | 180.0 | 180.0 | 180.0 | 180.0 | 180.0 | 180.0 | 180.0 | 180.0 |
| $S_{2, 4, 6}$ | 0.0 | 0.0 | 0.0 | 0.0 | 0.0 | 0.0 | 0.0 | 0.0 | 0.0 | 0.0 | 0.0 | 0.0 | 0.0 | 0.0 | 0.0 | 0.0 | 0.0 | 0.0 | 0.0 | 0.0 |
| $T_2$ | 1.8 | 2.1 | 2.3 | 1.6 | 1.8 | 2.1 | 2.4 | 1.6 | 1.9 | 2.1 | 2.4 | 1.7 | 1.9 | 2.2 | 2.4 | 1.7 | 2.0 | 2.2 | 2.5 | 1.7 |
| $\lambda_2$ | 3.2 | 275.6 | 187.1 | 86.0 | 357.9 | 269.5 | 181.0 | 79.1 | 350.0 | 260.8 | 171.4 | 69.0 | 339.5 | 250.2 | 161.1 | 59.0 | 330.3 | 241.8 | 153.6 | 52.4 |
| $\mu_2$ | 315.4 | 157.6 | 359.8 | 153.2 | 355.3 | 197.2 | 38.9 | 191.5 | 32.7 | 233.7 | 74.6 | 226.6 | 67.0 | 268.3 | 109.4 | 261.9 | 103.4 | 305.1 | 147.1 | 300.4 |
| $\nu_2$ | 131.2 | 62.1 | 353.0 | 248.2 | 179.0 | 109.6 | 40.0 | 294.5 | 224.4 | 154.1 | 83.7 | 337.5 | 267.0 | 196.7 | 126.4 | 20.7 | 311.0 | 241.4 | 172.1 | 67.3 |
| $\rho_1$ | 126.6 | 54.3 | 341.8 | 232.8 | 160.7 | 90.7 | 18.0 | 271.3 | 202.6 | 135.4 | 69.8 | 328.3 | 263.2 | 197.1 | 129.4 | 23.4 | 313.0 | 241.8 | 169.9 | 61.0 |
| $MK$ | 343.3 | 87.0 | 190.8 | 271.1 | 14.4 | 117.0 | 218.7 | 295.8 | 35.1 | 133.4 | 229.9 | 302.8 | 39.3 | 136.5 | 234.3 | 311.3 | 52.2 | 154.3 | 257.2 | 337.3 |
| $2MK$ | 308.6 | 148.8 | 349.5 | 140.3 | 341.0 | 181.8 | 22.8 | 174.5 | 16.6 | 219.4 | 62.8 | 216.8 | 60.6 | 263.8 | 106.3 | 258.4 | 99.6 | 300.6 | 141.3 | 292.8 |
| $MN$ | 138.6 | 252.7 | 7.0 | 59.3 | 173.2 | 286.7 | 39.2 | 90.6 | 202.8 | 314.5 | 66.0 | 115.6 | 226.9 | 338.5 | 90.4 | 140.9 | 253.7 | 6.9 | 120.6 | 172.8 |
| $MS$ | 337.2 | 78.6 | 180.1 | 257.1 | 358.7 | 99.6 | 200.5 | 276.8 | 17.2 | 117.5 | 217.6 | 293.2 | 33.3 | 133.4 | 233.8 | 309.9 | 50.6 | 151.6 | 252.8 | 329.8 |
| $2SM$ | 22.8 | 281.4 | 179.9 | 102.9 | 1.6 | 260.4 | 159.5 | 83.2 | 342.8 | 242.5 | 142.4 | 66.8 | 326.7 | 226.6 | 126.2 | 50.1 | 309.4 | 208.4 | 107.2 | 30.2 |
| $Mf$ | 212.8 | 117.1 | 22.3 | 313.8 | 218.3 | 121.8 | 23.6 | 309.7 | 206.4 | 99.9 | 350.3 | 265.3 | 154.4 | 45.8 | 300.4 | 224.7 | 125.2 | 27.7 | 291.6 | 222.8 |
| $MSf$ | 22.8 | 281.4 | 179.9 | 102.9 | 83.2 | 260.4 | 159.5 | 83.2 | 342.8 | 242.5 | 142.4 | 66.8 | 326.7 | 226.6 | 126.2 | 50.1 | 309.4 | 208.4 | 107.2 | 30.2 |
| $Mm$ | 175.8 | 264.5 | 353.2 | 95.0 | 183.7 | 272.4 | 1.2 | 103.0 | 191.7 | 280.4 | 9.1 | 110.9 | 199.6 | 288.4 | 17.1 | 118.9 | 207.6 | 296.3 | 25.0 | 126.8 |
| $Sa$ | 279.9 | 279.7 | 279.4 | 280.2 | 280.0 | 279.7 | 279.5 | 280.2 | 280.0 | 279.8 | 279.5 | 280.3 | 280.0 | 279.8 | 279.5 | 280.3 | 280.0 | 279.8 | 279.6 | 280.3 |
| $Ssa$ | 199.9 | 199.4 | 198.9 | 200.4 | 199.9 | 199.4 | 199.0 | 200.5 | 200.0 | 199.5 | 199.0 | 200.5 | 200.0 | 199.5 | 199.1 | 200.6 | 200.1 | 199.6 | 199.2 | 200.6 |

208

Table 15.—Equilibrium argument ($V_0 + u$) for meridian of Greenwich at beginning of each calendar year, 1850 to 2000—Con.

| Constituent | 1950 | 1951 | 1952 | 1953 | 1954 | 1955 | 1956 | 1957 | 1958 | 1959 | 1960 | 1961 | 1962 | 1963 | 1964 | 1965 | 1966 | 1967 | 1968 | 1969 |
|---|---|---|---|---|---|---|---|---|---|---|---|---|---|---|---|---|---|---|---|---|
| J₁ | 225.2 | 317.2 | 49.1 | 154.6 | 4.3 | 335.0 | 63.3 | 164.0 | 249.0 | 332.5 | 55.5 | 153.2 | 238.3 | 325.2 | 53.7 | 157.6 | 248.4 | 340.0 | 71.9 | 178.0 |
| K₁ | 9.8 | 12.1 | 14.2 | 17.0 | 18.3 | 18.7 | 18.2 | 17.6 | 14.9 | 11.4 | 7.7 | 5.3 | 2.7 | 1.2 | 0.9 | 2.4 | 3.8 | 5.6 | 7.8 | 11.2 |
| K₂ | 199.5 | 204.4 | 209.0 | 214.7 | 217.0 | 217.6 | 216.0 | 214.4 | 209.0 | 202.4 | 195.8 | 191.4 | 186.2 | 182.8 | 181.6 | 184.4 | 186.9 | 190.8 | 195.4 | 202.4 |
| L₂ | 86.6 | 298.5 | 147.4 | 306.1 | 118.5 | 322.8 | 165.4 | 338.9 | 153.8 | 342.4 | 179.5 | 5.4 | 191.0 | 2.4 | 194.4 | 30.4 | 230.4 | 26.6 | 215.8 | 57.9 |
| M₁ | 155.3 | 91.6 | 22.2 | 265.5 | 159.0 | 85.3 | 21.9 | 266.9 | 155.7 | 52.4 | 341.3 | 252.2 | 146.5 | 36.5 | 302.6 | 236.6 | 147.6 | 40.3 | 302.7 | 237.8 |
| M₂ | 71.3 | 172.8 | 274.2 | 351.0 | 92.1 | 192.9 | 293.5 | 9.4 | 109.6 | 209.7 | 309.6 | 25.5 | 125.7 | 226.0 | 326.7 | 43.1 | 144.2 | 245.5 | 346.9 | 64.0 |
| M₃ | 106.9 | 79.2 | 51.3 | 346.6 | 318.1 | 289.4 | 260.3 | 194.2 | 164.5 | 134.6 | 104.6 | 38.2 | 8.5 | 339.1 | 310.0 | 244.7 | 216.3 | 188.2 | 160.3 | 96.0 |
| M₄ | 142.6 | 345.5 | 188.3 | 342.1 | 184.2 | 25.8 | 227.0 | 18.9 | 219.3 | 59.4 | 259.5 | 50.9 | 251.3 | 92.1 | 293.3 | 86.2 | 288.4 | 130.9 | 333.8 | 128.0 |
| M₆ | 213.8 | 158.3 | 102.5 | 333.1 | 276.3 | 218.7 | 160.4 | 28.4 | 328.9 | 269.2 | 209.3 | 76.4 | 17.0 | 318.1 | 293.3 | 129.4 | 72.6 | 16.4 | 320.7 | 191.9 |
| M₈ | 285.1 | 331.1 | 16.7 | 324.2 | 8.4 | 51.6 | 93.9 | 37.8 | 78.6 | 118.3 | 159.0 | 101.8 | 142.6 | 184.2 | 226.6 | 172.5 | 216.8 | 261.9 | 307.6 | 255.9 |
| N₂ | 215.7 | 228.5 | 241.2 | 216.3 | 228.6 | 240.7 | 252.5 | 226.7 | 238.2 | 249.5 | 260.9 | 234.8 | 246.2 | 257.9 | 269.8 | 244.5 | 256.8 | 269.4 | 282.1 | 257.4 |
| 2N | 0.2 | 284.2 | 208.2 | 81.5 | 5.1 | 288.5 | 211.6 | 84.0 | 6.7 | 289.4 | 212.0 | 84.1 | 6.8 | 289.8 | 213.0 | 85.8 | 9.5 | 293.2 | 217.3 | 90.7 |
| O₁ | 61.6 | 159.8 | 258.0 | 331.2 | 70.3 | 170.1 | 271.1 | 348.2 | 92.2 | 197.1 | 303.2 | 23.0 | 126.8 | 229.0 | 329.8 | 44.1 | 143.1 | 241.6 | 339.8 | 52.6 |
| OO | 137.6 | 46.2 | 314.2 | 248.5 | 153.4 | 155.4 | 313.6 | 234.4 | 122.8 | 7.1 | 250.0 | 162.0 | 51.1 | 305.2 | 204.0 | 133.8 | 39.0 | 306.1 | 214.3 | 150.3 |
| P₁ | 349.9 | 350.2 | 350.4 | 349.6 | 349.6 | 350.1 | 350.3 | 349.6 | 349.9 | 350.1 | 350.3 | 349.6 | 349.8 | 350.1 | 350.3 | 349.6 | 349.8 | 350.1 | 350.3 | 349.5 |
| Q₁ | 206.1 | 215.5 | 225.0 | 196.4 | 206.8 | 217.9 | 230.2 | 205.4 | 260.1 | 237.2 | 254.4 | 232.3 | 247.4 | 260.7 | 272.9 | 245.5 | 255.8 | 265.5 | 275.0 | 246.6 |
| 2Q | 350.6 | 271.3 | 192.0 | 61.6 | 343.2 | 265.7 | 189.2 | 62.7 | 349.2 | 277.7 | 205.4 | 81.6 | 7.9 | 292.7 | 216.1 | 86.6 | 8.4 | 177.6 | 289.4 | 178.1 |
| R₂ | 178.0 | 177.7 | 177.5 | 178.2 | 178.0 | 177.7 | 177.4 | 178.2 | 177.9 | 177.7 | 177.4 | 178.1 | 177.9 | 177.6 | 177.4 | 178.1 | 177.8 | 177.6 | 177.3 | 178.1 |
| S₁ | 180.0 | 180.0 | 180.0 | 180.0 | 180.0 | 180.0 | 180.0 | 180.0 | 180.0 | 180.0 | 180.0 | 180.0 | 180.0 | 180.0 | 180.0 | 180.0 | 180.0 | 180.0 | 180.0 | 180.0 |
| S₂,₄,₆ | 0.0 | 0.0 | 0.0 | 0.0 | 0.0 | 0.0 | 0.0 | 0.0 | 0.0 | 0.0 | 0.0 | 0.0 | 0.0 | 0.4 | 0.0 | 0.0 | 0.0 | 0.0 | 0.0 | 0.0 |
| T₂ | 2.0 | 2.3 | 2.5 | 1.8 | 2.0 | 2.3 | 2.6 | 1.8 | 2.1 | 2.3 | 2.6 | 1.9 | 2.1 | 2.4 | 2.6 | 1.9 | 2.2 | 2.4 | 2.7 | 1.9 |
| λ | 324.4 | 236.4 | 148.3 | 47.0 | 318.6 | 229.9 | 141.0 | 38.8 | 309.5 | 320.1 | 130.7 | 28.2 | 299.0 | 209.9 | 121.0 | 19.3 | 290.9 | 202.7 | 114.7 | 13.6 |
| μ | 142.6 | 344.9 | 174.1 | 340.3 | 182.1 | 23.7 | 225.0 | 17.4 | 218.3 | 59.1 | 259.9 | 52.7 | 253.6 | 94.1 | 295.5 | 88.3 | 290.1 | 132.1 | 334.3 | 127.8 |
| ν | 358.2 | 289.2 | 220.1 | 115.1 | 46.6 | 335.4 | 266.0 | 160.1 | 89.8 | 19.3 | 308.8 | 202.7 | 132.4 | 62.2 | 352.3 | 246.9 | 177.0 | 108.2 | 39.1 | 294.4 |
| ρ₁ | 348.6 | 276.2 | 203.9 | 95.3 | 23.8 | 313.1 | 243.6 | 138.8 | 72.3 | 7.0 | 302.2 | 200.2 | 133.4 | 65.1 | 355.5 | 247.9 | 176.4 | 104.4 | 32.0 | 282.9 |
| MK | 81.0 | 184.9 | 288.4 | 8.1 | 110.4 | 211.6 | 311.7 | 27.0 | 124.6 | 221.2 | 317.4 | 30.7 | 128.4 | 227.3 | 327.5 | 45.6 | 148.0 | 251.1 | 354.7 | 75.1 |
| 2MK | 132.8 | 333.5 | 174.1 | 325.0 | 165.9 | 7.1 | 208.7 | 1.3 | 204.4 | 48.0 | 251.8 | 45.7 | 248.6 | 90.8 | 292.4 | 83.8 | 284.6 | 125.3 | 326.6 | 116.2 |
| MN | 287.0 | 41.3 | 155.4 | 207.3 | 320.7 | 73.6 | 186.0 | 236.2 | 347.8 | 99.2 | 210.6 | 260.7 | 11.9 | 24.0 | 236.4 | 287.6 | 41.0 | 154.8 | 269.0 | 321.4 |
| MS | 71.3 | 172.8 | 274.2 | 351.0 | 92.1 | 192.9 | 293.5 | 9.4 | 109.6 | 209.7 | 309.6 | 25.5 | 125.7 | 226.0 | 326.7 | 43.1 | 144.2 | 245.5 | 346.9 | 64.0 |
| 2SM | 288.7 | 187.2 | 85.8 | 9.0 | 267.9 | 167.1 | 66.5 | 350.6 | 250.4 | 150.3 | 50.2 | 334.5 | 234.3 | 134.0 | 33.3 | 316.9 | 215.8 | 114.5 | 13.1 | 296.0 |
| Mf | 128.0 | 33.2 | 298.1 | 228.6 | 131.6 | 32.7 | 291.3 | 213.1 | 105.3 | 354.8 | 243.4 | 159.5 | 52.1 | 308.1 | 207.1 | 134.8 | 38.0 | 302.3 | 207.2 | 138.9 |
| MSf | 288.7 | 187.2 | 85.8 | 9.0 | 2.3 | 167.1 | 66.5 | 350.6 | 250.4 | 150.3 | 50.2 | 334.5 | 234.3 | 134.0 | 33.3 | 316.9 | 215.8 | 114.5 | 13.1 | 296.0 |
| Mm | 215.6 | 304.3 | 33.0 | 134.8 | 2.5 | 312.2 | 41.0 | 142.7 | 231.5 | 320.2 | 48.9 | 150.7 | 239.4 | 328.1 | 56.9 | 158.6 | 247.4 | 336.1 | 64.8 | 166.6 |
| Sa | 280.1 | 279.8 | 279.6 | 280.4 | 279.9 | 279.7 | 279.6 | 280.4 | 280.1 | 280.0 | 279.7 | 280.4 | 280.1 | 279.9 | 279.7 | 280.4 | 280.2 | 280.4 | 279.7 | 280.5 |
| Ssa | 200.2 | 199.7 | 199.2 | 200.7 | 1.2 | 199.8 | 199.3 | 200.8 | 200.3 | 199.8 | 199.3 | 200.8 | 200.4 | 199.9 | 199.4 | 200.9 | 200.4 | 199.9 | 199.4 | 201.0 |

HARMONIC ANALYSIS AND PREDICTION OF TIDES

TABLE 15.—*Equilibrium argument* $(V_0 + u)$ *for meridian of Greenwich at beginning of each calendar year, 1850 to 2000*—Con.

| Constituent | 1970 | 1971 | 1972 | 1973 | 1974 | 1975 | 1976 | 1977 | 1978 | 1979 | 1980 | 1981 | 1982 | 1983 | 1984 | 1985 | 1986 | 1987 | 1988 | 1989 |
|---|---|---|---|---|---|---|---|---|---|---|---|---|---|---|---|---|---|---|---|---|
| J$_1$ | 270.0 | 1.8 | 93.1 | 197.4 | 286.6 | 14.3 | 100.2 | 198.6 | 281.8 | 4.9 | 89.1 | 188.9 | 276.5 | 5.6 | 95.8 | 201.0 | 292.7 | 24.7 | 116.8 | 222.8 |
| K$_1$ | 13.4 | 15.5 | 17.1 | 19.0 | 19.2 | 18.2 | 16.2 | 14.1 | 10.5 | 6.8 | 3.7 | 2.5 | 1.5 | 1.5 | 2.4 | 4.9 | 6.9 | 9.2 | 11.6 | 14.8 |
| K$_2$ | 207.2 | 211.5 | 214.8 | 218.4 | 218.2 | 215.8 | 211.5 | 207.6 | 200.9 | 194.2 | 188.3 | 185.6 | 183.1 | 182.6 | 184.2 | 189.3 | 193.5 | 198.3 | 203.3 | 210.0 |
| L$_2$ | 262.8 | 60.3 | 245.2 | 81.8 | 282.4 | 99.0 | 276.4 | 99.7 | 297.9 | 130.9 | 310.2 | 116.6 | 313.4 | 159.2 | 350.3 | 135.8 | 334.5 | 198.0 | 203.5 | 166.6 |
| M$_1$ | 155.2 | 48.6 | 307.8 | 234.8 | 158.7 | 51.8 | 302.5 | 198.6 | 135.7 | 41.8 | 291.5 | 170.8 | 87.9 | 30.0 | 291.6 | 171.7 | 82.0 | 31.2 | 299.1 | 179.8 |
| M$_2$ | 165.4 | 266.8 | 8.0 | 84.5 | 185.3 | 285.8 | 26.0 | 101.8 | 201.8 | 301.9 | 42.0 | 117.9 | 218.4 | 319.1 | 60.0 | 136.8 | 238.1 | 339.6 | 81.0 | 158.1 |
| M$_3$ | 68.1 | 40.2 | 11.9 | 306.8 | 277.9 | 248.6 | 219.0 | 152.7 | 122.7 | 92.8 | 63.0 | 356.8 | 327.6 | 298.6 | 270.0 | 205.2 | 177.2 | 149.3 | 121.6 | 57.1 |
| M$_4$ | 330.9 | 173.6 | 15.9 | 169.1 | 10.5 | 211.5 | 52.1 | 203.6 | 43.7 | 243.8 | 84.0 | 235.8 | 76.7 | 278.5 | 120.0 | 273.6 | 116.2 | 319.1 | 162.1 | 316.2 |
| M$_6$ | 136.3 | 80.4 | 23.9 | 263.6 | 195.8 | 137.2 | 78.1 | 305.3 | 245.5 | 185.7 | 126.0 | 353.7 | 295.1 | 237.2 | 180.0 | 50.3 | 354.3 | 298.7 | 243.1 | 114.2 |
| M$_8$ | 301.7 | 347.1 | 31.8 | 338.2 | 21.0 | 63.0 | 104.1 | 47.1 | 87.3 | 127.5 | 168.0 | 111.6 | 153.4 | 196.2 | 240.0 | 187.1 | 232.4 | 278.2 | 324.1 | 272.3 |
| N$_2$ | 270.1 | 282.7 | 295.2 | 270.0 | 282.0 | 293.8 | 305.3 | 297.3 | 290.4 | 301.9 | 313.3 | 287.4 | 299.2 | 311.2 | 323.2 | 298.4 | 310.0 | 323.7 | 336.4 | 311.7 |
| 2N | 14.8 | 298.7 | 222.4 | 95.4 | 18.7 | 301.8 | 224.6 | 96.8 | 19.4 | 302.0 | 224.7 | 97.0 | 20.0 | 303.2 | 226.7 | 100.0 | 23.8 | 307.8 | 231.9 | 105.3 |
| O$_1$ | 150.7 | 249.6 | 347.8 | 61.7 | 162.0 | 263.5 | 6.5 | 85.7 | 191.3 | 296.9 | 41.6 | 119.4 | 221.1 | 303.8 | 60.8 | 134.2 | 232.6 | 330.7 | 68.8 | 141.6 |
| OO | 58.7 | 326.4 | 232.6 | 163.9 | 64.3 | 321.1 | 212.6 | 126.5 | 9.8 | 253.0 | 139.2 | 71.6 | 347.4 | 213.4 | 117.6 | 51.0 | 318.5 | 227.0 | 135.6 | 71.2 |
| P$_1$ | 349.8 | 350.0 | 350.2 | 349.2 | 349.7 | 350.1 | 350.2 | 349.5 | 349.6 | 349.9 | 350.0 | 349.4 | 349.7 | 349.9 | 350.2 | 349.4 | 349.6 | 349.9 | 350.1 | 349.4 |
| Q$_1$ | 255.4 | 265.0 | 275.0 | 247.2 | 258.7 | 271.5 | 285.2 | 263.2 | 280.1 | 297.0 | 312.9 | 289.0 | 301.9 | 313.5 | 324.2 | 295.8 | 306.4 | 314.8 | 324.2 | 295.5 |
| 2Q | 0.1 | 281.0 | 202.2 | 72.6 | 355.4 | 279.5 | 205.1 | 80.7 | 8.8 | 297.0 | 224.3 | 98.5 | 22.7 | 305.6 | 227.5 | 97.4 | 305.4 | 219.6 | 89.9 | 88.9 |
| R$_2$ | 177.8 | 177.6 | 177.7 | 178.0 | 177.3 | 177.5 | 177.3 | 178.0 | 177.7 | 177.5 | 177.7 | 178.0 | 177.7 | 177.4 | 177.7 | 177.9 | 177.7 | 177.4 | 177.2 | 177.7 |
| S$_1$ | 180.0 | 180.0 | 180.0 | 180.0 | 180.0 | 180.0 | 180.0 | 180.0 | 180.0 | 180.0 | 180.0 | 180.0 | 180.0 | 180.0 | 180.0 | 180.0 | 180.0 | 180.0 | 180.0 | 180.0 |
| S$_2$, 4, 6 | 0.0 | 0.0 | 0.0 | 0.0 | 0.0 | 0.0 | 0.0 | 0.0 | 0.0 | 0.0 | 0.0 | 0.0 | 0.0 | 0.0 | 0.0 | 0.0 | 0.0 | 0.0 | 0.0 | 0.0 |
| T$_2$ | 2.2 | 2.4 | 2.7 | 2.0 | 2.2 | 2.5 | 2.7 | 2.0 | 2.3 | 2.5 | 2.8 | 2.0 | 2.3 | 2.6 | 2.8 | 2.1 | 2.3 | 2.6 | 2.8 | 2.1 |
| λ$_2$ | 285.6 | 197.1 | 109.1 | 7.6 | 278.8 | 189.7 | 100.6 | 358.5 | 268.8 | 179.4 | 90.3 | 347.8 | 258.8 | 170.2 | 81.4 | 340.0 | 251.9 | 163.9 | 75.9 | 334.8 |
| μ$_2$ | 330.0 | 172.1 | 14.0 | 167.0 | 8.4 | 209.7 | 50.7 | 202.8 | 43.6 | 244.4 | 85.3 | 237.6 | 78.8 | 280.2 | 121.9 | 275.1 | 117.2 | 319.4 | 161.6 | 315.0 |
| ν$_2$ | 225.3 | 156.1 | 86.8 | 341.5 | 271.7 | 201.7 | 131.4 | 25.3 | 314.9 | 244.4 | 174.0 | 68.0 | 358.0 | 288.5 | 218.6 | 113.5 | 44.3 | 335.2 | 266.2 | 161.4 |
| ρ$_1$ | 210.6 | 138.8 | 66.6 | 318.7 | 248.4 | 179.4 | 111.9 | 9.2 | 304.3 | 239.4 | 173.6 | 69.6 | 0.7 | 290.5 | 219.4 | 110.5 | 38.8 | 326.4 | 254.0 | 145.0 |
| MK | 178.9 | 282.3 | 25.1 | 103.6 | 204.4 | 304.0 | 42.2 | 115.9 | 212.3 | 308.7 | 45.7 | 120.4 | 219.8 | 320.5 | 62.4 | 141.7 | 245.0 | 348.8 | 92.6 | 172.8 |
| 2MK | 317.4 | 158.1 | 358.2 | 150.0 | 351.4 | 193.3 | 35.9 | 189.4 | 33.2 | 237.0 | 80.4 | 233.3 | 75.3 | 276.6 | 117.6 | 268.6 | 109.3 | 309.9 | 150.5 | 301.1 |
| MN | 75.6 | 189.5 | 303.2 | 354.5 | 107.2 | 219.5 | 331.3 | 203.7 | 132.4 | 243.8 | 355.3 | 45.3 | 157.5 | 242.2 | 23.4 | 75.1 | 189.1 | 303.2 | 150.5 | 316.2 |
| MS | 165.4 | 266.8 | 8.0 | 84.5 | 185.3 | 285.8 | 26.0 | 101.8 | 201.8 | 301.9 | 42.0 | 117.9 | 218.4 | 319.1 | 60.0 | 136.8 | 238.1 | 303.6 | 81.0 | 158.1 |
| 2SM | 194.6 | 93.2 | 352.0 | 275.5 | 174.7 | 74.2 | 334.0 | 258.2 | 158.2 | 58.1 | 318.0 | 242.1 | 141.6 | 40.9 | 300.0 | 223.2 | 121.9 | 20.4 | 279.0 | 201.9 |
| Mf | 44.0 | 308.7 | 212.4 | 141.1 | 41.3 | 298.8 | 193.0 | 110.4 | 359.3 | 248.0 | 138.8 | 59.0 | 216.2 | 216.2 | 118.4 | 48.4 | 313.0 | 218.1 | 123.4 | 54.8 |
| MSf | 194.0 | 93.2 | 352.0 | 275.5 | 174.7 | 74.2 | 352.0 | 258.2 | 158.2 | 58.1 | 334.0 | 242.1 | 141.6 | 40.9 | 300.0 | 223.2 | 121.9 | 20.4 | 279.0 | 201.9 |
| Mm | 255.3 | 344.0 | 72.8 | 174.6 | 263.3 | 352.0 | 80.7 | 182.5 | 271.2 | 0.0 | 88.7 | 190.5 | 279.2 | 7.9 | 96.6 | 198.4 | 287.1 | 15.9 | 104.6 | 206.4 |
| Sa | 280.2 | 280.0 | 279.8 | 280.1 | 280.3 | 280.0 | 279.8 | 280.2 | 280.3 | 280.1 | 279.8 | 280.0 | 280.3 | 280.1 | 279.8 | 280.1 | 280.4 | 280.1 | 279.9 | 280.6 |
| Ssa | 80.5 | 200.0 | 199.5 | 201.0 | 200.5 | 200.0 | 199.6 | 201.1 | 200.6 | 200.1 | 199.6 | 201.1 | 200.6 | 200.2 | 199.7 | 201.2 | 200.7 | 200.2 | 199.8 | 201.3 |

HARMONIC ANALYSIS AND PREDICTION OF TIDES    211

Table 15.—*Equilibrium argument* ($V_0 + u$) *for meridian of Greenwich at beginning of each calendar year, 1850 to 2000*—Con.

| Constituent | 1990 | 1991 | 1992 | 1993 | 1994 | 1995 | 1996 | 1997 | 1998 | 1999 | 2000 |
|---|---|---|---|---|---|---|---|---|---|---|---|
| $J_1$ | 314.4 | 45.2 | 135.2 | 237.9 | 325.0 | 50.2 | 134.0 | 231.1 | 322.5 | 39.3 | 125.8 |
| $K_1$ | 16.7 | 18.0 | 18.7 | 19.4 | 18.0 | 15.6 | 12.2 | 9.5 | 6.0 | 3.2 | 1.5 |
| $K_2$ | 213.6 | 216.6 | 217.6 | 218.5 | 215.4 | 210.3 | 204.0 | 199.2 | 192.7 | 187.2 | 183.4 |
| $L_2$ | 2.2 | 212.4 | 49.0 | 205.8 | 30.4 | 229.4 | 66.7 | 242.7 | 59.3 | 244.2 | 83.7 |
| $M_1$ | 85.9 | 33.5 | 305.0 | 194.1 | 79.3 | 4.4 | 293.5 | 176.9 | 64.8 | 319.9 | 251.4 |
| $M_2$ | 259.4 | 0.5 | 101.3 | 177.6 | 278.0 | 18.2 | 118.3 | 194.0 | 294.0 | 34.2 | 134.5 |
| $M_3$ | 29.1 | 0.7 | 332.0 | 266.4 | 237.0 | 207.3 | 177.4 | 110.9 | 81.0 | 51.3 | 21.8 |
| $M_4$ | 158.7 | 0.9 | 202.7 | 355.2 | 196.0 | 36.4 | 236.6 | 27.9 | 228.0 | 68.4 | 269.1 |
| $M_6$ | 58.1 | 1.4 | 304.0 | 172.8 | 114.0 | 54.6 | 354.9 | 221.9 | 162.1 | 102.6 | 43.6 |
| $M_8$ | 317.5 | 1.9 | 45.3 | 350.3 | 31.9 | 72.8 | 113.2 | 55.8 | 96.1 | 136.8 | 178.1 |
| $N_2$ | 324.3 | 336.7 | 348.8 | 323.3 | 334.9 | 346.4 | 357.8 | 331.7 | 83.0 | 354.5 | 6.1 |
| $2N$ | 29.2 | 312.8 | 236.3 | 108.9 | 31.9 | 314.6 | 237.3 | 109.4 | 32.0 | 314.7 | 237.6 |
| $O_1$ | 240.1 | 339.0 | 78.7 | 154.0 | 256.1 | 359.7 | 104.8 | 185.2 | 290.5 | 34.6 | 137.2 |
| $OO$ | 338.4 | 243.8 | 146.6 | 73.2 | 327.7 | 217.2 | 102.3 | 12.5 | 256.5 | 144.4 | 37.4 |
| $P_1$ | 349.6 | 349.8 | 350.1 | 349.3 | 349.6 | 349.8 | 350.1 | 349.3 | 349.6 | 349.8 | 350.0 |
| $Q_1$ | 305.0 | 315.2 | 326.1 | 299.7 | 313.0 | 327.9 | 344.3 | 322.9 | 339.5 | 354.9 | 8.8 |
| $2Q$ | 9.9 | 291.4 | 213.6 | 85.4 | 10.0 | 296.2 | 223.8 | 100.6 | 28.5 | 315.2 | 240.3 |
| $R_2$ | 177.6 | 177.4 | 177.1 | 177.8 | 177.6 | 177.3 | 177.1 | 177.8 | 177.5 | 177.3 | 177.0 |
| $S_1$ | 180.0 | 180.0 | 180.0 | 180.0 | 180.0 | 180.0 | 180.0 | 180.0 | 180.0 | 180.0 | 180.0 |
| $S_{2, 4, 6}$ | 0.0 | 0.0 | 0.0 | 0.0 | 0.0 | 0.0 | 0.0 | 0.0 | 0.0 | 0.0 | 0.0 |
| $T_2$ | 2.4 | 2.6 | 2.9 | 2.2 | 2.4 | 2.7 | 2.9 | 2.2 | 2.5 | 2.7 | 3.0 |
| $\lambda_2$ | 246.6 | 158.2 | 69.6 | 327.7 | 238.6 | 149.4 | 60.0 | 317.5 | 228.1 | 138.8 | 49.6 |
| $\mu_2$ | 157.1 | 358.9 | 200.5 | 353.2 | 194.3 | 35.4 | 236.1 | 28.2 | 229.2 | 69.9 | 271.0 |
| $\nu_2$ | 92.2 | 22.7 | 313.1 | 207.5 | 137.4 | 67.0 | 356.6 | 250.4 | 180.0 | 109.6 | 39.4 |
| $\rho_1$ | 72.9 | 1.3 | 290.4 | 183.9 | 115.4 | 48.6 | 343.1 | 241.6 | 176.5 | 110.1 | 42.1 |
| $MK$ | 276.0 | 18.5 | 120.0 | 197.0 | 296.0 | 33.8 | 130.5 | 203.4 | 300.0 | 37.4 | 136.0 |
| $2MK$ | 142.1 | 342.9 | 184.0 | 335.8 | 177.9 | 20.8 | 224.4 | 18.4 | 222.1 | 65.2 | 267.6 |
| $MN$ | 223.6 | 337.1 | 90.1 | 140.8 | 252.9 | 4.6 | 116.1 | 165.6 | 277.0 | 28.7 | 140.6 |
| $MS$ | 259.4 | 0.5 | 101.3 | 177.6 | 278.0 | 18.2 | 118.3 | 194.0 | 294.0 | 34.2 | 134.5 |
| $2SM$ | 100.6 | 359.5 | 258.7 | 182.4 | 82.0 | 341.8 | 241.7 | 166.0 | 66.0 | 325.8 | 225.5 |
| $Mf$ | 319.2 | 222.4 | 124.2 | 49.6 | 305.9 | 198.8 | 88.8 | 3.7 | 253.0 | 144.9 | 40.1 |
| $MSf$ | 100.6 | 359.5 | 258.7 | 182.4 | 82.0 | 341.8 | 241.7 | 166.0 | 66.0 | 325.8 | 225.5 |
| $Mm$ | 295.1 | 23.8 | 112.5 | 214.3 | 303.0 | 31.8 | 120.5 | 222.3 | 311.0 | 39.7 | 128.4 |
| $Sa$ | 280.4 | 280.2 | 279.9 | 280.7 | 280.4 | 280.2 | 279.9 | 280.7 | 280.4 | 280.2 | 280.0 |
| $Ssa$ | 200.8 | 200.3 | 199.8 | 201.3 | 200.8 | 200.4 | 199.9 | 201.4 | 200.9 | 200.4 | 200.0 |

212   U. S. COAST AND GEODETIC SURVEY

Table 16.—*Differences to adapt table 15 to beginning of each calendar month*

| Constituent | Month of year* | | | | | | | | | | | |
|---|---|---|---|---|---|---|---|---|---|---|---|---|
| | Jan. | Feb. | Mar. | Apr. | May | June | July | Aug. | Sept. | Oct. | Nov. | Dec. |
| | ° | ° | ° | ° | ° | ° | ° | ° | ° | ° | ° | ° |
| J₁ | 0.00 | 75.57 | 108.99 | 184.56 | 246.08 | 321.65 | 23.17 | 98.74 | 174.31 | 235.82 | 311.39 | 12.91 |
| K₁ | 0.00 | 30.56 | 58.15 | 88.71 | 118.28 | 148.83 | 178.40 | 208.96 | 239.51 | 269.08 | 299.64 | 329.21 |
| K₂ | 0.00 | 61.11 | 116.31 | 177.42 | 236.56 | 297.66 | 356.80 | 57.91 | 119.02 | 178.16 | 239.27 | 298.41 |
| L₂ | 0.00 | 9.19 | 52.33 | 61.54 | 82.02 | 91.21 | 111.71 | 120.90 | 130.09 | 160.59 | 159.78 | 180.29 |
| M₁,† | 0.00 | 345.54 | 7.32 | 352.86 | 350.48 | 336.02 | 333.64 | 319.18 | 304.72 | 302.34 | 287.88 | 285.50 |
| M₁ | 0.00 | 342.09 | 0.75 | 342.83 | 337.11 | 319.20 | 313.47 | 295.56 | 277.65 | 271.93 | 254.01 | 248.29 |
| M₂ | 0.00 | 324.17 | 1.49 | 325.66 | 314.22 | 278.39 | 266.95 | 231.12 | 195.30 | 183.85 | 148.02 | 136.58 |
| M₃ | 0.00 | 306.26 | 2.24 | 308.50 | 291.33 | 237.59 | 220.42 | 166.68 | 112.94 | 95.78 | 42.04 | 24.87 |
| M₄ | 0.00 | 288.35 | 2.98 | 291.33 | 268.44 | 196.79 | 173.90 | 102.24 | 30.59 | 7.70 | 296.05 | 273.16 |
| M₆ | 0.00 | 252.52 | 4.48 | 257.00 | 222.66 | 115.18 | 80.85 | 333.37 | 225.89 | 191.55 | 84.07 | 49.74 |
| M₈ | 0.00 | 216.69 | 5.97 | 222.66 | 176.88 | 33.58 | 347.80 | 204.49 | 61.18 | 15.40 | 232.10 | 186.32 |
| N₂ | 0.00 | 279.16 | 310.66 | 229.83 | 186.42 | 105.58 | 62.18 | 341.34 | 260.50 | 217.11 | 136.27 | 49.16 |
| 2N | 0.00 | 234.14 | 259.82 | 133.97 | 58.62 | 292.77 | 217.42 | 91.57 | 325.71 | 250.36 | 124.51 | 167.37 |
| O₁ | 0.00 | 293.62 | 303.34 | 236.96 | 195.94 | 129.56 | 88.55 | 22.16 | 315.78 | 274.77 | 208.39 | 131.04 |
| OO | 0.00 | 127.49 | 172.97 | 300.46 | 40.61 | 168.10 | 268.26 | 35.75 | 163.24 | 263.39 | 30.89 | 30.79 |
| P₁ | 0.00 | 329.44 | 301.85 | 271.29 | 241.72 | 211.17 | 181.60 | 151.04 | 120.49 | 90.92 | 60.36 | 30.79 |
| Q₁ | 0.00 | 248.60 | 252.50 | 141.11 | 68.14 | 143.93 | 243.78 | 132.39 | 20.99 | 308.03 | 196.63 | 123.67 |
| 2Q | 0.00 | 203.59 | 201.67 | 45.26 | 300.34 | 143.93 | 39.02 | 242.61 | 86.20 | 341.28 | 184.87 | 79.96 |
| R₂ | 0.00 | 30.56 | 58.15 | 88.71 | 118.28 | 148.83 | 178.40 | 208.96 | 239.51 | 269.08 | 299.64 | 329.21 |
| S₁ | 0.00 | 0.00 | 0.00 | 0.00 | 0.00 | 0.00 | 0.00 | 0.00 | 0.00 | 0.00 | 0.00 | 0.00 |
| T₂ | 0.00 | 329.44 | 301.85 | 271.29 | 241.72 | 211.17 | 181.60 | 151.04 | 120.49 | 90.92 | 60.36 | 30.79 |
| λ₂ | 0.00 | 314.98 | 309.16 | 264.15 | 232.20 | 187.19 | 155.24 | 110.22 | 65.21 | 33.26 | 348.24 | 316.29 |
| μ₂ | 0.00 | 288.35 | 2.98 | 291.33 | 268.44 | 196.79 | 173.90 | 102.24 | 30.59 | 7.70 | 296.05 | 273.16 |
| ν₂ | 0.00 | 333.36 | 53.82 | 27.18 | 36.24 | 9.60 | 18.66 | 352.02 | 325.38 | 334.44 | 307.81 | 316.87 |
| ρ₁ | 0.00 | 302.81 | 355.66 | 298.47 | 277.96 | 220.77 | 200.26 | 143.07 | 85.87 | 65.36 | 8.17 | 347.66 |
| MK | 0.00 | 354.73 | 59.64 | 54.37 | 72.50 | 67.23 | 85.35 | 80.08 | 74.81 | 92.93 | 87.66 | 105.79 |
| 2MK | 0.00 | 257.79 | 304.83 | 202.62 | 150.16 | 47.96 | 355.50 | 253.29 | 151.08 | 98.62 | 356.41 | 303.95 |
| MN | 0.00 | 243.33 | 312.15 | 195.48 | 140.64 | 23.97 | 329.13 | 212.47 | 95.80 | 40.96 | 284.29 | 229.45 |
| MS | 0.00 | 324.17 | 1.49 | 325.66 | 314.22 | 278.39 | 266.95 | 231.12 | 195.30 | 183.85 | 148.02 | 136.58 |
| 2SM | 0.00 | 35.83 | 358.51 | 34.34 | 45.78 | 81.61 | 93.05 | 128.88 | 164.70 | 176.15 | 211.98 | 223.42 |
| Mf | 0.00 | 96.94 | 114.82 | 211.75 | 282.34 | 19.27 | 89.86 | 186.79 | 283.73 | 354.31 | 91.25 | 161.83 |
| MSf | 0.00 | 35.83 | 358.51 | 34.34 | 45.78 | 81.61 | 93.05 | 128.88 | 164.70 | 176.15 | 211.98 | 223.42 |
| Mm | 0.00 | 45.02 | 50.84 | 95.85 | 127.80 | 172.81 | 204.76 | 249.78 | 294.79 | 326.74 | 11.76 | 43.71 |
| Sa | 0.00 | 30.56 | 58.15 | 88.71 | 118.28 | 148.83 | 178.40 | 208.96 | 239.51 | 269.08 | 299.64 | 329.21 |
| Ssa | 0.00 | 61.11 | 116.31 | 177.42 | 236.56 | 297.66 | 356.80 | 57.91 | 119.02 | 178.16 | 239.27 | 298.41 |

*This table was designed for direct use for common years. For a leap year the values given for the months of March to December, inclusive, apply to the last day of the preceding month, but may be used directly, provided an allowance is made in the day of month as indicated in the following table.

†The first line for constituent M₁ gives the difference as based upon the formula in table 2; the second line gives the differences as derived from the half speed of constituent M₂.

‡The differences for constituents S₁, S₃, S₄, S₆, etc., are each zero for every month.

## Table 17.—Differences to adapt table 15 to beginning of each day of month

| Constituent | Day of month* | | | | | | | | | | |
|---|---|---|---|---|---|---|---|---|---|---|---|
| | 1 | 2 | 3 | 4 | 5 | 6 | 7 | 8 | 9 | 10 | 11 |
| J₁ | 0.00 | 14.05 | 28.10 | 42.15 | 56.20 | 70.25 | 84.30 | 98.35 | 112.41 | 126.46 | 140.51 |
| K₁ | 0.00 | 0.99 | 1.97 | 2.96 | 3.94 | 4.93 | 5.91 | 6.90 | 7.88 | 8.87 | 9.86 |
| K₂ | 0.00 | 1.97 | 3.94 | 5.91 | 7.88 | 9.86 | 11.83 | 13.80 | 15.77 | 17.74 | 19.71 |
| L₂ | 0.00 | 348.68 | 337.37 | 326.05 | 314.73 | 303.42 | 292.10 | 280.78 | 269.47 | 258.15 | 246.84 |
| M₁† | 0.00 | 347.92 | 335.84 | 323.76 | 311.68 | 299.60 | 287.52 | 275.44 | 263.37 | 251.29 | 239.21 |
| M₂ | 0.00 | 347.81 | 335.62 | 323.43 | 311.24 | 299.05 | 286.86 | 274.66 | 262.47 | 250.28 | 238.09 |
| M₃ | 0.00 | 335.62 | 311.24 | 286.86 | 262.47 | 238.09 | 213.71 | 189.33 | 164.95 | 140.57 | 116.18 |
| M₄ | 0.00 | 335.62 | 311.24 | 286.86 | 262.47 | 238.09 | 213.71 | 189.33 | 164.95 | 140.57 | 116.18 |
| M₄ | 0.00 | 323.43 | 286.86 | 250.28 | 213.71 | 177.14 | 140.57 | 103.99 | 67.42 | 30.85 | 354.28 |
| M₆ | 0.00 | 311.24 | 262.47 | 213.71 | 164.95 | 116.18 | 67.42 | 18.66 | 329.90 | 281.13 | 232.37 |
| M₈ | 0.00 | 286.86 | 213.71 | 140.57 | 67.42 | 354.28 | 281.13 | 207.99 | 134.84 | 61.70 | 348.56 |
| N₂ | 0.00 | 322.55 | 285.11 | 247.66 | 210.21 | 172.77 | 135.32 | 97.88 | 60.43 | 22.98 | 345.54 |
| 2N | 0.00 | 309.49 | 258.98 | 208.47 | 157.95 | 107.44 | 56.93 | 6.42 | 315.91 | 265.40 | 214.89 |
| O₁ | 0.00 | 334.63 | 309.27 | 283.90 | 258.53 | 233.16 | 207.80 | 182.43 | 157.06 | 131.70 | 106.33 |
| OO | 0.00 | 27.34 | 54.68 | 82.02 | 109.35 | 136.69 | 164.03 | 191.37 | 218.71 | 246.05 | 273.38 |
| P₁ | 0.00 | 359.01 | 358.03 | 357.04 | 356.06 | 355.07 | 354.09 | 353.10 | 352.12 | 351.13 | 350.15 |
| | 0.00 | 359.01 | 358.03 | 357.04 | 356.06 | 355.07 | 354.09 | 353.10 | 352.12 | 351.13 | 350.15 |
| Q₁ | 0.00 | 321.57 | 283.14 | 244.70 | 206.27 | 167.84 | 129.41 | 90.98 | 52.54 | 14.11 | 335.68 |
| 2Q | 0.00 | 308.50 | 257.01 | 205.51 | 154.01 | 102.51 | 51.02 | 359.52 | 308.02 | 256.53 | 205.03 |
| R₂ | 0.00 | 0.99 | 1.97 | 2.96 | 3.94 | 4.93 | 5.91 | 6.90 | 7.88 | 8.87 | 9.86 |
| S₁ | 0.00 | 0.00 | 0.00 | 0.00 | 0.00 | 0.00 | 0.00 | 0.00 | 0.00 | 0.00 | 0.00 |
| T₂ | 0.00 | 359.01 | 358.03 | 357.04 | 356.06 | 355.07 | 354.09 | 353.10 | 352.12 | 351.13 | 350.15 |
| λ₂ | 0.00 | 346.93 | 333.87 | 320.80 | 307.74 | 294.68 | 281.61 | 268.54 | 255.48 | 242.42 | 229.35 |
| μ₂ | 0.00 | 311.24 | 262.47 | 213.71 | 164.95 | 116.18 | 67.42 | 18.66 | 329.90 | 281.13 | 232.37 |
| ν₂ | 0.00 | 324.30 | 288.60 | 252.91 | 217.21 | 181.51 | 145.81 | 110.11 | 74.42 | 38.72 | 3.02 |
| MK | 0.00 | 323.32 | 286.63 | 249.95 | 213.27 | 176.58 | 139.90 | 103.21 | 66.53 | 29.85 | 353.16 |
| 2MK | 0.00 | 336.60 | 313.21 | 289.81 | 266.42 | 243.02 | 219.62 | 196.23 | 172.83 | 149.44 | 126.04 |
| MN | 0.00 | 310.25 | 260.50 | 210.75 | 161.00 | 111.26 | 51.51 | 11.76 | 322.01 | 272.26 | 222.51 |
| MS | 0.00 | 298.17 | 236.34 | 174.52 | 112.69 | 50.86 | 349.03 | 287.20 | 225.38 | 163.55 | 101.72 |
| 2SM | 0.00 | 335.62 | 311.24 | 286.86 | 262.47 | 238.09 | 213.71 | 189.33 | 164.95 | 140.57 | 116.18 |
| Mf | 0.00 | 24.38 | 48.76 | 73.14 | 97.53 | 121.91 | 146.29 | 170.67 | 195.05 | 219.43 | 243.82 |
| MSf | 0.00 | 26.35 | 52.71 | 79.06 | 105.41 | 131.76 | 158.12 | 184.47 | 210.82 | 237.18 | 263.53 |
| Mm | 0.00 | 24.38 | 48.76 | 73.14 | 97.53 | 121.91 | 146.29 | 170.67 | 195.05 | 219.43 | 243.82 |
| Sa | 0.00 | 13.07 | 26.13 | 39.20 | 52.26 | 65.32 | 78.39 | 91.46 | 104.52 | 117.58 | 130.65 |
| Ssa | 0.00 | 0.99 | 1.97 | 2.96 | 3.94 | 4.93 | 5.91 | 6.90 | 7.88 | 8.87 | 9.86 |
| | 0.00 | 1.97 | 3.94 | 5.91 | 7.88 | 9.86 | 11.83 | 13.80 | 15.77 | 17.74 | 19.71 |

*The table is adapted directly for use with common years, but if the required date falls between Mar. 1 and Dec. 31, inclusive, in a leap year the day of month should be increased by one before entering the table.
†The first line for constituent M₁ gives the differences as based upon the formula in table 2, the second line gives the differences as derived from the half speed of constituent M₂.
‡The differences for constituents S₁, S₂, S₄, S₆, etc., are each zero for the beginning of every day.

TABLE 17.—*Differences to adapt table 15 to beginning of each day of month*—Continued

| Constituent | Day of month* | | | | | | | | | | |
|---|---|---|---|---|---|---|---|---|---|---|---|
| | 12 | 13 | 14 | 15 | 16 | 17 | 18 | 19 | 20 | 21 | 22 |
| $J_1$ | ° | ° | ° | ° | ° | ° | ° | ° | ° | ° | ° |
| | 154.56 | 168.61 | 182.66 | 196.71 | 210.76 | 224.81 | 238.86 | 252.91 | 266.96 | 281.01 | 295.06 |
| $K_1$ | 10.84 | 11.83 | 12.81 | 13.80 | 14.78 | 15.77 | 16.76 | 17.74 | 18.73 | 19.71 | 20.70 |
| $K_2$ | 21.68 | 23.66 | 25.63 | 27.60 | 29.57 | 31.54 | 33.51 | 35.48 | 37.46 | 39.43 | 41.40 |
| $L_2$ | 235.62 | 224.20 | 212.88 | 201.57 | 190.25 | 178.94 | 167.62 | 156.30 | 144.99 | 133.67 | 122.35 |
| $M_1$ | 227.13 | 215.05 | 202.97 | 190.89 | 178.81 | 166.73 | 154.65 | 142.57 | 130.49 | 118.41 | 106.33 |
| $M_2$ | 225.90 | 213.71 | 201.52 | 189.33 | 177.14 | 164.95 | 152.76 | 140.57 | 128.38 | 116.19 | 103.99 |
| $M_3$ | 91.80 | 67.42 | 43.04 | 18.66 | 354.28 | 329.90 | 305.52 | 281.13 | 256.75 | 232.37 | 207.99 |
| $M_4$ | 317.70 | 281.13 | 244.56 | 207.99 | 171.42 | 134.84 | 98.27 | 61.70 | 25.13 | 348.56 | 311.98 |
| $M_6$ | 183.61 | 134.84 | 86.08 | 37.32 | 348.56 | 299.79 | 251.03 | 202.27 | 153.50 | 104.74 | 55.98 |
| $M_8$ | 275.41 | 202.27 | 129.12 | 55.98 | 342.83 | 269.69 | 196.54 | 123.40 | 50.26 | 337.11 | 263.97 |
| $M_s$ | 7.21 | 269.69 | 172.16 | 74.64 | 337.11 | 239.58 | 142.06 | 44.53 | 307.01 | 209.48 | 111.95 |
| $N_2$ | 308.09 | 270.64 | 233.20 | 195.75 | 158.30 | 120.86 | 83.41 | 45.96 | 8.52 | 331.07 | 293.62 |
| $2N$ | 164.37 | 113.86 | 63.35 | 12.84 | 322.33 | 271.82 | 221.30 | 170.79 | 120.28 | 69.77 | 19.26 |
| $O$ | 80.96 | 55.59 | 30.23 | 4.86 | 339.49 | 314.13 | 288.76 | 263.39 | 238.02 | 212.66 | 187.29 |
| $OO$ | 300.72 | 328.06 | 355.40 | 22.74 | 50.08 | 77.42 | 104.75 | 132.09 | 159.43 | 186.77 | 214.11 |
| $P_1$ | −10.84 | −11.83 | −12.81 | −13.80 | −14.78 | −15.77 | −16.76 | −17.74 | −18.73 | −19.71 | −20.70 |
| $Q_1$ | 297.25 | 258.81 | 220.38 | 181.95 | 143.52 | 105.09 | 66.65 | 28.22 | 349.79 | 311.36 | 272.92 |
| $2Q$ | 153.53 | 102.03 | 50.54 | 359.04 | 307.54 | 256.05 | 204.55 | 153.05 | 101.55 | 50.06 | 358.56 |
| $R_2$ | 10.84 | 11.83 | 12.81 | 13.80 | 14.78 | 15.77 | 16.76 | 17.74 | 18.73 | 19.71 | 20.70 |
| $S_1$ | 0.00 | 0.00 | 0.00 | 0.00 | 0.00 | 0.00 | 0.00 | 0.00 | 0.00 | 0.00 | 0.00 |
| $T_2$ | −10.84 | −11.83 | −12.81 | −13.80 | −14.78 | −15.77 | −16.76 | −17.74 | −18.73 | −19.71 | −20.70 |
| $\lambda_2$ | 216.28 | 203.23 | 190.16 | 177.09 | 164.02 | 150.96 | 137.90 | 124.83 | 111.76 | 98.70 | 85.64 |
| $\mu_2$ | 183.61 | 134.84 | 86.08 | 37.32 | 348.56 | 299.79 | 251.03 | 202.27 | 153.50 | 104.74 | 55.98 |
| $\nu_2$ | 327.32 | 291.62 | 255.93 | 220.23 | 184.53 | 148.83 | 113.13 | 77.44 | 41.74 | 6.04 | 330.34 |
| $\rho_1$ | 316.48 | 279.80 | 243.11 | 206.43 | 169.75 | 133.06 | 96.38 | 59.69 | 23.01 | 346.33 | 309.64 |
| $MK$ | 102.65 | 79.25 | 55.85 | 32.46 | 9.06 | 345.67 | 322.27 | 298.88 | 275.48 | 252.08 | 228.69 |
| $2MK$ | 172.76 | 123.02 | 73.27 | 23.52 | 333.77 | 284.02 | 234.27 | 184.52 | 134.78 | 85.03 | 35.28 |
| $MN$ | 39.89 | 338.06 | 276.24 | 214.41 | 152.58 | 90.75 | 28.92 | 327.10 | 265.27 | 203.44 | 141.61 |
| $MS$ | 91.80 | 67.42 | 43.04 | 18.66 | 354.28 | 329.90 | 305.52 | 281.13 | 256.75 | 232.37 | 207.99 |
| $2SM$ | 268.20 | 292.58 | 316.96 | 341.34 | 5.72 | 30.10 | 54.48 | 78.87 | 103.25 | 127.63 | 152.01 |
| $Mf$ | 289.88 | 316.23 | 342.59 | 8.94 | 35.29 | 61.64 | 88.00 | 114.35 | 140.70 | 167.06 | 193.41 |
| $MSf$ | 268.20 | 292.58 | 316.96 | 341.34 | 5.72 | 30.10 | 54.48 | 78.87 | 103.25 | 127.63 | 152.01 |
| $Mm$ | 143.72 | 156.78 | 169.84 | 182.91 | 195.98 | 209.04 | 222.10 | 235.17 | 248.24 | 261.30 | 274.36 |
| $S_a$ | 10.84 | 11.83 | 12.81 | 13.80 | 14.78 | 15.77 | 16.76 | 17.74 | 18.73 | 19.71 | 20.70 |
| $Ssa$ | 21.68 | 23.66 | 25.63 | 27.60 | 29.57 | 31.54 | 33.51 | 35.48 | 37.46 | 39.43 | 41.40 |

*The table is adapted directly for use with common years, but if the required date falls between Mar. 1 and Dec. 31, inclusive, in a leap year the day of month should be increased by one before entering the table.

†The first line for constituent $M_1$ gives the differences as based upon the formula in table 2; the second line gives the differences as derived from the half speed of constituent $M_f$.

‡The differences for constituents $S_1$, $S_2$, $S_4$, $S_6$, etc., are each zero for the beginning of every day.

## Table 17.—Differences to adapt table 15 to beginning of each day of month—Continued

| Constituent | Day of month* | | | | | | | | | |
|---|---|---|---|---|---|---|---|---|---|---|
| | 23 | 24 | 25 | 26 | 27 | 28 | 29 | 30 | 31 | 32 |
| $J_1$ | 309.11 | 323.16 | 337.22 | 351.27 | 5.32 | 19.37 | 33.42 | 47.47 | 61.52 | 75.57 |
| $K_1$ | 21.68 | 22.67 | 23.66 | 24.64 | 25.63 | 26.61 | 27.60 | 28.58 | 29.57 | 30.56 |
| $K_2$ | 43.37 | 45.34 | 47.31 | 49.28 | 51.25 | 53.22 | 55.20 | 57.17 | 59.14 | 61.11 |
| $L_2$ | 111.04 | 99.72 | 88.40 | 77.09 | 65.77 | 54.45 | 43.14 | 31.82 | 20.50 | 9.19 |
| $M_1$† | 94.25 | 82.18 | 70.10 | 58.02 | 45.94 | 33.86 | 21.78 | 9.70 | 357.62 | 345.54 |
| $M_2$ | 91.80 | 79.61 | 67.42 | 55.23 | 43.04 | 30.85 | 18.66 | 6.47 | 354.28 | 342.09 |
| $M_3$ | 183.61 | 159.23 | 134.84 | 110.46 | 86.08 | 61.70 | 37.32 | 12.94 | 348.56 | 324.17 |
| $M_4$ | 275.41 | 238.84 | 202.27 | 165.69 | 129.12 | 92.55 | 55.98 | 19.40 | 342.83 | 306.26 |
| $M_6$ | 7.21 | 318.45 | 269.69 | 220.92 | 172.16 | 123.40 | 74.64 | 25.87 | 337.11 | 288.35 |
| $M_8$ | 190.82 | 117.68 | 44.53 | 331.39 | 258.24 | 185.10 | 111.95 | 38.81 | 325.66 | 252.52 |
| $N_2$ | 14.43 | 276.90 | 179.38 | 81.85 | 344.32 | 246.80 | 149.27 | 51.75 | 314.22 | 216.69 |
| $2N$ | 256.18 | 218.73 | 181.28 | 143.84 | 106.39 | 68.94 | 31.50 | 354.05 | 316.60 | 279.15 |
| $O_1$ | 328.75 | 278.24 | 227.72 | 177.21 | 126.70 | 76.19 | 25.68 | 335.17 | 284.66 | 234.15 |
| $OO$ | 161.92 | 136.56 | 111.19 | 85.82 | 60.45 | 35.09 | 9.72 | 344.35 | 318.99 | 293.62 |
| $P_1$ | 241.45 | 268.78 | 296.12 | 323.46 | 350.80 | 18.14 | 45.48 | 72.81 | 100.15 | 127.49 |
| $Q_1$ | −21.68 | −22.67 | −23.66 | −24.64 | −25.63 | −26.61 | −27.60 | −28.58 | −29.57 | −30.56 |
| $2Q$ | 234.49 | 196.06 | 157.63 | 119.20 | 80.76 | 42.33 | 3.90 | 325.47 | 287.04 | 248.60 |
| $R_2$ | 307.06 | 255.57 | 204.07 | 152.57 | 101.07 | 49.58 | 358.08 | 306.58 | 255.09 | 203.59 |
| $S_1$ | 21.68 | 22.67 | 23.66 | 24.64 | 25.63 | 26.61 | 27.60 | 28.58 | 29.57 | 30.56 |
| $T_2$ | 0.00 | 0.00 | 0.00 | 0.00 | 0.00 | 0.00 | 0.00 | 0.00 | 0.00 | 0.00 |
| $\lambda_2$ | −21.68 | −22.67 | −23.66 | −24.64 | −25.63 | −26.61 | −27.60 | −28.58 | −29.57 | −30.56 |
| $\mu_2$ | 72.57 | 59.50 | 46.44 | 33.38 | 20.31 | 7.24 | 354.18 | 341.12 | 328.05 | 314.98 |
| $\nu_2$ | 7.21 | 318.45 | 269.69 | 220.92 | 172.16 | 123.40 | 74.64 | 25.87 | 337.11 | 288.35 |
| $\rho_1$ | 294.64 | 258.95 | 223.25 | 187.55 | 151.85 | 116.15 | 80.46 | 44.76 | 9.06 | 333.36 |
| $MK$ | 272.96 | 236.28 | 199.59 | 162.91 | 126.22 | 89.54 | 52.86 | 16.17 | 339.49 | 302.81 |
| $2MK$ | 205.29 | 181.90 | 158.50 | 135.10 | 111.71 | 88.31 | 64.92 | 41.52 | 18.12 | 354.73 |
| $MN$ | 345.53 | 295.78 | 246.03 | 196.28 | 146.54 | 96.79 | 47.04 | 357.29 | 307.54 | 257.79 |
| $MS$ | 79.78 | 17.96 | 316.13 | 254.30 | 192.47 | 130.64 | 68.82 | 6.99 | 305.16 | 243.33 |
| $2SM$ | 183.61 | 159.23 | 134.84 | 110.46 | 86.08 | 61.70 | 37.32 | 12.94 | 348.56 | 324.17 |
| $Mf$ | 176.39 | 200.77 | 225.16 | 249.54 | 273.92 | 298.30 | 322.68 | 347.06 | 11.44 | 35.83 |
| $MSf$ | 219.76 | 246.11 | 272.47 | 298.82 | 325.17 | 351.52 | 17.88 | 44.23 | 70.58 | 96.64 |
| $Mm$ | 176.39 | 200.77 | 225.16 | 249.54 | 273.92 | 298.30 | 322.68 | 347.06 | 11.44 | 35.83 |
| $Sa$ | 287.43 | 300.50 | 313.56 | 326.62 | 339.69 | 352.76 | 5.82 | 18.88 | 31.95 | 45.02 |
| $Ssa$ | 21.68 | 22.67 | 23.66 | 24.64 | 25.63 | 26.61 | 27.60 | 28.58 | 29.57 | 30.56 |
| | 43.37 | 45.34 | 47.31 | 49.28 | 51.25 | 53.22 | 55.20 | 57.17 | 59.14 | 61.11 |

*The table is adapted directly for use with common years but if the required date falls between Mar. 1 and Dec. 31, inclusive, in a leap year the day of month should be increased by one before entering the table.
†The first line for constituent $M_1$ gives the differences as based upon the formula in Table 2; the second line gives the differences as derived from the half speed of constituent $M_2$.
‡The differences for constituents $S_1$, $S_2$, $S_4$, $S_6$, etc., are each zero for the beginning of every day.

216   U. S. COAST AND GEODETIC SURVEY

Table 18.—*Differences to adapt table 15 to beginning of each hour of day*

| Constituent | Hour of day | | | | | | | | | | | |
|---|---|---|---|---|---|---|---|---|---|---|---|---|
| | 0 | 1 | 2 | 3 | 4 | 5 | 6 | 7 | 8 | 9 | 10 | 11 |
| $J_1$ | 0.00 | 15.59 | 31.17 | 46.76 | 62.34 | 77.93 | 93.51 | 109.10 | 124.68 | 140.27 | 155.85 | 171.44 |
| $K_1$ | 0.00 | 15.04 | 30.08 | 45.12 | 60.16 | 75.21 | 90.25 | 105.29 | 120.33 | 135.37 | 150.41 | 165.45 |
| $K_2$ | 0.00 | 30.08 | 60.16 | 90.25 | 120.33 | 150.41 | 180.49 | 210.57 | 240.66 | 270.74 | 300.82 | 330.90 |
| $L_2$ | 0.00 | 29.53 | 59.06 | 88.59 | 118.11 | 147.64 | 177.17 | 206.70 | 236.23 | 265.76 | 295.28 | 324.81 |
| $M_1$† | 0.00 | 14.50 | 28.99 | 43.49 | 57.99 | 72.48 | 86.98 | 101.48 | 115.97 | 130.47 | 144.97 | 159.46 |
| $M_1$ | 0.00 | 14.49 | 28.98 | 43.48 | 57.97 | 72.46 | 86.95 | 101.44 | 115.94 | 130.43 | 144.92 | 159.41 |
| $M_2$ | 0.00 | 28.98 | 57.97 | 86.95 | 115.94 | 144.92 | 173.90 | 202.89 | 231.87 | 260.86 | 289.84 | 318.83 |
| $M_3$ | 0.00 | 43.48 | 86.95 | 130.43 | 173.90 | 217.38 | 260.86 | 304.33 | 347.81 | 31.29 | 74.76 | 118.24 |
| $M_4$ | 0.00 | 57.97 | 115.94 | 173.90 | 231.87 | 289.84 | 347.81 | 45.78 | 103.75 | 161.71 | 219.68 | 277.65 |
| $M_6$ | 0.00 | 86.95 | 173.90 | 260.86 | 347.81 | 74.76 | 161.71 | 248.67 | 335.62 | 62.57 | 149.52 | 236.48 |
| $M_8$ | 0.00 | 115.94 | 231.87 | 347.81 | 103.75 | 219.68 | 335.62 | 91.55 | 207.49 | 323.43 | 79.36 | 195.30 |
| $N_2$ | 0.00 | 28.44 | 56.88 | 85.32 | 113.76 | 142.20 | 170.64 | 199.08 | 227.52 | 255.96 | 284.40 | 312.84 |
| $2N$ | 0.00 | 27.90 | 55.79 | 83.69 | 111.58 | 139.48 | 167.37 | 195.27 | 223.16 | 251.06 | 278.95 | 306.85 |
| $O_1$ | 0.00 | 13.94 | 27.89 | 41.83 | 55.77 | 69.72 | 83.66 | 97.60 | 111.54 | 125.49 | 139.43 | 153.37 |
| $OO$ | 0.00 | 16.14 | 32.28 | 48.42 | 64.56 | 80.70 | 96.83 | 112.97 | 129.11 | 145.25 | 161.39 | 177.53 |
| $P_1$ | 0.00 | 14.96 | 29.92 | 44.88 | 59.84 | 74.79 | 89.75 | 104.71 | 119.67 | 134.63 | 149.59 | 164.55 |
| $Q_1$ | 0.00 | 13.40 | 26.80 | 40.20 | 53.59 | 66.99 | 80.39 | 93.79 | 107.19 | 120.59 | 133.99 | 147.39 |
| $2Q$ | 0.00 | 12.85 | 25.71 | 38.56 | 51.42 | 64.27 | 77.13 | 89.98 | 102.83 | 115.69 | 128.54 | 141.40 |
| $R_2$ | 0.00 | 30.04 | 60.08 | 90.12 | 120.16 | 150.21 | 180.25 | 210.29 | 240.33 | 270.37 | 300.41 | 330.45 |
| $S_1$ | 0.00 | 15.00 | 30.00 | 45.00 | 60.00 | 75.00 | 90.00 | 105.00 | 120.00 | 135.00 | 150.00 | 165.00 |
| $S_2$ | 0.00 | 30.00 | 60.00 | 90.00 | 120.00 | 150.00 | 180.00 | 210.00 | 240.00 | 270.00 | 300.00 | 330.00 |
| $S_4$ | 0.00 | 60.00 | 120.00 | 180.00 | 240.00 | 300.00 | 0.00 | 60.00 | 120.00 | 180.00 | 240.00 | 300.00 |
| $S_6$ | 0.00 | 90.00 | 180.00 | 270.00 | 0.00 | 90.00 | 180.00 | 270.00 | 0.00 | 90.00 | 180.00 | 270.00 |
| $T_2$ | 0.00 | 29.96 | 59.92 | 89.88 | 119.84 | 149.79 | 179.75 | 209.71 | 239.67 | 269.63 | 299.59 | 329.55 |
| $\lambda_2$ | 0.00 | 29.46 | 58.91 | 88.37 | 117.82 | 147.28 | 176.73 | 206.19 | 235.65 | 265.10 | 294.56 | 324.01 |
| $\mu_2$ | 0.00 | 27.97 | 55.94 | 83.90 | 111.87 | 139.84 | 167.81 | 195.78 | 223.75 | 251.71 | 279.68 | 307.65 |
| $\nu_2$ | 0.00 | 28.51 | 57.03 | 85.54 | 114.05 | 142.56 | 171.08 | 199.59 | 228.10 | 256.61 | 285.13 | 313.64 |
| $\rho_1$ | 0.00 | 13.47 | 26.94 | 40.41 | 53.89 | 67.36 | 80.83 | 94.30 | 107.77 | 121.24 | 134.72 | 148.19 |
| $MK$ | 0.00 | 44.03 | 88.05 | 132.08 | 176.10 | 220.13 | 264.15 | 308.18 | 352.20 | 36.23 | 80.25 | 124.28 |
| $2MK$ | 0.00 | 42.93 | 85.85 | 128.78 | 171.71 | 214.64 | 257.56 | 300.49 | 343.42 | 26.34 | 69.27 | 112.20 |
| $MN$ | 0.00 | 57.42 | 114.85 | 172.27 | 229.70 | 287.12 | 344.54 | 41.97 | 99.39 | 156.81 | 214.24 | 271.66 |
| $MS$ | 0.00 | 58.98 | 117.97 | 176.95 | 235.94 | 294.92 | 353.90 | 52.89 | 111.87 | 170.86 | 229.84 | 288.83 |
| $2SM$ | 0.00 | 31.02 | 62.03 | 93.05 | 124.06 | 155.08 | 186.10 | 217.11 | 248.13 | 279.14 | 310.16 | 341.17 |
| $Mf$ | 0.00 | 1.10 | 2.20 | 3.29 | 4.39 | 5.49 | 6.59 | 7.69 | 8.78 | 9.88 | 10.98 | 12.08 |
| $MSf$ | 0.00 | 1.02 | 2.03 | 3.05 | 4.06 | 5.08 | 6.10 | 7.11 | 8.13 | 9.14 | 10.16 | 11.17 |
| $Mm$ | 0.00 | 0.54 | 1.09 | 1.63 | 2.18 | 2.72 | 3.27 | 3.81 | 4.35 | 4.90 | 5.44 | 5.99 |
| $Sa$ | 0.00 | 0.04 | 0.08 | 0.12 | 0.16 | 0.21 | 0.25 | 0.29 | 0.33 | 0.37 | 0.41 | 0.45 |
| $Ssa$ | 0.00 | 0.08 | 0.16 | 0.25 | 0.33 | 0.41 | 0.49 | 0.57 | 0.66 | 0.74 | 0.82 | 0.90 |

†The first line for constituent $M_1$ gives the differences as based upon the formula in table 2; the second line gives the differences as derived from the half speed of constituent $M_2$.

Table 18.—*Differences to adapt table 15 to beginning of each hour of day*—Continued

| Constituent | Hour of day | | | | | | | | | | | |
|---|---|---|---|---|---|---|---|---|---|---|---|---|
| | 12 | 13 | 14 | 15 | 16 | 17 | 18 | 19 | 20 | 21 | 22 | 23 |
| J₁ | 187.03 | 202.61 | 218.20 | 233.78 | 249.37 | 264.95 | 280.54 | 296.12 | 311.71 | 327.29 | 342.88 | 358.47 |
| K₁ | 180.49 | 195.53 | 210.57 | 225.62 | 240.66 | 255.70 | 270.74 | 285.78 | 300.82 | 315.86 | 330.90 | 345.94 |
| K₂ | 0.99 | 31.07 | 61.15 | 91.23 | 121.31 | 151.40 | 181.48 | 211.56 | 241.64 | 271.72 | 301.81 | 331.89 |
| L₂ | 354.34 | 23.87 | 53.40 | 82.93 | 112.46 | 141.98 | 171.51 | 201.04 | 230.57 | 260.10 | 289.63 | 319.16 |
| M₁† | 173.96 | 188.46 | 202.95 | 217.45 | 231.95 | 246.44 | 260.94 | 275.44 | 289.93 | 304.43 | 318.93 | 333.42 |
| M₂ | 173.90 | 188.40 | 202.89 | 217.38 | 231.87 | 246.36 | 260.86 | 275.35 | 289.84 | 304.33 | 318.83 | 333.32 |
| M₃ | 347.81 | 16.79 | 45.78 | 74.76 | 103.75 | 132.73 | 161.71 | 190.70 | 219.68 | 248.67 | 277.65 | 306.63 |
| M₄ | 161.71 | 205.19 | 248.67 | 292.14 | 335.62 | 19.09 | 2.56 | 106.05 | 149.52 | 193.00 | 236.48 | 279.95 |
| M₆ | 335.62 | 33.59 | 91.55 | 149.52 | 207.49 | 265.46 | 323.43 | 381.40 | 79.36 | 137.33 | 195.30 | 253.27 |
| M₈ | 323.43 | 50.38 | 137.33 | 224.28 | 311.24 | 38.19 | 125.14 | 212.09 | 299.05 | 26.00 | 112.95 | 199.90 |
| Ma | 311.24 | 91.55 | 183.11 | 299.05 | 54.98 | 170.92 | 286.86 | 42.79 | 158.73 | 274.66 | 30.60 | 146.54 |
| N₂ | 341.28 | 9.72 | 38.16 | 58.43 | 86.33 | 114.22 | 151.92 | 180.35 | 208.79 | 237.23 | 265.67 | 294.11 |
| 2N | 334.74 | 2.64 | 30.53 | 58.43 | 86.33 | 114.22 | 142.12 | 170.01 | 197.91 | 225.80 | 253.70 | 281.59 |
| O₁ | 167.32 | 181.26 | 195.20 | 209.15 | 223.09 | 237.03 | 250.97 | 264.92 | 278.86 | 292.80 | 306.75 | 320.69 |
| OO | 193.67 | 209.81 | 225.95 | 242.09 | 258.23 | 274.36 | 290.50 | 306.64 | 322.78 | 338.92 | 355.06 | 11.20 |
| P₁ | 179.51 | 194.47 | 209.43 | 242.38 | 239.34 | 254.30 | 269.26 | 284.22 | 299.18 | 314.14 | 329.10 | 344.06 |
| Q₁ | 160.78 | 174.47 | 187.58 | 200.98 | 214.38 | 227.78 | 241.18 | 254.57 | 267.97 | 281.37 | 294.77 | 308.17 |
| 2Q | 154.25 | 167.11 | 179.96 | 192.81 | 205.67 | 218.52 | 231.38 | 244.23 | 257.09 | 269.94 | 282.79 | 295.65 |
| R₂ | 0.49 | 30.53 | 60.57 | 90.6 | 120.66 | 150.70 | 180.74 | 210.78 | 240.82 | 270.86 | 300.90 | 330.94 |
| S₁ | 180.00 | 195.00 | 210.00 | 225.00 | 240.00 | 255.00 | 270.00 | 285.00 | 300.00 | 315.00 | 330.00 | 345.00 |
| S₂ | 0.00 | 30.00 | 60.00 | 90.00 | 120.00 | 150.00 | 180.00 | 210.00 | 240.00 | 270.00 | 300.00 | 330.00 |
| S₄ | 0.00 | 60.00 | 120.00 | 180.00 | 240.00 | 300.00 | 0.00 | 60.00 | 120.00 | 180.00 | 240.00 | 300.00 |
| S₆ | 0.00 | 90.00 | 180.00 | 270.00 | 0.00 | 90.00 | 180.00 | 270.00 | 0.00 | 90.00 | 180.00 | 270.00 |
| T₂ | 359.51 | 29.47 | 59.43 | 89.38 | 119.34 | 149.30 | 179.26 | 209.22 | 239.18 | 269.14 | 299.10 | 329.06 |
| λ₂ | 353.47 | 22.92 | 52.38 | 81.83 | 111.29 | 140.75 | 170.20 | 199.66 | 229.11 | 258.57 | 288.02 | 317.48 |
| μ₂ | 335.62 | 3.59 | 31.55 | 59.52 | 87.49 | 115.46 | 143.43 | 171.40 | 199.36 | 227.33 | 255.30 | 283.27 |
| ν₂ | 342.15 | 10.66 | 39.18 | 7.69 | 96.20 | 124.71 | 153.23 | 181.74 | 210.25 | 238.76 | 267.28 | 295.79 |
| ρ₁ | 161.66 | 175.13 | 188.60 | 202.07 | 215.54 | 229.02 | 242.49 | 255.96 | 269.43 | 282.90 | 296.37 | 309.84 |
| MK | 168.30 | 212.33 | 256.35 | 300.38 | 344.40 | 28.43 | 72.45 | 116.48 | 160.50 | 204.53 | 248.55 | 292.58 |
| 2MK | 155.13 | 198.05 | 240.98 | 283.91 | 326.83 | 9.76 | 52.69 | 95.62 | 138.54 | 181.47 | 224.40 | 267.32 |
| MN | 329.09 | 26.51 | 83.93 | 141.36 | 198.78 | 256.21 | 313.63 | 11.05 | 68.48 | 125.90 | 183.32 | 240.75 |
| MS | 347.81 | 46.79 | 105.78 | 164.76 | 223.75 | 282.73 | 341.71 | 40.70 | 99.68 | 158.67 | 217.65 | 276.63 |
| 2SM | 12.19 | 43.21 | 74.22 | 105.24 | 136.25 | 167.27 | 198.29 | 229.30 | 260.32 | 291.33 | 322.35 | 353.37 |
| Mf | 13.18 | 14.27 | 15.37 | 16.47 | 17.57 | 18.67 | 19.76 | 20.86 | 21.96 | 23.06 | 24.16 | 25.25 |
| MSf | 12.19 | 13.21 | 14.22 | 15.24 | 16.25 | 17.27 | 18.29 | 19.30 | 20.32 | 21.33 | 22.35 | 23.37 |
| Mm | 6.53 | 7.08 | 7.62 | 8.17 | 8.71 | 9.25 | 9.80 | 10.34 | 10.89 | 11.43 | 11.98 | 12.52 |
| Sa | 0.49 | 0.53 | 0.57 | 0.62 | 0.66 | 0.70 | 0.74 | 0.78 | 0.82 | 0.86 | 0.90 | 0.94 |
| Ssa | 0.99 | 1.07 | 1.15 | 1.23 | 1.31 | 1.40 | 1.48 | 1.56 | 1.64 | 1.72 | 1.81 | 1.89 |

†The first line for constituent M₁ gives the differences as based upon the formula in table 2; the second line gives the differences as derived from the half speed of constituent M₂.

## Table 19.—*Products for Form 194*

[Multiplier=sin 15°=**0.259**]

|  | 0 | 1 | 2 | 3 | 4 | 5 | 6 | 7 | 8 | 9 |
|---|---|---|---|---|---|---|---|---|---|---|
| 0.00 | 0.000 | 0.259 | 0.518 | 0.777 | 1.036 | 1.295 | 1.554 | 1.813 | 2.072 | 2.331 |
| .01 | .003 | .262 | .521 | .780 | 1.039 | 1.298 | 1.557 | 1.816 | 2.075 | 2.334 |
| .02 | .005 | .264 | .523 | .782 | 1.041 | 1.300 | 1.559 | 1.818 | 2.077 | 2.336 |
| .03 | .008 | .267 | .526 | .785 | 1.044 | 1.303 | 1.562 | 1.821 | 2.080 | 2.339 |
| .04 | .010 | .269 | .528 | .787 | 1.046 | 1.305 | 1.564 | 1.823 | 2.082 | 2.341 |
| .05 | .013 | .272 | .531 | .790 | 1.049 | 1.308 | 1.567 | 1.826 | 2.085 | 2.344 |
| .06 | .016 | .275 | .534 | .793 | 1.052 | 1.311 | 1.570 | 1.829 | 2.088 | 2.347 |
| .07 | .018 | .277 | .536 | .795 | 1.054 | 1.313 | 1.572 | 1.831 | 2.090 | 2.349 |
| .08 | .021 | .280 | .539 | .798 | 1.057 | 1.316 | 1.575 | 1.834 | 2.093 | 2.352 |
| .09 | .023 | .282 | .541 | .800 | 1.059 | 1.318 | 1.577 | 1.836 | 2.095 | 2.354 |
| .10 | .026 | .285 | .544 | .803 | 1.062 | 1.321 | 1.580 | 1.839 | 2.098 | 2.357 |
| .11 | .028 | .287 | .546 | .805 | 1.064 | 1.323 | 1.582 | 1.841 | 2.100 | 2.359 |
| .12 | .031 | .290 | .549 | .808 | 1.067 | 1.326 | 1.585 | 1.844 | 2.103 | 2.362 |
| .13 | .034 | .293 | .552 | .811 | 1.070 | 1.329 | 1.588 | 1.847 | 2.106 | 2.365 |
| .14 | .036 | .295 | .554 | .813 | 1.072 | 1.331 | 1.590 | 1.849 | 2.108 | 2.367 |
| .15 | .039 | .298 | .557 | .816 | 1.075 | 1.334 | 1.593 | 1.852 | 2.111 | 2.370 |
| .16 | .041 | .300 | .559 | .818 | 1.077 | 1.336 | 1.595 | 1.854 | 2.113 | 2.372 |
| .17 | .044 | .303 | .562 | .821 | 1.080 | 1.339 | 1.598 | 1.857 | 2.116 | 2.375 |
| .18 | .047 | .306 | .565 | .824 | 1.083 | 1.342 | 1.601 | 1.860 | 2.119 | 2.378 |
| .19 | .049 | .308 | .567 | .826 | 1.085 | 1.344 | 1.603 | 1.862 | 2.121 | 2.380 |
| .20 | .052 | .311 | .570 | .829 | 1.088 | 1.347 | 1.606 | 1.865 | 2.124 | 2.383 |
| .21 | .054 | .313 | .572 | .831 | 1.090 | 1.349 | 1.608 | 1.867 | 2.126 | 2.385 |
| .22 | .057 | .316 | .575 | .834 | 1.093 | 1.352 | 1.611 | 1.870 | 2.129 | 2.388 |
| .23 | .060 | .319 | .578 | .837 | 1.096 | 1.355 | 1.614 | 1.873 | 2.132 | 2.391 |
| .24 | .062 | .321 | .580 | .839 | 1.098 | 1.357 | 1.616 | 1.875 | 2.134 | 2.393 |
| .25 | .065 | .324 | .583 | .842 | 1.101 | 1.360 | 1.619 | 1.878 | 2.137 | 2.396 |
| .26 | .067 | .326 | .585 | .844 | 1.103 | 1.362 | 1.621 | 1.880 | 2.139 | 2.398 |
| .27 | .070 | .329 | .588 | .847 | 1.106 | 1.365 | 1.624 | 1.883 | 2.142 | 2.401 |
| .28 | .073 | .332 | .591 | .850 | 1.109 | 1.368 | 1.627 | 1.886 | 2.145 | 2.404 |
| .29 | .075 | .334 | .593 | .852 | 1.111 | 1.370 | 1.629 | 1.888 | 2.147 | 2.406 |
| .30 | .078 | .337 | .596 | .855 | 1.114 | 1.373 | 1.632 | 1.891 | 2.150 | 2.409 |
| .31 | .080 | .339 | .598 | .857 | 1.116 | 1.375 | 1.634 | 1.893 | 2.152 | 2.411 |
| .32 | .083 | .342 | .601 | .860 | 1.119 | 1.378 | 1.637 | 1.896 | 2.155 | 2.414 |
| .33 | .085 | .344 | .603 | .862 | 1.121 | 1.380 | 1.639 | 1.898 | 2.157 | 2.416 |
| .34 | .088 | .347 | .606 | .865 | 1.124 | 1.383 | 1.642 | 1.901 | 2.160 | 2.419 |
| .35 | .091 | .350 | .609 | .868 | 1.127 | 1.386 | 1.645 | 1.904 | 2.163 | 2.422 |
| .36 | .093 | .352 | .611 | .870 | 1.129 | 1.388 | 1.647 | 1.906 | 2.165 | 2.424 |
| .37 | .096 | .355 | .614 | .873 | 1.132 | 1.391 | 1.650 | 1.909 | 2.168 | 2.427 |
| .38 | .098 | .357 | .616 | .875 | 1.134 | 1.393 | 1.652 | 1.911 | 2.170 | 2.429 |
| .39 | .101 | .360 | .619 | .878 | 1.137 | 1.396 | 1.655 | 1.914 | 2.173 | 2.432 |
| .40 | .104 | .363 | .622 | .881 | 1.140 | 1.399 | 1.658 | 1.917 | 2.176 | 2.435 |
| .41 | .106 | .365 | .624 | .883 | 1.142 | 1.401 | 1.660 | 1.919 | 2.178 | 2.437 |
| .42 | .109 | .368 | .627 | .886 | 1.145 | 1.404 | 1.663 | 1.922 | 2.181 | 2.440 |
| .43 | .111 | .370 | .629 | .888 | 1.147 | 1.406 | 1.665 | 1.924 | 2.183 | 2.442 |
| .44 | .114 | .373 | .632 | .891 | 1.150 | 1.409 | 1.668 | 1.927 | 2.186 | 2.445 |
| .45 | .117 | .376 | .635 | .894 | 1.153 | 1.412 | 1.671 | 1.930 | 2.189 | 2.448 |
| .46 | .119 | .378 | .637 | .896 | 1.155 | 1.414 | 1.673 | 1.932 | 2.191 | 2.450 |
| .47 | .122 | .381 | .640 | .899 | 1.158 | 1.417 | 1.676 | 1.935 | 2.194 | 2.453 |
| .48 | .124 | .383 | .642 | .901 | 1.160 | 1.419 | 1.678 | 1.937 | 2.196 | 2.455 |
| .49 | .127 | .386 | .645 | .904 | 1.163 | 1.422 | 1.681 | 1.940 | 2.199 | 2.458 |
| .50 | .130 | .388 | .648 | .906 | 1.166 | 1.424 | 1.684 | 1.942 | 2.202 | 2.460 |
|  | 0 | 1 | 2 | 3 | 4 | 5 | 6 | 7 | 8 | 9 |

HARMONIC ANALYSIS AND PREDICTION OF TIDES 219

Table 19.—*Products for Form 194*—Continued

[Multiplier=sin 15°=0.259]

|  | 0 | 1 | 2 | 3 | 4 | 5 | 6 | 7 | 8 | 9 |
|---|---|---|---|---|---|---|---|---|---|---|
| 0.50 | 0.130 | 0.388 | 0.648 | 0.906 | 1.166 | 1.424 | 1.684 | 1.942 | 2.202 | 2.460 |
| .51 | .132 | .391 | .650 | .909 | 1.168 | 1.427 | 1.686 | 1.945 | 2.204 | 2.463 |
| .52 | .135 | .394 | .653 | .912 | 1.171 | 1.430 | 1.689 | 1.948 | 2.207 | 2.466 |
| .53 | .137 | .396 | .655 | .914 | 1.173 | 1.432 | 1.691 | 1.950 | 2.209 | 2.468 |
| .54 | .140 | .399 | .658 | .917 | 1.176 | 1.435 | 1.694 | 1.953 | 2.212 | 2.471 |
| .55 | .142 | .401 | .660 | .919 | 1.178 | 1.437 | 1.696 | 1.955 | 2.214 | 2.473 |
| .56 | .145 | .404 | .663 | .922 | 1.181 | 1.440 | 1.699 | 1.958 | 2.217 | 2.476 |
| .57 | .148 | .407 | .666 | .925 | 1.184 | 1.443 | 1.702 | 1.961 | 2.220 | 2.479 |
| .58 | .150 | .409 | .668 | .927 | 1.186 | 1.445 | 1.704 | 1.963 | 2.222 | 2.481 |
| .59 | .153 | .412 | .671 | .930 | 1.189 | 1.448 | 1.707 | 1.966 | 2.225 | 2.484 |
| .60 | .155 | .414 | .673 | .932 | 1.191 | 1.450 | 1.709 | 1.968 | 2.227 | 2.486 |
| .61 | .158 | .417 | .676 | .935 | 1.194 | 1.453 | 1.712 | 1.971 | 2.230 | 2.489 |
| .62 | .161 | .420 | .679 | .938 | 1.197 | 1.456 | 1.715 | 1.974 | 2.233 | 2.492 |
| .63 | .163 | .422 | .681 | .940 | 1.199 | 1.458 | 1.717 | 1.976 | 2.235 | 2.494 |
| .64 | .166 | .425 | .684 | .943 | 1.202 | 1.461 | 1.720 | 1.979 | 2.238 | 2.497 |
| .65 | .168 | .427 | .686 | .945 | 1.204 | 1.463 | 1.722 | 1.981 | 2.240 | 2.499 |
| .66 | .171 | .430 | .689 | .948 | 1.207 | 1.466 | 1.725 | 1.984 | 2.243 | 2.502 |
| .67 | .174 | .433 | .692 | .951 | 1.210 | 1.469 | 1.728 | 1.987 | 2.246 | 2.505 |
| .68 | .176 | .435 | .694 | .953 | 1.212 | 1.471 | 1.730 | 1.989 | 2.248 | 2.507 |
| .69 | .179 | .438 | .697 | .956 | 1.215 | 1.474 | 1.733 | 1.992 | 2.251 | 2.510 |
| .70 | .181 | .440 | .699 | .958 | 1.217 | 1.476 | 1.735 | 1.994 | 2.253 | 2.512 |
| .71 | .184 | .443 | .702 | .961 | 1.220 | 1.479 | 1.738 | 1.997 | 2.256 | 2.515 |
| .72 | .186 | .445 | .704 | .963 | 1.222 | 1.481 | 1.740 | 1.999 | 2.258 | 2.517 |
| .73 | .189 | .448 | .707 | .966 | 1.225 | 1.484 | 1.743 | 2.002 | 2.261 | 2.520 |
| .74 | .192 | .451 | .710 | .969 | 1.228 | 1.487 | 1.746 | 2.005 | 2.264 | 2.523 |
| .75 | .194 | .453 | .712 | .971 | 1.230 | 1.489 | 1.748 | 2.007 | 2.266 | 2.525 |
| .76 | .197 | .456 | .715 | .974 | 1.233 | 1.492 | 1.751 | 2.010 | 2.269 | 2.528 |
| .77 | .199 | .458 | .717 | .976 | 1.235 | 1.494 | 1.753 | 2.012 | 2.271 | 2.530 |
| .78 | .202 | .461 | .720 | .979 | 1.238 | 1.497 | 1.756 | 2.015 | 2.274 | 2.533 |
| .79 | .205 | .464 | .723 | .982 | 1.241 | 1.500 | 1.759 | 2.018 | 2.277 | 2.536 |
| .80 | .207 | .466 | .725 | .984 | 1.243 | 1.502 | 1.761 | 2.020 | 2.279 | 2.538 |
| .81 | .210 | .469 | .728 | .987 | 1.246 | 1.505 | 1.764 | 2.023 | 2.282 | 2.541 |
| .82 | .212 | .471 | .730 | .989 | 1.248 | 1.507 | 1.766 | 2.025 | 2.284 | 2.543 |
| .83 | .215 | .474 | .733 | .992 | 1.251 | 1.510 | 1.769 | 2.028 | 2.287 | 2.546 |
| .84 | .218 | .477 | .736 | .995 | 1.254 | 1.513 | 1.772 | 2.031 | 2.290 | 2.549 |
| .85 | .220 | .479 | .738 | .997 | 1.256 | 1.515 | 1.774 | 2.033 | 2.292 | 2.551 |
| .86 | .223 | .482 | .741 | 1.000 | 1.259 | 1.518 | 1.777 | 2.036 | 2.295 | 2.554 |
| .87 | .225 | .484 | .743 | 1.002 | 1.261 | 1.520 | 1.779 | 2.038 | 2.297 | 2.556 |
| .88 | .228 | .487 | .746 | 1.005 | 1.264 | 1.523 | 1.782 | 2.041 | 2.300 | 2.559 |
| .89 | .231 | .490 | .749 | 1.008 | 1.267 | 1.526 | 1.785 | 2.044 | 2.303 | 2.562 |
| .90 | .233 | .492 | .751 | 1.010 | 1.269 | 1.528 | 1.787 | 2.046 | 2.305 | 2.564 |
| .91 | .236 | .495 | .754 | 1.013 | 1.272 | 1.531 | 1.790 | 2.049 | 2.308 | 2.567 |
| .92 | .238 | .497 | .756 | 1.015 | 1.274 | 1.533 | 1.792 | 2.051 | 2.310 | 2.569 |
| .93 | .241 | .500 | .759 | 1.018 | 1.277 | 1.536 | 1.795 | 2.054 | 2.313 | 2.572 |
| .94 | .243 | .502 | .761 | 1.020 | 1.279 | 1.538 | 1.797 | 2.056 | 2.315 | 2.574 |
| .95 | .246 | .505 | .764 | 1.023 | 1.282 | 1.541 | 1.800 | 2.059 | 2.318 | 2.577 |
| .96 | .249 | .508 | .767 | 1.026 | 1.285 | 1.544 | 1.803 | 2.062 | 2.321 | 2.580 |
| .97 | .251 | .510 | .769 | 1.028 | 1.287 | 1.546 | 1.805 | 2.064 | 2.323 | 2.582 |
| .98 | .254 | .513 | .772 | 1.031 | 1.290 | 1.549 | 1.808 | 2.067 | 2.326 | 2.585 |
| .99 | .256 | .515 | .774 | 1.033 | 1.292 | 1.551 | 1.810 | 2.069 | 2.328 | 2.587 |
| 1.00 | .259 | .518 | .777 | 1.036 | 1.295 | 1.554 | 1.813 | 2.072 | 2.331 | 2.590 |
|  | 0 | 1 | 2 | 3 | 4 | 5 | 6 | 7 | 8 | 9 |

Table 19.—*Products for Form 194*—Continued

[Multiplier=sin 30°=**0.500**]

|        | 0      | 1     | 2      | 3      | 4      | 5      | 6      | 7      | 8      | 9      |
|--------|--------|-------|--------|--------|--------|--------|--------|--------|--------|--------|
| 0.00   | 0.000  | 0.500 | 1.000  | 1.500  | 2.000  | 2.500  | 3.000  | 3.500  | 4.000  | 4.500  |
| .01    | .005   | .505  | 1.005  | 1.505  | 2.005  | 2.505  | 3.005  | 3.505  | 4.005  | 4.505  |
| .02    | .010   | .510  | 1.010  | 1.510  | 2.010  | 2.510  | 3.010  | 3.510  | 4.010  | 4.510  |
| .03    | .015   | .515  | 1.015  | 1.515  | 2.015  | 2.515  | 3.015  | 3.515  | 4.015  | 4.515  |
| .04    | .020   | .520  | 1.020  | 1.520  | 2.020  | 2.520  | 3.020  | 3.520  | 4.020  | 4.520  |
| .05    | .025   | .525  | 1.025  | 1.525  | 2.025  | 2.525  | 3.025  | 3.525  | 4.025  | 4.525  |
| .06    | .030   | .530  | 1.030  | 1.530  | 2.030  | 2.530  | 3.030  | 3.530  | 4.030  | 4.530  |
| .07    | .035   | .535  | 1.035  | 1.535  | 2.035  | 2.535  | 3.035  | 3.535  | 4.035  | 4.535  |
| .08    | .040   | .540  | 1.040  | 1.540  | 2.040  | 2.540  | 3.040  | 3.540  | 4.040  | 4.540  |
| .09    | .045   | .545  | 1.045  | 1.545  | 2.045  | 2.545  | 3.045  | 3.545  | 4.045  | 4.545  |
| .10    | .050   | .550  | 1.050  | 1.550  | 2.050  | 2.550  | 3.050  | 3.550  | 4.050  | 4.550  |
| .11    | .055   | .555  | 1.055  | 1.555  | 2.055  | 2.555  | 3.055  | 3.555  | 4.055  | 4.555  |
| .12    | .060   | .560  | 1.060  | 1.560  | 2.060  | 2.560  | 3.060  | 3.560  | 4.060  | 4.560  |
| .13    | .065   | .565  | 1.065  | 1.565  | 2.065  | 2.565  | 3.065  | 3.565  | 4.065  | 4.565  |
| .14    | .070   | .570  | 1.070  | 1.570  | 2.070  | 2.570  | 3.070  | 3.570  | 4.070  | 4.570  |
| .15    | .075   | .575  | 1.075  | 1.575  | 2.075  | 2.575  | 3.075  | 3.575  | 4.075  | 4.575  |
| .16    | .080   | .580  | 1.080  | 1.580  | 2.080  | 2.580  | 3.080  | 3.580  | 4.080  | 4.580  |
| .17    | .085   | .585  | 1.085  | 1.585  | 2.085  | 2.585  | 3.085  | 3.585  | 4.085  | 4.585  |
| .18    | .090   | .590  | 1.090  | 1.590  | 2.090  | 2.590  | 3.090  | 3.590  | 4.090  | 4.590  |
| .19    | .095   | .595  | 1.095  | 1.595  | 2.095  | 2.595  | 3.095  | 3.595  | 4.095  | 4.595  |
| .20    | .100   | .600  | 1.100  | 1.600  | 2.100  | 2.600  | 3.100  | 3.600  | 4.100  | 4.600  |
| .21    | .105   | .605  | 1.105  | 1.605  | 2.105  | 2.605  | 3.105  | 3.605  | 4.105  | 4.605  |
| .22    | .110   | .610  | 1.110  | 1.610  | 2.110  | 2.610  | 3.110  | 3.610  | 4.110  | 4.610  |
| .23    | .115   | .615  | 1.115  | 1.615  | 2.115  | 2.615  | 3.115  | 3.615  | 4.115  | 4.615  |
| .24    | .120   | .620  | 1.120  | 1.620  | 2.120  | 2.620  | 3.120  | 3.620  | 4.120  | 4.620  |
| .25    | .125   | .625  | 1.125  | 1.625  | 2.125  | 2.625  | 2.125  | 3.625  | 4.125  | 4.625  |
| .26    | .130   | .630  | 1.130  | 1.630  | 2.130  | 2.630  | 3.130  | 3.630  | 4.130  | 4.630  |
| .27    | .135   | .635  | 1.135  | 1.635  | 2.135  | 2.635  | 3.135  | 3.635  | 4.135  | 4.635  |
| .28    | .140   | .640  | 1.140  | 1.640  | 2.140  | 2.640  | 3.140  | 3.640  | 4.140  | 4.640  |
| .29    | .145   | .645  | 1.145  | 1.645  | 2.145  | 2.645  | 3.145  | 3.645  | 4.145  | 4.645  |
| .30    | .150   | .650  | 1.150  | 1.650  | 2.150  | 2.650  | 3.150  | 3.650  | 4.150  | 4.650  |
| .31    | .155   | .655  | 1.155  | 1.655  | 2.155  | 2.655  | 3.155  | 3.655  | 4.155  | 4.655  |
| .32    | .160   | .660  | 1.160  | 1.660  | 2.160  | 2.660  | 3.160  | 3.660  | 4.160  | 4.660  |
| .33    | .165   | .665  | 1.165  | 1.665  | 2.165  | 2.665  | 3.165  | 3.665  | 4.165  | 4.665  |
| .34    | .170   | .670  | 1.170  | 1.670  | 2.170  | 2.670  | 3.170  | 3.670  | 4.170  | 4.670  |
| .35    | .175   | .675  | 1.175  | 1.675  | 2.175  | 2.675  | 3.175  | 3.675  | 4.175  | 4.675  |
| .36    | .180   | .680  | 1.180  | 1.680  | 2.180  | 2.680  | 3.180  | 3.680  | 4.180  | 4.680  |
| .37    | .185   | .685  | 1.185  | 1.685  | 2.185  | 2.685  | 3.185  | 3.685  | 4.185  | 4.685  |
| .38    | .190   | .690  | 1.190  | 1.690  | 2.190  | 2.690  | 3.190  | 3.690  | 4.190  | 4.690  |
| .39    | .195   | .695  | 1.195  | 1.695  | 2.195  | 2.695  | 3.195  | 3.695  | 4.195  | 4.695  |
| .40    | .200   | .700  | 1.200  | 1.700  | 2.200  | 2.700  | 3.200  | 3.700  | 4.200  | 4.700  |
| .41    | .205   | .705  | 1.205  | 1.705  | 2.205  | 2.705  | 3.205  | 3.705  | 4.205  | 4.705  |
| .42    | .210   | .710  | 1.210  | 1.710  | 2.210  | 2.710  | 3.210  | 3.710  | 4.210  | 4.710  |
| .43    | .215   | .715  | 1.215  | 1.715  | 2.215  | 2.715  | 3.215  | 3.715  | 4.215  | 4.715  |
| .44    | .220   | .720  | 1.220  | 1.720  | 2.220  | 2.720  | 3.220  | 3.720  | 4.220  | 4.720  |
| .45    | .225   | .725  | 1.225  | 1.725  | 2.225  | 2.725  | 3.225  | 3.725  | 4.225  | 4.725  |
| .46    | .230   | .730  | 1.230  | 1.730  | 2.230  | 2.730  | 3.230  | 3.730  | 4.230  | 4.730  |
| .47    | .235   | .735  | 1.235  | 1.735  | 2.235  | 2.735  | 3.235  | 3.735  | 4.235  | 4.735  |
| .48    | .240   | .740  | 1.240  | 1.740  | 2.240  | 2.740  | 3.240  | 3.740  | 4.240  | 4.740  |
| .49    | .245   | .745  | 1.245  | 1.745  | 2.245  | 2.745  | 3.245  | 3.745  | 4.245  | 4.745  |
| .50    | .250   | .750  | 1.250  | 1.750  | 2.250  | 2.750  | 3.250  | 3.750  | 4.250  | 4.750  |
|        | 0      | 1     | 2      | 3      | 4      | 5      | 6      | 7      | 8      | 9      |

## Table 19.—Products for Form 194—Continued

[Multiplier=sin 30°=**0.500**]

| | 0 | 1 | 2 | 3 | 4 | 5 | 6 | 7 | 8 | 9 |
|---|---|---|---|---|---|---|---|---|---|---|
| 0.50 | 0.250 | 0.750 | 1.250 | 1.750 | 2.250 | 2.750 | 3.250 | 3.750 | 4.250 | 4.750 |
| .51 | .255 | .755 | 1.255 | 1.755 | 2.255 | 2.755 | 3.255 | 3.755 | 4.255 | 4.755 |
| .52 | .260 | .760 | 1.260 | 1.760 | 2.260 | 2.760 | 3.260 | 3.760 | 4.260 | 4.760 |
| .53 | .265 | .765 | 1.265 | 1.765 | 2.265 | 2.765 | 3.265 | 3.765 | 4.265 | 4.765 |
| .54 | .270 | .770 | 1.270 | 1.770 | 2.270 | 2.770 | 3.270 | 3.770 | 4.270 | 4.770 |
| .55 | .275 | .775 | 1.275 | 1.775 | 2.275 | 2.775 | 3.275 | 3.775 | 4.275 | 4.775 |
| .56 | .280 | .780 | 1.280 | 1.780 | 2.280 | 2.780 | 3.280 | 3.780 | 4.280 | 4.780 |
| .57 | .285 | .785 | 1.285 | 1.785 | 2.285 | 2.785 | 3.285 | 3.785 | 4.285 | 4.785 |
| .58 | .290 | .790 | 1.290 | 1.790 | 2.290 | 2.790 | 3.290 | 3.790 | 4.290 | 4.790 |
| .59 | .295 | .795 | 1.295 | 1.795 | 2.295 | 2.795 | 3.295 | 3.795 | 4.295 | 4.795 |
| .60 | .300 | .800 | 1.300 | 1.800 | 2.300 | 2.800 | 3.300 | 3.800 | 4.300 | 4.800 |
| .61 | .305 | .805 | 1.305 | 1.805 | 2.305 | 2.805 | 3.305 | 3.805 | 4.305 | 4.805 |
| .62 | .310 | .810 | 1.310 | 1.810 | 2.310 | 2.810 | 3.310 | 3.810 | 4.310 | 4.810 |
| .63 | .315 | .815 | 1.315 | 1.815 | 2.315 | 2.815 | 3.315 | 3.815 | 4.315 | 4.815 |
| .64 | .320 | .820 | 1.320 | 1.820 | 2.320 | 2.820 | 3.320 | 3.820 | 4.320 | 4.820 |
| .65 | .325 | .825 | 1.325 | 1.825 | 2.325 | 2.825 | 3.325 | 3.825 | 4.325 | 4.825 |
| .66 | .330 | .830 | 1.330 | 1.830 | 2.330 | 2.830 | 3.330 | 3.830 | 4.330 | 4.830 |
| .67 | .335 | .835 | 1.335 | 1.835 | 2.335 | 2.835 | 3.335 | 3.835 | 4.335 | 4.835 |
| .68 | .340 | .840 | 1.340 | 1.840 | 2.340 | 2.840 | 3.340 | 3.840 | 4.340 | 4.840 |
| .69 | .345 | .845 | 1.345 | 1.845 | 2.345 | 2.845 | 3.345 | 3.845 | 4.345 | 4.845 |
| .70 | .350 | .850 | 1.350 | 1.850 | 2.350 | 2.850 | 3.350 | 3.850 | 4.350 | 4.850 |
| .71 | .355 | .855 | 1.355 | 1.855 | 2.355 | 2.855 | 3.355 | 3.855 | 4.355 | 4.855 |
| .72 | .360 | .860 | 1.360 | 1.860 | 2.360 | 2.860 | 3.360 | 3.860 | 4.360 | 4.860 |
| .73 | .365 | .865 | 1.365 | 1.865 | 2.365 | 2.865 | 3.365 | 3.865 | 4.365 | 4.865 |
| .74 | .370 | .870 | 1.370 | 1.870 | 2.370 | 2.870 | 3.370 | 3.870 | 4.370 | 4.870 |
| .75 | .375 | .875 | 1.375 | 1.875 | 2.375 | 2.875 | 3.375 | 3.875 | 4.375 | 4.875 |
| .76 | .380 | .880 | 1.380 | 1.880 | 2.380 | 2.880 | 3.380 | 3.880 | 4.380 | 4.880 |
| .77 | .385 | .885 | 1.385 | 1.885 | 2.385 | 2.885 | 3.385 | 3.885 | 4.385 | 4.885 |
| .78 | .390 | .890 | 1.390 | 1.890 | 2.390 | 2.890 | 3.390 | 3.890 | 4.390 | 4.890 |
| .79 | .395 | .895 | 1.395 | 1.895 | 2.395 | 2.895 | 3.395 | 3.895 | 4.395 | 4.895 |
| .80 | .400 | .900 | 1.400 | 1.900 | 2.400 | 2.900 | 3.400 | 3.900 | 4.400 | 4.900 |
| .81 | .405 | .905 | 1.405 | 1.905 | 2.405 | 2.905 | 3.405 | 3.905 | 4.405 | 4.905 |
| .82 | .410 | .910 | 1.410 | 1.910 | 2.410 | 2.910 | 3.410 | 3.910 | 4.410 | 4.910 |
| .83 | .415 | .915 | 1.415 | 1.915 | 2.415 | 2.915 | 3.415 | 3.915 | 4.415 | 4.915 |
| .84 | .420 | .920 | 1.420 | 1.920 | 2.420 | 2.920 | 3.420 | 3.920 | 4.420 | 4.920 |
| .85 | .425 | .925 | 1.425 | 1.925 | 2.425 | 2.925 | 3.425 | 3.925 | 4.425 | 4.925 |
| .86 | .430 | .930 | 1.430 | 1.930 | 2.430 | 2.930 | 3.430 | 3.930 | 4.430 | 4.930 |
| .87 | .435 | .935 | 1.435 | 1.935 | 2.435 | 2.935 | 3.435 | 3.935 | 4.435 | 4.935 |
| .88 | .440 | .940 | 1.440 | 1.940 | 2.440 | 2.940 | 3.440 | 3.940 | 4.440 | 4.940 |
| .89 | .445 | .945 | 1.445 | 1.945 | 2.445 | 2.945 | 3.445 | 3.945 | 4.445 | 4.945 |
| .90 | .450 | .950 | 1.450 | 1.950 | 2.450 | 2.950 | 3.450 | 3.950 | 4.450 | 4.950 |
| .91 | .455 | .955 | 1.455 | 1.955 | 2.455 | 2.955 | 3.455 | 3.955 | 4.455 | 4.955 |
| .92 | .460 | .960 | 1.460 | 1.960 | 2.460 | 2.960 | 3.460 | 3.960 | 4.460 | 4.960 |
| .93 | .465 | .965 | 1.465 | 1.965 | 2.465 | 2.965 | 3.465 | 3.965 | 4.465 | 4.965 |
| .94 | .470 | .970 | 1.470 | 1.970 | 2.470 | 2.970 | 3.470 | 3.970 | 4.470 | 4.970 |
| .95 | .475 | .975 | 1.475 | 1.975 | 2.475 | 2.975 | 3.475 | 3.975 | 4.475 | 4.975 |
| .96 | .480 | .980 | 1.480 | 1.980 | 2.480 | 2.980 | 3.480 | 3.980 | 4.480 | 4.980 |
| .97 | .485 | .985 | 1.485 | 1.985 | 2.485 | 2.985 | 3.485 | 3.985 | 4.485 | 4.985 |
| .98 | .490 | .990 | 1.490 | 1.990 | 2.490 | 2.990 | 3.490 | 3.990 | 4.490 | 4.990 |
| .99 | .495 | .995 | 1.495 | 1.995 | 2.495 | 2.995 | 3.495 | 3.995 | 4.495 | 4.995 |
| 1.00 | .500 | 1.000 | 1.500 | 2.000 | 2.500 | 3.000 | 3.500 | 4.000 | 4.500 | 5.000 |
| | 0 | 1 | 2 | 3 | 4 | 5 | 6 | 7 | 8 | 9 |

U. S. COAST AND GEODETIC SURVEY

Table 19.—*Products for Form 194*—Continued

[Multiplier=sin 45°=**0.707**]

|       | 0     | 1     | 2     | 3     | 4     | 5     | 6     | 7     | 8     | 9     |
|-------|-------|-------|-------|-------|-------|-------|-------|-------|-------|-------|
| 0.00  | 0.000 | 0.707 | 1.414 | 2.121 | 2.828 | 3.535 | 4.242 | 4.949 | 5.656 | 6.363 |
| .01   | .007  | .714  | 1.421 | 2.128 | 2.835 | 3.542 | 4.249 | 4.956 | 5.663 | 6.370 |
| .02   | .014  | .721  | 1.428 | 2.135 | 2.842 | 3.549 | 4.256 | 4.963 | 5.670 | 6.377 |
| .03   | .021  | .728  | 1.435 | 2.142 | 2.849 | 3.556 | 4.263 | 4.970 | 5.677 | 6.384 |
| .04   | .028  | .735  | 1.442 | 2.149 | 2.856 | 3.563 | 4.270 | 4.977 | 5.684 | 6.391 |
| .05   | .035  | .742  | 1.449 | 2.156 | 2.863 | 3.570 | 4.277 | 4.984 | 5.691 | 6.398 |
| .06   | .042  | .749  | 1.456 | 2.163 | 2.870 | 3.577 | 4.284 | 4.991 | 5.698 | 6.405 |
| .07   | .049  | .756  | 1.463 | 2.170 | 2.877 | 3.584 | 4.291 | 4.998 | 5.705 | 6.412 |
| .08   | .057  | .764  | 1.471 | 2.178 | 2.885 | 3.592 | 4.299 | 5.006 | 5.713 | 6.420 |
| .09   | .064  | .771  | 1.478 | 2.185 | 2.892 | 3.599 | 4.306 | 5.013 | 5.720 | 6.427 |
| .10   | .071  | .778  | 1.485 | 2.192 | 2.899 | 3.606 | 4.313 | 5.020 | 5.727 | 6.434 |
| .11   | .078  | .785  | 1.492 | 2.199 | 2.906 | 3.613 | 4.320 | 5.027 | 5.734 | 6.441 |
| .12   | .085  | .792  | 1.499 | 2.206 | 2.913 | 3.620 | 4.327 | 5.034 | 5.741 | 6.448 |
| .13   | .092  | .799  | 1.506 | 2.213 | 2.920 | 3.627 | 4.334 | 5.041 | 5.748 | 6.455 |
| .14   | .099  | .806  | 1.513 | 2.220 | 2.927 | 3.634 | 4.341 | 5.048 | 5.755 | 6.462 |
| .15   | .106  | .813  | 1.520 | 2.227 | 2.934 | 3.641 | 4.348 | 5.055 | 5.762 | 6.469 |
| .16   | .113  | .820  | 1.527 | 2.234 | 2.941 | 3.648 | 4.355 | 5.062 | 5.769 | 6.476 |
| .17   | .120  | .827  | 1.534 | 2.241 | 2.948 | 3.655 | 4.362 | 5.069 | 5.776 | 6.483 |
| .18   | .127  | .834  | 1.541 | 2.248 | 2.955 | 3.662 | 4.369 | 5.076 | 5.783 | 6.490 |
| .19   | .134  | .841  | 1.548 | 2.255 | 2.962 | 3.669 | 4.376 | 5.083 | 5.790 | 6.497 |
| .20   | .141  | .848  | 1.555 | 2.262 | 2.969 | 3.676 | 4.383 | 5.090 | 5.797 | 6.504 |
| .21   | .148  | .855  | 1.562 | 2.269 | 2.976 | 3.683 | 4.390 | 5.097 | 5.804 | 6.511 |
| .22   | .156  | .863  | 1.570 | 2.277 | 2.984 | 3.691 | 4.398 | 5.105 | 5.812 | 6.519 |
| .23   | .163  | .870  | 1.577 | 2.284 | 2.991 | 3.698 | 4.405 | 5.112 | 5.819 | 6.526 |
| .24   | .170  | .877  | 1.584 | 2.291 | 2.998 | 3.705 | 4.412 | 5.119 | 5.826 | 6.533 |
| .25   | .177  | .884  | 1.591 | 2.298 | 3.005 | 3.712 | 4.419 | 5.126 | 5.833 | 6.540 |
| .26   | .184  | .891  | 1.598 | 2.305 | 3.012 | 3.719 | 4.426 | 5.133 | 5.840 | 6.547 |
| .27   | .191  | .898  | 1.605 | 2.312 | 3.019 | 3.726 | 4.433 | 5.140 | 5.847 | 6.554 |
| .28   | .198  | .905  | 1.612 | 2.319 | 3.026 | 3.733 | 4.440 | 5.147 | 5.854 | 6.561 |
| .29   | .205  | .912  | 1.619 | 2.326 | 3.033 | 3.740 | 4.447 | 5.154 | 5.861 | 6.568 |
| .30   | .212  | .919  | 1.626 | 2.333 | 3.040 | 3.747 | 4.454 | 5.161 | 5.868 | 6.575 |
| .31   | .219  | .926  | 1.633 | 2.340 | 3.047 | 3.754 | 4.461 | 5.168 | 5.875 | 6.582 |
| .32   | .226  | .933  | 1.640 | 2.347 | 3.054 | 3.761 | 4.468 | 5.175 | 5.882 | 6.589 |
| .33   | .233  | .940  | 1.647 | 2.354 | 3.061 | 3.768 | 4.475 | 5.182 | 5.889 | 6.596 |
| .34   | .240  | .947  | 1.654 | 2.361 | 3.068 | 3.775 | 4.482 | 5.189 | 5.896 | 6.603 |
| .35   | .247  | .954  | 1.661 | 2.368 | 3.075 | 3.782 | 4.489 | 5.196 | 5.903 | 6.610 |
| .36   | .255  | .962  | 1.669 | 2.376 | 3.083 | 3.790 | 4.497 | 5.204 | 5.911 | 6.618 |
| .37   | .262  | .969  | 1.676 | 2.383 | 3.090 | 3.797 | 4.504 | 5.211 | 5.918 | 6.625 |
| .38   | .269  | .976  | 1.683 | 2.390 | 3.097 | 3.804 | 4.511 | 5.218 | 5.925 | 6.632 |
| .39   | .276  | .983  | 1.690 | 2.397 | 3.104 | 3.811 | 4.518 | 5.225 | 5.932 | 6.639 |
| .40   | .283  | .990  | 1.697 | 2.404 | 3.111 | 3.818 | 4.525 | 5.232 | 5.939 | 6.646 |
| .41   | .290  | .997  | 1.704 | 2.411 | 3.118 | 3.825 | 4.532 | 5.239 | 5.946 | 6.653 |
| .42   | .297  | 1.004 | 1.711 | 2.418 | 3.125 | 3.832 | 4.539 | 5.246 | 5.953 | 6.660 |
| .43   | .304  | 1.011 | 1.718 | 2.425 | 3.132 | 3.839 | 4.546 | 5.253 | 5.960 | 6.667 |
| .44   | .311  | 1.018 | 1.725 | 2.432 | 3.139 | 3.846 | 4.553 | 5.260 | 5.967 | 6.674 |
| .45   | .318  | 1.025 | 1.732 | 2.439 | 3.146 | 3.853 | 4.560 | 5.267 | 5.974 | 6.681 |
| .46   | .325  | 1.032 | 1.739 | 2.446 | 3.153 | 3.860 | 4.567 | 5.274 | 5.981 | 6.688 |
| .47   | .332  | 1.039 | 1.746 | 2.453 | 3.160 | 3.867 | 4.574 | 5.281 | 5.988 | 6.695 |
| .48   | .339  | 1.046 | 1.753 | 2.460 | 3.167 | 3.874 | 4.581 | 5.288 | 5.995 | 6.702 |
| .49   | .346  | 1.053 | 1.760 | 2.467 | 3.174 | 3.881 | 4.588 | 5.295 | 6.002 | 6.709 |
| .50   | .354  | 1.060 | 1.768 | 2.474 | 3.182 | 3.888 | 4.596 | 5.302 | 6.010 | 6.716 |
|       | 0     | 1     | 2     | 3     | 4     | 5     | 6     | 7     | 8     | 9     |

## Table 19.—Products for Form 194—Continued

[Multiplier = sin 45° = **0.707**]

|       | 0     | 1     | 2     | 3     | 4     | 5     | 6     | 7     | 8     | 9     |
|-------|-------|-------|-------|-------|-------|-------|-------|-------|-------|-------|
| 0.50  | 0.354 | 1.060 | 1.768 | 2.474 | 3.182 | 3.888 | 4.596 | 5.302 | 6.010 | 6.716 |
| .51   | .361  | 1.068 | 1.775 | 2.482 | 3.189 | 3.896 | 4.603 | 5.310 | 6.017 | 6.724 |
| .52   | .368  | 1.075 | 1.782 | 2.489 | 3.196 | 3.903 | 4.610 | 5.317 | 6.024 | 6.731 |
| .53   | .375  | 1.082 | 1.789 | 2.496 | 3.203 | 3.910 | 4.617 | 5.324 | 6.031 | 6.738 |
| .54   | .382  | 1.089 | 1.796 | 2.503 | 3.210 | 3.917 | 4.624 | 5.331 | 6.038 | 6.745 |
| .55   | .389  | 1.096 | 1.803 | 2.510 | 3.217 | 3.924 | 4.631 | 5.338 | 6.045 | 6.752 |
| .56   | .396  | 1.103 | 1.810 | 2.517 | 3.224 | 3.931 | 4.638 | 5.345 | 6.052 | 6.759 |
| .57   | .403  | 1.110 | 1.817 | 2.524 | 3.231 | 3.938 | 4.645 | 5.352 | 6.059 | 6.766 |
| .58   | .410  | 1.117 | 1.824 | 2.531 | 3.238 | 3.945 | 4.652 | 5.359 | 6.066 | 6.773 |
| .59   | .417  | 1.124 | 1.831 | 2.538 | 3.245 | 3.952 | 4.659 | 5.366 | 6.073 | 6.780 |
| .60   | .424  | 1.131 | 1.838 | 2.545 | 3.252 | 3.959 | 4.666 | 5.373 | 6.080 | 6.787 |
| .61   | .431  | 1.138 | 1.845 | 2.552 | 3.259 | 3.966 | 4.673 | 5.380 | 6.087 | 6.794 |
| .62   | .438  | 1.145 | 1.852 | 2.559 | 3.266 | 3.973 | 4.680 | 5.387 | 6.094 | 6.801 |
| .63   | .445  | 1.152 | 1.859 | 2.566 | 3.273 | 3.980 | 4.687 | 5.394 | 6.101 | 6.808 |
| .64   | .452  | 1.159 | 1.866 | 2.573 | 3.280 | 3.987 | 4.694 | 5.401 | 6.108 | 6.815 |
| .65   | .460  | 1.167 | 1.874 | 2.581 | 3.288 | 3.995 | 4.702 | 5.409 | 6.116 | 6.823 |
| .66   | .467  | 1.174 | 1.881 | 2.588 | 3.295 | 4.002 | 4.709 | 5.416 | 6.123 | 6.830 |
| .67   | .474  | 1.181 | 1.888 | 2.595 | 3.302 | 4.009 | 4.716 | 5.423 | 6.130 | 6.837 |
| .68   | .481  | 1.188 | 1.895 | 2.602 | 3.309 | 4.016 | 4.723 | 5.430 | 6.137 | 6.844 |
| .69   | .488  | 1.195 | 1.902 | 2.609 | 3.316 | 4.023 | 4.730 | 5.437 | 6.144 | 6.851 |
| .70   | .495  | 1.202 | 1.909 | 2.616 | 3.323 | 4.030 | 4.737 | 5.444 | 6.151 | 6.858 |
| .71   | .502  | 1.209 | 1.916 | 2.623 | 3.330 | 4.037 | 4.744 | 5.451 | 6.158 | 6.865 |
| .72   | .509  | 1.216 | 1.923 | 2.630 | 3.337 | 4.044 | 4.751 | 5.458 | 6.165 | 6.872 |
| .73   | .516  | 1.223 | 1.930 | 2.637 | 3.344 | 4.051 | 4.758 | 5.465 | 6.172 | 6.879 |
| .74   | .523  | 1.230 | 1.937 | 2.644 | 3.351 | 4.058 | 4.765 | 5.472 | 6.179 | 6.886 |
| .75   | .530  | 1.237 | 1.944 | 2.651 | 3.358 | 4.065 | 4.772 | 5.479 | 6.186 | 6.893 |
| .76   | .537  | 1.244 | 1.951 | 2.658 | 3.365 | 4.072 | 4.779 | 5.486 | 6.193 | 6.900 |
| .77   | .544  | 1.251 | 1.958 | 2.665 | 3.372 | 4.079 | 4.786 | 5.493 | 6.200 | 6.907 |
| .78   | .551  | 1.258 | 1.965 | 2.672 | 3.379 | 4.086 | 4.793 | 5.500 | 6.207 | 6.914 |
| .79   | .559  | 1.266 | 1.973 | 2.680 | 3.387 | 4.094 | 4.801 | 5.508 | 6.215 | 6.922 |
| .80   | .566  | 1.273 | 1.980 | 2.687 | 3.394 | 4.101 | 4.808 | 5.515 | 6.222 | 6.929 |
| .81   | .573  | 1.280 | 1.987 | 2.694 | 3.401 | 4.108 | 4.815 | 5.522 | 6.229 | 6.936 |
| .82   | .580  | 1.287 | 1.994 | 2.701 | 3.408 | 4.115 | 4.822 | 5.529 | 6.236 | 6.943 |
| .83   | .587  | 1.294 | 2.001 | 2.708 | 3.415 | 4.122 | 4.829 | 5.536 | 6.243 | 6.950 |
| .84   | .594  | 1.301 | 2.008 | 2.715 | 3.422 | 4.129 | 4.836 | 5.543 | 6.250 | 6.957 |
| .85   | .601  | 1.308 | 2.015 | 2.722 | 3.429 | 4.136 | 4.843 | 5.550 | 6.257 | 6.964 |
| .86   | .608  | 1.315 | 2.022 | 2.729 | 3.436 | 4.143 | 4.850 | 5.557 | 6.264 | 6.971 |
| .87   | .615  | 1.322 | 2.029 | 2.736 | 3.443 | 4.150 | 4.857 | 5.564 | 6.271 | 6.978 |
| .88   | .622  | 1.329 | 2.036 | 2.743 | 3.450 | 4.157 | 4.864 | 5.571 | 6.278 | 6.985 |
| .89   | .629  | 1.336 | 2.043 | 2.750 | 3.457 | 4.164 | 4.871 | 5.578 | 6.285 | 6.992 |
| .90   | .636  | 1.343 | 2.050 | 2.757 | 3.464 | 4.171 | 4.878 | 5.585 | 6.292 | 6.999 |
| .91   | .643  | 1.350 | 2.057 | 2.764 | 3.471 | 4.178 | 4.885 | 5.592 | 6.299 | 7.006 |
| .92   | .650  | 1.357 | 2.064 | 2.771 | 3.478 | 4.185 | 4.892 | 5.599 | 6.306 | 7.013 |
| .93   | .658  | 1.365 | 2.072 | 2.779 | 3.486 | 4.193 | 4.900 | 5.607 | 6.314 | 7.021 |
| .94   | .665  | 1.372 | 2.079 | 2.786 | 3.493 | 4.200 | 4.907 | 5.614 | 6.321 | 7.028 |
| .95   | .672  | 1.379 | 2.086 | 2.793 | 3.500 | 4.207 | 4.914 | 5.621 | 6.328 | 7.035 |
| .96   | .679  | 1.386 | 2.093 | 2.800 | 3.507 | 4.214 | 4.921 | 5.628 | 6.335 | 7.042 |
| .97   | .686  | 1.393 | 2.100 | 2.807 | 3.514 | 4.221 | 4.928 | 5.635 | 6.342 | 7.049 |
| .98   | .693  | 1.400 | 2.107 | 2.814 | 3.521 | 4.228 | 4.935 | 5.642 | 6.349 | 7.056 |
| .99   | .700  | 1.407 | 2.114 | 2.821 | 3.528 | 4.235 | 4.942 | 5.649 | 6.356 | 7.063 |
| 1.00  | .707  | 1.414 | 2.121 | 2.828 | 3.535 | 4.242 | 4.949 | 5.656 | 6.363 | 7.070 |
|       | 0     | 1     | 2     | 3     | 4     | 5     | 6     | 7     | 8     | 9     |

Table 19.—*Products for Form 194*—Continued

[Multiplier=sin 60°=**0.866**]

|       | 0     | 1     | 2     | 3     | 4     | 5     | 6     | 7     | 8     | 9     |
|-------|-------|-------|-------|-------|-------|-------|-------|-------|-------|-------|
| 0.00  | 0.000 | 0.866 | 1.732 | 2.598 | 3.464 | 4.330 | 5.196 | 6.062 | 6.928 | 7.794 |
| .01   | .009  | .875  | 1.741 | 2.607 | 3.473 | 4.339 | 5.205 | 6.071 | 6.937 | 7.803 |
| .02   | .017  | .883  | 1.749 | 2.615 | 3.481 | 4.347 | 5.213 | 6.079 | 6.945 | 7.811 |
| .03   | .026  | .892  | 1.758 | 2.624 | 3.490 | 4.356 | 5.222 | 6.088 | 6.954 | 7.820 |
| .04   | .035  | .901  | 1.767 | 2.633 | 3.499 | 4.365 | 5.231 | 6.097 | 6.963 | 7.829 |
| .05   | .043  | .909  | 1.775 | 2.641 | 3.507 | 4.373 | 5.239 | 6.105 | 6.971 | 7.837 |
| .06   | .052  | .918  | 1.784 | 2.650 | 3.516 | 4.382 | 5.248 | 6.114 | 6.980 | 7.846 |
| .07   | .061  | .927  | 1.793 | 2.659 | 3.525 | 4.391 | 5.257 | 6.123 | 6.989 | 7.855 |
| .08   | .069  | .935  | 1.801 | 2.667 | 3.533 | 4.399 | 5.265 | 6.131 | 6.997 | 7.863 |
| .09   | .078  | .944  | 1.810 | 2.676 | 3.542 | 4.408 | 5.274 | 6.140 | 7.006 | 7.872 |
| .10   | .087  | .953  | 1.819 | 2.685 | 3.551 | 4.417 | 5.283 | 6.149 | 7.015 | 7.881 |
| .11   | .095  | .961  | 1.827 | 2.693 | 3.559 | 4.425 | 5.291 | 6.157 | 7.023 | 7.889 |
| .12   | .104  | .970  | 1.836 | 2.702 | 3.568 | 4.434 | 5.300 | 6.166 | 7.032 | 7.898 |
| .13   | .113  | .979  | 1.845 | 2.711 | 3.577 | 4.443 | 5.309 | 6.175 | 7.041 | 7.907 |
| .14   | .121  | .987  | 1.853 | 2.719 | 3.585 | 4.451 | 5.317 | 6.183 | 7.049 | 7.915 |
| .15   | .130  | .996  | 1.862 | 2.728 | 3.594 | 4.460 | 5.326 | 6.192 | 7.058 | 7.924 |
| .16   | .139  | 1.005 | 1.871 | 2.737 | 3.603 | 4.469 | 5.335 | 6.201 | 7.067 | 7.933 |
| .17   | .147  | 1.013 | 1.879 | 2.745 | 3.611 | 4.477 | 5.343 | 6.209 | 7.075 | 7.941 |
| .18   | .156  | 1.022 | 1.888 | 2.754 | 3.620 | 4.486 | 5.352 | 6.218 | 7.084 | 7.950 |
| .19   | .165  | 1.031 | 1.897 | 2.763 | 3.629 | 4.495 | 5.361 | 6.227 | 7.093 | 7.959 |
| .20   | .173  | 1.039 | 1.905 | 2.771 | 3.637 | 4.503 | 5.369 | 6.235 | 7.101 | 7.967 |
| .21   | .182  | 1.048 | 1.914 | 2.780 | 3.646 | 4.512 | 5.378 | 6.244 | 7.110 | 7.976 |
| .22   | .191  | 1.057 | 1.923 | 2.789 | 3.655 | 4.521 | 5.387 | 6.253 | 7.119 | 7.985 |
| .23   | .199  | 1.065 | 1.931 | 2.797 | 3.663 | 4.529 | 5.395 | 6.261 | 7.127 | 7.993 |
| .24   | .208  | 1.074 | 1.940 | 2.806 | 3.672 | 4.538 | 5.404 | 6.270 | 7.136 | 8.002 |
| .25   | .216  | 1.082 | 1.948 | 2.814 | 3.680 | 4.546 | 5.412 | 6.278 | 7.144 | 8.010 |
| .26   | .225  | 1.091 | 1.957 | 2.823 | 3.689 | 4.555 | 5.421 | 6.287 | 7.153 | 8.019 |
| .27   | .234  | 1.100 | 1.966 | 2.832 | 3.698 | 4.564 | 5.430 | 6.296 | 7.162 | 8.028 |
| .28   | .242  | 1.108 | 1.974 | 2.840 | 3.706 | 4.572 | 5.438 | 6.304 | 7.170 | 8.036 |
| .29   | .251  | 1.117 | 1.983 | 2.849 | 3.715 | 4.581 | 5.447 | 6.313 | 7.179 | 8.045 |
| .30   | .260  | 1.126 | 1.992 | 2.858 | 3.724 | 4.590 | 5.456 | 6.322 | 7.188 | 8.054 |
| .31   | .268  | 1.134 | 2.000 | 2.866 | 3.732 | 4.598 | 5.464 | 6.330 | 7.196 | 8.062 |
| .32   | .277  | 1.143 | 2.009 | 2.875 | 3.741 | 4.607 | 5.473 | 6.339 | 7.205 | 8.071 |
| .33   | .286  | 1.152 | 2.018 | 2.884 | 3.750 | 4.616 | 5.482 | 6.348 | 7.214 | 8.080 |
| .34   | .294  | 1.160 | 2.026 | 2.892 | 3.758 | 4.624 | 5.490 | 6.356 | 7.222 | 8.088 |
| .35   | .303  | 1.169 | 2.035 | 2.901 | 3.767 | 4.633 | 5.499 | 6.365 | 7.231 | 8.097 |
| .36   | .312  | 1.178 | 2.044 | 2.910 | 3.776 | 4.642 | 5.508 | 6.374 | 7.240 | 8.106 |
| .37   | .320  | 1.186 | 2.052 | 2.918 | 3.784 | 4.650 | 5.516 | 6.382 | 7.248 | 8.114 |
| .38   | .329  | 1.195 | 2.061 | 2.927 | 3.793 | 4.659 | 5.525 | 6.391 | 7.257 | 8.123 |
| .39   | .338  | 1.204 | 2.070 | 2.936 | 3.802 | 4.668 | 5.534 | 6.400 | 7.266 | 8.132 |
| .40   | .346  | 1.212 | 2.078 | 2.944 | 3.810 | 4.676 | 5.542 | 6.408 | 7.274 | 8.140 |
| .41   | .355  | 1.221 | 2.087 | 2.953 | 3.819 | 4.685 | 5.551 | 6.417 | 7.283 | 8.149 |
| .42   | .364  | 1.230 | 2.096 | 2.962 | 3.828 | 4.694 | 5.560 | 6.426 | 7.292 | 8.158 |
| .43   | .372  | 1.238 | 2.104 | 2.970 | 3.836 | 4.702 | 5.568 | 6.434 | 7.300 | 8.166 |
| .44   | .381  | 1.247 | 2.113 | 2.979 | 3.845 | 4.711 | 5.577 | 6.443 | 7.309 | 8.175 |
| .45   | .390  | 1.256 | 2.122 | 2.988 | 3.854 | 4.720 | 5.586 | 6.452 | 7.318 | 8.184 |
| .46   | .398  | 1.264 | 2.130 | 2.996 | 3.862 | 4.728 | 5.594 | 6.460 | 7.326 | 8.192 |
| .47   | .407  | 1.273 | 2.139 | 3.005 | 3.871 | 4.737 | 5.603 | 6.469 | 7.335 | 8.201 |
| .48   | .416  | 1.282 | 2.148 | 3.014 | 3.880 | 4.746 | 5.612 | 6.478 | 7.344 | 8.210 |
| .49   | .424  | 1.290 | 2.156 | 3.022 | 3.888 | 4.754 | 5.620 | 6.486 | 7.352 | 8.218 |
| .50   | .433  | 1.299 | 2.165 | 3.031 | 3.897 | 4.763 | 5.629 | 6.495 | 7.361 | 8.227 |
|       | 0     | 1     | 2     | 3     | 4     | 5     | 6     | 7     | 8     | 9     |

HARMONIC ANALYSIS AND PREDICTION OF TIDES

Table 19.—*Products for Form 194*—Continued

[Multiplier=sin 60°=**0.866**]

|      | 0     | 1     | 2     | 3     | 4     | 5     | 6     | 7     | 8     | 9     |
|------|-------|-------|-------|-------|-------|-------|-------|-------|-------|-------|
| 0.50 | 0.433 | 1.299 | 2.165 | 3.031 | 3.897 | 4.763 | 5.629 | 6.495 | 7.361 | 8.227 |
| .51  | .442  | 1.308 | 2.174 | 3.040 | 3.906 | 4.772 | 5.638 | 6.504 | 7.370 | 8.236 |
| .52  | .450  | 1.316 | 2.182 | 3.048 | 3.914 | 4.780 | 5.646 | 6.512 | 7.378 | 8.244 |
| .53  | .459  | 1.325 | 2.191 | 3.057 | 3.923 | 4.789 | 5.655 | 6.521 | 7.387 | 8.253 |
| .54  | .468  | 1.334 | 2.200 | 3.066 | 3.932 | 4.798 | 5.664 | 6.530 | 7.396 | 8.262 |
| .55  | .476  | 1.342 | 2.208 | 3.074 | 3.940 | 4.806 | 5.672 | 6.538 | 7.404 | 8.270 |
| .56  | .485  | 1.351 | 2.217 | 3.083 | 3.949 | 4.815 | 5.681 | 6.547 | 7.413 | 8.279 |
| .57  | .494  | 1.360 | 2.226 | 3.092 | 3.958 | 4.824 | 5.690 | 6.556 | 7.422 | 8.288 |
| .58  | .502  | 1.368 | 2.234 | 3.100 | 3.966 | 4.832 | 5.698 | 6.564 | 7.430 | 8.296 |
| .59  | .511  | 1.377 | 2.243 | 3.109 | 3.975 | 4.841 | 5.707 | 6.573 | 7.439 | 8.305 |
| .60  | .520  | 1.386 | 2.252 | 3.118 | 3.984 | 4.850 | 5.716 | 6.582 | 7.448 | 8.314 |
| .61  | .528  | 1.394 | 2.260 | 3.126 | 3.992 | 4.858 | 5.724 | 6.590 | 7.456 | 8.322 |
| .62  | .537  | 1.403 | 2.269 | 3.135 | 4.001 | 4.867 | 5.733 | 6.599 | 7.465 | 8.331 |
| .63  | .546  | 1.412 | 2.278 | 3.144 | 4.010 | 4.876 | 5.742 | 6.608 | 7.474 | 8.340 |
| .64  | .554  | 1.420 | 2.286 | 3.152 | 4.018 | 4.884 | 5.750 | 6.616 | 7.482 | 8.348 |
| .65  | .563  | 1.429 | 2.295 | 3.161 | 4.027 | 4.893 | 5.759 | 6.625 | 7.491 | 8.357 |
| .66  | .572  | 1.438 | 2.304 | 3.170 | 4.036 | 4.902 | 5.768 | 6.634 | 7.500 | 8.366 |
| .67  | .580  | 1.446 | 2.312 | 3.178 | 4.044 | 4.910 | 5.776 | 6.642 | 7.508 | 8.374 |
| .68  | .589  | 1.455 | 2.321 | 3.187 | 4.053 | 4.919 | 5.785 | 6.651 | 7.517 | 8.383 |
| .69  | .598  | 1.464 | 2.330 | 3.196 | 4.062 | 4.928 | 5.794 | 6.660 | 7.526 | 8.392 |
| .70  | .606  | 1.472 | 2.338 | 3.204 | 4.070 | 4.936 | 5.802 | 6.668 | 7.534 | 8.400 |
| .71  | .615  | 1.481 | 2.347 | 3.213 | 4.079 | 4.945 | 5.811 | 6.677 | 7.543 | 8.409 |
| .72  | .624  | 1.490 | 2.356 | 3.222 | 4.088 | 4.954 | 5.820 | 6.686 | 7.552 | 8.418 |
| .73  | .632  | 1.498 | 2.364 | 3.230 | 4.096 | 4.962 | 5.828 | 6.694 | 7.560 | 8.426 |
| .74  | .641  | 1.507 | 2.373 | 3.239 | 4.105 | 4.977 | 5.837 | 6.703 | 7.569 | 8.435 |
| .75  | .650  | 1.516 | 2.382 | 3.248 | 4.114 | 4.980 | 5.846 | 6.712 | 7.578 | 8.444 |
| .76  | .658  | 1.524 | 2.390 | 3.256 | 4.122 | 4.988 | 5.854 | 6.720 | 7.586 | 8.452 |
| .77  | .667  | 1.533 | 2.399 | 3.265 | 4.131 | 4.997 | 5.863 | 6.729 | 7.595 | 8.461 |
| .78  | .675  | 1.541 | 2.407 | 3.273 | 4.139 | 5.005 | 5.871 | 6.737 | 7.603 | 8.469 |
| .79  | .684  | 1.550 | 2.416 | 3.282 | 4.148 | 5.014 | 5.880 | 6.746 | 7.612 | 8.478 |
| .80  | .693  | 1.559 | 2.425 | 3.291 | 4.157 | 5.023 | 5.889 | 6.755 | 7.621 | 8.487 |
| .81  | .701  | 1.567 | 2.433 | 3.299 | 4.165 | 5.031 | 5.897 | 6.763 | 7.629 | 8.495 |
| .82  | .710  | 1.576 | 2.442 | 3.308 | 4.174 | 5.040 | 5.906 | 6.772 | 7.638 | 8.504 |
| .83  | .719  | 1.585 | 2.451 | 3.317 | 4.183 | 5.049 | 5.915 | 6.781 | 7.647 | 8.513 |
| .84  | .727  | 1.593 | 2.459 | 3.325 | 4.191 | 5.057 | 5.923 | 6.789 | 7.655 | 8.521 |
| .85  | .736  | 1.602 | 2.468 | 3.334 | 4.200 | 5.066 | 5.932 | 6.798 | 7.664 | 8.530 |
| .86  | .745  | 1.611 | 2.477 | 3.343 | 4.209 | 5.075 | 5.941 | 6.807 | 7.673 | 8.539 |
| .87  | .753  | 1.619 | 2.485 | 3.351 | 4.217 | 5.083 | 5.949 | 6.815 | 7.681 | 8.547 |
| .88  | .762  | 1.628 | 2.494 | 3.360 | 4.226 | 5.092 | 5.958 | 6.824 | 7.690 | 8.556 |
| .89  | .771  | 1.637 | 2.503 | 3.369 | 4.235 | 5.101 | 5.967 | 6.833 | 7.699 | 8.565 |
| .90  | .779  | 1.645 | 2.511 | 3.377 | 4.243 | 5.109 | 5.975 | 6.841 | 7.707 | 8.573 |
| .91  | .788  | 1.654 | 2.520 | 3.386 | 4.252 | 5.118 | 5.984 | 6.850 | 7.716 | 8.582 |
| .92  | .797  | 1.663 | 2.529 | 3.395 | 4.261 | 5.127 | 5.993 | 6.859 | 7.725 | 8.591 |
| .93  | .805  | 1.671 | 2.537 | 3.403 | 4.269 | 5.135 | 6.001 | 6.867 | 7.733 | 8.599 |
| .94  | .814  | 1.680 | 2.546 | 3.412 | 4.278 | 5.144 | 6.010 | 6.876 | 7.742 | 8.608 |
| .95  | .823  | 1.689 | 2.555 | 3.421 | 4.287 | 5.153 | 6.019 | 6.885 | 7.751 | 8.617 |
| .96  | .831  | 1.697 | 2.563 | 3.429 | 4.295 | 5.161 | 6.027 | 6.893 | 7.759 | 8.625 |
| .97  | .840  | 1.706 | 2.572 | 3.438 | 4.304 | 5.170 | 6.036 | 6.902 | 7.768 | 8.634 |
| .98  | .849  | 1.715 | 2.581 | 3.447 | 4.313 | 5.179 | 6.045 | 6.911 | 7.777 | 8.643 |
| .99  | .857  | 1.723 | 2.589 | 3.455 | 4.321 | 5.187 | 6.053 | 6.919 | 7.785 | 8.651 |
| 1.00 | .866  | 1.732 | 2.598 | 3.464 | 4.330 | 5.196 | 6.062 | 6.928 | 7.794 | 8.660 |
|      | 0     | 1     | 2     | 3     | 4     | 5     | 6     | 7     | 8     | 9     |

Table 19.—**Products for Form 194**—Continued

[Multiplier=sin 75°=**0.966**]

|  | 0 | 1 | 2 | 3 | 4 | 5 | 6 | 7 | 8 | 9 |
|---|---|---|---|---|---|---|---|---|---|---|
| 0.00 | 0.000 | 0.966 | 1.932 | 2.898 | 3.864 | 4.830 | 5.796 | 6.762 | 7.728 | 8.694 |
| .01 | .010 | .976 | 1.942 | 2.908 | 3.874 | 4.840 | 5.806 | 6.772 | 7.738 | 8.704 |
| .02 | .019 | .985 | 1.951 | 2.917 | 3.883 | 4.849 | 5.815 | 6.781 | 7.747 | 8.713 |
| .03 | .029 | .995 | 1.961 | 2.927 | 3.893 | 4.859 | 5.825 | 6.791 | 7.757 | 8.723 |
| .04 | .039 | 1.005 | 1.971 | 2.937 | 3.903 | 4.869 | 5.835 | 6.801 | 7.767 | 8.733 |
| .05 | .048 | 1.014 | 1.980 | 2.946 | 3.912 | 4.878 | 5.844 | 6.810 | 7.776 | 8.742 |
| .06 | .058 | 1.024 | 1.990 | 2.956 | 3.922 | 4.888 | 5.854 | 6.820 | 7.786 | 8.752 |
| .07 | .068 | 1.034 | 2.000 | 2.966 | 3.932 | 4.898 | 5.864 | 6.830 | 7.796 | 8.762 |
| .08 | .077 | 1.043 | 2.009 | 2.975 | 3.941 | 4.907 | 5.873 | 6.839 | 7.805 | 8.771 |
| .09 | .087 | 1.053 | 2.019 | 2.985 | 3.951 | 4.917 | 5.883 | 6.849 | 7.815 | 8.781 |
| .10 | .097 | 1.063 | 2.029 | 2.995 | 3.961 | 4.927 | 5.893 | 6.859 | 7.825 | 8.791 |
| .11 | .106 | 1.072 | 2.038 | 3.004 | 3.970 | 4.936 | 5.902 | 6.868 | 7.834 | 8.800 |
| .12 | .116 | 1.082 | 2.048 | 3.014 | 3.980 | 4.946 | 5.912 | 6.878 | 7.844 | 8.810 |
| .13 | .126 | 1.092 | 2.058 | 3.024 | 3.990 | 4.956 | 5.922 | 6.888 | 7.854 | 8.820 |
| .14 | .135 | 1.101 | 2.067 | 3.033 | 3.999 | 4.965 | 5.931 | 6.897 | 7.863 | 8.829 |
| .15 | .145 | 1.111 | 2.077 | 3.043 | 4.009 | 4.975 | 5.941 | 6.907 | 7.873 | 8.839 |
| .16 | .155 | 1.121 | 2.087 | 3.053 | 4.019 | 4.985 | 5.951 | 6.917 | 7.883 | 8.849 |
| .17 | .164 | 1.130 | 2.096 | 3.062 | 4.028 | 4.994 | 5.960 | 6.926 | 7.892 | 8.858 |
| .18 | .174 | 1.140 | 2.106 | 3.072 | 4.038 | 5.004 | 5.970 | 6.936 | 7.902 | 8.868 |
| .19 | .184 | 1.150 | 2.116 | 3.082 | 4.048 | 5.014 | 5.980 | 6.946 | 7.912 | 8.878 |
| .20 | .193 | 1.159 | 2.125 | 3.091 | 4.057 | 5.023 | 5.989 | 6.955 | 7.921 | 8.887 |
| .21 | .203 | 1.169 | 2.135 | 3.101 | 4.067 | 5.033 | 5.999 | 6.965 | 7.931 | 8.897 |
| .22 | .213 | 1.179 | 2.145 | 3.111 | 4.077 | 5.043 | 6.009 | 6.975 | 7.941 | 8.907 |
| .23 | .222 | 1.188 | 2.154 | 3.120 | 4.086 | 5.052 | 6.018 | 6.984 | 7.950 | 8.916 |
| .24 | .232 | 1.198 | 2.164 | 3.130 | 4.096 | 5.062 | 6.028 | 6.994 | 7.960 | 8.926 |
| .25 | .242 | 1.208 | 2.174 | 3.140 | 4.106 | 5.072 | 6.038 | 7.004 | 7.970 | 8.936 |
| .26 | .251 | 1.217 | 2.183 | 3.149 | 4.115 | 5.081 | 6.047 | 7.013 | 7.979 | 8.945 |
| .27 | .261 | 1.227 | 2.193 | 3.159 | 4.125 | 5.091 | 6.057 | 7.023 | 7.989 | 8.955 |
| .28 | .270 | 1.236 | 2.202 | 3.168 | 4.134 | 5.100 | 6.066 | 7.032 | 7.998 | 8.964 |
| .29 | .280 | 1.246 | 2.212 | 3.178 | 4.144 | 5.110 | 6.076 | 7.042 | 8.008 | 8.974 |
| .30 | .290 | 1.256 | 2.222 | 3.188 | 4.154 | 5.120 | 6.086 | 7.052 | 8.018 | 8.984 |
| .31 | .299 | 1.265 | 2.231 | 3.197 | 4.163 | 5.129 | 6.095 | 7.061 | 8.027 | 8.993 |
| .32 | .309 | 1.275 | 2.241 | 3.207 | 4.173 | 5.139 | 6.105 | 7.071 | 8.037 | 9.003 |
| .33 | .319 | 1.285 | 2.251 | 3.217 | 4.183 | 5.149 | 6.115 | 7.081 | 8.047 | 9.013 |
| .34 | .328 | 1.294 | 2.260 | 3.226 | 4.192 | 5.158 | 6.124 | 7.090 | 8.056 | 9.022 |
| .35 | .338 | 1.304 | 2.270 | 3.236 | 4.202 | 5.168 | 6.134 | 7.100 | 8.066 | 9.032 |
| .36 | .348 | 1.314 | 2.280 | 3.246 | 4.212 | 5.178 | 6.144 | 7.110 | 8.076 | 9.042 |
| .37 | .357 | 1.323 | 2.289 | 3.255 | 4.221 | 5.187 | 6.153 | 7.119 | 8.085 | 9.051 |
| .38 | .367 | 1.333 | 2.299 | 3.265 | 4.231 | 5.197 | 6.163 | 7.129 | 8.095 | 9.061 |
| .39 | .377 | 1.343 | 2.309 | 3.275 | 4.241 | 5.207 | 6.173 | 7.139 | 8.105 | 9.071 |
| .40 | .386 | 1.352 | 2.318 | 3.284 | 4.250 | 5.216 | 6.182 | 7.148 | 8.114 | 9.080 |
| .41 | .396 | 1.362 | 2.328 | 3.294 | 4.260 | 5.226 | 6.192 | 7.158 | 8.124 | 9.090 |
| .42 | .406 | 1.372 | 2.338 | 3.304 | 4.270 | 5.236 | 6.202 | 7.168 | 8.134 | 9.100 |
| .43 | .415 | 1.381 | 2.347 | 3.313 | 4.279 | 5.245 | 6.211 | 7.177 | 8.143 | 9.109 |
| .44 | .425 | 1.391 | 2.357 | 3.323 | 4.289 | 5.255 | 6.221 | 7.187 | 8.153 | 9.119 |
| .45 | .435 | 1.401 | 2.367 | 3.333 | 4.299 | 5.265 | 6.231 | 7.197 | 8.163 | 9.129 |
| .46 | .444 | 1.410 | 2.376 | 3.342 | 4.308 | 5.274 | 6.240 | 7.206 | 8.172 | 9.138 |
| .47 | .454 | 1.420 | 2.386 | 3.352 | 4.318 | 5.284 | 6.250 | 7.216 | 8.182 | 9.148 |
| .48 | .464 | 1.430 | 2.396 | 3.362 | 4.328 | 5.294 | 6.260 | 7.226 | 8.192 | 9.158 |
| .49 | .473 | 1.439 | 2.405 | 3.371 | 4.337 | 5.303 | 6.269 | 7.235 | 8.201 | 9.167 |
| .50 | .483 | 1.449 | 2.415 | 3.381 | 4.347 | 5.313 | 6.279 | 7.245 | 8.211 | 9.177 |
|  | 0 | 1 | 2 | 3 | 4 | 5 | 6 | 7 | 8 | 9 |

HARMONIC ANALYSIS AND PREDICTION OF TIDES 227

Table 19.—*Products for Form 194*—Continued

[Multiplier=sin 75°=**0.966**]

|  | 0 | 1 | 2 | 3 | 4 | 5 | 6 | 7 | 8 | 9 |
|---|---|---|---|---|---|---|---|---|---|---|
| 0.50 | 0.483 | 1.449 | 2.415 | 3.381 | 4.347 | 5.313 | 6.279 | 7.245 | 8.211 | 9.177 |
| .51 | .493 | 1.459 | 2.425 | 3.391 | 4.357 | 5.323 | 6.289 | 7.255 | 8.221 | 9.187 |
| .52 | .502 | 1.468 | 2.434 | 3.400 | 4.366 | 5.332 | 6.298 | 7.264 | 8.230 | 9.196 |
| .53 | .512 | 1.478 | 2.444 | 3.410 | 4.376 | 5.342 | 6.308 | 7.274 | 8.240 | 9.206 |
| .54 | .522 | 1.488 | 2.454 | 3.420 | 4.386 | 5.352 | 6.318 | 7.284 | 8.250 | 9.216 |
| .55 | .531 | 1.497 | 2.463 | 3.429 | 4.395 | 5.361 | 6.327 | 7.293 | 8.259 | 9.225 |
| .56 | .541 | 1.507 | 2.473 | 3.439 | 4.405 | 5.371 | 6.337 | 7.303 | 8.269 | 9.235 |
| .57 | .551 | 1.517 | 2.483 | 3.449 | 4.415 | 5.381 | 6.347 | 7.313 | 8.279 | 9.245 |
| .58 | .560 | 1.526 | 2.492 | 3.458 | 4.424 | 5.390 | 6.356 | 7.322 | 8.288 | 9.254 |
| .59 | .570 | 1.536 | 2.502 | 3.468 | 4.434 | 5.400 | 6.366 | 7.332 | 8.298 | 9.264 |
| .60 | .580 | 1.546 | 2.512 | 3.478 | 4.444 | 5.410 | 6.376 | 7.342 | 8.308 | 9.274 |
| .61 | .589 | 1.555 | 2.521 | 3.487 | 4.453 | 5.419 | 6.385 | 7.351 | 8.317 | 9.283 |
| .62 | .599 | 1.565 | 2.531 | 3.497 | 4.463 | 5.429 | 6.395 | 7.361 | 8.327 | 9.293 |
| .63 | .609 | 1.575 | 2.541 | 3.507 | 4.473 | 5.439 | 6.405 | 7.371 | 8.337 | 9.303 |
| .64 | .618 | 1.584 | 2.550 | 3.516 | 4.482 | 5.448 | 6.414 | 7.380 | 8.346 | 9.312 |
| .65 | .628 | 1.594 | 2.560 | 3.526 | 4.492 | 5.458 | 6.424 | 7.390 | 8.356 | 9.322 |
| .66 | .638 | 1.604 | 2.570 | 3.536 | 4.502 | 5.468 | 6.434 | 7.400 | 8.366 | 9.332 |
| .67 | .647 | 1.613 | 2.579 | 3.545 | 4.511 | 5.477 | 6.443 | 7.409 | 8.375 | 9.341 |
| .68 | .657 | 1.623 | 2.589 | 3.555 | 4.521 | 5.487 | 6.453 | 7.419 | 8.385 | 9.351 |
| .69 | .667 | 1.633 | 2.599 | 3.565 | 4.531 | 5.497 | 6.463 | 7.429 | 8.395 | 9.361 |
| .70 | .676 | 1.642 | 2.608 | 3.574 | 4.540 | 5.506 | 6.472 | 7.438 | 8.404 | 9.370 |
| .71 | .686 | 1.652 | 2.618 | 3.584 | 4.550 | 5.516 | 6.482 | 7.448 | 8.414 | 9.380 |
| .72 | .696 | 1.662 | 2.628 | 3.594 | 4.560 | 5.526 | 6.492 | 7.458 | 8.424 | 9.390 |
| .73 | .705 | 1.671 | 2.637 | 3.603 | 4.569 | 5.535 | 6.501 | 7.467 | 8.433 | 9.399 |
| .74 | .715 | 1.681 | 2.647 | 3.613 | 4.579 | 5.545 | 6.511 | 7.477 | 8.443 | 9.409 |
| .75 | .724 | 1.690 | 2.656 | 3.622 | 4.588 | 5.554 | 6.520 | 7.486 | 8.452 | 9.418 |
| .76 | .734 | 1.700 | 2.666 | 3.632 | 4.598 | 5.564 | 6.530 | 7.496 | 8.462 | 9.428 |
| .77 | .744 | 1.710 | 2.676 | 3.642 | 4.608 | 5.574 | 6.540 | 7.506 | 8.472 | 9.438 |
| .78 | .753 | 1.719 | 2.685 | 3.651 | 4.617 | 5.583 | 6.549 | 7.515 | 8.481 | 9.447 |
| .79 | .763 | 1.729 | 2.695 | 3.661 | 4.627 | 5.593 | 6.559 | 7.525 | 8.491 | 9.457 |
| .80 | .773 | 1.739 | 2.705 | 3.671 | 4.637 | 5.603 | 6.569 | 7.535 | 8.501 | 9.467 |
| .81 | .782 | 1.748 | 2.714 | 3.680 | 4.646 | 5.612 | 6.578 | 7.544 | 8.510 | 9.476 |
| .82 | .792 | 1.758 | 2.724 | 3.690 | 4.656 | 5.622 | 6.588 | 7.554 | 8.520 | 9.486 |
| .83 | .802 | 1.768 | 2.734 | 3.700 | 4.666 | 5.632 | 6.598 | 7.564 | 8.530 | 9.496 |
| .84 | .811 | 1.777 | 2.743 | 3.709 | 4.675 | 5.641 | 6.607 | 7.573 | 8.539 | 9.505 |
| .85 | .821 | 1.787 | 2.753 | 3.719 | 4.685 | 5.651 | 6.617 | 7.583 | 8.549 | 9.515 |
| .86 | .831 | 1.797 | 2.763 | 3.729 | 4.695 | 5.661 | 6.627 | 7.593 | 8.559 | 9.525 |
| .87 | .840 | 1.806 | 2.772 | 3.738 | 4.704 | 5.670 | 6.636 | 7.602 | 8.568 | 9.534 |
| .88 | .850 | 1.816 | 2.782 | 3.748 | 4.714 | 5.680 | 6.646 | 7.612 | 8.578 | 9.544 |
| .89 | .860 | 1.826 | 2.792 | 3.758 | 4.724 | 5.690 | 6.656 | 7.622 | 8.588 | 9.554 |
| .90 | .869 | 1.835 | 2.801 | 3.767 | 4.733 | 5.699 | 6.665 | 7.631 | 8.597 | 9.563 |
| .91 | .879 | 1.845 | 2.811 | 3.777 | 4.743 | 5.709 | 6.675 | 7.641 | 8.607 | 9.573 |
| .92 | .889 | 1.855 | 2.821 | 3.787 | 4.753 | 5.719 | 6.685 | 7.651 | 8.617 | 9.583 |
| .93 | .898 | 1.864 | 2.830 | 3.796 | 4.762 | 5.728 | 6.694 | 7.660 | 8.626 | 9.592 |
| .94 | .908 | 1.874 | 2.840 | 3.806 | 4.772 | 5.738 | 6.704 | 7.670 | 8.636 | 9.602 |
| .95 | .918 | 1.884 | 2.850 | 3.816 | 4.782 | 5.748 | 6.714 | 7.680 | 8.646 | 9.612 |
| .96 | .927 | 1.893 | 2.859 | 3.825 | 4.791 | 5.757 | 6.723 | 7.689 | 8.655 | 9.621 |
| .97 | .937 | 1.903 | 2.869 | 3.835 | 4.801 | 5.767 | 6.733 | 7.699 | 8.665 | 9.631 |
| .98 | .947 | 1.913 | 2.879 | 3.845 | 4.811 | 5.777 | 6.743 | 7.709 | 8.675 | 9.641 |
| .99 | .956 | 1.922 | 2.888 | 3.854 | 4.820 | 5.786 | 6.752 | 7.718 | 8.684 | 9.650 |
| 1.00 | .966 | 1.932 | 2.898 | 3.864 | 4.830 | 5.796 | 6.762 | 7.728 | 8.694 | 9.660 |
|  | 0 | 1 | 2 | 3 | 4 | 5 | 6 | 7 | 8 | 9 |

## Table 20.—*Augmenting factors*

SHORT-PERIOD CONSTITUENTS,* FORMULA (308)

| | Augmenting factor | Logarithm | Remarks |
|---|---|---|---|
| Diurnal $J_1$, $K_1$, $M_1$, $O_1$, $OO$, $P_1$, $Q_1$, $2Q$, $\rho_1$ | 1.0029 | 0.001241 | Each tabulated solar hourly height used once and once only in summation; group covers one constituent hour; constituent day represented by 24 means. |
| Semidurnal $K_2$, $L_2$, $M_2$, $N_2$, $2N$, $R_2$, $T_2$, $\lambda_2$, $\mu_2$, $\nu_2$, $2SM$. | 1.0115 | 0.004972 | |
| Terdiurnal $M_3$, $MK$, $2MK$ | 1.0262 | 0.011220 | |
| Quarter-diurnal $M_4$, $MN$, $MS$ | 1.0472 | 0.020029 | |
| Sixth-diurnal $M_6$ | 1.1107 | 0.045605 | |
| Eighth-diurnal $M_8$ | 1.2092 | 0.082498 | |

SHORT-PERIOD CONSTITUENTS,* FORMULA (309)

| | Augmenting factor | Logarithm | | Augmenting factor | Logarithm | Remarks |
|---|---|---|---|---|---|---|
| $J_1$ | 1.0031 | 0.00134 | $P_1$ | 1.0028 | 0.00123 | |
| $K_1$ | 1.0029 | 0.00125 | $Q_1$ | 1.0023 | 0.00099 | |
| $K_2$ | 1.0116 | 0.00500 | $2Q$ | 1.0021 | 0.00091 | |
| $L_2$ | 1.0112 | 0.00482 | $R_2$ | 1.0115 | 0.00499 | |
| $M_1$ | 1.0027 | 0.00116 | $T_2$ | 1.0115 | 0.00496 | |
| $M_2$ | 1.0107 | 0.00464 | $\lambda_2$ | 1.0111 | 0.00479 | Each constituent hour of observation period receives one and only one of solar hourly heights in the summation; group covers one solar hour; each constituent day represented by 24 means. |
| $M_3$ | 1.0244 | 0.01047 | $\mu_2$ | 1.0100 | 0.00432 | |
| $M_4$ | 1.0440 | 0.01868 | $\nu_2$ | 1.0104 | 0.00449 | |
| $M_6$ | 1.1028 | 0.04251 | $\rho_1$ | 1.0023 | 0.00100 | |
| $M_8$ | 1.1934 | 0.07680 | $MK$ | 1.0250 | 0.01074 | |
| $N_2$ | 1.0103 | 0.00447 | $2MK$ | 1.0238 | 0.01021 | |
| $2N$ | 1.0099 | 0.00430 | $MN$ | 1.0431 | 0.01833 | |
| $O_1$ | 1.0025 | 0.00107 | $MS$ | 1.0456 | 0.01935 | |
| $OO$ | 1.0033 | 0.00144 | $2SM$ | 1.0123 | 0.00532 | |

LONG-PERIOD CONSTITUENTS, FORMULA (403)

| | Augmenting factor | Logarithm | Remarks |
|---|---|---|---|
| $Mm$ | 1.0050 | 0.00218 | Daily sums used as units in the summation for the divisional means, and all daily sums used; constituent month for Mm, Mf, MSf, and constituent year for Sa and Ssa represented by 24 means. |
| $Mf$ | 1.0205 | 0.00880 | |
| $MSf$ | 1.0192 | 0.00825 | |
| $Sa$ | 1.0029 | 0.00124 | |
| $Ssa$ | 1.0115 | 0.00497 | |

ANNUAL AND SEMIANNUAL CONSTITUENTS, FORMULA (404)

| | Augmenting factor | Logarithm | Remarks |
|---|---|---|---|
| $Sa$ | 1.0115 | 0.00497 | For analysis of 12 monthly means. |
| $Ssa$ | 1.0472 | 0.02003 | |

*For constituents $S_1$, $S_2$, $S_3$, etc., augmenting factor is unity.

HARMONIC ANALYSIS AND PREDICTION OF TIDES 229

Table 21.—*Acceleration in epoch of* $K_1$ *due to* $P_1$

[Argument $h-\frac{1}{2}\nu'$ refers to beginning of series]

| $h-\frac{1}{2}\nu'$ | Series | 14 days | 29 days | 58 days | 87 days | 105 days | 134 days | 163 days | 192 days | 221 days | 250 days | 279 days | 297 days | 326 days |
|---|---|---|---|---|---|---|---|---|---|---|---|---|---|---|
| 0 | 180 | +6.5 | +11.4 | +14.6 | +12.6 | +10.1 | +5.1 | +0.9 | +0.2 | +2.4 | +3.9 | +3.9 | +3.3 | +1.6 |
| 10 | 190 | +13.9 | +16.4 | +16.0 | +12.0 | +8.8 | +3.4 | −0.1 | +0.7 | +3.0 | +4.1 | +3.6 | +2.7 | +0.9 |
| 20 | 200 | +17.0 | +18.5 | +15.9 | +9.9 | +6.4 | +1.6 | −1.0 | +0.9 | +3.2 | +3.0 | +2.5 | +1.5 | 0.0 |
| 30 | 210 | +9. | + | +2. | +6. | +3. | −1. | −1. | +1. | +2. | +3. | +1. | +0. | −1. |
| 40 | 220 | +17.6 | +15.2 | +9.4 | +2.8 | −0.5 | −3.8 | −2.1 | +1.0 | +2.4 | +2.0 | +0.3 | −0.8 | −1.7 |
| 50 | 230 | +14.8 | +11.5 | +5.2 | 1.2 | −4.1 | −5.4 | −2.9 | +0.8 | +0.9 | +0.8 | −1.1 | −2.0 | −2.0 |
| 60 | 240 | +0. | +7. | +0. | −5. | −7.2 | −6.0 | −1. | +0. | + | −0. | −2. | −2. | −2.1 |
| 70 | 250 | +6.4 | +2.7 | −3.9 | −8.7 | −9.3 | −5.8 | −1.5 | +0.4 | +0.1 | −1.5 | −3.2 | −3.4 | −1.9 |
| 80 | 260 | +1.5 | −2.2 | −8.2 | −11.3 | −10.2 | −5.0 | −1.0 | +0.1 | −0.8 | −2.6 | −3.8 | −3.3 | −1.5 |
| 90 | 270 | −3.5 | −6.9 | −12.0 | −12.5 | −9.7 | −3.8 | −0.5 | −0.1 | −1.6 | −3.5 | −3.9 | −2.9 | −1.1 |
| 100 | 280 | −8.2 | −11.3 | −14.7 | −12.1 | −8.1 | −2.3 | 0.0 | −0.4 | −2.3 | −4.0 | −3.5 | −2.2 | −0.5 |
| 110 | 290 | −12.5 | −14.9 | −16.0 | −10.1 | −5.6 | −0.7 | +0.6 | −0.6 | −2.8 | −5.6 | −2.6 | −1.8 | 0.0 |
| 120 | 300 | −16.0 | −17.5 | −15.2 | −6.9 | −2.7 | +1.0 | +1.1 | −0. | −3. | − | −1. | −0. | +0. |
| 130 | 310 | −18.4 | −18.3 | −12.2 | −2.9 | +0.5 | +2.6 | +1.5 | −1.0 | −3.1 | −2.5 | −0.3 | +0.7 | +1.1 |
| 140 | 320 | −18.9 | −16.7 | −7.2 | +1.3 | +3.6 | +4.1 | +1.9 | −1.9 | −2.5 | −1.5 | +0.9 | +1.6 | +1.6 |
| 150 | 330 | −16.8 | −12.1 | −1.0 | +5.4 | +6.4 | +5.2 | +2.1 | −0. | −1. | 0. | +2. | +2. | +1.9 |
| 160 | 340 | −11.3 | −4.7 | +5.4 | +8.9 | +8.6 | +5.9 | +2.0 | −0.7 | −0.1 | +1.9 | +3.1 | +3.1 | +2.1 |
| 170 | 350 | −2.5 | +3.8 | +10.9 | +11.9 | +10.0 | +5.9 | +1.7 | +0.3 | +1.8 | +3.9 | +3.9 | +3.8 | +2.0 |
| 180 | 360 | +6. | +11. | +14. | +12. | +10. | +5. | +0. | −0. | +2. | +3. | +3. | +3. | +. |

Table 22.—*Ratio of increase in amplitude of* $K_1$ *due to* $P_1$

[Argument $h-\frac{1}{2}\nu'$ refers to beginning of series]

| $h-\frac{1}{2}\nu'$ | Series | 14 days | 29 days | 58 days | 87 days | 105 days | 134 days | 163 days | 192 days | 221 days | 250 days | 279 days | 297 days | 326 days |
|---|---|---|---|---|---|---|---|---|---|---|---|---|---|---|
| 0 | 180 | −0.31 | −0.26 | −0.12 | +0.01 | +0.06 | +0.09 | +0.06 | +0.01 | −0.02 | 0.00 | +0.03 | +0.05 | +0.05 |
| 10 | 190 | −0.25 | −0.17 | −0.02 | +0.09 | +0.12 | +0.12 | +0.07 | +0.01 | 0.00 | +0.02 | +0.06 | +0.07 | +0.06 |
| 20 | 200 | −0.15 | −0.06 | +0.07 | +0.16 | +0.17 | +0.13 | +0.06 | +0.02 | +0.02 | +0.05 | +0.08 | +0.08 | +0.07 |
| 30 | 210 | −0.04 | +0.04 | +0.16 | +0.20 | +0.20 | +0.13 | +0.05 | +0.02 | +0.04 | +0.07 | +0.09 | +0.09 | +0.06 |
| 40 | 220 | +0.07 | +0.14 | +0.23 | +0.23 | +0.21 | +0.12 | +0.04 | +0.03 | +0.05 | +0.09 | +0.09 | +0.09 | +0.05 |
| 50 | 230 | +0.17 | +0.23 | +0.27 | +0.24 | +0.19 | +0.09 | +0.03 | +0.03 | +0.06 | +0.09 | +0.09 | +0.08 | +0.04 |
| 60 | 240 | +0.25 | +0.28 | +0.29 | +0.22 | +0.15 | +0.05 | +0.02 | +0.04 | +0.07 | +0.09 | +0.08 | +0.06 | +0.03 |
| 70 | 250 | +0.30 | +0.31 | +0.28 | +0.18 | +0.10 | +0.02 | +0.01 | +0.04 | +0.08 | +0.09 | +0.07 | +0.04 | +0.02 |
| 80 | 260 | +0.33 | +0.32 | +0.24 | +0.12 | +0.05 | −0.01 | 0.00 | +0.04 | +0.08 | +0.08 | +0.05 | +0.02 | +0.01 |
| 90 | 270 | +0.32 | +0.29 | +0.18 | +0.04 | −0.02 | −0.04 | 0.00 | +0.04 | +0.07 | +0.06 | +0.02 | 0.00 | 0.00 |
| 100 | 280 | +0.28 | +0.23 | +0.10 | −0.03 | −0.07 | −0.06 | −0.01 | +0.04 | +0.06 | +0.03 | 0.00 | −0.01 | 0.00 |
| 110 | 290 | +0.22 | +0.15 | +0.01 | −0.10 | −0.11 | −0.07 | 0.00 | +0.04 | +0.04 | +0.01 | −0.02 | −0.02 | −0.01 |
| 120 | 300 | +0.13 | +0.05 | −0.09 | −0.15 | −0.14 | −0.07 | 0.00 | +0.03 | +0.02 | −0.01 | −0.03 | −0.03 | 0.00 |
| 130 | 310 | +0.03 | −0.06 | −0.17 | −0.18 | −0.14 | −0.06 | +0.01 | +0.03 | 0.00 | −0.03 | −0.04 | −0.03 | 0.00 |
| 140 | 320 | −0.08 | −0.16 | −0.23 | −0.19 | −0.13 | −0.04 | +0.02 | +0.02 | −0.01 | −0.04 | −0.04 | −0.02 | +0.01 |
| 150 | 330 | −0.19 | −0.25 | −0.26 | −0.17 | −0.10 | −0.01 | +0.03 | +0.02 | −0.02 | −0.04 | −0.03 | −0.01 | +0.02 |
| 160 | 340 | −0.28 | −0.30 | −0.25 | −0.12 | −0.05 | +0.03 | +0.04 | +0.01 | −0.03 | −0.03 | −0.01 | +0.01 | +0.03 |
| 170 | 350 | −0.32 | −0.31 | −0.19 | −0.06 | 0.00 | +0.06 | +0.05 | +0.01 | −0.03 | −0.02 | +0.01 | +0.03 | +0.04 |
| 180 | 360 | −0.31 | −0.26 | −0.12 | +0.01 | +0.06 | +0.09 | +0.06 | +0.01 | −0.02 | 0.00 | +0.03 | +0.05 | +0.05 |

Table 23.—*Acceleration in epoch of $S_2$ due to $K_2$*

[Argument $h-v''$ refers to beginning of series]

| Series $h-v''$ | | 15 days | 29 days | 58 days | 87 days | 105 days | 134 days | 163 days | 192 days | 221 days | 250 days | 279 days | 297 days | 326 days |
|---|---|---|---|---|---|---|---|---|---|---|---|---|---|---|
| | | ° | ° | ° | ° | ° | ° | ° | ° | ° | ° | ° | ° | ° |
| 0 | 180 | +3.2 | +5.9 | +10.1 | +10.4 | +8.0 | +3.2 | +0.4 | +0.1 | +1.3 | +2.9 | +3.2 | +2.4 | +0.9 |
| 10 | 190 | +7.2 | +9.6 | +12.3 | +10.0 | +6.7 | +2.0 | 0.0 | +0.3 | +1.9 | +3.3 | +2.9 | +1.9 | +0.5 |
| 20 | 200 | +10.8 | +12.6 | +13.2 | +8.4 | +4.7 | −0.6 | −0.5 | +0.5 | +2.4 | +3.3 | +2.2 | +1.1 | 0.0 |
| 30 | 210 | +13.7 | +14.6 | +12.5 | +5.7 | +2.3 | −0.9 | −0.9 | +0.7 | +2.6 | +2.9 | +1.3 | +0.3 | −0.5 |
| 40 | 220 | +15.4 | +15.0 | +9.9 | +2.5 | −0.4 | −2.2 | −1.3 | +0.8 | +2.5 | +2.0 | +0.3 | −0.6 | −1.0 |
| 50 | 230 | +15.4 | +13.5 | +5.8 | −1.1 | −3.0 | −3.4 | −1.6 | +0.8 | +2.0 | +0.9 | −0.8 | −1.4 | −1.3 |
| 60 | 240 | +13.2 | +9.6 | +0.8 | −4.5 | −5.4 | −4.4 | −1.7 | +0.7 | +1.2 | −0.4 | −1.8 | −2.1 | −1.6 |
| 70 | 250 | +8.6 | +3.7 | −4.4 | −7.4 | −7.2 | −4.9 | −1.7 | +0.5 | +0.1 | −1.6 | −2.6 | −2.6 | −1.7 |
| 80 | 260 | +1.9 | −3.0 | −8.8 | −9.5 | −8.3 | −4.9 | −1.3 | +0.2 | −1.0 | −2.6 | −3.1 | −2.8 | −1.6 |
| 90 | 270 | −5.5 | −9.1 | −11.9 | −10.4 | −8.3 | −4.2 | −0.7 | −0.2 | −1.9 | −3.2 | −3.3 | −2.7 | −1.3 |
| 100 | 280 | −11.2 | −13.2 | −13.2 | −9.9 | −7.3 | −2.7 | 0.0 | −0.5 | −2.4 | −3.4 | −3.0 | −2.2 | −0.7 |
| 110 | 290 | −14.6 | −15.0 | −12.7 | −8.2 | −5.2 | −0.8 | +0.8 | −0.7 | −2.6 | −3.1 | −2.3 | −1.4 | 0.0 |
| 120 | 300 | −15.6 | −14.7 | −10.9 | −5.6 | −2.6 | +1.2 | +1.4 | −0.8 | −2.4 | −2.5 | −1.4 | −0.4 | +0.8 |
| 130 | 310 | −14.7 | −12.9 | −8.0 | −2.4 | +0.5 | +3.1 | +1.7 | −0.8 | −2.0 | −1.7 | −0.3 | +0.7 | +1.3 |
| 140 | 320 | −12.4 | −10.0 | −4.4 | +1.0 | +3.4 | +4.4 | +1.7 | −0.7 | −1.5 | −0.7 | +0.8 | +1.6 | +1.7 |
| 150 | 330 | −9.1 | −6.4 | −0.6 | +4.4 | +5.9 | +4.9 | +1.6 | −0.5 | −0.8 | +0.3 | +1.9 | +2.4 | +1.7 |
| 160 | 340 | −5.3 | −2.3 | +3.3 | +7.3 | +7.7 | +4.8 | +1.3 | −0.3 | −0.1 | +1.3 | +2.7 | +2.8 | +1.6 |
| 170 | 350 | −1.1 | +1.9 | +7.0 | +9.4 | +8.4 | +4.2 | +0.9 | −0.1 | +0.7 | +2.2 | +3.2 | +2.8 | +1.3 |
| 180 | 360 | +3.2 | +5.9 | +10.1 | +10.4 | +8.0 | +3.2 | +0.4 | +0.1 | +1.3 | +2.9 | +3.2 | +2.4 | +0.9 |

Table 24.—*Ratio of increase in amplitude of $S_2$ due to $K_2$*

[Argument $h-v''$ refers to beginning of series]

| Series $h-v''$ | | 15 days | 29 days | 58 days | 87 days | 105 days | 134 days | 163 days | 192 days | 221 days | 250 days | 279 days | 297 days | 326 days |
|---|---|---|---|---|---|---|---|---|---|---|---|---|---|---|
| 0 | 180 | +0.26 | +0.24 | +0.15 | +0.03 | −0.02 | −0.04 | −0.01 | +0.03 | +0.05 | +0.04 | +0.01 | 0.00 | 0.00 |
| 10 | 190 | +0.23 | +0.19 | +0.08 | −0.03 | −0.06 | −0.05 | −0.01 | +0.03 | +0.04 | +0.02 | 0.00 | −0.01 | −0.01 |
| 20 | 200 | +0.18 | +0.12 | 0.00 | −0.09 | −0.10 | −0.06 | −0.01 | +0.03 | +0.03 | 0.00 | −0.02 | −0.02 | −0.01 |
| 30 | 210 | +0.10 | +0.04 | −0.08 | −0.13 | −0.12 | −0.06 | 0.00 | +0.02 | +0.01 | −0.01 | −0.03 | −0.03 | −0.01 |
| 40 | 220 | +0.01 | −0.05 | −0.15 | −0.15 | −0.13 | −0.05 | 0.00 | +0.02 | 0.00 | −0.03 | −0.04 | −0.03 | 0.00 |
| 50 | 230 | −0.08 | −0.14 | −0.19 | −0.16 | −0.11 | −0.03 | +0.01 | +0.02 | −0.01 | −0.04 | −0.03 | −0.02 | 0.00 |
| 60 | 240 | −0.17 | −0.21 | −0.21 | −0.14 | −0.09 | −0.01 | +0.02 | +0.01 | −0.02 | −0.04 | −0.02 | −0.01 | +0.01 |
| 70 | 250 | −0.23 | −0.25 | −0.20 | −0.10 | −0.05 | +0.02 | +0.03 | +0.01 | −0.03 | −0.03 | −0.01 | 0.00 | +0.02 |
| 80 | 260 | −0.27 | −0.25 | −0.16 | −0.05 | 0.00 | +0.05 | +0.04 | 0.00 | −0.02 | −0.02 | 0.00 | +0.02 | +0.03 |
| 90 | 270 | −0.25 | −0.21 | −0.10 | +0.01 | +0.05 | +0.08 | +0.05 | 0.00 | −0.01 | 0.00 | +0.02 | +0.04 | +0.04 |
| 100 | 280 | −0.20 | −0.15 | −0.02 | +0.07 | +0.10 | +0.10 | +0.05 | +0.01 | 0.00 | +0.02 | +0.04 | +0.05 | +0.05 |
| 110 | 290 | −0.13 | −0.06 | +0.05 | +0.12 | +0.14 | +0.11 | +0.05 | +0.01 | +0.01 | +0.04 | +0.06 | +0.06 | +0.05 |
| 120 | 300 | −0.03 | +0.03 | +0.13 | +0.17 | +0.16 | +0.11 | +0.04 | +0.01 | +0.03 | +0.05 | +0.07 | +0.07 | +0.05 |
| 130 | 310 | +0.06 | +0.11 | +0.18 | +0.19 | +0.17 | +0.09 | +0.03 | +0.02 | +0.04 | +0.07 | +0.08 | +0.07 | +0.04 |
| 140 | 320 | +0.14 | +0.18 | +0.22 | +0.19 | +0.15 | +0.07 | +0.02 | +0.02 | +0.05 | +0.07 | +0.07 | +0.06 | +0.03 |
| 150 | 330 | +0.21 | +0.23 | +0.24 | +0.18 | +0.13 | +0.04 | +0.01 | +0.03 | +0.06 | +0.07 | +0.06 | +0.05 | +0.02 |
| 160 | 340 | +0.25 | +0.26 | +0.23 | +0.14 | +0.08 | +0.01 | 0.00 | +0.03 | +0.06 | +0.07 | +0.05 | +0.03 | +0.01 |
| 170 | 350 | +0.27 | +0.26 | +0.20 | +0.09 | +0.03 | −0.02 | 0.00 | +0.03 | +0.06 | +0.06 | +0.03 | +0.02 | 0.00 |
| 180 | 360 | +0.26 | +0.24 | +0.15 | +0.03 | −0.02 | −0.04 | −0.01 | +0.03 | +0.05 | +0.04 | +0.01 | 0.00 | 0.00 |

Table 25.—*Acceleration in epoch of $S_2$ due to $T_2$*

[Argument $h-p_1$ refers to beginning of series]

| $h-p_1$ | 15 days | 29 days | 58 days | 87 days | 105 days | 134 days | 163 days | 192 days | 221 days | 250 days | 279 days | 297 days | 326 days |
|---|---|---|---|---|---|---|---|---|---|---|---|---|---|
| 0 | −0.4 | −0.8 | −1.5 | −2.0 | −2.2 | −2.4 | −2.3 | −2.0 | −1.5 | −1.1 | −0.6 | −0.4 | −0.1 |
| 10 | −1.0 | −1.3 | −1.9 | −2.4 | −2.5 | −2.6 | −2.4 | −2.0 | −1.4 | −0.9 | −0.5 | −0.3 | −0.1 |
| 20 | −1.5 | −1.8 | −2.2 | −2.7 | −2.7 | −2.7 | −2.3 | −1.8 | −1.3 | −0.7 | −0.3 | −0.2 | 0.0 |
| 30 | −2.0 | −2.2 | −2.7 | −2.9 | −2.9 | −2.7 | −2.2 | −1.7 | −1.1 | −0.6 | −0.2 | 0.0 | +0.1 |
| 40 | −2.4 | −2.6 | −3.0 | −3.0 | −2.9 | −2.6 | −2.0 | −1.4 | −0.8 | −0.4 | 0.0 | +0.1 | +0.1 |
| 50 | −2.7 | −2.9 | −3.1 | −3.1 | −2.9 | −2.4 | −1.8 | −1.2 | −0.6 | −0.2 | +0.1 | +0.2 | +0.2 |
| 60 | −3.0 | −3.2 | −3.2 | −3.0 | −2.8 | −2.2 | −1.5 | −0.9 | −0.3 | +0.1 | +0.3 | +0.3 | +0.2 |
| 70 | −3.2 | −3.3 | −3.2 | −2.9 | −2.5 | −1.9 | −1.2 | −0.5 | 0.0 | +0.3 | +0.4 | +0.4 | +0.3 |
| 80 | −3.4 | −3.3 | −3.1 | −2.6 | −2.2 | −1.5 | −0.8 | −0.2 | +0.3 | +0.5 | +0.6 | +0.5 | +0.3 |
| 90 | −3.4 | −3.3 | −2.9 | −2.3 | −1.9 | −1.1 | −0.4 | +0.2 | +0.5 | +0.7 | +0.7 | +0.6 | +0.4 |
| 100 | −3.3 | −3.1 | −2.6 | −1.9 | −1.4 | −0.7 | 0.0 | +0.5 | +0.8 | +0.9 | +0.8 | +0.6 | +0.4 |
| 110 | −3.1 | −2.8 | −2.2 | −1.4 | −1.0 | −0.3 | +0.4 | +0.8 | +1.0 | +1.0 | +0.9 | +0.7 | +0.4 |
| 120 | −2.8 | −2.5 | −1.8 | −0.9 | −0.4 | +0.3 | +0.8 | +1.1 | +1.2 | +1.1 | +0.9 | +0.7 | +0.4 |
| 130 | −2.4 | −2.0 | −1.2 | −0.4 | +0.1 | +0.8 | +1.2 | +1.4 | +1.4 | +1.2 | +0.9 | +0.7 | +0.4 |
| 140 | −1.9 | −1.5 | −0.7 | +0.2 | +0.6 | +1.2 | +1.5 | +1.6 | +1.5 | +1.3 | +0.9 | +0.7 | +0.3 |
| 150 | −1.4 | −0.9 | −0.1 | +0.7 | +1.1 | +1.6 | +1.8 | +1.8 | +1.6 | +1.3 | +0.9 | +0.7 | +0.3 |
| 160 | −0.8 | −0.3 | +0.5 | +1.2 | +1.6 | +1.9 | +2.1 | +2.0 | +1.7 | +1.3 | +0.8 | +0.6 | +0.3 |
| 170 | −0.2 | +0.3 | +1.1 | +1.7 | +2.0 | +2.2 | +2.2 | +2.0 | +1.6 | +1.2 | +0.8 | +0.5 | +0.2 |
| 180 | +0.5 | +0.9 | +1.6 | +2.2 | +2.3 | +2.5 | +2.3 | +2.0 | +1.6 | +1.1 | +0.7 | +0.4 | +0.1 |
| 190 | +1.1 | +1.5 | +2.1 | +2.5 | +2.6 | +2.6 | +2.4 | +2.0 | +1.5 | +1.0 | +0.5 | +0.3 | +0.1 |
| 200 | +1.6 | +2.0 | +2.5 | +2.8 | +2.8 | +2.7 | +2.3 | +1.8 | +1.3 | +0.8 | +0.4 | +0.2 | 0.0 |
| 210 | +2.1 | +2.4 | +2.8 | +3.0 | +3.0 | +2.6 | +2.2 | +1.7 | +1.1 | +0.6 | +0.2 | 0.0 | −0.1 |
| 220 | +2.6 | +2.8 | +3.1 | +3.1 | +2.9 | +2.5 | +2.0 | +1.5 | +0.9 | +0.4 | 0.0 | −0.1 | −0.1 |
| 230 | +2.9 | +3.1 | +3.2 | +3.0 | +2.9 | +2.4 | +1.8 | +1.2 | +0.6 | +0.2 | −0.1 | −0.2 | −0.2 |
| 240 | +3.2 | +3.3 | +3.2 | +3.0 | +2.7 | +2.1 | +1.5 | +0.9 | +0.3 | −0.1 | −0.3 | −0.3 | −0.3 |
| 250 | +3.3 | +3.3 | +3.2 | +2.8 | +2.5 | +1.8 | +1.2 | +0.5 | 0.0 | −0.3 | −0.4 | −0.4 | −0.3 |
| 260 | +3.4 | +3.3 | +3.0 | +2.5 | +2.1 | +1.5 | +0.8 | +0.2 | −0.3 | −0.5 | −0.6 | −0.5 | −0.3 |
| 270 | +3.3 | +3.2 | +2.8 | +2.2 | +1.8 | +1.1 | +0.4 | −0.2 | −0.6 | −0.7 | −0.7 | −0.6 | −0.4 |
| 280 | +3.2 | +3.0 | +2.5 | +1.8 | +1.4 | +0.6 | 0.0 | −0.5 | −0.8 | −0.9 | −0.8 | −0.7 | −0.4 |
| 290 | +2.9 | +2.7 | +2.1 | +1.3 | +0.9 | +0.2 | −0.4 | −0.8 | −1.1 | −1.1 | −0.9 | −0.7 | −0.4 |
| 300 | +2.6 | +2.3 | +1.6 | +0.9 | +0.4 | −0.3 | −0.8 | −1.1 | −1.3 | −1.2 | −0.9 | −0.7 | −0.4 |
| 310 | +2.2 | +1.9 | +1.1 | +0.4 | −0.1 | −0.7 | −1.2 | −1.4 | −1.4 | −1.3 | −0.9 | −0.7 | −0.4 |
| 320 | +1.7 | +1.4 | +0.6 | −0.2 | −0.6 | −1.1 | −1.5 | −1.7 | −1.5 | −1.3 | −0.9 | −0.7 | −0.3 |
| 330 | +1.2 | +0.8 | +0.1 | −0.7 | −1.0 | −1.5 | −1.8 | −1.8 | −1.6 | −1.3 | −0.9 | −0.6 | −0.3 |
| 340 | +0.7 | +0.3 | −0.5 | −1.1 | −1.5 | −1.9 | −2.0 | −2.0 | −1.7 | −1.3 | −0.8 | −0.6 | −0.2 |
| 350 | +0.1 | −0.2 | −1.0 | −1.6 | −1.9 | −2.2 | −2.2 | −2.0 | −1.6 | −1.2 | −0.7 | −0.5 | −0.2 |
| 360 | −0.4 | −0.8 | −1.5 | −2.0 | −2.2 | −2.4 | −2.3 | −2.0 | −1.5 | −1.1 | −0.6 | −0.4 | −0.1 |

Table 26.—*Resultant amplitude of $S_2$ due to $T_2$*

[Argument $h-p_1$ refers to beginning of series]

| Series $h-p_1$ | 15 days | 29 days | 58 days | 87 days | 105 days | 134 days | 163 days | 192 days | 221 days | 250 days | 279 days | 297 days | 326 days |
|---|---|---|---|---|---|---|---|---|---|---|---|---|---|
| 0 | 1.06 | 1.06 | 1.05 | 1.04 | 1.03 | 1.02 | 1.01 | 1.00 | 0.99 | 0.99 | 0.99 | 0.99 | 0.99 |
| 10 | 1.06 | 1.05 | 1.05 | 1.03 | 1.03 | 1.01 | 1.00 | 0.99 | 0.99 | 0.99 | 0.99 | 0.99 | 0.99 |
| 20 | 1.05 | 1.05 | 1.04 | 1.03 | 1.02 | 1.00 | 0.99 | 0.99 | 0.98 | 0.98 | 0.99 | 0.99 | 0.99 |
| 30 | 1.05 | 1.04 | 1.03 | 1.02 | 1.01 | 1.00 | 0.99 | 0.98 | 0.98 | 0.98 | 0.98 | 0.99 | 0.99 |
| 40 | 1.04 | 1.03 | 1.02 | 1.01 | 1.00 | 0.99 | 0.98 | 0.98 | 0.98 | 0.98 | 0.98 | 0.99 | 0.99 |
| 50 | 1.03 | 1.03 | 1.01 | 1.00 | 0.99 | 0.98 | 0.97 | 0.97 | 0.97 | 0.98 | 0.98 | 0.99 | 1.00 |
| 60 | 1.02 | 1.02 | 1.00 | 0.99 | 0.98 | 0.97 | 0.97 | 0.97 | 0.97 | 0.98 | 0.99 | 0.99 | 1.00 |
| 70 | 1.01 | 1.01 | 0.99 | 0.98 | 0.97 | 0.97 | 0.96 | 0.97 | 0.97 | 0.98 | 0.99 | 0.99 | 1.00 |
| 80 | 1.00 | 1.00 | 0.98 | 0.97 | 0.97 | 0.96 | 0.96 | 0.97 | 0.97 | 0.98 | 0.99 | 0.99 | 1.00 |
| 90 | 1.00 | 0.99 | 0.97 | 0.96 | 0.96 | 0.96 | 0.96 | 0.97 | 0.97 | 0.98 | 0.99 | 0.99 | 1.00 |
| 100 | 0.99 | 0.98 | 0.97 | 0.96 | 0.96 | 0.96 | 0.96 | 0.97 | 0.98 | 0.98 | 0.99 | 1.00 | 1.00 |
| 110 | 0.98 | 0.97 | 0.96 | 0.95 | 0.95 | 0.95 | 0.96 | 0.97 | 0.98 | 0.99 | 0.99 | 1.00 | 1.00 |
| 120 | 0.97 | 0.96 | 0.95 | 0.95 | 0.95 | 0.95 | 0.96 | 0.97 | 0.98 | 0.99 | 1.00 | 1.00 | 1.00 |
| 130 | 0.96 | 0.95 | 0.95 | 0.95 | 0.95 | 0.96 | 9.97 | 0.98 | 0.99 | 0.99 | 1.00 | 1.00 | 1.00 |
| 140 | 0.95 | 0.95 | 0.94 | 0.95 | 0.95 | 0.96 | 0.97 | 0.98 | 0.99 | 0.99 | 1.00 | 1.00 | 1.00 |
| 150 | 0.95 | 0.94 | 0.94 | 0.95 | 0.95 | 0.96 | 0.97 | 0.99 | 1.00 | 1.00 | 1.01 | 1.01 | 1.01 |
| 160 | 0.94 | 0.94 | 0.94 | 0.95 | 0.96 | 0.97 | 0.98 | 0.99 | 1.00 | 1.01 | 1.01 | 1.01 | 1.01 |
| 170 | 0.94 | 0.94 | 0.95 | 0.96 | 0.96 | 0.97 | 0.99 | 1.00 | 1.01 | 1.01 | 1.01 | 1.01 | 1.01 |
| 180 | 0.94 | 0.94 | 0.95 | 0.96 | 0.97 | 0.98 | 0.99 | 1.00 | 1.01 | 1.01 | 1.01 | 1.01 | 1.01 |
| 190 | 0.95 | 0.95 | 0.96 | 0.97 | 0.98 | 0.99 | 1.00 | 1.01 | 1.01 | 1.02 | 1.02 | 1.01 | 1.01 |
| 200 | 0.95 | 0.95 | 0.96 | 0.98 | 0.99 | 1.00 | 1.01 | 1.02 | 1.02 | 1.02 | 1.02 | 1.01 | 1.01 |
| 210 | 0.96 | 0.96 | 0.97 | 0.99 | 0.99 | 1.01 | 1.02 | 1.02 | 1.02 | 1.02 | 1.02 | 1.01 | 1.01 |
| 220 | 0.96 | 0.97 | 0.98 | 1.00 | 1.01 | 1.02 | 1.03 | 1.03 | 1.03 | 1.02 | 1.02 | 1.01 | 1.01 |
| 230 | 0.97 | 0.98 | 0.99 | 1.00 | 1.01 | 1.02 | 1.03 | 1.03 | 1.03 | 1.02 | 1.02 | 1.01 | 1.01 |
| 240 | 0.98 | 0.99 | 1.00 | 1.01 | 1.02 | 1.03 | 1.03 | 1.03 | 1.03 | 1.02 | 1.02 | 1.01 | 1.01 |
| 250 | 0.99 | 1.00 | 1.01 | 1.02 | 1.03 | 1.04 | 1.04 | 1.03 | 1.02 | 1.02 | 1.01 | 1.01 |
| 260 | 1.00 | 1.01 | 1.02 | 1.03 | 1.04 | 1.04 | 1.04 | 1.03 | 1.02 | 1.01 | 1.01 | 1.00 |
| 270 | 1.01 | 1.02 | 1.03 | 1.04 | 1.04 | 1.04 | 1.04 | 1.03 | 1.02 | 1.01 | 1.01 | 1.00 |
| 280 | 1.02 | 1.03 | 1.04 | 1.04 | 1.05 | 1.05 | 1.04 | 1.04 | 1.03 | 1.02 | 1.01 | 1.01 | 1.00 |
| 290 | 1.03 | 1.03 | 1.04 | 1.05 | 1.05 | 1.05 | 1.04 | 1.03 | 1.02 | 1.01 | 1.01 | 1.00 | 1.00 |
| 300 | 1.04 | 1.04 | 1.05 | 1.05 | 1.05 | 1.05 | 1.04 | 1.03 | 1.02 | 1.01 | 1.00 | 1.00 | 1.00 |
| 310 | 1.04 | 1.05 | 1.05 | 1.05 | 1.05 | 1.05 | 1.04 | 1.03 | 1.02 | 1.01 | 1.00 | 1.00 | 1.00 |
| 320 | 1.05 | 1.05 | 1.06 | 1.05 | 1.05 | 1.04 | 1.03 | 1.02 | 1.01 | 1.00 | 1.00 | 1.00 | 1.00 |
| 330 | 1.06 | 1.06 | 1.06 | 1.05 | 1.05 | 1.04 | 1.03 | 1.02 | 1.01 | 1.00 | 1.00 | 1.00 | 1.00 |
| 340 | 1.06 | 1.06 | 1.06 | 1.05 | 1.05 | 1.03 | 1.02 | 1.01 | 1.00 | 1.00 | 0.99 | 0.99 | 1.00 |
| 350 | 1.06 | 1.06 | 1.06 | 1.05 | 1.04 | 1.03 | 1.02 | 1.00 | 1.00 | 0.99 | 0.99 | 0.99 | 1.00 |
| 360 | 1.06 | 1.06 | 1.05 | 1.04 | 1.03 | 1.02 | 1.01 | 1.00 | 0.99 | 0.99 | 0.99 | 0.99 | 0.99 |

HARMONIC ANALYSIS AND PREDICTION OF TIDES 233

Table 27.—*Critical logarithms for Form 245*

| Natural number | Logarithm | Natural number | Logarithm | Natural number | Logarithm | Natural number | Logarithm | Natural number | Logarithm |
|---|---|---|---|---|---|---|---|---|---|
| 0.000 | ---------- | 0.050 | 8.6947 | 0.100 | 8.9979 | 0.150 | 9.1747 | 0.200 | 9.3000 |
| .001 | 6.6990 | .051 | 8.7033 | .101 | 9.0022 | .151 | 9.1776 | .201 | 9.3022 |
| .002 | 7.1761 | .052 | 8.7119 | .102 | 9.0065 | .152 | 9.1805 | .202 | 9.3043 |
| .003 | 7.3980 | .053 | 8.7202 | .103 | 9.0108 | .153 | 9.1833 | .203 | 9.3065 |
| .004 | 7.5441 | .054 | 8.7284 | .104 | 9.0150 | .154 | 9.1862 | .204 | 9.3086 |
| .005 | 7.6533 | .055 | 8.7365 | .105 | 9.0192 | .155 | 9.1890 | .205 | 9.3107 |
| .006 | 7.7404 | .056 | 8.7443 | .106 | 9.0233 | .156 | 9.1918 | .206 | 9.3129 |
| .007 | 7.8130 | .057 | 8.7521 | .107 | 9.0274 | .157 | 9.1946 | .207 | 9.3150 |
| .008 | 7.8751 | .058 | 8.7597 | .108 | 9.0315 | .158 | 9.1973 | .208 | 9.3171 |
| .009 | 7.9295 | .059 | 8.7672 | .109 | 9.0355 | .159 | 9.2001 | .209 | 9.3192 |
| .010 | 7.9778 | .060 | 8.7746 | .110 | 9.0395 | .160 | 9.2028 | .210 | 9.3212 |
| .011 | 8.0212 | .061 | 8.7818 | .111 | 9.0434 | .161 | 9.2055 | .211 | 9.3233 |
| .012 | 8.0607 | .062 | 8.7889 | .112 | 9.0473 | .162 | 9.2082 | .212 | 9.3254 |
| .013 | 8.0970 | .063 | 8.7959 | .113 | 9.0512 | .163 | 9.2109 | .213 | 9.3274 |
| .014 | 8.1304 | .064 | 8.8028 | .114 | 9.0551 | .164 | 9.2136 | .214 | 9.3295 |
| .015 | 8.1614 | .065 | 8.8096 | .115 | 9.0589 | .165 | 9.2162 | .215 | 9.3315 |
| .016 | 8.1904 | .066 | 8.8163 | .116 | 9.0626 | .166 | 9.2189 | .216 | 9.3335 |
| .017 | 8.2175 | .067 | 8.8229 | .117 | 9.0664 | .167 | 9.2215 | .217 | 9.3355 |
| .018 | 8.2431 | .068 | 8.8294 | .118 | 9.0701 | .168 | 9.2241 | .218 | 9.3375 |
| .019 | 8.2672 | .069 | 8.8357 | .119 | 9.0738 | .169 | 9.2267 | .219 | 9.3395 |
| .020 | 8.2901 | .070 | 8.8420 | .120 | 9.0774 | .170 | 9.2292 | .220 | 9.3415 |
| .021 | 8.3118 | .071 | 8.8482 | .121 | 9.0810 | .171 | 9.2318 | .221 | 9.3435 |
| .022 | 8.3325 | .072 | 8.8544 | .122 | 9.0846 | .172 | 9.2343 | .222 | 9.3454 |
| .023 | 8.3522 | .073 | 8.8604 | .123 | 9.0882 | .173 | 9.2368 | .223 | 9.3474 |
| .024 | 8.3711 | .074 | 8.8663 | .124 | 9.0917 | .174 | 9.2394 | .224 | 9.3493 |
| .025 | 8.3892 | .075 | 8.8722 | .125 | 9.0952 | .175 | 9.2419 | .225 | 9.3513 |
| .026 | 8.4066 | .076 | 8.8780 | .126 | 9.0987 | .176 | 9.2443 | .226 | 9.3532 |
| .027 | 8.4233 | .077 | 8.8837 | .127 | 9.1021 | .177 | 9.2468 | .227 | 9.3551 |
| .028 | 8.4394 | .078 | 8.8894 | .128 | 9.1056 | .178 | 9.2493 | .228 | 9.3570 |
| .029 | 8.4549 | .079 | 8.8949 | .129 | 9.1090 | .179 | 9.2517 | .229 | 9.3589 |
| .030 | 8.4699 | .080 | 8.9004 | .130 | 9.1123 | .180 | 9.2541 | .230 | 9.3608 |
| .031 | 8.4843 | .081 | 8.9059 | .131 | 9.1157 | .181 | 9.2565 | .231 | 9.3627 |
| .032 | 8.4984 | .082 | 8.9112 | .132 | 9.1190 | .182 | 9.2589 | .232 | 9.3646 |
| .033 | 8.5119 | .083 | 8.9165 | .133 | 9.1223 | .183 | 9.2613 | .233 | 9.3665 |
| .034 | 8.5251 | .084 | 8.9217 | .134 | 9.1255 | .184 | 9.2637 | .234 | 9.3683 |
| .035 | 8.5379 | .085 | 8.9269 | .135 | 9.1288 | .185 | 9.2661 | .235 | 9.3702 |
| .036 | 8.5503 | .086 | 8.9320 | .136 | 9.1320 | .186 | 9.2684 | .236 | 9.3720 |
| .037 | 8.5623 | .087 | 8.9371 | .137 | 9.1352 | .187 | 9.2707 | .237 | 9.3739 |
| .038 | 8.5741 | .088 | 8.9421 | .138 | 9.1384 | .188 | 9.2731 | .238 | 9.3757 |
| .039 | 8.5855 | .089 | 8.9470 | .139 | 9.1415 | .189 | 9.2754 | .239 | 9.3775 |
| .040 | 8.5967 | .090 | 8.9519 | .140 | 9.1446 | .190 | 9.2777 | .240 | 9.3794 |
| .041 | 8.6075 | .091 | 8.9567 | .141 | 9.1477 | .191 | 9.2799 | .241 | 9.3812 |
| .042 | 8.6181 | .092 | 8.9615 | .142 | 9.1508 | .192 | 9.2822 | .242 | 9.3830 |
| .043 | 8.6284 | .093 | 8.9662 | .143 | 9.1539 | .193 | 9.2845 | .243 | 9.3848 |
| .044 | 8.6385 | .094 | 8.9709 | .144 | 9.1569 | .194 | 9.2867 | .244 | 9.3866 |
| .045 | 8.6484 | .095 | 8.9755 | .145 | 9.1599 | .195 | 9.2890 | .245 | 9.3883 |
| .046 | 8.6581 | .096 | 8.9801 | .146 | 9.1629 | .196 | 9.2912 | .246 | 9.3901 |
| .047 | 8.6675 | .097 | 8.9846 | .147 | 9.1659 | .197 | 9.2934 | .247 | 9.3919 |
| .048 | 8.6767 | .098 | 8.9891 | .148 | 9.1688 | .198 | 9.2956 | .248 | 9.3936 |
| .049 | 8.6858 | .099 | 8.9935 | .149 | 9.1718 | .199 | 9.2978 | .249 | 9.3954 |
| .050 | 8.6947 | .100 | 8.9979 | .150 | 9.1747 | .200 | 9.3000 | .250 | 9.3971 |

TABLE 28.—*Constituent speed differences* $(b-a)$ *and* $\log (b-a)$

DIURNAL CONSTITUENTS

| B \ A | $J_1$ | $K_1$ | $M_1$ | $O_1$ | $OO$ | $P_1$ | $Q_1$ | $2Q$ | $S_1$ | $\rho_1$ |
|---|---|---|---|---|---|---|---|---|---|---|
| $J_1$ | 0<br>0 | −0.544375<br>9.735898 | −1.088749<br>0.036928 | −1.642408<br>0.215481 | +0.553658<br>9.743242 | −0.626512<br>9.796929 | −2.186782<br>0.339806 | −2.731157<br>0.436347 | −0.585443<br>9.767485 | −2.113929<br>0.325090 |
| $K_1$ | +0.544375<br>9.735898 | 0<br>0 | −0.544375<br>9.735898 | −1.098033<br>0.040615 | +1.098033<br>0.040615 | −0.082137<br>8.914539 | −1.642408<br>0.215481 | −2.186782<br>0.339806 | −0.041069<br>8.613514 | −1.569554<br>0.195776 |
| $M_1$ | +1.088749<br>0.036928 | +0.544375<br>9.735898 | 0<br>0 | −0.553658<br>9.743242 | +1.642408<br>0.215481 | +0.462237<br>9.664865 | −1.098033<br>0.040615 | −1.642408<br>0.215481 | +0.503306<br>9.701832 | −1.025179<br>0.010800 |
| $O_1$ | +1.642408<br>0.215481 | +1.098033<br>0.040615 | +0.553658<br>9.743242 | 0<br>0 | +2.196066<br>0.341645 | +1.015896<br>0.006849 | −0.544375<br>9.735898 | −1.088749<br>0.036928 | +1.056964<br>0.024060 | −0.471521<br>9.673501 |
| $OO$ | −0.553658<br>9.743242 | −1.098033<br>0.040615 | −1.642408<br>0.215481 | −2.196066<br>0.341645 | 0<br>0 | −1.180170<br>0.071945 | −2.740441<br>0.437820 | −3.284816<br>0.516511 | −1.139102<br>0.056563 | −2.667587<br>0. 4219 |
| $P_1$ | +0.626512<br>9.796929 | +0.082137<br>8.914539 | −0.462237<br>9.664865 | −1.015896<br>0.006849 | +1.180170<br>0.071945 | 0<br>0 | −1.560270<br>0.193200 | −2.104645<br>0.323179 | +0.041069<br>8.613514 | −1.487417<br>0.172433 |
| $Q_1$ | +2.186782<br>0.339806 | +1.642408<br>0.215481 | +1.098033<br>0.040615 | +0.544375<br>9.735898 | +2.740441<br>0.437820 | +1.560270<br>0.193200 | 0<br>0 | −0.544375<br>9.735898 | +1.601339<br>0.204483 | +0.072854<br>8.862453 |
| $2Q$ | +2.731157<br>0.436347 | +2.186782<br>0.339806 | +1.642408<br>0.215481 | +1.088749<br>0.036928 | +3.284816<br>0.516511 | +2.104645<br>0.323179 | +0.544375<br>9.735898 | 0<br>0 | +2.145714<br>0.331572 | +0.617228<br>9.790446 |
| $S_1$ | +0.585443<br>9.767485 | +0.041069<br>8.613514 | −0.503306<br>9.701832 | −1.056964<br>0.024060 | +1.139102<br>0.056563 | −0.041069<br>8.613514 | −1.601339<br>0.204483 | −2.145714<br>0.331572 | 0<br>0 | −1.528486<br>0.184261 |
| $\rho_1$ | +2.113929<br>0.325090 | +1.569554<br>0.195776 | +1.025179<br>0.010800 | +0.471521<br>9.673501 | +2.667587<br>0.426119 | +1.487417<br>0.172433 | −0.072854<br>8.862453 | −0.617228<br>9.790446 | +1.528486<br>0.184261 | 0<br>0 |

# HARMONIC ANALYSIS AND PREDICTION OF TIDES 235

Table 28.—*Constituent speed differences (b—a) and log (b—a)*—Continued

SEMIDIURNAL CONSTITUENTS

| A \ B | K₂ | L₂ | M₂ | N₂ | 2N | R₂ | S₂ | T₂ | λ₂ | μ₂ | ν₂ | 2SM |
|---|---|---|---|---|---|---|---|---|---|---|---|---|
| K₂ | 0<br>0 | −0.553658<br>9.743242 | −1.098033<br>0.040615 | −1.642408<br>0.215481 | −2.186782<br>0.339806 | 0.041071<br>8.613535 | −0.082137<br>8.914539 | −0.123204<br>9.090625 | −0.625512<br>9.796929 | −2.113929<br>0.325090 | −1.569554<br>0.195776 | +0.369<br>9.970235 |
| L₂ | +0.553658<br>9.743242 | 0<br>0 | −0.544375<br>9.735898 | −1.088749<br>0.036928 | −1.633124<br>0.213019 | +0.512588<br>9.709768 | +0.471521<br>9.673501 | +0.430454<br>9.633927 | −0.072854<br>8.862453 | −1.560270<br>0.193200 | −1.015896<br>0.006849 | +1.487417<br>0.172433 |
| M₂ | +1.098033<br>0.040615 | +0.544375<br>9.735898 | 0<br>0 | −0.544375<br>9.735898 | −1.088749<br>0.036928 | +1.056962<br>0.024059 | +1.015896<br>0.006849 | +0.974829<br>9.988928 | +0.471521<br>9.673501 | −1.015896<br>0.006849 | −0.471521<br>9.673501 | +2.031792<br>0.307879 |
| N₂ | +1.642408<br>0.215481 | +1.088749<br>0.036928 | +0.544375<br>9.735898 | 0<br>0 | −0.544375<br>9.735898 | +1.601337<br>0.204483 | +1.560270<br>0.193200 | +1.519204<br>0.181616 | +1.015896<br>0.006849 | −0.471521<br>9.673501 | +0.072854<br>8.862453 | +2.576166<br>0.410974 |
| 2N | +2.186782<br>0.339806 | +1.633124<br>0.213019 | +1.088749<br>0.036928 | +0.544375<br>9.735898 | 0<br>0 | +2.145712<br>0.331571 | +2.104645<br>0.323179 | +2.063579<br>0.314621 | +1.560270<br>0.193200 | +0.072854<br>8.862453 | +0.617228<br>9.790446 | +3.120541<br>0.494230 |
| R₂ | −0.041071<br>8.613535 | −0.512588<br>9.709768 | −1.056962<br>0.024059 | −1.601337<br>0.204483 | −2.145712<br>0.331571 | 0<br>0 | −0.041067<br>8.613493 | −0.082133<br>8.914518 | −0.585441<br>9.767483 | −2.072858<br>0.316570 | −1.528484<br>0.184261 | +0.789<br>9.988928 |
| S₂ | +0.082137<br>8.914539 | −0.471521<br>9.673501 | −1.015896<br>0.006849 | −1.560270<br>0.193200 | −2.104645<br>0.323179 | +0.041067<br>8.613493 | 0<br>0 | −0.041067<br>8.613493 | −0.544375<br>9.735898 | −2.031792<br>0.307879 | −1.487417<br>0.172433 | +1.196<br>0.006849 |
| T₂ | +0.123204<br>9.090625 | −0.430454<br>9.633927 | −0.974829<br>9.988928 | −1.519204<br>0.181616 | −2.063579<br>0.314621 | +0.082133<br>8.914518 | +0.041067<br>8.613493 | 0<br>0 | −0.503308<br>9.701834 | −1.990725<br>0.299011 | −1.446350<br>0.160273 | +1.056962<br>0.024059 |
| λ₂ | +0.625512<br>9.796929 | +0.072854<br>8.862453 | −0.471521<br>9.673501 | −1.015896<br>0.006849 | −1.560270<br>0.193200 | +0.585441<br>9.767483 | +0.544375<br>9.735898 | +0.503308<br>9.701834 | 0<br>0 | −1.487417<br>0.172433 | −0.943042<br>9.974531 | +1.560270<br>0.193200 |
| μ₂ | +2.113929<br>0.325090 | +1.560270<br>0.193200 | +1.015896<br>0.006849 | +0.471521<br>9.673501 | −0.072854<br>8.862453 | +2.072858<br>0.316570 | +2.031792<br>0.307879 | +1.990725<br>0.299011 | +1.487417<br>0.172433 | 0<br>0 | +0.544375<br>9.735898 | +3.047687<br>0.483970 |
| ν₂ | +1.569554<br>0.195776 | +1.015896<br>0.006849 | +0.471521<br>9.673501 | −0.072854<br>8.862453 | −0.617228<br>9.790446 | +1.528484<br>0.184261 | +1.487417<br>0.172433 | +1.446350<br>0.160273 | +0.943042<br>9.974531 | −0.544375<br>9.735898 | 0<br>0 | +2.503313<br>0.398515 |
| 2SM | −0.933759<br>0.970235 | −1.487417<br>0.172433 | −2.031792<br>0.307879 | −2.576166<br>0.410974 | −3.120541<br>0.494230 | −0.974829<br>9.988928 | −1.015896<br>0.006849 | −1.056962<br>0.024059 | −1.560270<br>0.193200 | −3.047687<br>0.483970 | −2.503313<br>0.398515 | 0<br>0 |

## Table 29.—*Elimination factors*

[Upper line for each constituent gives the logarithms of the factors; middle line, corresponding natural numbers; lower line, angles in degrees]

### SERIES 14 DAYS. DIURNAL CONSTITUENTS

| Constituent sought (A) | Disturbing constituents (B, C, etc.) | | | | | | | | |
|---|---|---|---|---|---|---|---|---|---|
| | $J_1$ | $K_1$ | $M_1$ | $O_1$ | $OO$ | $P_1$ | $Q_1$ | $2Q$ | $S_1$ | $\rho_1$ |
| $J_1$ | ------ | 9.7968 .626 269 | 8.2015 .016 357 | 9.3150 .207 264 | 9.7890 .615 93 | 9.7203 .525 255 | 8.3017 .020 353 | 9.0913 .123 261 | 9.7607 .576 262 | 8.1357 .014 185 |
| $K_1$ | 9.7968 .626 91 | ------ | 9.7968 .626 269 | 8.3839 .024 356 | 8.3839 .024 4 | 9.9958 .990 346 | 9.3150 .207 264 | 8.3017 .020 353 | 9.9990 .998 353 | 9.3344 .216 276 |
| $M_1$ | 8.2015 .016 3 | 9.7968 .626 91 | ------ | 9.7890 .615 267 | 9.3150 .207 96 | 9.8578 .721 78 | 8.3839 .024 356 | 9.3150 .207 264 | 9.8290 .675 85 | 8.6530 .045 188 |
| $O_1$ | 9.3150 .207 96 | 8.3839 .024 4 | 9.7890 .615 93 | ------ | 8.3826 .024 9 | 8.7358 .054 171 | 9.7968 .626 269 | 8.2015 .016 357 | 8.1361 .014 178 | 9.8516 .711 281 |
| $OO$ | 9.7890 .615 267 | 8.3839 .024 356 | 9.3150 .207 264 | 8.3826 .024 351 | ------ | 8.9571 .091 342 | 9.0878 .122 260 | 8.3320 .021 348 | 8.7710 .059 349 | 9.1065 .128 272 |
| $P_1$ | 9.7203 .525 105 | 9.9958 .990 14 | 9.8578 .721 282 | 8.7358 .054 189 | 8.9571 .091 18 | ------ | 9.3355 .217 278 | 8.2581 .018 186 | 9.9990 .998 7 | 9.3331 .215 290 |
| $Q_1$ | 8.3017 .020 7 | 9.3150 .207 96 | 8.3839 .024 4 | 9.7968 .626 91 | 9.0878 .122 100 | 9.3355 .217 82 | ------ | 9.7968 .626 269 | 9.3283 .213 89 | 9.9967 .992 12 |
| $2Q$ | 9.0913 .123 99 | 8.3017 .020 7 | 9.3150 .207 96 | 8.2015 .016 3 | 8.3320 .021 174 | 8.2581 .018 91 | 9.7968 .626 91 | ------ | 7.1244 .001 0 | 9.7298 .537 104 |
| $S_1$ | 9.7607 .576 98 | 9.9990 .998 7 | 9.8290 .675 275 | 8.1361 .014 182 | 8.7710 .059 11 | 9.9990 .998 353 | 9.3283 .213 271 | 7.1244 .001 0 | ------ | 9.3369 .217 283 |
| $\rho_1$ | 8.1357 .014 175 | 9.3344 .216 84 | 8.6530 .045 172 | 9.8516 .711 79 | 9.1065 .128 88 | 9.3331 .215 70 | 9.9967 .992 348 | 9.7298 .537 256 | 9.3369 .217 77 | ------ |

HARMONIC ANALYSIS AND PREDICTION OF TIDES 237

Table 29.—*Elimination factors*—Continued

SERIES 15 DAYS. SEMIDIURNAL CONSTITUENTS

| Constituent sought ($A$) | Disturbing constituents ($B$, $C$, etc.) | | | | | | | | | | |
|---|---|---|---|---|---|---|---|---|---|---|---|
| | $K_2$ | $L_2$ | $M_2$ | $N_2$ | 2N | $R_2$ | $S_2$ | $T_2$ | $\lambda_2$ | $\mu_2$ | $\nu_2$ | 2SM |
| $K_2$ | ----- | 9.7534 .567 260 | 8.9437 .088 342 | 9.2424 .175 244 | 8.9063 .081 326 | 9.9986 .997 353 | 9.9950 .989 345 | 9.9892 .975 338 | 9.6707 .468 247 | 8.7223 .053 339 | 9.2966 .198 257 | 8.8476 .070 168 |
| $L_2$ | 9.7534 .567 100 | ----- | 9.7627 .579 262 | 8.9055 .080 344 | 9.2507 .178 246 | 9.7927 .620 92 | 9.8276 .672 85 | 9.8585 .722 77 | 9.9961 .991 347 | 9.3018 .200 259 | 8.1941 .016 357 | 9.3301 .214 88 |
| $M_2$ | 8.9437 .088 18 | 9.7627 .579 98 | ----- | 9.7627 .579 262 | 8.9055 .080 344 | 8.7291 .054 10 | 8.1941 .016 3 | 8.4114 .026 175 | 9.8276 .672 85 | 8.1941 .016 357 | 9.8276 .672 275 | 8.1935 .016 6 |
| $N_2$ | 9.2424 .175 116 | 8.9055 .080 16 | 9.7627 .579 98 | ----- | 9.7627 .579 262 | 9.2760 .189 108 | 9.3018 .200 101 | 9.3204 .209 93 | 8.1941 .016 3 | 9.8276 .672 275 | 9.9961 .991 13 | 9.0793 .120 104 |
| 2N | 8.9063 .081 34 | 9.2507 .178 114 | 8.9055 .080 16 | 9.7627 .579 98 | ----- | 8.8167 .066 26 | 8.6888 .049 19 | 8.4856 .031 11 | 9.3018 .200 101 | 9.9961 .991 13 | 9.6823 .481 111 | 8.5765 .038 22 |
| $R_2$ | 9.9986 .997 7 | 9.7927 .620 268 | 8.7291 .054 350 | 9.2760 .189 252 | 8.8167 .066 334 | ----- | 9.9987 .997 353 | 9.9950 .989 345 | 9.7195 .524 255 | 8.5420 .035 347 | 9.3168 .207 265 | 8.4114 .026 175 |
| $S_2$ | 9.9950 .989 15 | 9.8276 .672 275 | 8.1941 .016 357 | 9.3018 .200 259 | 8.6888 .049 341 | 9.9987 .997 7 | ----- | 9.9987 .997 353 | 9.7627 .579 262 | 8.1935 .016 354 | 9.3301 .214 272 | 8.1941 .016 3 |
| $T_2$ | 9.9892 .975 22 | 9.8585 .722 283 | 8.4114 .026 185 | 9.3204 .209 267 | 8.4856 .031 349 | 9.9950 .989 15 | 9.9987 .997 7 | ----- | 9.8010 .632 269 | 7.6684 .005 182 | 9.3364 .217 280 | 8.7291 .054 10 |
| $\lambda_2$ | 9.6707 .468 113 | 9.9961 .991 13 | 9.8276 .672 275 | 8.1941 .016 357 | 9.3018 .200 259 | 9.7195 .524 105 | 9.7627 .579 98 | 9.8010 .632 91 | ----- | 9.3301 .214 272 | 8.7786 .060 190 | 9.3018 .200 101 |
| $\mu_2$ | 8.7223 .053 21 | 9.3018 .200 101 | 8.9141 .016 3 | 9.8276 .672 85 | 9.9961 .991 347 | 8.5420 .035 13 | 8.1935 .016 6 | 7.6684 .005 178 | 9.3301 .214 88 | ----- | 9.7627 .579 98 | 8.1926 .016 9 |
| $\nu_2$ | 9.2966 .198 103 | 8.1941 .016 3 | 9.8276 .672 85 | 9.9961 .991 347 | 9.6823 .481 249 | 9.3168 .207 95 | 9.3301 .214 88 | 9.3364 .217 80 | 8.7786 .060 170 | 9.7627 .579 262 | ----- | 9.1043 .127 91 |
| 2SM | 8.8476 .070 192 | 9.3301 .214 272 | 8.1935 .016 354 | 9.0793 .120 256 | 8.5765 .038 338 | 8.4114 .026 185 | 8.1941 .016 357 | 8.7291 .054 350 | 9.3018 .200 259 | 8.1926 .016 351 | 9.1043 .127 269 | ----- |

Table 29.—*Elimination factors*—Continued

**SERIES 29 DAYS. DIURNAL CONSTITUENTS**

| Constituent sought (A) | Disturbing constituents (B, C, etc.) | | | | | | | | |
|---|---|---|---|---|---|---|---|---|---|
| | $J_1$ | $K_1$ | $M_1$ | $O_1$ | OO | $P_1$ | $Q_1$ | 2Q | $S_1$ | $\rho_1$ |
| $J_1$ | ----- | 8.6955 .050 351 | 8.6896 .049 341 | 8.7199 .052 328 | 8.8144 .065 13 | 9.2092 .162 322 | 8.6937 .049 319 | 8.6672 .046 310 | 9.0538 .113 336 | 8.3224 .021 344 |
| $K_1$ | 8.6955 .050 9 | ----- | 8.6955 .050 351 | 8.7517 .056 338 | 8.7517 .056 22 | 9.9818 .959 331 | 8.7199 .052 328 | 8.6937 .049 319 | 9.9954 .990 346 | 8.0542 .011 354 |
| $M_1$ | 8.6896 .049 19 | 8.6955 .050 9 | ----- | 8.8144 .065 347 | 8.7199 .052 32 | 9.0674 .117 161 | 8.7517 .056 338 | 8.7199 .052 328 | 8.4418 .028 175 | 7.9579 .009 183 |
| $O_1$ | 8.7199 .052 32 | 8.7517 .056 22 | 8.8144 .065 13 | ----- | 8.7185 .052 44 | 8.2616 .018 174 | 8.6955 .050 351 | 8.6896 .049 341 | 8.3262 .021 8 | 8.9810 .096 196 |
| OO | 8.8144 .065 347 | 8.7517 .056 338 | 8.7199 .052 328 | 8.7185 .052 316 | ----- | 9.0332 .108 309 | 8.6848 .048 306 | 8.6504 .045 297 | 8.9334 .086 324 | 8.4666 .029 332 |
| $P_1$ | 9.2092 .162 38 | 9.9818 .959 29 | 9.0674 .117 199 | 8.2616 .018 186 | 9.0332 .108 51 | ----- | 7.7378 .005 357 | 8.2260 .017 348 | 9.9954 .990 14 | 8.6248 .042 202 |
| $Q_1$ | 8.6937 .049 41 | 8.7199 .052 32 | 8.7517 .056 22 | 8.6955 .050 9 | 8.6848 .048 54 | 7.7378 .005 3 | ----- | 8.6955 .050 351 | 8.4846 .031 17 | 9.9857 .968 25 |
| 2Q | 8.6672 .046 50 | 8.6937 .049 41 | 8.7199 .052 32 | 8.6896 .049 19 | 8.6504 .045 63 | 8.2260 .017 12 | 8.6955 .050 9 | ----- | 8.5377 .034 27 | 9.1825 .152 35 |
| $S_1$ | 9.0538 .113 24 | 9.9954 .990 14 | 8.4418 .028 185 | 8.3262 .021 352 | 8.9334 .086 36 | 9.9954 .990 346 | 8.4846 .031 343 | 8.5377 .034 333 | ----- | 8.1807 .015 188 |
| $\rho_1$ | 8.3224 .021 16 | 8.0542 .011 6 | 7.9579 .009 177 | 8.9810 .096 164 | 8.4666 .029 28 | 8.6248 .042 158 | 9.9857 .968 335 | 9.1825 .152 325 | 8.1807 .015 172 | ----- |

HARMONIC ANALYSIS AND PREDICTION OF TIDES 239

Table 29.—*Elimination factors*—Continued

SERIES 29 DAYS. SEMIDIURNAL CONSTITUENTS

| Constituent sought (A) | Disturbing constituents (B, C, etc.) | | | | | | | | | | |
|---|---|---|---|---|---|---|---|---|---|---|---|
| | $K_2$ | $L_2$ | $M_2$ | $N_2$ | 2N | $R_2$ | $S_2$ | $T_2$ | $\lambda_2$ | $\mu_2$ | $\nu_2$ | 2SM |
| $K_2$ | | 8.8144 .065 347 | 8.7517 .056 338 | 8.7199 .052 328 | 8.6937 .049 319 | 9.9954 .990 346 | 9.9818 .959 331 | 9.9587 .909 317 | 9.2092 .162 322 | 8.3224 .021 344 | 8.0542 .011 354 | 9.0054 .101 145 |
| $L_2$ | 8.8144 .065 13 | | 8.6955 .050 351 | 8.6896 .049 341 | 8.6798 .048 332 | 7.9581 .009 178 | 8.9810 .096 164 | 9.2842 .192 150 | 9.9857 .968 335 | 7.7378 .005 357 | 8.2616 .018 186 | 8.6248 .042 158 |
| $M_2$ | 8.7517 .056 22 | 8.6955 .050 9 | | 8.6955 .050 351 | 8.6896 .049 341 | 8.3262 .021 8 | 8.2616 .018 174 | 8.7772 .060 159 | 8.9810 .096 164 | 8.2616 .018 186 | 8.9810 .096 196 | 8.2588 .018 167 |
| $N_2$ | 8.7199 .052 32 | 8.6896 .049 19 | 8.6955 .050 9 | | 8.6955 .050 351 | 8.4846 .031 17 | 7.7378 .005 3 | 8.3278 .021 169 | 8.2616 .018 174 | 8.9810 .096 196 | 9.9857 .968 25 | 7.5900 .004 177 |
| 2N | 8.6937 .049 41 | 8.6798 .048 28 | 8.6896 .049 19 | 8.6955 .050 9 | | 8.5377 .034 27 | 8.2260 .017 12 | 7.4179 .003 178 | 7.7378 .005 3 | 9.9857 .968 25 | 9.1825 .152 35 | 7.7379 .005 6 |
| $R_2$ | 9.9954 .990 14 | 7.9581 .009 182 | 8.3262 .021 352 | 8.4846 .031 343 | 8.5377 .034 333 | | 9.9954 .990 346 | 9.9818 .959 331 | 9.0538 .113 336 | 7.2754 .002 359 | 8.1807 .015 188 | 8.7772 .060 159 |
| $S_2$ | 9.9818 .959 29 | 8.9810 .096 196 | 8.2616 .018 186 | 7.7378 .005 357 | 8.2260 .017 348 | 9.9954 .990 14 | | 9.9954 .990 346 | 8.6955 .050 351 | 8.2588 .018 193 | 8.6248 .042 202 | 8.2616 .018 174 |
| $T_2$ | 9.9587 .909 43 | 9.2842 .192 210 | 8.7772 .060 201 | 8.3278 .021 191 | 7.4179 .003 182 | 9.9818 .959 29 | 9.9954 .990 14 | | 8.4418 .028 185 | 8.5780 .038 207 | 8.8324 .068 217 | 8.3262 .021 8 |
| $\lambda_2$ | 9.2092 .162 38 | 9.9857 .968 25 | 8.9810 .096 196 | 8.2616 .018 186 | 7.7378 .005 357 | 9.0538 .113 24 | 8.6955 .050 9 | 8.4418 .028 175 | | 8.6248 .042 202 | 8.9640 .092 212 | 7.7378 .005 3 |
| $\mu_2$ | 8.3224 .021 16 | 7.7378 .005 3 | 8.2616 .018 174 | 8.9810 .096 164 | 9.9857 .968 335 | 7.2754 .002 1 | 8.2588 .018 167 | 8.5780 .038 153 | 8.6248 .042 158 | | 8.6955 .050 9 | 8.2539 .018 161 |
| $\nu_2$ | 8.0542 .011 6 | 8.2616 .018 174 | 8.9810 .096 164 | 9.9857 .968 335 | 9.1825 .152 325 | 8.1807 .015 172 | 8.6248 .042 158 | 8.8324 .068 143 | 8.9640 .092 148 | 8.6955 .050 351 | | 8.5015 .032 151 |
| 2SM | 9.0054 .101 215 | 8.6248 .042 202 | 8.2588 .018 193 | 7.5900 .004 183 | 7.7379 .005 354 | 8.7772 .060 201 | 8.2616 .018 186 | 8.3262 .021 352 | 7.7378 .005 357 | 8.2539 .018 199 | 8.5015 .032 209 | |

Table 29.—*Elimination factors*—Continued

**SERIES 58 DAYS. DIURNAL CONSTITUENTS**

| Constituent sought (A) | Disturbing constituents (B, C. etc.) | | | | | | | | | |
|---|---|---|---|---|---|---|---|---|---|---|
| | $J_1$ | $K_1$ | $M_1$ | $O_1$ | OO | $P_1$ | $Q_1$ | 2Q | $S_1$ | $\rho_1$ |
| $J_1$ | ------ ------ --- | 8.6896 .049 341 | 8.6657 .046 322 | 8.6504 .045 297 | 8.8039 .064 25 | 9.1056 .128 284 | 8.5715 .037 278 | 8.4713 .030 259 | 9.0154 .104 313 | 8.3059 .020 329 |
| $K_1$ | 8.6896 .049 19 | ------ ------ --- | 8.6896 .049 341 | 8.7185 .052 316 | 8.7185 .052 44 | 9.9254 .842 303 | 8.6504 .045 297 | 8.5715 .037 278 | 9.9818 .959 331 | 8.0520 .011 348 |
| $M_1$ | 8.6657 .046 38 | 8.6896 .049 19 | ------ ------ --- | 8.8039 .064 335 | 8.6504 .045 63 | 9.0427 .110 142 | 8.7185 .052 316 | 8.6504 .045 297 | 8.4403 .028 170 | 7.9572 .009 186 |
| $O_1$ | 8.6504 .045 63 | 8.7185 .052 44 | 8.8039 .064 25 | ------ ------ --- | 8.5737 .037 88 | 8.2588 .018 167 | 8.6896 .049 341 | 8.6657 .046 322 | 8.3224 .021 16 | 8.9640 .092 212 |
| OO | 8.8039 .064 335 | 8.7185 .052 316 | 8.6504 .045 297 | 8.5737 .037 272 | ------ ------ --- | 8.8349 .068 259 | 8.4575 .029 253 | 8.3057 .020 234 | 8.8391 .069 287 | 8.4112 .026 303 |
| $P_1$ | 9.1056 .128 76 | 9.9254 .842 57 | 9.0427 .110 218 | 8.2588 .018 193 | 8.8349 .068 101 | ------ ------ --- | 7.7379 .005 354 | 8.2155 .016 335 | 9.9818 .959 29 | 8.5907 .039 225 |
| $Q_1$ | 8.5715 .037 82 | 8.6504 .045 63 | 8.7185 .052 44 | 8.6896 .049 19 | 8.4575 .029 107 | 7.7379 .005 6 | ------ ------ --- | 8.6896 .049 341 | 8.4645 .029 35 | 9.9418 .875 51 |
| 2Q | 8.4713 .030 101 | 8.5715 .037 82 | 8.6504 .045 63 | 8.6657 .046 38 | 8.3057 .020 126 | 8.2155 .016 25 | 8.6896 .049 19 | ------ ------ --- | 8.4887 .031 53 | 9.0969 .125 70 |
| $S_1$ | 9.0154 .104 47 | 9.9818 .959 29 | 8.4403 .028 190 | 8.3224 .021 344 | 8.8391 .069 73 | 9.9818 .959 331 | 8.4645 .029 325 | 8.4887 .031 307 | ------ ------ --- | 8.1761 .015 196 |
| $\rho_1$ | 8.3059 .020 34 | 8.0520 .011 12 | 7.9572 .009 174 | 8.9640 .092 148 | 8.4112 .026 57 | 8.5907 .039 135 | 9.9418 .875 309 | 9.0969 .125 290 | 8.1761 .015 164 | ------ ------ --- |

HARMONIC ANALYSIS AND PREDICTION OF TIDES 241

Table 29.—*Elimination factors*—Continued

SERIES 58 DAYS. SEMIDIURNAL CONSTITUENTS

| Constituent sought ($A$) | Disturbing constituents ($B$, $C$, etc.) | | | | | | | | | | |
|---|---|---|---|---|---|---|---|---|---|---|---|
| | $K_2$ | $L_2$ | $M_2$ | $N_2$ | $2N$ | $R_2$ | $S_2$ | $T_2$ | $\lambda_2$ | $\mu_2$ | $\nu_2$ | $2SM$ |
| $K_2$ | ----- | 8.8039 .064 335 | 8.7185 .052 316 | 8.6504 .045 297 | 8.5715 .037 278 | 9.9818 .959 331 | 9.9254 .842 303 | 9.8237 .666 274 | 9.1056 .128 284 | 8.3059 .020 329 | 8.0520 .011 348 | 8.9185 .083 110 |
| $L_2$ | 8.8039 .064 25 | ----- | 8.6896 .049 341 | 8.6657 .046 322 | 8.6244 .042 303 | 7.9579 .009 177 | 8.9640 .092 148 | 9.2209 .166 120 | 9.9418 .875 309 | 7.7379 .005 354 | 8.2588 .018 193 | 8.5907 .039 135 |
| $M_2$ | 8.7185 .052 44 | 8.6896 .049 19 | ----- | 8.6896 .049 341 | 8.6657 .046 322 | 8.3224 .021 16 | 8.2588 .018 167 | 8.7480 .056 138 | 9.9640 .092 148 | 8.2588 .018 193 | 9.9640 .092 212 | 8.2475 .018 154 |
| $N_2$ | 8.6504 .045 63 | 8.6657 .046 38 | 8.6896 .049 19 | ----- | 8.6896 .049 341 | 8.4645 .029 35 | 7.7379 .005 6 | 8.3193 .021 157 | 8.2588 .018 167 | 9.9640 .092 212 | 9.9418 .875 51 | 7.5898 .004 173 |
| $2N$ | 8.5715 .037 82 | 8.6244 .042 57 | 8.6657 .046 38 | 8.6896 .049 19 | ----- | 8.4887 .031 53 | 8.2155 .016 25 | 7.4165 .003 176 | 7.7379 .005 6 | 9.9418 .875 51 | 9.0969 .125 70 | 7.7356 .005 12 |
| $R_2$ | 9.9818 .959 29 | 7.9579 .009 183 | 8.3224 .021 344 | 8.4645 .029 325 | 8.4887 .031 307 | ----- | 9.9818 .959 331 | 9.9254 .842 303 | 9.0154 .104 313 | 7.2736 .002 357 | 8.1761 .015 196 | 8.7480 .056 138 |
| $S_2$ | 9.9254 .842 57 | 8.9640 .092 212 | 8.2588 .018 193 | 7.7379 .005 354 | 8.2155 .016 335 | 9.9818 .959 29 | ----- | 9.9818 .959 331 | 8.6896 .049 341 | 8.2475 .018 206 | 8.5907 .039 225 | 8.2588 .018 167 |
| $T_2$ | 9.8237 .666 86 | 9.2209 .166 240 | 8.7480 .056 222 | 8.3193 .021 203 | 7.4165 .003 184 | 9.9254 .842 57 | 9.9818 .959 29 | ----- | 8.4402 .028 190 | 8.5270 .034 234 | 8.7366 .055 253 | 8.3224 .021 16 |
| $\lambda_2$ | 9.1056 .128 76 | 9.9418 .875 51 | 8.9640 .092 212 | 8.2588 .018 193 | 7.7379 .005 354 | 9.0154 .104 47 | 8.6896 .049 19 | 8.4402 .028 170 | ----- | 8.5907 .039 225 | 8.8933 .078 244 | 7.7379 .005 6 |
| $\mu_2$ | 8.3059 .020 31 | 7.7379 .005 6 | 8.2588 .018 167 | 8.9640 .092 148 | 9.9418 .875 309 | 7.2736 .002 357 | 8.2475 .018 154 | 8.5270 .034 126 | 8.5907 .039 135 | ----- | 8.6896 .049 19 | 8.2286 .017 141 |
| $\nu_2$ | 8.0520 .011 12 | 8.2588 .018 167 | 9.9640 .092 148 | 9.9418 .875 309 | 9.0969 .125 290 | 8.1761 .015 164 | 8.5907 .039 135 | 8.7366 .055 107 | 8.8933 .078 116 | 8.6896 .049 341 | ----- | 8.4439 .028 122 |
| $2SM$ | 8.9185 .083 250 | 8.5907 .039 225 | 8.2475 .018 206 | 7.5898 .004 187 | 7.7356 .005 348 | 8.7480 .056 222 | 8.2588 .018 193 | 8.3224 .021 344 | 7.7379 .005 354 | 8.2286 .017 219 | 8.4439 .028 238 | ----- |

Table 29.—*Elimination factors*—Continued

**SERIES 87 DAYS. DIURNAL CONSTITUENTS**

| Constituent sought (A) | Disturbing constituents (B, C, etc.) | | | | | | | | | |
|---|---|---|---|---|---|---|---|---|---|---|
| | $J_1$ | $K_1$ | $M_1$ | $O_1$ | OO | $P_1$ | $Q_1$ | 2Q | $S_1$ | $\rho_1$ |
| $J_1$ | ------ ------ ---- | 8.6798 .048 332 | 8.6244 .042 303 | 8.5225 .033 265 | 8.7857 .061 38 | 8.9030 .080 246 | 8.3232 .021 237 | 7.9841 .010 209 | 8.9481 .089 289 | 8.2780 .019 313 |
| $K_1$ | 8.6798 .048 28 | ------ ------ ---- | 8.6798 .048 332 | 8.6607 .046 294 | 8.6607 .046 66 | 9.8237 .666 274 | 8.5225 .033 265 | 8.3232 .021 237 | 9.9587 .909 317 | 8.0476 .011 341 |
| $M_1$ | 8.6244 .042 57 | 8.6798 .048 28 | ------ ------ ---- | 8.7857 .061 322 | 8.5225 .033 95 | 9.0002 .100 123 | 8.6607 .046 294 | 8.5225 .033 265 | 8.4376 .027 165 | 7.9556 .009 190 |
| $O_1$ | 8.5225 .033 95 | 8.6607 .046 66 | 8.7857 .061 38 | ------ ------ ---- | 8.2641 .018 133 | 8.2539 .018 161 | 8.6798 .048 332 | 8.6244 .042 303 | 8.3155 .021 23 | 8.9351 .086 228 |
| OO | 8.7857 .061 322 | 8.6607 .046 294 | 8.5225 .033 265 | 8.2641 .018 227 | ------ ------ ---- | 8.3377 .022 208 | 7.8138 .007 199 | 7.4337 .003 351 | 8.6579 .045 251 | 8.3116 .020 275 |
| $P_1$ | 8.9030 .080 114 | 9.8237 .666 86 | 9.0002 .100 237 | 8.2539 .018 199 | 8.3377 .022 152 | ------ ------ ---- | 7.7367 .005 351 | 8.1982 .016 323 | 9.9587 .909 43 | 8.5315 .034 247 |
| $Q_1$ | 8.3232 .021 123 | 8.5225 .033 95 | 8.6607 .046 66 | 8.6798 .048 28 | 7.8138 .007 161 | 7.7367 .005 9 | ------ ------ ---- | [8.6798 .048 332 | 8.4303 .027 52 | 9.8640 .731 76 |
| 2Q | 7.9841 .010 151 | 8.3232 .021 123 | 8.5225 .033 95 | 8.6244 .042 57 | 7.4337 .003 9 | 8.1982 .016 37 | 8.6798 .048 28 | ------ ------ ---- | 8.4014 .025 80 | 8.9351 .086 104 |
| $S_1$ | 8.9481 .089 71 | 9.9587 .909 43 | 8.4376 .003 195 | 8.3155 .021 337 | 8.6579 .045 109 | 9.9587 .909 317 | 8.4303 .027 308 | 8.4014 .025 280 | ------ ------ ---- | 8.1689 .015 204 |
| $\rho_1$ | 8.2780 .019 47 | 8.0476 .011 19 | 7.9556 .009 170 | 8.9351 .086 132 | 8.3116 .020 85 | 8.5315 .034 113 | 9.8640 .731 284 | 8.9351 .086 256 | 8.1689 .015 156 | ------ ------ ---- |

# HARMONIC ANALYSIS AND PREDICTION OF TIDES

## Table 29.—*Elimination factors*—Continued

### SERIES 87 DAYS. SEMIDIURNAL CONSTITUENTS

| Constituent sought ($A$) | Disturbing constituents ($B$, $C$, etc.) | | | | | | | | | | |
|---|---|---|---|---|---|---|---|---|---|---|---|
| | $K_2$ | $L_2$ | $M_2$ | $N_2$ | $2N$ | $R_2$ | $S_2$ | $T_2$ | $\lambda_2$ | $\mu_2$ | $\nu_2$ | 2SM |
| $K_2$ | ----- | 8.7857 .061 322 | 8.6607 .046 294 | 8.5225 .033 265 | 8.3232 .021 237 | 9.9587 .909 317 | 9.8237 .666 274 | 9.5416 .348 231 | 8.9030 .080 246 | 8.2780 .019 313 | 8.0476 .011 341 | 8.7538 .057 75 |
| $L_2$ | 8.7857 .061 38 | ----- | 8.6798 .048 332 | 8.6244 .042 303 | 8.5247 .033 275 | 7.9576 .009 175 | 8.9351 .086 132 | 9.1055 .127 89 | 9.8640 .731 284 | 7.7367 .005 351 | 8.2539 .018 199 | 8.5315 .034 113 |
| $M_2$ | 8.6607 .046 66 | 8.6798 .048 28 | ----- | 8.6798 .048 332 | 8.6244 .042 303 | 8.3155 .021 23 | 8.2539 .018 161 | 8.6976 .050 118 | 8.9351 .086 132 | 8.2539 .018 199 | 8.9351 .086 228 | 8.2286 .017 141 |
| $N_2$ | 8.5225 .033 95 | 8.6244 .042 57 | 8.6798 .048 28 | ----- | 8.6798 .048 332 | 8.4303 .027 52 | 7.7367 .005 9 | 8.3068 .020 146 | 8.2539 .018 161 | 8.9351 .086 228 | 9.8640 .731 76 | 7.5883 .004 170 |
| $2N$ | 8.3232 .021 123 | 8.5247 .033 85 | 8.6244 .042 57 | 8.6798 .048 28 | ----- | 8.4014 .025 80 | 8.1982 .016 37 | 7.4165 .003 174 | 7.7367 .005 9 | 9.8640 .731 76 | 8.9351 .086 104 | 7.7314 .005 18 |
| $R_2$ | 9.9587 .909 43 | 7.9576 .009 185 | 8.3155 .021 337 | 8.4303 .027 308 | 8.4014 .025 280 | ----- | 9.9587 .909 317 | 9.8237 .666 274 | 8.9481 .089 289 | 7.2740 .002 356 | 8.1689 .015 204 | 8.6976 .050 118 |
| $S_2$ | 9.8237 .666 86 | 8.9351 .086 228 | 8.2539 .018 199 | 7.7367 .005 351 | 8.1982 .016 323 | 9.9587 .909 43 | ----- | 9.9587 .909 317 | 8.6798 .048 332 | 8.2286 .017 219 | 8.5315 .034 247 | 8.2539 .018 161 |
| $T_2$ | 9.5416 .348 129 | 9.1055 .127 271 | 8.6976 .050 242 | 8.3068 .020 214 | 7.4165 .003 186 | 9.8237 .666 86 | 9.9587 .909 43 | ----- | 8.4376 .027 195 | 8.4358 .027 262 | 8.5521 .036 290 | 8.3155 .021 23 |
| $\lambda_2$ | 8.9030 .080 114 | 9.8640 .731 76 | 8.9351 .086 228 | 8.2539 .018 199 | 7.7367 .005 351 | 8.9481 .089 71 | 8.6798 .048 28 | 8.4376 .027 165 | ----- | 8.5315 .034 247 | 8.7629 .058 275 | 7.7367 .005 9 |
| $\mu_2$ | 8.2780 .019 47 | 7.7367 .005 9 | 8.2539 .018 161 | 8.9351 .086 132 | 9.8640 .731 284 | 7.2740 .002 141 | 8.2286 .017 4 | 8.4358 .027 98 | 8.5315 .034 113 | ----- | 8.6798 .048 28 | 8.1849 .015 122 |
| $\nu_2$ | 8.0476 .011 19 | 8.2539 .018 161 | 8.9351 .086 132 | 9.8640 .731 284 | 8.9351 .086 256 | 8.1689 .015 156 | 8.5315 .034 113 | 8.5521 .036 70 | 8.7629 .058 85 | 8.6798 .048 332 | ----- | 8.3401 .022 93 |
| 2SM | 8.7538 .057 285 | 8.5315 .034 247 | 8.2286 .017 219 | 7.5883 .004 190 | 7.7314 .005 342 | 8.6976 .050 242 | 8.2539 .018 199 | 8.3155 .021 337 | 7.7367 .005 351 | 8.1849 .015 238 | 8.3401 .022 267 | ----- |

## Table 29.—*Elimination factors*—Continued

### SERIES 105 DAYS. DIURNAL CONSTITUENTS

| Constituent sought (A) | Disturbing constituents (B, C, etc.) | | | | | | | | | |
|---|---|---|---|---|---|---|---|---|---|---|
| | $J_1$ | $K_1$ | $M_1$ | $O_1$ | $OO$ | $P_1$ | $Q_1$ | $2Q$ | $S_1$ | $\rho_1$ |
| $J_1$ | ------ ------ --- | 8.6704 .047 214 | 8.5885 .039 248 | 8.4422 .028 271 | 8.4953 .031 158 | 8.8322 .068 291 | 8.2332 .017 308 | 7.7808 .006 339 | 8.3722 .024 342 | 8.1065 .013 216 |
| $K_1$ | 8.6704 .047 146 | ------ ------ --- | 8.6704 .047 214 | 8.5381 .035 236 | 8.5381 .035 124 | 9.7311 .538 257 | 8.4422 .028 271 | 8.2332 .017 305 | 9.9393 .870 308 | 7.0766 .001 182 |
| $M_1$ | 8.5885 .039 112 | 8.6704 .047 146 | ------ ------ --- | 8.4953 .031 202 | 8.4422 .028 89 | 8.8219 .066 42 | 8.5381 .035 236 | 8.4422 .028 271 | 8.9548 .090 94 | 8.3679 .023 328 |
| $O_1$ | 8.4422 .028 89 | 8.5381 .035 124 | 8.4953 .031 158 | ------ ------ --- | 8.2803 .019 67 | 8.1856 .015 20 | 8.6704 .047 214 | 8.5885 .039 248 | 8.6113 .041 72 | 8.8929 .078 306 |
| $OO$ | 8.4953 .031 202 | 8.5381 .035 236 | 8.4422 .028 271 | 8.2803 .019 293 | ------ ------ --- | 8.4500 .028 313 | 7.9556 .009 327 | 6.4362 .000 181 | 7.5174 .003 185 | 8.1640 .015 239 |
| $P_1$ | 8.8322 .068 69 | 9.7311 .538 103 | 8.8219 .066 318 | 8.1856 .015 340 | 8.4500 .028 47 | ------ ------ --- | 7.8500 .007 194 | 8.2067 .016 228 | 9.9393 .870 52 | 8.4685 .029 286 |
| $Q_1$ | 8.2332 .017 55 | 8.4422 .028 89 | 8.5381 .035 124 | 8.6704 .047 146 | 7.9556 .009 33 | 7.8500 .007 166 | ------ ------ --- | 8.6704 .047 214 | 8.2396 .017 38 | 9.7951 .624 92 |
| $2Q$ | 7.7808 .006 21 | 8.2332 .017 55 | 8.4422 .028 89 | 8.5885 .039 112 | 6.4362 .000 179 | 8.2067 .016 132 | 8.6704 .047 146 | ------ ------ --- | 7.1241 .001 4 | 8.7943 .062 58 |
| $S_1$ | 8.3722 .024 18 | 9.9393 .870 52 | 8.9548 .090 266 | 8.6113 .041 288 | 7.5174 .003 175 | 9.9393 .870 308 | 8.2396 .017 322 | 7.1241 .001 356 | ------ ------ --- | 8.3820 .024 234 |
| $\rho_1$ | 8.1065 .013 144 | 7.0766 .001 178 | 8.3679 .023 32 | 8.8929 .078 54 | 8.1640 .015 121 | 8.4685 .029 74 | 9.7951 .624 268 | 8.7943 .062 302 | 8.3820 .024 126 | ------ ------ --- |

Table 29.—*Elimination factors*—Continued

SERIES 105 DAYS. SEMIDIURNAL CONSTITUENTS

| Constituent sought (A) | Disturbing constituents (B, C, etc.) | | | | | | | | | | |
|---|---|---|---|---|---|---|---|---|---|---|---|
| | $K_2$ | $L_2$ | $M_2$ | $N_2$ | 2N | $R_2$ | $S_2$ | $T_2$ | $\lambda_2$ | $\mu_2$ | $\nu_2$ | 2SM |
| $K_2$ | ----- | 8.4953 .031 202 | 8.5381 .035 236 | 8.4422 .028 271 | 8.2332 .017 305 | 9.9392 .869 308 | 9.7311 .538 257 | 9.1892 .155 205 | 8.8322 .068 291 | 8,1065 .013 216 | 7.0766 .001 182 | 8.6847 .048 97 |
| $L_2$ | 8.4953 .031 158 | ----- | 8.6704 .047 214 | 8.5885 .039 248 | 8.4347 .027 282 | 8.9311 .085 106 | 8.8929 .078 54 | 7.6403 .004 2 | 9.7951 .624 268 | 7.8500 .007 194 | 8.1856 .015 340 | 8.4685 .029 74 |
| $M_2$ | 8.5381 .035 124 | 8.6704 .047 146 | ----- | 8.6704 .047 214 | 8.5885 .039 248 | 8.6113 .041 72 | 8.1856 .015 20 | 8.3896 .025 148 | 8.8929 .078 54 | 8.1856 .015 340 | 8.8929 .078 306 | 8.1585 .014 40 |
| $N_2$ | 8.4422 .028 89 | 8.5885 .039 112 | 8.6704 .047 146 | ----- | 8.6704 .047 214 | 8.2395 .017 38 | 7.8500 .007 166 | 8.4362 .027 114 | 8.1856 .015 20 | 8.8929 .078 306 | 9.7951 .624 92 | 7.2638 .002 6 |
| 2N | 8.2332 .017 55 | 8.4347 .027 78 | 8.5885 .039 112 | 8.6704 .047 146 | ----- | 7.1241 .001 4 | 8.2067 .016 132 | 8.3366 .022 80 | 7.8500 .007 166 | 9.7951 .624 92 | 8.7943 .062 58 | 7.8368 .007 152 |
| $R_2$ | 9.9392 .869 52 | 8.9311 .085 254 | 8.6113 .041 288 | 8.2395 .017 322 | 7.1241 .001 356 | ----- | 9.9392 .869 308 | 9.7311 .538 257 | 8.3722 .024 342 | 8.3410 .022 268 | 8.3820 .024 234 | 8.3896 .025 148 |
| $S_2$ | 9.7311 .538 103 | 8.8929 .078 306 | 8.1856 .015 340 | 7.8500 .007 194 | 8.2067 .016 228 | 9.9392 .869 52 | ----- | 9.9392 .869 308 | 8.6704 .047 214 | 8.1585 .014 320 | 8.4685 .029 286 | 8.1856 .015 20 |
| $T_2$ | 9.1892 .155 155 | 7.6403 .004 358 | 8.3896 .025 212 | 8.4362 .027 246 | 8.3366 .022 280 | 9.7311 .538 103 | 9.9392 .869 52 | ----- | 8.9548 .090 266 | 7.6654 .005 192 | 8.0785 .012 338 | 8.6113 .041 72 |
| $\lambda_2$ | 8.8322 .068 69 | 9.7951 .624 92 | 8.8929 .078 306 | 8.1856 .015 340 | 7.8500 .007 194 | 8.3722 .024 18 | 8.6704 .047 146 | 8.9548 .090 94 | ----- | 8.4685 .029 286 | 8.6609 .046 252 | 7.8500 .007 166 |
| $\mu_2$ | 8.1065 .013 144 | 7.8500 .007 166 | 8.1856 .015 20 | 8.8929 .078 54 | 9.7951 .624 268 | 8.3410 .022 92 | 8.1585 .014 40 | 7.6654 .005 168 | 8.4685 .029 74 | ----- | 8.6704 .047 146 | 8.1117 .013 60 |
| $\nu_2$ | 7.0766 .001 178 | 8.1856 .015 20 | 8.8929 .078 54 | 9.7951 .624 268 | 8.7943 .062 302 | 8.3820 .024 126 | 8.4685 .029 74 | 8.0785 .012 22 | 8.6609 .046 108 | 8.6704 .047 214 | ----- | 8.2581 .018 94 |
| 2SM | 8.6847 .048 263 | 8.4685 .029 286 | 8.1585 .014 320 | 7.2638 .002 354 | 7.8368 .007 208 | 8.3896 .025 212 | 8.1856 .015 340 | 8.6113 .041 288 | 7.8500 .007 194 | 8.1117 .013 300 | 8.2581 .018 266 | ----- |

Table 29.—*Elimination factors*—Continued

**SERIES 134 DAYS. DIURNAL CONSTITUENTS**

| Constituent sought (A) | Disturbing constituents (B, C, etc.) | | | | | | | | |
|---|---|---|---|---|---|---|---|---|---|
| | $J_1$ | $K_1$ | $M_1$ | $O_1$ | OO | $P_1$ | $Q_1$ | 2Q | $S_1$ | $\rho_1$ |
| $J_1$ | ------ | 8.4360 .027 205 | 8.3946 .025 229 | 8.2695 .019 239 | 8.0361 .011 170 | 8.7345 .054 253 | 8.2094 .016 264 | 8.0930 .012 288 | 8.6047 .040 319 | 7.7771 .006 201 |
| $K_1$ | 8.4360 .027 155 | ------ | 8.4360 .027 205 | 8.2628 .018 214 | 8.2628 .018 146 | 9.5078 .322 228 | 8.2695 .019 239 | 8.2094 .016 264 | 9.8992 .793 294 | 7.1819 .002 356 |
| $M_1$ | 8.3946 .025 131 | 8.4360 .027 155 | ------ | 8.0361 .011 190 | 8.2695 .019 121 | 8.4838 .030 23 | 8.2628 .018 214 | 8.2695 .019 239 | 8.8500 .071 89 | 8.2196 .017 332 |
| $O_1$ | 8.2695 .019 121 | 8.2628 .018 146 | 8.0361 .011 170 | ------ | 8.1796 .015 111 | 7.9151 .008 14 | 8.4360 .027 205 | 8.3946 .025 229 | 8.5206 .033 80 | 8.6697 .047 322 |
| OO | 8.0361 .011 190 | 8.2628 .018 214 | 8.2695 .019 239 | 8.1796 .015 249 | ------ | 8.4760 .030 262 | 8.1133 .013 273 | 7.9812 .010 298 | 8.2156 .016 328 | 7.8315 .007 211 |
| $P_1$ | 8.7345 .054 107 | 9.5078 .322 132 | 8.4838 .030 337 | 7.9151 .008 346 | 8.4760 .030 98 | ------ | 7.6424 .004 191 | 7.9951 .010 216 | 9.8992 .793 66 | 8.2746 .019 308 |
| $Q_1$ | 8.2094 .016 96 | 8.2695 .019 121 | 8.2628 .018 146 | 8.4360 .027 155 | 8.1133 .013 87 | 7.6424 .004 169 | ------ | 8.4360 .027 205 | 8.2605 .018 55 | 9.6387 .435 117 |
| 2Q | 8.0930 .012 72 | 8.2094 .016 96 | 8.2695 .019 121 | 8.3946 .025 131 | 7.9812 .010 62 | 7.9951 .010 144 | 8.4360 .027 155 | ------ | 7.9233 .008 30 | 8.7610 .058 92 |
| $S_1$ | 8.6047 .040 41 | 9.8992 .793 66 | 8.8500 .071 271 | 8.5206 .033 280 | 8.2156 .016 32 | 9.8992 .793 294 | 8.2605 .018 305 | 7.9233 .008 330 | ------ | 8.3143 .021 242 |
| $\rho_1$ | 7.7771 .006 159 | 7.1819 .002 4 | 8.2196 .017 28 | 8.6697 .047 38 | 7.8315 .007 149 | 8.2746 .019 52 | 9.6387 .435 243 | 8.7610 .058 268 | 8.3143 .021 118 | ------ |

HARMONIC ANALYSIS AND PREDICTION OF TIDES 247

Table 29.—*Elimination factors*—Continued

SERIES 134 DAYS. SEMIDIURNAL CONSTITUENTS

| Constituent sought (A) | Disturbing constituents (B, C, etc.) | | | | | | | | | | |
|---|---|---|---|---|---|---|---|---|---|---|---|
| | $K_2$ | $L_2$ | $M_2$ | $N_2$ | 2N | $R_2$ | $S_2$ | $T_2$ | $\lambda_2$ | $\mu_2$ | $\nu_2$ | 2SM |
| $K_2$ | ----- | 8.0361 .011 190 | 8.2628 .018 214 | 8.2695 .019 239 | 8.2094 .016 264 | 9.8992 .793 294 | 9.5078 .322 228 | 8.9538 .090 342 | 8.7345 .054 253 | 7.7771 .006 201 | 7.1819 .002 356 | 8.5254 .034 61 |
| $L_2$ | 8.0361 .011 170 | ----- | 8.4360 .027 205 | 8.3946 .025 229 | 8.3215 .021 254 | 8.8285 .067 104 | 8.6697 .047 38 | 8.5871 .039 152 | 9.6387 .435 243 | 7.6424 .004 191 | 7.9151 .008 346 | 8.2746 .019 52 |
| $M_2$ | 8.2628 .018 146 | 8.4360 .027 155 | ----- | 8.4360 .027 205 | 8.3946 .025 229 | 8.5206 .033 80 | 7.9151 .008 14 | 8.4622 .029 128 | 8.6697 .047 38 | 7.9151 .008 346 | 8.6697 .047 322 | 7.9208 .008 27 |
| $N_2$ | 8.2695 .019 121 | 8.3946 .025 131 | 8.4360 .027 155 | ----- | 8.4360 .027 205 | 8.2605 .018 55 | 7.6424 .004 169 | 8.3592 .023 103 | 7.9151 .008 14 | 8.6697 .047 322 | 9.6387 .435 117 | 6.7753 .001 2 |
| 2N | 8.2094 .016 96 | 8.3215 .021 106 | 8.3946 .025 131 | 8.4360 .027 155 | ----- | 7.9233 .008 30 | 7.9951 .010 144 | 8.2280 .017 78 | 7.6424 .004 169 | 9.6387 .435 117 | 8.7610 .058 92 | 7.6344 .004 158 |
| $R_2$ | 9.8992 .793 66 | 8.8285 .067 256 | 8.5206 .033 280 | 8.2605 .018 305 | 7.9233 .008 330 | ----- | 9.8992 .793 294 | 9.5079 .322 228 | 8.6047 .040 319 | 8.2346 .017 267 | 8.3143 .021 242 | 8.4622 .029 128 |
| $S_2$ | 9.5078 .322 132 | 8.6697 .047 322 | 7.9151 .008 346 | 7.6424 .004 191 | 7.9951 .010 216 | 9.8992 .793 66 | ----- | 9.8992 .793 294 | 8.4360 .027 205 | 7.9028 .008 333 | 8.2746 .019 308 | 7.9151 .008 14 |
| $T_2$ | 8.9538 .090 18 | 8.5871 .039 208 | 8.4622 .029 232 | 8.3592 .023 257 | 8.2280 .017 282 | 9.5079 .322 132 | 9.8992 .793 66 | ----- | 8.8500 .071 271 | 8.0509 .011 219 | 7.7834 .006 194 | 8.5206 .033 80 |
| $\lambda_2$ | 8.7345 .054 107 | 9.6387 .435 117 | 8.6697 .047 322 | 7.9151 .008 346 | 7.6424 .004 191 | 8.6047 .040 41 | 8.4360 .027 155 | 8.8500 .071 89 | ----- | 8.2746 .019 308 | 8.5650 .037 284 | 7.6424 .004 169 |
| $\mu_2$ | 7.7771 .006 159 | 7.6424 .004 169 | 7.9151 .008 14 | 8.6697 .047 38 | 9.6387 .435 243 | 8.2346 .017 93 | 7.9028 .008 27 | 8.0509 .011 141 | 8.2746 .019 52 | ----- | 8.4360 .027 155 | 7.8820 .008 41 |
| $\nu_2$ | 7.1819 .002 4 | 7.9151 .008 14 | 8.6697 .047 38 | 9.6387 .435 243 | 8.7610 .058 268 | 8.3143 .021 118 | 8.2746 .019 52 | 7.7834 .006 166 | 8.5650 .037 76 | 8.4360 .027 205 | ----- | 8.1118 .013 65 |
| 2SM | 8.5254 .034 299 | 8.2746 .019 308 | 7.9028 .008 333 | 6.7753 .001 358 | 7.6344 .004 202 | 8.4622 .029 232 | 7.9151 .008 346 | 8.5206 .033 280 | 7.6424 .004 191 | 7.8820 .008 319 | 8.1118 .013 295 | ----- |

Table 29.—*Elimination factors*—Continued

**SERIES 163 DAYS. DIURNAL CONSTITUENTS**

| Constituent sought (A) | Disturbing constituents (B, C, etc.) | | | | | | | | |
|---|---|---|---|---|---|---|---|---|---|
| | $J_1$ | $K_1$ | $M_1$ | $O_1$ | $OO$ | $P_1$ | $Q_1$ | $2Q$ | $S_1$ | $\rho_1$ |
| $J_1$ | ------ | 8.1495 .014 195 | 8.1341 .014 210 | 7.9150 .008 207 | 7.4365 .003 3 | 8.4234 .027 215 | 7.9579 .009 223 | 7.9582 .009 238 | 8.6570 .045 295 | 7.0948 .001 185 |
| $K_1$ | 8.1495 .014 165 | ------ | 8.1495 .014 195 | 7.7528 .006 192 | 7.7528 .006 168 | 9.0723 .118 199 | 7.9150 .008 207 | 7.9579 .009 223 | 9.8470 .703 280 | 7.5128 .003 350 |
| $M_1$ | 8.1341 .014 150 | 8.1495 .014 165 | ------ | 7.4365 .003 357 | 7.9150 .008 153 | 7.6604 .005 4 | 7.7528 .006 192 | 7.9150 .008 207 | 8.7629 .058 84 | 8.0859 .012 335 |
| $O_1$ | 7.9150 .008 153 | 7.7528 .006 168 | 7.4365 .003 3 | ------ | 7.7427 .006 156 | 7.5513 .004 7 | 8.1495 .014 195 | 8.1341 .014 210 | 8.4422 .028 87 | 8.3724 .024 338 |
| $OO$ | 7.4365 .003 357 | 7.7528 .006 192 | 7.9150 .008 207 | 7.7427 .006 204 | ------ | 8.1140 .013 212 | 7.8343 .007 220 | 7.8631 .007 235 | 8.3776 .024 292 | 6.6248 .000 182 |
| $P_1$ | 8.4234 .027 145 | 9.0723 .118 161 | 7.6604 .005 356 | 7.5513 .004 353 | 8.1140 .013 148 | ------ | 7.4230 .003 188 | 7.7409 .006 203 | 9.8470 .703 80 | 7.9852 .010 331 |
| $Q_1$ | 7.9579 .009 137 | 7.9150 .008 153 | 7.7528 .006 168 | 8.1495 .014 165 | 7.8343 .007 140 | 7.4230 .003 172 | ------ | 8.1495 .014 195 | 8.2410 .017 72 | 9.3888 .245 142 |
| $2Q$ | 7.9582 .009 122 | 7.9579 .009 137 | 7.9150 .008 153 | 8.1341 .014 150 | 7.8631 .007 125 | 7.7409 .006 157 | 8.1495 .014 165 | ------ | 8.0589 .011 57 | 8.5769 .038 127 |
| $S_1$ | 8.6570 .045 65 | 9.8470 .703 80 | 8.7629 .058 276 | 8.4422 .028 273 | 8.3776 .024 68 | 9.8470 .703 280 | 8.2410 .017 288 | 8.0589 .011 57 | ------ | 8.2562 .018 250 |
| $\rho_1$ | 7.0948 .001 175 | 7.5128 .003 10 | 8.0859 .012 25 | 8.3724 .024 22 | 6.6248 .000 178 | 7.9852 .010 29 | 9.3888 .245 218 | 8.5769 .038 233 | 8.2562 .018 110 | ------ |

HARMONIC ANALYSIS AND PREDICTION OF TIDES        249

Table 29.—*Elimination factors*—Continued

**SERIES 163 DAYS. SEMIDIURNAL CONSTITUENTS**

| Constituent sought ($A$) | Disturbing constituents ($B$, $C$, etc.) | | | | | | | | | |
|---|---|---|---|---|---|---|---|---|---|---|
| | $K_2$ | $L_2$ | $M_2$ | $N_2$ | $2N$ | $R_2$ | $S_2$ | $T_2$ | $\lambda_2$ | $\mu_2$ | $\nu_2$ | $2SM$ |
| $K_2$ | ----- | 7.4365 .003 ---- | 7.7528 .006 357 | 7.9150 .008 192 | 7.9579 .009 207 | 9.8470 .703 223 | 9.0723 .118 280 | 9.3179 .208 199 | 8.4234 .027 299 | 7.0948 .001 215 | 7.5128 .003 185 | 8.1450 .014 350 26 |
| $L_2$ | 7.4365 .003 3 | ----- | 8.1495 .014 195 | 8.1341 .014 210 | 8.1078 .013 226 | 8.7464 .056 103 | 8.3724 .024 22 | 8.7614 .058 122 | 9.3888 .245 218 | 7.4230 .003 188 | 7.5513 .004 353 | 7.9852 .010 29 |
| $M_2$ | 7.7528 .006 168 | 8.1495 .014 165 | ----- | 8.1495 .014 195 | 8.1341 .014 210 | 8.4422 .028 87 | 7.5513 .004 7 | 8.4590 .029 107 | 8.3724 .024 22 | 7.5513 .004 353 | 8.3724 .024 338 | 7.5483 .004 14 |
| $N_2$ | 7.9150 .008 153 | 8.1341 .014 150 | 8.1495 .014 165 | ----- | 8.1495 .014 195 | 8.2410 .017 72 | 7.4230 .003 172 | 8.2850 .019 92 | 7.5513 .004 7 | 8.3724 .024 338 | 9.3888 .245 142 | 6.3062 .000 179 |
| $2N$ | 7.9579 .009 137 | 8.1078 .013 134 | 8.1341 .014 150 | 8.1495 .014 165 | ----- | 8.0588 .011 57 | 7.7409 .006 157 | 8.1397 .014 76 | 7.4230 .003 172 | 9.3888 .245 142 | 8.5769 .038 127 | 7.4186 .003 164 |
| $R_2$ | 9.8470 .703 80 | 8.7464 .056 257 | 8.4422 .028 273 | 8.2410 .017 288 | 8.0588 .011 303 | ----- | 9.8470 .703 280 | 9.0725 .118 199 | 8.6570 .045 295 | 8.1488 .014 265 | 8.2563 .018 250 | 8.4590 .029 107 |
| $S_2$ | 9.0723 .118 161 | 8.3724 .024 338 | 7.5513 .004 353 | 7.4230 .003 188 | 7.7409 .006 203 | 9.8470 .703 80 | ----- | 9.8470 .703 280 | 8.1495 .014 195 | 7.5483 .004 346 | 7.9852 .010 331 | 7.5513 .004 7 |
| $T_2$ | 9.3179 .208 61 | 8.7614 .058 238 | 8.4590 .029 253 | 8.2850 .019 268 | 8.1397 .014 284 | 9.0725 .118 161 | 9.8470 .703 80 | ----- | 8.7629 .058 276 | 8.1668 .015 246 | 8.1966 .016 231 | 8.4422 .028 87 |
| $\lambda_2$ | 8.4234 .027 145 | 9.3888 .245 142 | 8.3724 .024 338 | 7.5513 .004 353 | 7.4230 .003 188 | 8.6570 .045 65 | 8.1495 .014 165 | 8.7629 .058 84 | ----- | 7.9852 .010 331 | 8.3386 .022 315 | 7.4230 .003 172 |
| $\mu_2$ | 7.0948 .001 175 | 7.4230 .003 172 | 7.5513 .004 7 | 8.3724 .024 22 | 9.3888 .245 218 | 8.1488 .014 95 | 7.5483 .004 14 | 8.1668 .015 114 | 7.9852 .010 29 | ----- | 8.1495 .014 165 | 7.5426 .003 21 |
| $\nu_2$ | 7.5128 .003 10 | 7.5513 .004 7 | 8.3724 .024 22 | 9.3888 .245 218 | 8.5769 .038 233 | 8.2563 .018 110 | 7.9852 .010 29 | 8.1966 .016 129 | 8.3336 .022 45 | 8.1495 .014 195 | ----- | 7.8424 .007 36 |
| $2SM$ | 8.1450 .014 334 | 7.9852 .010 331 | 7.5483 .004 346 | 6.3062 .000 181 | 7.4186 .003 196 | 8.4590 .029 253 | 7.5513 .004 353 | 8.4422 .028 273 | 7.4230 .003 188 | 7.5426 .003 339 | 7.8424 .007 324 | ----- |

Table 29.—*Elimination factors*—Continued

**SERIES 192 DAYS. DIURNAL CONSTITUENTS**

| Constituent sought (A) | Disturbing constituents (B, C, etc.) | | | | | | | | |
|---|---|---|---|---|---|---|---|---|---|
| | $J_1$ | $K_1$ | $M_1$ | $O_1$ | $OO$ | $P_1$ | $Q_1$ | $2Q$ | $S_1$ | $\rho_1$ |
| $J_1$ | ------ | 7.6613 .005 186 | 7.6591 .005 192 | 7.0355 .001 356 | 8.0828 .012 16 | 7.3819 .002 357 | 6.5151 .000 182 | 7.0698 .001 187 | 8.6281 .042 271 | 7.3308 .002 350 |
| $K_1$ | 7.6613 .005 174 | ------ | 7.6613 .005 186 | 7.5891 .004 350 | 7.5891 .004 10 | 8.6868 .049 351 | 7.0355 .001 356 | 6.5151 .000 182 | 9.7807 .604 265 | 7.6468 .004 344 |
| $M_1$ | 7.6591 .005 168 | 7.6613 .005 174 | ------ | 8.0828 .012 344 | 7.0355 .001 4 | 8.1441 .014 165 | 7.5891 .004 350 | 7.0355 .001 356 | 8.6866 .049 80 | 7.9586 .009 338 |
| $O_1$ | 7.0355 .001 4 | 7.5891 .004 10 | 8.0828 .012 16 | ------ | 7.5826 .004 20 | 6.4230 .000 1 | 7.6613 .005 186 | 7.6591 .005 192 | 8.3698 .023 95 | 7.7679 .006 354 |
| $OO$ | 8.0828 .012 344 | 7.5891 .004 350 | 7.0355 .001 356 | 7.5826 .004 340 | ------ | 7.8388 .007 341 | 7.3409 .002 346 | 7.0344 .001 352 | 8.3250 .021 256 | 7.6132 .004 334 |
| $P_1$ | 7.3819 .002 3 | 8.6868 .049 9 | 8.1441 .014 195 | 6.4230 .000 359 | 7.8388 .007 19 | ------ | 7.1547 .001 185 | 7.3491 .002 191 | 9.7807 .604 95 | 7.3097 .002 353 |
| $Q_1$ | 6.5151 .000 178 | 7.0355 .001 4 | 7.5891 .004 10 | 7.6613 .005 174 | 7.3409 .002 14 | 7.1547 .001 175 | ------ | 7.6613 .005 186 | 8.1911 .016 89 | 8.8560 .072 168 |
| $2Q$ | 7.0698 .001 173 | 6.5151 .000 178 | 7.0355 .001 4 | 7.6591 .005 168 | 7.0344 .001 8 | 7.3491 .002 169 | 7.6613 .005 174 | ------ | 8.0615 .012 84 | 8.0931 .012 162 |
| $S_1$ | 8.6281 .042 89 | 9.7807 .604 95 | 8.6866 .049 280 | 8.3698 .023 265 | 8.3250 .021 104 | 9.7807 .604 265 | 8.1911 .016 271 | 8.0615 .012 276 | ------ | 8.2024 .016 258 |
| $\rho_1$ | 7.3308 .002 10 | 7.6468 .004 16 | 7.9586 .009 22 | 7.7679 .006 6 | 7.6132 .004 26 | 7.3097 .002 7 | 8.8560 .072 192 | 8.0931 .012 198 | 8.2024 .016 102 | ------ |

HARMONIC ANALYSIS AND PREDICTION OF TIDES 251

Table 29.—*Elimination factors*—Continued

SERIES 192 DAYS. SEMIDIURNAL CONSTITUENTS

| Constituent sought (A) | Disturbing constituents (B, C, etc.) | | | | | | | | | | |
|---|---|---|---|---|---|---|---|---|---|---|---|
| | $K_2$ | $L_2$ | $M_2$ | $N_2$ | 2N | $R_2$ | $S_2$ | $T_2$ | $\lambda_2$ | $\mu_2$ | $\nu_2$ | 2SM |
| $K_2$ | | 8.0828 .012 344 | 7.5891 .004 350 | 7.0355 .001 356 | 6.5151 .000 182 | 9.7807 .604 265 | 8.6868 .049 351 | 9.2922 .196 256 | 7.3819 .002 357 | 7.3308 .002 350 | 7.6468 .004 344 | 7.6012 .004 171 |
| $L_2$ | 8.0828 .012 16 | | 7.6613 .005 186 | 7.6591 .005 192 | 7.6554 .005 197 | 8.6778 .048 101 | 7.7679 .006 6 | 8.7615 .058 92 | 8.8560 .072 192 | 7.1547 .001 185 | 6.4230 .000 359 | 7.3097 .002 7 |
| $M_2$ | 7.5891 .004 10 | 7.6613 .005 174 | | 7.6613 .005 186 | 7.6591 .005 192 | 8.3698 .023 95 | 6.4230 .000 1 | 8.4057 .025 86 | 7.7679 .006 6 | 6.4230 .000 359 | 7.7679 .006 354 | 6.4265 .000 1 |
| $N_2$ | 7.0355 .001 4 | 7.6591 .005 168 | 7.6613 .005 174 | | 7.6613 .005 186 | 8.1911 .016 89 | 7.1547 .001 175 | 8.2077 .016 80 | 6.4230 .000 1 | 7.7679 .006 354 | 8.8560 .072 168 | 6.8803 .001 175 |
| 2N | 6.5151 .000 178 | 7.6554 .005 163 | 7.6591 .005 168 | 7.6613 .005 174 | | 8.0615 .012 84 | 7.3491 .002 169 | 8.0649 .012 74 | 7.1547 .001 175 | 8.8560 .072 168 | 8.0931 .012 162 | 7.1525 .001 170 |
| $R_2$ | 9.7807 .604 95 | 8.6778 .048 259 | 8.3698 .023 265 | 8.1911 .016 271 | 8.0615 .012 276 | | 9.7807 .604 265 | 8.6863 .049 351 | 8.6281 .042 271 | 8.0768 .012 264 | 8.2024 .016 258 | 8.4057 .025 86 |
| $S_2$ | 8.6868 .049 9 | 7.7679 .006 354 | 6.4230 .000 359 | 7.1547 .001 185 | 7.3491 .002 191 | 9.7807 .604 95 | | 9.7807 .604 265 | 7.6613 .005 186 | 6.4265 .000 359 | 7.3097 .002 353 | 6.4230 .000 1 |
| $T_2$ | 9.2922 .196 104 | 8.7615 .058 268 | 8.4057 .025 274 | 8.2077 .016 280 | 8.0649 .012 286 | 8.6863 .049 9 | 9.7807 .604 95 | | 8.6866 .049 280 | 8.0959 .012 273 | 8.2350 .017 268 | 8.3698 .023 95 |
| $\lambda_2$ | 7.3819 .002 3 | 8.8560 .072 168 | 7.7679 .006 354 | 6.4230 .000 359 | 7.1547 .001 185 | 8.6281 .042 89 | 7.6613 .005 174 | 8.6866 .049 80 | | 7.3097 .002 353 | 7.7656 .006 347 | 7.1547 .001 175 |
| $\mu_2$ | 7.3308 .002 10 | 7.1547 .001 175 | 6.4230 .000 1 | 7.7679 .006 6 | 8.8560 .072 192 | 8.0768 .012 96 | 6.4265 .000 1 | 8.0959 .012 87 | 7.3097 .002 7 | | 7.6613 .005 174 | 6.4253 .000 2 |
| $\nu_2$ | 7.6468 .004 16 | 6.4230 .000 1 | 7.7679 .006 6 | 8.8560 .072 192 | 8.0931 .012 198 | 8.2024 .016 102 | 7.3097 .002 7 | 8.2350 .017 92 | 7.7656 .006 13 | 7.6613 .005 186 | | 7.1202 .001 8 |
| 2SM | 7.6012 .004 189 | 7.3097 .002 353 | 6.4265 .000 359 | 6.8803 .001 185 | 7.1525 .001 190 | 8.4057 .025 274 | 6.4230 .000 359 | 8.3698 .023 265 | 7.1547 .001 185 | 6.4253 .000 358 | 7.1202 .001 352 | |

## Table 29.—*Elimination factors*—Continued

### SERIES 221 DAYS. DIURNAL CONSTITUENTS

| Constituent sought (A) | Disturbing constituents (B, C, etc.) | | | | | | | | | |
|---|---|---|---|---|---|---|---|---|---|---|
| | $J_1$ | $K_1$ | $M_1$ | $O_1$ | OO | $P_1$ | $Q_1$ | 2Q | $S_1$ | $\rho_1$ |
| $J_1$ | ------ | 7.4061 .003 356 | 7.4052 .003 353 | 7.8848 .008 324 | 8.2672 .019 28 | 8.3590 .023 318 | 7.7969 .006 321 | 7.7322 .005 317 | 8.5324 .034 247 | 7.6535 .005 334 |
| $K_1$ | 7.4061 .003 4 | ------ | 7.4061 .003 356 | 8.0179 .010 328 | 8.0179 .010 32 | 9.2077 .161 322 | 7.8848 .008 324 | 7.7969 .006 321 | 9.6969 .498 251 | 7.7209 .005 338 |
| $M_1$ | 7.4052 .003 7 | 7.4061 .003 4 | ------ | 8.2672 .019 332 | 7.8848 .008 36 | 8.4189 .026 146 | 8.0179 .010 328 | 7.8848 .008 324 | 8.6172 .041 75 | 7.8313 .007 341 |
| $O_1$ | 7.8848 .008 36 | 8.0179 .010 32 | 8.2672 .019 28 | ------ | 7.9465 .009 64 | 7.3352 .002 174 | 7.4061 .003 356 | 7.4052 .003 353 | 8.2991 .020 103 | 7.8800 .008 190 |
| OO | 8.2672 .019 332 | 8.0179 .010 328 | 7.8848 .008 324 | 7.9465 .009 296 | ------ | 8.2351 .017 290 | 7.8628 .007 292 | 7.7946 .006 289 | 8.0778 .012 219 | 7.8188 .007 306 |
| $P_1$ | 8.3590 .023 42 | 9.2077 .161 38 | 8.4189 .026 214 | 7.3352 .002 186 | 8.2351 .017 70 | ------ | 6.7176 .001 182 | 6.4350 .000 358 | 9.6969 .498 109 | 7.5854 .004 195 |
| $Q_1$ | 7.7969 .006 39 | 7.8848 .008 36 | 8.0179 .010 32 | 7.4061 .003 4 | 7.8628 .007 68 | 6.7176 .001 178 | ------ | 7.4061 .003 356 | 8.1112 .013 107 | 8.8310 .068 13 |
| 2Q | 7.7322 .005 43 | 7.7969 .006 39 | 7.8848 .008 36 | 7.4052 .003 7 | 7.7946 .006 71 | 6.4350 .000 2 | 7.4061 .003 4 | ------ | 7.9748 .009 110 | 8.0073 .010 17 |
| $S_1$ | 8.5324 .034 113 | 9.6969 .498 109 | 8.6172 .041 285 | 8.2991 .020 257 | 8.0778 .012 141 | 9.6969 .498 251 | 8.1112 .013 253 | 7.9748 .009 250 | ------ | 8.1495 .014 266 |
| $\rho_1$ | 7.6535 .005 26 | 7.7209 .005 22 | 7.8313 .007 19 | 7.8800 .008 170 | 7.8188 .007 54 | 7.5854 .004 165 | 8.8310 .068 347 | 8.0073 .010 343 | 8.1495 .014 94 | ------ |

HARMONIC ANALYSIS AND PREDICTION OF TIDES 253

Table 29.—*Elimination factors*—Continued

SERIES 221 DAYS. SEMIDIURNAL CONSTITUENTS

| Constituent sought (A) | Disturbing constituents (B, C, etc.) | | | | | | | | | | |
|---|---|---|---|---|---|---|---|---|---|---|---|
| | $K_2$ | $L_2$ | $M_2$ | $N_2$ | 2N | $R_2$ | $S_2$ | $T_2$ | $\lambda_2$ | $\mu_2$ | $\nu_2$ | 2SM |
| $K_2$ | ----- ----- ----- | 8.2872 .019 332 | 8.0179 .010 328 | 7.8848 .008 324 | 7.7969 .006 321 | 9.6969 .498 251 | 9.2077 .161 322 | 8.9831 .096 213 | 8.3590 .023 318 | 7.6535 .005 334 | 7.7209 .005 338 | 8.2035 .016 136 |
| $L_2$ | 8.2672 .019 28 | ----- ----- ----- | 7.4061 .003 356 | 7.4052 .003 353 | 7.4037 .003 349 | 8.6189 .042 99 | 7.8800 .008 170 | 8.6448 .044 62 | 8.3310 .068 347 | 6.7176 .001 182 | 7.3352 .002 186 | 7.5854 .004 165 |
| $M_2$ | 8.0179 .010 32 | 7.4061 .003 4 | ----- ----- ----- | 7.4061 .003 356 | 7.4052 .003 353 | 8.2991 .020 103 | 7.3352 .002 174 | 8.3038 .020 65 | 7.8800 .008 170 | 7.3352 .002 186 | 7.8800 .008 190 | 7.3333 .002 168 |
| $N_2$ | 7.8848 .008 36 | 7.4052 .003 7 | 7.4061 .003 4 | ----- ----- ----- | 7.4061 .003 356 | 8.1112 .013 107 | 6.7176 .001 178 | 8.1229 .013 69 | 7.3352 .002 174 | 7.8800 .008 190 | 8.8310 .068 13 | 7.0677 .001 172 |
| 2N | 7.7969 .006 39 | 7.4037 .003 11 | 7.4052 .003 7 | 7.4061 .003 4 | ----- ----- ----- | 7.9748 .009 110 | 6.4350 .000 2 | 7.9996 .010 73 | 6.7176 .001 178 | 8.8310 .068 13 | 8.0073 .010 17 | 6.7183 .001 176 |
| $R_2$ | 9.6969 .498 109 | 8.6189 .042 261 | 8.2991 .020 257 | 8.1112 .013 253 | 7.9748 .009 250 | ----- ----- ----- | 9.6970 .498 251 | 9.2076 .161 322 | 8.5324 .034 247 | 8.0145 .010 263 | 8.1495 .014 266 | 8.3038 .020 65 |
| $S_2$ | 9.2077 .161 38 | 7.8800 .008 190 | 7.3352 .002 186 | 6.7176 .001 182 | 6.4350 .000 358 | 9.6970 .498 109 | ----- ----- ----- | 9.6970 .498 251 | 7.4061 .003 356 | 7.3333 .002 192 | 7.5854 .004 195 | 7.3352 .002 174 |
| $T_2$ | 8.9831 .096 147 | 8.6448 .044 298 | 8.3038 .020 295 | 8.1229 .013 291 | 7.9996 .010 287 | 9.2076 .161 38 | 9.6970 .498 109 | ----- ----- ----- | 8.6172 .041 285 | 7.9704 .009 301 | 8.0914 .012 304 | 8.2991 .020 103 |
| $\lambda_2$ | 8.3590 .023 42 | 8.8310 .068 13 | 7.8800 .008 190 | 7.3352 .002 186 | 6.7176 .001 182 | 8.5324 .034 113 | 7.4061 .003 4 | 8.6172 .041 75 | ----- ----- ----- | 7.5854 .004 195 | 7.8738 .007 199 | 6.7176 .001 178 |
| $\mu_2$ | 7.6535 .005 26 | 6.7176 .001 178 | 7.3352 .002 174 | 7.8800 .008 170 | 8.3310 .068 347 | 8.0145 .010 97 | 7.3333 .002 168 | 7.9704 .009 59 | 7.5854 .004 165 | ----- ----- ----- | 7.4061 .003 4 | 7.3294 .002 162 |
| $\nu_2$ | 7.7209 .005 22 | 7.3352 .002 174 | 7.8800 .008 170 | 8.8310 .068 347 | 8.0073 .010 343 | 8.1495 .014 94 | 7.5854 .004 165 | 8.0914 .012 56 | 7.8738 .007 161 | 7.4061 .003 356 | ----- ----- ----- | 7.4945 .003 159 |
| 2SM | 8.2035 .016 224 | 7.5854 .004 195 | 7.3333 .002 192 | 7.0677 .001 188 | 6.7183 .001 184 | 8.3038 .020 295 | 7.3352 .002 186 | 8.2991 .020 257 | 6.7176 .001 182 | 7.3294 .002 198 | 7.4945 .003 201 | ----- ----- ----- |

246037—41——17

Table 29.—*Elimination factors*—Continued

**SERIES 250 DAYS. DIURNAL CONSTITUENTS**

| Constituent sought (A) | Disturbing constituents (B, C, etc.) | | | | | | | | |
|---|---|---|---|---|---|---|---|---|---|
| | $J_1$ | $K_1$ | $M_1$ | $O_1$ | OO | $P_1$ | $Q_1$ | 2Q | $S_1$ | $\rho_1$ |
| $J_1$ | ------ | 7.9011 .008 347 | 7.8898 .008 334 | 8.0302 .011 293 | 8.3544 .023 41 | 8.4768 .030 280 | 7.9350 .009 280 | 7.8438 .007 267 | 8.3527 .023 224 | 7.7796 .006 318 |
| $K_1$ | 7.9011 .008 13 | ------ | 7.9011 .008 347 | 8.1489 .014 306 | 8.1489 .014 54 | 9.3286 .213 294 | 8.0302 .011 293 | 7.9350 .009 280 | 9.5900 .389 237 | 7.7661 .006 331 |
| $M_1$ | 7.8898 .008 26 | 7.9011 .008 13 | ------ | 8.3544 .023 319 | 8.0302 .011 67 | 8.5201 .033 127 | 8.1489 .014 306 | 8.0302 .011 293 | 8.5519 .036 70 | 7.6982 .005 344 |
| $O_1$ | 8.0302 .011 67 | 8.1489 .014 54 | 8.3544 .023 41 | ------ | 7.9171 .008 108 | 7.6029 .004 168 | 7.9011 .008 347 | 7.8898 .008 334 | 8.2274 .017 111 | 8.2405 .017 205 |
| OO | 8.3544 .023 319 | 8.1489 .014 306 | 8.0302 .011 293 | 7.9171 .008 252 | ------ | 8.1443 .014 239 | 7.7748 .006 239 | 7.6181 .004 226 | 6.8959 .001 183 | 7.8514 .007 277 |
| $P_1$ | 8.4768 .030 80 | 9.3286 .213 66 | 8.5201 .033 233 | 7.6029 .004 192 | 8.1443 .014 121 | ------ | 6.2382 .000 359 | 7.3397 .002 346 | 9.5900 .389 123 | 7.8955 .008 218 |
| $Q_1$ | 7.9350 .009 80 | 8.0302 .011 67 | 8.1489 .014 54 | 7.9011 .008 13 | 7.7748 .006 121 | 6.2382 .000 1 | ------ | 7.9011 .008 347 | 7.9950 .010 124 | 9.2133 .163 39 |
| 2Q | 7.8438 .007 93 | 7.9350 .009 80 | 8.0302 .011 67 | 7.8898 .008 26 | 7.6181 .004 134 | 7.3397 .002 14 | 7.9011 .008 13 | ------ | 7.7821 .006 137 | 8.3852 .024 52 |
| $S_1$ | 8.3527 .023 136 | 9.5900 .389 123 | 8.5519 .036 290 | 8.2274 .017 249 | 6.8959 .001 177 | 9.5900 .389 237 | 7.9950 .010 236 | 7.7821 .006 223 | ------ | 8.0954 .012 275 |
| $\rho_1$ | 7.7796 .006 42 | 7.7661 .006 29 | 7.6982 .005 16 | 8.2405 .017 155 | 7.8514 .007 83 | 7.8955 .008 142 | 9.2133 .163 321 | 8.3852 .024 308 | 8.0954 .012 85 | ------ |

Table 29.—*Elimination factors*—Continued

SERIES 250 DAYS. SEMIDIURNAL CONSTITUENTS

| Constituent sought (A) | Disturbing constituents (B, C, etc.) | | | | | | | | | | | |
|---|---|---|---|---|---|---|---|---|---|---|---|---|
| | $K_2$ | $L_2$ | $M_2$ | $N_2$ | 2N | $R_2$ | $S_2$ | $T_2$ | $\lambda_2$ | $\mu_2$ | $\nu_2$ | 2SM |
| $K_2$ | ----- | 8.3544 .023 319 | 8.1489 .014 306 | 8.0302 .011 293 | 7.9350 .009 280 | 9.5900 .389 237 | 9.3286 .213 294 | 8.4129 .026 350 | 8.4768 .030 280 | 7.7796 .006 318 | 7.7661 .006 331 | 8.3023 .020 101 |
| $L_2$ | 8.3544 .023 41 | ----- | 7.9011 .008 347 | 7.8898 .008 334 | 7.8703 .007 321 | 8.5672 .037 98 | 8.2405 .017 155 | 8.3634 .023 31 | 9.2133 .163 321 | 6.2382 .000 359 | 7.6029 .004 192 | 7.8955 .008 143 |
| $M_2$ | 8.1489 .014 54 | 7.9011 .008 13 | ----- | 7.9011 .008 347 | 7.8898 .008 334 | 8.2274 .017 111 | 7.6029 .004 168 | 8.1376 .014 44 | 8.2405 .017 155 | 7.6029 .004 192 | 8.2405 .017 205 | 7.5928 .004 155 |
| $N_2$ | 8.0302 .011 67 | 7.8898 .008 26 | 7.9011 .008 13 | ----- | 7.9011 .008 347 | 7.9950 .010 124 | 6.2382 .000 1 | 8.0259 .011 58 | 7.6029 .004 168 | 8.2405 .017 205 | 9.2133 .163 39 | 7.1697 .001 168 |
| 2N | 7.9350 .009 80 | 7.8703 .007 39 | 7.8898 .008 26 | 7.9011 .008 13 | ----- | 7.7821 .006 137 | 7.3397 .002 14 | 7.9414 .009 71 | 6.2382 .000 1 | 9.2133 .163 39 | 8.3852 .024 52 | 6.2381 .000 2 |
| $R_2$ | 9.5900 .389 123 | 8.5672 .037 262 | 8.2274 .017 249 | 7.9950 .010 236 | 7.7821 .006 223 | ----- | 9.5901 .389 237 | 9.3286 .213 294 | 8.3528 .023 224 | 7.9596 .009 261 | 8.0954 .012 275 | 8.1376 .014 44 |
| $S_2$ | 9.3286 .213 66 | 8.2405 .017 205 | 7.6029 .004 192 | 6.2382 .000 359 | 7.3397 .002 346 | 9.5901 .389 123 | ----- | 9.5901 .389 237 | 7.9011 .008 347 | 7.5928 .004 205 | 7.8955 .008 218 | 7.6029 .004 168 |
| $T_2$ | 8.4129 .026 10 | 8.3634 .023 329 | 8.1376 .014 316 | 8.0259 .011 302 | 7.9414 .009 289 | 9.3286 .213 66 | 9.5901 .389 123 | ----- | 8.5519 .036 290 | 7.7084 .005 328 | 7.6345 .004 341 | 8.2274 .017 111 |
| $\lambda_2$ | 8.4768 .030 80 | 9.2133 .163 39 | 8.2405 .017 205 | 7.6029 .004 192 | 6.2382 .000 359 | 8.3528 .023 136 | 7.9011 .008 13 | 8.5519 .036 70 | ----- | 7.8955 .008 218 | 8.1962 .016 231 | 6.2382 .000 1 |
| $\mu_2$ | 7.7796 .006 42 | 6.2382 .000 1 | 7.6029 .004 168 | 8.2405 .017 155 | 9.2133 .163 321 | 7.9596 .009 99 | 7.5928 .004 155 | 7.7084 .005 32 | 7.8955 .008 142 | ----- | 7.9011 .008 13 | 7.5759 .004 143 |
| $\nu_2$ | 7.7661 .006 29 | 7.6029 .004 168 | 8.2405 .017 155 | 9.2133 .163 321 | 8.3852 .024 308 | 8.0954 .012 85 | 7.8955 .008 142 | 7.6345 .004 19 | 8.1962 .016 129 | 7.9011 .008 347 | ----- | 7.7671 .006 130 |
| 2SM | 8.3023 .020 259 | 7.8955 .008 218 | 7.5928 .004 205 | 7.1697 .001 192 | 6.2381 .000 358 | 8.1376 .014 316 | 7.6029 .004 192 | 8.2274 .017 249 | 6.2382 .000 359 | 7.5759 .004 217 | 7.7671 .006 230 | ----- |

Table 29.—*Elimination factors*—Continued

**SERIES 279 DAYS. DIURNAL CONSTITUENTS**

| Constituent sought (A) | Disturbing constituents (B, C, etc.) | | | | | | | | | |
|---|---|---|---|---|---|---|---|---|---|---|
| | $J_1$ | $K_1$ | $M_1$ | $O_1$ | OO | $P_1$ | $Q_1$ | 2Q | $S_1$ | $\rho_1$ |
| $J_1$ | ------ <br> ------ <br> --- | 8.0816 <br> .012 <br> 337 | 8.0469 <br> .011 <br> 315 | 8.0127 <br> .010 <br> 261 | 8.3961 <br> .025 <br> 54 | 8.3841 <br> .024 <br> 242 | 7.8250 <br> .007 <br> 239 | 7.5672 <br> .004 <br> 216 | 7.9987 <br> .010 <br> 200 | 7.8339 <br> .007 <br> 303 |
| $K_1$ | 8.0816 <br> .012 <br> 23 | ------ <br> ------ <br> --- | 8.0816 <br> .012 <br> 337 | 8.1800 <br> .015 <br> 284 | 8.1800 <br> .015 <br> 76 | 9.3172 <br> .208 <br> 265 | 8.0127 <br> .010 <br> 261 | 7.8250 <br> .007 <br> 239 | 9.4495 <br> .282 <br> 222 | 7.7947 <br> .006 <br> 325 |
| $M_1$ | 8.0469 <br> .011 <br> 45 | 8.0816 <br> .012 <br> 23 | ------ <br> ------ <br> --- | 8.3961 <br> .025 <br> 306 | 8.0127 <br> .010 <br> 99 | 8.5477 <br> .035 <br> 108 | 8.1800 <br> .015 <br> 284 | 8.0127 <br> .010 <br> 261 | 8.4890 <br> .031 <br> 65 | 7.5510 <br> .004 <br> 348 |
| $O_1$ | 8.0127 <br> .010 <br> 99 | 8.1800 <br> .015 <br> 76 | 8.3961 <br> .025 <br> 54 | ------ <br> ------ <br> --- | 7.5571 <br> .004 <br> 152 | 7.7343 <br> .005 <br> 161 | 8.0816 <br> .012 <br> 337 | 8.0469 <br> .011 <br> 315 | 8.1523 <br> .014 <br> 119 | 8.3798 <br> .024 <br> 221 |
| OO | 8.3961 <br> .025 <br> 306 | 8.1800 <br> .015 <br> 284 | 8.0127 <br> .010 <br> 261 | 7.5571 <br> .004 <br> 208 | ------ <br> ------ <br> --- | 7.3456 <br> .002 <br> 189 | 6.7358 <br> .001 <br> 185 | 7.1964 <br> .002 <br> 342 | 7.9211 <br> .008 <br> 326 | 7.7771 <br> .006 <br> 249 |
| $P_1$ | 8.3841 <br> .024 <br> 118 | 9.3172 <br> .208 <br> 95 | 8.5477 <br> .035 <br> 252 | 7.7343 <br> .005 <br> 199 | 7.3456 <br> .002 <br> 171 | ------ <br> ------ <br> --- | 6.8592 <br> .001 <br> 356 | 7.5574 <br> .004 <br> 334 | 9.4495 <br> .282 <br> 138 | 7.9990 <br> .010 <br> 240 |
| $Q_1$ | 7.8250 <br> .007 <br> 121 | 8.0127 <br> .010 <br> 99 | 8.1800 <br> .015 <br> 76 | 8.0816 <br> .012 <br> 23 | 6.7358 <br> .001 <br> 175 | 6.8592 <br> .001 <br> 4 | ------ <br> ------ <br> --- | 8.0816 <br> .012 <br> 337 | 7.8251 <br> .007 <br> 141 | 9.3242 <br> .211 <br> 64 |
| 2Q | 7.5672 <br> .004 <br> 144 | 7.8250 <br> .007 <br> 121 | 8.0127 <br> .010 <br> 99 | 8.0469 <br> .011 <br> 45 | 7.1964 <br> .002 <br> 18 | 7.5574 <br> .004 <br> 26 | 8.0816 <br> .012 <br> 23 | ------ <br> ------ <br> --- | 7.3460 <br> .002 <br> 164 | 8.4421 <br> .028 <br> 86 |
| $S_1$ | 7.9987 <br> .010 <br> 160 | 9.4495 <br> .282 <br> 138 | 8.4890 <br> .031 <br> 295 | 8.1523 <br> .014 <br> 241 | 7.9211 <br> .008 <br> 34 | 9.4495 <br> .282 <br> 222 | 7.8251 <br> .007 <br> 219 | 7.3460 <br> .002 <br> 196 | ------ <br> ------ <br> --- | 8.0384 <br> .011 <br> 283 |
| $\rho_1$ | 7.8339 <br> .007 <br> 57 | 7.7947 <br> .006 <br> 35 | 7.5510 <br> .004 <br> 12 | 8.3798 <br> .024 <br> 139 | 7.7771 <br> .006 <br> 111 | 7.9990 <br> .010 <br> 120 | 9.3242 <br> .211 <br> 296 | 8.4421 <br> .028 <br> 274 | 8.0384 <br> .011 <br> 77 | ------ <br> ------ <br> --- |

HARMONIC ANALYSIS AND PREDICTION OF TIDES 257

Table 29.—*Elimination factors*—Continued

SERIES 279 DAYS. SEMIDIURNAL CONSTITUENTS

| Constituent sought ($A$) | Disturbing constituents ($B$, $C$, etc.) | | | | | | | | | | |
|---|---|---|---|---|---|---|---|---|---|---|---|
| | $K_2$ | $L_2$ | $M_2$ | $N_2$ | $2N$ | $R_2$ | $S_2$ | $T_2$ | $\lambda_2$ | $\mu_2$ | $\nu_2$ | $2SM$ |
| $K_2$ | ------ | 8.3961 .025 306 | 8.1800 .015 284 | 8.0127 .010 261 | 7.8250 .007 239 | 9.4494 .281 222 | 9.3172 .208 265 | 9.0421 .110 308 | 8.3841 .024 242 | 7.8339 .007 303 | 7.7947 .006 325 | 8.2246 .017 66 |
| $L_2$ | 8.3961 .025 54 | ------ | 8.0816 .012 337 | 8.0469 .011 315 | 7.9866 .010 292 | 8.5211 .033 96 | 8.3798 .024 139 | 6.9057 .001 1 | 9.3242 .211 296 | 6.8592 .001 356 | 7.7343 .005 199 | 7.9990 .010 120 |
| $M_2$ | 8.1800 .015 76 | 8.0816 .012 23 | ------ | 8.0816 .012 337 | 8.0469 .011 315 | 8.1523 .014 119 | 7.7343 .005 161 | 7.8491 .007 24 | 8.3798 .024 139 | 7.7343 .005 199 | 8.3798 .024 221 | 7.7105 .005 142 |
| $N_2$ | 8.0127 .010 99 | 8.0469 .011 45 | 8.0816 .012 23 | ------ | 8.0816 .012 337 | 7.8251 .007 141 | 6.8592 .001 4 | 7.9108 .008 46 | 7.7343 .005 161 | 8.3798 .024 221 | 9.3242 .211 64 | 7.2354 .002 165 |
| $2N$ | 7.8250 .007 121 | 7.9866 .010 68 | 8.0469 .011 45 | 8.0816 .012 23 | ------ | 7.3463 .002 164 | 7.8721 .007 66 | 7.8885 .008 69 | 6.8592 .001 4 | 9.3242 .211 64 | 8.4421 .028 86 | 6.8588 .001 8 |
| $R_2$ | 9.4494 .281 138 | 8.5211 .033 264 | 8.1523 .014 241 | 7.8251 .007 219 | 7.3463 .002 164 | ------ | 9.4496 .282 223 | 9.3172 .208 265 | 7.9987 .010 200 | 7.9102 .008 260 | 8.0384 .011 283 | 7.8491 .007 24 |
| $S_2$ | 9.3172 .208 95 | 8.3798 .024 221 | 7.7343 .005 199 | 6.8592 .001 356 | 7.8721 .007 294 | 9.4496 .282 137 | ------ | 9.4496 .282 223 | 8.0816 .012 237 | 7.7105 .005 218 | 7.9990 .010 240 | 7.7343 .005 161 |
| $T_2$ | 9.0421 .110 52 | 6.9057 .001 359 | 7.8491 .007 336 | 7.9108 .008 314 | 7.8885 .008 291 | 9.3172 .208 95 | 9.4496 .282 137 | ------ | 8.4891 .031 295 | 6.8703 .001 355 | 7.5541 .004 198 | 8.1523 .014 119 |
| $\lambda_2$ | 8.3841 .024 118 | 9.3242 .211 64 | 8.3798 .024 221 | 7.7343 .005 199 | 6.8592 .001 356 | 7.9987 .010 160 | 8.0816 .012 123 | 8.4891 .031 65 | ------ | 7.9990 .010 240 | 8.2553 .018 263 | 6.8592 .001 4 |
| $\mu_2$ | 7.8339 .007 57 | 6.8592 .001 4 | 7.7343 .005 161 | 8.3798 .024 139 | 9.3242 .211 296 | 7.9102 .008 100 | 7.7105 .005 142 | 6.8703 .001 5 | 7.9990 .010 120 | ------ | 8.0816 .012 23 | 7.6697 .005 124 |
| $\nu_2$ | 7.7947 .006 35 | 7.7343 .005 161 | 8.3798 .024 139 | 9.3242 .211 296 | 8.4421 .028 274 | 8.0384 .011 77 | 7.9990 .010 120 | 7.5541 .004 162 | 8.2553 .018 97 | 8.0816 .012 337 | ------ | 7.8266 .007 101 |
| $2SM$ | 8.2246 .017 294 | 7.9990 .010 240 | 7.7105 .005 218 | 7.2354 .002 195 | 6.8588 .001 352 | 7.8491 .007 336 | 7.7343 .005 199 | 8.1523 .014 241 | 6.8592 .001 356 | 7.6697 .005 236 | 7.8266 .007 259 | ------ |

Table 29.—*Elimination factors*—Continued

**SERIES 297 DAYS. DIURNAL CONSTITUENTS**

| Constituent sought (A) | Disturbing constituents (B, C, etc.) | | | | | | | | | |
|---|---|---|---|---|---|---|---|---|---|---|
| | $J_1$ | $K_1$ | $M_1$ | $O_1$ | OO | $P_1$ | $Q_1$ | 2Q | $S_1$ | $\rho_1$ |
| $J_1$ | ------ | 8.2770 .019 220 | 8.1622 .015 260 | 7.9899 .010 266 | 7.5338 .003 173 | 8.3896 .025 287 | 7.7726 .006 306 | 7.1486 .001 346 | 8.4204 .026 253 | 7.5223 .003 206 |
| $K_1$ | 8.2770 .019 140 | ------ | 8.2770 .019 220 | 8.0269 .011 227 | 8.0269 .011 133 | 9.2565 .181 247 | 7.9899. .010 266 | 7.7726 .006 306 | 9.3360 .217 214 | 7.3907 .002 346 |
| $M_1$ | 8.1622 .015 100 | 8.2770 .019 140 | ------ | 7.5338 .003 187 | 7.9899 .010 94 | 8.2044 .016 27 | 8.0269 .011 227 | 7.9899 .010 266 | 7.5392 .003 174 | 8.1019 .013 306 |
| $O_1$ | 7.9899 .010 94 | 8.0269 .011 133 | 7.5338 .003 173 | ------ | 7.8638 .007 87 | 7.7467 .006 21 | 8.2770 .019 220 | 8.1622 .015 260 | 7.5336 .003 167 | 8.4724 .030 300 |
| OO | 7.5338 .003 187 | 8.0269 .011 227 | 7.9899 .010 266 | 7.8638 .007 273 | ------ | 8.0954 .012 294 | 7.6320 .004 313 | 6.7805 .001 353 | 8.1433 .014 260 | 7.5129 .003 213 |
| $P_1$ | 8.3896 .025 73 | 9.2565 .181 113 | 8.2044 .016 333 | 7.7467 .006 339 | 8.0954 .012 66 | ------ | 7.5300 .003 199 | 7.8163 .007 239 | 9.3360 .217 146 | 8.0286 .011 279 |
| $Q_1$ | 7.7726 .006 54 | 7.9899 .010 94 | 8.0269 .011 133 | 8.2770 .019 140 | 7.6320 .004 47 | 7.5300 .003 161 | ------ | 8.2770 .019 220 | 7.9031 .008 127 | 9.3366 .217 80 |
| 2Q | 7.1486 .001 14 | 7.7726 .006 54 | 7.9899 .010 94 | 8.1622 .015 100 | 6.7805 .001 7 | 7.8163 .007 121 | 8.2770 .019 140 | ------ | 7.8741 .007 87 | 8.2220 .017 40 |
| $S_1$ | 8.4204 .026 107 | 9.3360 .217 146 | 7.5392 .003 186 | 7.5336 .003 193 | 8.1433 .014 100 | 9.3360 .217 214 | 7.9031 .008 233 | 7.8741 .007 273 | ------ | 7.8897 .008 312 |
| $\rho_1$ | 7.5223 .003 154 | 7.3907 .002 14 | 8.1019 .013 54 | 8.4724 .030 60 | 7.5129 .003 147 | 8.0286 .011 81 | 9.3366 .217 280 | 8.2220 .017 320 | 7.8897 .008 48 | ------ |

HARMONIC ANALYSIS AND PREDICTION OF TIDES 259

Table 29.—*Elimination factors*—Continued

SERIES 297 DAYS. SEMIDIURNAL CONSTITUENTS

| Constituent sought ($A$) | Disturbing constituents ($B$, $C$, etc.) | | | | | | | | | | |
|---|---|---|---|---|---|---|---|---|---|---|---|
| | $K_2$ | $L_2$ | $M_2$ | $N_2$ | $2N$ | $R_2$ | $S_2$ | $T_2$ | $\lambda_2$ | $\mu_2$ | $\nu_2$ | $2SM$ |
| $K_2$ | ----- | 7.5338 .003 187 | 8.0269 .011 227 | 7.9899 .010 266 | 7.7726 .006 306 | 9.3359 .217 214 | 9.2565 .181 247 | 9.1076 .128 281 | 8.3896 .025 287 | 7.5223 .003 206 | 7.3907 .002 346 | 8.2357 .017 88 |
| $L_2$ | 7.5338 .003 173 | ----- | 8.2770 .019 220 | 8.1622 .015 260 | 7.9326 .009 300 | 8.1514 .014 27 | 8.4724 .030 60 | 8.5711 .037 94 | 9.3366 .217 280 | 7.5300 .003 199 | 7.7467 .006 339 | 8.0286 .011 81 |
| $M_2$ | 8.0269 .011 133 | 8.2770 .019 140 | ----- | 8.2770 .019 220 | 8.1622 .015 260 | 7.5339 .003 167 | 7.7467 .006 21 | 8.1268 .013 54 | 8.4724 .030 60 | 7.7467 .006 339 | 8.4724 .030 300 | 7.7179 .005 41 |
| $N_2$ | 7.9899 .010 94 | 8.1622 .015 100 | 8.2770 .019 140 | ----- | 8.2770 .019 220 | 7.9031 .008 127 | 7.5300 .003 161 | 7.4214 .003 14 | 7.7467 .006 21 | 8.4724 .030 300 | 9.3366 .217 80 | 6.2014 .000 1 |
| $2N$ | 7.7726 .006 54 | 7.9326 .009 60 | 8.1622 .015 100 | 8.2770 .019 140 | ----- | 7.8741 .007 87 | 7.8163 .007 121 | 7.5240 .003 155 | 7.5300 .003 161 | 9.3366 .217 80 | 8.2220 .017 40 | 7.5050 .003 142 |
| $R_2$ | 9.3359 .217 146 | 8.1514 .014 333 | 7.5339 .003 193 | 7.9031 .008 233 | 7.8741 .007 273 | ----- | 9.3362 .217 214 | 9.2566 .181 247 | 8.4204 .026 253 | 7.0150 .001 352 | 7.8897 .008 312 | 8.1268 .013 54 |
| $S_2$ | 9.2565 .181 113 | 8.4724 .030 300 | 7.7467 .006 339 | 7.5300 .003 199 | 7.8163 .007 239 | 9.3362 .217 146 | ----- | 9.3362 .217 214 | 8.2770 .019 220 | 7.7179 .005 319 | 8.0286 .011 279 | 7.7467 .006 21 |
| $T_2$ | 9.1076 .128 79 | 8.5711 .037 266 | 8.1268 .013 306 | 7.4214 .003 346 | 7.5240 .003 205 | 9.2566 .181 113 | 9.3362 .217 146 | ----- | 7.5385 .003 186 | 7.8920 .008 285 | 8.0039 .010 245 | 7.5339 .003 167 |
| $\lambda_2$ | 8.3896 .025 73 | 9.3366 .217 80 | 8.4724 .030 300 | 7.7467 .006 339 | 7.5300 .003 199 | 8.4204 .026 107 | 8.2770 .019 140 | 7.5385 .003 174 | ----- | 8.0286 .011 279 | 8.1647 .015 239 | 7.5300 .003 161 |
| $\mu_2$ | 7.5223 .003 154 | 7.5300 .003 161 | 7.7467 .006 21 | 8.4724 .030 60 | 9.3366 .217 280 | 7.0150 .001 8 | 7.7179 .005 41 | 7.8920 .008 75 | 8.0286 .011 81 | ----- | 8.2770 .019 140 | 7.6680 .005 62 |
| $\nu_2$ | 7.3907 .002 14 | 7.7467 .006 21 | 8.4724 .030 60 | 9.3366 .217 280 | 8.2220 .017 320 | 7.8897 .008 48 | 8.0286 .011 81 | 8.0039 .010 115 | 8.1647 .015 121 | 8.2770 .019 220 | ----- | 7.7984 .006 102 |
| $2SM$ | 8.2357 .017 272 | 8.0286 .011 279 | 7.7179 .005 319 | 6.2014 .000 359 | 7.5050 .003 218 | 8.1268 .013 306 | 7.7467 .006 339 | 7.5339 .003 193 | 7.5300 .003 199 | 7.6680 .005 298 | 7.7984 .006 258 | ----- |

Table 29.—*Elimination factors*—Continued

**SERIES 326 DAYS. DIURNAL CONSTITUENTS**

| Constituent sought (A) | Disturbing constituents (B, C. etc.) | | | | | | | | |
|---|---|---|---|---|---|---|---|---|---|
| | $J_1$ | $K_1$ | $M_1$ | $O_1$ | OO | $P_1$ | $Q_1$ | 2Q | $S_1$ | $\rho_1$ |
| $J_1$ | ------ ------ ------ | 8.1340 .014 210 | 8.0698 .012 241 | 7.8631 .007 235 | 7.4352 .003 6 | 8.3392 .022 249 | 7.8244 .007 265 | 7.6841 .005 296 | 8.2809 .019 230 | 7.0934 .001 190 |
| $K_1$ | 8.1340 .014 150 | ------ ------ ------ | 8.1340 .014 210 | 7.7427 .006 204 | 7.7427 .006 156 | 9.0470 .111 219 | 7.8631 .007 235 | 7.8244 .007 265 | 9.0723 .118 199 | 7.5061 .003 340 |
| $M_1$ | 8.0698 .012 119 | 8.1340 .014 150 | ------ ------ ------ | 7.4352 .003 354 | 7.8631 .007 125 | 7.6587 .005 8 | 7.7427 .006 204 | 7.8631 .007 235 | 7.7472 .006 169 | 8.0423 .011 310 |
| $O_1$ | 7.8631 .007 125 | 7.7427 .006 156 | 7.4352 .003 6 | ------ ------ ------ | 7.7018 .005 131 | 7.5483 .004 14 | 8.1340 .014 210 | 8.0698 .012 241 | 7.0956 .001 175 | 8.3386 .022 315 |
| OO | 7.4352 .003 354 | 7.7427 .006 204 | 7.8631 .007 235 | 7.7018 .005 229 | ------ ------ ------ | 8.0443 .011 243 | 7.7204 .005 259 | 7.6227 .004 290 | 7.9496 .009 224 | 6.6245 .000 184 |
| $P_1$ | 8.3392 .022 111 | 9.0470 .111 141 | 7.6587 .005 352 | 7.5483 .004 346 | 8.0443 .011 117 | ------ ------ ------ | 7.4186 .003 196 | 7.7040 .005 227 | 9.0723 .118 161 | 7.9254 .008 301 |
| $Q_1$ | 7.8244 .007 95 | 7.8631 .007 125 | 7.7427 .006 156 | 8.1340 .014 150 | 7.7204 .005 101 | 7.4186 .003 164 | ------ ------ ------ | 8.1340 .014 210 | 7.7258 .005 144 | 9.2882 .194 105 |
| 2Q | 7.6841 .005 64 | 7.8244 .007 95 | 7.8631 .007 125 | 8.0698 .012 119 | 7.6227 .004 70 | 7.7040 .005 133 | 8.1340 .014 150 | ------ ------ ------ | 7.7948 .006 114 | 8.3594 .023 75 |
| $S_1$ | 8.2809 .019 130 | 9.0723 .118 161 | 7.7472 .006 191 | 7.0956 .001 185 | 7.9496 .009 136 | 9.0723 .118 199 | 7.7258 .005 216 | 7.7948 .006 246 | ------ ------ ------ | 7.7844 .006 321 |
| $\rho_1$ | 7.0934 .001 170 | 7.5061 .003 20 | 8.0423 .011 50 | 8.3386 .022 45 | 6.6245 .000 176 | 7.9254 .008 59 | 9.2882 .194 255 | 8.3594 .023 285 | 7.7844 .006 39 | ------ ------ ------ |

HARMONIC ANALYSIS AND PREDICTION OF TIDES 261

Table 29.—*Elimination factors*—Continued

SERIES 326 DAYS. SEMIDIURNAL CONSTITUENTS

| Constituent sought ($A$) | Disturbing constituents ($B$, $C$, etc.) | | | | | | | | | | |
|---|---|---|---|---|---|---|---|---|---|---|---|
| | $K_2$ | $L_2$ | $M_2$ | $N_2$ | 2N | $R_2$ | $S_2$ | $T_2$ | $\lambda_2$ | $\mu_2$ | $\nu_2$ | 2SM |
| $K_2$ | | 7.4352 | 7.7427 | 7.8631 | 7.8244 | 9.0720 | 9.0470 | 9.0037 | 8.3392 | 7.0934 | 7.5061 | 8.0971 |
| | | .003 | .006 | .007 | .007 | .118 | .111 | .101 | .022 | .001 | .003 | .013 |
| | | | 354 | 204 | 235 | 265 | 199 | 219 | 238 | 249 | 190 | 340 | 53 |
| $L_2$ | 7.4352 | | 8.1340 | 8.0698 | 7.9526 | 8.0858 | 8.3386 | 8.4852 | 9.2882 | 7.4186 | 7.5483 | 7.9254 |
| | .003 | | | .014 | .012 | .009 | .012 | .022 | .031 | .194 | .003 | .004 | .008 |
| | 6 | | | 210 | 241 | 271 | 25 | 45 | 64 | 255 | 196 | 346 | 59 |
| $M_2$ | 7.7427 | 8.1340 | | 8.1340 | 8.0698 | 7.0956 | 7.5483 | 7.9190 | 8.3386 | 7.5483 | 8.3386 | 7.5347 |
| | .006 | .014 | | | .014 | .012 | .001 | .004 | .008 | .022 | .004 | .022 | .003 |
| | 156 | 150 | | | 210 | 241 | 175 | 14 | 34 | 45 | 346 | 315 | 28 |
| $N_2$ | 7.8631 | 8.0698 | 8.1340 | | 8.1340 | 7.7259 | 7.4186 | 6.7213 | 7.5483 | 8.3386 | 9.2882 | 6.3062 |
| | .007 | .012 | .014 | | | .014 | .005 | .003 | .001 | .004 | .022 | .194 | .000 |
| | 125 | 119 | 150 | | | 210 | 144 | 164 | 3 | 14 | 315 | 105 | 178 |
| 2N | 7.8244 | 7.9526 | 8.0698 | 8.1340 | | 7.7948 | 7.7040 | 7.5123 | 7.4186 | 9.2882 | 8.3594 | 7.4010 |
| | .007 | .009 | .012 | .014 | | | .006 | .005 | .003 | .003 | .194 | .023 | .003 |
| | 95 | 89 | 119 | 150 | | | 114 | 133 | 153 | 164 | 105 | 75 | 148 |
| $R_2$ | 9.0720 | 8.0858 | 7.0956 | 7.7259 | 7.7948 | | 9.0725 | 9.0472 | 8.2809 | 7.0444 | 7.7843 | 7.9190 |
| | .118 | .012 | .001 | .005 | .006 | | | .118 | .112 | .019 | .001 | .006 | .008 |
| | 161 | 335 | 216 | 246 | | | 199 | 219 | 230 | 351 | 321 | 34 |
| $S_2$ | 9.0470 | 8.3386 | 7.5483 | 7.4186 | 7.7040 | 9.0725 | | 9.0725 | 8.1340 | 7.5347 | 9.2254 | 7.5483 |
| | .111 | .022 | .004 | .003 | .005 | .118 | | | .118 | .014 | .003 | .008 | .004 |
| | 141 | 315 | 346 | 196 | 227 | 161 | | | 199 | 210 | 332 | 301 | 14 |
| $T_2$ | 9.0037 | 8.4852 | 7.9190 | 6.7213 | 7.5123 | 9.0472 | 9.0725 | | 7.7468 | 7.7359 | 7.9960 | 7.0956 |
| | .101 | .031 | .008 | .001 | .003 | .112 | .118 | | | .006 | .005 | .010 | .001 |
| | 122 | 296 | 326 | 357 | 207 | 141 | 161 | | | 191 | 312 | 282 | 175 |
| $\lambda_2$ | 8.3392 | 9.2882 | 8.3386 | 7.5483 | 7.4186 | 8.2809 | 8.1340 | 7.7568 | | 7.9254 | 8.1912 | 7.4186 |
| | .022 | .194 | .022 | .004 | .003 | .019 | .014 | .006 | | | .008 | .016 | .003 |
| | 111 | 105 | 315 | 346 | 196 | 130 | 150 | 169 | | | 301 | 271 | 164 |
| $\mu_2$ | 7.0934 | 7.4186 | 7.5483 | 8.3386 | 9.2882 | 7.0444 | 7.4347 | 7.7359 | 7.9254 | | 8.1340 | 7.5118 |
| | .001 | .003 | .004 | .022 | .194 | .001 | .003 | .005 | .008 | | | .014 | .003 |
| | 170 | 164 | 14 | 45 | 255 | 9 | 28 | 48 | 59 | | | 150 | 43 |
| $\nu_2$ | 7.5061 | 7.5483 | 8.3386 | 9.2882 | 8.3594 | 7.7843 | 7.9254 | 7.9960 | 8.1912 | 8.1340 | | 7.7477 |
| | .003 | .004 | .022 | .194 | .023 | .006 | .008 | .010 | .016 | .014 | | | .006 |
| | 20 | 14 | 45 | 255 | 285 | 39 | 59 | 78 | 89 | 210 | | | 73 |
| 2SM | 8.0971 | 7.9254 | 7.5347 | 6.3062 | 7.4010 | 7.9190 | 7.5483 | 7.0956 | 7.4186 | 7.5118 | 7.7477 | |
| | .013 | .008 | .003 | .000 | .003 | .008 | .004 | .001 | .003 | .003 | .006 | |
| | 307 | 301 | 332 | 182 | 212 | 326 | 346 | 185 | 196 | 317 | 287 | |

Table 29.—*Elimination factors*—Continued

**SERIES 355 DAYS. DIURNAL CONSTITUENTS**

| Constituent sought (A) | Disturbing constituents (B, C, etc.) | | | | | | | | |
|---|---|---|---|---|---|---|---|---|---|
| | $J_1$ | $K_1$ | $M_1$ | $O_1$ | $OO$ | $P_1$ | $Q_1$ | $2Q$ | $S_1$ | $\rho_1$ |
| $J_1$ | ------ | 7.9464 .009 201 | 7.9167 .008 222 | 7.5111 .003 203 | 7.8888 .008 19 | 8.0444 .011 211 | 7.6331 .004 224 | 7.6506 .004 245 | 8.0032 .010 206 | 6.7724 .001 355 |
| $K_1$ | 7.9464 .009 159 | ------ | 7.9464 .009 201 | 6.7064 .001 182 | 6.7064 .001 178 | 8.4581 .029 190 | 7.5111 .003 203 | 7.6331 .004 224 | 8.4598 .029 185 | 7.5794 .004 334 |
| $M_1$ | 7.9167 .008 138 | 7.9464 .009 159 | ------ | 7.8888 .008 341 | 7.5111 .003 157 | 7.7393 .005 169 | 6.7064 .001 182 | 7.5111 .003 203 | 7.8651 .007 164 | 7.9839 .010 313 |
| $O_1$ | 7.5111 .003 157 | 6.7064 .001 178 | 7.8888 .008 19 | ------ | 6.7060 .001 175 | 7.2500 .002 8 | 7.9464 .009 201 | 7.9167 .008 222 | 6.7729 .001 3 | 8.1364 .014 331 |
| $OO$ | 7.8888 .008 341 | 6.7064 .001 182 | 7.5111 .003 203 | 6.7060 .001 185 | ------ | 7.3914 .002 192 | 7.3284 .002 206 | 7.4740 .003 227 | 7.1838 .002 187 | 7.3105 .002 336 |
| $P_1$ | 8.0444 .011 149 | 8.4581 .029 170 | 7.7393 .005 191 | 7.2500 .002 352 | 7.3914 .002 168 | ------ | 7.2957 .002 193 | 7.5554 .004 214 | 8.4598 .029 175 | 7.7296 .005 324 |
| $Q_1$ | 7.6331 .004 136 | 7.5111 .003 157 | 6.7064 .001 178 | 7.9464 .009 159 | 7.3284 .002 167 | 7.2957 .002 | ------ | 7.9464 .009 201 | 7.4212 .003 162 | 9.1482 .141 130 |
| $2Q$ | 7.6506 .004 115 | 7.6331 .004 136 | 7.5111 .003 157 | 7.9167 .008 138 | 7.4740 .003 133 | 7.5554 .004 146 | 7.9464 .009 159 | ------ | 7.5984 .004 141 | 8.3129 .021 109 |
| $S_1$ | 8.0032 .010 154 | 8.4598 .029 175 | 7.8651 .007 196 | 6.7729 .001 357 | 7.1838 .002 173 | 8.4598 .029 185 | 7.4212 .003 198 | 7.5984 .004 219 | ------ | 7.6607 .005 329 |
| $\rho_1$ | 6.7724 .001 5 | 7.5794 .004 26 | 7.9839 .010 47 | 8.1364 .014 29 | 7.3105 .002 24 | 7.7296 .005 36 | 9.1482 .141 230 | 8.3129 .021 251 | 7.6607 .005 31 | ------ |

HARMONIC ANALYSIS AND PREDICTION OF TIDES 263

Table 29.—*Elimination factors*—Continued

SERIES 355 DAYS. SEMIDIURNAL CONSTITUENTS

| Constituent sought (A) | Disturbing constituents (B, C, etc.) | | | | | | | | | | |
|---|---|---|---|---|---|---|---|---|---|---|---|
| | K₂ | L₂ | M₂ | N₂ | 2N | R₂ | S₂ | T₂ | λ₂ | μ₂ | ν₂ | 2SM |
| K₂ | ----- | 7.8888 .008 341 | 6.7064 .001 182 | 7.5111 .003 203 | 7.6331 .004 224 | 8.4589 .029 185 | 8.4581 .029 190 | 8.4553 .029 195 | 8.0444 .011 211 | 6.7724 .001 355 | 7.5794 .004 334 | 7.6440 .004 18 |
| L₂ | 7.8888 .008 19 | ----- | 7.9464 .009 201 | 7.9167 .008 222 | 7.8652 .007 243 | 8.0217 .011 24 | 8.1364 .014 29 | 8.2393 .017 34 | 9.1482 .141 230 | 7.2957 .002 193 | 7.2500 .002 352 | 7.7296 .005 36 |
| M₂ | 6.7064 .001 178 | 7.9464 .009 159 | ----- | 7.9464 .009 201 | 7.9167 .008 222 | 6.7712 .001 3 | 7.2500 .002 8 | 7.4842 .003 13 | 8.1364 .014 29 | 7.2500 .002 352 | 8.1364 .014 331 | 7.2458 .002 15 |
| N₂ | 7.5111 .003 157 | 7.9167 .008 138 | 7.9464 .009 159 | ----- | 7.9464 .009 201 | 7.4212 .003 162 | 7.2957 .002 167 | 7.1008 .001 172 | 7.2500 .002 8 | 8.1364 .014 331 | 9.1482 .141 130 | 6.7017 .001 174 |
| 2N | 7.6331 .004 136 | 7.8652 .007 117 | 7.9167 .008 138 | 7.9464 .009 159 | ----- | 7.5985 .004 141 | 7.5554 .004 146 | 7.5017 .003 151 | 7.2957 .002 167 | 9.1482 .141 130 | 8.3129 .021 109 | 7.2840 .002 154 |
| R₂ | 8.4589 .029 175 | 8.0217 .011 336 | 6.7712 .001 357 | 7.4212 .003 198 | 7.5985 .004 219 | ----- | 8.4598 .029 185 | 8.4586 .029 190 | 8.0034 .010 206 | 7.0678 .001 350 | 7.6606 .005 329 | 7.4842 .003 13 |
| S₂ | 8.4581 .029 170 | 8.1364 .014 331 | 7.2500 .002 352 | 7.2957 .002 193 | 7.5554 .004 214 | 8.4598 .029 175 | ----- | 8.4598 .029 185 | 7.9464 .009 201 | 7.2458 .002 345 | 7.7296 .005 324 | 7.2500 .002 8 |
| T₂ | 8.4553 .029 165 | 8.2393 .017 326 | 7.4842 .003 347 | 7.1008 .001 188 | 7.5017 .003 209 | 8.4586 .029 170 | 8.4598 .029 175 | ----- | 7.8648 .007 196 | 7.3738 .002 340 | 7.7893 .006 319 | 6.7712 .001 3 |
| λ₂ | 8.0444 .011 149 | 9.1482 .141 130 | 8.1364 .014 331 | 7.2500 .002 352 | 7.2957 .002 193 | 8.0034 .010 154 | 7.9464 .009 159 | 7.8648 .007 164 | ----- | 7.7296 .005 324 | 8.0795 .012 303 | 7.2957 .002 167 |
| μ₂ | 6.7724 .001 5 | 7.2957 .002 167 | 7.2500 .002 8 | 8.1364 .014 29 | 9.1482 .141 230 | 7.0678 .001 10 | 7.2458 .002 15 | 7.3738 .002 20 | 7.7296 .005 36 | ----- | 7.9464 .009 159 | 7.2393 .002 23 |
| ν₂ | 7.5794 .004 26 | 7.2500 .002 8 | 8.1364 .014 29 | 9.1482 .141 230 | 8.3129 .021 251 | 7.6606 .005 31 | 7.7296 .005 36 | 7.7893 .006 41 | 8.0795 .012 57 | 7.9464 .009 201 | ----- | 7.5728 .004 44 |
| 2SM | 7.6440 .004 342 | 7.7296 .005 324 | 7.2458 .002 345 | 6.7017 .001 186 | 7.2840 .002 206 | 7.4842 .003 347 | 7.2500 .002 352 | 6.7712 .002 357 | 7.2957 .001 193 | 7.2393 .002 337 | 7.5728 .004 316 | ----- |

Table 29.—*Elimination factors*—Continued

**SERIES 369 DAYS. DIURNAL CONSTITUENTS**

| Constituent sought (A) | Disturbing constituents (B, C, etc.) | | | | | | | | | |
|---|---|---|---|---|---|---|---|---|---|---|
| | $J_1$ | $K_1$ | $M_1$ | $O_1$ | OO | $P_1$ | $Q_1$ | 2Q | $S_1$ | $\rho_1$ |
| $J_1$ | ------ ------ ---- | 8.3503 .022 290 | 7.8740 .007 219 | 7.8760 .008 287 | 8.3371 .022 112 | 8.2982 .020 286 | 7.5509 .004 217 | 7.4182 .003 326 | 8.3235 .021 288 | 5.7099 .000 0 |
| $K_1$ | 8.3503 .022 70 | ------ ------ ---- | 8.3503 .022 290 | 6.6332 .000 358 | 6.6332 .000 2 | 8.0072 .010 356 | 7.8760 .008 287 | 7.5509 .004 217 | 8.0074 .010 358 | 7.8892 .008 250 |
| $M_1$ | 7.8740 .007 141 | 8.3503 .022 70 | ------ ------ ---- | 8.3371 .022 248 | 7.8760 .008 73 | 8.4104 .026 67 | 6.6332 .000 358 | 7.8760 .008 287 | 8.3792 .024 69 | 7.9045 .008 321 |
| $O_1$ | 7.8760 .008 73 | 6.6332 .000 2 | 8.3371 .022 112 | ------ ------ ---- | 6.6329 .000 4 | 6.5537 .000 178 | 8.3503 .022 290 | 7.8740 .007 219 | 5.7100 .000 0 | 8.4169 .026 252 |
| OO | 8.3371 .022 248 | 6.6332 .000 358 | 7.8760 .008 287 | 6.6329 .000 356 | ------ ------ ---- | 7.0438 .001 354 | 7.6584 .005 285 | 7.3523 .002 215 | 6.8924 .001 356 | 7.6527 .004 248 |
| $P_1$ | 8.2982 .020 74 | 8.0072 .010 4 | 8.4104 .026 293 | 6.5537 .000 182 | 7.0438 .001 6 | ------ ------ ---- | 7.8885 .008 291 | 7.6024 .004 221 | 8.0074 .010 2 | 7.9217 .008 254 |
| $Q_1$ | 7.5509 .004 143 | 7.8760 .008 73 | 6.6332 .000 2 | 8.3503 .022 70 | 7.6584 .005 75 | 7.8885 .008 69 | ------ ------ ---- | 8.3503 .022 290 | 7.8824 .008 71 | 9.0329 .108 143 |
| 2Q | 7.4182 .003 34 | 7.5509 .004 143 | 7.8760 .008 73 | 8.3740 .007 141 | 7.3523 .002 145 | 7.6024 .004 139 | 8.3503 .022 70 | ------ ------ ---- | 7.5772 .004 141 | 8.0586 .011 33 |
| $S_1$ | 8.3235 .021 72 | 8.0074 .010 2 | 8.3792 .024 291 | 5.7100 .000 0 | 6.8924 .001 4 | 8.0074 .010 358 | 7.8824 .008 289 | 7.5772 .004 219 | ------ ------ ---- | 7.9055 .008 252 |
| $\rho_1$ | 5.7099 .000 0 | 7.8892 .008 110 | 7.9045 .008 39 | 8.4169 .026 108 | 7.6527 .004 112 | 7.9217 .008 106 | 9.0329 .108 217 | 8.0586 .011 327 | 7.9055 .008 108 | ------ ------ ---- |

HARMONIC ANALYSIS AND PREDICTION OF TIDES

Table 29.—*Elimination factors*—Continued

SERIES 369 DAYS. SEMIDIURNAL CONSTITUENTS

| Constituent sought (A) | Disturbing constituents (B, C, etc.) | | | | | | | | | | |
|---|---|---|---|---|---|---|---|---|---|---|---|
| | K₂ | L₂ | M₂ | N₂ | 2N | R₂ | S₂ | T₂ | λ₂ | μ₂ | ν₂ | 2SM |
| K₂ | ----- | 8.3371 .022 248 | 6.6332 .000 358 | 7.8760 .008 287 | 7.5509 .004 217 | 8.0074 .010 358 | 8.0072 .010 356 | 8.0076 .010 354 | 8.2982 .020 286 | 5.7099 .000 0 | 7.8892 .008 250 | 7.1088 .001 175 |
| L₂ | 8.3371 .022 112 | ----- | 8.3503 .022 290 | 7.8740 .007 219 | 7.6166 .004 329 | 8.3758 .024 110 | 8.4169 .026 108 | 8.4607 .029 106 | 9.0329 .108 217 | 7.8885 .008 291 | 6.5537 .000 182 | 7.9217 .008 106 |
| M₂ | 6.6332 .000 2 | 8.3503 .022 70 | ----- | 8.3503 .022 290 | 7.8740 .007 219 | 5.7100 .000 0 | 6.5537 .000 178 | 6.9049 .001 177 | 8.4169 .026 108 | 6.5537 .000 182 | 8.4169 .026 252 | 6.5549 .000 177 |
| N₂ | 7.8760 .008 73 | 7.8740 .007 141 | 8.3503 .022 70 | ----- | 8.3503 .022 290 | 7.8824 .008 71 | 7.8885 .008 69 | 7.8944 .008 67 | 6.5537 .000 178 | 8.4169 .026 252 | 9.0329 .108 143 | 7.6658 .005 67 |
| 2N | 7.5509 .004 143 | 7.6166 .004 31 | 7.8740 .007 141 | 8.3503 .022 70 | ----- | 7.5772 .004 141 | 7.6024 .004 139 | 7.6268 .004 138 | 7.8885 .008 69 | 9.0329 .108 143 | 8.0586 .011 33 | 7.4452 .003 138 |
| R₂ | 8.0074 .010 2 | 8.3758 .024 250 | 5.7100 .000 0 | 7.8824 .008 289 | 7.5772 .004 219 | ----- | 8.0074 .010 358 | 8.0072 .010 356 | 8.3235 .021 288 | 6.1771 .000 181 | 7.9055 .008 252 | 6.9049 .001 177 |
| S₂ | 8.0072 .010 4 | 8.4169 .026 252 | 6.5537 .000 182 | 7.8885 .008 291 | 7.6024 .004 221 | 8.0074 .010 2 | ----- | 8.0074 .010 358 | 8.3503 .022 290 | 6.5549 .000 183 | 7.9217 .008 254 | 6.5537 .000 178 |
| T₂ | 8.0076 .010 6 | 8.4607 .029 254 | 6.9049 .001 183 | 7.8944 .008 293 | 7.6268 .004 222 | 8.0072 .010 4 | 8.0074 .010 2 | ----- | 8.3792 .024 291 | 6.7601 .001 185 | 7.9377 .009 256 | 5.7100 .000 0 |
| λ₂ | 8.2982 .020 74 | 9.0329 .108 143 | 8.4164 .026 252 | 6.5537 .000 182 | 7.8885 .008 291 | 8.3235 .021 72 | 8.3503 .022 70 | 8.3792 .024 69 | ----- | 7.9217 .008 254 | 7.9044 .008 324 | 7.8885 .008 69 |
| μ₂ | 5.7099 .000 0 | 7.8885 .008 69 | 6.5537 .000 178 | 8.4169 .026 108 | 9.0329 .108 217 | 6.1771 .000 179 | 6.5549 .000 177 | 6.7601 .001 175 | 7.9217 .008 106 | ----- | 8.3503 .022 70 | 6.5542 .000 175 |
| ν₂ | 7.8892 .008 110 | 6.5537 .000 178 | 8.4169 .026 108 | 9.0329 .108 217 | 8.0586 .011 327 | 7.9055 .008 108 | 7.9217 .008 106 | 7.9377 .009 104 | 7.9044 .008 36 | 8.3503 .022 290 | ----- | 7.6990 .005 105 |
| 2SM | 7.1088 .001 185 | 7.9217 .008 254 | 6.5549 .000 183 | 7.6658 .005 293 | 7.4452 .003 222 | 6.9049 .001 183 | 6.5537 .000 182 | 5.7100 .000 0 | 7.8885 .008 291 | 6.5542 .000 185 | 7.6990 .005 255 | ----- |

Table 30.—*Products of amplitudes and angular functions for Form 245*

| ° | 1 | | 2 | | 3 | | 4 | | 5 | | ° |
|---|---|---|---|---|---|---|---|---|---|---|---|
| | Sin | Cos | Sin | Cos | Sin | Cos | Sin | Cos | Sin | Cos | |
| 0 | 0.000 | 1.000 | 0.000 | 2.000 | 0.000 | 3.000 | 0.000 | 4.000 | 0.000 | 5.000 | 90 |
| 1 | .017 | 1.000 | .035 | 2.000 | .052 | 3.000 | .070 | 3.999 | .087 | 4.999 | 89 |
| 2 | .035 | 0.999 | .070 | 1.999 | .105 | 2.998 | .140 | 3.998 | .174 | 4.997 | 88 |
| 3 | .052 | .999 | .105 | 1.997 | .157 | 2.996 | .209 | 3.995 | .262 | 4.993 | 87 |
| 4 | .070 | .998 | .140 | 1.995 | .209 | 2.993 | .279 | 3.990 | .349 | 4.988 | 86 |
| 5 | .087 | .996 | .174 | 1.992 | .261 | 2.989 | .349 | 3.985 | .436 | 4.981 | 85 |
| 6 | .105 | .995 | .209 | 1.989 | .314 | 2.984 | .418 | 3.978 | .523 | 4.973 | 84 |
| 7 | .122 | .993 | .244 | 1.985 | .366 | 2.978 | .487 | 3.970 | .609 | 4.963 | 83 |
| 8 | .139 | .990 | .278 | 1.981 | .418 | 2.971 | .557 | 3.961 | .696 | 4.951 | 82 |
| 9 | .156 | .988 | .313 | 1.975 | .469 | 2.963 | .626 | 3.951 | .782 | 4.938 | 81 |
| 10 | .174 | .985 | .347 | 1.970 | .521 | 2.954 | .695 | 3.939 | .868 | 4.924 | 80 |
| 11 | .191 | .982 | .382 | 1.963 | .572 | 2.945 | .763 | 3.927 | .954 | 4.908 | 79 |
| 12 | .208 | .978 | .416 | 1.956 | .624 | 2.934 | .832 | 3.913 | 1.040 | 4.891 | 78 |
| 13 | .225 | .974 | .450 | 1.949 | .675 | 2.923 | .900 | 3.897 | 1.125 | 4.872 | 77 |
| 14 | .242 | .970 | .484 | 1.941 | .726 | 2.911 | .968 | 3.881 | 1.210 | 4.852 | 76 |
| 15 | .259 | .966 | .518 | 1.932 | .776 | 2.898 | 1.035 | 3.864 | 1.294 | 4.830 | 75 |
| 16 | .276 | .961 | .551 | 1.923 | .827 | 2.884 | 1.103 | 3.845 | 1.378 | 4.806 | 74 |
| 17 | .292 | .956 | .585 | 1.913 | .877 | 2.869 | 1.169 | 3.825 | 1.462 | 4.782 | 73 |
| 18 | .309 | .951 | .618 | 1.902 | .927 | 2.853 | 1.236 | 3.804 | 1.545 | 4.755 | 72 |
| 19 | .326 | .946 | .651 | 1.891 | .977 | 2.837 | 1.302 | 3.782 | 1.628 | 4.728 | 71 |
| 20 | .342 | .940 | .684 | 1.879 | 1.026 | 2.819 | 1.368 | 3.759 | 1.710 | 4.698 | 70 |
| 21 | .358 | .934 | .717 | 1.867 | 1.075 | 2.801 | 1.433 | 3.734 | 1.792 | 4.668 | 69 |
| 22 | .375 | .927 | .749 | 1.854 | 1.124 | 2.782 | 1.498 | 3.709 | 1.873 | 4.636 | 68 |
| 23 | .391 | .920 | .781 | 1.841 | 1.172 | 2.762 | 1.563 | 3.682 | 1.954 | 4.602 | 67 |
| 24 | .407 | .914 | .813 | 1.827 | 1.220 | 2.741 | 1.627 | 3.654 | 2.034 | 4.568 | 66 |
| 25 | .423 | .906 | .845 | 1.813 | 1.268 | 2.719 | 1.690 | 3.625 | 2.113 | 4.532 | 65 |
| 26 | .438 | .899 | .877 | 1.798 | 1.315 | 2.696 | 1.753 | 3.595 | 2.192 | 4.494 | 64 |
| 27 | .454 | .891 | .908 | 1.782 | 1.362 | 2.673 | 1.816 | 3.564 | 2.270 | 4.455 | 63 |
| 28 | .469 | .883 | .939 | 1.766 | 1.408 | 2.649 | 1.878 | 3.532 | 2.347 | 4.415 | 62 |
| 29 | .485 | .875 | .970 | 1.749 | 1.454 | 2.624 | 1.939 | 3.498 | 2.424 | 4.373 | 61 |
| 30 | .500 | .866 | 1.000 | 1.732 | 1.500 | 2.598 | 2.000 | 3.464 | 2.500 | 4.330 | 60 |
| 31 | .515 | .857 | 1.030 | 1.714 | 1.545 | 2.572 | 2.060 | 3.429 | 2.575 | 4.286 | 59 |
| 32 | .530 | .848 | 1.060 | 1.696 | 1.590 | 2.544 | 2.120 | 3.392 | 2.650 | 4.240 | 58 |
| 33 | .545 | .839 | 1.089 | 1.677 | 1.634 | 2.516 | 2.179 | 3.355 | 2.723 | 4.193 | 57 |
| 34 | .559 | .829 | 1.118 | 1.658 | 1.678 | 2.487 | 2.237 | 3.316 | 2.796 | 4.145 | 56 |
| 35 | .574 | .819 | 1.147 | 1.638 | 1.721 | 2.457 | 2.294 | 3.277 | 2.868 | 4.096 | 55 |
| 36 | .588 | .809 | 1.176 | 1.618 | 1.763 | 2.427 | 2.351 | 3.236 | 2.939 | 4.045 | 54 |
| 37 | .602 | .799 | 1.204 | 1.597 | 1.805 | 2.396 | 2.407 | 3.195 | 3.009 | 3.993 | 53 |
| 38 | .616 | .788 | 1.231 | 1.576 | 1.847 | 2.364 | 2.463 | 3.152 | 3.078 | 3.940 | 52 |
| 39 | .629 | .777 | 1.259 | 1.554 | 1.888 | 2.331 | 2.517 | 3.109 | 3.147 | 3.886 | 51 |
| 40 | .643 | .766 | 1.286 | 1.532 | 1.928 | 2.298 | 2.571 | 3.064 | 3.214 | 3.830 | 50 |
| 41 | .656 | .755 | 1.312 | 1.509 | 1.968 | 2.264 | 2.624 | 3.019 | 3.280 | 3.774 | 49 |
| 42 | .669 | .743 | 1.338 | 1.486 | 2.007 | 2.229 | 2.677 | 2.973 | 3.346 | 3.716 | 48 |
| 43 | .682 | .731 | 1.364 | 1.463 | 2.046 | 2.194 | 2.728 | 2.925 | 3.410 | 3.657 | 47 |
| 44 | .695 | .719 | 1.389 | 1.439 | 2.084 | 2.158 | 2.779 | 2.877 | 3.473 | 3.597 | 46 |
| 45 | 0.707 | 0.707 | 1.414 | 1.414 | 2.121 | 2.121 | 2.828 | 2.828 | 3.536 | 3.536 | 45 |
| | Cos | Sin | Cos | Sin | Cos | Sin | Cos | Sin | Cos | Sin | |
| | 1 | | 2 | | 3 | | 4 | | 5 | | |

HARMONIC ANALYSIS AND PREDICTION OF TIDES 267

Table 30.—*Products of amplitudes and angular functions for Form 245*—Continued

| ° | 6 | | 7 | | 8 | | 9 | | ° |
|---|---|---|---|---|---|---|---|---|---|
| | Sin | Cos | Sin | Cos | Sin | Cos | Sin | Cos | |
| 0 | 0.000 | 6.000 | 0.000 | 7.000 | 0.000 | 8.000 | 0.000 | 9.000 | 90 |
| 1 | .105 | 5.999 | .122 | 6.999 | .140 | 7.999 | .157 | 8.999 | 89 |
| 2 | .209 | 5.996 | .244 | 6.996 | .279 | 7.995 | .314 | 8.995 | 88 |
| 3 | .314 | 5.992 | .366 | 6.990 | .419 | 7.989 | .471 | 8.988 | 87 |
| 4 | .419 | 5.985 | .488 | 6.983 | .558 | 7.980 | .628 | 8.978 | 86 |
| 5 | .523 | 5.977 | .610 | 6.973 | .697 | 7.970 | .784 | 8.966 | 85 |
| 6 | .627 | 5.967 | .732 | 6.962 | .836 | 7.956 | .941 | 8.951 | 84 |
| 7 | .731 | 5.955 | .853 | 6.948 | .975 | 7.940 | 1.097 | 8.933 | 83 |
| 8 | .835 | 5.942 | .974 | 6.932 | 1.113 | 7.922 | 1.253 | 8.912 | 82 |
| 9 | .939 | 5.926 | 1.095 | 6.914 | 1.251 | 7.902 | 1.408 | 8.889 | 81 |
| 10 | 1.042 | 5.909 | 1.216 | 6.894 | 1.389 | 7.878 | 1.563 | 8.863 | 80 |
| 11 | 1.145 | 5.890 | 1.336 | 6.871 | 1.526 | 7.853 | 1.717 | 8.835 | 79 |
| 12 | 1.247 | 5.869 | 1.455 | 6.847 | 1.663 | 7.825 | 1.871 | 8.803 | 78 |
| 13 | 1.350 | 5.846 | 1.575 | 6.821 | 1.800 | 7.795 | 2.025 | 8.769 | 77 |
| 14 | 1.452 | 5.822 | 1.693 | 6.792 | 1.935 | 7.762 | 2.177 | 8.733 | 76 |
| 15 | 1.553 | 5.796 | 1.812 | 6.762 | 2.071 | 7.727 | 2.329 | 8.693 | 75 |
| 16 | 1.654 | 5.768 | 1.929 | 6.729 | 2.205 | 7.690 | 2.481 | 8.651 | 74 |
| 17 | 1.754 | 5.738 | 2.047 | 6.694 | 2.339 | 7.650 | 2.631 | 8.607 | 73 |
| 18 | 1.854 | 5.706 | 2.163 | 6.657 | 2.472 | 7.608 | 2.781 | 8.560 | 72 |
| 19 | 1.953 | 5.673 | 2.279 | 6.619 | 2.605 | 7.564 | 2.930 | 8.510 | 71 |
| 20 | 2.052 | 5.638 | 2.394 | 6.578 | 2.736 | 7.518 | 3.078 | 8.457 | 70 |
| 21 | 2.150 | 5.601 | 2.509 | 6.535 | 2.867 | 7.469 | 3.225 | 8.402 | 69 |
| 22 | 2.248 | 5.563 | 2.622 | 6.490 | 2.997 | 7.417 | 3.371 | 8.345 | 68 |
| 23 | 2.344 | 5.523 | 2.735 | 6.444 | 3.126 | 7.364 | 3.517 | 8.284 | 67 |
| 24 | 2.440 | 5.481 | 2.847 | 6.395 | 3.254 | 7.308 | 3.661 | 8.222 | 66 |
| 25 | 2.536 | 5.438 | 2.958 | 6.344 | 3.381 | 7.250 | 3.804 | 8.157 | 65 |
| 26 | 2.630 | 5.393 | 3.069 | 6.292 | 3.507 | 7.190 | 3.945 | 8.089 | 64 |
| 27 | 2.724 | 5.346 | 3.178 | 6.237 | 3.632 | 7.128 | 4.086 | 8.019 | 63 |
| 28 | 2.817 | 5.298 | 3.286 | 6.181 | 3.756 | 7.064 | 4.225 | 7.947 | 62 |
| 29 | 2.909 | 5.248 | 3.394 | 6.122 | 3.878 | 6.997 | 4.363 | 7.872 | 61 |
| 30 | 3.000 | 5.196 | 3.500 | 6.062 | 4.000 | 6.928 | 4.500 | 7.794 | 60 |
| 31 | 3.090 | 5.143 | 3.605 | 6.000 | 4.120 | 6.857 | 4.635 | 7.715 | 59 |
| 32 | 3.180 | 5.088 | 3.709 | 5.936 | 4.239 | 6.784 | 4.769 | 7.632 | 58 |
| 33 | 3.268 | 5.032 | 3.812 | 5.871 | 4.357 | 6.709 | 4.902 | 7.548 | 57 |
| 34 | 3.355 | 4.974 | 3.914 | 5.803 | 4.474 | 6.632 | 5.033 | 7.461 | 56 |
| 35 | 3.441 | 4.915 | 4.015 | 5.734 | 4.589 | 6.553 | 5.162 | 7.372 | 55 |
| 36 | 3.527 | 4.854 | 4.115 | 5.663 | 4.702 | 6.472 | 5.290 | 7.281 | 54 |
| 37 | 3.611 | 4.792 | 4.213 | 5.590 | 4.815 | 6.389 | 5.416 | 7.188 | 53 |
| 38 | 3.694 | 4.728 | 4.310 | 5.516 | 4.925 | 6.304 | 5.541 | 7.092 | 52 |
| 39 | 3.776 | 4.663 | 4.405 | 5.440 | 5.035 | 6.217 | 5.664 | 6.994 | 51 |
| 40 | 3.857 | 4.596 | 4.500 | 5.362 | 5.142 | 6.128 | 5.785 | 6.894 | 50 |
| 41 | 3.936 | 4.528 | 4.592 | 5.283 | 5.248 | 6.038 | 5.905 | 6.792 | 49 |
| 42 | 4.015 | 4.459 | 4.684 | 5.202 | 5.353 | 5.945 | 6.022 | 6.688 | 48 |
| 43 | 4.092 | 4.388 | 4.774 | 5.119 | 5.456 | 5.851 | 6.138 | 6.582 | 47 |
| 44 | 4.168 | 4.316 | 4.863 | 5.035 | 5.557 | 5.755 | 6.252 | 6.474 | 46 |
| 45 | 4.243 | 4.243 | 4.950 | 4.950 | 5.657 | 5.657 | 6.364 | 6.364 | 45 |
| | Cos | Sin | Cos | Sin | Cos | Sin | Cos | Sin | |
| | 6 | | 7 | | 8 | | 9 | | |

Table 31.—*For construction of primary stencils*

| Difference | | Constituent 2Q | | | | | | | | | | | | | | | | | |
|---|---|---|---|---|---|---|---|---|---|---|---|---|---|---|---|---|---|---|---|
| Hour 0 | | d. | h. | d. | h. | d. | h. | d. | h. | d. | h. | d. | h. | d. | h. | d. | h. | d. | h. | d. | h. |
| +23 | −1 | 1 | 0 | 7 | 21 | 14 | 21 | 21 | 20* | 28 | 20* | 35 | 20 | 42 | 20 | 49 | 19* | 56 | 19* | 63 | 19 |
| +22 | −2 | | 4 | 8 | 4 | 15 | 4 | 22 | 3* | 29 | 3* | 36 | 3 | 43 | 3 | 50 | 2* | 57 | 2* | 64 | 2 |
| +21 | −3 | | 11 | | 11 | | 11 | | 10* | | 10* | | 10 | | 10 | | 9* | | 9* | | 9 |
| +20 | −4 | | 18 | | 18 | | 18 | | 17* | | 17* | | 17 | | 17 | | 16* | | 16* | | 16 |
| +19 | −5 | 2 | 1 | 9 | 1 | 16 | 1 | 23 | 0* | 30 | 0* | 37 | 0 | 44 | 0 | | 23* | | 23* | | 23 |
|  |  |  | 8 |  | 8 |  | 8 |  | 7* |  | 7* |  | 7 |  | 7 | 51 | 6* | 58 | 6* | 65 | 6 |
| +18 | −6 |  | 15 |  | 15 |  | 15 |  | 14* |  | 14* |  | 14 |  | 14 |  | 13* |  | 13* |  | 13 |
| +17 | −7 |  | 22 |  | 22 |  | 21* |  | 21* |  | 21* |  | 21 |  | 21 |  | 20* |  | 20* |  | 20 |
| +16 | −8 | 3 | 5 | 10 | 5 | 17 | 4* | 24 | 4* | 31 | 4* | 38 | 4 | 45 | 4 | 52 | 3* | 59 | 3* | 66 | 3 |
| +15 | −9 |  | 12 |  | 12 |  | 11* |  | 11* |  | 11* |  | 11 |  | 11 |  | 10* |  | 10* |  | 10 |
| +14 | −10 |  | 19 |  | 19 |  | 18* |  | 18* |  | 18* |  | 18 |  | 18 |  | 17* |  | 17* |  | 17 |
| +13 | −11 | 4 | 2 | 11 | 2 | 18 | 1* | 25 | 1* | 32 | 1* | 39 | 1 | 46 | 1 | 53 | 0* | 60 | 0* | 67 | 0 |
| +12 | −12 |  | 9 |  | 9 |  | 8* |  | 8* |  | 8 |  | 8 |  | 8 |  | 7* |  | 7* |  | 7 |
| +11 | −13 |  | 16 |  | 16 |  | 15* |  | 15* |  | 15 |  | 15 |  | 15 |  | 14* |  | 14* |  | 14 |
| +10 | −14 |  | 23 |  | 23 |  | 22* |  | 22* |  | 22 |  | 22 |  | 22 |  | 21* |  | 21* |  | 21 |
| +9 | −15 | 5 | 6 | 12 | 6 | 19 | 5* | 26 | 5* | 33 | 5 | 40 | 5 | 47 | 5 | 54 | 4* | 61 | 4* | 68 | 4 |
| +8 | −16 |  | 13 |  | 13 |  | 12* |  | 12* |  | 12 |  | 12 |  | 12 |  | 11* |  | 11* |  | 11 |
| +7 | −17 |  | 20 |  | 20 |  | 19* |  | 19* |  | 19 |  | 19 |  | 19 |  | 18* |  | 18* |  | 18 |
| +6 | −18 | 6 | 3 | 13 | 3 | 20 | 2* | 27 | 2* | 34 | 2 | 41 | 2 | 48 | 1* | 55 | 1* | 62 | 1* | 69 | 1 |
| +5 | −19 |  | 10 |  | 10 |  | 9* |  | 9* |  | 9 |  | 9 |  | 8* |  | 8* |  | 8* |  | 8 |
| +4 | −20 |  | 17 |  | 17 |  | 16* |  | 16* |  | 16 |  | 16 |  | 15* |  | 15* |  | 15* |  | 15 |
| +3 | −21 | 7 | 0 | 14 | 0 |  | 23* |  | 23* |  | 23 |  | 23 |  | 22* |  | 22* |  | 22* |  | 22 |
| +2 | −22 |  | 7 |  | 7 | 21 | 6* | 28 | 6* | 35 | 6 | 42 | 6 | 49 | 5* | 56 | 5* | 63 | 5* | 70 | 5 |
| +1 | −23 |  | 14 |  | 14 |  | 13* |  | 13* |  | 13 |  | 13 |  | 12* |  | 12* |  | 12* |  | 12 |

| Difference | | Constituent 2Q | | | | | | | | | | | | | | | | | |
|---|---|---|---|---|---|---|---|---|---|---|---|---|---|---|---|---|---|---|---|
| Hour 0 | | d. | h. | d. | h. | d. | h. | d. | h. | d. | h. | d. | h. | d. | h. | d. | h. | d. | h. | d. | h. |
| +23 | −1 | 70 | 19 | 77 | 19 | 84 | 18* | 91 | 18* | 98 | 18 | 105 | 18 | 112 | 17* | 119 | 17* | 126 | 17 | 133 | 17 |
| +22 | −2 | 71 | 2 | 78 | 2 | 85 | 1* | 92 | 1* | 99 | 1 | 106 | 1 | 113 | 0* | 120 | 0* | 127 | 0 | 134 | 0 |
| +21 | −3 |  | 9 |  | 9 |  | 8* |  | 8* |  | 8 |  | 8 |  | 7* |  | 7* |  | 7 |  | 7 |
| +20 | −4 |  | 16 |  | 16 |  | 15* |  | 15* |  | 15 |  | 15 |  | 14* |  | 14* |  | 14 |  | 14 |
| +19 | −5 |  | 23 |  | 23 |  | 22* |  | 22* |  | 22 |  | 22 |  | 21* |  | 21* |  | 21 |  | 21 |
|  |  | 72 | 6 | 79 | 5* | 86 | 5* | 93 | 5* | 100 | 5 | 107 | 5 | 114 | 4* | 121 | 4* | 128 | 4 | 135 | 4 |
| +18 | −6 |  | 13 |  | 12* |  | 12* |  | 12* |  | 12 |  | 12 |  | 11* |  | 11* |  | 11 |  | 11 |
| +17 | −7 |  | 20 |  | 19* |  | 19* |  | 19* |  | 19 |  | 19 |  | 18* |  | 18* |  | 18 |  | 18 |
| +16 | −8 | 73 | 3 | 80 | 2* | 87 | 2* | 94 | 2* | 101 | 2 | 108 | 2 | 115 | 1* | 122 | 1* | 129 | 1 | 136 | 1 |
| +15 | −9 |  | 10 |  | 9* |  | 9* |  | 9* |  | 9 |  | 9 |  | 8* |  | 8* |  | 8 |  | 8 |
| +14 | −10 |  | 17 |  | 16* |  | 16* |  | 16* |  | 16 |  | 16 |  | 15* |  | 15* |  | 15 |  | 15 |
| +13 | −11 | 74 | 0 |  | 23* |  | 23* |  | 23 |  | 23 |  | 23 |  | 22* |  | 22* |  | 22 |  | 22 |
| +12 | −12 |  | 7 | 81 | 6* | 88 | 6* | 95 | 6 | 102 | 6 | 109 | 6 | 116 | 5* | 123 | 5* | 130 | 5 | 137 | 5 |
| +11 | −13 |  | 14 |  | 13* |  | 13* |  | 13 |  | 13 |  | 13 |  | 12* |  | 12* |  | 12 |  | 12 |
| +10 | −14 |  | 21 |  | 20* |  | 20* |  | 20 |  | 20 |  | 20 |  | 19* |  | 19* |  | 19 |  | 19 |
| +9 | −15 | 75 | 4 | 82 | 3* | 89 | 3* | 96 | 3 | 103 | 3 | 110 | 3 | 117 | 2* | 124 | 2* | 131 | 2 | 138 | 2 |
| +8 | −16 |  | 11 |  | 10* |  | 10* |  | 10 |  | 10 |  | 10 |  | 9* |  | 9* |  | 9 |  | 9 |
| +7 | −17 |  | 18 |  | 17* |  | 17* |  | 17 |  | 17 |  | 16* |  | 16* |  | 16* |  | 16 |  | 16 |
| +6 | −18 | 76 | 1 | 83 | 0* | 90 | 0* | 97 | 0 | 104 | 0 |  | 23* |  | 23* |  | 23* |  | 23 |  | 23 |
| +5 | −19 |  | 8 |  | 7* |  | 7* |  | 7 |  | 7 | 111 | 6* | 118 | 6* | 125 | 6* | 132 | 6 | 139 | 6 |
| +4 | −20 |  | 15 |  | 14* |  | 14* |  | 14 |  | 14 |  | 13* |  | 13* |  | 13* |  | 13 |  | 13 |
| +3 | −21 |  | 22 |  | 21* |  | 21* |  | 21 |  | 21 |  | 20* |  | 20* |  | 20* |  | 20 |  | 20 |
| +2 | −22 | 77 | 5 | 84 | 4* | 91 | 4* | 98 | 4 | 105 | 4 | 112 | 3* | 119 | 3* | 126 | 3 | 133 | 3 | 140 | 3 |
| +1 | −23 |  | 12 |  | 11* |  | 11* |  | 11 |  | 11 |  | 10* |  | 10* |  | 10 |  | 10 |  | 10 |

HARMONIC ANALYSIS AND PREDICTION OF TIDES 269

TABLE 31.—*For construction of primary stencils*—Continued

| Difference | | Constituent 2Q | | | | | | | | |
|---|---|---|---|---|---|---|---|---|---|---|
| Hour | | d. h. | d. h. | d. h. | d. h. | d. h. | d. h. | d. h. | d. h. | d. h. |
| 0 | | 140 17 | 147 16* | 154 16* | 161 16 | 168 16 | 175 15* | 182 15* | 189 15 | 196 15 | 203 15 |
| +23 | −1 | 141 0 | 23* | 23* | 23 | 23 | 22* | 22* | 22 | 22 | 22 |
| +22 | −2 | 7 | 148 6* | 155 6* | 162 6 | 169 6 | 176 5* | 183 5* | 190 5 | 197 5 | 204 4* |
| +21 | −3 | 14 | 13* | 13* | 13 | 13 | 12* | 12* | 12 | 12 | 11* |
| +20 | −4 | 20* | 20* | 20* | 20 | 20 | 19* | 19 | 19 | 19 | 18* |
| +19 | −5 | 142 3* | 149 3* | 156 3* | 163 3 | 170 3 | 177 2* | 184 2* | 191 2 | 198 2 | 205 1* |
| +18 | −6 | 10* | 10* | 10* | 10 | 10 | 9* | 9* | 9 | 9 | 8* |
| +17 | −7 | 17* | 17* | 17* | 17 | 17 | 16* | 16* | 16 | 16 | 15* |
| +16 | −8 | 143 0* | 150 0* | 157 0* | 164 0 | 171 0 | 23* | 23* | 23 | 23 | 22* |
| +15 | −9 | 7* | 7* | 7* | 7 | 7 | 178 6* | 185 6* | 192 6 | 199 6 | 206 5* |
| +14 | −10 | 14* | 14* | 14 | 14 | 14 | 13* | 13* | 13 | 13 | 12* |
| +13 | −11 | 21* | 21* | 21 | 21 | 21 | 20* | 20* | 20 | 20 | 19* |
| +12 | −12 | 144 4* | 151 4* | 158 4 | 165 4 | 172 4 | 179 3* | 186 3* | 193 3 | 200 3 | 207 2* |
| +11 | −13 | 11* | 11* | 11 | 11 | 11 | 10* | 10* | 10 | 10 | 9* |
| +10 | −14 | 18* | 18* | 18 | 18 | 18 | 17* | 17* | 17 | 17 | 16* |
| +9 | −15 | 145 1* | 152 1* | 159 1 | 166 1 | 173 0* | 180 0* | 187 0* | 194 0 | 201 0 | 23* |
| +8 | −16 | 8* | 8* | 8 | 8 | 7* | 7* | 7* | 7 | 7 | 208 6* |
| +7 | −17 | 15* | 15* | 15 | 15 | 14* | 14* | 14* | 14 | 14 | 13* |
| +6 | −18 | 22* | 22* | 22 | 22 | 21* | 21* | 21* | 21 | 21 | 20* |
| +5 | −19 | 146 5* | 153 5* | 160 5 | 167 5 | 174 4* | 181 4* | 188 4* | 195 4 | 202 4 | 209 3* |
| +4 | −20 | 12* | 12* | 12 | 12 | 11* | 11* | 11* | 11 | 11 | 10* |
| +3 | −21 | 19* | 19* | 19 | 19 | 18* | 18* | 18 | 18 | 18 | 17* |
| +2 | −22 | 147 2* | 154 2* | 161 2 | 168 2 | 175 1* | 182 1* | 189 1 | 196 1 | 203 1 | 210 0* |
| +1 | −23 | 9* | 9* | 9 | 9 | 8* | 8* | 8 | 8 | 8 | 7* |

| Difference | | Constituent 2Q | | | | | | | | |
|---|---|---|---|---|---|---|---|---|---|---|
| Hour | | d. h. | d. h. | d. h. | d. h. | d. h. | d. h. | d. h. | d. h. | d. h. |
| 0 | | 210 14* | 217 14* | 224 14 | 231 14 | 238 13* | 245 13* | 252 13 | 259 13 | 266 13 | 273 12* |
| +23 | −1 | 21* | 21* | 21 | 21 | 20* | 20* | 20 | 20 | 19* | 19* |
| +22 | −2 | 211 4* | 218 4* | 225 4 | 232 4 | 239 3* | 246 3* | 253 3 | 260 3 | 267 2* | 274 2* |
| +21 | −3 | 11* | 11* | 11 | 11 | 10* | 10* | 10 | 10 | 9* | 9* |
| +20 | −4 | 18* | 18* | 18 | 18 | 17* | 17* | 17 | 17 | 16* | 16* |
| +19 | −5 | 212 1* | 219 1* | 226 1 | 233 1 | 240 0* | 247 0* | 254 0 | 261 0 | 23* | 23* |
| +18 | −6 | 8* | 8* | 8 | 8 | 7* | 7* | 7 | 7 | 268 6* | 275 6* |
| +17 | −7 | 15* | 15* | 15 | 15 | 14* | 14* | 14 | 14 | 13* | 13* |
| +16 | −8 | 22* | 22 | 22 | 22 | 21* | 21* | 21 | 21 | 20* | 20* |
| +15 | −9 | 213 5* | 220 5 | 227 5 | 234 5 | 241 4* | 248 4* | 255 4 | 262 4 | 269 3* | 276 3* |
| +14 | −10 | 12* | 12 | 12 | 12 | 11* | 11* | 11 | 11 | 10* | 10* |
| +13 | −11 | 19* | 19 | 19 | 19 | 18* | 18* | 18 | 18 | 17* | 17* |
| +12 | −12 | 214 2* | 221 2 | 228 2 | 235 2 | 242 1* | 249 1* | 256 1 | 263 1 | 270 0* | 277 0* |
| +11 | −13 | 9* | 9 | 9 | 9 | 8* | 8* | 8 | 8 | 7* | 7* |
| +10 | −14 | 16* | 16 | 16 | 16 | 15* | 15* | 15 | 15 | 14* | 14* |
| +9 | −15 | 23* | 23 | 23 | 23 | 22* | 22* | 22 | 22 | 21* | 21* |
| +8 | −16 | 215 6* | 222 6 | 229 6 | 236 5* | 243 5* | 250 5* | 257 5 | 264 5 | 271 4* | 278 4* |
| +7 | −17 | 13* | 13 | 13 | 12* | 12* | 12* | 12 | 12 | 11* | 11* |
| +6 | −18 | 20* | 20 | 20 | 19* | 19* | 19* | 19 | 19 | 18* | 18* |
| +5 | −19 | 216 3* | 223 3 | 230 3 | 237 2* | 244 2* | 251 2 | 258 2 | 265 2 | 272 1* | 279 1* |
| +4 | −20 | 10* | 10 | 10 | 9* | 9* | 9 | 9 | 9 | 8* | 8* |
| +3 | −21 | 17* | 17 | 17 | 16* | 16* | 16 | 16 | 16 | 15* | 15* |
| +2 | −22 | 217 0* | 224 0 | 231 0 | 23* | 23* | 23 | 23 | 23 | 22* | 22* |
| +1 | −23 | 7* | 7 | 7 | 238 6* | 245 6* | 252 6 | 259 6 | 266 6 | 273 5* | 280 5* |

246037—41——18

# 270    U. S. COAST AND GEODETIC SURVEY

## Table 31.—*For construction of primary stencils*—Continued

| Difference | | Constituent 2Q | | | | | | | | |
|---|---|---|---|---|---|---|---|---|---|---|
| Hour | | d. h. | d. h. | d. h. | d. h. | d. h. | d. h. | d. h. | d. h. | d. h. | d. h. |
| 0 | | 280 12* | 287 12 | 294 12 | 301 11* | 308 11* | 315 11 | 322 11 | 329 10* | 336 10* | 343 10* |
| +23 | −1 | 19* | 19 | 19 | 18* | 18* | 18 | 18 | 17* | 17* | 17* |
| +22 | −2 | 281 2* | 288 2 | 295 2 | 302 1* | 309 1* | 316 1 | 323 1 | 330 0* | 337 0* | 344 0* |
| +21 | −3 | 9* | 9 | 9 | 8* | 8* | 8 | 8 | 7* | 7* | 7* |
| +20 | −4 | 16* | 16 | 16 | 15* | 15* | 15 | 15 | 14* | 14* | 14* |
| +19 | −5 | 23* | 23 | 23 | 22* | 22* | 22 | 22 | 22* | 21* | 21 |
| +18 | −6 | 282 6* | 289 6 | 296 6 | 303 5* | 303 5* | 317 5 | 324 5 | 331 4* | 338 4* | 345 4 |
| +17 | −7 | 13 | 13 | 13 | 12* | 12* | 12 | 12 | 11* | 11* | 11 |
| +16 | −8 | 20 | 20 | 20 | 19* | 19* | 19 | 19 | 18* | 18* | 18 |
| +15 | −9 | 283 3 | 290 3 | 297 3 | 304 2* | 311 2* | 318 2 | 325 2 | 332 1* | 339 1* | 346 1 |
| +14 | −10 | 10 | 10 | 10 | 9* | 9* | 9 | 9 | 8* | 8* | 8 |
| +13 | −11 | 17 | 17 | 17 | 16* | 16* | 16 | 16 | 15* | 15* | 15 |
| +12 | −12 | 284 0 | 291 0 | 23* | 23* | 23* | 23 | 23 | 22* | 22* | 22 |
| +11 | −13 | 7 | 7 | 298 6* | 305 6* | 312 6* | 319 6 | 326 6 | 333 5* | 340 5* | 347 5 |
| +10 | −14 | 14 | 14 | 13* | 13* | 13* | 13 | 13 | 12* | 12* | 12 |
| +9 | −15 | 21 | 21 | 20* | 20* | 20* | 20 | 20 | 19* | 19* | 19 |
| +8 | −16 | 285 4 | 292 4 | 299 3* | 306 3* | 313 3* | 320 3 | 327 3 | 334 2* | 341 2* | 318 2 |
| +7 | −17 | 11 | 11 | 10* | 10* | 10* | 10 | 10 | 9* | 9* | 9 |
| +6 | −18 | 18 | 18 | 17* | 17* | 17 | 17 | 17 | 16* | 16* | 16 |
| +5 | −19 | 286 1 | 293 1 | 300 0* | 307 0* | 314 0 | 321 0 | 328 0 | 23* | 23* | 23 |
| +4 | −20 | 8 | 8 | 7* | 7* | 7 | 7 | 7 | 335 6* | 342 6* | 349 6 |
| +3 | −21 | 15 | 15 | 14* | 14* | 14 | 14 | 14 | 13* | 13* | 13 |
| +2 | −22 | 22 | 22 | 21* | 21* | 21 | 21 | 21 | 20* | 20* | 20 |
| +1 | −23 | 287 5 | 294 5 | 301 4* | 308 4* | 315 4 | 322 4 | 329 4 | 336 3* | 343 3* | 350 3 |

| Difference | | Constituent 2Q | | | Constituent Q | | | | | | |
|---|---|---|---|---|---|---|---|---|---|---|---|
| Hour | | d. h. | d. h. | d. h. | d. h. | d. h. | d. h. | d. h. | d. h. | d. h. | d. h. |
| 0 | | 350 10 | 357 10 | 364 9* | 1 0 | 10 5 | 19 13* | 28 22* | 38 7* | 47 16 | 57 1 |
| +23 | −1 | 17 | 17 | 16* | 5* | 14 | 23 | 29 8 | 16* | 48 1* | 10* |
| +22 | −2 | 351 0 | 358 0 | 23* | 15 | 23* | 20 8* | 17 | 39 2 | 11 | 19* |
| +21 | −3 | 7 | 7 | 365 6* | 2 0 | 11 9 | 18 | 30 2* | 11* | 20 | 58 5 |
| +20 | −4 | 14 | 14 | 13* | 9* | 18* | 21 3 | 12 | 21 | 49 5* | 14*. |
| +19 | −5 | 21 | 21 | 20* | 19 | 12 3* | 12* | 21* | 40 6 | 15 | 59 0 |
| +18 | −6 | 352 4 | 359 4 | 366 3* | 3 4* | 13 | 22 | 31 6* | 15* | 50 0* | 9 |
| +17 | −7 | 11 | 11 | 10* | 13* | 22* | 22 7* | 16 | 41 1 | 9* | 18* |
| +16 | −8 | 18 | 18 | 17* | 23 | 13 8 | 16* | 32 1* | 16* | 19 | 60 4 |
| +15 | −9 | 353 1 | 360 1 | 367 0* | 4 8* | 17 | 23 2 | 11 | 19* | 51 4* | 13 |
| +14 | −10 | 8 | 8 | 7* | 17* | 14 2* | 11* | 20 | 42 5 | 14 | 22* |
| +13 | −11 | 15 | 15 | 14* | 5 3 | 12 | 20* | 33 5* | 14* | 23 | 61 8 |
| +12 | −12 | 22 | 21* | 21* | 12* | 21* | 24 6 | 15 | 23* | 52 8* | 17* |
| +11 | −13 | 354 5 | 361 4* | 368 4* | 22 | 15 6* | 15* | 34 0* | 43 9 | 18 | 62 2* |
| +10 | −14 | 12 | 11* | 11* | 6 7 | 16 | 25 1 | 9* | 18* | 53 3* | 12 |
| +9 | −15 | 19 | 18* | 18* | 16* | 16 1* | 10 | 19 | 44 4 | 12* | 21* |
| +8 | −16 | 355 2 | 362 1* | 369 1* | 7 2 | 11 | 19* | 35 4* | 13 | 22 | 63 7 |
| +7 | −17 | 9 | 8* | 8* | 11* | 20 | 26 5 | 13* | 22* | 54 7* | 16 |
| +6 | −18 | 16 | 15* | 15* | 20* | 17 5* | 14* | 23 | 45 8 | 16* | 64 1* |
| +5 | −19 | 23 | 22* | 22* | 8 6 | 15 | 23* | 36 8* | 17* | 55 2 | 11 |
| +4 | −20 | 356 6 | 363 5* | 370 5* | 15* | 18 0 | 27 9 | 18 | 46 2* | 11* | 20* |
| +3 | −21 | 13 | 12* | | 9 1 | 9* | 18* | 37 3 | 12 | 21 | 65 5* |
| +2 | −22 | 20 | 19* | | 10 | 19 | 28 4 | 12* | 21* | 56 6 | 15 |
| +1 | −23 | 357 3 | 364 2* | | 19* | 19 4* | 13 | 22 | 47 7 | 15* | 66 0* |

HARMONIC ANALYSIS AND PREDICTION OF TIDES

TABLE 31.—*For construction of primary stencils*—Continued

| Difference | | Constituent Q | | | | | | | | | |
|---|---|---|---|---|---|---|---|---|---|---|---|
| Hour 0 | | d. h.<br>66 9* | d. h.<br>75 18* | d. h.<br>85 3* | d. h.<br>94 12 | d. h.<br>103 21 | d. h.<br>113 6 | d. h.<br>122 14* | d. h.<br>131 23* | d. h.<br>141 8 | d. h.<br>150 17 |
| +23 | −1 | 19 | 76 4 | 12* | 21* | 104 6* | 15 | 123 0 | 132 9 | 17* | 151 2* |
| +22 | −2 | 67 4* | 13* | 22 | 95 7 | 15* | 114 0* | 9* | 18 | 142 3 | 12 |
| +21 | −3 | 14 | 22* | 86 7* | 16* | 105 1 | 10 | 18* | 133 3* | 12* | 21 |
| +20 | −4 | 23 | 77 8 | 17 | 96 1* | 10* | 19* | 124 4 | 13 | 21* | 152 6* |
| +19 | −5 | 68 8* | 17* | 87 2 | 11 | 20 | 115 4* | 13* | 22* | 143 7 | 16 |
| +18 | −6 | 18 | 78 3 | 11* | 20* | 106 5 | 14 | 23 | 134 7* | 16* | 153 1* |
| +17 | −7 | 69 3* | 12 | 21 | 97 6 | 14* | 23* | 125 8 | 17 | 144 2 | 10* |
| +16 | −8 | 12* | 21* | 88 6* | 15 | 107 0 | 116 8* | 17* | 135 2* | 11 | 20 |
| +15 | −9 | 22 | 79 7 | 15* | 98 0* | 9* | 18 | 126 3 | 11* | 20* | 154 5* |
| +14 | −10 | 70 7* | 16 | 89 1 | 10 | 18* | 117 3* | 12* | 21 | 145 6 | 14* |
| +13 | −11 | 17 | 80 1* | 10* | 19 | 108 4 | 13 | 21* | 136 6* | 15* | 155 0 |
| +12 | −12 | 71 2 | 11 | 20 | 99 4* | 13* | 22 | 127 7 | 16 | 146 0* | 9* |
| +11 | −13 | 11* | 20* | 90 5 | 14 | 23 | 118 7* | 16* | 137 1 | 10 | 19 |
| +10 | −14 | 21 | 81 5* | 14* | 23* | 109 8 | 17 | 128 2 | 10* | 19* | 156 4 |
| +9 | −15 | 72 6* | 15 | 91 0 | 100 8* | 17* | 119 2* | 11 | 20 | 147 5 | 13* |
| +8 | −16 | 15* | 82 0* | 9 | 18 | 110 3 | 11* | 20* | 138 5* | 14 | 23 |
| +7 | −17 | 73 1 | 10 | 18* | 101 3* | 12 | 21 | 129 6 | 14* | 23* | 157 8* |
| +6 | −18 | 10* | 19 | 92 4 | 13 | 21* | 120 6* | 15 | 139 0 | 148 9 | 17* |
| +5 | −19 | 19* | 83 4* | 13* | 22 | 111 7 | 16 | 130 0* | 9* | 18 | 158 3 |
| +4 | −20 | 74 5 | 14 | 22* | 102 7* | 16* | 121 1 | 10 | 19 | 149 3* | 12* |
| +3 | −21 | 14* | 23* | 93 8 | 17 | 112 1* | 10* | 19* | 140 4 | 13 | 22 |
| +2 | −22 | 75 0 | 84 8* | 17* | 103 2* | 11 | 20 | 131 4* | 13* | 22* | 159 7 |
| +1 | −23 | 9 | 18 | 94 3 | 11* | 20* | 122 5* | 14 | 23 | 150 7* | 16* |

| Difference | | Constituent Q | | | | | | | | | |
|---|---|---|---|---|---|---|---|---|---|---|---|
| Hour 0 | | d. h.<br>160 2 | d. h.<br>169 10* | d. h.<br>178 19* | d. h.<br>188 4* | d. h.<br>197 13 | d. h.<br>206 22 | d. h.<br>216 6* | d. h.<br>225 15* | d. h.<br>235 0* | d. h.<br>244 9 |
| +23 | −1 | 11 | 20 | 179 5 | 13* | 22* | 207 7* | 16 | 226 1 | 9* | 18* |
| +22 | −2 | 20* | 170 5* | 14 | 23 | 198 8 | 16* | 217 1* | 10* | 19 | 245 4 |
| +21 | −3 | 161 6 | 15 | 23* | 189 8* | 17 | 208 2 | 11 | 19* | 236 4* | 13* |
| +20 | −4 | 15* | 171 0 | 180 9 | 18 | 199 2* | 11* | 20 | 227 5 | 14 | 22* |
| +19 | −5 | 162 0* | 9* | 18* | 190 3 | 12 | 21 | 218 5* | 14* | 23 | 246 8 |
| +18 | −6 | 10 | 19 | 181 3* | 12* | 21* | 209 6 | 15 | 228 0 | 237 8* | 17* |
| +17 | −7 | 19* | 172 4 | 13 | 22 | 200 6* | 15* | 219 0* | 9 | 18 | 247 2* |
| +16 | −8 | 163 5 | 13* | 22* | 191 7 | 16 | 210 1 | 9* | 18* | 238 3* | 12 |
| +15 | −9 | 14 | 23 | 182 8 | 16* | 201 1* | 10 | 19 | 229 4 | 12* | 21* |
| +14 | −10 | 23* | 173 8* | 17 | 192 2 | 11 | 19 | 220 4* | 13 | 22 | 248 7 |
| +13 | −11 | 164 9 | 17* | 183 2* | 11* | 20 | 211 5 | 14 | 22* | 239 7* | 16 |
| +12 | −12 | 18* | 174 3 | 12 | 20* | 202 5* | 14* | 23 | 230 8 | 17 | 249 1* |
| +11 | −13 | 165 3* | 12* | 21* | 193 6 | 15 | 23* | 221 8* | 17* | 240 2 | 11 |
| +10 | −14 | 13 | 22 | 184 6* | 15* | 203 0* | 212 9 | 18 | 231 2* | 11* | 20* |
| +9 | −15 | 22* | 175 7 | 16 | 194 1 | 9* | 18* | 222 3 | 12 | 21 | 250 5* |
| +8 | −16 | 166 7* | 16* | 185 1* | 10 | 19 | 213 4 | 12* | 21* | 241 6 | 15 |
| +7 | −17 | 17 | 176 2 | 10* | 19* | 204 4* | 13 | 22 | 232 7 | 15* | 251 0* |
| +6 | −18 | 167 2* | 11* | 20 | 195 5 | 13* | 22* | 223 7* | 16 | 242 1 | 10 |
| +5 | −19 | 12 | 20* | 186 5* | 14* | 23 | 214 8 | 16* | 233 1* | 10* | 19 |
| +4 | −20 | 21 | 177 6 | 15 | 23* | 205 8* | 17* | 224 2 | 11 | 19* | 252 4* |
| +3 | −21 | 168 6* | 15* | 187 0 | 196 9 | 18 | 215 2* | 11* | 20* | 243 5 | 14 |
| +2 | −22 | 16 | 178 1 | 9* | 18* | 206 3 | 12 | 21 | 234 5* | 14* | 23* |
| +1 | −23 | 169 1* | 10 | 19 | 197 3* | 12* | 21* | 225 6 | 15 | 244 0 | 253 8* |

TABLE 31.—*For construction of primary stencils*—Continued

| Difference | | Constituent Q | | | | | | | | | |
|---|---|---|---|---|---|---|---|---|---|---|---|
| Hour 0 | | d. h.<br>253 18 | d. h.<br>263 3 | d. h.<br>272 11* | d. h.<br>281 20* | d. h.<br>291 5 | d. h.<br>300 14 | d. h.<br>309 23 | d. h.<br>319 7* | −d. h.<br>328 16* | d. h.<br>338 1* |
| +23 | −1 | 254 3* | 12 | 21 | 282 6 | 14* | 23* | 310 8 | 17 | 329 2 | 10* |
| +22 | −2 | 12* | 21* | 273 6* | 15 | 292 0 | 301 9 | 17* | 320 2* | 11 | 20 |
| +21 | −3 | 22 | 264 7 | 15* | 283 0* | 9* | 18 | 311 3 | 12 | 20* | 339 5* |
| +20 | −4 | 255 7* | 16* | 274 1 | 10 | 18* | 302 3* | 12* | 21 | 330 6 | 15 |
| +19 | −5 | 17 | 265 1* | 10* | 19* | 293 4 | 13 | 21* | 321 6* | 15* | 340 0 |
| +18 | −6 | 256 2 | 11 | 20 | 284 4* | 13* | 22 | 312 7 | 16 | 331 0* | 9* |
| +17 | −7 | 11* | 20* | 275 5 | 14 | 23 | 303 7* | 16* | 322 1 | 10 | 19 |
| +16 | −8 | 21 | 266 5* | 14* | 23* | 294 8 | 17 | 313 2 | 10* | 19* | 341 4 |
| +15 | −9 | 257 6* | 15 | 276 0 | 285 8* | 17* | 304 2* | 11 | 20 | 332 5 | 13* |
| +14 | −10 | 15* | 267 0* | 9* | 18 | 295 3 | 11* | 323 5* | 14 | 23 |
| +13 | −11 | 258 1 | 10 | 18* | 286 3* | 12* | 21 | 314 6 | 14* | 23* | 342 8* |
| +12 | −12 | 10* | 19 | 277 4 | 13 | 21* | 305 6* | 15* | 324 0 | 333 9 | 17* |
| +11 | −13 | 20 | 268 4* | 13* | 22 | 296 7 | 16 | 315 0* | 9* | 18* | 343 3 |
| +10 | −14 | 259 5 | 14 | 23 | 287 7* | 16* | 306 1 | 10 | 19 | 334 3* | 12* |
| +9 | −15 | 14* | 23* | 278 8 | 17 | 297 1* | 10* | 19* | 325 4 | 13 | 22 |
| +8 | −16 | 260 0 | 269 8* | 17* | 288 2* | 11 | 20 | 316 4* | 13* | 22* | 344 7 |
| +7 | −17 | 9 | 18 | 279 3 | 11* | 20* | 307 5* | 14 | 23 | 335 7* | 16* |
| +6 | −18 | 18* | 270 3* | 12 | 21 | 298 6 | 14* | 23* | 326 8* | 17 | 345 2 |
| +5 | −19 | 261 4 | 13 | 21* | 289 6* | 15 | 308 0 | 317 9 | 17* | 336 2* | 11* |
| +4 | −20 | 13* | 22 | 280 7 | 16 | 299 0* | 9 | 18 | 327 3 | 12 | 21* |
| +3 | −21 | 22* | 271 7* | 16* | 290 1 | 10 | 19* | 318 3* | 12* | 21 | 346 6 |
| +2 | −22 | 262 8 | 17 | 281 1* | 10* | 19* | 309 4 | 13 | 21* | 337 6* | 15* |
| +1 | −23 | 17* | 272 2 | 11 | 20 | 300 4* | 13* | 22* | 328 7 | 16 | 347 0* |

| Difference | | Constituent Q | | | Constituent ρ | | | | | | |
|---|---|---|---|---|---|---|---|---|---|---|---|
| Hour 0 | | d. h.<br>347 10 | d. h.<br>356 19 | d. h.<br>366 3* | d. h.<br>1 0 | d. h.<br>10 15* | d. h.<br>20 11 | d. h.<br>30 6* | d. h.<br>40 2 | d. h.<br>49 21* | d. h.<br>59 17 |
| +23 | −1 | 19* | 357 4* | 13 | 5* | 11 1 | 20* | 16 | 12 | 50 7* | 60 3 |
| +22 | −2 | 348 5 | 13* | 22* | 15* | 11 | 21 6* | 31 2 | 21* | 17 | 12* |
| +21 | −3 | 14 | 23 | 367 8 | 2 1* | 21 | 16* | 12 | 41 7* | 51 3 | 22* |
| +20 | −4 | 23* | 358 8* | 17 | 11 | 12 6* | 22 2 | 21* | 17 | 12* | 61 8* |
| +19 | −5 | 349 9 | 18 | 368 2* | 21 | 16* | 12 | 32 7* | 42 3 | 22* | 18 |
| +18 | −6 | 18* | 359 3 | 12 | 3 6* | 13 2* | 22 | 17* | 13 | 52 8* | 62 4 |
| +17 | −7 | 350 3* | 12* | 21* | 16* | 12 | 23 7* | 33 3 | 22* | 18 | 13* |
| +16 | −8 | 13 | 22 | 369 6* | 4 2* | 22 | 17* | 13 | 43 8* | 53 4 | 23* |
| +15 | −9 | 22* | 360 7 | 16 | 12 | 14 7* | 24 3 | 22* | 18* | 14 | 63 9* |
| +14 | −10 | 351 8 | 16* | 370 1* | 22 | 17* | 13 | 34 8* | 44 4 | 23* | 19 |
| +13 | −11 | 17 | 361 2 | ---------- | 5 8 | 15 3* | 23 | 18* | 14 | 54 9* | 64 5 |
| +12 | −12 | 352 2* | 11* | ---------- | 17* | 13 | 25 8* | 35 4 | 23* | 19 | 15 |
| +11 | −13 | 12 | 20* | ---------- | 6 3* | 23 | 18* | 14 | 45 9* | 55 5 | 65 0* |
| +10 | −14 | 21 | 362 6 | ---------- | 13 | 16 9 | 26 4* | 36 0 | 19* | 15 | 10* |
| +9 | −15 | 353 6* | 15* | ---------- | 23 | 18* | 14 | 9* | 46 5 | 56 0* | 20 |
| +8 | −16 | 16 | 363 1 | ---------- | 7 9 | 17 4* | 27 0 | 19* | 15 | 10* | 66 6 |
| +7 | −17 | 354 1* | 10 | ---------- | 18* | 14 | 9* | 37 5* | 47 1 | 20* | 16 |
| +6 | −18 | 10* | 19* | ---------- | 8 4* | 18 0 | 19* | 15 | 10* | 57 6 | 67 1* |
| +5 | −19 | 20 | 364 5 | ---------- | 14* | 10 | 28 5* | 38 1 | 20* | 16 | 11* |
| +4 | −20 | 355 5* | 14* | ---------- | 9 0 | 19* | 15 | 10* | 48 6 | 58 2 | 21* |
| +3 | −21 | 15 | 23* | ---------- | 10 | 19 5* | 29 1 | 20* | 16 | 11* | 68 7 |
| +2 | −22 | 356 0 | 365 9 | ---------- | 19* | 15* | 11 | 39 6* | 49 2 | 21* | 17 |
| +1 | −23 | 9* | 18* | ---------- | 10 5* | 20 1 | 20* | 16 | 11* | 59 7 | 69 2* |

## HARMONIC ANALYSIS AND PREDICTION OF TIDES

Table 31.—*For construction of primary stencils*—Continued

| Difference | | Constituent ρ | | | | | | | | | |
|---|---|---|---|---|---|---|---|---|---|---|---|
| Hour | | d. h. | d. h. | d. h. | d. h. | d. h. | d. h. | d. h. | d. h. | d. h. | d. h. |
| 0 | | 69 12* | 79 8 | 89 3* | 98 23 | 108 18* | 118 14 | 128 9* | 138 5 | 148 1 | 157 20* |
| +23 | −1 | 22* | 18 | 13* | 99 9 | 109 4* | 119 0 | 19* | 15 | 10* | 158 6 |
| +22 | −2 | 70 8 | 80 3* | 23 | 18* | 14* | 10 | 129 5* | 139 1 | 20* | 16 |
| +21 | −3 | 18 | 13* | 90 9 | 100 4* | 110 0 | 19* | 15 | 10* | 149 6 | 159 1* |
| +20 | −4 | 71 4 | 23* | 19 | 14* | 10 | 120 5* | 130 1 | 20* | 16 | 11* |
| +19 | −5 | 13* | 81 9 | 91 4* | 101 0 | 19* | 15 | 11 | 140 6* | 150 2 | 21* |
| +18 | −6 | 23* | 19 | 14* | 10 | 111 5* | 121 1 | 20* | 16 | 11* | 160 7 |
| +17 | −7 | 72 9 | 82 5 | 92 0* | 20 | 15* | 11 | 131 6* | 141 2 | 21* | 17 |
| +16 | −8 | 19 | 14* | 10 | 102 5* | 112 1 | 20* | 16 | 11* | 151 7* | 161 3 |
| +15 | −9 | 73 5 | 83 0* | 20 | 15* | 11 | 122 6* | 132 2 | 21* | 17 | 12* |
| +14 | −10 | 14* | 10 | 93 5* | 103 1* | 21 | 16* | 12 | 142 7* | 152 3 | 22* |
| +13 | −11 | 74 0* | 20 | 15* | 11 | 113 6* | 123 2 | 21* | 17 | 12* | 162 8 |
| +12 | −12 | 10* | 84 6 | 94 1* | 21 | 16* | 12 | 133 7* | 143 3 | 22* | 18 |
| +11 | −13 | 20 | 15* | 11 | 104 6* | 114 2 | 21* | 17* | 13 | 153 8* | 163 4 |
| +10 | −14 | 75 6 | 85 1* | 21 | 16* | 12 | 124 7* | 134 3 | 22* | 18 | 13* |
| +9 | −15 | 15* | 11* | 95 7 | 105 2* | 22 | 17* | 13 | 144 8* | 154 4 | 23* |
| +8 | −16 | 76 1* | 21 | 16* | 12 | 115 7* | 125 3 | 22* | 18 | 14 | 164 9* |
| +7 | −17 | 11* | 86 7 | 96 2* | 22 | 17* | 13 | 135 8* | 145 4 | 23* | 19 |
| +6 | −18 | 21 | 16* | 12 | 106 8 | 116 3* | 23 | 18* | 14 | 155 9* | 165 5 |
| +5 | −19 | 77 7 | 87 2* | 22 | 17* | 13 | 126 8* | 136 4 | 23* | 19 | 14* |
| +4 | −20 | 17 | 12* | 97 8 | 107 3* | 23 | 18* | 14 | 146 9* | 156 5 | 166 0* |
| +3 | −21 | 78 2* | 22 | 17* | 13 | 117 8* | 127 4* | 137 0 | 19* | 15 | 10* |
| +2 | −22 | 12* | 88 8 | 98 3* | 23 | 18* | 14 | 9* | 147 5 | 157 0* | 20 |
| +1 | −23 | 22 | 18 | 13* | 108 9 | 118 4* | 128 0 | 19* | 15 | 10* | 167 6 |

| Difference | | Constituent ρ | | | | | | | | | |
|---|---|---|---|---|---|---|---|---|---|---|---|
| Hour | | d. h. | d. h. | d. h. | d. h. | d. h. | d. h. | d. h. | d. h. | d. h. | d. h. |
| 0 | | 167 16 | 177 11* | 187 7 | 197 2* | 206 22 | 216 17* | 226 13 | 236 8* | 246 4 | 255 23* |
| +23 | −1 | 168 1* | 21 | 16* | 12 | 207 7* | 217 3* | 23 | 18* | 14 | 256 9* |
| +22 | −2 | 11* | 178 7 | 188 2* | 22 | 17* | 13 | 227 8* | 237 4 | 23* | 19 |
| +21 | −3 | 21 | 17 | 12* | 198 8 | 208 3* | 23 | 18* | 14 | 247 9* | 257 5 |
| +20 | −4 | 169 7 | 179 2* | 22 | 17* | 13 | 218 8* | 228 4 | 238 0 | 19* | 15 |
| +19 | −5 | 17 | 12* | 189 8 | 199 3* | 23 | 18* | 14 | 9* | 248 5 | 258 0* |
| +18 | −6 | 170 2* | 22 | 17* | 13* | 209 9 | 219 4* | 229 0 | 19* | 15 | 10* |
| +17 | −7 | 12* | 180 8 | 190 3* | 23 | 18* | 14 | 9* | 239 5 | 249 0* | 20 |
| +16 | −8 | 22* | 18 | 13* | 200 9 | 210 4* | 220 0 | 19* | 15 | 10* | 259 6 |
| +15 | −9 | 171 8 | 181 3* | 23 | 18* | 14 | 10 | 230 5* | 240 1 | 20* | 16 |
| +14 | −10 | 18 | 13* | 191 9 | 201 4* | 211 0 | 19* | 15 | 10* | 250 6 | 260 1* |
| +13 | −11 | 172 4 | 23* | 19 | 14* | 10 | 221 5* | 231 1 | 20* | 16 | 11* |
| +12 | −12 | 13* | 182 9 | 192 4* | 202 0 | 19* | 15 | 10* | 241 6* | 251 2 | 21* |
| +11 | −13 | 23* | 19 | 14* | 10 | 212 5* | 222 1 | 20* | 16 | 11* | 261 7 |
| +10 | −14 | 173 9 | 183 4* | 193 0* | 20 | 15* | 11 | 232 6* | 242 2 | 21* | 17 |
| +9 | −15 | 19 | 14* | 10 | 203 5* | 213 1 | 20* | 16 | 11* | 252 7 | 262 3 |
| +8 | −16 | 174 5 | 184 0* | 20 | 15* | 11 | 223 6* | 233 2 | 21* | 17 | 12* |
| +7 | −17 | 14* | 10 | 194 5* | 204 1 | 20* | 16* | 12 | 243 7* | 253 3 | 22* |
| +6 | −18 | 175 0* | 20 | 11 | 214 6* | 224 2 | 21* | 17 | 12* | 263 8 | |
| +5 | −19 | 10* | 185 6 | 195 1* | 21 | 16* | 12 | 234 7* | 244 3 | 22* | 18 |
| +4 | −20 | 20 | 15* | 11 | 205 6* | 215 2 | 21* | 17 | 13 | 254 8* | 264 4 |
| +3 | −21 | 176 6 | 186 1* | 16* | 12 | 225 7* | 235 3 | 22* | 18 | 13* | |
| +2 | −22 | 15* | 11 | 196 7 | 206 2* | 22 | 17* | 13 | 245 8* | 255 4 | 23* |
| +1 | −23 | 177 1* | 21 | 16* | 12 | 216 7* | 226 3 | 22* | 18 | 13* | 265 9* |

Table 31.—*For construction of primary stencils*—Continued

| Difference | | Constituent ρ | | | | | | | | | |
|---|---|---|---|---|---|---|---|---|---|---|---|
| Hour | 0 | d. h. | d. h. | d. h. | d. h. | d. h. | d. h. | d. h. | d. h. | d. h. | d. h. |
| 0 | | 265 19 | 275 14* | 285 10 | 295 5* | 305 1 | 314 20* | 324 16 | 334 12 | 344 7* | 354 3 |
| +23 | −1 | 266 5 | 276 0* | 20 | 15* | 11 | 315 6* | 325 2 | 21* | 17 | 12* |
| +22 | −2 | 14* | 10 | 286 6 | 296 1* | 21 | 16* | 12 | 355 7* | 345 3 | 22* |
| +21 | −3 | 267 0* | 20 | 15* | 11 | 306 6* | 316 2 | 21* | 17 | 12* | 355 8* |
| +20 | −4 | 10* | 277 6 | 287 1* | 21 | 16* | 12 | 326 7* | 336 3 | 22* | 18 |
| +19 | −5 | 20 | 15* | 11 | 297 6* | 307 2* | 22 | 17* | 13 | 346 8* | 356 4 |
| +18 | −6 | 268 6 | 278 1* | 21 | 16* | 12 | 317 7* | 327 3 | 22* | 18 | 13* |
| +17 | −7 | 16 | 11* | 288 7 | 298 2* | 22 | 17* | 13 | 337 8* | 347 4 | 23* |
| +16 | −8 | 269 1* | 21 | 16* | 12 | 308 7* | 318 3 | 23 | 18* | 14 | 357 9* |
| +15 | −9 | 11* | 279 7 | 289 2* | 22 | 17* | 13 | 328 8* | 338 4 | 23* | 19 |
| +14 | −10 | 21 | 16* | 12* | 299 8 | 309 3* | 23 | 18* | 14 | 348 9* | 358 5 |
| +13 | −11 | 270 7 | 280 2* | 22 | 17* | 13 | 319 8* | 329 4 | 23* | 19 | 15 |
| +12 | −12 | 17 | 12* | 290 8 | 300 3* | 23 | 18* | 14 | 339 9* | 349 5 | 359 0* |
| +11 | −13 | 271 2* | 22 | 17* | 13 | 310 9 | 320 4* | 330 0 | 19* | 15 | 10* |
| +10 | −14 | 12* | 281 8 | 291 3* | 23 | 311 4* | 321 0 | 9* | 340 5 | 350 0* | 20 |
| +9 | −15 | 22* | 18 | 13* | 301 9 | 311 4* | 321 0 | 19* | 15 | 10* | 360 6 |
| +8 | −16 | 272 8 | 282 3* | 23 | 18* | 14 | 9* | 331 5* | 341 1 | 20* | 16 |
| +7 | −17 | 18 | 13* | 292 9 | 302 4* | 312 0 | 19* | 15 | 10* | 351 6 | 361 1* |
| +6 | −18 | 273 3* | 23* | 19 | 14* | 10 | 322 5* | 332 1 | 20* | 16 | 11* |
| +5 | −19 | 13* | 283 9 | 293 4* | 303 0 | 19* | 15 | 10* | 342 6 | 352 2 | 21* |
| +4 | −20 | 23* | 19 | 14* | 10 | 313 5* | 323 1 | 20* | 16 | 11* | 362 7 |
| +3 | −21 | 274 9 | 284 4* | 294 0 | 19* | 15* | 11 | 333 6* | 343 2 | 21* | 17 |
| +2 | −22 | 19 | 14* | 10 | 304 5* | 314 1 | 20* | 16 | 11* | 353 7 | 363 2* |
| +1 | −23 | 275 5 | 285 0* | 20 | 15* | 11 | 324 6* | 334 2 | 21* | 17 | 12* |

| Difference | ρ | Constituent O | | | | | | | | | |
|---|---|---|---|---|---|---|---|---|---|---|---|
| Hour | 0 | d. h. | d. h. | d. h. | d. h. | d. h. | d. h. | d. h. | d. h. | d. h. | d. h. |
| 0 | | 363 22* | 1 0 | 14 22* | 29 3 | 43 7* | 57 12 | 71 16* | 85 21 | 100 2 | 114 6* |
| +23 | −1 | 364 8 | 8 | 15 12* | 17 | 21* | 58 2 | 72 7 | 86 11* | 16 | 20* |
| +22 | −2 | 18 | 22 | 16 2* | 30 7 | 44 12 | 16* | 21 | 87 1* | 101 6 | 115 11 |
| +21 | −3 | 365 4 | 2 12 | 17 | 21* | 45 2 | 59 6* | 73 11 | 16 | 20* | 116 1 |
| +20 | −4 | 13* | 3 2* | 17 7 | 31 11* | 16 | 21 | 74 1* | 88 6 | 102 10* | 15 |
| +19 | −5 | 23* | 16* | 21 | 32 2 | 46 6* | 60 11 | 15* | 20 | 103 1 | 117 5* |
| +18 | −6 | 366 9 | 4 7 | 18 11* | 16 | 20* | 61 1 | 75 6 | 89 10* | 15 | 19* |
| +17 | −7 | 19 | 21 | 19 1* | 33 6 | 47 11 | 15* | 20 | 90 0* | 104 5 | 118 10 |
| +16 | −8 | 367 5 | 5 11 | 16 | 20* | 48 1 | 62 5* | 76 10 | 15 | 19* | 119 0 |
| +15 | −9 | 14* | 6 1* | 20 6 | 34 10* | 15 | 20 | 77 0* | 91 5 | 105 9* | 14 |
| +14 | −10 | 368 0* | 15* | 20 | 35 1 | 49 5* | 63 10 | 14* | 19 | 106 0 | 120 4* |
| +13 | −11 | 10* | 7 6 | 21 10* | 15 | 19* | 64 0 | 78 5 | 92 9* | 14 | 18* |
| +12 | −12 | 20 | 20 | 22 0* | 36 5 | 50 9* | 14* | 19 | 23* | 107 4 | 121 8* |
| +11 | −13 | 369 6 | 8 10 | 14* | 19* | 51 0 | 65 4* | 79 9 | 93 13* | 18* | 23 |
| +10 | −14 | 15* | 9 0* | 23 5 | 37 9* | 14 | 18* | 23* | 94 4 | 108 8* | 122 13 |
| +9 | −15 | 370 1* | 14* | 19 | 23* | 52 4* | 66 9 | 80 13* | 18 | 22* | 123 3* |
| +8 | −16 | -------- | 10 4* | 24 9* | 38 14 | 18* | 23 | 81 3* | 95 8* | 109 13 | 17* |
| +7 | −17 | -------- | 19 | 23* | 39 4 | 53 8* | 67 13* | 18 | 22* | 110 3 | 124 7* |
| +6 | −18 | -------- | 11 9 | 25 13* | 18* | 23 | 68 3* | 82 8 | 96 12* | 17* | 22 |
| +5 | −19 | -------- | 23* | 26 4 | 40 8* | 54 13 | 17* | 22* | 97 3 | 111 7* | 125 12 |
| +4 | −20 | -------- | 12 13* | 18 | 22* | 55 3* | 69 8 | 83 12* | 17 | 21* | 126 2* |
| +3 | −21 | -------- | 13 3* | 27 8* | 41 13 | 17* | 22 | 84 2* | 98 7* | 112 12 | 16* |
| +2 | −22 | -------- | 18 | 22* | 42 3 | 56 7* | 70 12* | 17 | 21* | 113 2 | 127 6* |
| +1 | −23 | -------- | 14 8 | 28 12* | 17* | 22 | 71 2* | 85 7 | 99 11* | 16 | 21 |

HARMONIC ANALYSIS AND PREDICTION OF TIDES 275

Table 31.—*For construction of primary stencils*—Continued

| Difference | | Constituent O | | | | | | | | |
|---|---|---|---|---|---|---|---|---|---|---|
| Hours | | d. h. | d. h. | d. h. | d. h. | d. h. | d. h. | d. h. | d. h. | d. h. |
| 0 | | 128 11 | 142 15* | 156 20 | 171 1 | 185 5* | 199 10 | 213 14* | 227 19 | 242 0 | 256 4* |
| +23 | −1 | 129 1 | 143 6 | 157 10* | 15 | 19* | 200 0 | 214 5 | 228 9* | 14 | 18* |
| +22 | −2 | 15* | 20 | 158 0* | 172 5 | 186 10 | 14* | 19 | 23* | 243 4 | 257 9 |
| +21 | −3 | 130 5* | 144 10 | 15 | 19* | 187 0 | 201 4* | 215 9 | 229 14 | 18* | 23 |
| +20 | −4 | 20 | 145 0* | 159 5 | 173 9* | 14 | 19 | 23* | 230 4 | 244 8* | 258 13 |
| +19 | −5 | 131 10 | 14* | 19 | 174 0 | 188 4* | 202 9 | 216 13* | 18 | 23 | 259 3* |
| +18 | −6 | 132 0 | 146 5 | 160 9* | 14 | 18* | 23 | 217 4 | 231 8* | 245 13 | 17* |
| +17 | −7 | 14* | 19 | 23* | 175 4 | 189 9 | 203 13* | 18 | 22* | 246 3 | 260 8 |
| +16 | −8 | 133 4* | 147 9 | 161 14 | 18* | 23 | 204 3* | 218 8 | 232 13 | 17* | 22 |
| +15 | −9 | 19 | 23* | 162 4 | 176 8* | 190 13 | 18 | 22* | 233 3 | 247 7* | 261 12 |
| +14 | −10 | 134 9 | 148 13* | 18 | 22* | 191 3* | 205 8 | 219 12* | 17 | 21* | 262 2* |
| +13 | −11 | 23 | 149 3* | 163 8* | 177 13 | 17* | 22 | 220 2* | 234 7* | 248 12 | 16* |
| +12 | −12 | 135 13* | 18 | 22* | 178 3 | 192 7* | 206 12* | 17 | 21* | 249 2 | 263 6* |
| +11 | −13 | 136 3* | 150 8 | 164 12* | 17* | 22 | 207 2* | 221 7 | 235 11* | 16* | 21 |
| +10 | −14 | 17* | 22* | 165 3 | 179 7* | 193 12 | 16* | 21* | 236 2 | 250 6* | 264 11 |
| +9 | −15 | 137 8 | 151 12* | 17 | 21* | 194 2* | 208 7 | 222 11* | 16 | 20* | 265 1* |
| +8 | −16 | 22 | 152 2* | 166 7* | 180 12 | 16* | 21 | 223 1* | 237 6* | 251 11 | 15* |
| +7 | −17 | 138 12* | 17 | 21* | 181 2 | 195 6* | 209 11* | 16 | 20* | 252 1 | 266 5* |
| +6 | −18 | 139 2* | 153 7 | 167 11* | 16* | 21 | 210 1* | 224 6 | 238 10* | 15* | 20 |
| +5 | −19 | 16* | 21* | 168 2 | 182 6* | 196 11 | 15* | 20* | 239 1 | 253 5* | 267 10 |
| +4 | −20 | 140 7 | 154 11* | 16 | 20* | 197 1* | 211 6 | 225 10* | 15 | 19* | 268 0* |
| +3 | −21 | 21 | 155 1* | 169 6* | 183 11 | 15* | 20 | 226 0* | 240 5 | 254 10 | 14* |
| +2 | −22 | 141 11* | 16 | 20* | 184 1 | 198 5* | 212 10 | 15 | 19* | 255 0 | 269 4* |
| +1 | −23 | 142 1* | 156 6 | 170 10* | 15 | 20 | 213 0* | 227 5 | 241 9* | 14 | 19 |

| Difference | | Constituent O | | | | | | | | Component 2N | |
|---|---|---|---|---|---|---|---|---|---|---|---|
| Hour | | d. h | d. h. | d. h. | d. h. | d. h. | d. h. | d. h. | d. h. | d. h. | d. h. |
| 0 | | 270 9 | 284 13* | 298 18 | 312 23 | 327 3* | 341 8 | 355 12* | 369 17 | 1 0 | 14 23* |
| +23 | −1 | 23 | 285 4 | 299 8* | 313 13 | 17* | 22 | 356 3 | 370 7* | 8 | 15 14 |
| +22 | −2 | 271 13* | 18 | 22* | 314 3 | 328 8 | 342 12* | 17 | ---------- | 22 | 16 4 |
| +21 | −3 | 272 3* | 286 8 | 300 13 | 17* | 22 | 343 2* | 357 7 | ---------- | 2 12* | 18* |
| +20 | −4 | 18 | 22* | 301 3 | 315 7* | 329 12 | 17 | 21* | ---------- | 3 2* | 17 8* |
| +19 | −5 | 273 8 | 287 12* | 17 | 22 | 330 2* | 344 7 | 358 11* | ---------- | 17 | 23 |
| +18 | −6 | 22 | 288 3 | 302 7* | 316 12 | 16* | 21 | 359 2 | ---------- | 4 7 | 18 13 |
| +17 | −7 | 274 12* | 17 | 21* | 317 2 | 331 7 | 345 11* | 16 | ---------- | 21* | 19 3* |
| +16 | −8 | 275 2* | 289 7 | 303 11* | 16* | 21 | 346 1* | 360 6 | ---------- | 5 11* | 18 |
| +15 | −9 | 16* | 21* | 304 2 | 318 6* | 332 11 | 15* | 20* | ---------- | 6 2 | 20 8 |
| +14 | −10 | 276 7 | 290 11* | 16 | 20* | 333 1* | 347 6* | 361 10* | ---------- | 16 | 22* |
| +13 | −11 | 21 | 291 1* | 305 6* | 319 11 | 15* | 20 | 362 0* | ---------- | 7 6* | 21 12* |
| +12 | −12 | 277 11* | 16 | 20* | 320 1 | 334 5* | 348 10* | 15 | ---------- | 20* | 22 3 |
| +11 | −13 | 278 1* | 292 6 | 306 10* | 15* | 20 | 349 0* | 363 5 | ---------- | 8 11 | 17 |
| +10 | −14 | 15* | 20* | 307 1 | 321 5* | 335 10 | 14* | 19* | ---------- | 9 1 | 23 7* |
| +9 | −15 | 279 6 | 293 10* | 15 | 19* | 336 0* | 350 5 | 364 9* | ---------- | 15* | 21* |
| +8 | −16 | 20 | 294 0* | 308 5* | 322 10 | 14* | 19 | 23* | ---------- | 10 5* | 24 12 |
| +7 | −17 | 280 10* | 15 | 19* | 323 0 | 337 4* | 351 9* | 365 14 | ---------- | 20 | 25 2 |
| +6 | −18 | 281 0* | 295 5 | 309 9* | 14 | 19 | 23* | 366 4 | ---------- | 11 10 | 16* |
| +5 | −19 | 14* | 19* | 310 0 | 324 4* | 338 9 | 352 13* | 18 | ---------- | 12 0* | 26 6* |
| +4 | −20 | 282 5 | 296 9* | 14 | 18* | 23 | 353 4 | 367 8* | ---------- | 14* | 21 |
| +3 | −21 | 19 | 23* | 311 4 | 325 9 | 339 13* | 18 | 22* | ---------- | 13 5 | 27 11 |
| +2 | −22 | 283 9 | 297 14 | 18* | 23 | 340 3* | 354 8 | 368 13 | ---------- | 19 | 28 1* |
| +1 | −23 | 23* | 298 4 | 312 8* | 326 13 | 18 | 22* | 369 3 | ---------- | 14 9* | 15* |

Table 31.—*For construction of primary stencils*—Continued

| Difference | | Constituent 2N | | | | | | | | |
|---|---|---|---|---|---|---|---|---|---|---|
| Hour | | d. h. | d. h. | d. h. | d. h. | d. h. | d. h. | d. h. | d. h. | d. h. |
| 0 | | 29 6 | 43 12 | 57 18 | 72 0 | 86 6 | 100 12* | 114 18* | 129 0* | 143 6* | 157 12* |
| +23 | −1 | 20 | 44 2 | 58 8* | 14* | 20* | 101 2* | 115 8* | 15 | 21 | 158 3 |
| +22 | −2 | 30 10* | 16* | 22* | 73 4* | 87 10* | 17 | 23 | 130 5 | 144 11 | 17 |
| +21 | −3 | 31 0* | 45 6* | 59 13 | 19 | 88 1 | 102 7 | 116 13 | 19* | 145 1* | 159 7* |
| +20 | −4 | 15 | 21 | 60 3 | 74 9 | 15 | 21* | 117 3* | 131 9* | 15* | 21* |
| +19 | −5 | 32 5 | 46 11 | 17 | 23* | 89 5* | 103 11* | 17* | 132 0 | 146 6 | 160 12 |
| +18 | −6 | 19* | 47 1* | 61 7* | 75 13* | 20 | 104 2 | 118 8 | 14 | 20 | 161 2* |
| +17 | −7 | 33 9 | 15* | 22 | 76 4 | 90 10 | 16 | 22 | 133 4* | 147 10* | 16* |
| +16 | −8 | 34 0 | 48 6 | 62 12 | 18 | 91 0* | 105 6* | 119 12* | 18* | 148 0* | 162 7 |
| +15 | −9 | 14 | 20 | 63 2* | 77 8* | 14* | 20* | 120 2* | 134 9 | 15 | 21 |
| +14 | −10 | 35 4* | 49 10* | 16* | 22* | 92 5 | 106 11 | 17 | 23 | 149 5 | 163 11* |
| +13 | −11 | 18* | 50 0* | 64 7 | 78 13 | 19 | 107 1 | 121 7 | 135 13* | 19* | 164 1* |
| +12 | −12 | 36 9 | 15 | 21 | 79 3 | 93 9* | 15* | 21* | 136 3* | 150 9* | 16 |
| +11 | −13 | 23 | 51 5 | 65 11* | 17* | 23* | 108 5* | 122 11* | 18 | 151 0 | 165 6 |
| +10 | −14 | 37 13* | 19* | 66 1* | 80 7* | 94 14 | 20 | 123 2 | 137 8 | 14 | 20* |
| +9 | −15 | 38 3* | 52 9* | 16 | 22 | 95 4 | 109 10 | 16 | 22* | 152 4* | 166 10* |
| +8 | −16 | 18 | 53 0 | 67 6 | 81 12 | 18* | 110 0* | 124 6* | 138 12* | 18* | 167 1 |
| +7 | −17 | 39 8 | 14 | 20* | 82 2* | 96 8* | 14* | 20* | 139 3 | 153 9 | 15 |
| +6 | −18 | 22* | 54 4* | 68 10* | 16* | 23 | 111 5 | 125 11 | 17 | 23 | 168 5* |
| +5 | −19 | 40 12* | 19 | 69 1 | 83 7 | 97 13 | 19 | 126 1* | 140 7* | 154 13* | 19* |
| +4 | −20 | 41 3 | 55 9 | 15 | 21 | 98 3* | 112 9 | 15* | 21* | 155 3* | 169 10 |
| +3 | −21 | 17 | 23* | 70 5* | 84 11* | 17* | 23 | 127 6 | 141 12 | 18 | 170 0 |
| +2 | −22 | 42 7* | 56 13* | 19* | 85 1* | 99 8 | 113 14 | 20 | 142 2 | 156 8 | 14* |
| +1 | −23 | 21* | 57 4 | 71 10 | 16 | 22 | 114 4 | 128 10* | 16* | 22* | 171 4* |

| Difference | | Constituent 2N | | | | | | | | |
|---|---|---|---|---|---|---|---|---|---|---|
| Hour | | d. h. | d. h. | d. h. | d. h. | d. h. | d. h. | d. h. | d. h. | d. h. |
| 0 | | 171 19 | 186 1 | 200 7 | 214 13 | 228 19 | 243 1* | 257 7* | 271 13* | 285 19* | 300 1* |
| +23 | −1 | 172 9 | 15 | 21* | 215 3* | 229 9* | 15* | 21* | 272 4 | 286 10 | 16 |
| +22 | −2 | 23* | 187 5* | 201 11* | 17* | 23* | 244 6 | 258 12 | 18 | 287 0 | 301 6 |
| +21 | −3 | 173 13* | 19* | 202 2 | 216 8 | 230 14 | 20 | 259 2 | 273 8* | 14* | 20* |
| +20 | −4 | 174 4 | 188 10 | 16 | 22 | 231 4 | 245 10* | 16* | 22* | 288 4* | 302 10* |
| +19 | −5 | 18 | 189 0 | 203 6* | 217 12* | 18* | 246 0* | 260 6* | 274 13 | 19 | 303 1 |
| +18 | −6 | 175 8* | 14* | 20* | 218 2* | 232 9 | 15 | 21 | 275 3 | 289 9 | 15* |
| +17 | −7 | 22* | 190 4* | 204 11 | 17 | 23 | 247 5 | 261 11 | 17* | 23* | 304 5* |
| +16 | −8 | 176 13 | 19 | 205 1 | 219 7 | 233 13* | 19* | 262 1* | 276 7* | 290 13* | 20 |
| +15 | −9 | 177 3 | 191 9 | 15* | 21* | 234 3* | 248 9* | 15* | 22 | 291 4 | 305 10 |
| +14 | −10 | 17* | 23* | 206 5* | 220 11* | 18 | 249 0 | 263 6 | 277 12 | 18 | 306 0* |
| +13 | −11 | 178 7* | 192 13* | 20 | 221 2 | 235 8 | 14 | 20 | 278 2* | 292 8* | 14* |
| +12 | −12 | 22 | 193 4 | 207 10 | 16 | 22* | 250 4* | 264 16* | 16* | 22* | 307 5 |
| +11 | −13 | 179 12 | 18 | 208 0* | 222 6* | 236 12* | 18* | 265 0* | 279 7 | 293 13 | 19 |
| +10 | −14 | 180 2* | 194 8* | 14* | 20* | 237 3 | 251 9 | 15 | 21 | 294 3 | 308 9* |
| +9 | −15 | 16* | 22* | 209 5 | 223 11 | 17 | 23 | 266 5 | 280 11* | 17* | 23* |
| +8 | −16 | 181 7 | 195 13 | 19 | 224 1 | 238 7* | 252 13* | 19* | 281 1* | 295 7* | 309 14 |
| +7 | −17 | 21 | 196 3 | 210 9* | 15* | 21* | 253 3* | 267 10 | 16 | 22 | 310 4 |
| +6 | −18 | 182 11* | 17* | 23* | 225 5* | 239 12 | 18 | 268 0 | 282 6 | 296 12 | 18* |
| +5 | −19 | 183 1* | 197 8 | 211 14 | 20 | 240 2 | 254 8 | 14* | 20* | 297 2* | 311 8* |
| +4 | −20 | 16 | 22 | 212 4 | 226 10 | 16* | 269 4* | 283 10* | 16* | 23 |
| +3 | −21 | 184 6 | 198 12* | 18* | 227 0* | 241 6* | 255 12* | 19 | 284 1 | 298 7 | 312 13 |
| +2 | −22 | 20* | 199 2* | 213 8* | 14* | 21 | 256 3 | 270 9 | 15 | 21 | 313 3* |
| +1 | −23 | 185 10* | 17 | 23 | 228 5 | 242 11 | 17 | 23* | 285 5* | 299 11* | 17* |

HARMONIC ANALYSIS AND PREDICTION OF TIDES 277

Table 31.—*For construction of primary stencils*—Continued

| Difference | | Constituent 2N | | | | Constituent μ | | | |
|---|---|---|---|---|---|---|---|---|---|
| Hour | | d. h. | d. h. | d. h. | d. h. | d. h. | d. h. | d. h. | d. h. | d. h. |
| 0 | | 314 8 | 328 14 | 342 20 | 357 2 | 1 0 | 15 11* | 30 6 | 45 0* | 59 19 | 74 13 |
| +23 | −1 | 22 | 329 4 | 343 10* | 16* | 8 | 16 2* | 21 | 15 | 60 9* | 75 4 |
| +22 | −2 | 315 12* | 18* | 344 0* | 358 6* | 23 | 17* | 31 11* | 46 6 | 61 0* | 18* |
| +21 | −3 | 316 2* | 330 8* | 15 | 21 | 2 13* | 17 8 | 32 2* | 21 | 15 | 76 9* |
| +20 | −4 | 17 | 23 | 345 5 | 359 11 | 3 4* | 23 | 17 | 47 11* | 62 6 | 77 0* |
| +19 | −5 | 317 7 | 331 13 | 19* | 360 1* | 19 | 18 13* | 33 8 | 48 2* | 20* | 15 |
| +18 | −6 | 21* | 332 3* | 346 9* | 15* | 4 10 | 19 4* | 22* | 17 | 63 11* | 78 6 |
| +17 | −7 | 318 11* | 17* | 347 0 | 361 6 | 5 0* | 19 | 34 13* | 49 8 | 64 2 | 20* |
| +16 | −8 | 319 2 | 333 8 | 14 | 20 | 15* | 20 10 | 35 4 | 22* | 17 | 79 11* |
| +15 | −9 | 16 | 22 | 348 4* | 362 10* | 6 6* | 21 0* | 19 | 50 13* | 65 7* | 80 2* |
| +14 | −10 | 320 6* | 334 12* | 18* | 363 0* | 21 | 15* | 36 10 | 51 4 | 22* | 17 |
| +13 | −11 | 20* | 335 2* | 349 9 | 15 | 7 12 | 22 6 | 37 0* | 19 | 66 13* | 81 7* |
| +12 | −12 | 321 11 | 17 | 23 | 364 5 | 8 2* | 21 | 15* | 52 9* | 67 4 | 22* |
| +11 | −13 | 322 1 | 336 7 | 350 13* | 19* | 17* | 23 11* | 38 6 | 53 0* | 19 | 82 13 |
| +10 | −14 | 15* | 21* | 351 3* | 365 9* | 9 8 | 24 2* | 21 | 15 | 68 9* | 83 4 |
| +9 | −15 | 323 5* | 337 11* | 18 | 366 0 | 23 | 17 | 39 11* | 54 6 | 69 0* | 18* |
| +8 | −16 | 20 | 338 2 | 352 8 | 14 | 10 13* | 25 8 | 40 2* | 20* | 15 | 84 9 |
| +7 | −17 | 324 10 | 16* | 22* | 367 4* | 11 4* | 22* | 17 | 55 11* | 70 6 | 85 0 |
| +6 | −18 | 325 0* | 339 6* | 353 12* | 18* | 19 | 26 13* | 41 8 | 56 2 | 20* | 15 |
| +5 | −19 | 14* | 21 | 354 3 | 368 9 | 12 10 | 27 4* | 22* | 17 | 71 11* | 86 5* |
| +4 | −20 | 326 5 | 340 11 | 17 | 23 | 13 0* | 19 | 42 13* | 57 8 | 72 2 | 20* |
| +3 | −21 | 19 | 341 1* | 355 7* | 369 13* | 15* | 28 10 | 43 4 | 22* | 17 | 87 11* |
| +2 | −22 | 327 9* | 15* | 21* | 370 3* | 14 6 | 29 0* | 19 | 58 13* | 73 7* | 88 2 |
| +1 | −23 | 23* | 342 6 | 356 12 | ------ | 21 | 15* | 44 9* | 59 4 | 22* | 17 |

| Difference | | Constituent μ | | | | | | | | |
|---|---|---|---|---|---|---|---|---|---|---|
| Hour | | d. h. | d. h. | d. h. | d. h. | d. h. | d. h. | d. h. | d. h. | d. h. |
| 0 | | 89 7* | 104 2 | 118 20* | 133 14* | 148 9 | 163 3* | 177 22 | 192 16 | 207 10* | 222 5 |
| +23 | −1 | 22* | 16* | 119 11 | 134 5* | 149 0 | 18 | 178 12* | 193 7 | 208 1* | 19* |
| +22 | −2 | 90 13 | 105 7* | 120 2 | 20 | 14* | 164 9 | 179 3* | 21* | 16 | 223 10* |
| +21 | −3 | 91 4 | 22 | 16* | 135 11 | 150 5* | 23* | 18 | 194 12* | 209 7 | 224 1 |
| +20 | −4 | 18* | 106 13 | 121 7* | 136 1* | 20 | 165 14* | 180 9 | 195 3 | 21* | 16 |
| +19 | −5 | 92 9* | 107 4 | 22 | 16* | 151 11 | 166 5 | 23* | 18 | 210 12* | 225 6* |
| +18 | −6 | 93 0 | 18* | 122 13 | 137 7* | 152 1* | 20 | 181 14* | 196 8* | 211 3 | 21* |
| +17 | −7 | 15 | 108 9* | 123 3* | 22 | 16* | 167 11 | 182 5 | 23* | 18 | 226 12 |
| +16 | −8 | 94 5* | 109 0 | 18* | 138 13 | 153 7 | 168 1* | 20 | 197 14* | 212 8* | 227 3 |
| +15 | −9 | 20* | 15 | 124 9 | 139 3* | 22 | 16* | 183 10* | 198 5 | 23* | 18 |
| +14 | −10 | 95 11 | 110 5* | 125 0 | 18* | 154 12* | 169 7 | 184 1* | 20 | 213 14 | 228 8* |
| +13 | −11 | 96 2 | 20* | 14* | 140 9 | 155 3* | 22 | 16 | 199 10* | 214 5 | 23* |
| +12 | −12 | 17 | 111 11 | 126 5* | 141 0 | 18 | 170 12* | 185 7 | 200 1* | 19* | 229 14 |
| +11 | −13 | 97 7* | 112 2 | 20* | 14* | 156 9 | 171 3* | 21* | 16 | 215 10* | 230 5 |
| +10 | −14 | 22* | 16* | 127 11 | 142 5* | 157 0 | 18 | 186 12* | 201 7 | 216 1 | 19* |
| +9 | −15 | 98 13 | 113 7* | 128 2 | 20 | 14* | 172 9 | 187 3 | 21* | 16 | 231 10* |
| +8 | −16 | 99 4 | 22 | 16* | 143 11 | 158 5* | 23* | 18 | 202 12* | 217 7 | 232 1 |
| +7 | −17 | 18* | 114 13 | 129 7* | 144 1* | 20 | 173 14* | 188 9 | 203 3 | 21* | 16 |
| +6 | −18 | 100 9* | 115 3* | 18 | 22* | 159 11 | 174 5 | 23* | 18 | 218 12* | 233 6* |
| +5 | −19 | 101 0 | 18* | 130 13 | 145 7 | 160 1* | 20 | 189 14* | 204 8* | 219 3 | 21* |
| +4 | −20 | 15 | 116 9 | 131 3* | 22 | 16* | 175 10* | 190 5 | 23* | 18 | 234 12 |
| +3 | −21 | 102 5* | 117 0 | 18* | 146 12* | 161 7 | 176 1* | 20 | 205 14 | 220 8* | 235 3 |
| +2 | −22 | 20* | 15 | 132 9* | 147 3* | 22 | 16 | 191 10* | 206 5 | 23* | 17* |
| +1 | −23 | 103 11 | 118 5* | 133 0 | 18* | 162 12* | 177 7 | 192 1* | 19* | 221 14 | 236 8* |

278                    U. S. COAST AND GEODETIC SURVEY

Table 31.—*For construction of primary stencils*—Continued

| Difference | | Constituent μ | | | | | | | | | |
|---|---|---|---|---|---|---|---|---|---|---|---|
| Hour 0 | | d. h. | d. h. | d. h. | d. h. | d. h. | d. h. | d. h. | d. h. | d. h. | d. h. |
|  |  | 236 23 | 251 17* | 266 12 | 281 6* | 296 0* | 310 19 | 325 13* | 340 8 | 355 2 | 369 20* |
| +23 | −1 | 237 14 | 252 8* | 267 2* | 21 | 15* | 311 10 | 326 4 | 22* | 17 | 370 11* |
| +22 | −2 | 238 5 | 23 | 17* | 282 12 | 297 6 | 312 0* | 19 | 341 13* | 356 7* | |
| +21 | −3 | 19* | 253 14 | 268 8* | 283 2* | 21 | 15* | 327 9* | 342 4 | 22* | |
| +20 | −4 | 239 10* | 254 4* | 23 | 17* | 298 12 | 313 6 | 328 0* | 19 | 357 13 | |
| +19 | −5 | 240 1 | 19 | 269 14 | 284 8 | 299 2* | 21 | 15* | 343 9* | 358 4 | |
| +18 | −6 | 16 | 255 10 | 270 4* | 23 | 17* | 314 11* | 329 6 | 344 0* | 19 | |
| +17 | −7 | 241 6* | 256 1 | 19* | 285 13* | 300 8 | 315 2* | 21 | 15 | 359 9* | |
| +16 | −8 | 21* | 15* | 271 10 | 286 4* | 23 | 17 | 330 11* | 345 6 | 360 0* | |
| +15 | −9 | 242 12 | 257 6* | 272 1 | 19 | 301 13* | 316 8 | 331 2* | 20* | 15 | |
| +14 | −10 | 243 3 | 21* | 15* | 287 10 | 302 4* | 22* | 17 | 346 11* | 361 6 | |
| +13 | −11 | 17* | 258 12 | 273 6* | 288 1 | 19 | 317 13* | 332 8 | 347 2 | 20* | |
| +12 | −12 | 244 8* | 259 3 | 21 | 15* | 303 10 | 318 4* | 22 | 17 | 362 11* | |
| +11 | −13 | 23 | 17* | 274 12 | 289 6* | 304 0* | 19 | 333 13* | 348 8 | 363 2 | |
| +10 | −14 | 245 14 | 260 8* | 275 2* | 21 | 15* | 319 10 | 334 4 | 22* | 17 | |
| +9 | −15 | 246 4* | 23 | 17* | 290 12 | 305 6 | 320 0* | 19 | 349 13* | 364 7* | |
| +8 | −16 | 19* | 261 14 | 276 8 | 291 2* | 21 | 15* | 335 9* | 350 4 | 22* | |
| +7 | −17 | 247 10 | 262 4* | 23 | 17* | 306 11* | 321 6 | 336 0* | 19 | 365 13 | |
| +6 | −18 | 248 1 | 19* | 277 13* | 292 8 | 307 2* | 21 | 15 | 351 9* | 366 4 | |
| +5 | −19 | 16 | 263 10 | 278 4* | 23 | 17 | 322 11* | 337 6 | 352 0* | 18* | |
| +4 | −20 | 249 6* | 264 1 | 19* | 293 13* | 308 8 | 323 2* | 20* | 15 | 367 9* | |
| +3 | −21 | 21* | 15* | 279 10 | 294 4* | 23 | 17 | 338 11* | 353 6 | 368 0 | |
| +2 | −22 | 250 12 | 265 6* | 280 1 | 19 | 309 13* | 324 8 | 339 2* | 20* | 15 | |
| +1 | −23 | 251 3 | 21 | 15* | 295 10 | 310 4* | 22* | 17 | 354 11* | 369 6 | |

| Difference | | Constituent N | | | | | | | | | |
|---|---|---|---|---|---|---|---|---|---|---|---|
| Hour 0 | | d. h. | d. h. | d. h. | d. h. | d. h. | d. h. | d. h. | d. h. | d. h. | d. h. |
|  |  | 1 0 | 19 20* | 39 2 | 58 7* | 77 13 | 96 18* | 116 0 | 135 5* | 154 11 | 173 16* |
| +23 | −1 | 10* | 20 16 | 21* | 59 2* | 78 8 | 97 13* | 19 | 136 0* | 155 6 | 174 11* |
| +22 | −2 | 2 5* | 21 11 | 40 16* | 22 | 79 3* | 98 9 | 117 14* | 20 | 156 1* | 175 6* |
| +21 | −3 | 3 1 | 22 6* | 41 11* | 60 17 | 22* | 99 4 | 118 9* | 137 15 | 20* | 176 2 |
| +20 | −4 | 20 | 23 1* | 42 7 | 61 12* | 80 18 | 23* | 119 5 | 138 10* | 157 15* | 21 |
| +19 | −5 | 4 15* | 20* | 43 2 | 62 7* | 81 13 | 100 18* | 120 0 | 139 5* | 158 11 | 177 16* |
| +18 | −6 | 5 10* | 24 16 | 21* | 63 3 | 82 8* | 101 14 | 19* | 140 0* | 159 6 | 178 11* |
| +17 | −7 | 6 5* | 25 11 | 44 16* | 22 | 83 3* | 102 9 | 121 14* | 20 | 160 1* | 179 7 |
| +16 | −8 | 7 1 | 26 6* | 45 12 | 64 17* | 23 | 103 4 | 122 9* | 141 15 | 20* | 180 2 |
| +15 | −9 | 20 | 27 1* | 46 7 | 65 12* | 84 18 | 23* | 123 5 | 142 10* | 161 16 | 21* |
| +14 | −10 | 8 15* | 21 | 47 2* | 66 8 | 85 13 | 104 18* | 124 0 | 143 5* | 162 11 | 181 16 |
| +13 | −11 | 9 10* | 28 16 | 21* | 67 3 | 86 8* | 105 14 | 19* | 144 1 | 163 6* | 182 12 |
| +12 | −12 | 10 6 | 29 11* | 48 17 | 22 | 87 3* | 106 9 | 125 14* | 20 | 164 1* | 183 7 |
| +11 | −13 | 11 1 | 30 6* | 49 12 | 68 17* | 23 | 107 4* | 126 10 | 145 15* | 21 | 184 2 |
| +10 | −14 | 20* | 31 2 | 50 7 | 69 12* | 88 18 | 23* | 127 5 | 146 10* | 165 16 | 21* |
| +9 | −15 | 12 15* | 21 | 51 2* | 70 8 | 89 13* | 108 19 | 128 0* | 147 6 | 166 11 | 185 16* |
| +8 | −16 | 13 11 | 32 16 | 21* | 71 3 | 90 8* | 109 14 | 19* | 148 1 | 167 6* | 186 12 |
| +7 | −17 | 14 6 | 33 11* | 52 17 | 22* | 91 4 | 110 9* | 129 15 | 20 | 168 1* | 187 7 |
| +6 | −18 | 15 1 | 34 6* | 53 12 | 72 17 | 23 | 111 4* | 130 10 | 149 15* | 21 | 188 2* |
| +5 | −19 | 20* | 35 2 | 54 7* | 73 13 | 92 18* | 23* | 131 5 | 150 10* | 169 16 | 21* |
| +4 | −20 | 16 15* | 21 | 55 2* | 74 8 | 93 13* | 112 19 | 132 0* | 151 6* | 170 11* | 189 17 |
| +3 | −21 | 17 11 | 36 16* | 22 | 75 3* | 94 8* | 113 14 | 19* | 152 1 | 171 6* | 190 12 |
| +2 | −22 | 18 6 | 37 11* | 56 17 | 22* | 95 4 | 114 9* | 133 15 | 20* | 172 2 | 191 7* |
| +1 | −23 | 19 1* | 38 7 | 57 12* | 76 17* | 23 | 115 4* | 134 10 | 153 15* | 21 | 192 2* |

HARMONIC ANALYSIS AND PREDICTION OF TIDES 279

Table 31.—*For construction of primary stencils*—Continued

| Difference | | Constituent N | | | | | | | | | | | |
|---|---|---|---|---|---|---|---|---|---|---|---|---|---|
| Hour | | d. h. | d. h. | d. h. | d. h. | d. h. | d. h. | d. h. | d. h. | d. h. | d. h. | d. h. | d. h. |
| 0 | | 192 21* | 212 3 | 231 8* | 250 14 | 269 19* | 289 1 | 308 6* | 327 12 | 346 17* | 365 23 |
| +23 | −1 | 193 17 | | 22* | 232 4 | 251 9* | 270 15 | | 20 | 309 1* | 328 7 | 347 12* | 366 18 |
| +22 | −2 | 194 12 | 213 17* | | 23 | 252 4* | 271 10 | 290 15* | | 21 | 329 2* | 348 8 | 367 13* |
| +21 | −3 | 195 7* | 214 13 | 233 18* | 253 0 | 272 5 | 291 10* | 310 16 | | 21* | 349 3 | 368 8* |
| +20 | −4 | 196 2* | 215 8 | 234 13* | | 19 | 273 0* | 292 6 | 311 11* | 330 17 | | 22* | 369 4 |
| +19 | −5 | | 22 | 216 3* | 235 9 | 254 14 | | 19* | 293 1 | 312 6* | 331 12 | 350 17* | | 23 |
| +18 | −6 | 197 17 | | 22* | 236 4 | 255 9* | 274 15 | | 20* | 313 2 | 332 7* | 351 13 | 370 18 |
| +17 | −7 | 198 12* | 217 18 | | 23 | 256 4* | 275 10 | 294 15* | | 21 | 333 2* | 352 8 | ........ |
| +16 | −8 | 199 7* | 218 13 | 237 18* | 257 0 | 276 5* | 295 11 | 314 16* | | 21* | 353 3 | ........ |
| +15 | −9 | 200 3 | 219 8 | 238 13* | | 19 | 277 0* | 296 6 | 315 11* | 334 17 | | 22* | ........ |
| +14 | −10 | | 22 | 220 3* | 239 9 | 258 14* | | 20 | 297 1* | 316 6* | 335 12 | 354 17* | ........ |
| +13 | −11 | 201 17 | | 22* | 240 4 | 259 9* | 278 15 | | 20* | 317 2 | 336 7* | 355 13 | ........ |
| +12 | −12 | 202 12* | 221 18 | | 23* | 260 5 | 279 10* | 298 15* | | 21 | 337 2* | 356 8 | ........ |
| +11 | −13 | 203 7* | 222 13 | 241 18* | 261 0 | 280 5* | 299 11 | 318 16* | | 22 | 357 3* | ........ |
| +10 | −14 | 204 3 | 223 8* | 242 14 | | 19* | 281 0* | 300 6 | 319 11* | 338 17 | | 22* | ........ |
| +9 | −15 | | 22 | 224 3* | 243 9 | 262 14* | | 20 | 301 1* | 320 7 | 339 12* | 358 18 | ........ |
| +8 | −16 | 205 17* | | 23 | 244 4* | 263 9* | 282 15 | | 20* | 321 2 | 340 7* | 359 13 | ........ |
| +7 | −17 | 206 12* | 225 18 | | 23* | 264 5 | 283 10* | 302 16 | | 21* | 341 3 | 360 8* | ........ |
| +6 | −18 | 207 8 | 226 13* | 245 18* | 265 0 | 284 5* | 303 11 | 322 16* | | 22 | 361 3* | ........ |
| +5 | −19 | 208 3 | 227 8* | 246 14 | | 19* | 285 1 | 304 6* | 323 12 | 342 17* | | 22* | ........ |
| +4 | −20 | | 22* | 228 3* | 247 9 | 266 14* | | 20 | 305 1* | 324 7 | 343 12* | 362 18 | ........ |
| +3 | −21 | 209 17* | | 23 | 248 4* | 267 10 | 286 15* | | 21 | 325 2* | 344 7* | 363 13 | ........ |
| +2 | −22 | 210 12* | 229 18 | | 23* | 268 5 | 287 10* | 306 16 | | 21* | 345 3 | 364 8* | ........ |
| +1 | −23 | 211 8 | 230 13* | 249 19 | 269 0* | 288 6 | 307 11 | 326 16* | | 22 | 365 3* | ........ |

| Difference | | Constituent ν | | | | | | | | | | | |
|---|---|---|---|---|---|---|---|---|---|---|---|---|---|
| Hour | | d. h. | d. h. | d. h. | d. h. | d. h. | d. h. | d. h. | d. h. | d. h. | d. h. | d. h. |
| 0 | | 1 0 | 20 18* | 40 23 | 61 3 | 81 7 | 101 11 | 121 15 | 141 19 | 161 23 | 182 3 |
| +23 | −1 | | 11 | 21 15 | 41 19 | | 23 | 82 3 | 102 7 | 122 11 | 142 15* | 162 19* | | 23* |
| +22 | −2 | 2 7 | 22 11 | 42 15 | 62 19 | | 23 | 103 3* | 123 7* | 143 11* | 163 15* | 183 19* |
| +21 | −3 | 3 3 | 23 7 | 43 11* | 63 15* | 83 19* | | 23* | 124 3* | 144 7* | 164 11* | 184 15* |
| +20 | −4 | | 23* | 24 3* | 44 7* | 64 11* | 84 15* | 104 19* | | 23* | 145 4 | 165 8 | 185 12 |
| +19 | −5 | 4 19* | | 23* | 45 3* | 65 7* | 85 12 | 105 16 | 125 20 | 146 0 | 166 4 | 186 8 |
| +18 | −6 | 5 15* | 25 19* | 46 0 | 66 4 | 86 8 | 106 12 | 126 16 | | 20 | 167 0 | 187 4 |
| +17 | −7 | 6 12 | 26 16 | | 20 | 67 0 | 87 4 | 107 8 | 127 12 | 147 16* | | 20* | 188 0* |
| +16 | −8 | 7 8 | 27 12 | 47 16 | | 20 | 88 0* | 108 4* | 128 8* | 148 12* | 168 16* | | 20* |
| +15 | −9 | 8 4 | 28 8 | 48 12* | 68 16* | | 20* | 109 0* | 129 4* | 149 8* | 169 12* | 189 16* |
| +14 | −10 | 9 0* | 29 4* | 49 8* | 69 12* | 89 16* | | 20* | 130 0* | 150 5 | 170 9 | 190 13 |
| +13 | −11 | | 20* | 30 0* | 50 4* | 70 8* | 90 13 | 110 17 | | 21 | 151 1 | 171 5 | 191 9 |
| +12 | −12 | 10 16* | | 21 | 51 1 | 71 5 | 91 9 | 111 13 | 131 17 | | 21 | 172 1 | 192 5 |
| +11 | −13 | 11 13 | 31 17 | | 21 | 72 1 | 92 5 | 112 9 | 132 13 | 152 17* | | 21* | 193 1* |
| +10 | −14 | 12 9 | 32 13 | 52 17 | | 21 | 93 1* | 113 5* | 133 9* | 153 13* | 173 17* | | 21* |
| +9 | −15 | 13 5 | 33 9* | 53 13* | 73 17* | | 21* | 114 1* | 134 5* | 154 9* | 174 13* | 194 17* |
| +8 | −16 | 14 1* | 34 5* | 54 9* | 74 13* | 94 17* | | 21* | 135 1* | 155 6 | 175 10 | 195 14 |
| +7 | −17 | | 21* | 35 1* | 55 5* | 75 9* | 95 14 | 115 18 | | 22 | 156 2 | 176 6 | 196 10 |
| +6 | −18 | 15 17* | | 22 | 56 2 | 76 6 | 96 10 | 116 14 | 136 18 | | 22 | 177 2 | 197 6* |
| +5 | −19 | 16 14 | 36 18 | | 22 | 77 2 | 97 6 | 117 10 | 137 14 | 157 18* | | 22* | 198 2* |
| +4 | −20 | 17 10 | 37 14 | 57 18 | | 22 | 98 2* | 118 6* | 138 10* | 158 14* | 178 18* | | 22* |
| +3 | −21 | 18 6 | 38 10* | 58 14* | 78 18* | | 22* | 119 2* | 139 6* | 159 10* | 179 14* | 199 19 |
| +2 | −22 | 19 2* | 39 6* | 59 10* | 79 14* | 99 18* | | 22* | 140 3 | 160 7 | 180 11 | 200 15 |
| +1 | −23 | | 22* | 40 2* | 60 6* | 80 10* | 100 15 | 120 19 | | 23 | 161 3 | 181 7 | 201 11 |

Table 31.—*For construction of primary stencils*—Continued

| Difference | | Constituent ν | | | | | | | | |
|---|---|---|---|---|---|---|---|---|---|---|
| Hour | | d. h. | d. h. | d. h. | d. h. | d. h. | d. h. | d. h. | d. h. | d. h. |
| 0 | | 202 7* | 222 11* | 242 15* | 262 19* | 282 23* | 303 3 | 323 7* | 343 11* | 363 16 |
| +23 | −1 | 203 3* | 223 7* | 243 11* | 263 15* | 283 19* | 23 | 324 4 | 344 8 | 364 12 |
| +22 | −2 | 23* | 224 3* | 244 7* | 264 12 | 284 16 | 304 20 | 325 0 | 345 4 | 365 8 |
| +21 | −3 | 204 20 | 225 0 | 245 4 | 265 8 | 285 12 | 305 16 | 20 | 346 0 | 366 4* |
| +20 | −4 | 205 16 | 20 | 246 0 | 266 4 | 286 8 | 306 12* | 326 16* | 20* | 367 0* |
| +19 | −5 | 206 12 | 226 16 | 20 | 267 0* | 287 4* | 307 8* | 327 12* | 347 16* | 20* |
| +18 | −6 | 207 8* | 227 12* | 247 16* | 20* | 288 0* | 308 4* | 328 8* | 348 12* | 368 17 |
| +17 | −7 | 208 4* | 228 8* | 248 12* | 268 16* | 20* | 309 1 | 329 5 | 349 9 | 369 13 |
| +16 | −8 | 209 0* | 229 4* | 249 8* | 269 13 | 289 17 | 21 | 330 1 | 350 5 | 370 9 |
| +15 | −9 | 21 | 230 1 | 250 5 | 270 9 | 290 13 | 310 17 | 21 | 351 1 | |
| +14 | −10 | 210 17 | 21 | 251 1 | 271 5 | 291 9 | 311 13* | 331 17* | 21* | |
| +13 | −11 | 211 13 | 231 17 | 21* | 272 1* | 292 5* | 312 9* | 332 13* | 352 17* | |
| +12 | −12 | 212 9* | 232 13* | 252 17* | 21* | 293 1* | 313 5* | 333 9* | 353 13* | |
| +11 | −13 | 213 5* | 233 9* | 253 13* | 273 17* | 21* | 314 2 | 334 6 | 354 10 | |
| +10 | −14 | 214 1* | 234 5* | 254 10 | 274 14 | 294 18 | 22 | 335 2 | 355 6 | |
| +9 | −15 | 22 | 235 2 | 255 6 | 275 10 | 295 14 | 315 18 | 22 | 356 2 | |
| +8 | −16 | 215 18 | 22 | 256 2 | 276 6 | 296 10 | 316 14* | 336 18* | 22* | |
| +7 | −17 | 216 14 | 236 18 | 22* | 277 2* | 297 6* | 317 10* | 337 14* | 357 18* | |
| +6 | −18 | 217 10* | 237 14* | 257 18* | 22* | 298 2* | 318 6* | 338 10* | 358 14* | |
| +5 | −19 | 218 6* | 238 10* | 258 14* | 278 18* | 22* | 319 3 | 339 7 | 359 11 | |
| +4 | −20 | 219 2* | 239 6* | 259 11 | 279 15 | 299 19 | 23 | 340 3 | 360 7 | |
| +3 | −21 | 23 | 240 3 | 260 7 | 280 11 | 300 15 | 320 19 | 23 | 361 3 | |
| +2 | −22 | 220 19 | 23 | 261 3 | 281 7 | 301 11 | 321 15* | 341 19* | 23* | |
| +1 | −23 | 221 15 | 241 19 | 23* | 282 3 | 302 7* | 322 11* | 342 15* | 362 19* | |

| Difference | | Constituent 2 MK | | | | | | | | |
|---|---|---|---|---|---|---|---|---|---|---|
| Hour | | d. h. | d. h. | d. h. | d. h. | d. h. | d. h. | d. h. | d. h. | d. h. |
| 0 | | 1 0 | 22 7 | 44 0 | 65 17 | 87 10 | 109 3 | 130 20 | 152 13 | 174 6 | 195 23 |
| +23 | −1 | 11* | 23 4* | 21* | 66 14* | 88 7* | 110 0* | 131 17* | 153 10* | 175 4 | 196 21 |
| +22 | −2 | 2 9* | 24 2* | 45 19* | 67 12* | 89 5* | 22* | 132 15* | 154 8* | 176 1* | 197 18* |
| +21 | −3 | 3 7 | 25 0 | 46 17 | 68 10 | 90 3 | 111 20 | 133 13 | 155 6 | 23 | 198 16 |
| +20 | −4 | 4 4* | 22 | 47 15 | 69 8 | 91 1 | 112 18 | 134 11 | 156 4 | 177 21 | 199 14 |
| +19 | −5 | 5 2* | 26 19* | 48 12* | 70 5* | 22* | 113 15* | 135 8* | 157 1* | 178 18* | 200 11* |
| +18 | −6 | 6 0 | 27 17 | 49 10 | 71 3 | 92 20 | 114 13 | 136 6* | 23* | 179 16* | 201 9* |
| +17 | −7 | 22 | 28 15 | 50 8 | 72 1 | 93 18 | 115 11 | 137 4 | 158 21 | 180 14 | 202 7 |
| +16 | −8 | 7 19* | 29 12* | 51 5* | 22* | 94 15* | 116 8* | 138 1* | 159 18* | 181 11* | 203 4* |
| +15 | −9 | 8 17* | 30 10* | 52 3* | 73 20* | 95 13* | 117 6* | 23* | 160 16* | 182 9* | 204 2* |
| +14 | −10 | 9 15 | 31 8 | 53 1 | 74 18 | 96 11 | 118 4 | 139 21 | 161 14 | 183 7 | 205 0 |
| +13 | −11 | 10 12* | 32 5* | 22* | 75 16 | 97 9 | 119 2 | 140 19 | 162 12 | 184 5 | 22 |
| +12 | −12 | 11 10* | 33 3* | 54 20* | 76 13* | 98 6* | 23* | 141 16* | 163 9* | 185 2* | 206 19* |
| +11 | −13 | 12 8 | 34 1 | 55 18 | 77 11 | 99 4 | 120 21 | 142 14 | 164 7 | 186 0* | 207 17* |
| +10 | −14 | 13 6 | 23 | 56 16 | 78 9 | 100 2 | 121 19 | 143 12 | 165 5 | 22 | 208 15 |
| +9 | −15 | 14 3* | 35 20* | 57 13* | 79 6* | 23* | 122 16* | 144 9* | 166 2* | 187 19* | 209 12* |
| +8 | −16 | 15 1 | 36 18* | 58 11* | 80 4* | 101 21* | 123 14* | 145 7* | 167 0* | 188 17* | 210 10* |
| +7 | −17 | 23 | 37 16 | 59 9 | 81 2 | 102 19 | 124 12 | 146 5 | 22 | 189 15 | 211 8 |
| +6 | −18 | 16 20* | 38 13* | 60 6* | 23* | 103 16* | 125 10 | 147 3 | 168 20 | 190 13 | 212 6 |
| +5 | −19 | 17 18* | 39 11* | 61 4* | 82 21* | 104 14* | 126 7* | 148 0* | 169 17* | 191 10* | 213 3* |
| +4 | −20 | 18 16 | 40 9 | 62 2 | 83 19 | 105 12 | 127 5 | 22 | 170 15 | 192 8 | 214 1 |
| +3 | −21 | 19 14 | 41 7 | 63 0 | 84 17 | 106 10 | 128 3 | 149 20 | 171 13 | 193 6 | 23 |
| +2 | −22 | 20 11* | 42 4* | 21* | 85 14* | 107 7* | 129 0* | 150 17* | 172 10* | 194 3* | 215 20* |
| +1 | −23 | 21 9 | 43 2 | 64 19 | 86 12* | 108 5* | 22* | 151 15* | 173 8* | 195 1* | 216 18* |

## HARMONIC ANALYSIS AND PREDICTION OF TIDES

Table 31.—*For construction of primary stencils*—Continued

| Difference | | Constituent 2MK | | | | | | | | Constituent MN | |
|---|---|---|---|---|---|---|---|---|---|---|---|
| Hour | | d. h. | d. h. | d. h. | d. h. | d. h. | d. h. | d. h. | d. h. | d. h. | d. h. |
| 0 | | 217 16 | 239 9 | 261 2 | 282 19 | 304 12 | 326 5 | 347 22 | 369 15 | 1 0 | 23 20 |
| +23 | −1 | 218 14 | 240 7 | 262 0 | 283 17 | 305 10 | 327 3 | 348 20 | 370 13 | 12* | 24 19* |
| +22 | −2 | 219 11* | 241 4* | 21* | 284 14* | 306 7* | 328 0* | 349 17* | | 2 11 | 25 18* |
| +21 | −3 | 220 9 | 242 2 | 263 19* | 285 12* | 307 5* | 22* | 350 15* | | 3 11 | 26 18 |
| +20 | −4 | 221 7 | 243 0 | 264 17 | 286 10 | 308 3 | 329 20 | 351 13 | | 4 10* | 27 17 |
| +19 | −5 | 222 4* | 21* | 265 14* | 287 7* | 309 0* | 330 17* | 352 10* | | 5 9* | 28 16* |
| +18 | −6 | 223 2* | 244 19* | 266 12* | 288 5* | 22* | 331 15* | 353 8* | | 6 9 | 29 16 |
| +17 | −7 | 224 0 | 245 17 | 267 10 | 289 3 | 310 20 | 332 13 | 354 6 | | 7 8 | 30 15 |
| +16 | −8 | 22 | 246 15 | 268 8 | 290 1 | 311 18 | 333 11 | 355 4 | | 8 7* | 31 14* |
| +15 | −9 | 225 19* | 247 12* | 269 5* | 22* | 312 15* | 334 8* | 356 1* | | 9 6* | 32 13* |
| +14 | −10 | 226 17 | 248 10 | 270 3 | 291 20 | 313 13* | 335 6* | 23* | | 10 6 | 33 13 |
| +13 | −11 | 227 15 | 249 8 | 271 1 | 292 18 | 314 11 | 336 4 | 357 21 | | 11 5* | 34 12* |
| +12 | −12 | 228 12* | 250 5* | 22* | 293 15* | 315 8* | 337 1* | 358 18* | | 12 4* | 35 11* |
| +11 | −13 | 229 10* | 251 3* | 272 20* | 294 13* | 316 6* | 23* | 359 16* | | 13 4 | 36 11 |
| +10 | −14 | 230 8 | 252 1 | 273 18 | 295 11 | 317 4 | 338 21 | 360 14 | | 14 3 | 37 10 |
| +9 | −15 | 231 5* | 22* | 274 16 | 296 9 | 318 2 | 339 19 | 361 12 | | 15 2* | 38 9* |
| +8 | −16 | 232 3* | 253 20* | 275 13* | 297 6* | 23* | 340 16* | 362 9* | | 16 2 | 39 8* |
| +7 | −17 | 233 1 | 254 18 | 276 11 | 298 4 | 319 21 | 341 14 | 363 7* | | 17 1 | 40 8 |
| +6 | −18 | 23 | 255 16 | 277 9 | 299 2 | 320 19 | 342 12 | 364 5 | | 18 0* | 41 7* |
| +5 | −19 | 234 20* | 256 13* | 278 6* | 23* | 321 16* | 343 9* | 365 2* | | 23* | 42 6* |
| +4 | −20 | 235 18* | 257 11* | 279 4* | 300 21* | 322 14* | 344 7* | 366 0* | | 19 23 | 43 6 |
| +3 | −21 | 236 16 | 258 9 | 280 2 | 301 19 | 323 12 | 345 5 | 22 | | 20 22 | 44 5 |
| +2 | −22 | 237 13* | 259 6* | 23* | 302 16* | 324 10 | 346 3 | 367 20 | | 21 21* | 45 4* |
| +1 | −23 | 238 11* | 260 4* | 281 21* | 303 14* | 325 7* | 347 0* | 368 17* | | 22 21 | 46 4 |

| Difference | | Constituent MN | | | | | | | | | |
|---|---|---|---|---|---|---|---|---|---|---|---|
| Hour | | d. h. | d. h. | d. h. | d. h. | d. h. | d. h. | d. h. | d. h. | d. h. | d. h. |
| 0 | | 47 3 | 70 10 | 93 17 | 117 0 | 140 7 | 163 14 | 186 21 | 210 4 | 233 11 | 256 18 |
| +23 | −1 | 48 2* | 71 9* | 94 16* | 23 | 141 6 | 164 13 | 187 20 | 211 3 | 234 10 | 257 17 |
| +22 | −2 | 49 1* | 72 8* | 95 15* | 118 22* | 142 5* | 165 12* | 188 19* | 212 2* | 235 9* | 258 16* |
| +21 | −3 | 50 1 | 73 8 | 96 15 | 119 22 | 143 5 | 166 12 | 189 18* | 213 1* | 236 8* | 259 15* |
| +20 | −4 | 51 0 | 74 7 | 97 14 | 120 21 | 144 4 | 167 11 | 190 18 | 214 1 | 237 8 | 260 15 |
| +19 | −5 | 23* | 75 6* | 98 13* | 121 20* | 145 3* | 168 10* | 191 17* | 215 0* | 238 7* | 261 14 |
| +18 | −6 | 52 23 | 76 6 | 99 12* | 122 19* | 146 2* | 169 9* | 192 16* | 23* | 239 6* | 262 13* |
| +17 | −7 | 53 22 | 77 5 | 100 12 | 123 19 | 147 2 | 170 9 | 193 16 | 216 23 | 240 6 | 263 13 |
| +16 | −8 | 54 21* | 78 4* | 101 11* | 124 18* | 148 1 | 171 8 | 194 15 | 217 22 | 241 5 | 264 12 |
| +15 | −9 | 55 20* | 79 3* | 102 10* | 125 17* | 149 0* | 172 7* | 195 14* | 218 21* | 242 4* | 265 11* |
| +14 | −10 | 56 20 | 80 3 | 103 10 | 126 17 | 150 0 | 173 7 | 196 14 | 219 20* | 243 3* | 266 10* |
| +13 | −11 | 57 19 | 81 2 | 104 9 | 127 16 | 23 | 174 6 | 197 13 | 220 20 | 244 3 | 267 10 |
| +12 | −12 | 58 18* | 82 1* | 105 8* | 128 15* | 151 22* | 175 5* | 198 12* | 221 19* | 245 2* | 268 9* |
| +11 | −13 | 59 18 | 83 1 | 106 8 | 129 14* | 152 21* | 176 4* | 199 11* | 222 18* | 246 1* | 269 8* |
| +10 | −14 | 60 17 | 84 0 | 107 7 | 130 14 | 153 21 | 177 4 | 200 11 | 223 18 | 247 1 | 270 8 |
| +9 | −15 | 61 16* | 23* | 108 6* | 131 13* | 154 20* | 178 3 | 201 10 | 224 17 | 248 0 | 271 7 |
| +8 | −16 | 62 15* | 85 22* | 109 5* | 132 12* | 155 19* | 179 2* | 202 9* | 225 16* | 23* | 272 6* |
| +7 | −17 | 63 15 | 86 22 | 110 5 | 133 12 | 156 19 | 180 2 | 203 9 | 226 16 | 249 22* | 273 5* |
| +6 | −18 | 64 14* | 87 21 | 111 4 | 134 11 | 157 18* | 181 1 | 204 8 | 227 15 | 250 22 | 274 5 |
| +5 | −19 | 65 13* | 88 20* | 112 3* | 135 10* | 158 17* | 182 0* | 205 7* | 228 14* | 251 21* | 275 4* |
| +4 | −20 | 66 13 | 89 20 | 113 3 | 136 10 | 159 16* | 23* | 206 6* | 229 13* | 252 20* | 276 3* |
| +3 | −21 | 67 12 | 90 19 | 114 2 | 137 9 | 160 16 | 183 23 | 207 6 | 230 13 | 253 20 | 277 3 |
| +2 | −22 | 68 11* | 91 18* | 115 1* | 138 8* | 161 15* | 184 22* | 208 5* | 231 12 | 254 19 | 278 2 |
| +1 | −23 | 69 10* | 92 17 | 116 0* | 139 7* | 162 14* | 185 21* | 209 4* | 232 11* | 255 18* | 279 1* |

## Table 31.—*For construction of primary stencils*—Continued

| Difference | | Constituent MN | | | | Constituent M | | | | | |
|---|---|---|---|---|---|---|---|---|---|---|---|
| Hour 0 | | d. h. | d. h. | d. h. | d. h. | d. h. | d. h. | d. h. | d. h. | d. h. | d. h. |
| +23 | −1 | 280 0* | 303 7* | 326 14* | 349 21* | 1 0 | 29 22* | 59 11* | 89 0 | 118 13 | 148 1* |
| +22 | −2 | 281 0 | 304 7 | 327 14 | 350 21 | 15* | 31 4 | 60 17 | 90 5* | 119 18* | 149 7 |
| +21 | −3 | 23* | 305 6* | 328 13* | 351 20 | 2 21 | 32 10 | 61 22* | 91 11 | 121 0 | 150 12* |
| +20 | −4 | 282 22* | 306 5* | 329 12* | 352 19* | 4 2* | 33 15* | 63 4 | 92 17 | 122 5* | 151 18 |
| +19 | −5 | 283 22 | 307 5 | 330 12 | 353 19 | 5 8 | 34 21 | 64 9* | 93 22* | 123 11 | 153 0 |
|  |  | 284 21 | 308 4 | 331 11 | 354 18 | 6 13* | 36 2* | 65 15 | 95 4 | 124 16* | 154 5* |
| +18 | −6 | 285 20* | 309 3* | 332 10* | 355 17* | 7 19 | 37 8 | 66 20* | 96 9* | 125 22 | 155 11 |
| +17 | −7 | 286 20 | 310 2* | 333 9* | 356 16* | 9 0* | 38 13* | 68 2 | 97 15 | 127 3* | 156 16* |
| +16 | −8 | 287 19 | 311 2 | 334 9 | 357 16 | 10 6 | 39 19 | 69 7* | 98 20* | 128 9 | 157 22 |
| +15 | −9 | 288 18* | 312 1* | 335 8* | 358 15* | 11 12 | 41 0* | 70 13 | 100 2 | 129 14* | 159 3* |
| +14 | −10 | 289 17* | 313 0* | 336 7* | 359 14* | 12 17* | 42 6 | 71 19 | 101 7* | 130 20 | 160 9 |
| +13 | −11 | 290 17 | 314 0 | 337 7 | 360 14 | 13 23 | 43 11* | 73 0* | 102 13 | 132 2 | 161 14* |
| +12 | −12 | 291 16 | 23 | 338 6 | 361 13 | 15 4* | 44 17 | 74 6 | 103 18* | 133 7* | 162 20 |
| +11 | −13 | 292 15* | 315 22* | 339 5* | 362 12* | 16 10 | 45 22* | 75 11* | 105 0 | 134 13 | 164 1* |
| +10 | −14 | 293 15 | 316 22 | 340 5 | 363 11* | 17 15* | 47 4 | 76 17 | 106 5* | 135 18* | 165 7 |
| +9 | −15 | 294 14 | 317 21 | 341 4 | 364 11 | 18 21 | 48 9* | 77 22* | 107 11 | 137 0 | 166 12* |
| +8 | −16 | 295 13* | 318 20* | 342 3* | 365 10* | 20 2* | 49 15 | 79 4 | 108 16* | 138 5* | 167 18 |
| +7 | −17 | 296 12* | 319 19* | 343 2* | 366 9* | 21 8 | 50 20* | 80 9* | 109 22 | 139 11 | 168 23* |
| +6 | −18 | 297 12 | 320 19 | 344 2 | 367 9 | 22 13* | 52 2* | 81 15 | 111 3* | 140 16* | 170 5 |
| +5 | −19 | 298 11* | 321 18 | 345 1 | 368 8 | 23 19 | 53 8 | 82 20* | 112 9* | 141 22 | 171 10* |
| +4 | −20 | 299 10* | 322 17* | 346 0* | 369 7* | 25 0* | 54 13* | 84 2 | 113 15 | 143 3* | 172 16* |
| +3 | −21 | 300 10 | 323 17 | 347 0 | 370 7 | 26 6 | 55 19 | 85 -7* | 114 20* | 144 9 | 173 22 |
| +2 | −22 | 301 9 | 324 16 | 23 |  | 27 11* | 57 0* | 86 13 | 116 2 | 145 14* | 175 3* |
| +1 | −23 | 302 8* | 325 15* | 348 22* |  | 28 17 | 58 6 | 87 18* | 117 7* | 146 20 | 176 9 |

| Difference | | Constituent M | | | | | | | Constituent MK | | |
|---|---|---|---|---|---|---|---|---|---|---|---|
| Hour 0 | | d. h. | d. h. | d. h. | d. h. | d. h. | d. h. | d. h. | d. h. | d. h. | d. h. |
| +23 | −1 | 177 14* | 207 3 | 236 16 | 266 4* | 295 17* | 325 6 | 354 19 | 1 0 | 46 5* | 92 9* |
| +22 | −2 | 178 20 | 208 8* | 237 21* | 267 10 | 296 23 | 326 11* | 356 0* | 2 0 | 48 3* | 94 7* |
| +21 | −3 | 180 1* | 209 14 | 239 3 | 268 15* | 298 4* | 327 17 | 357 6 | 3 22 | 50 2 | 96 6 |
| +20 | −4 | 181 7 | 210 19* | 240 8* | 269 21 | 299 10 | 328 22* | 358 11* | 5 20 | 52 0 | 98 4 |
| +19 | −5 | 182 12* | 212 1 | 241 14 | 271 2* | 300 15* | 330 4 | 359 17 | 7 18* | 53 22 | 100 2 |
|  |  | 183 18 | 213 7 | 242 19* | 272 8 | 301 21 | 331 9* | 360 22* | 9 16* | 55 20* | 102 0* |
| +18 | −6 | 184 23* | 214 12* | 244 1 | 273 14 | 303 2* | 332 15 | 362 4 | 11 14* | 57 18* | 103 22* |
| +17 | −7 | 186 5 | 215 18 | 245 6* | 274 19* | 304 8 | 333 21 | 363 9* | 13 13 | 59 16* | 105 20* |
| +16 | −8 | 187 10* | 216 23* | 246 12 | 276 1 | 305 13* | 335 2* | 364 15 | 15 11 | 61 15 | 107 18* |
| +15 | −9 | 188 16 | 218 5 | 247 17* | 277 6* | 306 19 | 336 8 | 365 20* | 17 9 | 63 13 | 109 17 |
| +14 | −10 | 189 21* | 219 10* | 248 23 | 278 12 | 308 0* | 337 13* | 367 2 | 19 7* | 65 11 | 111 15 |
| +13 | −11 | 191 3 | 220 16 | 250 4* | 279 17* | 309 6 | 338 19 | 368 7* | 21 5* | 67 9* | 113 13 |
| +12 | −12 | 192 9 | 221 21* | 251 10 | 280 23 | 310 11* | 340 0* | 369 13 | 23 3* | 69 7* | 115 11* |
| +11 | −13 | 193 14* | 223 3 | 252 16 | 282 4* | 311 17 | 341 6 | 370 18* | 25 2 | 71 5* | 117 9* |
| +10 | −14 | 194 20 | 224 8* | 253 21* | 283 10 | 312 23 | 342 11* |  | 27 0 | 73 4 | 119 7* |
| +9 | −15 | 196 1* | 225 14 | 255 3 | 284 15* | 314 4* | 343 17 |  | 28 22 | 75 2 | 121 6 |
| +8 | −16 | 197 7 | 226 19* | 256 8* | 285 21 | 315 10 | 344 22* |  | 30 20* | 77 0 | 123 4 |
| +7 | −17 | 198 12* | 228 1 | 257 14 | 287 2* | 316 15* | 346 4 |  | 32 18* | 78 22* | 125 2 |
| +6 | −18 | 199 18 | 229 6* | 258 19* | 288 8 | 317 21 | 347 9* |  | 34 16* | 80 20* | 127 0* |
| +5 | −19 | 200 23* | 230 12 | 260 1 | 289 13* | 319 2* | 348 15 |  | 36 14* | 82 18* | 128 22* |
| +4 | −20 | 202 5 | 231 17* | 261 6* | 290 19 | 320 8 | 349 20* |  | 38 13 | 84 17 | 130 20* |
| +3 | −21 | 203 10* | 232 23* | 262 12 | 292 0* | 321 13* | 351 2 |  | 40 11 | 86 15 | 132 19 |
| +2 | −22 | 204 16 | 234 5 | 263 17* | 293 6* | 322 19 | 352 7* |  | 42 9 | 88 13 | 134 17 |
| +1 | −23 | 205 21* | 235 10* | 264 23 | 294 12 | 324 0* | 353 13* |  | 44 7* | 90 11* | 136 15 |

HARMONIC ANALYSIS AND PREDICTION OF TIDES 283

Table 31.—*For construction of primary stencils*—Continued

| Difference | | Constituent MK | | | | | | Constituent λ | | | | | |
|---|---|---|---|---|---|---|---|---|---|---|---|---|---|
| Hour | | d. h. | d. h. | d. h. | d. h. | d. h. | d. h. | d. h. | d. h. | d. h. | d. h. |
| 0 | | 138 13* | 184 17 | 230 21 | 277 1 | 323 5 | 369 9 | 1 0 | 55 0 | 110 2* | 165 5 |
| +23 | −1 | 140 11* | 186 15* | 232 19* | 278 23 | 325 3 | 371 7 | 2 4* | 57 7 | 112 9* | 167 12 |
| +22 | −2 | 142 9* | 188 13* | 234 17* | 280 21* | 327 1 | ------- | 4 11* | 59 14 | 114 16* | 169 19* |
| +21 | −3 | 144 8 | 190 11* | 236 15* | 282 19* | 328 23* | ------- | 6 18* | 61 21 | 117 0 | 172 2* |
| +20 | −4 | 146 6 | 192 10 | 238 14 | 284 17* | 330 21* | ------- | 9 1* | 64 4* | 119 7 | 174 9* |
| +19 | −5 | 148 4 | 194 8 | 240 12 | 286 16 | 332 19* | ------- | 11 8* | 66 11* | 121 14 | 176 16* |
| +18 | −6 | 150 2* | 196 6 | 242 10 | 288 14 | 334 18 | ------- | 13 16 | 68 18* | 123 21 | 178 23* |
| +17 | −7 | 152 0* | 198 4* | 244 8 | 290 12 | 336 16 | ------- | 15 23 | 71 1* | 126 4 | 181 7 |
| +16 | −8 | 153 22* | 200 2* | 246 6* | 292 10* | 338 14 | ------- | 18 6 | 73 8* | 128 11* | 183 14 |
| +15 | −9 | 155 21 | 202 0* | 248 4* | 294 8* | 340 12* | ------- | 20 13 | 75 16 | 130 18* | 185 21 |
| +14 | −10 | 157 19 | 203 23 | 250 2* | 296 6* | 342 10* | ------- | 22 20* | 77 23 | 133 1* | 188 4 |
| +13 | −11 | 159 17 | 205 21 | 252 1 | 298 5 | 344 8* | ------- | 25 3* | 80 6 | 135 8* | 190 11* |
| +12 | −12 | 161 15* | 207 19 | 253 23 | 300 3 | 346 7 | ------- | 27 10* | 82 13 | 137 15* | 192 18* |
| +11 | −13 | 163 13* | 209 17* | 255 21 | 302 1 | 348 5 | ------- | 28 17* | 84 20 | 139 23 | 195 1* |
| +10 | −14 | 165 11* | 211 15* | 257 19* | 303 23* | 350 3 | ------- | 32 0* | 87 3* | 142 6 | 197 8* |
| +9 | −15 | 167 10 | 213 13* | 259 17* | 305 21* | 352 1* | ------- | 34 8 | 89 10* | 144 13 | 199 15* |
| +8 | −16 | 169 8 | 215 12 | 261 15* | 307 19* | 353 23* | ------- | 36 15 | 91 17* | 146 20 | 201 23 |
| +7 | −17 | 171 6 | 217 10 | 263 14 | 309 18 | 355 21* | ------- | 38 22 | 94 0* | 149 3* | 204 6 |
| +6 | −18 | 173 4* | 219 8 | 265 12 | 311 16 | 357 20 | ------- | 41 5 | 96 8 | 151 10* | 206 13 |
| +5 | −19 | 175 2* | 221 6* | 267 10 | 313 14 | 359 18 | ------- | 43 12* | 98 15 | 153 17* | 208 20 |
| +4 | −20 | 177 0* | 223 4* | 269 8* | 315 12 | 361 16 | ------- | 45 19* | 100 22 | 156 0* | 211 3 |
| +3 | −21 | 178 22* | 225 2* | 271 6* | 317 10* | 363 14* | ------- | 48 2* | 103 5 | 158 7* | 213 10* |
| +2 | −22 | 180 21 | 227 1 | 273 4* | 319 8* | 365 12* | ------- | 50 9* | 105 12 | 160 15 | 215 17* |
| +1 | −23 | 182 19 | 228 23 | 275 3 | 321 6* | 367 10* | ------- | 52 16* | 107 19* | 162 22 | 218 0* |

| Difference | | Constituent λ | | | Constituent MS | | | | | | | |
|---|---|---|---|---|---|---|---|---|---|---|---|---|
| Hour | | d. h. | d. h. | d. h. | d. h. | d. h. | d. h. | d. h. | d. h. | d. h. | d. h. |
| 0 | | 220 7* | 275 10* | 330 13 | 1 0 | 58 20* | 117 22 | 176 23* | 236 1 | 295 2* | 354 4 |
| +23 | −1 | 222 15 | 277 17* | 332 20 | 2 6* | 61 7* | 120 9 | 179 10* | 238 12 | 297 13* | 356 15 |
| +22 | −2 | 224 22 | 280 0* | 335 3 | 4 17* | 63 19 | 122 20* | 181 21* | 240 23 | 300 0* | 359 2 |
| +21 | −3 | 227 5 | 282 7* | 337 10 | 7 4* | 66 6 | 125 7* | 184 9 | 243 10* | 302 11* | 361 13 |
| +20 | −4 | 229 12 | 284 14* | 339 17* | 9 15* | 68 17 | 127 18* | 186 20 | 245 21* | 304 23 | 364 0* |
| +19 | −5 | 231 19 | 286 22 | 342 0* | 12 2* | 71 4 | 130 5* | 189 7 | 248 8* | 307 10 | 366 11* |
| +18 | −6 | 234 2* | 289 5 | 344 7* | 14 13* | 73 15 | 132 16* | 191 18 | 250 19* | 309 21 | 368 22* |
| +17 | −7 | 236 9* | 291 12 | 346 14* | 17 0* | 76 2 | 135 3* | 194 5 | 253 6* | 312 8 | 371 9* |
| +16 | −8 | 238 16* | 293 19 | 348 22 | 19 11* | 78 13 | 137 14* | 196 16 | 255 17* | 314 19 | ------- |
| +15 | −9 | 240 23* | 296 2* | 351 5 | 21 23 | 81 0 | 140 1* | 199 3 | 258 4* | 317 6 | ------- |
| +14 | −10 | 243 7 | 298 9* | 353 12 | 24 10 | 83 11* | 142 13 | 201 14 | 260 15* | 319 17 | ------- |
| +13 | −11 | 245 14 | 300 16* | 355 19 | 26 21 | 85 22* | 145 0 | 204 1* | 263 3 | 322 4 | ------- |
| +12 | −12 | 247 21 | 302 23* | 358 2 | 29 8 | 88 9* | 147 11 | 206 12* | 265 14 | 324 15* | ------- |
| +11 | −13 | 250 4 | 305 6* | 360 9* | 31 19 | 90 20* | 149 22 | 208 23* | 268 1 | 327 2* | ------- |
| +10 | −14 | 252 11 | 307 14 | 362 16* | 34 6 | 93 7* | 152 9 | 211 10* | 270 12 | 329 13* | ------- |
| +9 | −15 | 254 18* | 309 21 | 364 23* | 36 17 | 95 18* | 154 20 | 213 21* | 272 23 | 332 0* | ------- |
| +8 | −16 | 257 1* | 312 4 | 367 6* | 39 4 | 98 5* | 157 7 | 216 8* | 275 10 | 334 11* | ------- |
| +7 | −17 | 259 8* | 314 11 | 369 14 | 41 15* | 100 16* | 159 18 | 218 19* | 277 21 | 336 22* | ------- |
| +6 | −18 | 261 15* | 316 18* | 371 21 | 44 2* | 103 4 | 162 5* | 221 6* | 280 8 | 339 9* | ------- |
| +5 | −19 | 263 22* | 319 1* | ------- | 46 13* | 105 15 | 164 16* | 223 18 | 282 19* | 341 20* | ------- |
| +4 | −20 | 266 6 | 321 8* | ------- | 49 0* | 108 2 | 167 3* | 226 5 | 285 6* | 344 8 | ------- |
| +3 | −21 | 268 13 | 323 15* | ------- | 51 11* | 110 13 | 169 14* | 228 16 | 287 17* | 346 19 | ------- |
| +2 | −22 | 270 20 | 325 22* | ------- | 53 22* | 113 0 | 172 1* | 231 3 | 290 4* | 349 6 | ------- |
| +1 | −23 | 273 3 | 328 6 | ------- | 56 9* | 115 11 | 174 12* | 233 14 | 292 15* | 351 17 | ------- |

Table 31.—*For construction of primary stencils*—Continued

| Difference | | Constituent L | | | | | | Constituent P | | Constituent T |
|---|---|---|---|---|---|---|---|---|---|---|
| | | d. h. | d. h. | d. h. | d. h. | d. h. | d. h. | d. h. | d. h. | d. h. |
| Hour 0 | | 1 0 | 63 8 | 126 23 | 190 14 | 254 5 | 317 20 | 1 0 | 358 16 | 1 0 |
| +23 | −1 | 2 8* | 65 23* | 129 14* | 193 5* | 256 20* | 320 11* | 8 15* | 373 21 | 16 6 |
| +22 | −2 | 5 0 | 68 15 | 132 6 | 195 21 | 259 12 | 323 3 | 23 20* | | 46 16* |
| +21 | −3 | 7 16 | 71 7 | 134 22 | 198 12* | 262 3* | 325 18* | 39 2 | | 77 3 |
| +20 | −4 | 10 7* | 73 22* | 137 13* | 201 4* | 264 19* | 328 10* | 54 7 | | 107 13* |
| +19 | −5 | 12 23 | 76 14 | 140 5 | 203 20 | 267 11 | 331 2 | 69 12* | | 138 0 |
| +18 | −6 | 15 14* | 79 5* | 142 20* | 206 11* | 270 2* | 333 17* | 84 17* | | 168 10* |
| +17 | −7 | 18 6* | 81 21* | 145 12* | 209 3 | 272 18 | 336 9 | 99 23 | | 198 21 |
| +16 | −8 | 20 22 | 84 13 | 148 4 | 211 19 | 275 10 | 339 1 | 115 4 | | 229 7* |
| +15 | −9 | 23 13* | 87 4* | 150 19* | 214 10* | 278 1* | 341 16* | 130 9* | | 259 18 |
| +14 | −10 | 26 5 | 89 20 | 153 11 | 217 2 | 280 17 | 344 8 | 145 14* | | 290 4* |
| +13 | −11 | 28 21 | 92 12 | 156 2* | 219 17* | 283 8* | 346 23* | 160 20 | | 320 15 |
| +12 | −12 | 31 12* | 95 3* | 158 18* | 222 9* | 286 0* | 349 15* | 176 1 | | 351 1* |
| +11 | −13 | 34 4 | 97 19 | 161 10 | 225 1 | 288 16 | 352 7 | 191 6* | | 381 12 |
| +10 | −14 | 36 19* | 100 10* | 164 1* | 227 16* | 291 7* | 354 22* | 206 11* | | |
| +9 | −15 | 39 11* | 103 2* | 166 17 | 230 8 | 293 23 | 357 14 | 221 17 | | |
| +8 | −16 | 42 3 | 105 18 | 169 9 | 233 0 | 296 15 | 360 6 | 236 22 | | |
| +7 | −17 | 44 18* | 108 9* | 172 0* | 235 15* | 299 6* | 362 21* | 252 3* | | |
| +6 | −18 | 47 10 | 111 1 | 174 16 | 238 7 | 301 22 | 365 13 | 267 8* | | |
| +5 | −19 | 50 2 | 113 17 | 177 7* | 240 22* | 304 13* | 368 4* | 282 13* | | |
| +4 | −20 | 52 17* | 116 8* | 179 23* | 243 14* | 307 5* | 370 20* | 297 19 | | |
| +3 | −21 | 55 9 | 119 0 | 182 15 | 246 6 | 309 21 | | 313 0 | | |
| +2 | −22 | 58 0* | 121 15* | 185 6* | 248 21* | 312 12* | | 328 5* | | |
| +1 | −23 | 60 16* | 124 7* | 187 22 | 251 13 | 315 4 | | 343 10* | | |

| Difference | | Constituent R | Constituent K | | Constituent 2SM | | | | | | |
|---|---|---|---|---|---|---|---|---|---|---|---|
| | | d. h. | d. h. | d. h. | d. h. | d. h. | d. h. | d. h. | d. h. | d. h. | d. h. |
| Hour 0 | | 1 0 | 1 0 | 358 16 | 1 0 | 29 22* | 59 11* | 89 0 | 118 13 | 148 1* | 177 14* |
| +1 | −23 | 16 6 | 8 15* | 373 21 | 15* | 31 4 | 60 17 | 90 5* | 119 18* | 149 7 | 178 20 |
| +2 | −22 | 46 16* | 23 20* | | 2 21 | 32 10 | 61 22* | 91 11 | 121 0 | 150 12* | 180 1* |
| +3 | −21 | 77 3 | 39 2 | | 4 2* | 33 15* | 63 4 | 92 17 | 122 5* | 151 18 | 181 7 |
| +4 | −20 | 107 13* | 54 7 | | 5 8 | 34 21 | 64 9* | 93 22* | 123 11 | 153 0 | 182 12* |
| +5 | −19 | 138 0 | 69 12* | | 6 13* | 36 2* | 65 15 | 95 4 | 124 16* | 154 5* | 183 18 |
| +6 | −18 | 168 10* | 84 17* | | 7 19 | 37 8 | 66 20* | 96 9* | 125 22 | 155 11 | 184 23* |
| +7 | −17 | 198 21 | 99 23 | | 9 0* | 38 13* | 68 2 | 97 15 | 127 3* | 156 16* | 186 5 |
| +8 | −16 | 229 7* | 115 4 | | 10 6 | 39 19 | 69 7* | 98 20* | 128 9 | 157 22 | 187 10* |
| +9 | −15 | 259 18 | 130 9* | | 11 12 | 41 0* | 70 13 | 100 2 | 129 14* | 159 3* | 188 16 |
| +10 | −14 | 290 4* | 145 14* | | 12 17* | 42 6 | 71 19 | 101 7* | 130 20 | 160 9 | 189 21* |
| +11 | −13 | 320 15 | 160 20 | | 13 23 | 43 11* | 73 0* | 102 13 | 132 2 | 161 14* | 191 3 |
| +12 | −12 | 351 1* | 176 1 | | 15 4* | 44 17 | 74 6 | 103 18* | 133 7* | 162 20 | 192 9 |
| +13 | −11 | 381 12 | 191 6* | | 16 10 | 45 22* | 75 11* | 105 0 | 134 13 | 164 1* | 193 14* |
| +14 | −10 | | 206 11* | | 17 15* | 47 4 | 76 17 | 106 5* | 135 18* | 165 7 | 194 20 |
| +15 | −9 | | 221 17 | | 18 21 | 48 9* | 77 22* | 107 11 | 137 0 | 166 12* | 196 1* |
| +16 | −8 | | 236 22 | | 20 2* | 49 15 | 79 4 | 108 16* | 138 5* | 167 18 | 197 7 |
| +17 | −7 | | 252 3* | | 21 8 | 50 20* | 80 9* | 109 22 | 139 11 | 168 23* | 198 12* |
| +18 | −6 | | 267 8* | | 22 13* | 52 2* | 81 15 | 111 3* | 140 16* | 170 5 | 199 18 |
| +19 | −5 | | 282 13* | | 23 19 | 53 8 | 82 20* | 112 9* | 141 22 | 171 10* | 200 23* |
| +20 | −4 | | 297 19 | | 25 0* | 54 13* | 84 2 | 113 15 | 143 3* | 172 16* | 202 5 |
| +21 | −3 | | 313 0 | | 26 6 | 55 19 | 85 7* | 114 20* | 144 9 | 173 22 | 203 10* |
| +22 | −2 | | 328 5* | | 27 11* | 57 0* | 86 13 | 116 2 | 145 14* | 175 3* | 204 16 |
| +23 | −1 | | 343 10* | | 28 17 | 58 6 | 87 18* | 117 7* | 146 20 | 176 9 | 205 21* |

HARMONIC ANALYSIS AND PREDICTION OF TIDES

Table 31.—*For construction of primary stencils*—Continued

| Difference | | Constituent 2SM | | | | | | Constituent J | | | |
|---|---|---|---|---|---|---|---|---|---|---|---|
| Hour | | d. h. | d. h. | d. h. | d. h. | d. h. | d. h. | d. h. | d. h. | d. h. | d. h. |
| 0 | | 207 3 | 236 16 | 266 4* | 295 17* | 325 6 | 354 19 | 1 0 | 26 3 | 51 18 | 77 8* |
| +1 | −23 | 208 8* | 237 21* | 267 10 | 296 23 | 326 11* | 356 0* | 13* | 27 4* | 52 19* | 78 10* |
| +2 | −22 | 209 14 | 239 3 | 268 15* | 298 4* | 327 17 | 357 6 | 2 15 | 28 6 | 53 21 | 79 12 |
| +3 | −21 | 210 19* | 240 8* | 269 21 | 299 10 | 328 22* | 358 11* | 3 17 | 29 7* | 54 22* | 80 13* |
| +4 | −20 | 212 1 | 241 14 | 271 2* | 300 15* | 330 4 | 359 17 | 4 18* | 30 9* | 56 0* | 81 15 |
| +5 | −19 | 213 7 | 242 19* | 272 8 | 301 21 | 331 9* | 360 22* | 5 20 | 31 11 | 57 2 | 82 17 |
| +6 | −18 | 214 12* | 244 1 | 273 14 | 303 2* | 332 15 | 362 4 | 6 21* | 32 12* | 58 3* | 83 18* |
| +7 | −17 | 215 18 | 245 6* | 274 19* | 304 8 | 333 21 | 363 9* | 7 23* | 33 14 | 59 5 | 84 20 |
| +8 | −16 | 216 23* | 246 12 | 276 1 | 305 13* | 335 2* | 364 15 | 9 1 | 34 16 | 60 6* | 85 21* |
| +9 | −15 | 218 5 | 247 17* | 277 6* | 306 19 | 336 8 | 365 20* | 10 2* | 35 17* | 61 8* | 86 23* |
| +10 | −14 | 219 10* | 248 23 | 278 12 | 308 0* | 337 13* | 367 2 | 11 4 | 36 19 | 62 10 | 88 1 |
| +11 | −13 | 220 16 | 250 4* | 279 17* | 309 6 | 338 19 | 368 7* | 12 6 | 37 20* | 63 11* | 89 2* |
| +12 | −12 | 221 21* | 251 10 | 280 23 | 310 11* | 340 0* | 369 13 | 13 7* | 38 22* | 64 13 | 90 4 |
| +13 | −11 | 223 3 | 252 16 | 282 4* | 311 17 | 341 6 | 370 18* | 14 9 | 40 0 | 65 15 | 91 6 |
| +14 | −10 | 224 8* | 253 21* | 283 10 | 312 23 | 342 11* | | 15 10* | 41 1* | 66 16* | 92 7* |
| +15 | −9 | 225 14 | 255 3 | 284 15* | 314 4* | 343 17 | | 16 12* | 42 3 | 67 18 | 93 9 |
| +16 | −8 | 226 19* | 256 8* | 285 21 | 315 10 | 344 22* | | 17 14 | 43 5 | 68 19* | 94 10* |
| +17 | −7 | 228 1 | 257 14 | 287 2* | 316 15* | 346 4 | | 18 15* | 44 6* | 69 21* | 95 12* |
| +18 | −6 | 229 6* | 258 19* | 288 8 | 317 21 | 347 9* | | 19 17 | 45 8 | 70 23 | 96 14 |
| +19 | −5 | 230 12 | 260 1 | 289 13* | 319 2* | 348 15 | | 20 18* | 46 9* | 72 0* | 97 15* |
| +20 | −4 | 231 17* | 261 6* | 290 19 | 320 8 | 349 20* | | 21 20* | 47 11* | 73 2 | 98 17 |
| +21 | −3 | 232 23* | 262 12 | 292 0* | 321 13* | 351 2 | | 22 22 | 48 13 | 74 4 | 99 18* |
| +22 | −2 | 234 5 | 263 17* | 293 6* | 322 19 | 352 7* | | 23 23* | 49 14* | 75 5* | 100 20* |
| +23 | −1 | 235 10* | 264 23 | 294 12 | 324 0* | 353 13* | | 25 1 | 50 16 | 76 7 | 101 22 |

| Difference | | Constituent J | | | | | | | | | |
|---|---|---|---|---|---|---|---|---|---|---|---|
| Hour | | d. h. | d. h. | d. h. | d. h. | d. h. | d. h. | d. h. | d. h. | d. h. | d. h. |
| 0 | | 102 23* | 128 14* | 154 5* | 179 20* | 205 11* | 231 2 | 256 17 | 282 8 | 307 23 | 333 14 |
| +1 | −23 | 104 1 | 129 16 | 155 7 | 180 22 | 206 13 | 232 4 | 257 18* | 283 9* | 309 0* | 334 15* |
| +2 | −22 | 105 3 | 130 18 | 156 8* | 181 23* | 207 14* | 233 5* | 258 20* | 284 11* | 310 2 | 335 17 |
| +3 | −21 | 106 4* | 131 19* | 157 10* | 183 1 | 208 16 | 234 7 | 259 22. | 285 13 | 311 4 | 336 18* |
| +4 | −20 | 107 6 | 132 21 | 158 12 | 184 3 | 209 18 | 235 8* | 260 23* | 286 14* | 312 5* | 337 20* |
| +5 | −19 | 108 7* | 133 22* | 159 13* | 185 4* | 210 19* | 236 10* | 262 1 | 287 16 | 313 7 | 338 22 |
| +6 | −18 | 109 9* | 135 0* | 160 15 | 186 6 | 211 21 | 237 12 | 263 3 | 288 18 | 314 8* | 339 23* |
| +7 | −17 | 110 11 | 136 2 | 161 17 | 187 7* | 212 22* | 238 13* | 264 4* | 289 19* | 315 10* | 341 1 |
| +8 | −16 | 111 12* | 137 3* | 162 18* | 188 9* | 214 0* | 239 15 | 265 6 | 290 21 | 316 12 | 342 3 |
| +9 | −15 | 112 14 | 138 5 | 163 20 | 189 11 | 215 2 | 240 17 | 266 7* | 291 22* | 317 13* | 343 4* |
| +10 | −14 | 113 16 | 139 6* | 164 21* | 190 12* | 216 3* | 241 18* | 267 9* | 293 0* | 318 15 | 344 6 |
| +11 | −13 | 114 17* | 140 8* | 165 23* | 191 14 | 217 5 | 242 20 | 268 11 | 294 2 | 319 17 | 345 7* |
| +12 | −12 | 115 19 | 141 10 | 167 1 | 192 16 | 218 6* | 243 21* | 269 12* | 295 3* | 320 18* | 346 9* |
| +13 | −11 | 116 20* | 142 11* | 168 2* | 193 17* | 219 8* | 244 23* | 270 14 | 296 5 | 321 20 | 347 11 |
| +14 | −10 | 117 22* | 143 13 | 169 4 | 194 19 | 220 10 | 246 1 | 271 16 | 297 6* | 322 21* | 348 12* |
| +15 | −9 | 119 0 | 144 15 | 170 6 | 195 20* | 221 11* | 247 2* | 272 17* | 298 8* | 323 23* | 349 14 |
| +16 | −8 | 120 1* | 145 16* | 171 7* | 196 22* | 222 13 | 248 4 | 273 19 | 299 10 | 325 1 | 350 16 |
| +17 | −7 | 121 3 | 146 18 | 172 9 | 198 0 | 223 15 | 249 6 | 274 20* | 300 11* | 326 2* | 351 17* |
| +18 | −6 | 122 5 | 147 19* | 173 10* | 199 1* | 224 16* | 250 7* | 275 22* | 301 13 | 327 4 | 352 19 |
| +19 | −5 | 123 6* | 148 21* | 174 12* | 200 3 | 225 18 | 251 9 | 277 0 | 302 15 | 328 6 | 353 20* |
| +20 | −4 | 124 8 | 149 23 | 175 14 | 201 5 | 226 19* | 252 10* | 278 1* | 303 16* | 329 7* | 354 22* |
| +21 | −3 | 125 9* | 151 0* | 176 15* | 202 6* | 227 21* | 253 12* | 279 3 | 304 18 | 330 9 | 356 0 |
| +22 | −2 | 126 11* | 152 2 | 177 17 | 203 8 | 228 23 | 254 14 | 280 5 | 305 19* | 331 10* | 357 1* |
| +23 | −1 | 127 13 | 153 4 | 178 18* | 204 9* | 230 0* | 255 15* | 281 6* | 306 21* | 332 12* | 358 3 |

246037—41——19

Table 31.—*For construction of primary stencils*—Continued

| Difference | | Con. J | Constituent OO | | | | | | | | |
|---|---|---|---|---|---|---|---|---|---|---|---|
| Hour 0 | | d. h. | d. h. | d. h. | d. h. | d. h. | d. h. | d. h. | d. h. | d. h. | d. h. |
| 0 | | 359 5 | 1 0 | 13 22 | 27 2 | 40 6* | 53 10* | 66 14* | 79 18* | 92 22* | 106 2* |
| +1 | −23 | 360 6* | 7* | 14 11* | 15* | 19* | 23* | 67 3* | 80 7* | 93 11* | 15* |
| +2 | −22 | 361 8 | 20* | 15 0* | 28 4* | 41 8* | 54 12* | 16* | 20* | 94 1 | 107 5 |
| +3 | −21 | 362 9* | 2 9* | 13* | 17* | 22 | 55 2 | 68 6 | 81 10 | 14 | 18 |
| +4 | −20 | 363 11* | 23 | 16 3 | 29 7 | 42 11 | 15 | 19 | 23 | 95 3 | 108 7 |
| +5 | −19 | 364 13 | 3 12 | 16 | 20 | 43 0 | 56 4 | 69 8 | 82 12 | 16* | 20* |
| +6 | −18 | 365 14* | 4 1 | 17 5 | 30 9* | 13* | 17* | 21* | 83 1* | 96 5* | 109 9* |
| +7 | −17 | 366 16 | 14* | 18* | 22* | 44 2* | 57 6* | 70 10* | 14* | 18* | 22* |
| +8 | −16 | 367 18 | 5 3* | 18 7* | 31 11* | 15* | 19* | 23* | 84 3* | 97 8 | 110 12 |
| +9 | −15 | 368 19* | 16* | 20* | 32 1 | 45 5 | 58 9 | 71 13 | 17 | 21 | 111 1 |
| +10 | −14 | 369 21 | 6 6 | 19 10 | 14 | 18 | 22 | 72 2 | 85 6 | 98 10 | 14 |
| +11 | −13 | 370 22* | 19 | 23 | 33 3 | 46 7 | 59 11 | 15 | 19 | 23* | 112 3* |
| +12 | −12 | -------- | 7 8 | 20 12 | 16* | 20* | 60 0* | 73 4* | 86 8* | 99 12* | 16* |
| +13 | −11 | -------- | 21* | 21 1* | 34 5* | 47 9* | 13* | 17* | 21* | 100 1* | 113 5* |
| +14 | −10 | -------- | 8 10* | 14* | 18* | 22* | 61 2* | 74 6* | 87 11 | 15 | 19 |
| +15 | −9 | -------- | 23* | 22 3* | 35 8 | 48 12 | 16 | 20 | 88 0 | 101 4 | 114 8 |
| +16 | −8 | -------- | 9 13 | 17 | 21 | 49 1 | 62 5 | 75 9 | 13 | 17 | 21 |
| +17 | −7 | -------- | 10 2 | 23 6 | 36 10 | 14 | 18 | 22 | 89 2* | 102 6* | 115 10* |
| +18 | −6 | -------- | 15 | 19 | 23* | 50 3* | 63 7* | 76 11* | 15* | 19* | 23* |
| +19 | −5 | -------- | 11 4* | 24 8* | 37 12* | 16* | 20* | 77 0* | 90 4* | 103 8* | 116 12* |
| +20 | −4 | -------- | 17* | 21* | 38 1* | 51 5* | 64 9* | 13* | 18 | 22 | 117 2 |
| +21 | −3 | -------- | 12 6* | 25 10* | 15 | 19 | 23 | 78 3 | 91 7 | 104 11 | 15 |
| +22 | −2 | -------- | 20 | 26 0 | 39 4 | 52 8 | 65 12 | 16 | 20 | 105 0 | 118 4 |
| +23 | −1 | -------- | 13 9 | 13 | 17 | 21 | 66 1 | 79 5 | 92 9* | 13* | 17* |

| Difference | | Constituent OO | | | | | | | | | |
|---|---|---|---|---|---|---|---|---|---|---|---|
| Hour 0 | | d. h. | d. h. | d. h. | d. h. | d. h. | d. h. | d. h. | d. h. | d. h. | d. h. |
| 0 | | 119 6* | 132 10* | 145 14* | 158 18* | 171 22* | 185 2* | 198 6* | 211 11 | 224 15 | 237 19 |
| +1 | −23 | 19* | 23* | 146 4 | 159 8 | 172 12 | 16 | 20 | 212 0 | 225 4 | 238 8 |
| +2 | −22 | 120 9 | 133 13 | 17 | 21 | 173 1 | 186 5 | 199 9 | 13 | 17 | 21 |
| +3 | −21 | 22 | 134 2 | 147 6 | 160 10 | 14 | 18 | 22 | 213 2* | 226 6* | 239 10* |
| +4 | −20 | 121 11 | 15 | 19* | 23* | 174 3* | 187 7* | 200 11* | 15* | 19* | 23* |
| +5 | −19 | 122 0* | 135 4* | 148 8* | 161 12* | 16* | 20* | 201 0* | 214 4* | 227 8* | 240 12* |
| +6 | −18 | 13* | 17* | 21* | 162 1* | 175 5* | 188 9* | 14 | 18 | 22 | 241 2 |
| +7 | −17 | 123 2* | 136 6* | 149 11 | 15 | 19 | 23 | 202 3 | 215 7 | 228 11 | 15 |
| +8 | −16 | 16 | 20 | 150 0 | 163 4 | 176 8 | 189 12 | 16 | 20 | 229 0 | 242 4 |
| +9 | −15 | 124 5 | 137 9 | 13 | 17 | 21 | 190 1 | 203 5* | 216 9* | 13* | 17* |
| +10 | −14 | 18 | 22 | 151 2* | 164 6* | 177 10* | 14* | 18* | 22* | 230 2* | 243 6* |
| +11 | −13 | 125 7* | 138 11* | 15* | 19* | 23* | 191 3* | 204 7* | 217 11* | 15* | 19* |
| +12 | −12 | 20* | 139 0* | 152 4* | 165 8* | 178 12* | 16* | 21 | 218 1 | 231 5 | 244 9 |
| +13 | −11 | 126 9* | 13* | 18 | 22 | 179 2 | 192 6 | 205 10 | 14 | 18 | 22 |
| +14 | −10 | 23 | 140 3 | 153 7 | 166 11 | 15 | 19 | 23 | 219 3 | 232 7 | 245 11 |
| +15 | −9 | 127 12 | 16 | 20 | 167 0 | 180 4 | 193 8 | 206 12* | 16* | 20* | 246 0* |
| +16 | −8 | 128 1 | 141 5 | 154 9* | 13* | 17* | 21* | 207 1* | 220 5* | 233 9* | 13* |
| +17 | −7 | 14* | 18* | 22* | 168 2* | 181 6* | 194 10* | 14* | 18* | 22* | 247 2* |
| +18 | −6 | 129 3* | 142 7* | 155 11* | 15* | 19* | 23* | 208 4 | 221 8 | 234 12 | 16 |
| +19 | −5 | 16* | 20* | 156 1 | 169 5 | 182 9 | 195 13 | 17 | 21 | 235 1 | 248 5 |
| +20 | −4 | 130 6 | 143 10 | 14 | 18 | 22 | 196 2 | 209 6 | 222 10 | 14 | 18 |
| +21 | −3 | 19 | 23 | 157 3 | 170 7 | 183 11 | 15 | 19* | 23* | 236 3* | 249 7* |
| +22 | −2 | 131 8 | 144 12* | 16* | 20* | 184 0* | 197 4* | 210 8* | 223 12* | 16* | 20* |
| +23 | −1 | 21* | 145 1* | 158 5* | 171 9* | 13* | 17* | 21* | 224 1* | 237 5* | 250 9* |

HARMONIC ANALYSIS AND PREDICTION OF TIDES 287

Table 31.—*For construction of primary stencils*—Continued

| Difference | | Constituent OO | | | | | | | | | |
|---|---|---|---|---|---|---|---|---|---|---|---|
| Hour 0 | | d. h.<br>250 23 | d. h.<br>264 3 | d. h.<br>277 7 | d. h.<br>290 11 | d. h.<br>303 15 | d. h.<br>316 19 | d. h.<br>329 23 | d. h.<br>343 3 | d. h.<br>356 7 | d. h.<br>369 11 |
| +1 | −23 | 251 12 | 16 | 20 | 291 0 | 304 4 | 317 8* | 330 12* | 16* | 20* | 370 0* |
| +2 | −22 | 252 1 | 265 5* | 278 9* | 13* | 17* | 21* | 331 1* | 344 5* | 357 9* | |
| +3 | −21 | 14* | 18* | 22* | 292 2* | 305 6* | 318 10* | 14* | 18* | 22* | |
| +4 | −20 | 253 3* | 266 7* | 279 11* | 15* | 19* | 319 0 | 332 4 | 345 8 | 358 12 | |
| +5 | −19 | 16* | 21 | 280 1 | 293 5 | 306 9 | 13 | 17 | 21 | 359 1 | |
| +6 | −18 | 254 6 | 267 10 | 14 | 18 | 22 | 320 2 | 333 6 | 346 10 | 14 | |
| +7 | −17 | 19 | 23 | 281 3 | 294 7 | 307 11 | 15* | 19* | 23* | 360 3* | |
| +8 | −16 | 255 8 | 268 12* | 16* | 20* | 308 0* | 321 4* | 334 8* | 347 12* | 16* | |
| +9 | −15 | 21* | 269 1* | 282 5* | 295 9* | 13* | 17* | 21* | 348 1* | 361 5* | |
| +10 | −14 | 256 10* | 14* | 18* | 22* | 309 2* | 322 7 | 335 11 | 15 | 19 | |
| +11 | −13 | 23* | 270 4 | 283 8 | 296 12 | 16 | 20 | 336 0 | 349 4 | 362 8 | |
| +12 | −12 | 257 13 | 17 | 21 | 297 1 | 310 5 | 323 9 | 13 | 17 | 21 | |
| +13 | −11 | 258 2 | 271 6 | 284 10 | 14 | 18 | 22* | 337 2* | 350 6* | 363 10* | |
| +14 | −10 | 15* | 19* | 23* | 298 3* | 311 7* | 324 11* | 15* | 19* | 23* | |
| +15 | −9 | 259 4* | 272 8* | 285 12* | 16* | 20* | 325 0* | 338 4* | 351 8* | 364 12* | |
| +16 | −8 | 17* | 21* | 286 1* | 299 5* | 312 9* | 14 | 18 | 22 | 365 2 | |
| +17 | −7 | 260 7 | 273 11 | 15 | 19 | 23 | 326 3 | 339 7 | 352 11 | 15 | |
| +18 | −6 | 20 | 274 0 | 287 4 | 300 8 | 313 12 | 16 | 20 | 353 0 | 366 4 | |
| +19 | −5 | 261 9 | 13 | 17 | 21 | 314 1 | 327 5* | 340 9* | 13* | 17* | |
| +20 | −4 | 22* | 275 2* | 288 6* | 301 10* | 14* | 18* | 22* | 354 2* | 367 6* | |
| +21 | −3 | 262 11* | 15* | 19* | 23* | 315 3* | 328 7* | 341 11* | 15* | 19* | |
| +22 | −2 | 263 0* | 276 4* | 289 8* | 302 12* | 17 | 21 | 342 1 | 355 5 | 368 9 | |
| +23 | −1 | 14 | 18 | 22 | 303 2 | 316 6 | 329 10 | 14 | 18 | 22 | |

Table 32.—*Divisors for primary stencil sums*

CONSTITUENT J

| Series | 29 | 58 | 87 | 105 | 134 | 163 | 192 | 221 | 250 | 279 | 297 | 326 | 355 | 369 |
|---|---|---|---|---|---|---|---|---|---|---|---|---|---|---|
| Hour | | | | | | | | | | | | | | |
| 0 | 30 | 59 | 87 | 106 | 134 | 164 | 192 | 221 | 250 | 279 | 298 | 326 | 355 | 370 |
| 1 | 31 | 59 | 89 | 106 | 135 | 164 | 193 | 222 | 250 | 280 | 298 | 327 | 356 | 369 |
| 2 | 28 | 58 | 86 | 104 | 134 | 162 | 192 | 220 | 250 | 278 | 296 | 326 | 354 | 369 |
| 3 | 30 | 59 | 88 | 106 | 135 | 165 | 192 | 222 | 251 | 280 | 299 | 326 | 356 | 370 |
| 4 | 29 | 59 | 88 | 104 | 135 | 163 | 193 | 222 | 250 | 280 | 297 | 327 | 355 | 369 |
| 5 | 28 | 59 | 87 | 105 | 134 | 163 | 193 | 221 | 251 | 278 | 297 | 326 | 355 | 370 |
| 6 | 30 | 57 | 88 | 106 | 134 | 165 | 192 | 222 | 250 | 280 | 298 | 326 | 356 | 369 |
| 7 | 28 | 58 | 87 | 104 | 134 | 163 | 193 | 221 | 250 | 279 | 297 | 327 | 354 | 369 |
| 8 | 29 | 58 | 88 | 106 | 134 | 164 | 193 | 222 | 251 | 279 | 298 | 326 | 356 | 371 |
| 9 | 29 | 57 | 87 | 105 | 134 | 163 | 192 | 222 | 250 | 280 | 297 | 326 | 355 | 369 |
| 10 | 28 | 58 | 86 | 104 | 134 | 162 | 193 | 220 | 250 | 278 | 297 | 326 | 354 | 368 |
| 11 | 30 | 59 | 88 | 107 | 134 | 164 | 193 | 223 | 251 | 280 | 299 | 327 | 357 | 370 |
| 12 | 29 | 57 | 87 | 104 | 134 | 162 | 191 | 221 | 250 | 279 | 296 | 326 | 354 | 368 |
| 13 | 28 | 58 | 85 | 104 | 133 | 162 | 191 | 220 | 250 | 278 | 297 | 325 | 354 | 368 |
| 14 | 30 | 58 | 88 | 106 | 134 | 164 | 192 | 223 | 250 | 280 | 297 | 327 | 356 | 369 |
| 15 | 29 | 58 | 87 | 105 | 135 | 162 | 192 | 220 | 251 | 279 | 296 | 327 | 355 | 369 |
| 16 | 28 | 58 | 86 | 105 | 133 | 163 | 191 | 220 | 250 | 279 | 297 | 325 | 355 | 369 |
| 17 | 30 | 57 | 87 | 105 | 134 | 163 | 192 | 221 | 250 | 280 | 296 | 326 | 355 | 368 |
| 18 | 28 | 58 | 86 | 104 | 134 | 162 | 192 | 220 | 250 | 278 | 296 | 325 | 355 | 369 |
| 19 | 29 | 58 | 87 | 106 | 133 | 163 | 191 | 221 | 249 | 280 | 297 | 325 | 356 | 369 |
| 20 | 29 | 57 | 87 | 104 | 134 | 162 | 191 | 220 | 249 | 279 | 296 | 326 | 354 | 368 |
| 21 | 28 | 58 | 85 | 104 | 133 | 162 | 191 | 219 | 249 | 277 | 296 | 325 | 354 | 369 |
| 22 | 30 | 58 | 88 | 106 | 134 | 164 | 192 | 222 | 249 | 279 | 298 | 326 | 356 | 369 |
| 23 | 28 | 57 | 86 | 104 | 134 | 161 | 191 | 219 | 249 | 277 | 295 | 325 | 353 | 368 |

CONSTITUENT K

| Series | 14 | 29 | 58 | 87 | 105 | 134 | 163 | 192 | 221 | 250 | 279 | 297 | 326 | 355 | 369 |
|---|---|---|---|---|---|---|---|---|---|---|---|---|---|---|---|
| Hour | | | | | | | | | | | | | | | |
| 0 | 15 | 30 | 59 | 88 | 106 | 135 | 164 | 193 | 221 | 250 | 279 | 297 | 326 | 355 | 369 |
| 1 | 14 | 30 | 59 | 88 | 106 | 135 | 164 | 193 | 222 | 251 | 279 | 297 | 326 | 355 | 369 |
| 2 | 14 | 29 | 59 | 88 | 106 | 135 | 164 | 193 | 222 | 251 | 280 | 298 | 327 | 355 | 369 |
| 3 | 14 | 29 | 59 | 88 | 106 | 135 | 164 | 193 | 222 | 251 | 280 | 298 | 327 | 356 | 370 |
| 4 | 14 | 29 | 57 | 87 | 105 | 134 | 163 | 192 | 221 | 250 | 279 | 297 | 326 | 355 | 369 |
| 5 | 14 | 29 | 58 | 88 | 105 | 134 | 163 | 192 | 221 | 250 | 279 | 297 | 326 | 355 | 369 |
| 6 | 14 | 29 | 58 | 87 | 106 | 135 | 163 | 192 | 221 | 250 | 279 | 297 | 326 | 355 | 369 |
| 7 | 14 | 29 | 58 | 87 | 105 | 135 | 164 | 193 | 221 | 250 | 279 | 296 | 325 | 354 | 368 |
| 8 | 14 | 29 | 58 | 87 | 105 | 135 | 164 | 193 | 222 | 251 | 280 | 298 | 327 | 355 | 369 |
| 9 | 14 | 29 | 58 | 87 | 105 | 134 | 164 | 193 | 222 | 251 | 280 | 298 | 327 | 356 | 370 |
| 10 | 14 | 29 | 57 | 86 | 104 | 133 | 163 | 192 | 221 | 250 | 279 | 297 | 326 | 355 | 369 |
| 11 | 14 | 29 | 58 | 87 | 105 | 133 | 162 | 192 | 221 | 250 | 279 | 297 | 326 | 355 | 369 |
| 12 | 14 | 29 | 58 | 87 | 105 | 134 | 163 | 192 | 221 | 250 | 279 | 297 | 326 | 355 | 369 |
| 13 | 14 | 29 | 58 | 87 | 105 | 134 | 163 | 192 | 221 | 250 | 279 | 297 | 326 | 355 | 369 |
| 14 | 14 | 29 | 58 | 87 | 105 | 134 | 163 | 192 | 222 | 251 | 280 | 297 | 326 | 355 | 369 |
| 15 | 13 | 28 | 57 | 86 | 104 | 133 | 162 | 191 | 220 | 250 | 279 | 297 | 326 | 355 | 368 |
| 16 | 14 | 29 | 58 | 86 | 104 | 133 | 162 | 191 | 220 | 249 | 279 | 297 | 326 | 355 | 369 |
| 17 | 14 | 29 | 58 | 87 | 105 | 133 | 162 | 191 | 220 | 249 | 279 | 297 | 326 | 355 | 369 |
| 18 | 14 | 29 | 58 | 87 | 105 | 134 | 163 | 191 | 220 | 249 | 278 | 297 | 326 | 355 | 369 |
| 19 | 14 | 29 | 58 | 87 | 105 | 134 | 163 | 192 | 221 | 250 | 278 | 297 | 326 | 355 | 369 |
| 20 | 14 | 29 | 58 | 87 | 105 | 134 | 163 | 192 | 221 | 250 | 279 | 297 | 326 | 355 | 369 |
| 21 | 14 | 28 | 57 | 86 | 104 | 133 | 162 | 191 | 220 | 249 | 278 | 296 | 325 | 355 | 369 |
| 22 | 14 | 29 | 58 | 86 | 104 | 133 | 162 | 191 | 220 | 249 | 278 | 296 | 325 | 355 | 369 |
| 23 | 14 | 29 | 58 | 87 | 105 | 134 | 162 | 191 | 220 | 249 | 278 | 296 | 325 | 354 | 369 |

HARMONIC ANALYSIS AND PREDICTION OF TIDES 289

Table 32.—*Divisors for primary stencil sums*—Continued

CONSTITUENT L

| Series | 29 | 58 | 87 | 105 | 134 | 163 | 192 | 221 | 250 | 279 | 297 | 326 | 355 | 369 |
|---|---|---|---|---|---|---|---|---|---|---|---|---|---|---|
| Hour | | | | | | | | | | | | | | |
| 0 | 29 | 59 | 87 | 105 | 133 | 163 | 191 | 221 | 250 | 279 | 297 | 326 | 355 | 369 |
| 1 | 29 | 59 | 87 | 106 | 134 | 164 | 192 | 222 | 251 | 279 | 297 | 326 | 355 | 369 |
| 2 | 29 | 58 | 87 | 106 | 134 | 163 | 192 | 221 | 250 | 280 | 298 | 326 | 356 | 370 |
| 3 | 30 | 58 | 87 | 105 | 134 | 163 | 192 | 221 | 250 | 279 | 298 | 326 | 356 | 370 |
| 4 | 30 | 58 | 88 | 106 | 135 | 164 | 192 | 222 | 250 | 279 | 297 | 326 | 355 | 370 |
| 5 | 29 | 58 | 88 | 106 | 134 | 164 | 192 | 222 | 250 | 280 | 298 | 327 | 356 | 369 |
| 6 | 29 | 57 | 86 | 105 | 133 | 163 | 191 | 221 | 249 | 279 | 297 | 325 | 355 | 368 |
| 7 | 30 | 59 | 88 | 106 | 135 | 164 | 193 | 222 | 250 | 279 | 298 | 326 | 356 | 369 |
| 8 | 30 | 58 | 88 | 105 | 135 | 164 | 193 | 221 | 251 | 280 | 298 | 327 | 357 | 370 |
| 9 | 29 | 57 | 87 | 104 | 133 | 163 | 191 | 221 | 250 | 279 | 296 | 326 | 355 | 369 |
| 10 | 30 | 58 | 87 | 105 | 134 | 164 | 192 | 221 | 249 | 279 | 296 | 326 | 354 | 368 |
| 11 | 29 | 58 | 87 | 105 | 134 | 162 | 192 | 222 | 250 | 280 | 297 | 326 | 355 | 369 |
| 12 | 29 | 58 | 87 | 104 | 134 | 162 | 192 | 221 | 250 | 279 | 297 | 326 | 355 | 369 |
| 13 | 29 | 58 | 88 | 105 | 135 | 163 | 192 | 220 | 250 | 279 | 296 | 326 | 354 | 368 |
| 14 | 29 | 58 | 88 | 105 | 134 | 163 | 193 | 221 | 250 | 280 | 297 | 327 | 355 | 370 |
| 15 | 28 | 58 | 86 | 105 | 134 | 163 | 192 | 221 | 250 | 279 | 297 | 327 | 355 | 369 |
| 16 | 28 | 58 | 86 | 104 | 134 | 162 | 191 | 220 | 249 | 278 | 296 | 325 | 353 | 367 |
| 17 | 28 | 57 | 86 | 104 | 134 | 162 | 192 | 220 | 250 | 278 | 297 | 326 | 355 | 369 |
| 18 | 29 | 58 | 87 | 105 | 134 | 162 | 192 | 220 | 250 | 278 | 296 | 326 | 355 | 369 |
| 19 | 29 | 58 | 87 | 105 | 135 | 163 | 192 | 221 | 250 | 279 | 297 | 326 | 354 | 369 |
| 20 | 28 | 58 | 86 | 105 | 134 | 163 | 192 | 221 | 250 | 279 | 297 | 327 | 355 | 369 |
| 21 | 28 | 58 | 86 | 104 | 132 | 162 | 191 | 219 | 249 | 277 | 296 | 324 | 354 | 368 |
| 22 | 29 | 58 | 87 | 105 | 134 | 163 | 193 | 221 | 251 | 279 | 297 | 325 | 355 | 369 |
| 23 | 29 | 58 | 87 | 105 | 134 | 163 | 193 | 221 | 251 | 279 | 298 | 326 | 355 | 370 |

CONSTITUENT M

| Series | 15 | 29 | 58 | 87 | 105 | 134 | 163 | 192 | 221 | 250 | 279 | 297 | 326 | 355 | 369 |
|---|---|---|---|---|---|---|---|---|---|---|---|---|---|---|---|
| Hour | | | | | | | | | | | | | | | |
| 0 | 15 | 29 | 59 | 87 | 105 | 135 | 164 | 192 | 222 | 250 | 279 | 297 | 325 | 355 | 369 |
| 1 | 15 | 29 | 57 | 87 | 105 | 134 | 163 | 192 | 221 | 250 | 279 | 296 | 326 | 354 | 369 |
| 2 | 15 | 28 | 58 | 86 | 105 | 134 | 162 | 192 | 221 | 250 | 279 | 296 | 325 | 354 | 369 |
| 3 | 16 | 29 | 59 | 88 | 107 | 135 | 165 | 193 | 222 | 251 | 281 | 299 | 328 | 357 | 371 |
| 4 | 16 | 30 | 58 | 87 | 106 | 135 | 164 | 193 | 222 | 251 | 280 | 297 | 326 | 355 | 370 |
| 5 | 15 | 28 | 57 | 86 | 104 | 134 | 163 | 192 | 221 | 250 | 278 | 296 | 325 | 354 | 368 |
| 6 | 15 | 29 | 58 | 87 | 106 | 134 | 163 | 192 | 222 | 250 | 280 | 297 | 326 | 355 | 369 |
| 7 | 16 | 29 | 58 | 87 | 105 | 134 | 163 | 192 | 221 | 250 | 279 | 296 | 326 | 354 | 369 |
| 8 | 16 | 29 | 59 | 87 | 106 | 135 | 164 | 193 | 221 | 251 | 280 | 298 | 326 | 355 | 370 |
| 9 | 15 | 29 | 58 | 87 | 106 | 135 | 165 | 193 | 223 | 251 | 280 | 298 | 327 | 357 | 371 |
| 10 | 15 | 29 | 57 | 87 | 105 | 134 | 163 | 192 | 221 | 250 | 279 | 296 | 326 | 354 | 368 |
| 11 | 15 | 28 | 57 | 86 | 104 | 133 | 162 | 192 | 221 | 250 | 278 | 296 | 325 | 354 | 369 |
| 12 | 15 | 29 | 58 | 87 | 105 | 133 | 162 | 191 | 220 | 250 | 280 | 297 | 326 | 355 | 368 |
| 13 | 15 | 30 | 59 | 88 | 105 | 134 | 163 | 192 | 221 | 250 | 279 | 298 | 327 | 355 | 369 |
| 14 | 15 | 29 | 58 | 87 | 105 | 134 | 163 | 192 | 220 | 250 | 278 | 297 | 326 | 356 | 369 |
| 15 | 14 | 29 | 58 | 87 | 104 | 134 | 163 | 192 | 222 | 250 | 279 | 298 | 326 | 356 | 369 |
| 16 | 15 | 29 | 57 | 87 | 104 | 133 | 162 | 191 | 220 | 249 | 278 | 296 | 326 | 354 | 368 |
| 17 | 15 | 29 | 59 | 87 | 105 | 134 | 162 | 192 | 220 | 250 | 279 | 298 | 326 | 355 | 369 |
| 18 | 14 | 29 | 58 | 87 | 105 | 133 | 163 | 191 | 220 | 249 | 278 | 297 | 326 | 355 | 368 |
| 19 | 15 | 30 | 58 | 88 | 105 | 135 | 163 | 192 | 221 | 250 | 279 | 297 | 326 | 356 | 369 |
| 20 | 14 | 28 | 57 | 86 | 103 | 133 | 162 | 191 | 220 | 249 | 277 | 296 | 325 | 354 | 368 |
| 21 | 14 | 29 | 58 | 87 | 105 | 133 | 162 | 192 | 221 | 250 | 280 | 298 | 327 | 356 | 369 |
| 22 | 15 | 30 | 59 | 88 | 105 | 134 | 163 | 192 | 221 | 249 | 279 | 298 | 327 | 355 | 369 |
| 23 | 15 | 29 | 58 | 87 | 105 | 134 | 163 | 192 | 220 | 250 | 278 | 296 | 325 | 355 | 369 |

Table 32.—*Divisors for primary stencil sums*—Continued

CONSTITUENT N

| Series | 15 | 29 | 58 | 87 | 105 | 134 | 163 | 192 | 221 | 250 | 279 | 297 | 326 | 355 | 369 |
|---|---|---|---|---|---|---|---|---|---|---|---|---|---|---|---|
| Hour | | | | | | | | | | | | | | | |
| 0 | 16 | 29 | 58 | 87 | 105 | 134 | 163 | 191 | 220 | 250 | 279 | 297 | 327 | 356 | 370 |
| 1 | 16 | 29 | 58 | 88 | 106 | 135 | 165 | 194 | 223 | 252 | 281 | 299 | 327 | 357 | 370 |
| 2 | 15 | 29 | 57 | 87 | 105 | 133 | 162 | 191 | 220 | 248 | 278 | 296 | 324 | 354 | 367 |
| 3 | 16 | 30 | 58 | 88 | 106 | 134 | 163 | 192 | 221 | 249 | 279 | 297 | 326 | 355 | 370 |
| 4 | 16 | 30 | 58 | 87 | 105 | 135 | 164 | 193 | 223 | 252 | 282 | 299 | 328 | 357 | 371 |
| 5 | 15 | 30 | 59 | 88 | 106 | 134 | 164 | 192 | 222 | 250 | 279 | 297 | 326 | 355 | 369 |
| 6 | 15 | 29 | 58 | 87 | 105 | 133 | 163 | 191 | 221 | 249 | 278 | 296 | 324 | 354 | 367 |
| 7 | 15 | 29 | 58 | 87 | 105 | 133 | 163 | 191 | 220 | 250 | 279 | 298 | 326 | 357 | 370 |
| 8 | 14 | 29 | 58 | 88 | 107 | 135 | 164 | 194 | 223 | 251 | 281 | 299 | 327 | 356 | 370 |
| 9 | 15 | 30 | 58 | 88 | 105 | 134 | 163 | 192 | 221 | 249 | 279 | 297 | 325 | 354 | 368 |
| 10 | 15 | 30 | 58 | 88 | 105 | 134 | 163 | 191 | 221 | 249 | 279 | 297 | 326 | 356 | 370 |
| 11 | 15 | 30 | 58 | 86 | 106 | 135 | 165 | 193 | 224 | 252 | 281 | 299 | 328 | 357 | 371 |
| 12 | 15 | 28 | 59 | 87 | 106 | 134 | 164 | 192 | 220 | 250 | 278 | 297 | 325 | 355 | 368 |
| 13 | 15 | 28 | 58 | 86 | 104 | 133 | 161 | 191 | 219 | 249 | 277 | 295 | 324 | 354 | 368 |
| 14 | 14 | 28 | 57 | 86 | 104 | 133 | 161 | 191 | 220 | 250 | 279 | 297 | 326 | 354 | 369 |
| 15 | 14 | 29 | 58 | 88 | 105 | 135 | 164 | 194 | 222 | 251 | 280 | 298 | 327 | 355 | 370 |
| 16 | 15 | 29 | 58 | 86 | 104 | 134 | 162 | 191 | 220 | 249 | 277 | 295 | 325 | 353 | 368 |
| 17 | 15 | 28 | 58 | 86 | 104 | 134 | 162 | 191 | 220 | 249 | 278 | 296 | 327 | 355 | 370 |
| 18 | 15 | 28 | 58 | 87 | 105 | 134 | 164 | 193 | 222 | 252 | 280 | 298 | 327 | 356 | 371 |
| 19 | 15 | 29 | 59 | 87 | 105 | 134 | 163 | 192 | 220 | 250 | 278 | 296 | 325 | 354 | 367 |
| 20 | 14 | 28 | 57 | 86 | 104 | 133 | 161 | 191 | 219 | 249 | 277 | 295 | 325 | 353 | 367 |
| 21 | 14 | 28 | 57 | 86 | 103 | 133 | 161 | 192 | 220 | 249 | 279 | 297 | 326 | 354 | 368 |
| 22 | 16 | 30 | 59 | 88 | 106 | 137 | 165 | 194 | 223 | 252 | 281 | 298 | 328 | 356 | 370 |
| 23 | 15 | 29 | 58 | 86 | 104 | 133 | 162 | 191 | 220 | 249 | 277 | 295 | 325 | 353 | 367 |

CONSTITUENT 2N

| Series | 29 | 58 | 87 | 105 | 134 | 163 | 192 | 221 | 250 | 279 | 297 | 326 | 355 | 369 |
|---|---|---|---|---|---|---|---|---|---|---|---|---|---|---|
| Hour | | | | | | | | | | | | | | |
| 0 | 28 | 58 | 86 | 105 | 135 | 163 | 193 | 222 | 251 | 280 | 299 | 327 | 357 | 371 |
| 1 | 30 | 58 | 88 | 106 | 135 | 165 | 194 | 223 | 252 | 281 | 299 | 329 | 357 | 371 |
| 2 | 28 | 58 | 87 | 105 | 134 | 164 | 193 | 222 | 250 | 279 | 297 | 325 | 353 | 368 |
| 3 | 30 | 59 | 88 | 106 | 136 | 164 | 193 | 221 | 251 | 280 | 298 | 326 | 356 | 370 |
| 4 | 29 | 57 | 86 | 104 | 132 | 161 | 190 | 220 | 249 | 278 | 295 | 325 | 353 | 368 |
| 5 | 28 | 58 | 86 | 105 | 134 | 163 | 192 | 222 | 251 | 280 | 298 | 326 | 356 | 369 |
| 6 | 30 | 58 | 88 | 106 | 135 | 164 | 194 | 222 | 252 | 281 | 298 | 328 | 356 | 370 |
| 7 | 29 | 59 | 88 | 106 | 135 | 165 | 193 | 223 | 251 | 280 | 297 | 325 | 354 | 368 |
| 8 | 29 | 59 | 88 | 106 | 135 | 163 | 192 | 220 | 249 | 279 | 296 | 325 | 365 | 368 |
| 9 | 29 | 57 | 86 | 104 | 133 | 162 | 191 | 220 | 250 | 278 | 296 | 326 | 354 | 360 |
| 10 | 29 | 58 | 87 | 106 | 135 | 164 | 193 | 223 | 251 | 280 | 298 | 327 | 357 | 370 |
| 11 | 29 | 58 | 87 | 100 | 135 | 164 | 194 | 222 | 251 | 280 | 298 | 326 | 355 | 369 |
| 12 | 29 | 58 | 88 | 106 | 135 | 165 | 193 | 221 | 249 | 277 | 295 | 325 | 354 | 368 |
| 13 | 29 | 59 | 88 | 105 | 134 | 162 | 190 | 219 | 248 | 278 | 296 | 325 | 354 | 368 |
| 14 | 29 | 57 | 86 | 104 | 133 | 161 | 191 | 220 | 250 | 278 | 297 | 326 | 355 | 370 |
| 15 | 29 | 58 | 87 | 105 | 133 | 163 | 192 | 222 | 250 | 280 | 298 | 327 | 356 | 370 |
| 16 | 29 | 58 | 87 | 104 | 134 | 163 | 192 | 221 | 251 | 279 | 297 | 325 | 354 | 368 |
| 17 | 29 | 58 | 88 | 105 | 134 | 163 | 192 | 220 | 249 | 278 | 296 | 326 | 355 | 369 |
| 18 | 29 | 59 | 87 | 104 | 132 | 161 | 189 | 219 | 248 | 278 | 296 | 325 | 354 | 368 |
| 19 | 29 | 57 | 86 | 103 | 133 | 161 | 191 | 220 | 249 | 278 | 297 | 326 | 355 | 369 |
| 20 | 30 | 59 | 88 | 106 | 134 | 164 | 193 | 222 | 251 | 280 | 299 | 328 | 357 | 371 |
| 21 | 28 | 58 | 87 | 104 | 134 | 163 | 192 | 221 | 250 | 279 | 297 | 325 | 354 | 367 |
| 22 | 30 | 58 | 87 | 106 | 135 | 163 | 192 | 220 | 249 | 278 | 296 | 326 | 355 | 369 |
| 23 | 28 | 56 | 85 | 103 | 131 | 161 | 189 | 219 | 248 | 277 | 295 | 325 | 354 | 368 |

## Table 32.—*Divisors for primary stencil sums*—Continued

### CONSTITUENT O

| Series | 14 | 29 | 58 | 87 | 105 | 134 | 163 | 192 | 221 | 250 | 279 | 297 | 326 | 355 | 369 |
|---|---|---|---|---|---|---|---|---|---|---|---|---|---|---|---|
| *Hours* | | | | | | | | | | | | | | | |
| 0 | 13 | 29 | 58 | 87 | 106 | 135 | 164 | 192 | 222 | 251 | 279 | 298 | 327 | 355 | 369 |
| 1 | 14 | 29 | 59 | 88 | 105 | 134 | 164 | 192 | 221 | 251 | 280 | 298 | 327 | 355 | 369 |
| 2 | 14 | 28 | 57 | 86 | 105 | 133 | 162 | 192 | 221 | 250 | 279 | 296 | 325 | 354 | 368 |
| 3 | 14 | 30 | 57 | 87 | 105 | 134 | 164 | 193 | 221 | 251 | 280 | 297 | 326 | 356 | 370 |
| 4 | 14 | 29 | 58 | 87 | 106 | 135 | 163 | 193 | 222 | 250 | 280 | 297 | 325 | 354 | 369 |
| 5 | 14 | 29 | 59 | 87 | 105 | 135 | 163 | 192 | 222 | 251 | 280 | 297 | 326 | 355 | 369 |
| 6 | 14 | 29 | 58 | 87 | 105 | 134 | 164 | 193 | 222 | 251 | 279 | 297 | 326 | 355 | 369 |
| 7 | 14 | 28 | 58 | 87 | 105 | 135 | 164 | 192 | 222 | 251 | 280 | 297 | 327 | 355 | 369 |
| 8 | 14 | 29 | 58 | 88 | 106 | 134 | 164 | 193 | 221 | 250 | 278 | 296 | 325 | 355 | 368 |
| 9 | 15 | 30 | 58 | 87 | 106 | 134 | 163 | 193 | 221 | 250 | 279 | 297 | 326 | 355 | 369 |
| 10 | 14 | 29 | 58 | 87 | 105 | 135 | 164 | 193 | 221 | 249 | 279 | 297 | 326 | 356 | 370 |
| 11 | 14 | 30 | 59 | 88 | 107 | 136 | 164 | 192 | 222 | 251 | 280 | 299 | 327 | 356 | 370 |
| 12 | 14 | 29 | 59 | 87 | 105 | 135 | 163 | 192 | 221 | 250 | 279 | 297 | 327 | 355 | 369 |
| 13 | 13 | 28 | 57 | 87 | 104 | 132 | 161 | 189 | 219 | 248 | 276 | 295 | 324 | 353 | 367 |
| 14 | 14 | 29 | 58 | 87 | 105 | 133 | 163 | 192 | 220 | 250 | 279 | 297 | 327 | 356 | 369 |
| 15 | 14 | 29 | 58 | 87 | 105 | 133 | 162 | 192 | 221 | 250 | 280 | 297 | 326 | 355 | 370 |
| 16 | 14 | 30 | 58 | 87 | 104 | 134 | 163 | 192 | 221 | 250 | 279 | 298 | 325 | 355 | 369 |
| 17 | 14 | 29 | 58 | 86 | 104 | 133 | 161 | 191 | 220 | 248 | 277 | 296 | 325 | 354 | 369 |
| 18 | 13 | 28 | 58 | 87 | 104 | 134 | 163 | 191 | 221 | 250 | 279 | 297 | 327 | 355 | 369 |
| 19 | 14 | 29 | 58 | 88 | 104 | 133 | 163 | 192 | 221 | 250 | 278 | 297 | 326 | 355 | 368 |
| 20 | 15 | 29 | 58 | 87 | 105 | 134 | 163 | 192 | 220 | 250 | 279 | 296 | 326 | 355 | 369 |
| 21 | 14 | 29 | 58 | 87 | 105 | 133 | 163 | 192 | 220 | 249 | 279 | 297 | 326 | 356 | 370 |
| 22 | 15 | 30 | 57 | 86 | 105 | 134 | 162 | 191 | 221 | 250 | 279 | 298 | 326 | 355 | 369 |
| 23 | 14 | 28 | 58 | 86 | 104 | 134 | 162 | 192 | 221 | 249 | 279 | 297 | 326 | 355 | 369 |

### CONSTITUENT OO

| Series | 29 | 58 | 87 | 105 | 134 | 163 | 192 | 221 | 250 | 279 | 297 | 326 | 355 | 369 |
|---|---|---|---|---|---|---|---|---|---|---|---|---|---|---|
| *Hours* | | | | | | | | | | | | | | |
| 0 | 29 | 58 | 86 | 104 | 134 | 163 | 192 | 221 | 250 | 280 | 298 | 326 | 355 | 369 |
| 1 | 30 | 60 | 88 | 107 | 136 | 164 | 193 | 221 | 250 | 279 | 297 | 326 | 355 | 369 |
| 2 | 29 | 57 | 86 | 103 | 133 | 162 | 192 | 220 | 250 | 280 | 297 | 327 | 355 | 369 |
| 3 | 31 | 60 | 89 | 107 | 137 | 166 | 194 | 223 | 251 | 281 | 298 | 327 | 355 | 370 |
| 4 | 30 | 58 | 87 | 104 | 132 | 162 | 191 | 220 | 249 | 278 | 297 | 326 | 355 | 369 |
| 5 | 29 | 59 | 88 | 106 | 135 | 166 | 194 | 223 | 251 | 280 | 298 | 326 | 355 | 369 |
| 6 | 28 | 58 | 86 | 104 | 132 | 161 | 190 | 219 | 249 | 277 | 297 | 325 | 355 | 368 |
| 7 | 29 | 58 | 88 | 105 | 135 | 165 | 194 | 223 | 251 | 280 | 298 | 327 | 355 | 369 |
| 8 | 29 | 59 | 88 | 105 | 134 | 163 | 191 | 220 | 249 | 278 | 297 | 326 | 355 | 369 |
| 9 | 29 | 58 | 87 | 105 | 134 | 163 | 193 | 223 | 251 | 280 | 298 | 326 | 355 | 369 |
| 10 | 28 | 58 | 87 | 105 | 133 | 162 | 191 | 219 | 248 | 277 | 296 | 325 | 355 | 368 |
| 11 | 28 | 57 | 87 | 104 | 134 | 162 | 193 | 222 | 251 | 280 | 299 | 327 | 355 | 369 |
| 12 | 29 | 58 | 88 | 106 | 135 | 163 | 193 | 221 | 250 | 278 | 296 | 326 | 355 | 369 |
| 13 | 29 | 57 | 88 | 104 | 133 | 162 | 192 | 221 | 250 | 280 | 297 | 328 | 356 | 370 |
| 14 | 30 | 59 | 88 | 107 | 135 | 164 | 192 | 222 | 250 | 278 | 296 | 325 | 354 | 369 |
| 15 | 28 | 57 | 85 | 104 | 132 | 162 | 191 | 220 | 250 | 279 | 297 | 327 | 356 | 369 |
| 16 | 29 | 58 | 87 | 106 | 135 | 164 | 193 | 222 | 251 | 279 | 297 | 326 | 354 | 369 |
| 17 | 29 | 57 | 86 | 104 | 133 | 161 | 190 | 220 | 249 | 278 | 296 | 326 | 355 | 369 |
| 18 | 30 | 58 | 88 | 106 | 135 | 165 | 193 | 224 | 252 | 281 | 298 | 327 | 356 | 370 |
| 19 | 28 | 57 | 85 | 104 | 132 | 161 | 189 | 218 | 248 | 277 | 295 | 323 | 354 | 368 |
| 20 | 29 | 58 | 87 | 106 | 135 | 164 | 193 | 222 | 252 | 280 | 298 | 326 | 356 | 369 |
| 21 | 28 | 58 | 86 | 104 | 133 | 161 | 190 | 218 | 248 | 277 | 295 | 324 | 354 | 368 |
| 22 | 29 | 57 | 87 | 105 | 135 | 163 | 193 | 222 | 251 | 281 | 298 | 327 | 356 | 370 |
| 23 | 29 | 58 | 87 | 105 | 134 | 163 | 191 | 220 | 249 | 278 | 295 | 325 | 354 | 369 |

Table 32.—*Divisors for primary stencil sums*—Continued

CONSTITUENT P

| Series | 29 | 58 | 87 | 105 | 134 | 163 | 192 | 221 | 250 | 279 | 297 | 326 | 355 | 369 |
|---|---|---|---|---|---|---|---|---|---|---|---|---|---|---|
| *Hour* | | | | | | | | | | | | | | |
| 0 | 29 | 58 | 87 | 105 | 135 | 164 | 193 | 222 | 251 | 280 | 298 | 327 | 356 | 369 |
| 1 | 29 | 58 | 87 | 105 | 134 | 163 | 192 | 221 | 250 | 279 | 297 | 297 | 354 | 368 |
| 2 | 29 | 58 | 87 | 105 | 134 | 163 | 192 | 222 | 251 | 280 | 298 | 327 | 355 | 369 |
| 3 | 29 | 59 | 88 | 106 | 135 | 164 | 193 | 222 | 251 | 280 | 298 | 327 | 356 | 370 |
| 4 | 29 | 58 | 87 | 105 | 134 | 164 | 193 | 222 | 251 | 280 | 297 | 326 | 355 | 369 |
| 5 | 29 | 58 | 87 | 105 | 134 | 163 | 192 | 221 | 250 | 279 | 296 | 325 | 354 | 368 |
| 6 | 29 | 58 | 87 | 105 | 134 | 163 | 192 | 221 | 251 | 279 | 297 | 326 | 355 | 369 |
| 7 | 29 | 58 | 88 | 106 | 135 | 164 | 193 | 222 | 251 | 279 | 297 | 326 | 356 | 370 |
| 8 | 29 | 58 | 87 | 105 | 134 | 163 | 192 | 221 | 249 | 278 | 296 | 325 | 354 | 368 |
| 9 | 29 | 58 | 87 | 105 | 134 | 164 | 193 | 221 | 250 | 279 | 297 | 326 | 355 | 369 |
| 10 | 29 | 58 | 87 | 105 | 134 | 163 | 192 | 220 | 249 | 279 | 297 | 326 | 355 | 369 |
| 11 | 29 | 58 | 88 | 106 | 135 | 164 | 192 | 221 | 250 | 279 | 297 | 326 | 356 | 370 |
| 12 | 29 | 58 | 87 | 105 | 134 | 163 | 191 | 220 | 249 | 278 | 296 | 325 | 354 | 368 |
| 13 | 29 | 58 | 87 | 105 | 134 | 162 | 192 | 221 | 250 | 279 | 297 | 326 | 355 | 369 |
| 14 | 30 | 59 | 88 | 106 | 135 | 163 | 192 | 221 | 250 | 280 | 298 | 327 | 356 | 370 |
| 15 | 29 | 58 | 87 | 105 | 133 | 162 | 191 | 220 | 249 | 278 | 296 | 325 | 354 | 368 |
| 16 | 29 | 58 | 87 | 106 | 134 | 163 | 192 | 221 | 250 | 279 | 297 | 326 | 355 | 370 |
| 17 | 29 | 58 | 87 | 104 | 133 | 162 | 192 | 221 | 250 | 279 | 297 | 326 | 355 | 369 |
| 18 | 30 | 59 | 87 | 105 | 134 | 163 | 192 | 221 | 250 | 279 | 298 | 327 | 356 | 370 |
| 19 | 29 | 58 | 86 | 104 | 133 | 162 | 191 | 220 | 249 | 278 | 296 | 325 | 354 | 368 |
| 20 | 29 | 57 | 86 | 104 | 134 | 163 | 192 | 221 | 250 | 279 | 297 | 326 | 355 | 369 |
| 21 | 29 | 57 | 86 | 104 | 133 | 162 | 191 | 221 | 250 | 279 | 297 | 326 | 355 | 369 |
| 22 | 28 | 57 | 86 | 104 | 133 | 162 | 191 | 220 | 249 | 278 | 296 | 325 | 354 | 368 |
| 23 | 28 | 58 | 87 | 105 | 134 | 163 | 192 | 221 | 250 | 279 | 298 | 327 | 356 | 370 |

CONSTITUENT Q

| Series | 29 | 58 | 87 | 105 | 134 | 163 | 192 | 221 | 250 | 279 | 297 | 326 | 355 | 369 |
|---|---|---|---|---|---|---|---|---|---|---|---|---|---|---|
| *Hour* | | | | | | | | | | | | | | |
| 0 | 29 | 59 | 88 | 106 | 136 | 164 | 194 | 222 | 250 | 280 | 297 | 326 | 355 | 368 |
| 1 | 29 | 58 | 86 | 104 | 133 | 162 | 191 | 221 | 250 | 280 | 298 | 327 | 357 | 370 |
| 2 | 29 | 59 | 88 | 106 | 135 | 165 | 193 | 222 | 251 | 280 | 298 | 326 | 354 | 368 |
| 3 | 28 | 56 | 86 | 103 | 132 | 161 | 190 | 220 | 249 | 278 | 297 | 326 | 354 | 369 |
| 4 | 30 | 59 | 89 | 107 | 136 | 166 | 195 | 225 | 253 | 282 | 299 | 328 | 356 | 370 |
| 5 | 29 | 58 | 87 | 105 | 133 | 162 | 191 | 219 | 249 | 277 | 296 | 325 | 254 | 369 |
| 6 | 30 | 59 | 88 | 107 | 136 | 165 | 195 | 224 | 254 | 281 | 300 | 328 | 356 | 371 |
| 7 | 30 | 58 | 87 | 104 | 133 | 162 | 191 | 219 | 248 | 277 | 295 | 324 | 354 | 369 |
| 8 | 28 | 58 | 87 | 105 | 135 | 164 | 194 | 223 | 251 | 280 | 298 | 326 | 356 | 369 |
| 9 | 29 | 59 | 88 | 106 | 135 | 163 | 192 | 221 | 248 | 278 | 296 | 325 | 355 | 369 |
| 10 | 28 | 58 | 86 | 104 | 134 | 163 | 192 | 221 | 250 | 280 | 298 | 327 | 355 | 370 |
| 11 | 29 | 58 | 88 | 106 | 134 | 164 | 192 | 220 | 249 | 277 | 295 | 324 | 353 | 368 |
| 12 | 29 | 57 | 86 | 104 | 133 | 163 | 191 | 220 | 250 | 279 | 297 | 327 | 356 | 370 |
| 13 | 30 | 59 | 89 | 107 | 136 | 165 | 192 | 221 | 250 | 278 | 296 | 325 | 354 | 369 |
| 14 | 29 | 58 | 87 | 104 | 133 | 161 | 191 | 220 | 250 | 279 | 297 | 327 | 356 | 371 |
| 15 | 30 | 59 | 88 | 107 | 136 | 164 | 194 | 222 | 251 | 280 | 297 | 326 | 355 | 368 |
| 16 | 29 | 58 | 86 | 104 | 133 | 161 | 190 | 219 | 248 | 278 | 296 | 325 | 355 | 368 |
| 17 | 29 | 59 | 87 | 106 | 135 | 164 | 193 | 223 | 251 | 280 | 298 | 326 | 355 | 368 |
| 18 | 28 | 57 | 85 | 103 | 131 | 160 | 188 | 218 | 247 | 277 | 295 | 324 | 354 | 367 |
| 19 | 29 | 58 | 87 | 105 | 134 | 164 | 193 | 223 | 252 | 281 | 299 | 328 | 356 | 370 |
| 20 | 29 | 57 | 86 | 104 | 132 | 161 | 190 | 218 | 247 | 276 | 294 | 324 | 353 | 367 |
| 21 | 30 | 57 | 87 | 105 | 134 | 164 | 193 | 222 | 252 | 281 | 299 | 328 | 356 | 370 |
| 22 | 29 | 58 | 87 | 105 | 134 | 162 | 191 | 220 | 249 | 277 | 295 | 325 | 354 | 368 |
| 23 | 27 | 56 | 85 | 103 | 133 | 162 | 192 | 221 | 251 | 280 | 298 | 327 | 357 | 370 |

HARMONIC ANALYSIS AND PREDICTION OF TIDES

Table 32.—*Divisors for primary stencil sums*—Continued

CONSTITUENT 2Q

| Series | 29 | 58 | 87 | 105 | 134 | 163 | 192 | 221 | 250 | 279 | 297 | 326 | 355 | 369 |
|---|---|---|---|---|---|---|---|---|---|---|---|---|---|---|
| *Hour* | | | | | | | | | | | | | | |
| 0 | 25 | 50 | 83 | 113 | 142 | 167 | 192 | 217 | 242 | 279 | 309 | 334 | 359 | 371 |
| 1 | 25 | 59 | 101 | 116 | 141 | 166 | 191 | 216 | 255 | 293 | 308 | 333 | 358 | 370 |
| 2 | 36 | 77 | 102 | 117 | 142 | 167 | 192 | 233 | 269 | 293 | 309 | 334 | 359 | 371 |
| 3 | 39 | 64 | 89 | 104 | 129 | 154 | 196 | 230 | 255 | 279 | 295 | 320 | 345 | 366 |
| 4 | 25 | 50 | 75 | 90 | 115 | 159 | 192 | 217 | 242 | 266 | 282 | 307 | 355 | 370 |
| 5 | 25 | 50 | 75 | 90 | 136 | 167 | 192 | 217 | 242 | 266 | 282 | 332 | 358 | 370 |
| 6 | 25 | 50 | 83 | 113 | 142 | 167 | 192 | 217 | 241 | 277 | 309 | 334 | 358 | 370 |
| 7 | 25 | 60 | 102 | 117 | 142 | 167 | 192 | 217 | 254 | 293 | 309 | 334 | 358 | 370 |
| 8 | 36 | 76 | 101 | 116 | 141 | 166 | 191 | 232 | 267 | 292 | 308 | 333 | 357 | 369 |
| 9 | 39 | 64 | 89 | 104 | 129 | 154 | 197 | 230 | 255 | 280 | 296 | 320 | 345 | 365 |
| 10 | 25 | 50 | 75 | 90 | 115 | 159 | 191 | 215 | 240 | 265 | 281 | 305 | 353 | 369 |
| 11 | 25 | 50 | 75 | 90 | 136 | 167 | 192 | 216 | 241 | 266 | 283 | 331 | 358 | 370 |
| 12 | 25 | 50 | 83 | 113 | 142 | 167 | 192 | 216 | 241 | 277 | 307 | 332 | 357 | 369 |
| 13 | 25 | 60 | 101 | 117 | 142 | 167 | 191 | 216 | 254 | 293 | 308 | 333 | 358 | 370 |
| 14 | 37 | 77 | 101 | 117 | 142 | 167 | 191 | 231 | 268 | 293 | 308 | 333 | 358 | 370 |
| 15 | 38 | 63 | 87 | 103 | 128 | 153 | 194 | 229 | 254 | 279 | 294 | 319 | 344 | 364 |
| 16 | 25 | 50 | 74 | 90 | 115 | 159 | 191 | 216 | 241 | 266 | 281 | 306 | 354 | 370 |
| 17 | 25 | 49 | 74 | 90 | 136 | 165 | 190 | 215 | 240 | 265 | 280 | 330 | 357 | 369 |
| 18 | 25 | 49 | 81 | 113 | 142 | 166 | 191 | 216 | 241 | 278 | 308 | 333 | 358 | 370 |
| 19 | 25 | 58 | 101 | 117 | 142 | 166 | 191 | 216 | 254 | 292 | 307 | 332 | 357 | 369 |
| 20 | 36 | 76 | 101 | 117 | 141 | 166 | 191 | 231 | 268 | 293 | 308 | 333 | 358 | 370 |
| 21 | 37 | 62 | 87 | 103 | 127 | 152 | 194 | 229 | 254 | 279 | 294 | 319 | 344 | 364 |
| 22 | 24 | 49 | 74 | 90 | 114 | 158 | 191 | 216 | 241 | 266 | 281 | 306 | 354 | 370 |
| 23 | 24 | 49 | 74 | 90 | 135 | 166 | 191 | 216 | 241 | 266 | 281 | 331 | 358 | 370 |

CONSTITUENT R

| Series | 29 | 58 | 87 | 105 | 134 | 163 | 192 | 221 | 250 | 279 | 297 | 326 | 355 | 369 |
|---|---|---|---|---|---|---|---|---|---|---|---|---|---|---|
| *Hour* | | | | | | | | | | | | | | |
| 0 | 30 | 59 | 88 | 106 | 135 | 164 | 193 | 222 | 251 | 280 | 298 | 327 | 356 | 370 |
| 1 | 29 | 59 | 88 | 106 | 135 | 164 | 193 | 222 | 251 | 280 | 298 | 326 | 355 | 369 |
| 2 | 29 | 58 | 88 | 106 | 135 | 164 | 193 | 222 | 251 | 279 | 297 | 326 | 355 | 369 |
| 3 | 29 | 58 | 87 | 105 | 135 | 164 | 193 | 221 | 250 | 279 | 297 | 326 | 355 | 369 |
| 4 | 29 | 58 | 87 | 105 | 134 | 163 | 192 | 221 | 250 | 279 | 297 | 326 | 355 | 369 |
| 5 | 29 | 58 | 86 | 104 | 133 | 162 | 192 | 221 | 250 | 279 | 297 | 326 | 355 | 369 |
| 6 | 28 | 57 | 86 | 104 | 133 | 162 | 191 | 221 | 250 | 279 | 297 | 326 | 355 | 369 |
| 7 | 29 | 58 | 87 | 105 | 134 | 163 | 192 | 221 | 251 | 280 | 298 | 327 | 356 | 370 |
| 8 | 29 | 58 | 87 | 105 | 134 | 163 | 192 | 221 | 250 | 280 | 298 | 327 | 356 | 370 |
| 9 | 29 | 58 | 87 | 105 | 134 | 163 | 192 | 221 | 250 | 279 | 298 | 327 | 356 | 370 |
| 10 | 29 | 58 | 87 | 105 | 134 | 163 | 192 | 221 | 250 | 279 | 297 | 327 | 356 | 370 |
| 11 | 29 | 58 | 87 | 105 | 134 | 163 | 192 | 221 | 250 | 279 | 297 | 326 | 356 | 370 |
| 12 | 29 | 58 | 87 | 105 | 134 | 163 | 192 | 221 | 250 | 279 | 297 | 326 | 354 | 368 |
| 13 | 29 | 58 | 87 | 105 | 134 | 163 | 192 | 221 | 250 | 279 | 296 | 325 | 354 | 368 |
| 14 | 29 | 58 | 87 | 105 | 134 | 163 | 192 | 221 | 249 | 278 | 296 | 325 | 354 | 368 |
| 15 | 29 | 58 | 87 | 105 | 134 | 163 | 191 | 220 | 249 | 278 | 296 | 325 | 354 | 368 |
| 16 | 29 | 58 | 87 | 105 | 133 | 162 | 191 | 220 | 249 | 278 | 296 | 325 | 354 | 368 |
| 17 | 29 | 57 | 86 | 104 | 133 | 162 | 191 | 220 | 249 | 278 | 296 | 325 | 354 | 368 |
| 18 | 29 | 58 | 87 | 105 | 134 | 163 | 192 | 221 | 250 | 279 | 297 | 326 | 355 | 369 |
| 19 | 29 | 58 | 87 | 105 | 134 | 163 | 192 | 221 | 250 | 279 | 297 | 326 | 355 | 369 |
| 20 | 29 | 58 | 87 | 105 | 134 | 163 | 192 | 221 | 250 | 279 | 297 | 326 | 355 | 369 |
| 21 | 29 | 58 | 87 | 105 | 134 | 163 | 192 | 221 | 250 | 279 | 297 | 326 | 355 | 369 |
| 22 | 29 | 58 | 87 | 105 | 134 | 163 | 192 | 221 | 250 | 279 | 297 | 326 | 355 | 369 |
| 23 | 29 | 58 | 87 | 105 | 134 | 163 | 192 | 221 | 250 | 279 | 297 | 326 | 355 | 369 |

## Table 32.—*Divisors for primary stencil sums*—Continued

### CONSTITUENT T

| Series | 29 | 58 | 87 | 105 | 134 | 163 | 192 | 221 | 250 | 279 | 297 | 326 | 355 | 369 |
|---|---|---|---|---|---|---|---|---|---|---|---|---|---|---|
| *Hour* | | | | | | | | | | | | | | |
| 0 | 29 | 58 | 88 | 106 | 135 | 164 | 193 | 222 | 251 | 280 | 298 | 327 | 356 | 370 |
| 1 | 29 | 58 | 87 | 105 | 134 | 163 | 192 | 221 | 250 | 279 | 297 | 326 | 355 | 369 |
| 2 | 29 | 58 | 87 | 105 | 134 | 163 | 192 | 221 | 250 | 279 | 297 | 326 | 355 | 369 |
| 3 | 29 | 58 | 87 | 105 | 134 | 163 | 192 | 221 | 250 | 279 | 297 | 326 | 355 | 369 |
| 4 | 29 | 58 | 87 | 105 | 134 | 163 | 193 | 222 | 251 | 280 | 298 | 328 | 357 | 371 |
| 5 | 30 | 59 | 88 | 106 | 135 | 164 | 193 | 222 | 251 | 280 | 298 | 327 | 356 | 370 |
| 6 | 29 | 58 | 87 | 105 | 134 | 163 | 192 | 221 | 250 | 279 | 297 | 326 | 355 | 369 |
| 7 | 29 | 58 | 87 | 105 | 134 | 163 | 192 | 221 | 250 | 279 | 297 | 326 | 355 | 369 |
| 8 | 29 | 58 | 87 | 105 | 134 | 163 | 192 | 221 | 250 | 279 | 297 | 326 | 355 | 369 |
| 9 | 29 | 58 | 87 | 105 | 135 | 164 | 193 | 222 | 251 | 281 | 299 | 328 | 357 | 371 |
| 10 | 29 | 58 | 87 | 105 | 134 | 163 | 192 | 221 | 250 | 279 | 297 | 326 | 355 | 369 |
| 11 | 29 | 58 | 87 | 105 | 134 | 163 | 192 | 221 | 250 | 279 | 297 | 326 | 355 | 369 |
| 12 | 29 | 58 | 87 | 105 | 134 | 163 | 192 | 221 | 250 | 279 | 297 | 326 | 354 | 368 |
| 13 | 29 | 58 | 87 | 105 | 134 | 163 | 192 | 221 | 250 | 279 | 297 | 325 | 355 | 369 |
| 14 | 29 | 58 | 59 | 88 | 106 | 135 | 164 | 193 | 223 | 252 | 281 | 298 | 327 | 356 | 370 |
| 15 | 29 | 58 | 87 | 105 | 134 | 163 | 192 | 221 | 250 | 278 | 296 | 325 | 354 | 368 |
| 16 | 29 | 58 | 87 | 105 | 134 | 163 | 192 | 221 | 249 | 278 | 296 | 325 | 354 | 368 |
| 17 | 29 | 58 | 87 | 105 | 134 | 163 | 192 | 220 | 249 | 278 | 296 | 325 | 354 | 368 |
| 18 | 29 | 58 | 87 | 105 | 134 | 163 | 191 | 220 | 249 | 278 | 297 | 326 | 355 | 369 |
| 19 | 29 | 58 | 87 | 105 | 134 | 163 | 192 | 221 | 250 | 279 | 297 | 326 | 355 | 369 |
| 20 | 29 | 58 | 87 | 105 | 133 | 162 | 191 | 220 | 249 | 278 | 296 | 325 | 354 | 368 |
| 21 | 29 | 58 | 86 | 104 | 133 | 162 | 191 | 220 | 249 | 278 | 296 | 325 | 354 | 368 |
| 22 | 29 | 57 | 86 | 104 | 133 | 162 | 191 | 220 | 249 | 278 | 296 | 325 | 354 | 368 |
| 23 | 28 | 57 | 86 | 104 | 133 | 162 | 191 | 220 | 250 | 279 | 297 | 326 | 355 | 369 |

### CONSTITUENT λ

| Series | 29 | 58 | 87 | 105 | 134 | 163 | 192 | 221 | 250 | 279 | 297 | 326 | 355 | 369 |
|---|---|---|---|---|---|---|---|---|---|---|---|---|---|---|
| *Hour* | | | | | | | | | | | | | | |
| 0 | 29 | 58 | 89 | 107 | 135 | 164 | 194 | 223 | 252 | 280 | 298 | 330 | 358 | 372 |
| 1 | 29 | 57 | 87 | 106 | 134 | 162 | 191 | 221 | 250 | 278 | 296 | 325 | 355 | 369 |
| 2 | 29 | 57 | 86 | 104 | 134 | 162 | 191 | 219 | 250 | 278 | 296 | 324 | 354 | 369 |
| 3 | 31 | 59 | 88 | 105 | 136 | 165 | 194 | 222 | 252 | 282 | 300 | 328 | 357 | 371 |
| 4 | 31 | 59 | 88 | 105 | 134 | 164 | 193 | 221 | 250 | 279 | 298 | 326 | 355 | 369 |
| 5 | 29 | 59 | 88 | 105 | 134 | 162 | 193 | 221 | 250 | 278 | 297 | 326 | 355 | 369 |
| 6 | 29 | 58 | 88 | 105 | 134 | 162 | 193 | 221 | 250 | 278 | 296 | 326 | 355 | 369 |
| 7 | 29 | 57 | 88 | 106 | 135 | 163 | 192 | 223 | 252 | 280 | 298 | 326 | 355 | 371 |
| 8 | 29 | 57 | 86 | 104 | 134 | 162 | 191 | 219 | 250 | 278 | 296 | 324 | 354 | 368 |
| 9 | 30 | 58 | 87 | 104 | 135 | 163 | 192 | 220 | 250 | 279 | 297 | 325 | 354 | 367 |
| 10 | 31 | 60 | 88 | 106 | 135 | 166 | 195 | 223 | 252 | 282 | 301 | 329 | 358 | 371 |
| 11 | 28 | 59 | 87 | 105 | 134 | 162 | 193 | 221 | 249 | 278 | 297 | 326 | 355 | 368 |
| 12 | 28 | 57 | 87 | 105 | 134 | 162 | 191 | 221 | 249 | 278 | 296 | 326 | 355 | 368 |
| 13 | 28 | 57 | 87 | 105 | 134 | 162 | 190 | 221 | 249 | 278 | 296 | 324 | 354 | 368 |
| 14 | 28 | 57 | 85 | 105 | 134 | 163 | 191 | 220 | 251 | 280 | 298 | 326 | 355 | 371 |
| 15 | 29 | 58 | 86 | 104 | 134 | 163 | 191 | 220 | 249 | 279 | 296 | 325 | 353 | 367 |
| 16 | 30 | 59 | 87 | 105 | 133 | 164 | 192 | 221 | 249 | 280 | 297 | 326 | 354 | 368 |
| 17 | 28 | 60 | 88 | 106 | 134 | 164 | 194 | 223 | 251 | 280 | 299 | 329 | 357 | 371 |
| 18 | 28 | 57 | 87 | 105 | 133 | 162 | 191 | 221 | 249 | 278 | 295 | 326 | 354 | 368 |
| 19 | 28 | 57 | 86 | 105 | 133 | 162 | 190 | 221 | 249 | 278 | 295 | 324 | 354 | 368 |
| 20 | 28 | 57 | 85 | 104 | 133 | 162 | 190 | 219 | 249 | 278 | 295 | 324 | 353 | 368 |
| 21 | 29 | 58 | 86 | 104 | 134 | 164 | 192 | 221 | 250 | 281 | 298 | 327 | 355 | 370 |
| 22 | 30 | 59 | 87 | 105 | 133 | 164 | 192 | 221 | 249 | 279 | 297 | 326 | 354 | 368 |
| 23 | 28 | 58 | 87 | 105 | 133 | 163 | 192 | 221 | 249 | 277 | 296 | 326 | 354 | 368 |

HARMONIC ANALYSIS AND PREDICTION OF TIDES 295

Table 32.—*Divisors for primary stencil sums*—Continued

CONSTITUENT $\mu$

| Series | 29 | 58 | 87 | 105 | 134 | 163 | 192 | 221 | 250 | 279 | 297 | 326 | 355 | 369 |
|---|---|---|---|---|---|---|---|---|---|---|---|---|---|---|
| *Hour* | | | | | | | | | | | | | | |
| 0 | 29 | 59 | 89 | 105 | 135 | 163 | 192 | 223 | 252 | 280 | 299 | 326 | 356 | 369 |
| 1 | 30 | 61 | 88 | 107 | 135 | 164 | 194 | 223 | 252 | 282 | 298 | 327 | 356 | 369 |
| 2 | 30 | 57 | 88 | 105 | 134 | 164 | 193 | 221 | 250 | 280 | 296 | 326 | 355 | 369 |
| 3 | 27 | 57 | 87 | 104 | 134 | 163 | 190 | 219 | 249 | 276 | 296 | 325 | 353 | 368 |
| 4 | 30 | 60 | 87 | 106 | 135 | 162 | 192 | 222 | 249 | 279 | 298 | 326 | 356 | 369 |
| 5 | 30 | 57 | 86 | 106 | 133 | 163 | 193 | 220 | 250 | 280 | 297 | 327 | 356 | 370 |
| 6 | 27 | 56 | 86 | 103 | 133 | 163 | 190 | 220 | 250 | 278 | 297 | 326 | 354 | 369 |
| 7 | 29 | 59 | 88 | 106 | 136 | 163 | 193 | 223 | 251 | 280 | 299 | 326 | 355 | 369 |
| 8 | 30 | 59 | 87 | 107 | 134 | 164 | 194 | 222 | 250 | 279 | 296 | 326 | 356 | 369 |
| 9 | 29 | 57 | 87 | 104 | 134 | 163 | 191 | 219 | 249 | 279 | 296 | 326 | 354 | 369 |
| 10 | 28 | 58 | 88 | 105 | 134 | 163 | 191 | 221 | 251 | 278 | 298 | 326 | 355 | 369 |
| 11 | 30 | 59 | 86 | 105 | 134 | 162 | 192 | 222 | 249 | 279 | 297 | 326 | 356 | 369 |
| 12 | 29 | 57 | 86 | 105 | 133 | 163 | 193 | 220 | 250 | 280 | 297 | 327 | 356 | 369 |
| 13 | 28 | 57 | 87 | 104 | 134 | 164 | 191 | 221 | 251 | 279 | 297 | 326 | 353 | 369 |
| 14 | 29 | 59 | 88 | 106 | 136 | 163 | 193 | 222 | 250 | 278 | 298 | 325 | 355 | 368 |
| 15 | 30 | 59 | 87 | 107 | 134 | 164 | 193 | 222 | 250 | 280 | 297 | 327 | 357 | 370 |
| 16 | 29 | 57 | 87 | 104 | 133 | 162 | 191 | 219 | 249 | 278 | 296 | 326 | 354 | 369 |
| 17 | 28 | 57 | 86 | 103 | 133 | 162 | 190 | 220 | 249 | 277 | 297 | 325 | 354 | 368 |
| 18 | 29 | 59 | 86 | 106 | 135 | 163 | 193 | 222 | 250 | 280 | 298 | 327 | 356 | 369 |
| 19 | 30 | 57 | 87 | 106 | 134 | 164 | 193 | 221 | 251 | 281 | 298 | 327 | 356 | 370 |
| 20 | 27 | 57 | 87 | 104 | 134 | 163 | 191 | 221 | 250 | 277 | 296 | 325 | 353 | 369 |
| 21 | 30 | 60 | 88 | 106 | 136 | 164 | 193 | 222 | 250 | 279 | 297 | 326 | 356 | 369 |
| 22 | 30 | 58 | 87 | 105 | 132 | 162 | 192 | 220 | 249 | 279 | 295 | 325 | 356 | 369 |
| 23 | 28 | 56 | 85 | 101 | 131 | 161 | 190 | 219 | 249 | 278 | 295 | 325 | 352 | 369 |

CONSTITUENT $\nu$

| Series | 29 | 58 | 87 | 105 | 134 | 163 | 192 | 221 | 250 | 279 | 297 | 326 | 355 | 369 |
|---|---|---|---|---|---|---|---|---|---|---|---|---|---|---|
| *Hour* | | | | | | | | | | | | | | |
| 0 | 31 | 59 | 86 | 103 | 135 | 165 | 193 | 221 | 249 | 283 | 300 | 327 | 355 | 368 |
| 1 | 28 | 56 | 83 | 103 | 134 | 161 | 189 | 216 | 250 | 278 | 295 | 322 | 351 | 367 |
| 2 | 28 | 56 | 89 | 107 | 135 | 162 | 190 | 224 | 252 | 280 | 297 | 324 | 358 | 371 |
| 3 | 28 | 60 | 89 | 106 | 134 | 161 | 195 | 222 | 250 | 278 | 295 | 329 | 357 | 370 |
| 4 | 33 | 62 | 90 | 107 | 135 | 167 | 196 | 223 | 251 | 280 | 302 | 329 | 357 | 370 |
| 5 | 30 | 57 | 85 | 102 | 135 | 164 | 192 | 219 | 249 | 281 | 299 | 326 | 354 | 367 |
| 6 | 28 | 55 | 83 | 104 | 134 | 161 | 189 | 217 | 250 | 277 | 295 | 322 | 352 | 369 |
| 7 | 28 | 55 | 90 | 107 | 135 | 162 | 191 | 224 | 252 | 279 | 296 | 325 | 358 | 371 |
| 8 | 28 | 61 | 89 | 106 | 134 | 161 | 195 | 222 | 250 | 277 | 295 | 329 | 357 | 370 |
| 9 | 34 | 62 | 90 | 107 | 134 | 169 | 197 | 224 | 252 | 282 | 302 | 330 | 358 | 371 |
| 10 | 29 | 56 | 84 | 101 | 134 | 162 | 190 | 217 | 248 | 279 | 296 | 324 | 351 | 365 |
| 11 | 28 | 55 | 85 | 107 | 134 | 162 | 190 | 219 | 251 | 278 | 295 | 323 | 353 | 371 |
| 12 | 28 | 56 | 89 | 106 | 133 | 161 | 190 | 223 | 251 | 278 | 295 | 326 | 356 | 370 |
| 13 | 29 | 62 | 90 | 107 | 134 | 164 | 195 | 223 | 251 | 278 | 298 | 330 | 357 | 371 |
| 14 | 32 | 60 | 88 | 105 | 134 | 167 | 194 | 222 | 250 | 281 | 300 | 328 | 355 | 369 |
| 15 | 27 | 55 | 83 | 101 | 133 | 161 | 188 | 216 | 247 | 278 | 295 | 323 | 350 | 365 |
| 16 | 27 | 55 | 86 | 107 | 134 | 162 | 189 | 221 | 250 | 278 | 295 | 323 | 355 | 371 |
| 17 | 27 | 58 | 88 | 106 | 133 | 161 | 192 | 223 | 250 | 278 | 295 | 327 | 356 | 369 |
| 18 | 30 | 62 | 89 | 107 | 134 | 165 | 195 | 223 | 250 | 278 | 299 | 330 | 357 | 370 |
| 19 | 31 | 59 | 86 | 103 | 134 | 166 | 193 | 221 | 248 | 282 | 299 | 327 | 354 | 367 |
| 20 | 27 | 55 | 82 | 101 | 133 | 161 | 188 | 216 | 249 | 278 | 295 | 322 | 350 | 366 |
| 21 | 27 | 55 | 87 | 106 | 134 | 162 | 189 | 222 | 250 | 278 | 295 | 322 | 356 | 369 |
| 22 | 27 | 59 | 88 | 105 | 133 | 160 | 193 | 223 | 250 | 278 | 295 | 328 | 357 | 370 |
| 23 | 31 | 62 | 89 | 106 | 134 | 165 | 195 | 223 | 250 | 279 | 300 | 328 | 356 | 369 |

Table 32.—*Divisors for primary stencil sums*—Continued

CONSTITUENT p

| Series | 29 | 58 | 87 | 105 | 134 | 163 | 192 | 221 | 250 | 279 | 297 | 326 | 355 | 369 |
|---|---|---|---|---|---|---|---|---|---|---|---|---|---|---|
| *Hour* | | | | | | | | | | | | | | |
| 0 | 30 | 59 | 89 | 107 | 135 | 164 | 193 | 222 | 251 | 279 | 297 | 326 | 355 | 369 |
| 1 | 29 | 59 | 88 | 105 | 134 | 164 | 193 | 222 | 251 | 280 | 299 | 328 | 356 | 372 |
| 2 | 28 | 57 | 87 | 105 | 134 | 162 | 191 | 220 | 249 | 278 | 295 | 324 | 353 | 367 |
| 3 | 29 | 58 | 87 | 106 | 135 | 165 | 194 | 224 | 252 | 281 | 299 | 328 | 357 | 371 |
| 4 | 30 | 58 | 87 | 106 | 135 | 163 | 192 | 221 | 250 | 279 | 297 | 326 | 355 | 370 |
| 5 | 29 | 58 | 87 | 106 | 134 | 163 | 192 | 220 | 250 | 278 | 296 | 325 | 354 | 368 |
| 6 | 28 | 58 | 87 | 106 | 136 | 165 | 193 | 223 | 252 | 282 | 299 | 328 | 357 | 371 |
| 7 | 29 | 58 | 87 | 104 | 134 | 163 | 192 | 221 | 249 | 278 | 296 | 325 | 354 | 369 |
| 8 | 29 | 58 | 87 | 105 | 135 | 165 | 194 | 222 | 250 | 280 | 298 | 326 | 355 | 370 |
| 9 | 29 | 58 | 87 | 105 | 133 | 162 | 192 | 222 | 251 | 280 | 298 | 327 | 357 | 371 |
| 10 | 29 | 57 | 86 | 104 | 133 | 162 | 191 | 220 | 249 | 278 | 296 | 325 | 353 | 366 |
| 11 | 28 | 58 | 87 | 105 | 134 | 163 | 193 | 223 | 252 | 280 | 298 | 327 | 356 | 369 |
| 12 | 29 | 58 | 86 | 104 | 133 | 162 | 190 | 219 | 249 | 279 | 297 | 326 | 355 | 369 |
| 13 | 30 | 59 | 88 | 106 | 135 | 163 | 192 | 221 | 250 | 280 | 297 | 326 | 355 | 368 |
| 14 | 29 | 58 | 87 | 104 | 134 | 163 | 193 | 221 | 250 | 280 | 299 | 328 | 357 | 370 |
| 15 | 28 | 57 | 86 | 104 | 132 | 162 | 190 | 219 | 248 | 276 | 295 | 324 | 354 | 367 |
| 16 | 30 | 59 | 88 | 106 | 135 | 164 | 193 | 222 | 250 | 279 | 298 | 327 | 355 | 369 |
| 17 | 28 | 57 | 86 | 104 | 133 | 161 | 191 | 220 | 250 | 279 | 298 | 327 | 356 | 370 |
| 18 | 29 | 58 | 87 | 104 | 133 | 162 | 191 | 220 | 248 | 277 | 295 | 325 | 354 | 367 |
| 19 | 30 | 58 | 87 | 106 | 135 | 164 | 193 | 222 | 251 | 280 | 297 | 326 | 356 | 369 |
| 20 | 28 | 57 | 86 | 104 | 132 | 161 | 190 | 219 | 249 | 277 | 295 | 325 | 354 | 368 |
| 21 | 30 | 59 | 87 | 105 | 134 | 163 | 191 | 220 | 249 | 278 | 296 | 324 | 353 | 368 |
| 22 | 29 | 58 | 88 | 105 | 135 | 164 | 193 | 222 | 251 | 281 | 298 | 328 | 357 | 371 |
| 23 | 29 | 58 | 86 | 104 | 133 | 162 | 191 | 219 | 248 | 277 | 295 | 323 | 352 | 367 |

CONSTITUENT MK

| Series | 29 | 58 | 87 | 105 | 134 | 163 | 192 | 221 | 250 | 279 | 297 | 326 | 355 | 369 |
|---|---|---|---|---|---|---|---|---|---|---|---|---|---|---|
| *Hour.* | | | | | | | | | | | | | | |
| 0 | 30 | 59 | 88 | 105 | 135 | 164 | 192 | 222 | 251 | 279 | 297 | 325 | 355 | 368 |
| 1 | 29 | 58 | 88 | 106 | 135 | 163 | 192 | 222 | 251 | 280 | 298 | 327 | 356 | 369 |
| 2 | 29 | 58 | 88 | 105 | 134 | 164 | 192 | 221 | 249 | 278 | 297 | 326 | 355 | 369 |
| 3 | 30 | 58 | 86 | 104 | 133 | 163 | 191 | 221 | 250 | 278 | 296 | 325 | 355 | 368 |
| 4 | 29 | 58 | 87 | 106 | 134 | 163 | 192 | 222 | 251 | 280 | 298 | 326 | 356 | 369 |
| 5 | 29 | 58 | 87 | 105 | 134 | 164 | 192 | 220 | 250 | 279 | 298 | 327 | 356 | 369 |
| 6 | 30 | 58 | 86 | 105 | 134 | 164 | 193 | 221 | 250 | 279 | 297 | 326 | 356 | 369 |
| 7 | 30 | 58 | 87 | 105 | 133 | 163 | 192 | 221 | 251 | 279 | 297 | 326 | 355 | 369 |
| 8 | 30 | 59 | 88 | 106 | 135 | 164 | 193 | 221 | 251 | 280 | 299 | 327 | 355 | 369 |
| 9 | 28 | 57 | 86 | 105 | 134 | 164 | 192 | 220 | 249 | 278 | 296 | 325 | 354 | 369 |
| 10 | 29 | 58 | 86 | 104 | 133 | 163 | 192 | 221 | 251 | 279 | 297 | 326 | 355 | 369 |
| 11 | 29 | 59 | 88 | 106 | 134 | 162 | 192 | 221 | 250 | 279 | 298 | 326 | 354 | 369 |
| 12 | 28 | 58 | 86 | 105 | 134 | 163 | 192 | 220 | 250 | 279 | 297 | 326 | 355 | 369 |
| 13 | 29 | 58 | 86 | 105 | 133 | 162 | 192 | 221 | 250 | 278 | 297 | 326 | 355 | 369 |
| 14 | 29 | 59 | 88 | 106 | 134 | 163 | 193 | 222 | 250 | 280 | 297 | 326 | 355 | 370 |
| 15 | 29 | 59 | 87 | 106 | 135 | 163 | 192 | 221 | 250 | 280 | 297 | 327 | 355 | 369 |
| 16 | 28 | 57 | 86 | 105 | 133 | 162 | 192 | 220 | 249 | 278 | 296 | 326 | 355 | 369 |
| 17 | 29 | 59 | 87 | 104 | 134 | 163 | 193 | 222 | 250 | 279 | 296 | 325 | 354 | 369 |
| 18 | 29 | 58 | 87 | 105 | 134 | 162 | 191 | 220 | 249 | 279 | 296 | 326 | 354 | 368 |
| 19 | 28 | 57 | 87 | 105 | 135 | 163 | 193 | 221 | 250 | 279 | 297 | 327 | 356 | 370 |
| 20 | 29 | 57 | 86 | 103 | 133 | 162 | 191 | 221 | 249 | 279 | 296 | 325 | 354 | 369 |
| 21 | 29 | 58 | 88 | 105 | 134 | 162 | 191 | 221 | 250 | 280 | 297 | 326 | 355 | 369 |
| 22 | 28 | 57 | 87 | 105 | 135 | 163 | 191 | 221 | 250 | 280 | 297 | 327 | 355 | 369 |
| 23 | 29 | 57 | 87 | 104 | 134 | 163 | 192 | 221 | 249 | 278 | 297 | 325 | 355 | 370 |

HARMONIC ANALYSIS AND PREDICTION OF TIDES 297

Table 32.—*Divisors for primary stencil sums*—Continued

CONSTITUENT 2MK

| Series | 29 | 58 | 87 | 105 | 134 | 163 | 192 | 221 | 250 | 279 | 297 | 326 | 355 | 369 |
|---|---|---|---|---|---|---|---|---|---|---|---|---|---|---|
| *Hour* | | | | | | | | | | | | | | |
| 0 | 30 | 59 | 87 | 106 | 134 | 164 | 193 | 221 | 251 | 280 | 298 | 326 | 355 | 369 |
| 1 | 29 | 59 | 87 | 106 | 134 | 164 | 193 | 221 | 250 | 280 | 298 | 327 | 356 | 370 |
| 2 | 29 | 59 | 87 | 106 | 134 | 163 | 192 | 221 | 250 | 279 | 297 | 325 | 355 | 368 |
| 3 | 29 | 58 | 86 | 105 | 133 | 163 | 193 | 222 | 251 | 280 | 298 | 326 | 355 | 369 |
| 4 | 30 | 59 | 87 | 105 | 134 | 164 | 192 | 221 | 250 | 279 | 297 | 326 | 355 | 369 |
| 5 | 29 | 59 | 88 | 106 | 135 | 164 | 192 | 222 | 251 | 279 | 298 | 327 | 357 | 370 |
| 6 | 29 | 58 | 86 | 104 | 134 | 163 | 191 | 221 | 250 | 278 | 297 | 325 | 354 | 368 |
| 7 | 30 | 59 | 87 | 105 | 135 | 164 | 192 | 221 | 251 | 280 | 299 | 327 | 356 | 370 |
| 8 | 30 | 58 | 87 | 105 | 134 | 163 | 192 | 221 | 250 | 279 | 296 | 326 | 355 | 368 |
| 9 | 29 | 57 | 87 | 104 | 134 | 164 | 192 | 221 | 251 | 279 | 297 | 326 | 356 | 369 |
| 10 | 30 | 58 | 88 | 105 | 135 | 164 | 192 | 221 | 251 | 279 | 297 | 326 | 355 | 369 |
| 11 | 30 | 59 | 88 | 106 | 135 | 164 | 193 | 222 | 251 | 280 | 297 | 327 | 356 | 370 |
| 12 | 29 | 57 | 87 | 105 | 134 | 162 | 192 | 221 | 249 | 278 | 296 | 326 | 356 | 369 |
| 13 | 29 | 57 | 87 | 104 | 134 | 162 | 191 | 221 | 249 | 278 | 296 | 325 | 354 | 368 |
| 14 | 29 | 57 | 86 | 104 | 133 | 162 | 191 | 220 | 249 | 279 | 297 | 326 | 355 | 369 |
| 15 | 29 | 58 | 87 | 105 | 134 | 162 | 192 | 221 | 249 | 279 | 296 | 326 | 355 | 369 |
| 16 | 28 | 57 | 87 | 105 | 135 | 163 | 192 | 222 | 250 | 279 | 297 | 326 | 354 | 369 |
| 17 | 28 | 57 | 86 | 104 | 133 | 162 | 191 | 220 | 249 | 278 | 296 | 325 | 353 | 368 |
| 18 | 29 | 59 | 88 | 106 | 135 | 163 | 193 | 222 | 250 | 280 | 298 | 328 | 356 | 371 |
| 19 | 28 | 58 | 87 | 105 | 134 | 162 | 192 | 221 | 249 | 278 | 296 | 325 | 354 | 368 |
| 20 | 28 | 57 | 87 | 104 | 133 | 162 | 191 | 220 | 250 | 279 | 297 | 326 | 355 | 369 |
| 21 | 29 | 58 | 87 | 105 | 133 | 163 | 192 | 220 | 250 | 279 | 297 | 326 | 354 | 369 |
| 22 | 28 | 58 | 87 | 106 | 134 | 163 | 193 | 221 | 250 | 279 | 297 | 326 | 355 | 370 |
| 23 | 28 | 57 | 87 | 104 | 133 | 162 | 191 | 219 | 249 | 278 | 296 | 325 | 354 | 368 |

CONSTITUENT MN

| Series | 29 | 58 | 87 | 105 | 134 | 163 | 192 | 221 | 250 | 279 | 297 | 326 | 355 | 269 |
|---|---|---|---|---|---|---|---|---|---|---|---|---|---|---|
| *Hour* | | | | | | | | | | | | | | |
| 0 | 28 | 56 | 85 | 104 | 134 | 163 | 193 | 223 | 253 | 283 | 301 | 328 | 356 | 370 |
| 1 | 30 | 60 | 90 | 109 | 139 | 166 | 194 | 222 | 250 | 277 | 296 | 325 | 355 | 370 |
| 2 | 28 | 56 | 84 | 101 | 129 | 158 | 188 | 218 | 248 | 278 | 297 | 326 | 355 | 369 |
| 3 | 30 | 60 | 91 | 109 | 139 | 168 | 198 | 226 | 254 | 281 | 299 | 326 | 355 | 370 |
| 4 | 30 | 59 | 87 | 104 | 132 | 159 | 187 | 217 | 248 | 277 | 296 | 325 | 355 | 369 |
| 5 | 28 | 57 | 88 | 106 | 136 | 165 | 195 | 225 | 256 | 283 | 301 | 328 | 356 | 369 |
| 6 | 30 | 61 | 90 | 109 | 136 | 164 | 192 | 220 | 247 | 277 | 295 | 325 | 355 | 369 |
| 7 | 28 | 56 | 83 | 101 | 130 | 160 | 190 | 221 | 250 | 280 | 298 | 327 | 355 | 368 |
| 8 | 30 | 61 | 90 | 109 | 138 | 168 | 196 | 224 | 251 | 279 | 296 | 325 | 355 | 369 |
| 9 | 29 | 57 | 84 | 102 | 129 | 157 | 187 | 218 | 247 | 277 | 295 | 325 | 355 | 369 |
| 10 | 30 | 60 | 89 | 108 | 137 | 167 | 198 | 228 | 255 | 283 | 300 | 328 | 356 | 369 |
| 11 | 31 | 61 | 89 | 107 | 134 | 162 | 190 | 218 | 247 | 277 | 295 | 325 | 356 | 370 |
| 12 | 28 | 55 | 84 | 102 | 132 | 162 | 193 | 222 | 252 | 282 | 299 | 327 | 355 | 368 |
| 13 | 30 | 59 | 89 | 107 | 137 | 165 | 193 | 220 | 248 | 276 | 294 | 324 | 355 | 369 |
| 14 | 28 | 55 | 83 | 100 | 128 | 159 | 189 | 218 | 248 | 278 | 296 | 327 | 355 | 368 |
| 15 | 30 | 59 | 89 | 107 | 137 | 168 | 198 | 225 | 253 | 281 | 298 | 326 | 356 | 370 |
| 16 | 30 | 58 | 86 | 103 | 131 | 159 | 187 | 216 | 246 | 276 | 294 | 325 | 355 | 369 |
| 17 | 28 | 56 | 86 | 104 | 135 | 165 | 195 | 224 | 254 | 282 | 299 | 327 | 355 | 368 |
| 18 | 29 | 59 | 89 | 107 | 135 | 163 | 190 | 218 | 246 | 276 | 295 | 325 | 354 | 369 |
| 19 | 27 | 55 | 83 | 100 | 131 | 161 | 190 | 220 | 250 | 280 | 299 | 328 | 355 | 369 |
| 20 | 29 | 59 | 89 | 108 | 138 | 168 | 195 | 223 | 251 | 279 | 296 | 325 | 354 | 369 |
| 21 | 28 | 56 | 84 | 101 | 129 | 157 | 186 | 216 | 246 | 276 | 295 | 325 | 354 | 369 |
| 22 | 28 | 58 | 88 | 107 | 137 | 167 | 196 | 226 | 254 | 282 | 299 | 327 | 354 | 368 |
| 23 | 29 | 59 | 88 | 105 | 133 | 161 | 188 | 216 | 246 | 276 | 295 | 325 | 354 | 369 |

Table 32.—*Divisors for primary stencil sums*—Continued

CONSTITUENT MS

| Series | 29 | 58 | 87 | 105 | 134 | 163 | 192 | 221 | 250 | 279 | 297 | 326 | 355 | 369 |
|---|---|---|---|---|---|---|---|---|---|---|---|---|---|---|
| Hour | | | | | | | | | | | | | | |
| 0 | 30 | 59 | 88 | 106 | 135 | 164 | 192 | 222 | 250 | 279 | 297 | 326 | 354 | 369 |
| 1 | 30 | 58 | 88 | 106 | 134 | 164 | 192 | 221 | 250 | 279 | 296 | 326 | 355 | 369 |
| 2 | 29 | 58 | 87 | 105 | 134 | 163 | 192 | 222 | 250 | 280 | 298 | 327 | 355 | 369 |
| 3 | 30 | 58 | 88 | 107 | 135 | 165 | 194 | 223 | 252 | 281 | 298 | 328 | 356 | 370 |
| 4 | 29 | 58 | 87 | 105 | 134 | 163 | 191 | 221 | 249 | 279 | 296 | 325 | 354 | 368 |
| 5 | 30 | 58 | 88 | 105 | 134 | 164 | 192 | 221 | 250 | 280 | 297 | 327 | 256 | 370 |
| 6 | 29 | 58 | 88 | 105 | 135 | 164 | 194 | 223 | 252 | 281 | 299 | 328 | 356 | 371 |
| 7 | 30 | 58 | 87 | 105 | 134 | 162 | 192 | 220 | 250 | 278 | 296 | 326 | 354 | 368 |
| 8 | 29 | 58 | 87 | 104 | 134 | 162 | 191 | 220 | 249 | 278 | 296 | 326 | 355 | 369 |
| 9 | 29 | 57 | 87 | 104 | 133 | 162 | 192 | 220 | 250 | 279 | 297 | 326 | 355 | 369 |
| 10 | 30 | 59 | 88 | 106 | 136 | 164 | 193 | 222 | 251 | 279 | 298 | 327 | 355 | 370 |
| 11 | 30 | 58 | 88 | 105 | 134 | 163 | 192 | 220 | 250 | 278 | 296 | 326 | 354 | 368 |
| 12 | 28 | 58 | 87 | 104 | 134 | 162 | 192 | 221 | 250 | 279 | 298 | 326 | 356 | 370 |
| 13 | 28 | 57 | 86 | 105 | 134 | 163 | 193 | 221 | 251 | 279 | 297 | 325 | 355 | 369 |
| 14 | 29 | 59 | 87 | 105 | 135 | 163 | 192 | 221 | 250 | 278 | 297 | 325 | 354 | 369 |
| 15 | 29 | 58 | 87 | 105 | 134 | 163 | 192 | 220 | 250 | 278 | 296 | 325 | 355 | 369 |
| 16 | 28 | 58 | 86 | 104 | 134 | 163 | 193 | 222 | 251 | 280 | 299 | 327 | 356 | 371 |
| 17 | 29 | 58 | 87 | 106 | 135 | 163 | 193 | 221 | 251 | 279 | 297 | 326 | 355 | 369 |
| 18 | 28 | 58 | 86 | 104 | 133 | 162 | 190 | 220 | 248 | 278 | 296 | 324 | 354 | 367 |
| 19 | 28 | 57 | 86 | 104 | 132 | 162 | 191 | 220 | 249 | 279 | 297 | 326 | 356 | 369 |
| 20 | 29 | 59 | 87 | 106 | 134 | 164 | 192 | 222 | 250 | 279 | 298 | 326 | 355 | 369 |
| 21 | 29 | 58 | 86 | 105 | 133 | 162 | 191 | 220 | 249 | 278 | 296 | 325 | 354 | 367 |
| 22 | 28 | 58 | 86 | 104 | 133 | 162 | 190 | 220 | 248 | 278 | 296 | 325 | 355 | 368 |
| 23 | 28 | 57 | 86 | 105 | 133 | 163 | 192 | 221 | 250 | 280 | 297 | 326 | 356 | 369 |

CONSTITUENT 2SM

| Series | 29 | 58 | 87 | 105 | 134 | 163 | 192 | 221 | 250 | 279 | 297 | 326 | 355 | 369 |
|---|---|---|---|---|---|---|---|---|---|---|---|---|---|---|
| Hour | | | | | | | | | | | | | | |
| 0 | 28 | 57 | 87 | 106 | 136 | 164 | 192 | 220 | 250 | 280 | 299 | 329 | 356 | 369 |
| 1 | 30 | 60 | 88 | 106 | 133 | 163 | 193 | 223 | 252 | 280 | 297 | 325 | 355 | 370 |
| 2 | 28 | 56 | 86 | 105 | 135 | 165 | 193 | 220 | 249 | 279 | 297 | 327 | 356 | 370 |
| 3 | 30 | 60 | 89 | 107 | 135 | 163 | 193 | 223 | 253 | 281 | 298 | 326 | 355 | 370 |
| 4 | 28 | 55 | 84 | 103 | 133 | 163 | 191 | 219 | 247 | 276 | 294 | 324 | 353 | 367 |
| 5 | 30 | 60 | 90 | 107 | 135 | 163 | 193 | 223 | 253 | 282 | 299 | 326 | 355 | 370 |
| 6 | 28 | 56 | 84 | 103 | 133 | 163 | 192 | 220 | 248 | 277 | 295 | 325 | 355 | 370 |
| 7 | 30 | 60 | 90 | 109 | 137 | 164 | 193 | 223 | 253 | 283 | 300 | 328 | 356 | 370 |
| 8 | 29 | 57 | 85 | 103 | 133 | 163 | 193 | 221 | 249 | 277 | 295 | 325 | 355 | 370 |
| 9 | 29 | 59 | 89 | 108 | 137 | 165 | 193 | 222 | 252 | 282 | 300 | 328 | 356 | 370 |
| 10 | 30 | 59 | 86 | 104 | 133 | 163 | 193 | 223 | 251 | 279 | 295 | 325 | 355 | 370 |
| 11 | 28 | 57 | 87 | 106 | 136 | 164 | 192 | 220 | 249 | 279 | 298 | 327 | 355 | 369 |
| 12 | 30 | 59 | 87 | 104 | 132 | 161 | 191 | 221 | 249 | 277 | 295 | 323 | 353 | 367 |
| 13 | 28 | 57 | 87 | 105 | 135 | 164 | 191 | 219 | 249 | 279 | 298 | 328 | 356 | 369 |
| 14 | 30 | 60 | 88 | 105 | 133 | 162 | 192 | 222 | 251 | 279 | 297 | 325 | 355 | 369 |
| 15 | 27 | 55 | 85 | 103 | 133 | 163 | 191 | 219 | 247 | 277 | 296 | 326 | 355 | 368 |
| 16 | 30 | 60 | 89 | 106 | 134 | 162 | 192 | 222 | 252 | 281 | 298 | 326 | 355 | 369 |
| 17 | 28 | 56 | 84 | 102 | 132 | 162 | 191 | 219 | 247 | 276 | 295 | 325 | 355 | 368 |
| 18 | 30 | 60 | 90 | 108 | 135 | 163 | 192 | 222 | 252 | 282 | 300 | 328 | 355 | 369 |
| 19 | 29 | 57 | 85 | 102 | 132 | 162 | 192 | 220 | 248 | 276 | 295 | 325 | 355 | 369 |
| 20 | 29 | 59 | 89 | 107 | 135 | 163 | 190 | 220 | 250 | 280 | 298 | 326 | 354 | 367 |
| 21 | 29 | 57 | 85 | 102 | 132 | 162 | 192 | 221 | 249 | 276 | 295 | 325 | 355 | 369 |
| 22 | 28 | 58 | 88 | 106 | 135 | 163 | 191 | 220 | 250 | 280 | 299 | 327 | 355 | 368 |
| 23 | 30 | 58 | 86 | 103 | 132 | 162 | 192 | 222 | 250 | 278 | 295 | 325 | 355 | 369 |

HARMONIC ANALYSIS AND PREDICTION OF TIDES 299

Table 33.—*For construction of secondary stencils*

| Constituent A. | | | | S | | | | L | | | |
|---|---|---|---|---|---|---|---|---|---|---|---|
| Constituent B. | OO | | 2SM | | K and P | | R and T | | MS | | λ | |
| Page | J hours | Difference, hours | J hours | Difference, hours | S hours | Difference, hours | S hours | Difference, hours | L hours | Difference, hours | L hours | Difference, hours |
|  |  | + |  | − |  | ± |  | ± |  | − |  | − |
| 1 | 0–23 | 3 | 0–23 | 0 | 0–23 | 0 | 0–23 | 0 | 0–23 | 0 | 0–23 | 0 |
| 2 | 10– 3 | 9 | 0–23 | 1 | 0–23 | 1 | 0–23 | 0 | 0–23 | 0 | 0–23 | 1 |
| 3 | 16– 4 | 15 | 0–23 | 2 | 0–23 | 1 | 0–23 | 1 | 17–21 | 0 | 0–23 | 1 |
| 4 | 23– 5 | 21 | 0–23 | 3 | 0–23 | 2 | 0–23 | 1 | 0–23 | 1 | 0–23 | 1 |
| 5 | 5– 6 | 3 | 0–23 | 4 | 0–23 | 2 | 0–23 | 1 | 0–23 | 1 | 0–23 | 2 |
| 6 | 0–23 | 10 | 0–23 | 5 | 0– 1 | 2 | 0–23 | 1 | 0–23 | 1 | 0–23 | 2 |
| 7 | 19–12 | 16 | 0–23 | 6 | 0–23 | 3 | 0–15 | 1 | 0–23 | 1 | 0–23 | 3 |
| 8 | 1–12 | 22 | 1–11 | 6 | 0–23 | 3 | 0–23 | 2 | 0–23 | 2 | 0–23 | 3 |
| 9 | 8–13 | 4 | 0–23 | 7 | 0–23 | 4 | 0–23 | 2 | 0–23 | 2 | 0–23 | 3 |
| 10 | 14 | 10 | 0–23 | 8 | 0–23 | 4 | 0–23 | 2 | 0–23 | 2 | 0–23 | 4 |
| 11 | 0–23 | 17 | 0–23 | 9 | 0–23 | 5 | 0–23 | 2 | 0–23 | 2 | 0–23 | 4 |
| 12 | 3–20 | 23 | 0–23 | 10 | 0–23 | 5 | 0–23 | 3 | 0–23 | 2 | 0–23 | 5 |
| 13 | 10–21 | 5 | 0–23 | 11 | 0–23 | 6 | 0–23 | 3 | 0–23 | 3 | 0–23 | 5 |
| 14 | 16–22 | 11 | 0–23 | 12 | 0–23 | 6 | 0–23 | 3 | 0–23 | 3 | 12–20 | 5 |
| 15 | 23 | 17 | 0–23 | 13 | 0–23 | 7 | 0–23 | 3 | 0–23 | 3 | 0–23 | 6 |
| 16 | 6– 4 | 0 | 6– 3 | 13 | 0–23 | 7 | 0–23 | 4 | 0–23 | 3 | 0–23 | 6 |
| 17 | 12– 5 | 6 | 0–23 | 14 | 0–23 | 8 | 0–23 | 4 | 0–23 | 3 | 0–23 | 7 |
| 18 | 19– 6 | 12 | 0–23 | 15 | 0–23 | 8 | 0–23 | 4 | 0–23 | 4 | 0–23 | 7 |
| 19 | 1– 7 | 18 | 0–23 | 16 | 0– 8 | 8 | 0–23 | 4 | 0–23 | 4 | 0–23 | 8 |
| 20 | 8 | 0 | 0–23 | 17 | 0–23 | 9 | 0–23 | 4 | 0–23 | 4 | 0–23 | 8 |
| 21 | 0–23 | 7 | 0–23 | 18 | 0–23 | 9 | 0–23 | 5 | 0–23 | 4 | 0–23 | 8 |
| 22 | 21–14 | 13 | 0–23 | 19 | 0–23 | 10 | 0–23 | 5 | 0–23 | 4 | 0–23 | 9 |
| 23 | 4–14 | 19 | 4– 9 | 19 | 0–23 | 10 | 0–23 | 5 | 0–23 | 5 | 0–23 | 9 |
| 24 | 10–15 | 1 | 0–23 | 20 | 0–23 | 11 | 0–23 | 5 | 0–23 | 5 | 0–23 | 10 |
| 25 | 0–23 | 8 | 0–23 | 21 | 0–23 | 11 | 0–23 | 6 | 0–23 | 5 | 0–23 | 10 |
| 26 | 23–21 | 14 | 0–23 | 22 | 0–23 | 12 | 0–23 | 6 | 0–23 | 5 | 0–23 | 10 |
| 27 | 6–22 | 20 | 0–23 | 23 | 0–23 | 12 | 0–23 | 6 | 0–23 | 5 | 0–23 | 11 |
| 28 | 12–23 | 2 | 0–23 | 0 | 0–23 | 13 | 0–23 | 6 | 0–23 | 6 | 0–23 | 11 |
| 29 | 19– 0 | 8 | 0–23 | 1 | 0–23 | 13 | 0–23 | 7 | 0–23 | 6 | 0–23 | 12 |
| 30 | 1 | 14 | 0–23 | 2 | 0–23 | 14 | 0–23 | 7 | 0–23 | 6 | 0–23 | 12 |
| 31 | 8– 6 | 21 | 8– 1 | 2 | 0–23 | 14 | 0–23 | 7 | 0–23 | 6 | 0–23 | 12 |
| 32 | 15– 7 | 3 | 0–23 | 3 | 0–15 | 14 | 0–23 | 7 | 0–23 | 6 | 0–23 | 13 |
| 33 | 21– 8 | 9 | 0–23 | 4 | 0–23 | 15 | 0–23 | 7 | 0–23 | 7 | 0–23 | 13 |
| 34 | 4– 9 | 15 | 0–23 | 5 | 0–23 | 15 | 0–23 | 8 | 0–23 | 7 | 0–23 | 14 |
| 35 | 0–23 | 22 | 0–23 | 6 | 0–23 | 16 | 0–23 | 8 | 0–23 | 7 | 0–23 | 14 |
| 36 | 17–15 | 4 | 0–23 | 7 | 0–23 | 16 | 0–23 | 8 | 0–23 | 7 | 2–21 | 14 |
| 37 | 23–15 | 10 | 0–23 | 8 | 0–23 | 17 | 0–23 | 8 | 0–23 | 7 | 0–23 | 15 |
| 38 | 6–16 | 16 | 6– 8 | 8 | 0–23 | 17 | 0–23 | 9 | 0–23 | 8 | 0–23 | 15 |
| 39 | 12–17 | 22 | 0–23 | 9 | 0–23 | 18 | 0–23 | 9 | 0–23 | 8 | 0–23 | 16 |
| 40 | 0–23 | 5 | 0–23 | 10 | 0–23 | 18 | 0–23 | 9 | 0–23 | 8 | 0–23 | 16 |
| 41 | 2–23 | 11 | 0–23 | 11 | 0–23 | 19 | 0–23 | 9 | 0–23 | 8 | 13–15 | 16 |
| 42 | 8– 0 | 17 | 0–23 | 12 | 0–23 | 19 | 0–23 | 10 | 0–23 | 8 | 0–23 | 17 |
| 43 | 15– 1 | 23 | 0–23 | 13 | 0–23 | 20 | 0–23 | 10 | 0–23 | 9 | 0–23 | 17 |
| 44 | 21– 2 | 5 | 0–23 | 14 | 0–23 | 20 | 0–23 | 10 | 0–23 | 9 | 0–23 | 18 |
| 45 | 0–23 | 12 | 0–23 | 15 | 0–23 | 20 | 0–23 | 10 | 0–23 | 9 | 0–23 | 18 |
| 46 | 10– 8 | 18 | 10–23 | 15 | 0–23 | 21 | 0–23 | 10 | 0–23 | 9 | 0–23 | 19 |
| 47 | 17– 9 | 0 | 0–23 | 16 | 0–23 | 21 | 0–23 | 11 | 0–23 | 9 | 0–23 | 19 |
| 48 | 23–10 | 6 | 0–23 | 17 | 0–23 | 22 | 0–23 | 11 | 0–23 | 10 | 0–23 | 19 |
| 49 | 6–11 | 12 | 0–23 | 18 | 0–23 | 22 | 0–23 | 11 | 0–23 | 10 | 0–23 | 20 |
| 50 | 0–23 | 19 | 0–23 | 19 | 0–23 | 23 | 0–23 | 11 | 0–23 | 10 | 0–23 | 20 |
| 51 | 19–16 | 1 | 0–23 | 20 | 0–23 | 23 | 0–23 | 12 | 0–23 | 10 | 0–23 | 21 |
| 52 | 2–17 | 7 | 0–23 | 21 | 0–23 | 0 | 0–23 | 12 | 8–16 | 10 | 0–23 | 21 |
| (53) | 7–14 | 12 | 0–23 | 21 | 0–23 | 0 | 0–23 | 12 | 0–23 | 11 | 0–23 | 21 |

300  U. S. COAST AND GEODETIC SURVEY

Table 33.—*For construction of secondary stencils*—Continued

| Constituent A | L | | M | | N | | N | | O | | O | Di... |
|---|---|---|---|---|---|---|---|---|---|---|---|---|
| Constituent B | MK | | MN | | 2MK | | ν | | μ | | 2N | |
| Page | L hours | Difference, hours | M hours | Difference, hours | N hours | Difference, hours | N hours | Difference, hours | O hours | Difference, hours | O hours | Di... en ho |
|  |  | − |  | − |  | + |  | + |  | + |  | + |
| 1 | 23–10 | 0 | 0–23 | 1 | 20– 7 | 0 | 0–23 | 0 | 0–23 | 0 | 0–23 | |
| 2 | 20– 8 | 1 | 0–23 | 2 | 11–23 | 1 | 0–23 | 1 | 0–23 | 1 | 0–23 | |
| 3 | 17– 5 | 2 | 0–23 | 4 | 2–14 | 2 | 0–23 | 1 | 0–23 | 1 | 0–23 | |
| 4 | 15– 3 | 3 | 0–23 | 5 | 17– 6 | 3 | 0–23 | 1 | 0–23 | 2 | 0–23 | |
| 5 | 12– 1 | 4 | 0–23 | 7 | 9–21 | 4 | 0–23 | 2 | 0–23 | 2 | 0–23 | |
| 6 | 9–22 | 5 | 0–23 | 8 | 0–13 | 5 | 0–23 | 2 | 7– 8 | 2 | 0–23 | |
| 7 | 7–20 | 6 | 0–23 | 10 | 15– 4 | 6 | 0–23 | 3 | 0–23 | 3 | 0–23 | |
| 8 | 4–17 | 7 | 5– 0 | 11 | 6–19 | 7 | 0–23 | 3 | 0–23 | 3 | 0–23 | |
| 9 | 2–15 | 8 | 0–23 | 13 | 22–11 | 8 | 0–23 | 3 | 0–23 | 4 | 0–23 | |
| 10 | 23–12 | 9 | 18– 8 | 14 | 13– 2 | 9 | 0–23 | 4 | 0–23 | 4 | 0–23 | |
| 11 | 20–10 | 10 | 0–23 | 16 | 4–18 | 10 | 0–23 | 4 | 0–23 | 5 | 0–23 | |
| 12 | 18– 7 | 11 | 7–15 | 17 | 20– 9 | 11 | 0–23 | 5 | 0–23 | 5 | 0–23 | |
| 13 | 15– 5 | 12 | 0–23 | 19 | 11– 1 | 12 | 0–23 | 5 | 0–23 | 6 | 0–23 | |
| 14 | 12– 2 | 13 | 19–22 | 20 | 2–16 | 13 | 2–10 | 5 | 0–23 | 6 | 0–23 | |
| 15 | 10– 0 | 14 | 0–23 | 22 | 17– 2 | 14 | 0–23 | 6 | 0–23 | 7 | 0–23 | |
| 16 | 7–21 | 15 | 0–23 | 0 | 9–23 | 15 | 0–23 | 6 | 0–23 | 7 | 0–23 | |
| 17 | 4–19 | 16 | 0–23 | 1 | 0–15 | 16 | 0–23 | 7 | 0–23 | 8 | 0–23 | |
| 18 | 2–17 | 17 | 0–23 | 3 | 15– 6 | 17 | 0–23 | 7 | 0–23 | 8 | 0–23 | |
| 19 | 23–14 | 18 | 0–23 | 4 | 6–21 | 18 | 0–23 | 8 | 21– 5 | 8 | 0–23 | |
| 20 | 21–12 | 19 | 0–23 | 6 | 22–13 | 19 | 0–23 | 8 | 0–23 | 9 | 0–23 | |
| 21 | 18– 9 | 20 | 0–23 | 7 | 13– 4 | 20 | 0–23 | 8 | 0–23 | 9 | 0–23 | |
| 22 | 15– 7 | 21 | 0–23 | 9 | 4–20 | 21 | 0–23 | 9 | 0–23 | 10 | 0–23 | |
| 23 | 13– 5 | 22 | 0–23 | 10 | 19–11 | 22 | 0–23 | 9 | 0–23 | 10 | 0–23 | |
| 24 | 10– 2 | 23 | 0–23 | 12 | 11– 3 | 23 | 0–23 | 10 | 0–23 | 11 | 0–25 | |
| 25 | 7– 0 | 0 | 0–23 | 13 | 2–18 | 0 | 0–23 | 10 | 0–23 | 11 | 0–23 | |
| 26 | 5–21 | 1 | 0–23 | 15 | 17–10 | 1 | 0–23 | 10 | 0–23 | 12 | 0–23 | |
| 27 | 2–19 | 2 | 0–23 | 16 | 8– 1 | 2 | 0–23 | 11 | 0–23 | 12 | 0–23 | |
| 28 | 23–16 | 3 | 0–23 | 18 | 0–16 | 3 | 0–23 | 11 | 0–23 | 13 | 0–23 | |
| 29 | 21–14 | 4 | 6– 0 | 19 | 15– 8 | 4 | 0–23 | 12 | 0–23 | 13 | 0–23 | |
| 30 | 18–11 | 5 | 0–23 | 21 | 6–23 | 5 | 0–23 | 12 | 0–23 | 14 | 0–23 | |
| 31 | 15– 9 | 6 | 19– 7 | 22 | 22–15 | 6 | 0–23 | 12 | 0–23 | 14 | 0–23 | |
| 32 | 13– 6 | 7 | 0–23 | 0 | 13– 6 | 7 | 0–23 | 13 | 11– 4 | 14 | 0–23 | |
| 33 | 10– 4 | 8 | 7–15 | 1 | 4–22 | 8 | 0–23 | 13 | 0–23 | 15 | 0–23 | |
| 34 | 8– 1 | 9 | 0–23 | 3 | 19–13 | 9 | 0–23 | 14 | 0–23 | 15 | 0–23 | |
| 35 | 5–23 | 10 | 20–22 | 4 | 11– 5 | 10 | 0–23 | 14 | 0–23 | 16 | 0–23 | |
| 36 | 2–21 | 11 | 0–23 | 6 | 2–20 | 11 | 2–20 | 14 | 0–23 | 16 | 0–23 | |
| 37 | 0–18 | 12 | 0–23 | 8 | 17–12 | 12 | 0–23 | 15 | 0–23 | 17 | 0–23 | |
| 38 | 21–16 | 13 | 0–23 | 9 | 8– 3 | 13 | 0–23 | 15 | 0–23 | 17 | 0–23 | |
| 39 | 18–13 | 14 | 0–23 | 11 | 0–18 | 14 | 0–23 | 16 | 0–23 | 18 | 0–23 | |
| 40 | 16–11 | 15 | 0–23 | 12 | 15–10 | 15 | 0–23 | 16 | 0–23 | 18 | 0–23 | |
| 41 | 13– 8 | 16 | 0–23 | 14 | 6– 1 | 16 | 6–9 | 16 | 0–23 | 19 | 0–23 | |
| 42 | 10– 6 | 17 | 0–23 | 15 | 21–17 | 17 | 0–23 | 17 | 0–23 | 19 | 0–23 | |
| 43 | 8– 4 | 18 | 0–23 | 17 | 13– 8 | 18 | 0–23 | 17 | 0–23 | 20 | 0–23 | |
| 44 | 5– 1 | 19 | 0–23 | 18 | 4– 0 | 19 | 0–23 | 18 | 0–23 | 20 | 0–23 | |
| 45 | 3–23 | 20 | 0–23 | 20 | 19–15 | 20 | 0–23 | 18 | 1–23 | 20 | 0–23 | |
| 46 | 0–20 | 21 | 0–23 | 21 | 10– 6 | 21 | 0–23 | 19 | 0–23 | 21 | 0–23 | |
| 47 | 21–18 | 22 | 0–23 | 23 | 2–22 | 22 | 0–23 | 19 | 0–23 | 21 | 0–23 | |
| 48 | 19–15 | 23 | 18–16 | 0 | 17–13 | 23 | 0–23 | 19 | 0–23 | 22 | 0–23 | |
| 49 | 16–13 | 0 | 0–23 | 2 | 8– 5 | 0 | 0–23 | 20 | 0–23 | 22 | 0–23 | |
| 50 | 13–10 | 1 | 6–23 | 3 | 0–20 | 1 | 0–23 | 20 | 0–23 | 23 | 0–23 | |
| 51 | 11– 8 | 2 | 0–23 | 5 | 15–12 | 2 | 0–23 | 21 | 0–23 | 23 | 0–23 | |
| 52 | 8– 5 | 3 | 19– 7 | 6 | 6– 3 | 3 | 0–23 | 21 | 0–23 | 0 | 0–23 | |
| (53) | 0–23 | 4 | 0–23 | 8 | 0–23 | 4 | 0–23 | 21 | 0–23 | 0 | 0–23 | |

HARMONIC ANALYSIS AND PREDICTION OF TIDES

Table 33.—*For construction of secondary stencils*—Continued

| Constituent A | O | | | | | | | | |
|---|---|---|---|---|---|---|---|---|---|
| Constituent B | ρ | | Q | | | | 2Q | | |
| Page | O hours | Difference, hours | O hours | Difference, hours | O hours | Difference, hours | O hours | Difference, hours | O hours | Difference, hours |

| Page | O hours | Diff. hours | O hours | Diff. hours | O hours | Diff. hours | O hours | Diff. hours | O hours | Diff. hours |
|---|---|---|---|---|---|---|---|---|---|---|
| 1 | 18- 1 | 2 | 0-23 | 3 | 18-22 | 5 | 23-11 | 6 | 12-17 | 7 |
| 2 | 0-23 | 8 | 6- 3 | 9 | 6- 8 | 17 | 9-20 | 18 | 21- 5 | 19 |
| 3 | 18-15 | 13 | 18-12 | 15 | 18- 6 | 6 | 7-17 | 7 | | |
| 4 | 7-19 | 18 | 7-22 | 21 | 7-16 | 18 | 17- 5 | 19 | 6 | 20 |
| 5 | 19-22 | 23 | 19- 8 | 3 | 19- 1 | 6 | 2-14 | 7 | 15-18 | 8 |
| 6 | 0-23 | 5 | 7-18 | 9 | 7-11 | 18 | 12- 0 | 19 | 1- 6 | 20 |
| 7 | 19-12 | 10 | 19- 3 | 15 | 19-21 | 6 | 22-10 | 7 | 11-18 | 8 |
| 8 | 7-16 | 15 | 7-13 | 21 | 7-19 | 19 | 20- 6 | 20 | | |
| 9 | 19 | 20 | 19-23 | 3 | 19- 5 | 7 | 0-18 | 8 | | |
| 10 | 8- 5 | 2 | 8 | 9 | 8-15 | 19 | 16- 3 | 20 | 4- 7 | 21 |
| 11 | 20- 9 | 7 | 0-23 | 16 | 20- 0 | 7 | 1-13 | 8 | 14-19 | 9 |
| 12 | 8-13 | 12 | 8- 5 | 22 | 8-10 | 19 | 11-23 | 20 | 0- 7 | 21 |
| 13 | 0-23 | 18 | 20-15 | 4 | 20 | 7 | 21- 8 | 8 | 9-19 | 9 |
| 14 | 8- 2 | 23 | 8- 1 | 10 | 8-18 | 20 | 19- 7 | 21 | | |
| 15 | 20- 6 | 4 | 20-10 | 16 | 20- 4 | 8 | 5-17 | 9 | 18-19 | 10 |
| 16 | 9 | 9 | 9-20 | 22 | 9-14 | 20 | 15- 2 | 21 | 3- 8 | 22 |
| 17 | 21-19 | 15 | 21- 6 | 4 | 21-23 | 8 | 0-12 | 9 | 13-20 | 10 |
| 18 | 9-23 | 20 | 9-15 | 10 | 9 | 20 | 10-22 | 21 | 23- 8 | 22 |
| 19 | 21- 3 | 1 | 21- 1 | 16 | 21- 7 | 9 | 8-20 | 10 | | |
| 20 | 0-23 | 7 | 9-11 | 22 | 9-17 | 21 | 18- 6 | 22 | 7- 8 | 23 |
| 21 | 21-16 | 12 | 0-23 | 5 | 21- 3 | 9 | 4-16 | 10 | 17-20 | 11 |
| 22 | 10-20 | 17 | 10- 8 | 11 | 10-12 | 21 | 13- 1 | 22 | 2- 9 | 23 |
| 23 | 22- 0 | 22 | 22-17 | 17 | 22 | 9 | 23-11 | 10 | 12-21 | 11 |
| 24 | 0-23 | 4 | 10- 3 | 23 | 10-21 | 22 | 22- 9 | 23 | | |
| 25 | 22-13 | 9 | 22-13 | 5 | 22- 6 | 10 | 7-19 | 11 | 20-21 | 12 |
| 26 | 10-17 | 14 | 10-22 | 11 | 10-16 | 22 | 17- 5 | 23 | 6- 9 | 0 |
| 27 | 22- 6 | 19 | 22- 8 | 17 | 22- 2 | 10 | 3-14 | 11 | 15-21 | 12 |
| 28 | 10- 6 | 1 | 10-18 | 23 | 10-11 | 22 | 12- 0 | 23 | 1- 9 | 0 |
| 29 | 23-10 | 6 | 23- 3 | 5 | 23-10 | 11 | 11-22 | 12 | | |
| 30 | 11-14 | 11 | 11-13 | 11 | 11-19 | 23 | 20- 8 | 0 | 9-10 | 1 |
| 31 | 0-23 | 17 | 23 | 17 | 23- 5 | 11 | 6-18 | 12 | 19-22 | 13 |
| 32 | 11- 3 | 22 | 0-23 | 0 | 11-15 | 23 | 16- 4 | 0 | 5-10 | 10 |
| 33 | 23- 7 | 3 | 23-20 | 6 | 23- 0 | 11 | 1-13 | 12 | 14-22 | 13 |
| 34 | 11 | 8 | 11- 5 | 12 | 11-23 | 0 | 0-10 | 1 | | |
| 35 | 0-20 | 14 | 0-15 | 18 | 0- 9 | 12 | 10-21 | 13 | 22-23 | 14 |
| 36 | 12- 0 | 19 | 12- 1 | 0 | 12-18 | 0 | 19- 7 | 1 | 8-11 | 2 |
| 37 | 0- 4 | 0 | 0-10 | 6 | 0- 4 | 12 | 5-17 | 13 | 18-23 | 14 |
| 38 | 0-23 | 6 | 12-20 | 12 | 12-14 | 0 | 15- 2 | 1 | 3-11 | 2 |
| 39 | 0-17 | 11 | 0- 6 | 18 | 0-12 | 13 | 13-23 | 14 | | |
| 40 | 12-21 | 16 | 12-15 | 0 | 12-22 | 1 | 23-11 | 2 | | |
| 41 | 1 | 21 | 1 | 6 | 1- 7 | 13 | 8-20 | 14 | 21- 0 | 15 |
| 42 | 13-10 | 3 | 0-23 | 13 | 13-17 | 1 | 18- 6 | 2 | 7-12 | 3 |
| 43 | 1-14 | 8 | 1-22 | 19 | 1- 3 | 13 | 4-16 | 14 | 17- 0 | 15 |
| 44 | 13-16 | 13 | 13- 8 | 1 | 13 | 1 | 14- 1 | 2 | 2-12 | 3 |
| 45 | 0-23 | 19 | 1-17 | 7 | 1-11 | 14 | 12- 0 | 15 | | |
| 46 | 13- 7 | 0 | 13- 3 | 13 | 13-21 | 2 | 22- 9 | 3 | 10-12 | 4 |
| 47 | 2-11 | 5 | 2-13 | 19 | 2- 6 | 14 | 7-19 | 15 | 20- 1 | 16 |
| 48 | 14-15 | 10 | 14-22 | 1 | 14-16 | 2 | 17- 5 | 3 | 6-13 | 4 |
| 49 | 2- 0 | 16 | 2- 8 | 7 | 2 | 14 | 3-15 | 15 | 16- 1 | 16 |
| 50 | 14- 4 | 21 | 14-18 | 3 | 14- 0 | 3 | 1-13 | 4 | | |
| 51 | 2- 8 | 2 | 2- 3 | 19 | 2-10 | 15 | 11-23 | 16 | 0- 1 | 17 |
| 52 | 0-23 | 8 | 0-23 | 2 | 14-20 | 3 | 21- 8 | 4 | 9-13 | 5 |
| (53) | 4-16 | 12 | 4-23 | 7 | 4 | 13 | 5-16 | 14 | 17- 3 | 15 |

Table 34.—*For summation of long-period constituents*

ASSIGNMENT OF DAILY SUMS FOR CONSTITUENT Mf

| Constituent division | Days of series |
|---|---|
| 0 | 1  28  55  82*  110  137  164*  192  219  246  274  301  328  356 |
| 1 | 2  29  56  84  111  138  166  193  220  248*  275  302  330*  357 |
| 2 | 3  30  57*  85  112  139  167  194  221  249  276  303  331  358 |
| 3 | 4  31  59  86  113  141*  168  195  223*  250  277  304*  332  359 |
| 4 | 5  32  60  87  114  142  169  196  224  251  278  306  333  360 |
| 5 | 6  34*  61  88  115*  143  170  197*  225  252  279  307  334  361 |
| 6 | 7  35  62  89  117  144  171  199  226  253  281*  308  335  363* |
| 7 | 8*  36  63  90*  118  145  172  200  227  254  282  309  336  364 |
| 8 | 10  37  64  92  119  146  174*  201  228  256*  283  310  337*  365 |
| 9 | 11  38  65  93  120  147  175  202  229  257  284  311  339  366 |
| 10 | 12  39  67*  94  121  149*  176  203  230*  258  285  312*  340  367 |
| 11 | 13  40  68  95  122  150  177  204  232  259  286  314  341  368 |
| 12 | 14  42*  69  96  123*  151  178  205*  233  260  287  315  342  369 |
| 13 | 15  43  70  97  125  152  179  207  234  261  289*  316  343  ____ |
| 14 | 16*  44  71  98  126  153  180  208  235  262  290  317  344  ____ |
| 15 | 18  45  72  100*  127  154  182*  209  236  263*  291  318  345*  ____ |
| 16 | 19  46  73  101  128  155  183  210  237  265  292  319  347  ____ |
| 17 | 20  47  75*  102  129  156*  184  211  238*  266  293  320  348  ____ |
| 18 | 21  48  76  103  130  158  185  212  240  267  294  322*  349  ____ |
| 19 | 22  49*  77  104  131*  159  186  213  241  268  295  323  350  ____ |
| 20 | 23  51  78  105  133  160  187  215*  242  269  297*  324  351  ____ |
| 21 | 24  52  79  106  134  161  188  216  243  270  298  325  352  ____ |
| 22 | 26*  53  80  108*  135  162  189*  217  244  271*  299  326  353  ____ |
| 23 | 27  54  81  109  136  163  191  218  245  273  300  327  355*  ____ |

ASSIGNMENT OF DAILY SUMS FOR CONSTITUENT MSf

| Constituent division | Days of series |
|---|---|
| 0 | 1  30  60*  89  119  148  178  207  237  266  296  325  355 |
| 1 | 2  31  61  90  120  149*  179  208*  238  268*  297  327*  356 |
| 2 | 3  33*  62  92*  121  151  180  210  239  269  298  328  357 |
| 3 | 4  34  63  93  122  152  181*  211  240*  270  300*  329  359* |
| 4 | 5*  35  65*  94  124*  153  183  212  242  271  301  330  360 |
| 5 | 7  36  66  95  125  154  184  213*  243  272*  302  332*  361 |
| 6 | 8  37*  67  96*  126  156*  185  215  244  274  303  333  362 |
| 7 | 9  39  68  98  127  157  186  216  245  275  304*  334  364* |
| 8 | 10  40  69*  99  128*  158  188*  217  247*  276  306  335  365 |
| 9 | 12*  41  71  100  130  159  189  218  248  277  307  336*  366 |
| 10 | 13  42  72  101*  131  160*  190  220*  249  279*  308  338  367 |
| 11 | 14  44*  73  103  132  162  191  221  250  280  309  339  368* |
| 12 | 15  45  74  104  133*  163  192*  222  252*  281  311*  340  ____ |
| 13 | 17*  46  76*  105  135  164  194  223  253  282  312  341  ____ |
| 14 | 18  47  77  106  136  165*  195  224*  254  284*  313  343*  ____ |
| 15 | 19  49*  78  108*  137  167  196  226  255  285  314  344  ____ |
| 16 | 20  50  79  109  138  168  197*  227  256*  286  316*  345  ____ |
| 17 | 21*  51  80*  110  140*  169  199  228  258  287  317  346  ____ |
| 18 | 23  52  82  111  141  170  200  229  259  288*  318  348*  ____ |
| 19 | 24  53*  83  112*  142  172*  201  231*  260  290  319  349  ____ |
| 20 | 25  55  84  114  143  173  202  232  261  291  320*  350  ____ |
| 21 | 26  56  85*  115  144*  174  204*  233  263*  292  322  351  ____ |
| 22 | 28*  57  87  116  146  175  205  234  264  293  323  352*  ____ |
| 23 | 29  58  88  117*  147  176*  206  236*  265  295*  324  354  ____ |

Table 34.—*For summation of long-period constituents*—Continued

ASSIGNMENT OF DAILY SUMS FOR CONSTITUENT Mm

| Constituent division | Days of series |
|---|---|
| 0 | 1 28 56 83 111 138 166 193 221 249* 276 304 331 359 |
| 1 | 2 29 57 84 112 139* 167 195* 222 250 277 305 332 360 |
| 2 | 3 30 58 85* 113 141 168 196 223 251 278 306 333* 361 |
| 3 | 4 32* 59 87 114 142 169 197 224 252 280* 307 335 362 |
| 4 | 5 33 60 88 115 143 170* 198 226* 253 281 308 336 363 |
| 5 | 6 34 61 89 116* 144 172 199 227 254 282 309 337 364* |
| 6 | 7 35 63* 90 118 145 173 200 228 255 283 311* 338 366 |
| 7 | 9* 36 64 91 119 146 174 201* 229 257* 284 312 339 367 |
| 8 | 10 37 65 92 120 147* 175 203 230 258 285 313 340 368 |
| 9 | 11 38 66 94* 121 149 176 204 231 259 286 314 342* 369 |
| 10 | 12 40* 67 95 122 150 177 205 232* 260 288* 315 343 ___ |
| 11 | 13 41 68 96 123 151 178* 206 234 261 289 316 344 ___ |
| 12 | 14 42 69 97 125* 152 180 207 235 262 290 317 345 ___ |
| 13 | 15* 43 71* 98 126 153 181 208 236 263* 291 319* 346 ___ |
| 14 | 17 44 72 99 127 154 182 200* 237 265 292 320 347 ___ |
| 15 | 18 45 73 100 128 156* 183 211 238 266 293 321 348 ___ |
| 16 | 19 46* 74 102* 129 157 184 212 239 267 294* 322 350* ___ |
| 17 | 20 48 75 103 130 158 185 213 240* 268 296 323 351 ___ |
| 18 | 21 49 76 104 131 159 187* 214 242 269 297 324 352 ___ |
| 19 | 22 50 77* 105 133* 160 188 215 243 270 298 325* 353 ___ |
| 20 | 23* 51 79 106 134 161 189 216 244 271* 299 327 354 ___ |
| 21 | 25 52 80 107 135 162 190 218* 245 273 300 328 355 ___ |
| 22 | 26 53 81 108* 136 164* 191 219 246 274 301 329 356* ___ |
| 23 | 27 54* 82 110 137 165 192 220 247 275 302* 330 358 ___ |

ASSIGNMENT OF DAILY SUMS FOR CONSTITUENT Sa

| Constituent division | Days of series | Constituent division | Days of series |
|---|---|---|---|
| 0 | { 1- 8 / 359-369 | 12 | 176-190 |
| 1 | 9- 23 | 13 | 191-205 |
| 2 | 24- 38 | 14 | 206-221 |
| 3 | 39- 53 | 15 | 222-236 |
| 4 | 54- 69 | 16 | 237-251 |
| 5 | 70- 84 | 17 | 252-266 |
| 6 | 85- 99 | 18 | 267-282 |
| 7 | 100-114 | 19 | 283-297 |
| 8 | 115-129 | 20 | 298-312 |
| 9 | 130-145 | 21 | 313-327 |
| 10 | 146-160 | 22 | 328-342 |
| 11 | 161-175 | 23 | 343-358 |

Table 35.—**Products** $\left(a\dfrac{S}{15}\right)$ **for Form 444**

| Constituent | Time meridian in hours = S÷15 | | | | | | | |
|---|---|---|---|---|---|---|---|---|
| | 1.000 | 2.000 | 3.000 | 4.000 | 5.000 | 5.500 | 6.000 | 6.500 |
| | Products, in degrees | | | | | | | |
| M₂ | 28.98 | 57.97 | 86.95 | 115.94 | 144.92 | 159.41 | 173.90 | 188.40 |
| S₂ | 30.00 | 60.00 | 90.00 | 120.00 | 150.00 | 165.00 | 180.00 | 195.00 |
| N₂ | 28.44 | 58.88 | 85.32 | 113.76 | 142.20 | 156.42 | 170.64 | 184.86 |
| K₁ | 15.04 | 30.08 | 45.12 | 60.16 | 75.21 | 82.73 | 90.25 | 97.77 |
| M₄ | 57.97 | 115.94 | 173.90 | 231.87 | 289.84 | 318.83 | 347.81 | 16.79 |
| O₁ | 13.94 | 27.89 | 41.83 | 55.77 | 69.72 | 76.69 | 83.66 | 90.63 |
| M₆ | 86.95 | 173.90 | 260.86 | 347.81 | 74.76 | 118.24 | 161.71 | 205.19 |
| (MK)₃ | 44.03 | 88.05 | 132.08 | 176.10 | 220.13 | 242.14 | 264.15 | 286.16 |
| S₄ | 60.00 | 120.00 | 180.00 | 240.00 | 300.00 | 330.00 | 0.00 | 30.00 |
| (MN)₄ | 57.42 | 114.85 | 172.27 | 229.70 | 287.12 | 315.83 | 344.54 | 13.25 |
| ν₂ | 28.51 | 57.03 | 85.54 | 114.05 | 142.56 | 156.82 | 171.08 | 185.33 |
| S₆ | 90.00 | 180.00 | 270.00 | 0.00 | 90.00 | 135.00 | 180.00 | 225.00 |
| μ₂ | 27.97 | 55.94 | 83.90 | 111.87 | 139.84 | 153.83 | 167.81 | 181.79 |
| (2N)₂ | 27.90 | 55.79 | 83.69 | 111.58 | 139.48 | 153.42 | 167.37 | 181.32 |
| (OO)₁ | 16.14 | 32.28 | 48.42 | 64.56 | 80.70 | 88.77 | 96.83 | 104.90 |
| λ₂ | 29.46 | 58.91 | 88.37 | 117.82 | 147.28 | 162.01 | 176.73 | 191.46 |
| S₁ | 15.00 | 30.00 | 45.00 | 60.00 | 75.00 | 82.50 | 90.00 | 97.50 |
| M₁ | 14.50 | 28.99 | 43.49 | 57.99 | 72.48 | 79.73 | 86.98 | 94.23 |
| J₁ | 15.59 | 31.17 | 46.76 | 62.34 | 77.93 | 85.72 | 93.51 | 101.31 |
| Mm | 0.54 | 1.09 | 1.63 | 2.18 | 2.72 | 2.99 | 3.27 | 3.54 |
| Ssa | 0.08 | 0.16 | 0.25 | 0.33 | 0.41 | 0.45 | 0.49 | 0.53 |
| Sa | 0.04 | 0.08 | 0.12 | 0.16 | 0.21 | 0.23 | 0.25 | 0.27 |
| MSf | 1.02 | 2.03 | 3.05 | 4.06 | 5.08 | 5.59 | 6.10 | 6.60 |
| Mf | 1.10 | 2.20 | 3.29 | 4.39 | 5.49 | 6.04 | 6.59 | 7.14 |
| ρ₁ | 13.47 | 26.94 | 40.41 | 53.89 | 67.36 | 74.09 | 80.83 | 87.56 |
| Q₁ | 13.40 | 26.80 | 40.20 | 53.59 | 66.99 | 73.69 | 80.39 | 87.09 |
| T₂ | 29.96 | 59.92 | 89.88 | 119.84 | 149.79 | 164.77 | 179.75 | 194.73 |
| R₂ | 30.04 | 60.08 | 90.12 | 120.16 | 150.21 | 165.23 | 180.25 | 195.27 |
| (2Q)₁ | 12.85 | 25.71 | 38.56 | 51.42 | 64.27 | 70.70 | 77.13 | 83.55 |
| P₁ | 14.96 | 29.92 | 44.88 | 59.84 | 74.80 | 82.27 | 89.75 | 97.23 |
| (2SM)₂ | 31.02 | 62.03 | 93.05 | 124.06 | 155.08 | 170.59 | 186.10 | 201.60 |
| M₃ | 43.48 | 86.95 | 130.43 | 173.90 | 217.38 | 239.12 | 260.86 | 282.60 |
| L₂ | 29.53 | 59.06 | 88.59 | 118.11 | 147.64 | 162.41 | 177.17 | 191.94 |
| (2MK)₃ | 42.93 | 85.85 | 128.78 | 171.71 | 214.64 | 236.10 | 257.56 | 279.03 |
| K₂ | 30.08 | 60.16 | 90.25 | 120.33 | 150.41 | 165.45 | 180.49 | 195.53 |
| M₈ | 115.94 | 231.87 | 347.81 | 103.75 | 219.68 | 277.65 | 335.62 | 33.59 |
| (MS)₄ | 58.98 | 117.97 | 176.95 | 235.94 | 294.92 | 324.41 | 353.90 | 23.40 |

Table 35.—Products $\left(a\dfrac{S}{15}\right)$ for Form 444—Continued

| Constituent | Time meridian in hours = $S \div 15$ | | | | | | | |
|---|---|---|---|---|---|---|---|---|
| | 7.000 | 8.000 | 9.000 | 10.000 | 10.500 | 11.000 | 11.500 | 12.000 |
| | Products in degrees | | | | | | | |
| $M_2$ | 202.89 | 231.87 | 260.86 | 289.84 | 304.33 | 318.83 | 333.32 | 347.81 |
| $S_2$ | 210.00 | 240.00 | 270.00 | 300.00 | 315.00 | 330.00 | 345.00 | 0.00 |
| $N_2$ | 199.08 | 227.52 | 255.96 | 284.40 | 298.62 | 312.84 | 327.06 | 341.28 |
| $K_1$ | 105.29 | 120.33 | 135.37 | 150.41 | 157.93 | 165.45 | 172.97 | 180.49 |
| $M_4$ | 45.78 | 103.75 | 161.71 | 219.68 | 248.67 | 277.65 | 306.63 | 335.62 |
| $O_1$ | 97.60 | 111.54 | 125.49 | 139.43 | 146.40 | 153.37 | 160.34 | 167.32 |
| $M_6$ | 248.67 | 335.62 | 62.57 | 149.52 | 193.00 | 236.48 | 279.95 | 323.43 |
| $(MK)_3$ | 308.18 | 352.20 | 36.23 | 80.25 | 102.26 | 124.28 | 146.29 | 168.30 |
| $S_4$ | 60.00 | 120.00 | 180.00 | 240.00 | 270.00 | 300.00 | 330.00 | 0.00 |
| $(MN)_4$ | 41.97 | 99.39 | 156.81 | 214.24 | 242.95 | 271.66 | 300.37 | 329.09 |
| $\nu_2$ | 199.59 | 228.10 | 256.61 | 285.13 | 299.38 | 313.64 | 327.89 | 342.15 |
| $S_6$ | 270.00 | 0.00 | 90.00 | 180.00 | 225.00 | 270.00 | 315.00 | 0.00 |
| $\mu_2$ | 195.78 | 223.75 | 251.71 | 279.68 | 293.67 | 307.65 | 321.63 | 335.62 |
| $(2N)_2$ | 195.27 | 223.16 | 251.06 | 278.95 | 292.90 | 306.85 | 320.80 | 334.74 |
| $(OO)_1$ | 112.97 | 129.11 | 145.25 | 161.39 | 169.46 | 177.53 | 185.60 | 193.67 |
| $\lambda_2$ | 206.19 | 235.65 | 265.10 | 294.56 | 309.28 | 324.01 | 338.74 | 353.47 |
| $S_1$ | 105.00 | 120.00 | 135.00 | 150.00 | 157.50 | 165.00 | 172.50 | 180.00 |
| $M_1$ | 101.48 | 115.97 | 130.47 | 144.97 | 152.22 | 159.46 | 166.71 | 173.96 |
| $J_1$ | 109.10 | 124.68 | 140.27 | 155.85 | 163.65 | 171.44 | 179.23 | 187.03 |
| Mm | 3.81 | 4.35 | 4.90 | 5.44 | 5.72 | 5.99 | 6.26 | 6.53 |
| Ssa | 0.57 | 0.66 | 0.74 | 0.82 | 0.86 | 0.90 | 0.94 | 0.99 |
| Sa | 0.29 | 0.33 | 0.37 | 0.41 | 0.43 | 0.45 | 0.47 | 0.49 |
| MSf | 7.11 | 8.13 | 9.14 | 10.16 | 10.67 | 11.17 | 11.68 | 12.19 |
| Mf | 7.69 | 8.78 | 9.88 | 10.98 | 11.53 | 12.08 | 12.63 | 13.18 |
| $\rho_1$ | 94.30 | 107.77 | 121.24 | 134.72 | 141.45 | 148.19 | 154.92 | 161.66 |
| $Q_1$ | 93.79 | 107.19 | 120.59 | 133.99 | 140.69 | 147.39 | 154.08 | 160.78 |
| $T_2$ | 209.71 | 239.67 | 269.63 | 299.59 | 314.57 | 329.55 | 344.53 | 359.51 |
| $R_2$ | 210.29 | 240.33 | 270.37 | 300.41 | 315.43 | 330.45 | 345.47 | 0.49 |
| $(2Q)_1$ | 89.98 | 102.83 | 115.69 | 128.54 | 134.97 | 141.40 | 147.82 | 154.25 |
| $P_1$ | 104.71 | 119.67 | 134.63 | 149.59 | 157.07 | 164.55 | 172.03 | 179.51 |
| $(2SM)_2$ | 217.11 | 248.13 | 279.14 | 310.16 | 325.67 | 341.17 | 356.68 | 12.19 |
| $M_3$ | 304.33 | 347.81 | 31.29 | 74.76 | 96.50 | 118.24 | 139.98 | 161.71 |
| $L_2$ | 206.70 | 236.23 | 265.76 | 295.28 | 310.05 | 324.81 | 339.58 | 354.34 |
| $(2MK)_3$ | 300.49 | 343.42 | 26.34 | 69.27 | 90.73 | 112.20 | 133.66 | 155.13 |
| $K_2$ | 210.57 | 240.66 | 270.74 | 300.82 | 315.86 | 330.90 | 345.94 | 0.99 |
| $M_8$ | 91.55 | 207.49 | 323.43 | 79.36 | 137.33 | 195.30 | 253.27 | 311.24 |
| $(MS)_4$ | 52.89 | 111.87 | 170.86 | 229.84 | 259.33 | 288.83 | 318.32 | 347.81 |

Table 36.—*Angle differences for Form 445*

| Constituent | Jan. 1, $0^h$, to Feb. 1, $0^h$ | | Feb. 1, $0^h$, to Dec. 31, $24^h$ | | | | Jan. 1, $0^h$, to Dec. 31, $24^h$ | | | |
|---|---|---|---|---|---|---|---|---|---|---|
| | | | Common year | | Leap year | | Common year | | Leap year | |
| | ° | ° | ° | ° | ° | ° | ° | ° | ° | ° |
| $M_2$ | +324.2 | −35.8 | +136.6 | −223.4 | +112.2 | −247.8 | +100.8 | −259.2 | +76.4 | −283.6 |
| $S_2$ | 0 | 0 | 0 | 0 | 0 | 0 | 0 | 0 | 0 | 0 |
| $N_2$ | +279 | −81 | +93 | −267 | +56 | −304 | +12 | −348 | +335 | −25 |
| $K_1$ | +31 | −329 | +329 | −31 | +330 | −30 | 0 | 0 | +1 | −359 |
| $M_4$ | +288 | −72 | +274 | −86 | +225 | −135 | +202 | −158 | +153 | −207 |
| $O_1$ | +294 | −66 | +167 | −193 | +142 | −218 | +101 | −259 | +76 | −284 |
| $M_6$ | +253 | −107 | +49 | −311 | +336 | −24 | +302 | −58 | +229 | −131 |
| $(MK)_3$ | +355 | −5 | +106 | −254 | +82 | −278 | +101 | −259 | +77 | −283 |
| $S_4$ | 0 | 0 | 0 | 0 | 0 | 0 | 0 | 0 | 0 | 0 |
| $(MN)_4$ | +243 | −117 | +230 | −130 | +168 | −192 | +113 | −247 | +51 | −309 |
| $\nu_2$ | +333 | −27 | +317 | −43 | +281 | −79 | +290 | −70 | +254 | −106 |
| $S_6$ | 0 | 0 | 0 | 0 | 0 | 0 | 0 | 0 | 0 | 0 |
| $\mu_2$ | +288 | −72 | +274 | −86 | +225 | −135 | +202 | −158 | +153 | −207 |
| $(2N)_2$ | +234 | −126 | +49 | −311 | +359 | −1 | +283 | −77 | +233 | −127 |
| $(OO)_1$ | +127 | −233 | +131 | −229 | +159 | −201 | +258 | −102 | +286 | −74 |
| $\lambda_2$ | +315 | −45 | +316 | −44 | +303 | −57 | +271 | −89 | +258 | −102 |
| $S_1$ | 0 | 0 | 0 | 0 | 0 | 0 | 0 | 0 | 0 | 0 |
| $M_1$ | +342 | −18 | +248 | −112 | +236 | −124 | +230 | −130 | +218 | −142 |
| $J_1$ | +76 | −284 | +12 | −348 | +27 | −333 | +88 | −272 | +103 | −257 |
| $Mm$ | +45 | −315 | +44 | −316 | +57 | −303 | +89 | −271 | +102 | −258 |
| $Ssa$ | +61 | −299 | +299 | −61 | +300 | −60 | 0 | 0 | 1 | −359 |
| $Sa$ | +31 | −329 | +329 | −31 | +330 | −30 | 0 | 0 | +1 | −359 |
| $MSf$ | +36 | −324 | +223 | −137 | +248 | −112 | +259 | −101 | +284 | −76 |
| $Mf$ | +97 | −263 | +162 | −198 | +188 | −172 | +259 | −101 | +285 | −75 |
| $\rho_1$ | +303 | −57 | +348 | −12 | +311 | −49 | +291 | −69 | +254 | −106 |
| $Q_1$ | +249 | −111 | +123 | −237 | +85 | −275 | +12 | −348 | +334 | −26 |
| $T_2$ | +329 | −31 | +31 | −329 | +30 | −330 | 0 | 0 | +259 | −1 |
| $R_2$ | +31 | −329 | +329 | −31 | +330 | −30 | 0 | 0 | +1 | −359 |
| $(2Q)_1$ | +204 | −156 | +80 | −280 | +28 | −332 | +284 | −76 | +232 | −128 |
| $P_1$ | +329 | −31 | +31 | −329 | +30 | −330 | 0 | 0 | +359 | −1 |
| $(2SM)_2$ | +36 | −324 | +223 | −137 | +248 | −112 | +259 | −101 | +284 | −76 |
| $M_3$ | +306 | −54 | +25 | −335 | +348 | −12 | +331 | −29 | +294 | −66 |
| $L_2$ | +9 | −351 | +180 | −180 | +169 | −191 | +189 | −171 | +178 | −182 |
| $(2MK)_3$ | +258 | −102 | +304 | −56 | +254 | −106 | +202 | −158 | +152 | −208 |
| $K_2$ | +61 | −299 | +299 | −61 | +300 | −60 | 0 | 0 | +1 | −359 |
| $M_8$ | +217 | −143 | +186 | −174 | +89 | −271 | +43 | −317 | +306 | −54 |
| $(MS)_4$ | +324 | −36 | +137 | −223 | +112 | −248 | +101 | −259 | +76 | −284 |

# HARMONIC ANALYSIS AND PREDICTION OF TIDES 307

Table 37.—*U. S. Coast and Geodetic Survey tide-predicting machine No. 2*

## GENERAL GEARS AND CONNECTING SHAFTS

| Gears and Shafts | Face or diameter | Number of teeth | Pitch | Period of rotation | Remarks |
|---|---|---|---|---|---|
| | *Inches* | | | *Dial hours* | |
| S-1 | 0.56 | | | 4 | Hand crank shaft for operating machine. |
| G-1 | 0.56 | 40 | 24 | 4 | Spur gear on shaft 1. |
| G-2 | 0.50 | 120 | 24 | 12 | Spur-stud gear. |
| G-3 | 0.50 | 120 | 24 | 12 | Spur gear on shaft 2. |
| S-2 | 0.50 | | | 12 | Short horizontal shaft. |
| G-4 | 0.41 | 72 | 24 | 12 | Bevel gear on shaft 2. |
| G-5 | 0.41 | 72 | 24 | 12 | Bevel gear on shaft 3. |
| S-3 | 0.50 | | | 12 | Diagonal shaft connecting with middle section. |
| G-6 | 0.38 | 75 | 30 | 12 | Bevel gear on shaft 3. |
| G-7 | 0.38 | 75 | 30 | 12 | Bevel gear on shaft 4. |
| S-4 | 0.38 | | | 12 | Short vertical shaft through desk top. |
| G-8 | 0.38 | 75 | 30 | 12 | Bevel gear on shaft 4. |
| G-9 | 0.38 | 75 | 30 | 12 | Bevel gear on shaft 5. |
| S-5 | 0.38 | | | 12 | Short horizontal shaft. |
| G-10 | 0.27 | 75 | 30 | 12 | Bevel gear on shaft 5. |
| G-11 | 0.27 | 75 | 30 | 12 | Bevel gear on shaft 6. |
| S-6 | 0.38 | | | 12 | Main vertical shaft of dial case. |
| G-12 | 0.17 | 60 | 48 | 12 | Releasable bevel gear on shaft 6. |
| G-13 | 0.17 | 120 | 48 | 24 | Bevel gear on shaft 7. |
| S-7 | 0.15 | | | 24 | Intermediate shaft to hour hand. |
| G-14 | 0.17 | 84 | 48 | 24 | Bevel gear on shaft 7. |
| G-15 | 0.17 | 84 | 48 | 24 | Bevel gear on shaft 8. |
| S-8 | 0.15 | | | 24 | Hour-hand shaft. |
| G-16 | 0.17 | 180 | 48 | 12 | Releasable bevel gear on shaft 6. |
| G-17 | 0.17 | 60 | 48 | 4 | Bevel gear on shaft 9. |
| S-9 | 0.15 | | | 4 | Intermediate shaft to minute hand. |
| G-18 | 0.17 | 240 | 48 | 4 | Bevel gear on shaft 9. |
| G-19 | 0.17 | 60 | 48 | 1 | Bevel gear on shaft 10. |
| S-10 | 0.15 | | | 1 | Minute-hand shaft. |
| G-20 | 0.17 | 60 | 48 | 12 | Releasable bevel gear on shaft 6. |
| G-21 | 0.17 | 120 | 48 | 24 | Bevel gear on shaft 11. |
| S-11 | 0.15 | | | 24 | Intermediate shaft to day dial. |
| G-22 | | 1 | | 24 | Worm screw, 0.56 inch diameter, 18 threads to inch on shaft 11. |
| G-23 | | 366 | | 24×366 | Worm wheel, 6.47 inch diameter, on shaft 12. |
| S-12 | 0.31 | | | 24×366 | Day dial shaft. |
| G-24 | 0.25 | 46 | 40 | 12 | Spur gear at top of shaft 6. |
| G-25 | 0.25 | 60 | 40 | | Spur-stud gear. |
| G-26 | 0.25 | 60 | 40 | | Spur-stud gear connected with gear 25 by ratchet wheel and pawl. |
| G-27 | 0.25 | 46 | 40 | 12 | Spur gear at lower end of feeding roller. |
| G-28 | 0.41 | 72 | 24 | 12 | Bevel gear on shaft 3. |
| G-29 | 0.41 | 72 | 24 | 12 | Bevel gear on shaft 13. |
| S-13 | 0.44 | | | 12 | Main vertical shaft of middle section. |
| G-30 | 0.38 | 110 | 30 | 12 | Spur gear on shaft 13. |
| G-31 | 0.38 | 110 | 30 | 12 | Spur stud gear. |
| G-32 | 0.38 | 110 | 30 | 12 | Spur stud gear on shaft 14. |
| S-14 | 0.38 | | | 12 | Front vertical shaft of rear section. |
| G-33 | 0.28 | 75 | 30 | 12 | Bevel gear on shaft 14. |
| G-34 | 0.28 | 75 | 30 | 12 | Bevel gear on shaft 15. |
| S-15 | 0.50 | | | 12 | Main connecting horizontal shaft of rear section. |
| G-35 | 0.28 | 75 | 30 | 12 | Bevel gear on shaft 15. |
| G-36 | 0.28 | 75 | 30 | 12 | Bevel gear on shaft 16. |
| S-16 | 0.38 | | | 12 | Rear vertical shaft of rear section. |

Table 38.—*U. S. Coast and Geodetic Survey tide-predicting machine No. 2*

**CONSTITUENT GEARS AND MAXIMUM AMPLITUDE SETTINGS**

| Constituents | Theoretical speed per hour | Teeth in gear wheels | | | | Gear speed per dial hour | Error per year | Maximum amplitude settings of cranks |
|---|---|---|---|---|---|---|---|---|
| | | Vertical shafts I | Intermediate shafts II | III | Crank shafts IV | | | |
| | ° | | | | | ° | ° | Units |
| $J_1$ | 15.5854433 | 107 | 90 | 52 | 119 | 15.5854342 | −0.08 | 1.4 |
| $K_1$ | 15.0410686 | 61 | 73 | 51 | 85 | 15.0410959 | +.24 | 11.0 |
| $K_2$ | 30.0821372 | 122 | 80 | 96 | 146 | 30.0821918 | +.48 | 3.9 |
| $L_2$ | 29.5284788 | 104 | 61 | 56 | 97 | 29.5284773 | −.01 | 2.4 |
| *$M_1$ | 14.4920521 | 103 | 85 | 59 | 148 | 14.4920509 | −.01 | 1.0 |
| $M_2$ | 28.9841042 | 103 | 74 | 59 | 85 | 28.9841017 | −.02 | 20.0 |
| $M_3$ | 43.4761563 | 86 | 62 | 70 | 67 | 43.4761675 | +.10 | 1.4 |
| $M_4$ | 57.9682084 | 118 | 74 | 103 | 85 | 57.9682035 | −.04 | 4.0 |
| $M_6$ | 86.9523126 | 140 | 62 | 86 | 67 | 86.9523351 | +.20 | 1.0 |
| $M_8$ | 115.9364168 | 118 | 37 | 103 | 85 | 115.9364070 | +.09 | 0.4 |
| $N_2$ | 28.4397296 | 65 | 46 | 53 | 79 | 28.4397358 | +.05 | 6.0 |
| $2N$ | 27.8953548 | 68 | 58 | 46 | 58 | 27.8953627 | +.07 | 1.0 |
| $O_1$ | 13.9430356 | 92 | 89 | 58 | 129 | 13.9430363 | +.01 | 9.0 |
| $OO$ | 16.1391016 | 134 | 131 | 71 | 135 | 16.1391009 | −.01 | 0.8 |
| $P_1$ | 14.9589314 | 91 | 73 | 50 | 125 | 14.9589041 | −.24 | 4.8 |
| $Q_1$ | 13.3986609 | 84 | 88 | 51 | 109 | 13.3986656 | +.04 | 3.0 |
| $2Q$ | 12.8542862 | 127 | 114 | 50 | 130 | 12.8542510 | −.31 | 0.6 |
| $R_2$ | 30.0410686 | 85 | 50 | 43 | 73 | 30.0410959 | +.24 | 0.4 |
| $S_1$ | 15.0000000 | 63 | 75 | 50 | 84 | 15.0000000 | .00 | 2.0 |
| $S_2$ | 30.0000000 | 70 | 70 | 70 | 70 | 30.0000000 | .00 | 9.8 |
| $S_4$ | 60.0000000 | 75 | 45 | 60 | 50 | 60.0000000 | .00 | 1.0 |
| $S_6$ | 90.0000000 | 90 | 48 | 80 | 50 | 90.0000000 | .00 | 0.4 |
| $T_2$ | 29.9589314 | 81 | 50 | 45 | 73 | 29.9589041 | −.24 | 1.0 |
| $\lambda_2$ | 29.4556254 | 131 | 65 | 57 | 117 | 29.4556213 | −.04 | 0.4 |
| $\mu_2$ | 27.9682084 | 125 | 82 | 74 | 121 | 27.9681516 | −.50 | 1.2 |
| $\nu_2$ | 28.5125830 | 89 | 69 | 70 | 95 | 28.5125858 | +.02 | 2.0 |
| $\rho_1$ | 13.4715144 | 69 | 70 | 41 | 90 | 13.4714286 | −.75 | 0.8 |
| $MK$ | 44.0251728 | 120 | 81 | 105 | 106 | 44.0251572 | −.14 | 1.9 |
| $2MK$ | 42.9271398 | 81 | 52 | 79 | 86 | 42.9271020 | −.33 | 1.4 |
| $MN$ | 57.4238338 | 135 | 42 | 53 | 89 | 57.4237560 | −.68 | 0.7 |
| $MS$ | 58.9841042 | 118 | 61 | 62 | 61 | 58.9841440 | −.35 | 2.0 |
| $2SM$ | 31.0158958 | 69 | 47 | 50 | 71 | 31.0158825 | −.12 | 1.4 |
| $Mf$ | 1.0980330 | 84 | 45 | 1 | 51 | 1.0980392 | +.05 | 4.0 |
| $MSf$ | 1.0158958 | 149 | 80 | 1 | 55 | 1.0159091 | +.12 | 2.0 |
| $Mm$ | 0.5443747 | 93 | 41 | 1 | 125 | 0.5443902 | +.14 | 3.0 |
| $Sa$ | 0.0410686 | 51 | { 149 / 125 } | { 1 / 60 } | 120 | 0.0410738 | +.05 | 8.0 |
| $Ssa$ | 0.0821372 | 51 | 149 | 1 | 125 | 0.0821477 | +.09 | 3.0 |

*Designed for one-half of speed of $M_2$.

HARMONIC ANALYSIS AND PREDICTION OF TIDES 309

Table 39.—*Synodic periods of constituents*

DIURNAL CONSTITUENTS

|  | $J_1$ | $K_1$ | $M_1$ | $O_1$ | $OO$ | $P_1$ | $Q_1$ | $2Q$ | $S_1$ |
|---|---|---|---|---|---|---|---|---|---|
|  | Days. | Days. | Days. | Days. | Days. | Days. | Days. | Days. | Days. |
| $K_1$ | 27.555 | | | | | | | | |
| $M_1$ | 13.777 | 27.555 | | | | | | | |
| $O_1$ | 9.133 | 13.661 | 27.555 | | | | | | |
| $OO$ | 27.093 | 13.661 | 9.133 | 6.830 | | | | | |
| $P_1$ | 23.942 | 182.621 | 32.451 | 14.765 | 12.710 | | | | |
| $Q_1$ | 6.859 | 9.133 | 13.661 | 27.555 | 5.474 | 9.614 | | | |
| $2Q$ | 5.492 | 6.859 | 9.133 | 13.777 | 4.566 | 7.127 | 27.555 | | |
| $S_1$ | 25.622 | 365.243 | 29.803 | 14.192 | 13.168 | 365.243 | 9.367 | 6.991 | |
| $\rho_1$ | 7.096 | 9.557 | 14.632 | 31.812 | 5.623 | 10.085 | 205.892 | 24.302 | 9.814 |

SEMIDIURNAL CONSTITUENTS

|  | $K_2$ | $L_2$ | $M_2$ | $N_2$ | $2N$ | $R_2$ | $S_2$ | $T_2$ | $\lambda_2$ | $\mu_2$ | $\nu_2$ |
|---|---|---|---|---|---|---|---|---|---|---|---|
|  | Days. | Days. | Days. | Days. | Days. | Days. | Days. | Days. | Days. | Days. | Days. |
| $L_2$ | 27.093 | | | | | | | | | | |
| $M_2$ | 13.661 | 27.555 | | | | | | | | | |
| $N_2$ | 9.133 | 13.777 | 27.555 | | | | | | | | |
| $2N$ | 6.859 | 9.185 | 13.777 | 27.555 | | | | | | | |
| $R_2$ | 365.225 | 29.263 | 14.192 | 9.367 | 6.991 | | | | | | |
| $S_2$ | 182.621 | 31.812 | 14.765 | 9.614 | 7.127 | 365.259 | | | | | |
| $T_2$ | 121.748 | 34.847 | 15.387 | 9.874 | 7.269 | 182.630 | 365.259 | | | | |
| $\lambda_2$ | 23.942 | 205.892 | 31.812 | 14.765 | 9.614 | 25.622 | 27.555 | 29.803 | | | |
| $\mu_2$ | 7.096 | 9.614 | 14.765 | 31.812 | 205.892 | 7.236 | 7.383 | 7.535 | 10.085 | | |
| $\nu_2$ | 9.557 | 14.765 | 31.812 | 205.892 | 24.302 | 9.814 | 10.085 | 10.371 | 15.906 | 27.555 | |
| $2SM$ | 16.064 | 10.085 | 7.383 | 5.823 | 4.807 | 15.387 | 14.765 | 14.192 | 9.614 | 4.922 | 5.992 |

Table 40.—*Day of the common year corresponding to day of month*

[For leap year increase all numbers after February 29 by 1 day]

| Day of month. | Jan. | Feb. | Mar. | Apr. | May | June | July | Aug. | Sept. | Oct. | Nov. | Dec. |
|---|---|---|---|---|---|---|---|---|---|---|---|---|
| 1 | 1 | 32 | 60 | 91 | 121 | 152 | 182 | 213 | 244 | 274 | 305 | 335 |
| 2 | 2 | 33 | 61 | 92 | 122 | 153 | 183 | 214 | 245 | 275 | 306 | 336 |
| 3 | 3 | 34 | 62 | 93 | 123 | 154 | 184 | 215 | 246 | 276 | 307 | 337 |
| 4 | 4 | 35 | 63 | 94 | 124 | 155 | 185 | 216 | 247 | 277 | 308 | 338 |
| 5 | 5 | 36 | 64 | 95 | 125 | 156 | 186 | 217 | 248 | 278 | 309 | 339 |
| 6 | 6 | 37 | 65 | 96 | 126 | 157 | 187 | 218 | 249 | 279 | 310 | 340 |
| 7 | 7 | 38 | 66 | 97 | 127 | 158 | 188 | 219 | 250 | 280 | 311 | 341 |
| 8 | 8 | 39 | 67 | 98 | 128 | 159 | 189 | 220 | 251 | 281 | 312 | 342 |
| 9 | 9 | 40 | 68 | 99 | 129 | 160 | 190 | 221 | 252 | 282 | 313 | 343 |
| 10 | 10 | 41 | 69 | 100 | 130 | 161 | 191 | 222 | 253 | 283 | 314 | 344 |
| 11 | 11 | 42 | 70 | 101 | 131 | 162 | 192 | 223 | 254 | 284 | 315 | 345 |
| 12 | 12 | 43 | 71 | 102 | 132 | 163 | 193 | 224 | 255 | 285 | 316 | 346 |
| 13 | 13 | 44 | 72 | 103 | 133 | 164 | 194 | 225 | 256 | 286 | 317 | 347 |
| 14 | 14 | 45 | 73 | 104 | 134 | 165 | 195 | 226 | 257 | 287 | 318 | 348 |
| 15 | 15 | 46 | 74 | 105 | 135 | 166 | 196 | 227 | 258 | 288 | 319 | 349 |
| 16 | 16 | 47 | 75 | 106 | 136 | 167 | 197 | 228 | 259 | 289 | 320 | 350 |
| 17 | 17 | 48 | 76 | 107 | 137 | 168 | 198 | 229 | 260 | 290 | 321 | 351 |
| 18 | 18 | 49 | 77 | 108 | 138 | 169 | 199 | 230 | 261 | 291 | 322 | 352 |
| 19 | 19 | 50 | 78 | 109 | 139 | 170 | 200 | 231 | 262 | 292 | 323 | 353 |
| 20 | 20 | 51 | 79 | 110 | 140 | 171 | 201 | 232 | 263 | 293 | 324 | 354 |
| 21 | 21 | 52 | 80 | 111 | 141 | 172 | 202 | 233 | 264 | 294 | 325 | 355 |
| 22 | 22 | 53 | 81 | 112 | 142 | 173 | 203 | 234 | 265 | 295 | 326 | 356 |
| 23 | 23 | 54 | 82 | 113 | 143 | 174 | 204 | 235 | 266 | 296 | 327 | 357 |
| 24 | 24 | 55 | 83 | 114 | 144 | 175 | 205 | 236 | 267 | 297 | 328 | 358 |
| 25 | 25 | 56 | 84 | 115 | 145 | 176 | 206 | 237 | 268 | 298 | 329 | 359 |
| 26 | 26 | 57 | 85 | 116 | 146 | 177 | 207 | 238 | 269 | 299 | 330 | 360 |
| 27 | 27 | 58 | 86 | 117 | 147 | 178 | 208 | 239 | 270 | 300 | 331 | 361 |
| 28 | 28 | 59 | 87 | 118 | 148 | 179 | 209 | 240 | 271 | 301 | 332 | 362 |
| 29 | 29 | | 88 | 119 | 149 | 180 | 210 | 241 | 272 | 302 | 333 | 363 |
| 30 | 30 | | 89 | 120 | 150 | 181 | 211 | 242 | 273 | 303 | 334 | 364 |
| 31 | 31 | | 90 | | 151 | | 212 | 243 | | 304 | | 365 |

Table 41.—*Values of h in formula* $h=(1+r^2+2r\cos x)^{\frac{1}{2}}$

| x | r | | | | | | | | | | | x |
|---|---|---|---|---|---|---|---|---|---|---|---|---|
| | 0.0 | 0.1 | 0.2 | 0.3 | 0.4 | 0.5 | 0.6 | 0.7 | 0.8 | 0.9 | 1.0 | |
| ° | | | | | | | | | | | | ° |
| 0 | 1. | 1.100 | 1.200 | 1.300 | 1 400 | 1.500 | 1.600 | 1.700 | 1.800 | 1.900 | 2.000 | 360 |
| 10 | 1.000 | 1.099 | 1.197 | 1.296 | 1.396 | 1.495 | 1.594 | 1.694 | 1.793 | 1.893 | 1.992 | 350 |
| 20 | 1.000 | 1.095 | 1.190 | 1.286 | 1.383 | 1.480 | 1.577 | 1.675 | 1.773 | 1.871 | 1.970 | 340 |
| 30 | 1. | 1.088 | 1.177 | 1.269 | 1.361 | 1.455 | 1.549 | 1.644 | 1.739 | 1.835 | 1.932 | 330 |
| 40 | 1.000 | 1.079 | 1.160 | 1.245 | 1.331 | 1.420 | 1.510 | 1.601 | 1.693 | 1.786 | 1.879 | 320 |
| 50 | 1.000 | 1.067 | 1.139 | 1.215 | 1.294 | 1.376 | 1.460 | 1.546 | 1.634 | 1.723 | 1.813 | 310 |
| 60 | 1. | 1.054 | 1.114 | 1.179 | 1.249 | 1.323 | 1.400 | 1.480 | 1.562 | 1.646 | 1.732 | 300 |
| 70 | 1.000 | 1.038 | 1.085 | 1.138 | 1.197 | 1.262 | 1.331 | 1.403 | 1.479 | 1.557 | 1.638 | 290 |
| 80 | 1.000 | 1.022 | 1.053 | 1.093 | 1.140 | 1.193 | 1.252 | 1.316 | 1.385 | 1.457 | 1.532 | 280 |
| 90 | 1.000 | 1.005 | 1.020 | 1.044 | 1.077 | 1.118 | 1.166 | 1.221 | 1.281 | 1.345 | 1.414 | 270 |
| 100 | 1.000 | 0.988 | 0.985 | 0.993 | 1.010 | 1.037 | 1.073 | 1.117 | 1.167 | 1.224 | 1.286 | 260 |
| 110 | 1.000 | 0.970 | 0.950 | 0.941 | 0.941 | 0.953 | 0.974 | 1.006 | 1.045 | 1.093 | 1.147 | 250 |
| 120 | 1.000 | 0.954 | 0.917 | 0.889 | 0.872 | 0.866 | 0.872 | 0.889 | 0.917 | 0.954 | 1.000 | 240 |
| 130 | 1.000 | 0.939 | 0.885 | 0.839 | 0.804 | 0.779 | 0.767 | 0.768 | 0.782 | 0.808 | 0.845 | 230 |
| 140 | 1.000 | 0.926 | 0.856 | 0.794 | 0.740 | 0.696 | 0.664 | 0.646 | 0.644 | 0.657 | 0.684 | 220 |
| 150 | 1.000 | 0.915 | 0.833 | 0.755 | 0.684 | 0.620 | 0.566 | 0.527 | 0.504 | 0.501 | 0.518 | 210 |
| 160 | 1.0 | 0.907 | 0.815 | 0.725 | 0.639 | 0.557 | 0.482 | 0.418 | 0.369 | 0.344 | 0.347 | 200 |
| 170 | 1.0 | 0.902 | 0.804 | 0.706 | 0.610 | 0.515 | 0.422 | 0.334 | 0.254 | 0.193 | 0.174 | 190 |
| 180 | 1.000 | 0.900 | 0.800 | 0.700 | 0.600 | 0.500 | 0.400 | 0.300 | 0.200 | 0.100 | 0.000 | 180 |

Table 42.—*Values of k in formula* $k=\tan^{-1}\dfrac{r\sin x}{1+r\cos x}$

[When x is between 180° and 360°, tabular values are negative]

| x | r | | | | | | | | | | | x |
|---|---|---|---|---|---|---|---|---|---|---|---|---|
| | 0.0 | 0.1 | 0.2 | 0.3 | 0.4 | 0.5 | 0.6 | 0.7 | 0.8 | 0.9 | 1.0 | |
| ° | ° | ° | ° | ° | ° | ° | ° | ° | ° | ° | ° | ° |
| 0 | 0.00 | 0.00 | 0.00 | 0.00 | 0.00 | 0.00 | 0.00 | 0.00 | 0.00 | 0.00 | 0.00 | 360 |
| 10 | 0.00 | 0.90 | 1.66 | 2.30 | 2.85 | 3.33 | 3.75 | 4.12 | 4.45 | 4.73 | 5.00 | 350 |
| 20 | 0.00 | 1.78 | 3.30 | 4.58 | 5.68 | 6.63 | 7.47 | 8.22 | 8.88 | 9.47 | 10.00 | 340 |
| 30 | 0.00 | 2.63 | 4.87 | 6.78 | 8.45 | 9.90 | 11.17 | 12.30 | 13.30 | 14.20 | 15.00 | 330 |
| 40 | 0.00 | 3.42 | 6.36 | 8.92 | 11.13 | 13.08 | 14.80 | 16.82 | 17.68 | 18.90 | 20.00 | 320 |
| 50 | 0.00 | 4.12 | 7.73 | 10.90 | 13.70 | 16.17 | 18.35 | 20.30 | 22.03 | 23.60 | 25.00 | 310 |
| 60 | 0.00 | 4.72 | 8.95 | 12.73 | 16.10 | 19.10 | 21.78 | 24.18 | 26.33 | 28.27 | 30.00 | 300 |
| 70 | 0.00 | 5.20 | 9.98 | 14.33 | 18.30 | 21.87 | 25.07 | 27.95 | 30.55 | 32.88 | 35.00 | 290 |
| 80 | 0.00 | 5.53 | 10.78 | 15.68 | 20.22 | 24.37 | 28.15 | 31.58 | 34.67 | 37.47 | 40.00 | 280 |
| 90 | 0.00 | 5.71 | 11.31 | 16.70 | 21.80 | 26.57 | 30.96 | 34.99 | 38.66 | 41.99 | 45.00 | 270 |
| 100 | 0.00 | 5.72 | 11.53 | 17.82 | 22.95 | 28.33 | 33.42 | 38.12 | 42.45 | 46.42 | 50.00 | 260 |
| 110 | 0.00 | 5.55 | 11.41 | 17.43 | 23.53 | 29.55 | 35.35 | 40.85 | 45.98 | 50.70 | 55.00 | 250 |
| 120 | 0.00 | 5.22 | 10.89 | 17.00 | 23.42 | 30.00 | 36.58 | 43.00 | 49.10 | 54.78 | 60.00 | 240 |
| 130 | 0.00 | 4.68 | 9.97 | 15.90 | 22.42 | 29.45 | 36.80 | 44.27 | 51.60 | 58.57 | 65.00 | 230 |
| 140 | 0.00 | 3.98 | 8.63 | 14.05 | 20.33 | 27.52 | 35.52 | 44.13 | 53.02 | 61.77 | 70.00 | 220 |
| 150 | 0.00 | 3.13 | 6.90 | 11.45 | 17.02 | 23.80 | 31.98 | 41.63 | 52.48 | 63.88 | 75.00 | 210 |
| 160 | 0.00 | 2.17 | 4.81 | 8.13 | 12.37 | 17.88 | 25.20 | 34.98 | 47.82 | 63.38 | 80.00 | 200 |
| 170 | 0.00 | 1.10 | 2.48 | 4.23 | 6.53 | 9.70 | 14.28 | 21.37 | 33.22 | 53.98 | 85.00 | 190 |
| 180 | 0.00 | 0.00 | 0.00 | 0.00 | 0.00 | 0.00 | 0.00 | 0.00 | 0.00 | 0.00 | -------- | 180 |

# EXPLANATION OF SYMBOLS USED IN THIS BOOK

Although the following list is fairly comprehensive, some of the symbols given may at times be used in the text to represent other quantities not listed below, but such application will be made clear by the context.

$A$    (1) General symbol for a tidal constituent or its amplitude. It is sometimes written with a subscript to indicate the species of the constituent (par. 52).
(2) General symbol with an identifying subscript for a constituent term in the development of the lunar tide-producing force (par. 66).
(3) The particular tidal constituent being cleared by the elimination process (par. 245).
(4) Azimuth of tide-producing body reckoned from the south through the west (par. 80).
(5) Azimuth of horizontal component of force in any given direction (par. 85).

$a$    (1) Speed or rate of change in argument of constituent $A$.
(2) Mean radius of earth.

$B$    (1) Tidal constituent following constituent $A$ in a series.
(2) General symbol with an identifying subscript for a constituent term in the development of the solar tide-producing force (par. 117).
(3) General symbol for disturbing constituents in elimination process (par. 245).

$b$    Speed or rate of change in argument of constituent $B$.

$C$    (1) Mean constituent coefficient (par. 74).
(2) General symbol for coefficients of cosine terms in Fourier series (par. 187).

$c$    Reciprocal of mean value of $1/d$.

$c_1$    Reciprocal of mean value of $1/d_1$.

$D$    Declination of moon or sun.

$d$    Distance from center of earth to center of moon.

$d_1$    Distance from center of earth to center of sun.

$E$    (1) Mass of earth.
(2) Argument of tidal constituent (same as $V+u$).

$e$    Eccentricity of moon's orbit.

$e_1$    Eccentricity of earth's orbit.

$F$    Reduction factor, reciprocal of node factor $f$ (par. 78).

$F_a$    Horizontal component of tide-producing force in azimuth $A$. When numerals are annexed, the first digit (3 or 4) signifies the power of the parallax of the moon or sun involved in the development and the second digit (0, 1, 2, or 3) indicates the species of the terms included in the group. Thus $F_{a30}$ represents that part of the horizontal component in azimuth $A$ that comprises the long-period terms depending upon the cube of the parallax.

$F_s$    South horizontal component of tide-producing force. (See $F_a$ for explanation of annexed numerals.)

$F_v$    Vertical component of tide-producing force. (See $F_a$ for explanation of annexed numerals.)

$F_w$    West horizontal component of tide-producing force. (See $F_a$ for explanation of annexed numerals.)

$f$    Node factor (par. 77).

$G$    (1) Greenwich epoch or phase lag of a tidal constituent (par. 226).
(2) Gear ratio of predicting machine (par. 396).

$g$    (1) Mean acceleration of gravity on earth's surface.
(2) Modified epoch of tidal constituent, same as $\kappa'$ (par. 225).

$H$    Mean amplitude of a tidal constituent (par. 143).

| | |
|---|---|
| $H_0$ | Mean water level above datum used for tabulation. |
| $h$ | (1) Mean longitude of sun; also rate of change in same.<br>(2) Height of tide at any time. |
| $h_3$ | Height of equilibrium tide involving cube of moon's parallax. A second digit in the subscript limits the height to that due to terms of the single species indicated by this digit (pars. 97 and 101). |
| $h_4$ | Height of equilibrium tide involving 4th power of moon's parallax. A second digit in the subscript has the same significance as in the case of $h_3$. |
| $I$ | Obliquity of lunar orbit with respect to earth's equator. |
| $i$ | Inclination of lunar orbit to the ecliptic. |
| $J_1$ | Tidal constituent. |
| $j$ | Longitude of moon in its orbit reckoned from lunar intersection. |

$K_1$, $K_2$ Tidal constituents.
$KJ_2$, $KP_1$, $KQ_1$ Tidal constituents.

| | |
|---|---|
| $k$ | Difference between mean and true longitude of moon (par. 59). |
| $L$ | Longitude of place; positive for west longitude, negative for east longitude. |

$L_2$, $LP_1$ Tidal constituents.

| | |
|---|---|
| $M$ | Mass of moon. |

$M_1$, $M_2$, $M_3$, $M_4$, $M_6$, $M_8$ Tidal constituents.

| | |
|---|---|
| Mf | Tidal constituent. |

$MK_3$, $2MK_3$, $MK_4$ Tidal constituents.

| | |
|---|---|
| Mm | Tidal constituent. |

$MN_4$, $2MN_6$, $MNS_2$ Tidal constituents.

| | |
|---|---|
| $MP_1$ | Tidal constituent. |

$2MS_2$, $MS_4$, $2MS_6$, $3MS_8$, $2(MS)_8$ Tidal constituents.

| | |
|---|---|
| MSf | Tidal constituent. |

$MSN_6$, $2MSN_8$ Tidal constituents.

| | |
|---|---|
| $m$ | Ratio of mean motion of sun to that of moon (par. 62). |
| $N$ | Longitude of moon's node; also rate of change in same. |

$N_2$, $2N_2$, $NJ_1$ Tidal constituents.
$O_1$, $OO_1$ Tidal constituents.

| | |
|---|---|
| $P$ | Mean longitude of lunar perigee reckoned from lunar intersection (par. 122). |
| $P_1$ | Tidal constituent. |
| $p$ | (1) Mean longitude of lunar perigee; also rate of change in same.<br>(2) Numeral indicating species of constituent, frequently written as the subscript of the constituent symbol. In special case used with long-period constituents to show number of periods in month or year. |
| $p_1$ | Mean longitude of solar perigee; also rate of change in same. |
| $Q$ | Term in argument of constituent $M_1$ (par. 123). |
| $Q_a$ | Factor in amplitude of constituent $M_1$ (par. 122). |
| $Q_u$ | Term in argument of constituent $M_1$ (par. 122). |

$Q_1$, $2Q_1$ Tidal constituents.

| | |
|---|---|
| $R$ | (1) Amplitude of constituent pertaining to a particular time (par. 143).<br>(2) Term in argument of constituent $L_2$ (par. 129). |
| $R_a$ | Factor in amplitude of constituent $L_2$ (par. 129). |

$R_2$, $RP_1$ Tidal constituents.

| | |
|---|---|
| $r$ | Distance of any point from center of earth. |
| $S$ | (1) Mass of sun.<br>(2) Longitude of time meridian; positive for west longitude, negative for east longitude.<br>(3) General symbol for coefficients of sine terms in Fourier series (par. 187).<br>(4) Working scale factor of predicting machine. |
| $S'$ | Solar factor $U_1/U$ (par. 118). |

$S_1$, $S_2$, $S_3$, $S_4$, $S_6$, $S_8$ Tidal constituents.

| | |
|---|---|
| Sa | Tidal constituent. |
| $SK_3$ | Tidal constituent. |
| $2SM_6$ | Tidal constituent. |

HARMONIC ANALYSIS AND PREDICTION OF TIDES    313

$SO_1$, $SO_3$    Tidal constituents.
Ssa    Tidal constituent.
$s$    Mean longitude of moon; also rate of change in same.
$s'$    True longitude of moon in orbit referred to equinox (par. 59).
$T$    (1) Number of Julian centuries reckoned from Greenwich mean noon, December 31, 1899.
(2) Hour angle of mean sun.
(3) Time expressed in degrees of constituent reckoned from phase zero of Greenwich argument (par. 439).
$T_2$    Tidal constituent.
$t$    (1) Hour angle of tide-producing body.
(2) Time reckoned from beginning of tidal series.
$U$    Basic factor $(M/E)(a/c)^3$.
$U_1$    Factor $(S/E)(a/c_1)^3$.
$u$    Part of constituent argument depending upon variations in obliquity of lunar orbit (par. 71).
$V$    (1) Principal portion of constitute argument (par. 71).
(2) Velocity of current (par. 330).
$(V+u)$    Constituent argument at any time.
$(V_o+u)$    Constituent argument at beginning of a tidal series.
$V_g$    Potential due to gravity at earth's surface (par. 96).
$V_3$    Tide-producing potential involving cube of moon's parallax (par. 94).
$V_4$    Tide-producing potential involving 4th power of moon's parallax (par. 94.)
$X$    Longitude of observer reckoned in celestial equator from lunar intersection.
$Y$    Latitude of observer. When combined with a subscript consisting of a letter and numerals, it represents the latitude factor to be used with the tidal force component similarly marked (par. 79).
$z$    Geocentric zenith distance of tide-producing body.
$\alpha$ (Alpha)    General symbol for the initial phase of tidal constituent $A$.
$\beta$ (Beta)    Initial phase of constituent $B$.
$\gamma$ (Gamma)    Initial phase of constituent $C$.
$\delta$ (Delta)    Initial phase of constituent $D$.
$\epsilon$ (Epsilon)    Initial phase of constituent $E$.
$\zeta$ (Zeta)    The explement of the initial phase of a constituent (par. 221).
$\theta_1$ (Theta)    Tidal constituent, same as $\lambda 0_1$.
$\kappa$ (Kappa)    Local phase lag or epoch of tidal constituent (par. 144).
$\kappa'$    Modified epoch of tidal constituent (par. 225).
$\lambda_2$ (Lambda)    Tidal constituent.
$\mu$ (Mu)    Attraction of gravitation between unit masses at unit distance.
$\mu_2$    Tidal constituent, same as $2MS_2$.
$\nu$ (Nu)    Right ascension of lunar intersection (par. 24).
$\nu'$    Term in argument of lunisolar constituent $K_1$ (par. 133).
$2\nu''$    Term in argument of lunisolar constituent $K_2$ (par. 135).
$\nu_2$    Tidal constituent.
$\xi$ (Xi)    Longitude in moon's orbit of lunar intersection (par. 24).
$\pi$ (Pi)    An angle of 3.14159 radians or 180°.
$\pi_1$    Tidal constituent, same as $TK_1$.
$\rho_1$ (Rho)    Tidal constituent, same as $\nu K_1$.
$\sigma_1$ (Sigma)    Tidal constituent, same as $\nu J_1$.
$\tau$ (Tau)    Length of series in mean solar hours (par. 248).
$\varphi_1$ (Phi)    Tidal constituent, same as $KP_1$.
$\chi_1$ (Chi)    Tidal constituent, same as $LP_1$.
$\psi_1$ (Psi)    Tidal constituent, same as $RP_1$.
$\omega$ (Omega)    Obliquity of ecliptic.
♈    Vernal equinox.
☊    Moon's ascending node.

# INDEX

## A

| | Page |
|---|---|
| Adams, J. C. | 1 |
| Airy, George B. | 1 |
| Amplitude of constituent | 2, 49 |
| Analysis of high and low waters | 100 |
| Analysis of monthly sea level | 98, 114 |
| Analysis of observations | 49 |
| Analysis of tidal currents | 118 |
| Anomalistic month, year | 4 |
| Approximation, degree of | 8 |
| Argument. (See Equilibrium argument.) | |
| Astres fictifs | 23 |
| Astronomical data | 3, 153, 162 |
| Astronomical day | 3 |
| Astronomical periods | 163 |
| Astronomical tide | 30 |
| Augmenting factors | 71, 91, 157, 228 |

## B

| | |
|---|---|
| Basic factor | 24 |

## C

| | |
|---|---|
| Calendars | 4 |
| Civil day | 3 |
| Coefficients | 24 |
| Component. (See Constituent tides.) | |
| Component of force, horizontal | 26 |
| Component of force, vertical | 15 |
| Compound tides | 47, 167 |
| Constituent day | 3 |
| Constituent hour | 4 |
| Constituent tides | 2, 16, 87 |
| Formulas | 21, 35, 39 |
| Tables | 153, 164, 167 |
| Currents, analysis | 118 |
| Currents, prediction | 147 |

## D

| | |
|---|---|
| Darwin, G. H. | 1 |
| Datum for prediction | 124, 144 |
| Day, several kinds | 3 |
| Day of year, table | 309 |
| Declinational factor | 17 |
| Degree of approximation | 8 |
| Development of tide-producing force | 10 |
| Diurnal constituents | 16 |

## E

| | |
|---|---|
| Eccentricity of orbit | 4 |
| Eclipse year | 4 |
| Elimination | 84, 116, 158, 236 |
| Elliptic factor | 24 |
| Epoch of constituent | 49, 75 |
| Equations of moon's motion | 19 |

| | Page |
|---|---|
| Equilibrium argument | 22, 50, 75, 108, 124, 157, 204 |
| Equilibrium theory | 28 |
| Equilibrium tide | 28, 38 |
| Equinox | 6 |
| Eudoxas | 1 |
| Evection | 4 |
| Explanation of tables | 153 |
| Explanation of tidal movement | 2 |
| Extreme equilibrium tide | 33 |
| Extreme tide-producing force | 13 |

## F

| | |
|---|---|
| Factor F. (See Reduction factor.) | |
| Factor f. (See Node factor.) | |
| Ferrel, William | 1, 127 |
| Forms for analysis of tides | 104 |
| Forms for predicting machine | 143 |
| Fourier series | 62 |
| Fourth power of moon's parallax | 34 |
| Fundamental astronomical data | 153, 162 |
| Fundamental formulas | 10 |

## G

| | |
|---|---|
| General coefficient | 24 |
| General explanation, tidal movement | 2 |
| Gravitational tide | 30 |
| Greatest equilibrium tide | 33 |
| Greatest tide-producing force | 13 |
| Greenwich argument | 76 |
| Greenwich epoch | 77 |
| Gregorian calendar | 4 |

## H

| | |
|---|---|
| Harmonic analysis | 3, 49, 112 |
| Harmonic constants | 3, 49, 143 |
| Harmonic prediction | 3, 123 |
| Harris, Rollin A. | 1 |
| High and low water analysis | 100 |
| Historical statement | 1 |
| Horizontal component, tide-producing force | 26, 37 |
| Hour, several kinds | 4 |
| Hourly heights | 104 |
| Hydraulic current | 148 |
| Hydrographic datum | 144 |

## I

| | |
|---|---|
| Inclination of moon's orbit | 6, 155, 173 |
| Inference of constants | 78, 114 |

315

# INDEX

## J
| | Page |
|---|---|
| Julian calendar | 4 |

## K
| | Page |
|---|---|
| $K_1$ and $K_2$ tides | 44 |
| Kelvin, Lord | 1, 126 |

## L
| | Page |
|---|---|
| Laplace | 1 |
| Latitude | 6 |
| Latitude factors | 17, 24, 154, 168 |
| Length of series | 51 |
| Lesser lunar constituents | 35 |
| Lesser solar constituents | 40 |
| Lesser tide-producing force | 34, 40 |
| Longitude | 6 |
| Longitude, lunar and solar elements | 162, 170 |
| Long-period constituents | 16, 87, 302 |
| $L_2$-tide | 43, 156, 177, 192 |
| Lunar constituents | 21, 35 |
| Lunar day | 3 |
| Lunar hour | 4 |
| Lunar intersection | 6 |
| Lunar node | 6, 8 |
| Lunisolar tides | 44 |

## M
| | Page |
|---|---|
| Mean constituent coefficient | 24 |
| Mean longitude | 7 |
| Meteorological tides | 46 |
| Month, several kinds | 4 |
| Monthly sea-level analysis | 98, 114 |
| Moon's motion, equations | 19 |
| Moon's node | 6, 8 |
| Moon's parallax, 4th power | 34 |
| $M_1$-tide | 41, 156, 179, 192 |

## N
| | Page |
|---|---|
| Node, lunar | 6, 8 |
| Node factor | 25 |
|   Compound tides | 47 |
|   Constituent $K_1$ | 45 |
|   Constituent $K_2$ | 46 |
|   Constituent $L_2$ | 44 |
|   Constituent $M_1$ | 43 |
|   Lesser tide-producing force | 36 |
|   Predictions | 124 |
|   Table | 199 |
| Nodical month | 4 |

## O
| | Page |
|---|---|
| Obliquity factor | 24 |
| Obliquity of ecliptic | 6 |
| Obliquity of moon's orbit | 6 |
| Observational data | 50 |
| Overtides | 47 |

## P
| | Page |
|---|---|
| Period of constituent | 3 |
| Periods, astronomical | 163 |
| Phase lag | 49, 75 |
| Phase of constituent | 2 |
| Poor, Charles Lane | 1 |
| Potential | 30 |
| Predicting machine. (See Tide-predicting machine.) | |
| Prediction of tidal currents | 147 |
| Prediction of tides | 123 |
| Principal lunar constituents | 21 |
| Principal solar constituents | 39 |

## R
| | Page |
|---|---|
| Record of observations | 50 |
| Reduction factor | 25, 111, 156, 186 |

## S
| | Page |
|---|---|
| Secondary stencils | 57, 159, 299 |
| Semidiurnal constituents | 16 |
| Settings for tide-predicting machine | 145, 306 |
| Shallow-water constituents | 46, 167 |
| Shidy, L. P | 53 |
| Sidereal day | 3 |
| Sidereal hour, month, year | 4 |
| Solar day | 3 |
| Solar factor | 40 |
| Solar hour | 4 |
| Solar tides | 39 |
| South component, tide-producing force | 26, 37 |
| Species of constituent | 16 |
| Speed of constituent | 3, 23 |
| Stationary wave | 2 |
| Stencil sums | 107 |
| Stencils | 53, 106, 158, 268 |
| Summarized formulas: | |
|   Equilibrium tide | 33 |
|   Lesser tide-producing force | 36 |
|   Principal tide-producing force | 33 |
| Summation for analysis | 52 |
| Surface of equilibrium | 30, 32 |
| Symbols used in book | 311 |
| Synodical month | 4 |
| Synodic periods of constituents | 161, 309 |

## T
| | Page |
|---|---|
| Tables | 162 |
|   Explanation | 153 |
| Terdiurnal constituents | 34 |
| Thomson, Sir William | 1, 126 |
| Tidal currents | 118, 147 |
| Tidal movement | 2 |
| Tide-predicting machine | 126 |
|   Adjustments | 139 |
|   Automatic stopping device | 135 |
|   Base | 127 |
|   Constituent cranks | 130 |
|   Constituent dials | 131 |
|   Constituent pulleys | 132 |
|   Constituent sliding frames | 131 |
|   Datum of heights | 141 |
|   Day dial | 128 |
|   Dial hour | 128 |
|   Dimensions | 127 |
|   Doubling gears | 132 |
|   Forms used | 143 |
|   Gear speeds | 129 |
|   Gearing | 128, 160, 307 |
|   Graph scale | 137 |

| | |
|---|---|
| Tide-predicting machine—Con. Page | |
| Height formula | 126 |
| Height predictions | 134 |
| Height scale | 134, 141 |
| Height side | 128 |
| High and low water marking device | 139 |
| Hour marking device | 139 |
| Marigram gears | 137, 141 |
| Marigram scale | 137 |
| Nonreversing ratchet | 136 |
| Operation of machine | 142 |
| Paper | 136, 142 |
| Pens | 138, 142 |
| Plane of reference | 141 |
| Positive and negative directions | 131 |
| Predicting | 142 |
| Releasable gears | 130 |
| Scale, amplitude settings | 132, 140 |
| Scale, height dial | 134, 141 |
| Scale, marigram | 137 |
| Scale, table | 138 |
| Scale, working | 135 |
| Setting machine | 140 |
| Stopping device | 135 |
| Summation chains | 133 |
| Tide-predicting machine—Con. Page | |
| Summation wheels | 133 |
| Tide curve | 136 |
| Time dials | 128 |
| Time formula | 126, 132 |
| Time prediction | 135 |
| Time side | 128 |
| Verification of settings | 142 |
| Tide-producing force | 10 |
| Tide-producing potential | 31 |
| Tropical month, year | 4 |

**V**

| | |
|---|---|
| Variation inequality | 4 |
| Vernal equinox | 6 |
| Vertical component, tide-producing force | 15, 34 |

**W**

| | |
|---|---|
| West component, tide-producing force | 26, 37 |

**Y**

| | |
|---|---|
| Year, several kinds | 4 |
| Young, Thomas | 1 |

O

Lightning Source UK Ltd.
Milton Keynes UK
UKHW02f1533250818
327772UK00005B/548/P